Markus Reiher and Alexander Wolf

Relativistic Quantum Chemistry

Further Reading

P. Comba, T. W. Hambley, B. Martin

Molecular Modelling of Inorganic Compounds

2009
ISBN: 978-3-527-31799-8

H.-D. Höltje, W. Sippl, D. Rognan, G. Folkers

Molecular Modeling

Basic Principles and Applications

2008
ISBN: 978-3-527-31568-0

Markus Reiher and Alexander Wolf

Relativistic Quantum Chemistry

The Fundamental Theory of Molecular Science

WILEY-
VCH

WILEY-VCH Verlag GmbH & Co. KGaA

The Authors

Prof. Dr. Markus Reiher
Dr. Alexander Wolf
ETH Zuerich
Laboratory for Physical Chemistry
Hoenggerberg Campus
Wolfgang-Pauli-Strasse 10
8093 Zuerich
Switzerland

Library of Congress Card No.:
applied for

British Library Cataloguing-in-Publication Data
A catalogue record for this book is available from the British Library.

Bibliographic information published by the Deutsche Nationalbibliothek
The Deutsche Nationalbibliothek lists this publication in the Deutsche Nationalbibliografie; detailed bibliographic data are available on the Internet at http://dnb.d-nb.de.

Printing betz-druck GmbH, Darmstadt
Binding Litges & Dopf GmbH, Heppenheim

ISBN: 978-3-527-31292-4

Contents

Relativistic Quantum Chemistry. Markus Reiher and Alexander Wolf

ISBN: 978-3-527-31292-4

Preface

A relativistic consistent quantum-theoretical description of electronic bound states in atoms was first introduced in atomic physics as early as the late 1920s and has been pushed forward since that time. It was believed, however, that effects stemming from Einstein's theory of relativity were of little or even no importance to chemistry. This changed in the 1970s when it was recognized by Pyykkö, Pitzer, Desclaux, Grant and others that several 'unusual' features in heavy-element chemistry and spectroscopy can only be explained in terms of so-called relativistic effects. *Relativistic effects* denote the deviation of results obtained in a theoretical framework which is in accordance with Einstein's theory of special relativity from nonrelativistic Schrödinger quantum mechanics. Since then, the development of quantum chemical methods for the description of *relativistic* electronic structures has made huge progress — particularly since the late 1980s.

Current relativistic electronic structure theory is now in a mature and well-developed state. We are in possession of sufficiently detailed knowledge on relativistic approximations and relativistic Hamiltonian operators which will be demonstrated in the course of this book. Once a relativistic Hamiltonian has been chosen, the electronic wave function can be constructed using methods well known from nonrelativistic quantum chemistry, and the calculation of molecular properties can be performed in close analogy to the standard nonrelativistic framework. In addition, the derivation and efficient implementation of quantum chemical methods based on (quasi-)relativistic Hamiltonians have facilitated a very large amount of computational studies in heavy element chemistry over the last two decades. Relativistic effects are now well understood, and many problems in contemporary relativistic quantum chemistry are technical rather than fundamental in nature.

We aim to present coherently all its essential aspects in textbook form using a homogeneous notation throughout the book. The greatest challenge of this endeavor is to give a description of the whole theory ranging from the fundamental physical concepts to the final application of the theory to issues of chemical relevance like molecular structure, energetics, and properties. The

Relativistic Quantum Chemistry. Markus Reiher and Alexander Wolf

ISBN: 978-3-527-31292-4

presentation will be concise and focus on the essential ideas as well as on analytical results rather than on too many of the unavoidable technical details, which might blur the view on the physics and concepts behind the calculations. To illustrate these important points in more detail:

(i) It is the nature of approximate relativistic many-electron theories that a large number of effective Hamiltonians may be deduced (defining thus a plethora of different relativistic quantum chemical approaches), though this is neither advantageous nor desirable because of the huge amount and variety of numerical data whose accuracy may be difficult to assess. Instead a rather small number of well-justified approximate Hamiltonians should suffice as we shall see.

(ii) In a similar manner, so-called relativistic effects are discussed *in extenso* in chemistry although these effects are, of course, purely artificial in nature since any fundamental physical theory has to be based on the principles of special relativity. The errors introduced by a nonrelativistic approximate description, which do not occur in a relativistic framework and which cannot in principle be measured in experiments, are called relativistic effects. However, this definition of relativistic effects may only be useful to address surprising observations in the chemistry of homologous compounds when the properties of a heavy-element compound deviate from those of its lighter homologs as in the most prominent cases of the *liquid* state of mercury at ambient conditions and of the *yellowish* color of gold.

Nevertheless, technical details of implementations as well as a moderate amount of discussion of these relativistic effects will be covered by this book as these are needed in practice when actual calculations are to be performed and their results interpreted. In addition, technical and implementational issues are incorporated to demonstrate how the relativistic many-particle equations are actually solved and what effort is required for this.

Because of the wide range of topics covered by this book it was appropriate to start with an introductory chapter, in order to prepare the ground for a convenient perception of structure and material presented in this book. While writing the book we realized that there was no space to present various aspects of our topic in sufficient depth. Still, our main goal was to present in great detail all essential ideas and how they are connected. Since this has always been the guideline for our own research, we may have put the focus too much on what we always have considered to be important throughout the past decade. At least we tried to derive as much as possible of the material presented from scratch to make this book as original as possible. In order to compensate for deficiencies that certainly result from our specific choice of material and its presentation we provide references to selected papers from the original literature. Although we included a considerable number of such references, it is hardly possible to provide a complete list of references as this would comprise several thousands of papers. Fortunately, Pekka Pyykkö made a huge effort to set up a complete data base of references, first published as a series of books [1–3] and since the mid-1990s also available online on the internet [4]. And, we may draw the reader's attention to reviews by leading experts in the field as collected by Schwerdtfeger [5, 6] and by Hess [7]. Of course, this book provides all the basic vocabulary and knowledge required to dig more deeply into the literature.

Finally, it is a pleasure to mention that our view of relativistic many-electron theory has been shaped over a period of more than a decade in which we had the opportunity to sharpen our understanding by comparison with the views of colleagues who shared their knowledge to various extents with us; these are (in alphabetical order): PD Dr. D. Andrae, D. Dath, Prof. E. Eliav, PD Dr. T. Fleig, Prof. L. Gagliardi, Prof. B. A. Hess, Prof. H.-J. Himmel, Prof. J. Hinze, Prof. J. Hutter, Prof. H. J. Å. Jensen, Prof. G. Jeschke, Prof. U. Kaldor, Dr. D. Kędziera, Prof. B. Kirchner, Dr. T. Koch (it took only 13 years), Dr. A. Landau, Prof. R. Lindh, Prof. P.-Å. Malmqvist, Prof. B. Meier, Prof. F. Merkt, Prof. U. Müller-Herold, Prof. F. Neese, Dr. J. Neugebauer, Prof. P. Pyykkö, Prof. M. Quack, Prof. R. Riek, Prof. B. O. Roos, Prof. K. Ruud, Prof. V. Sandoghdar, Dr. T. Saue, Prof. W. Scherer, Prof. W. H. E. Schwarz, Prof. P. Schwerdtfeger, Prof. H. Siedentop, Dr. R. Szmytkowski, Prof. W. van Gunsteren, Prof. C. van Wüllen, Prof. L. Visscher, and Prof. M. Westerhausen.

We thank G. Eickerling, R. Mastalerz, and S. Schenk for help in preparing the cover picture and the figures in section 16.6 and especially R. Mastalerz and S. Luber for help gathering classic books and old papers via the ETH library. Moreover, MR is indebted to Romy Isenegger for keeping many of the administrative duties of present-day university life away from him so that he could spend sufficient time on writing this book. MR would also like to thank all members of his group (Dr. G. Eickerling, Dr. C. Herrmann, Dr. C. Jacob, K. Kiewisch, S. Luber, Dr. I. Malkin, K. Marti, R. Mastalerz, Dr. G. Moritz, Dr. J. Neugebauer, M. Podewitz, Dr. S. Schenk, Dr. M. Stiebritz, L. Yu) for their dedicated work. AW is deeply indebted to Barbara Pfeiffer for her continuous support and patience in the course of preparing this manuscript. Last but not least, it is a pleasure to thank Dr. Elke Maase and Dr. Rainer Münz of Wiley-VCH for help with all publishing issues of this book.

Markus Reiher and Alexander Wolf *Zürich, June 2008*

1
Introduction

This first chapter provides a short reader's guide, which may help to make the material presented in this book more easily accessible. Moreover, it serves to highlight the philosophy and intention of this book.

1.1
Philosophy of this Book

Relativistic quantum chemistry is the relativistic formulation of quantum mechanics applied to many-electron systems, i.e., to atoms, molecules and solids. It combines the principles of Einstein's theory of special relativity, which have to be obeyed by any fundamental physical theory, with the basic rules of quantum mechanics. By construction, it represents the most fundamental theory of all molecular sciences, which describes matter by the action, interaction, and motion of the elementary particles of the theory. In this sense it is important for physicists, chemists, material scientists, and biologists with a molecular view of the world. It is important to note that the energy range relevant to the molecular sciences allows us to operate with a reduced and idealized set of "elementary" particles. "Elementary" to chemistry are atomic nuclei and electrons. In most cases, neither the structure of the nuclei nor the *explicit* description of photons is required for the theory of molecular processes. Of course, this elementary level is not always the most appropriate one if it comes to the investigation of very large nanometer-sized molecular systems. Nevertheless it has two very convenient features:

(i) As a fundamental theory it does not require any experimental information (other than that needed to formulate the basic axioms of the theory) as input and is thus a so-called *first-principles* or *ab initio* theory, which can be deduced completely from the postulates of relativistic quantum mechanics.

(ii) As the theory describes the motion of elementary particles solely on the basis of the laws of quantum mechanics and special relativity, it can be

Relativistic Quantum Chemistry. Markus Reiher and Alexander Wolf

ISBN: 978-3-527-31292-4

used for the derivation of more approximate models for large molecules without falling back on experimental data. This can be achieved by integrating out those degrees of freedom that are deemed irrelevant, thus arriving at effective *ab initio* or even *classical* model theories which describe certain chemical processes sufficiently well.

For all issues relevant to the chemistry and physics of atoms, molecules, clusters, and solids only electromagnetic and — to a negligible extent — weak interactions, which are responsible for the radioactive β-decay and the non-conservation of parity, contribute. The internal structure of hadrons, i.e., protons and neutrons built up by quarks governed by strong interactions and also gravitational forces, do not play any role and are thus not covered by this presentation.

A main purpose of this book is to provide a structured and self-contained presentation of relativistic quantum chemistry as a semi-classical theory. We deem this necessary as there hardly exists any such contiguous and detailed presentation. The main reason appears to be the fact that quantum electrodynamics was developed in the 1930s and 1940s shortly after the advent of the new quantum theory. It was already clear in those days that quantum electrodynamics represents the fundamental theory of light and electrons, and hence the semi-classical theory was almost instantaneously abandoned. However, this most elegant and sophisticated theory of quantum electrodynamics is too abstract to be grasped immediately. The connection to classical physics, which has always been a guiding principle in quantum theory and even for the development of quantum electrodynamics, is seldom made. Instead, especially modern accounts start directly with the field-quantized formulation although the semi-classical theory is sufficient for chemistry as countless numerical studies in quantum chemistry demonstrated. This issue had already been noted by Dirac in his famous lectures on quantum theory, but he nevertheless changed the presentation of the electromagnetic quantum field theory in one of the later editions and gave up the semi-classical theory. However, because of the paramount importance of the semi-classical theory to chemistry, we derive this theory from the very basis, i.e., from classical electrodynamics. Of course, the transition to quantum electrodynamics is also made in this book, but it plays a minor role. This transition can be much better understood once the problems with the classical and the semi-classical theory have been worked out. Only then can one fully appreciate the emergence of concepts like *retarded electromagnetic interactions, magnetic spin–spin coupling* or *orbit–other-orbit interaction*.

Because of the limited space available and the vast number of relativistic studies on atoms and molecules, we chose to accompany the derivation of the theory with a multitude of references to the original research literature in order to provide the reader with a topical overview of results for electronic systems.

Relativity adds a new dimension to quantum chemistry, which is the choice of the Hamiltonian operator. While the Hamiltonian of a molecule is exactly known in nonrelativistic quantum mechanics (if one focuses on the dominating electrostatic monopole interactions to be considered as being transmitted instantaneously), this is no longer the case for the relativistic formulation. Numerical results obtained by many researchers over the past decades have shown how Hamiltonians which capture most of the (numerical) effect of relativity on physical observables can be derived. Relativistic quantum chemistry thus comes in various flavors, which are more or less well rooted in fundamental physical theory and whose relation to one another will be described in detail in this book. The new dimension of relativistic Hamiltonians makes the presentation of the relativistic many-electron theory very complicated, and the degree of complexity is far greater than for nonrelativistic quantum chemistry. However, the relativistic theory provides the consistent approach toward the description of nature: molecular structures containing heavy atoms can *only* be treated correctly within a relativistic framework. Prominent examples known to everyone are the color of gold and the liquid state of mercury at room temperature. Moreover, it must be understood that relativistic quantum chemistry provides universal theoretical means that are applicable to any element from the periodic table or to any molecule — not only to heavy-element compounds.

It is the nature of the subject that makes its presentation rather formal and requires some basic, mainly conceptual knowledge in mathematics and physics. However, only standard mathematical techniques (like differential and integral calculus, matrix algebra) are required. More advanced subjects like complex analysis and tensor calculus are occasionally also used. Furthermore, also basic knowledge of classical Newtonian mechanics and electrodynamics will be helpful to more quickly understand the concise but short review of these matters in the second chapter of this book.

Many (pseudo-)relativistic quantum chemical approaches provide methods which can be implemented in computer programs in a very efficient manner — an aspect which may be called a boundary condition of computational chemistry imposed on theoretical chemistry. Most of these approaches demand only as much computing time as their nonrelativistic analogs (or are more expensive by a constant but small factor). The quantum chemistry community has developed a certain working knowledge of the reliability of these relativistic methods, but their relation to one another sometimes remains unclear. In the light of the importance of relativistic methods, we therefore derive all methods from first principles and highlight their development to sophisticated computational tools in chemistry. In doing so, we shall understand which aspects of the rigorous theory of relativistic quantum *mechanics* survive in relativistic quantum *chemistry*; we learn and have to accept that relativistic effects in extended molecules are about the art of efficiently correcting numerical results like energies for many-electron systems in an essentially nonrelativistic framework.

1.2
Short Reader's Guide

A book on a theoretical topic, which is based on fundamental physical theory but extends to the realm of experimental chemistry — like the present one does — always faces difficulties, as it might be too formal for chemists while it could be too technical and specialized for non-experts and even physicists. Our main goal is the presentation of an almost complete derivation of the relativistic theory for many-electron systems as the fundamental theory of the molecular sciences. For this purpose, we have tried to introduce all essential concepts and ideas and derive all basic equations explicitly. As a consequence, parts of the book — like the solution of the Dirac hydrogen atom — seem to be lengthy. However, many derivations cannot be easily found in such detail elsewhere in the literature (if at all). Also, the number of misconceptions can be rather large regarding issues of relativistic quantum chemistry, which is the reason why we try to provide a self-contained presentation of the theory. The resulting equations for many-particle systems cannot be solved analytically, and the derivation of working equations is strongly driven by the need to be able to solve them on modern computer hardware. Although it is not possible to delve deeply into how this is achieved in every detail, the reader will be given sufficient hints and information so that the equations derived do not appear like mathematical deadwood but are designed to be actually solvable, the essential boundary condition for quantum chemistry. Often, however, only one way for solving these equations is chosen in our presentation, but references to other possibilities are included.

In order to grasp all the essential ideas of the book the reader is strongly advised to go through it from the very beginning. However, each chapter is designed to represent a single unit which may be read with little knowledge of the contents of other chapters. This will help readers to use the book also as a quick reference (in combination with the index). However, especially the later chapters make frequent reference to results derived in previous parts of the book.

Our guiding principles for the preparation of this book were the goals to make all derivations comprehensible and to elaborate on all basic principles relevant for each topic. Each chapter is headed by a brief summary of the essential ideas. Of course, it is hardly possible to present all steps of a derivation, but we aim at a reasonable number of steps, which should allow for a quick re-derivation of the equations given. Needless to say, the more understanding of physical theories and quantum chemical methods the reader already possesses, the easier it will be to understand the presentation. If the reader finds the extensive mathematical formalism of relativistic quantum chemistry too involved and complicated, we ask for patience: it should still be possible to access the basic ideas. Furthermore, all equations are in general presented in a way such that it is possible to easily verify intermediate derivation steps not explicitly given. If such steps are particularly easy, the reader will find comments like "it can be easily verified" or "it follows immediately" with additional references to equations. The purpose of these comments is to let the reader know that the derivation is indeed easy so that he or she can instantly realize that the ansatz is wrong from the start if a re-derivation turns out to be too complicated.

The topics of the individual chapters are well separated and the division of the book into five major parts emphasizes this structure. Part I contains all material, which is essential for understanding the physical ideas behind the merging of classical mechanics, principles of special relativity, and quantum mechanics to the complex field of relativistic quantum chemistry. However, one or all of these three chapters may be skipped by the experienced reader. As is good practice in theoretical physics (and even in textbooks on physical chemistry), *exact* treatments of the relativistic theory of the electron as well as analytically solvable problems like the Dirac electron in a central field (i.e., the Dirac hydrogen atom) are contained in part II.

Chemistry and the molecular sciences start with the many-particle theories of physics; part III of the book deals with these many-electron extensions of the theoretical framwork, which have their foundations in the one-electron framework presented in part II. The first chapter in part III is on the most general many-electron theory known in physics: quantum electrodynamics (QED). From the point of view of physics this is the fundamental theory of chemistry although far too complicated to be used for calculations on systems with more than a few electrons. Standard chemistry does not require all features covered by QED (like pair creation), and so neither does a basic and at the same time practical theory of chemistry. Three subsequent chapters describe the suitable approximations, which provide a first-quantized theory for many-electron systems with a, basically, fixed number of particles. A major result from this discussion is the fact that this successful model is still plagued by practical as well as by conceptual difficulties. As a consequence further 'simplifications' are introduced, which eliminate the conceptual difficulties; these 'simplifications' are discussed in part IV.

The reader who proceeds stepwise will realize at this point that the theories of relativistic quantum chemistry are approximate anyhow and do not obey the basic formal principles of relativity, though they capture most of the effect on the numerical values of physical observables relevant to chemistry. After this stringent development of relativistic quantum chemical methods for many-electron systems, we are in a position to finally discuss molecular properties and relativistic effects in chemistry in part V.

1.3 Notational Conventions and Choice of Units

Several conventions are used throughout this book. We may summarize some basic principles of notation here.

Cartesian vectors with three components, so-called 3-vectors, are indicated by boldface type, e.g., the position vector $\boldsymbol{r}$. Relativistic four-component vectors, so-called 4-vectors, are denoted by normal type. If not otherwise stated, all vectors irrespective of their dimension are assumed to be column vectors. The corresponding row vector is given by the transposed quantity, e.g., $\boldsymbol{r}^T = (x, y, z)$. For the sake of brevity, however, the transposition is not explicitly denoted in ubiquitous scalar products like $\boldsymbol{\sigma} \cdot \boldsymbol{p}$ or $\boldsymbol{\alpha} \cdot \boldsymbol{p}$, for example.

A dot on top of a symbol for a physical quantity denotes the *total* time derivative of that quantity, e.g., $\dot{\boldsymbol{r}} = \mathrm{d}\boldsymbol{r}/\mathrm{d}t$. Partial time derivatives are symbolized by $\partial/\partial t$. The spatial derivatives in one- or three-dimensional space are symbolized by $A'(x)$ or $\nabla\phi$ and $\nabla \cdot \boldsymbol{A}(\boldsymbol{r})$, respectively.

The same symbols will be used for classical quantities as well as for the corresponding operators of the quantized formulation. Consequently, $\boldsymbol{p}$ might symbolize classical momentum as well as the quantum mechanical momentum operator $\boldsymbol{p} = -i\hbar\nabla$ for example. The detailed meaning of symbols will become obvious from the context. Occasionally one might encounter a hat on top of a symbol chosen in order to emphasize that this symbol denotes an operator. However, a hat on top of a vector may also denote the corresponding unit vector pointing in the direction of the vector, e.g., the position vector might thus be expressed as $\boldsymbol{r} = r\,\hat{\boldsymbol{r}}$.

Most quantities require extensive use of indices. In principle, we may distinguish the following sets of indices:

(i) General coordinates: In relativistic or covariant equations lower-case Latin indices $i, j, k, \ldots$ generally run over the three spatial coordinates, usually denoted as x, y, z or x_1, x_2, x_3. Greek indices $\alpha, \beta, \gamma, \delta, \ldots$ as well as $\mu, \nu, \varrho, \tau, \ldots$ run over the four space-time coordinates ranging from 0 to 3, where $x^0 = ct$ represents the time-like coordinate and x^i the space-like coordinates.

(ii) Particle coordinates in molecular systems: For molecular systems nuclear coordinates are denoted by capital letter indices $I, J, K, \ldots$, while electronic coordinates are labeled by lower-case Latin indices $i, j, k, \ldots$.

(iii) Basis functions: One-electron functions (i.e., orbitals and spinors) are distinguished via Latin indices. We use $i, j, k, \ldots$ for occupied orbitals and $a, b, c, \ldots$ for virtuals. Basis functions used for the representation of the one-electron functions are labelled by Greek indices $\mu, \nu, \kappa, \lambda, \ldots$.

(iv) One-electron quantities are distinguished from those of the total system by lower-case versus capital letters. Examples are the one-electron Hamiltonian h and the many-particle Hamiltonian H or the orbital ψ and the total wave function Ψ. Of course, if a system contains only one electron, this distinction becomes meaningless. Only in these cases may the lower- and upper-case symbols be used synonymously.

In relativistic equations the position of indices (subscripts or superscripts) is determined by the transformation property of the corresponding quantity under Lorentz transformations. In nonrelativistic equations indices will always be chosen as subscripts.

Almost all equations in this book are given in so-called *Gaussian units*, i.e., the dielectric constant of the vacuum is dimensionless with $4\pi\varepsilon_0 = 1$. The whole existence of Gaussian units is based on the (trivial) fact that we are free to choose the proportionality constant in Coulomb's force law to be 1 (Gaussian units) instead of $1/4\pi\epsilon_0$ (SI units). The positive elementary charge e, the electron (rest) mass m_e, and the reduced Planck constant $\hbar = h/2\pi$ are explicitly taken into account. For the sake of convenience, a few equations, however, are given in *Hartree atomic units*, where these fundamental physical constants all adopt a numerical value of one, since in most accounts on quantum chemistry these units are employed. In Hartree atomic units (often abbreviated as 'a.u.'), a charge is then measured in multiples of the elementary charge e and a mass is given in terms of multiples of the electron's rest mass m_e. The speed of light, however, is approximately 137.037 a.u. and depends on experimental measurements [8] (in contrast to the SI unit system where the value of c is fixed). The fine-structure constant, which is $\alpha = e^2/(4\pi\epsilon_0\hbar c)$ in SI units or $\alpha = e^2/(\hbar c)$ in Gaussian units, takes the simple form $\alpha = 1/c$ in Hartree atomic units, so that expansions in terms of the fine-structure constant become equivalent to expansions in $1/c$. It is important to emphasize that the choice of $1/(4\pi\epsilon_0) = 1$ in Hartree atomic units automatically implies that Gaussian units are the basis of this atomic-units system. In the literature, one encounters a third choice of units — often called *natural units* and applied in quantum field theory — where the speed of light c and the reduced Planck constant $\hbar$ are set equal to unity, while the electron's rest mass m_e and the elementary charge e differ from unity. Hence, only those fundamental physical constants which are independent of specific elementary particles are chosen to be unity in this system of units.

The use of SI units bears a catch. This is the fact that the dielectric constant of the vacuum ϵ_0 is connected with the speed of light c via the permittivity μ_0, $\epsilon_0\mu_0 = c^{-2}$. As a consequence, equations containing electromagnetic potentials may carry a c or not depending on the system of units chosen [9]. The important point to understand is that magnetic fields $\boldsymbol{B}$ are suppressed by a factor of $1/c$ when compared to electric fields $\boldsymbol{E}$ in *any* system of units. This fact is most easily seen in Gaussian units rather than SI units, which is the reason why we abandon the latter. Furthermore, in contrast to the SI system of units, only in Gaussian units electric ($\boldsymbol{E}$) and magnetic ($\boldsymbol{B}$) fields, or analogously, scalar (ϕ) and vector ($\boldsymbol{A}$) potentials feature the same units, clearly exhibiting their physical nature and indicating their transformation properties into one another.

Relativistic theories often employ series expansions in powers of $1/c$ in order to study cases of small velocities or the nonrelativistic limit. The physics, however, must not change by the choice of units and we recall that all quantities in, for example, a force law need to be expressed in the same units (e.g., the electrical charges also take different values in different systems of units).

Finally, the logical symbols $\forall$ ('for all') and $\exists$ ('there exists') are employed as abbreviations rather than as precise logical quantors. For truncated series expansions, we use the Landau notation $O(...)$ to denote the leading order in the expansion parameter of the truncated terms.

Part I

FUNDAMENTALS

2
Elements of Classical Mechanics and Electrodynamics

All phenomena of classical nonrelativistic mechanics are solely based on Newton's laws of motion, which are valid in any inertial frame of reference. The natural symmetry operations of classical mechanics are the Galilean transformations, mediating the transition from one inertial coordinate system to another. The fundamental laws of classical mechanics can equally well be formulated applying the elegant Lagrangian and Hamiltonian descriptions based on Hamilton's action principle. Maxwell's equations for electric and magnetic fields are introduced as the basic laws of classical electrodynamics.

2.1
Elementary Newtonian Mechanics

In this chapter basic concepts and formal structures of classical nonrelativistic mechanics and electrodynamics are presented. In this context the term *classical* is used in order to draw the distinction with respect to the corresponding quantum theories which will be dealt with in later chapters of this book. In this chapter we will exclusively cover classical nonrelativistic mechanics or Newtonian mechanics and will thus often apply the word *nonrelativistic* for the sake of brevity. The focus of the discussion is on those aspects and formal structures of the theory which will be modified by the transition to the relativistic formulation in chapter 3.

2.1.1
Newton's Laws of Motion

The subject of classical mechanics is the description of the motion of material bodies under the influence of given forces. All phenomena of classical nonrelativistic mechanics can be deduced from three basic axioms or laws of motion, which were first presented by Sir Isaac Newton in 1687 in his work *Philosophiae Naturalis Principia Mathematica* [10]. In modern language they can be formulated as:

Relativistic Quantum Chemistry. Markus Reiher and Alexander Wolf

ISBN: 978-3-527-31292-4

1. Existence of inertial systems:
   There are inertial systems (IS) or inertial frames of reference, in which the forceless motion of a particle is described by a constant velocity,

$$\dot{\boldsymbol{r}}(t) = \text{const.} \tag{2.1}$$

2. Equation of motion in IS:
   In inertial systems the motion of a single particle under the influence of a resulting force $\boldsymbol{F}$ is described by the equation of motion,

$$\dot{\boldsymbol{p}} = m\ddot{\boldsymbol{r}} = \boldsymbol{F} \tag{2.2}$$

   which implicitly requires a constant (inertial) mass, i.e., $\dot{m}\dot{\boldsymbol{r}} = 0$. The resulting force $\boldsymbol{F}$ is the vectorial superposition of all acting forces: $\boldsymbol{F} = \sum_i \boldsymbol{F}_i$.

3. Principle of *actio equals reactio*:
   For every action force $\boldsymbol{F}_{12}$, with which a particle 1 is acting on another particle 2 or its environment there is an equal and opposite reaction force $\boldsymbol{F}_{21}$, with which the other particle 2 or the environment is acting on particle 1,

$$\boldsymbol{F}_{12} = -\boldsymbol{F}_{21} \tag{2.3}$$

   Furthermore, the internal forces between two particles are always acting along the line connecting them: $\boldsymbol{F}_{12} \times (\boldsymbol{r}_1 - \boldsymbol{r}_2) = 0$.

Newton's first law is often denoted as the principle of inertia. However, the formulation presented above constitutes a much stronger statement. It is also an existence theorem for very special coordinate systems, the inertial systems or inertial frames of reference, in which the laws of nature take a particularly *simple* form. Especially, due to Eq. (2.1) free particles move in straight lines with constant velocity in inertial systems. Here, of course, $\boldsymbol{r}$ denotes the spatial coordinate vector

$$\boldsymbol{r} = x\boldsymbol{e}_x + y\boldsymbol{e}_y + z\boldsymbol{e}_z = \sum_{i=1}^{3} x_i\boldsymbol{e}_i \longrightarrow \begin{pmatrix} x \\ y \\ z \end{pmatrix} = (x_i) = \boldsymbol{r} \tag{2.4}$$

which is represented by the column vector of its *Cartesian* components. We emphasize that the whole presentation of this section with its particularly simple expressions refers only to flat Cartesian frames of reference, where the basis vectors $\boldsymbol{e}_i$ are mutually orthogonal and depend neither on time nor on the position of the particle. They establish a global frame of reference. Experimentally one finds that inertial frames of reference are at rest with respect to the fixed stars background of the universe or move uniformly against it.

In the following we will freely choose any of the representations of $\boldsymbol{r}$ given by Eq. (2.4), where it is assumed that Latin indices range from 1 to 3 if not otherwise stated. In particular, we do not distinguish between a vector and its components with respect to a given basis, but use the same symbol $\boldsymbol{r}$ to denote both the vector itself and the set of its components. For simplicity the components of all vectors are consequently denoted with subscripts throughout this chapter, since in nonrelativistic physics the positioning of indices does not have any significance at all — in contrast to the relativistic situation where it labels the transformation property of the corresponding object under Lorentz transformations as will be discussed in chapter 3.

The second axiom defines both the mass m of a particle and the force $\boldsymbol{F}$ acting upon it as experimental measurands and establishes the dynamical law of Eq. (2.2) between those quantities. For given initial conditions $\boldsymbol{r}(t_1)$ and $\dot{\boldsymbol{r}}(t_1)$ the trajectory $\boldsymbol{r} = \boldsymbol{r}(t)$ of a particle in IS is thereby determined uniquely. In Newtonian mechanics the (inertial) mass m of the particle is assumed to be constant and in particular independent of the velocity

$$\boldsymbol{v}(t) = \dot{\boldsymbol{r}}(t) = \frac{\mathrm{d}\boldsymbol{r}(t)}{\mathrm{d}t} \tag{2.5}$$

of the particle. We may thus always identify m in Eq. (2.2) with the rest mass of the particle (later denoted m_0), i.e., its resistance against acceleration when it is at rest. The particularly simple form of Eq. (2.2) of the equation of motion is only valid for inertial systems. In other, more general frames of reference like rotating or accelerated coordinate systems, the equation of motion is more complicated and contains additional terms due to inertial forces. The quantity

$$\boldsymbol{p} = m\boldsymbol{v} = m\dot{\boldsymbol{r}} \tag{2.6}$$

is called the *linear* or *kinematical* or *mechanical momentum* of the particle, whereas

$$\boldsymbol{l} = \boldsymbol{r} \times \boldsymbol{p} = m\,\boldsymbol{r} \times \dot{\boldsymbol{r}} \tag{2.7}$$

defines the *angular momentum* of the particle. Its ith Cartesian component may conveniently be expressed as

$$l_i = \sum_{j,k=1}^{3} \varepsilon_{ijk} x_j p_k \tag{2.8}$$

where the totally antisymmetric *Levi-Cività* symbol ε_{ijk} is defined by

$$\varepsilon_{ijk} = \left\{ \begin{array}{lll} +1 & & (ijk) \text{ is an even permutation of (123)} \\ -1 & \text{if} & (ijk) \text{ is an odd permutation of (123)} \\ 0 & & \text{else} \end{array} \right\} \tag{2.9}$$

Newton's third law, the principle of *actio equals reactio* of Eq. (2.3) is a general principle valid for all areas of physics. Without the amendment after Eq. (2.3) a closed system consisting of two particles with fixed distance could arbitrarily increase its total angular momentum by accelerating its rotational motion around its center of mass by itself, which has never been observed.

2.1.2
Galilean Transformations

2.1.2.1 Relativity Principle of Galilei

The Newtonian laws of motion have been formulated for inertial frames of reference only, but no special IS has been singled out so far, since classical nonrelativistic mechanics relies on the *Galilean principle of relativity*:

1. All inertial frames are equivalent.

2. Newton's laws of motion have the same form in all inertial frames.

These two brief statements need some explanation. The term *equivalent* in the first statement means that no IS has to be favored over all other IS. In particular no experiment may in principle be performed which singles out one specific IS. As a consequence, an observer can freely choose any IS for the description of a physical *event* E

$$\text{event E}: \quad (t, \boldsymbol{r}) \;=\; (t, \boldsymbol{r}(t)) \;=\; \big(t, x_i(t)\big) \tag{2.10}$$

which may be the collision of two particles or the emission of a photon by an excited atom, for example. As soon as a coordinate system IS has been chosen, each event E is completely determined by the specification of the time coordinate t and the three spatial coordinates $\boldsymbol{r}$. In this context one should be aware of the meaning of *time* in classical nonrelativistic physics: it is simply an independent parameter describing the propagation of the particle in three-dimensional configuration space $\mathbb{R}^3$, i.e., its trajectory $\boldsymbol{r} = \boldsymbol{r}(t)$. A trajectory in IS has to be interpreted as a sequence of events.

The very same physical event E may also be described with reference to another Cartesian inertial frame IS′ yielding the coordinates $(t', \boldsymbol{r}')$. The coordinates of E for both inertial frames IS and IS′ are related to each other by a so-called *Galilei transformation*,

$$\big(t, x_i\big) \text{ in IS} \;\longrightarrow\; \big(t', x_i'\big) \text{ in IS}' \tag{2.11}$$

Here the viewpoint of *passive* transformations has been taken, i.e., one physical event or system is described with reference to two different inertial frames IS and IS′, cf. Figure 2.1.

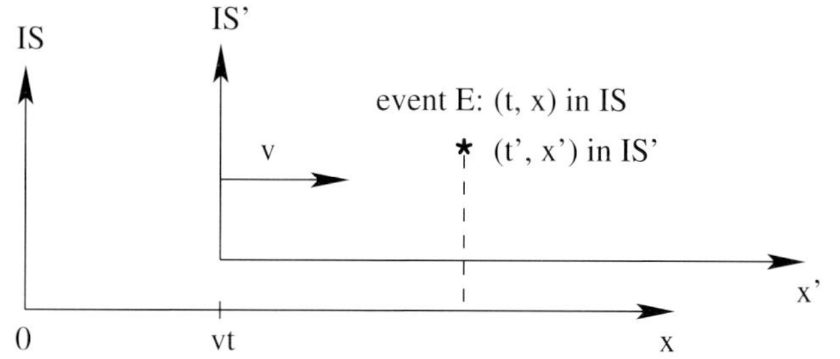

Figure 2.1 Illustration of a Galilei boost in x-direction. For $t = t' = 0$ the two inertial frames IS and IS′ coincide, and for later times the coordinates of the event E with respect to IS and IS′ are related to each other according to Eq. (2.15).

After these preliminaries we can discuss the second statement of Galilei's relativity principle. It states that the *form* of Newton's laws does not change under Galilean transformations relating two inertial frames, i.e., Newton's laws are supposed to be *covariant* or *invariant in form* under arbitrary Galilean transformations. In other words, Newton's laws are valid in all inertial frames. They are, according to the Galilean principle of relativity, *relativistic* and thus fundamental laws of nature. In non-inertial frames, e.g., in rotating or accelerated frames of reference, Newton's laws will have a different form since they are modified by centrifugal or inertia terms. We note in passing that the relativity principle is, of course, only valid for non-dissipative forces excluding friction, since otherwise there would be a special IS, where the particle is at rest and no friction occurs.

2.1.2.2 **General Galilean Transformations and Boosts**

We now exploit the relativity (in the sense of Galilei) of Newton's laws for the determination of the most general Galilean transformations. Newtonian physics crucially relies on the concept of *absolute time*: the time difference $\mathrm{d}t$ between two events is the same in all inertial frames. The time shown by a clock is in particular *independent* of the state of motion of the clock. As a consequence the most general relation between the time t in IS and the time t' in IS′ is given by

$$t' = t - t_0 \qquad \Longleftrightarrow \qquad \mathrm{d}t' = \mathrm{d}t \tag{2.12}$$

where t_0 describes a constant shift of the origin of the time axes, which does not have any influence on Newton's laws. Furthermore, since Newton's laws contain only second derivatives of the trajectory $\boldsymbol{r} = \boldsymbol{r}(t)$, the spatial coordinates of event E in IS and IS′ are only related via transformations like

$$x'_i = \sum_{j=1}^{3} R_{ij} x_j - v_i t - a_i \qquad \Longleftrightarrow \qquad \boldsymbol{r}' = R\boldsymbol{r} - \boldsymbol{v}t - \boldsymbol{a} \tag{2.13}$$

A more general coordinate transformation relating two inertial frames, e.g., a transformation which contains quadratic terms, would not leave the form of Newton's laws invariant. Equations (2.12) and (2.13) establish thus the most general Galilean transformation, which contains a constant temporal shift described by t_0, constant spatial shifts described by a_i and uniform spatial shifts described by $v_i t$, and a constant spatial rotation described by the orthogonal transformation R. If R were not orthogonal, the transition to another inertial frame of reference would change the length of objects, or more generally, the distance between events. The invariance of spatial distances, however, is the most important and defining property of Galilean transformations, as will be discussed below. Since orthogonal matrices with $RR^T = \mathbf{1}$ are completely specified by three parameters, which may be chosen as the Euler angles for example, the general Galilean transformation depends on ten independent parameters, which are constants that do *not* depend on the space and time coordinates. The set of all Galilean transformations represents a non-Abelian 10-parameter group which is called the *Galilei group*.

As a direct consequence of Eq. (2.13) we find the general Galilean formula for the transformation of velocities,

$$\dot{\boldsymbol{r}}' = R\dot{\boldsymbol{r}} - \boldsymbol{v} \tag{2.14}$$

i.e., a rotation and the familiar addition of velocities.

The discussion in this book will mostly be restricted to very special Galilean transformations for which the relative velocity v in x-direction between IS and IS′ is the only parameter, cf. Figure 2.1. This transformation is the *Galilei boost* in x-direction and given by

$$x' = x - vt\,, \qquad y' = y\,, \qquad z' = z\,, \text{ and } \quad t' = t \tag{2.15}$$

It is accompanied by the simple Galilean formula for the addition of velocities, $\dot{x}' = \dot{x} - v$, which is a special case of Eq. (2.14) for vanishing rotation of the two inertial frames.

2.1.2.3 Galilei Covariance of Newton's Laws

Covariance of equations shall denote their invariance in form under coordinate transformations. Covariance of Newton's first law, $\ddot{x}_i = 0$ $(i = 1, 2, 3)$, under general Galilean transformations given by Eqs. (2.12) and (2.13) is obvious, since

$$\ddot{x}'_i(t') = \frac{d^2 x'_i}{dt'^2} \overset{(2.12)}{=} \frac{d^2 x'_i}{dt^2} \overset{(2.13)}{=} \sum_{j=1}^{3} R_{ij}\ddot{x}_j(t) \tag{2.16}$$

Newton's first law is thus valid in one specific IS if and only if it is valid in all other inertial frames related to IS via Galilean transformations.

Newton's second law in IS is in general given by

$$m\ddot{x}_i(t) = F_i(\boldsymbol{r}, \dot{\boldsymbol{r}}, t) \tag{2.17}$$

where the force F_i might also depend on the velocity $\dot{\boldsymbol{r}}$ of the particle (like in cases where frictional or Lorentz forces are present). Note that this setting also includes the ordinary cases where the force does not depend on the particle's velocity or where it is even constant in all space.

If we rename i as j, multiply Eqs. (2.17) by R_{ij}, and sum over j, we arrive at Newton's second law in IS$'$ by virtue of Eq. (2.16),

$$m\ddot{x}'_i(t') = \sum_{j=1}^{3} R_{ij} F_j(\boldsymbol{r}, \dot{\boldsymbol{r}}, t) \equiv F'_i(\boldsymbol{r}', \dot{\boldsymbol{r}}', t') \tag{2.18}$$

Equations (2.17) and (2.18) have exactly the same form, i.e., Newton's equation of motion is indeed covariant under Galilean transformations. These two equations describe the same physical situation with respect to two different inertial frames of reference. Though the physical vectorial force is of course the same in both frames of reference, $\boldsymbol{F} = \boldsymbol{F}'$, its components F_i and F'_i are in general different functions of their arguments. This relationship is given by the second equality of Eq. (2.18).

2.1.2.4 **Scalars, Vectors, and Tensors in Three-Dimensional Space**

Newtonian mechanics is set up in flat and Euclidean three-dimensional space $\mathbb{R}^3$ equipped with a linear time axis. The most general symmetry transformations of this theory, the Galilean transformations, are identified as those which leave the *length* of objects, or more generally the spatial distance

$$r_{12} = |\boldsymbol{r}_2 - \boldsymbol{r}_1| = \sqrt{(\boldsymbol{r}_2 - \boldsymbol{r}_1)^2} \tag{2.19}$$

between two events E_1 and E_2 invariant, $r_{12} = r'_{12}$. Here both events have to be described with reference to the same inertial system, which may be either IS or IS$'$. In this sense also space (or, more precisely, spatial distances) have an *absolute* meaning in Newtonian mechanics. As a consequence of Eq. (2.13) the mixing of different components of the position vector $\boldsymbol{r}$ under Galilean transformations is only due to rotations. We will thus focus on the rotations R only in order to define scalars, vectors, and tensors by their transformation properties under Galilean transformations. Rotations in $\mathbb{R}^3$ are described by orthogonal matrices

$$R = (R_{ij}) = \begin{pmatrix} R_{11} & R_{12} & R_{13} \\ R_{21} & R_{22} & R_{23} \\ R_{31} & R_{32} & R_{33} \end{pmatrix} \tag{2.20}$$

For example, a rotation by an angle ϑ about the z-axis is then given by

$$R_z(\vartheta) = \begin{pmatrix} \cos\vartheta & -\sin\vartheta & 0 \\ \sin\vartheta & \cos\vartheta & 0 \\ 0 & 0 & 1 \end{pmatrix} \tag{2.21}$$

where the rotational direction is in accordance with the so-called "right-hand rule", cf. Figure 2.2 in section 2.2.1.

The entity of all nine (real-valued) components R_{ij} defines the rotation R, where the first or left index (i) labels the row and the second or right index (j) labels the column of the corresponding entry of the matrix. Orthogonal matrices are defined by the requirement that the inverse transformation is given by the transposed matrix,

$$R^{-1} = R^T \tag{2.22}$$

which is identical to the conditions

$$RR^T = R^T R = \mathbf{1} \iff \sum_{k=1}^{3} R_{ik}(R^T)_{kj} = \sum_{k=1}^{3} R_{ik}R_{jk} = \sum_{k=1}^{3} R_{ki}R_{kj} = \delta_{ij} \tag{2.23}$$

and immediately implies that $\det R = \pm 1$. We will mainly focus on proper rotations with $\det R = +1$ in this book and do not concern ourselves with the inversions with $\det R = -1$. The set of all proper rotations R constitutes the non-Abelian special orthogonal group or rotation group of dimension three, $SO(3)$. The entries δ_{ij} of the symmetric three-dimensional Kronecker symbol or unity operator or identity operator are defined by

$$\delta = (\delta_{ij}) = \begin{pmatrix} 1 & 0 & 0 \\ 0 & 1 & 0 \\ 0 & 0 & 1 \end{pmatrix} = \mathbf{1} = id \tag{2.24}$$

and the components of the inverse rotation R^{-1} are given by

$$\left(R^{-1}\right)_{ij} = \left(R^T\right)_{ij} = R_{ji} \tag{2.25}$$

The squared distance between two infinitesimal neighboring events may thus be expressed as

$$\mathrm{d}r^2 = \sum_{i,j=1}^{3} \delta_{ij}\mathrm{d}x_i\mathrm{d}x_j = \mathrm{d}x^2 + \mathrm{d}y^2 + \mathrm{d}z^2 \tag{2.26}$$

We are now in the position to precisely define scalars, vectors, and tensors under Galilean transformations. Each quantity, which features the same transformation property like the position vector x_i,

$$x_i' = \sum_{j=1}^{3} R_{ij}x_j \qquad \iff \qquad x_j = \sum_{i=1}^{3} R_{ij}x_i' \tag{2.27}$$

is called a *Galilei vector* or just a vector. Thus, according to Eq. (2.18) the force F_i is also a vector. Furthermore, since due to Eq. (2.27) the components of R may be expressed as derivatives of the new coordinates with respect to the old ones or vice versa,

$$R_{ij} = \frac{\partial x'_i}{\partial x_j} = \frac{\partial x_j}{\partial x'_i} \tag{2.28}$$

also the gradient ∂_i is a vector,

$$\partial'_i = \frac{\partial}{\partial x'_i} = \sum_{j=1}^{3} \frac{\partial x_j}{\partial x'_i} \frac{\partial}{\partial x_j} \overset{(2.28)}{=} R_{ij} \partial_j \tag{2.29}$$

Thus, in three-dimensional Euclidean space, coordinates and derivatives with respect to those coordinates, i.e., gradients, as well as all other vectors transform in exactly the same way. The origin of this simple transformation behavior is solely rooted in the relation of Eq. (2.22), which makes redundant the need for a distinction between different types of vectors in nonrelativistic theories. This is no longer true for a relativistic regime where one has to distinguish between vectors and co-vectors, cf. chapter 3.

Any n-index quantity T with 3^n components $T_{i_1 i_2 \ldots i_n}$ is called a tensor of the nth rank if its components transform like an n-fold product of vectors under Galilean transformations,

$$T'_{i_1 i_2 \ldots i_n} = \sum_{j_1=1}^{3} \sum_{j_2=1}^{3} \cdots \sum_{j_n=1}^{3} R_{i_1 j_1} R_{i_2 j_2} \cdot \cdots \cdot R_{i_n j_n} T_{j_1 j_2 \ldots j_n} \tag{2.30}$$

According to this definition, a tensor of the first rank is simply a vector. As examples of second rank tensors within classical mechanics one might think of the inertia tensor $\theta = (\theta_{ij})$ describing the rotational motion of a rigid body, or the unity tensor δ defined by Eq. (2.24). A tensor of the second rank can always be expressed as a matrix. Note, however, that not each matrix is a tensor. Any tensor is uniquely defined within *one* given inertial system IS, and its components may be transformed to another coordinate system IS$'$. This transition to another coordinate system is described by orthogonal transformation matrices R, which are thus *not* tensors at all but mediate the change of coordinates. The matrices R are not defined with respect to one specific IS, but relate two inertial systems IS and IS$'$.

Quantities without any indices like the mass m, which are not only covariant but invariant under Galilean transformations, are called *Galilei scalars* or zero rank tensors. They have exactly the same value in all inertial frames of reference.

2.1.3
Basic Conservation Laws for One Particle in Three Dimensions

As a direct consequence of Newton's laws one finds the elementary conservation laws for linear and angular momentum and energy. Consider again one particle with mass m in three-dimensional space. If the particle is moving freely, i.e., the resulting force acting on it vanishes, $\boldsymbol{F} = 0$, the linear momentum $\boldsymbol{p}$ is conserved due to Eq. (2.2).

Vectorial multiplication of Eq. (2.2) with $\boldsymbol{r}$ from the left yields

$$\boldsymbol{r} \times m\ddot{\boldsymbol{r}} \;=\; \boldsymbol{r} \times \dot{\boldsymbol{p}} \;=\; \boldsymbol{r} \times \boldsymbol{F} \;\equiv\; \boldsymbol{M} \tag{2.31}$$

with $\boldsymbol{M}$ being the *torque* or *moment of force*. Due to the identity

$$\dot{\boldsymbol{l}} \overset{(2.7)}{=} \frac{\mathrm{d}}{\mathrm{d}t}(\boldsymbol{r} \times \boldsymbol{p}) \;=\; \dot{\boldsymbol{r}} \times \boldsymbol{p} + \boldsymbol{r} \times \dot{\boldsymbol{p}} \overset{(2.6)}{=} \boldsymbol{r} \times \dot{\boldsymbol{p}} \overset{(2.31)}{=} \boldsymbol{M} \tag{2.32}$$

this leads immediately to the conservation theorem for angular momentum: If the total torque $\boldsymbol{M}$ vanishes, angular momentum $\boldsymbol{l}$ will be conserved. In particular this is always guaranteed for central force problems, where $\boldsymbol{F}$ is always parallel to $\boldsymbol{r}$.

For the discussion of energy conservation we first need the general concept of *conservative force*. A force $\boldsymbol{F}$ is said to be conservative if there exists a time-independent scalar potential energy $U = U(\boldsymbol{r})$ such that

$$\boldsymbol{F} \cdot \dot{\boldsymbol{r}} \;=\; -\frac{\mathrm{d}U(\boldsymbol{r})}{\mathrm{d}t} \;=\; -\dot{\boldsymbol{r}} \cdot \mathrm{grad}\, U(\boldsymbol{r}) \;=\; -\dot{\boldsymbol{r}} \cdot \nabla U(\boldsymbol{r}) \tag{2.33}$$

where the multi-dimensional chain rule for $\mathrm{d}U(\boldsymbol{r}(t))/\mathrm{d}t$ has been applied. Eq. (2.33) assumes that there are no dissipative forces like friction. This immediately implies the most general form of a conservative force,

$$\boldsymbol{F} \;=\; \boldsymbol{F}(\boldsymbol{r}, \dot{\boldsymbol{r}}, t) \;=\; -\,\mathrm{grad}\, U(\boldsymbol{r}) + \dot{\boldsymbol{r}} \times \boldsymbol{Y}(\boldsymbol{r}, t) \tag{2.34}$$

with $\boldsymbol{Y}$ being an arbitrary vector field. In order to understand that the solution given by Eq. (2.34) indeed fulfills Eq. (2.33), we note that the second term of the right hand side of Eq. (2.34) yields a vanishing scalar product with $\dot{\boldsymbol{r}}$ as the resulting vector of the vector product $\dot{\boldsymbol{r}} \times \boldsymbol{Y}(\boldsymbol{r}, t)$ is perpendicular to $\dot{\boldsymbol{r}}$.

The most prominent example of such a velocity-dependent conservative force is the Lorentz force of electrodynamics, cf. Eq. (2.102) in section 2.4. After multiplication by $\dot{\boldsymbol{r}}$, Newton's second law in Eq. (2.2) yields for conservative forces

$$\frac{\mathrm{d}}{\mathrm{d}t}\left(\frac{1}{2} m \dot{\boldsymbol{r}}^2\right) \;=\; \boldsymbol{F} \cdot \dot{\boldsymbol{r}} \overset{(2.33)}{=} -\frac{\mathrm{d}U(\boldsymbol{r})}{\mathrm{d}t} \tag{2.35}$$

If we *define* the total energy E of a particle as

$$E \;=\; \tfrac{1}{2} m \dot{\boldsymbol{r}}^2 + U(\boldsymbol{r}) \;=\; T(\dot{\boldsymbol{r}}) + U(\boldsymbol{r}) \tag{2.36}$$

the conservation theorem for the energy thus reads: If all forces acting on a particle are conservative in the sense of Eq. (2.34), the total energy E will be conserved. Conservative forces do thus conserve the energy, which is the origin of their name. The quantities T and U are called the *kinetic* and *potential* energy of the particle, respectively, and both feature the dimension of an energy. Nevertheless, U is also often simply called the potential.

In the absence of velocity-dependent potentials any conservative force defines a vector field without vortices,

$$\boldsymbol{F}(\boldsymbol{r}) = -\mathrm{grad}\, U(\boldsymbol{r}) \quad \Longleftrightarrow \quad \mathrm{curl}\, \boldsymbol{F}(\boldsymbol{r}) = \nabla \times \boldsymbol{F}(\boldsymbol{r}) = 0 \tag{2.37}$$

From Stokes' theorem (cf. appendix A.1.3) the work W done around closed orbits ∂A vanishes for those conservative forces

$$W = -\oint_{\partial A} \boldsymbol{F} \cdot \mathrm{d}\boldsymbol{r} \stackrel{(A.20)}{=} -\int_A (\mathrm{curl}\, \boldsymbol{F}) \cdot \mathrm{d}\boldsymbol{\sigma} \stackrel{(2.37)}{=} 0 \tag{2.38}$$

where ∂A denotes the boundary of the plane A and $\boldsymbol{\sigma}$ is the outer normal vector perpendicular to the plane.

2.1.4
Collection of N Particles

All results of classical mechanics of 1 particle in 3 dimensions presented so far do directly transform to the situation of N particles. For a system of N particles in 3 dimensions the total kinematical momentum $\boldsymbol{P}$ is given as the vectorial sum of all N one-particle momenta,

$$\boldsymbol{P} = \sum_{i=1}^{N} \boldsymbol{p}_i = \sum_{i=1}^{N} m_i \dot{\boldsymbol{r}}_i \tag{2.39}$$

Accordingly, the total angular momentum is given by

$$\boldsymbol{L} = \sum_{i=1}^{N} \boldsymbol{l}_i = \sum_{i=1}^{N} \boldsymbol{r}_i \times \boldsymbol{p}_i \tag{2.40}$$

In the case of non-interacting particles, the (conservative) force acting on particle number i ($i = 1, \ldots, N$) is given as the gradient of the potential energy with respect to the coordinates of the particle,

$$\boldsymbol{F}_i = -\mathrm{grad}_i\, U(\boldsymbol{r}_1, \ldots, \boldsymbol{r}_i, \ldots, \boldsymbol{r}_N) \tag{2.41}$$

It is important to realize that the potential energy U is still a scalar, i.e., real-valued function, but depends on all $3N$ Cartesian coordinates of the N-particle system.

2.2 Lagrangian Formulation

All aspects of Newtonian mechanics can equally well be formulated within the more general Lagrangian framework based on a single scalar function, the Lagrangian. These formal developments are essential prerequisites for the later discussion of relativistic mechanics and relativistic quantum field theories. As a matter of fact the importance of the Lagrangian formalism for contemporary physics cannot be overestimated as it has strongly contributed to the development of every branch of modern theoretical physics. We will thus briefly discuss its most central formal aspects within the framework of classical Newtonian mechanics.

2.2.1 Generalized Coordinates and Constraints

For many situations of interest elementary Newtonian mechanics is not directly applicable, since the system might be subject to a set of constraints. These constraints and the corresponding *a priori* unknown forces of constraint hamper both the setting up and the solution of the Newtonian equations of motion. For example, consider a planar pendulum of mass m with fixed length l oscillating in the two-dimensional xy-plane as sketched in Figure 2.2. Experimentally the fixed length of the pendulum may be realized by an iron rod of length l pivoted around the z-axis. At every moment the total force, i.e., the vectorial sum of the gravitational force $\boldsymbol{F}_{\mathrm{g}}$ and the force of constraint $\boldsymbol{F}_{\mathrm{c}}$ directed toward the origin will be tangential to the circle of radius l. This pendulum can, of course, be described using Cartesian coordinates. Since its motion, however, is restricted or constrained to a one-dimensional curve, the Cartesian coordinates $\boldsymbol{r} = (x, y)$ are not independent of each other. They have to satisfy the constraint

$$\boldsymbol{r}^2(t) = x^2(t) + y^2(t) = l^2 \qquad \forall t \tag{2.42}$$

which reduces the number of degrees of freedom by one.

Accordingly, the position of the mass is completely determined by specifying the angle ϑ, which may assume any value and is hence not subject to a constraint. This angular coordinate is much better adapted to the symmetry of the system since it automatically respects the constraint. Unconstrained coordinates of this type are called *generalized coordinates*. This new coordinate ϑ and the Cartesian coordinates x and y are uniquely related to one another according to

$$x(t) = l \sin(\vartheta(t)) \quad \text{and} \quad y(t) = -l \cos(\vartheta(t)) \tag{2.43}$$

Any mechanical system of N particles in $\mathbb{R}^3$, which is subject to R constraints, features in general $f = 3N - R$ degrees of freedom represented by

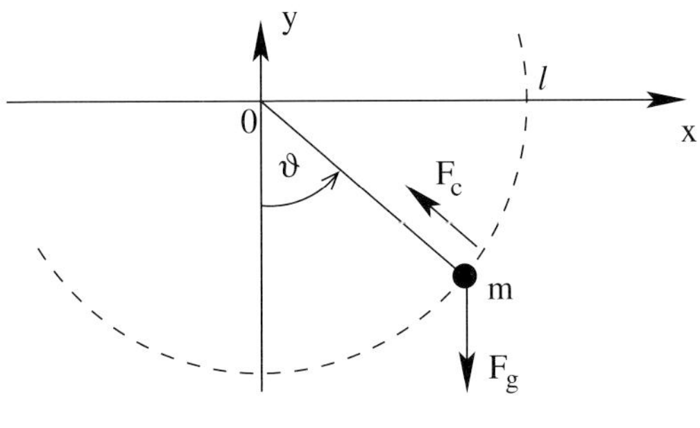

Figure 2.2 Planar pendulum of mass m and fixed length l in the xy-plane. The motion of the mass is restricted to the dashed one-dimensional circle by action of an *a priori* unknown force of constraint F_c, and thus completely described by a single generalized coordinate ϑ.

generalized coordinates $q_1, q_2, \ldots, q_f$ compatible with all constraints. These generalized coordinates may assume any value independently of each other, and the position of all N particles is uniquely determined by the f generalized coordinates,

$$x_i = x_i(q_1, q_2, \ldots, q_f, t) \qquad (i = 1, 2, \ldots, 3N) \tag{2.44}$$

For a given physical system the choice of generalized coordinates is not unique, and often there exists another set of generalized coordinates Q_i $(i = 1, \ldots, f)$ which is equally well suited for the description of the system. It is a matter of personal taste and experience which set to choose. Obviously, generalized coordinates do not necessarily have to feature the dimension of length.

2.2.2
Hamiltonian Principle and Euler–Lagrange Equations

2.2.2.1 Discrete System of Point Particles

As soon as suitable generalized coordinates q have been chosen, the Lagrangian formalism immediately permits the derivation of the equations of motion following a general algorithm based on the Hamiltonian principle. The first step is to express the kinetic and potential energy of the system in terms of the generalized coordinates. This yields the Lagrangian

$$L(q, \dot{q}, t) = T(q, \dot{q}, t) - U(q, \dot{q}, t) \tag{2.45}$$

of the system under investigation. Here the abbreviation q denotes the set of all f generalized coordinates q_i and $\dot{q}$ the set of all corresponding f generalized velocities $\dot{q}_i$. The Lagrangian is a scalar function of the generalized coordinates q, the generalized velocities $\dot{q}$, and the time t, and comprises the complete information about the system. The coordinates q and the velocities $\dot{q}$ are not independent of each other since the latter are just the time derivatives of the former. At each moment in time the instantaneous configuration of the system is thus described by the set of f values for the generalized coordinates q, i.e., by one point in an f-dimensional hyperspace denoted as *configuration space*. As time goes by the system will move along a trajectory in configuration space, which is called the path of motion of the system.

The time integral over the Lagrangian between two fixed times t_1 and t_2 defines the action

$$S = S[q] = \int_{t_1}^{t_2} \mathrm{d}t\, L(q, \dot{q}, t) \tag{2.46}$$

which is a *functional* of the f-dimensional trajectory $q = q(t)$. Each trajectory in configuration space with fixed boundary conditions $q(t_1) = q_1$ and $q(t_2) = q_2$ yields a certain value for the action S. According to the *Hamiltonian principle* the actual trajectory of the system is such that the action is stationary, i.e., S features either an extremum (minimum or maximum) or a stationary point for the actual path of motion of the system. This means, out of all infinitely many possible paths connecting the fixed starting and end points q_1 and q_2 in configuration space, exactly this path $q(t)$ is realized for which the one-dimensional action integral in Eq. (2.46) is stationary with respect to all infinitesimal variations δq of the path,

$$\delta S = \delta S[q] = \delta \int_{t_1}^{t_2} \mathrm{d}t\, L(q, \dot{q}, t) = 0 \qquad \forall \delta q \tag{2.47}$$

Note that only those variations of the path are considered which leave the starting and end points invariant, i.e., there is no variation at the boundaries of the path,

$$\delta q(t_1) = 0 \quad \text{and} \quad \delta q(t_2) = 0 \tag{2.48}$$

This situation is illustrated in Figure 2.3. As in most cases of practical interest the actual path minimizes the action, the Hamiltonian principle defined by Eq. (2.47) is sometimes also denoted as the *principle of least action*. It might be considered as the most fundamental law of nature, valid in all areas of physics, and all other fundamental laws or equations can be deduced from this principle.

We will now explicitly execute these infinitesimal variations

$$q \longrightarrow q + \delta q \quad \Longleftrightarrow \quad q_i \longrightarrow q_i + \delta q_i \qquad \forall i = 1, 2, \ldots, f \tag{2.49}$$

of the path in order to derive the equations of motion the actual path has to satisfy. Applying a Taylor series expansion of the Lagrangian around the

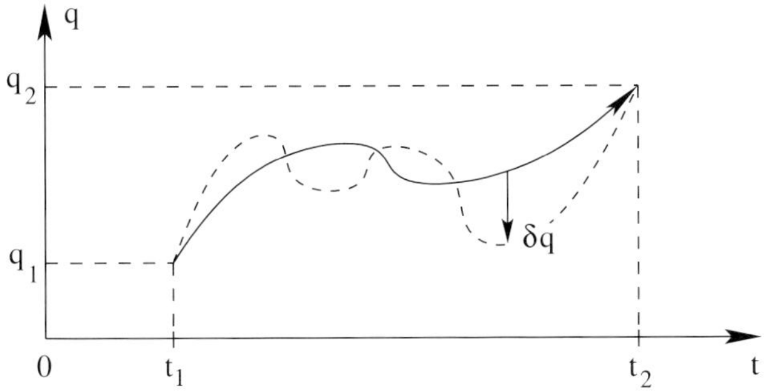

Figure 2.3 Trajectories $q = q(t)$ with fixed boundary conditions $q(t_1) = q_1$ and $q(t_2) = q_2$. The actual path (solid line) is characterized by stationary action, $\delta S = 0$, whereas for any other path connecting the fixed starting and end points (dashed line) the Hamiltonian principle (2.47) is not satisfied.

actual path, the corresponding variation of the action is thus given by

$$\delta S \;=\; S[q+\delta q] - S[q] \;=\; \int_{t_1}^{t_2} \mathrm{d}t \left\{ \sum_{i=1}^{f} \left(\frac{\partial L}{\partial q_i}\delta q_i + \frac{\partial L}{\partial \dot{q}_i}\delta \dot{q}_i \right) \right\} \tag{2.50}$$

As the variation of the velocities is determined by those of the coordinates, $\delta \dot{q}_i = \frac{\mathrm{d}}{\mathrm{d}t}(\delta q_i)$, the variation of the action may be reformulated using integration by parts (p.I.),

$$\begin{aligned} \delta S \;&\overset{\text{(p.I.)}}{=}\; \sum_{i=1}^{f} \left\{ \int_{t_1}^{t_2} \mathrm{d}t \left(\frac{\partial L}{\partial q_i} - \frac{\mathrm{d}}{\mathrm{d}t}\frac{\partial L}{\partial \dot{q}_i} \right) \delta q_i \right\} \;+\; \sum_{i=1}^{f} \frac{\partial L}{\partial \dot{q}_i}\delta q_i \bigg|_{t_1}^{t_2} \\ &=\; \sum_{i=1}^{f} \left\{ \int_{t_1}^{t_2} \mathrm{d}t \left(\frac{\partial L}{\partial q_i} - \frac{\mathrm{d}}{\mathrm{d}t}\frac{\partial L}{\partial \dot{q}_i} \right) \delta q_i \right\} \end{aligned} \tag{2.51}$$

where the surface term has been neglected due to the vanishing variation in Eq. (2.48) at the boundaries. Since all generalized coordinates and thus all infinitesimal variations δq_i are independent of each other, the Hamiltonian principle given by Eq. (2.47) is then exactly satisfied for all variations if and only if the integrand in parenthesis vanishes. This yields the equations of motion or the so-called *Euler–Lagrange equations* for the actual path,

$$\frac{\mathrm{d}}{\mathrm{d}t}\frac{\partial L}{\partial \dot{q}_i} - \frac{\partial L}{\partial q_i} = 0 \qquad \forall i = 1, 2, \ldots, f \tag{2.52}$$

The solution of this set of f second-order differential equations is uniquely determined by $2f$ integrational constants, which may either be incorporated via the fixed boundary conditions of the variational procedure $\big(q(t_1), q(t_2)\big)$

or via the initial conditions $\big(q(t_1), \dot{q}(t_1)\big)$. The quantity

$$p_i = \frac{\partial L}{\partial \dot{q}_i} \tag{2.53}$$

is called the *canonical* or *generalized momentum* conjugated to the generalized coordinate q_i. It must not be confused with the linear momentum given by Eq. (2.6) though, in the absence of vector potentials, both quantities coincide. In order to include also velocity-dependent potentials and electromagnetic forces into the Lagrangian framework, it is convenient to introduce the concept of *generalized forces* defined by

$$G_i = -\frac{\partial U(q, \dot{q}, t)}{\partial q_i} + \frac{\mathrm{d}}{\mathrm{d}t}\frac{\partial U(q, \dot{q}, t)}{\partial \dot{q}_i} \tag{2.54}$$

acting on the ith degree of freedom. G_i might be interpreted as the force in direction i. For velocity-independent potentials this expression reduces to the ith component of the familiar negative gradient of U.

2.2.2.2 **Explicit Example: Planar Pendulum**

In order to illustrate these concepts we will explicitly discuss the planar pendulum of Figure 2.2. In original Cartesian coordinates, which are not generalized coordinates, the Lagrangian is given as

$$L = \tfrac{1}{2} m \left(\dot{x}^2 + \dot{y}^2\right) - mgy. \tag{2.55}$$

Exploiting the generalized coordinate ϑ and Eq. (2.43), the Lagrangian may be expressed in terms of the generalized coordinate only,

$$L = L(\vartheta, \dot{\vartheta}) = \tfrac{1}{2} m l^2 \dot{\vartheta}^2 + mgl \cos\vartheta \tag{2.56}$$

According to Eq. (2.53) the canonical momentum is thus given as $p_\vartheta = ml^2\dot{\vartheta}$, the generalized tangential force is given by $G_\vartheta = -mgl \sin\vartheta$, and the Euler–Lagrange equation (2.52) may be rearranged to yield

$$\ddot{\vartheta} + \frac{g}{l} \sin\vartheta = 0 \tag{2.57}$$

This is the Newtonian equation of motion for a two-dimensional pendulum of fixed length subject to the gravitational field.

2.2.2.3 **Continuous Systems of Fields**

In practical applications the independent variables of the system are often not discrete but described by a set of continuous functions, so-called fields. For simplicity we may assume here that our system is specified by f scalar and real-valued fields $\phi_i = \phi_i(\boldsymbol{r}, t)$ which each depend on space and time. Set-ups of this kind can be found in electrodynamics and the mechanics of liquids and continuous media, for example. In field theory the role of space and time has been modified as compared to the mechanics of discrete particles discussed above. The time coordinate is no longer singled out as simply parametrizing the spatial motion of the particles but is now a coordinate completely equivalent to the spatial ones. This may be seen as a first glance toward the inevitable necessity of relativistic formulations, and, indeed, the paramountcy of field theories for the development of modern physics is inseparably connected to their covariant relativistic structure.

The Lagrangian L is given as the spatial integral over the Lagrangian density $\mathcal{L}$ over all available space V,

$$L = \int_V d^3r\, \mathcal{L} = \int_V d^3r\, \mathcal{L}(\phi, \dot{\phi}, \nabla\phi, \boldsymbol{r}, t) \tag{2.58}$$

where ϕ is again a shorthand notation for the set of all f fields ϕ_i, which shall be the generalized variables of the system. Since these variables are functions of space and time the Lagrangian density depends on both temporal and spatial derivatives of these variables, and in the most general case also explicitly on space and time itself. At each point of space the fields ϕ and their derivatives with respect to time ($\dot{\phi}$) and space ($\nabla\phi$) uniquely determine the local value of the Lagrangian density $\mathcal{L}$. The action is still given by Eq. (2.46) and thus reads

$$S = S[\phi] = \int_{t_1}^{t_2} dt \int_V d^3r\, \mathcal{L}(\phi, \dot{\phi}, \nabla\phi, \boldsymbol{r}, t) \tag{2.59}$$

Varying the generalized variables in exactly the same way as before for discrete systems, the Euler–Lagrange equations of field theory follow from the Hamiltonian principle and read

$$\frac{\partial\mathcal{L}}{\partial\phi_i} - \frac{\partial}{\partial t}\frac{\partial\mathcal{L}}{\partial\dot{\phi}_i} - \nabla\cdot\frac{\partial\mathcal{L}}{\partial(\nabla\phi_i)} = 0 \qquad \forall i = 1, 2, \ldots f \tag{2.60}$$

Obviously, the scalar product between the nabla operator and the partial derivative of $\mathcal{L}$ has to be interpreted as the divergence of the corresponding vector field. These equations of motion determine the actual form of the fields. Their solution is, however, much more involved than for discrete systems. The canonical momentum field conjugate to the field ϕ_i is analogously defined by

$$\pi_i = \frac{\partial\mathcal{L}}{\partial\dot{\phi}_i} \tag{2.61}$$

2.2.3
Symmetries and Conservation Laws

2.2.3.1 Gauge Transformations of the Lagrangian

We have introduced the Lagrangian L as a theoretical quantity that — after a suitable set of generalized coordinates has been chosen — may be used to yield the correct equations of motion. However, the Lagrangian itself has no physical meaning at all and it is *not* an observable in contrast to the kinetic and potential energy, which are available by experiment. The Lagrangian is thus not unique but bears some degree of ambiguity. In this sense all Lagrangians are said to be *equivalent* to each other if they give rise to the same Euler–Lagrange equations. Due to the Hamiltonian principle $\delta S = 0$ this is guaranteed for two Lagrangians L and L', defined both in terms of the same generalized coordinates, if they differ only by a total time derivative of a function F depending on the coordinates and time,

$$L'(q, \dot{q}, t) = L(q, \dot{q}, t) + \frac{\mathrm{d}}{\mathrm{d}t} F(q, t) \tag{2.62}$$

It is immediately verified that L and L' yield the same equations of motion by inserting L' into Eq. (2.52). Transformations of the form of Eq. (2.62) are called *gauge transformations* of the Lagrangian. As an example consider a free particle in three dimensions. The standard Lagrangian in Cartesian coordinates reads

$$L(\boldsymbol{r}, \dot{\boldsymbol{r}}, t) = T(\dot{\boldsymbol{r}}) = \frac{1}{2} m \dot{\boldsymbol{r}}^2 \tag{2.63}$$

and is modified by a general Galilean transformation of the form of Eq. (2.13) to yield

$$L'(\boldsymbol{r}, \dot{\boldsymbol{r}}, t) = \frac{1}{2} m \left(R\dot{\boldsymbol{r}} - \boldsymbol{v} \right)^2 = L(\boldsymbol{r}, \dot{\boldsymbol{r}}, t) + \frac{\mathrm{d}}{\mathrm{d}t} \left(\frac{1}{2} m v^2 t - m R \boldsymbol{r} \cdot \boldsymbol{v} \right) \tag{2.64}$$

where R and $\boldsymbol{v}$ are the orthogonal transformation and the relative velocity of the two systems of reference, respectively, while $\dot{\boldsymbol{r}}$ is the velocity of the particle. Galilean transformations of the coordinates are thus accompanied by gauge transformations of the Lagrangian.

2.2.3.2 Energy and Momentum Conservation

The Lagrangian formalism permits a very general discussion of conservation laws and corresponding constants of motion. Consider a system with f degrees of freedom and corresponding Lagrangian $L = L(q, \dot{q}, t)$. If L does not explicitly depend on the generalized coordinate q_i, i.e., $\partial L / \partial q_i = 0$, q_i is called a *cyclic* coordinate of the problem. Due to the equations of motion in Eq. (2.52) the corresponding canonical momentum p_i is conserved,

$$\frac{\partial L}{\partial q_i} = 0 \quad \Longrightarrow \quad p_i = \text{const.} \tag{2.65}$$

An elementary discussion of energy conservation within the Lagrangian formalism is a little bit more subtle. As it has been indicated by Eq. (2.45) the kinetic energy expressed in terms of the generalized coordinates q does in general depend on these coordinates q, velocities $\dot{q}$, and time t, and so does the potential U. For this general set-up the identity

$$\frac{\mathrm{d}}{\mathrm{d}t}\left(\sum_{i=1}^{f} \frac{\partial L}{\partial \dot{q}_i}\dot{q}_i - L\right) = -\frac{\partial L}{\partial t} \tag{2.66}$$

which is due to the Euler–Lagrange equations (2.52), yields the conservation theorem

$$\frac{\partial L}{\partial t} = 0 \quad \Longrightarrow \quad \sum_{i=1}^{f} \frac{\partial L}{\partial \dot{q}_i}\dot{q}_i - L = \text{const.} \tag{2.67}$$

This general conservation theorem is related to energy conservation as follows: If the constraints do not explicitly depend on time, i.e., the coordinate transformation in Eq. (2.44) does not explicitly depend on time, $\partial x_i/\partial t = 0$, and the potential is static and independent of the generalized velocities, $U = U(q)$, then the expression

$$\sum_{i=1}^{f} \frac{\partial L}{\partial \dot{q}_i}\dot{q}_i - L \stackrel{(B.8)}{=} T(q,\dot{q}) + U(q) \stackrel{(2.36)}{=} E \tag{2.68}$$

equals the energy E of the system, which will later become the Hamiltonian function H. In appendix B a detailed mathematical proof of this relation is presented. For these systems the energy E is thus conserved if the Lagrangian does not explicitly depend on time, $\partial L/\partial t = 0$.

2.2.3.3 General Space–Time Symmetries

The conservation theorems for energy and momentum presented above are special cases of the much more general *Noether theorem* which relates each continuous symmetry transformation of the system with a conserved quantity. It can thus be shown that the energy is conserved if the system is translationally invariant in time, i.e., the Lagrangian L is mapped to an equivalent Lagrangian L' by time translations $t \to t + \varepsilon$. This result holds for any branch of physics: energy conservation is a consequence of the homogeneity of time. Similarly, momentum conservation is related to spatial translational invariance of the system, i.e., it is due to the homogeneity of space. If this homogeneity of space is broken by the presence of forces acting on the system, momentum will not be conserved. The total angular momentum is a constant of the motion if the system is rotationally invariant, i.e., it is due to the isotropy of space. For example this is always guaranteed if the potential only depends on the distance between particles and not on their position in space.

Based on the discussion so far we already understand in the context of classical Newtonian mechanics that the investigation of fundamental symmetries of space, time and fundamental physical equations of motion provides deep insight into physical truth. We will come back to the question of the correct relativistic covariance principle valid for all physics in chapter 3. It may be anticipated here that the study of symmetry and covariance requirements will finally provide the means to construct the basic equations of motion for a freely moving electron in chapter 5 and for an electron in external scalar and vector fields in chapter 6.

2.3 Hamiltonian Mechanics

2.3.1 Hamiltonian Principle and Canonical Equations

2.3.1.1 System of Point Particles

The third equivalent formulation of classical mechanics to be briefly discussed here is the Hamiltonian formalism. Its main practical importance especially for molecular simulations lies in the solution of practical problems for processes that can be adequately described by classical mechanics despite their intrinsically quantum mechanical character (like protein folding processes). However, more important for our purposes here is that it can serve as a useful starting point for the transition to quantum theory. The basic idea of the Hamiltonian formalism is to eliminate the f generalized velocities $\dot{q}_i$ in favor of the canonical momenta p_i defined by Eq. (2.53). This is achieved by a Legendre transformation of the Lagrangian with respect to the velocities,

$$H = H(q,p,t) = \sum_{i=1}^{f} p_i \dot{q}_i(q,p,t) - L\big(q, \dot{q}(q,p,t), t\big) \tag{2.69}$$

H is called the *Hamiltonian* of the system and depends only on the coordinates q, on the canonical momenta p conjugate to these coordinates, and on the time t. The $2f$ variables q and p are also denoted as *canonical variables* of the system. They are treated as being independent of each other within the Hamiltonian framework and specify the state of the system completely. The $2f$-dimensional space spanned by the canonical variables is called the *phase space* or Γ*–space*. Each point in phase space uniquely determines one state of the system, and the time evolution of the system is given as a trajectory in phase space.

Since the $2f$ canonical variables are independent of each other the Hamiltonian principle has to be modified. The variation of the action

$$S \;=\; S[q,p] \;=\; \int_{t_1}^{t_2} \mathrm{d}t\, L \;\stackrel{(2.69)}{=}\; \int_{t_1}^{t_2} \mathrm{d}t \left(\sum_{i=1}^{f} p_i \dot{q}_i - H(q,p,t) \right) \tag{2.70}$$

has now to be stationary with respect to all $2f$ infinitesimal variations of the variables q and p with fixed boundary conditions,

$$\delta q(t_1) \;=\; \delta q(t_2) \;=\; 0 \quad \text{and} \quad \delta p(t_1) \;=\; \delta p(t_2) \;=\; 0 \tag{2.71}$$

Executing this variation, integrating by parts, and reordering the terms yields for the variation of the action

$$\delta S[q,p] \;=\; \sum_{i=1}^{f} \left\{ \int_{t_1}^{t_2} \mathrm{d}t \left[\left(\dot{q}_i - \frac{\partial H}{\partial p_i} \right) \delta p_i \;-\; \left(\dot{p}_i + \frac{\partial H}{\partial q_i} \right) \delta q_i \right] \right\} \tag{2.72}$$

This variation vanishes exactly then *for all* independent variations δq_i and δp_i $(i = 1, \ldots, f)$ if and only if both expressions in parentheses vanish. This condition yields the *Hamiltonian* or *canonical* equations,

$$\dot{q}_i \;=\; \frac{\partial H}{\partial p_i} \quad \text{and} \quad \dot{p}_i \;=\; -\frac{\partial H}{\partial q_i} \qquad \forall i = 1, 2, \ldots, f \tag{2.73}$$

of the system. Equation (2.73) is a set of $2f$ differential first-order equations instead of the set of f second-order differential equations occurring in the Lagrangian formalism. Alternatively, the canonical equations could equally well be derived by calculating the total time derivative of the Hamiltonian H and exploiting the Euler–Lagrange equations (2.52). Taking advantage of the canonical equations (2.73) it is obvious that both the total and the partial time derivative of the Hamiltonian are given by

$$\frac{\mathrm{d}H}{\mathrm{d}t} \;=\; \frac{\partial H}{\partial t} \;=\; -\frac{\partial L}{\partial t} \tag{2.74}$$

Hence the Hamiltonian is conserved if it does not explicitly depend on time. Furthermore, if the constraints on the Cartesian coordinates do not explicitly depend on time and the potential is independent of the velocities the Hamiltonian function equals the energy,

$$H = T + U \tag{2.75}$$

as already seen in Eq. (2.68).

2.3.1.2 Continuous System of Fields

Similarly to the Lagrangian formulation the Hamiltonian for continuous systems described by fields is given as the spatial integral over the Hamiltonian density $\mathcal{H}$,

$$\mathcal{H} = \mathcal{H}(\phi, \nabla\phi, \pi, \boldsymbol{r}, t) = \sum_{i=1}^{f} \pi_i \dot{\phi}_i - \mathcal{L}(\phi, \dot{\phi}, \nabla\phi, \boldsymbol{r}, t) \tag{2.76}$$

which is the Legendre transform of $\mathcal{L}$ with respect to $\dot{\phi}$. Variation of the independent variables ϕ and π and application of the Hamiltonian principle $\delta S[\phi, \pi] = 0$ gives the canonical equations for field theory,

$$\dot{\phi}_i = \frac{\partial \mathcal{H}}{\partial \pi_i} \quad \text{and} \quad \dot{\pi}_i = -\frac{\partial \mathcal{H}}{\partial \phi_i} + \nabla \cdot \frac{\partial \mathcal{H}}{\partial (\nabla \phi_i)} \qquad \forall i = 1, 2, \ldots, f \tag{2.77}$$

2.3.2 Poisson Brackets and Conservation Laws

A very important object of the Hamiltonian formalism for both the investigation of conservation theorems and the transition towards quantum theory are the Poisson brackets. Consider two arbitrary *phase space functions* $u = u(q, p, t)$ and $v = v(q, p, t)$ depending only on the variables of phase space q and p and the time t. The Poisson bracket between u and v is defined as

$$\{u, v\} = \{u, v\}_{q,p} = \sum_{i=1}^{f} \left(\frac{\partial u}{\partial q_i} \frac{\partial v}{\partial p_i} - \frac{\partial u}{\partial p_i} \frac{\partial v}{\partial q_i} \right) \tag{2.78}$$

It is always defined with respect to the given set of canonical variables which is sometimes symbolized by suitable indices. The result, i.e., the Poisson bracket itself is, in general, another phase space function. In anticipation of the next section 2.3.3, all Poisson brackets are invariant under canonical transformations, i.e., their *value* does not depend on the choice of canonical variables. The indices q and p attached to the Poisson brackets are thus often suppressed.

Since also the canonical variables q and p are phase space functions, we may calculate their Poisson brackets. They are given by

$$\{q_i, q_j\} = 0 \quad , \quad \{p_i, p_j\} = 0 \quad , \quad \{q_i, p_j\} = \delta_{ij} \qquad \forall i, j = 1, 2, \ldots, f \tag{2.79}$$

and are denoted as the fundamental Poisson brackets. We briefly mention some obvious but important algebraic properties of Poisson brackets:

$$\{u, v\} = -\{v, u\} \quad \text{and} \quad \{u, u\} = 0 \tag{2.80}$$

$$\{u, vw\} = \{u, v\}\, w + v\, \{u, w\} \tag{2.81}$$

$$\{u, \{v, w\}\} + \{v, \{w, u\}\} + \{w, \{u, v\}\} = 0 \tag{2.82}$$

The last Eq. (2.82) is called *Jacobi's identity*; it is simple but quite tedious to prove. We note in passing that the Poisson brackets of classical mechanics presented here will be promoted to the commutators of quantum mechanics (scaled by $-\mathrm{i}/\hbar$) by the transition to quantum theory in chapter 4.

Poisson brackets might be employed to cast the time dependence of an arbitrary phase space function $u = u(q, p, t)$ in a more compact form, which might be useful for the exhibition of conservation laws. Since the total time derivative of u is given as

$$\dot{u} = \frac{\mathrm{d}u}{\mathrm{d}t} = \sum_{i=1}^{f} \left(\frac{\partial u}{\partial q_i} \dot{q}_i + \frac{\partial u}{\partial p_i} \dot{p}_i \right) + \frac{\partial u}{\partial t} \tag{2.83}$$

it might be rewritten by exploitation of the canonical equations (2.73) and the definition of Poisson brackets in Eq. (2.78) to yield the important formula

$$\frac{\mathrm{d}u}{\mathrm{d}t} = \{u, H\} + \frac{\partial u}{\partial t} \tag{2.84}$$

Inserting $u = H$ immediately re-derives the left equality of Eq. (2.74), and choosing u as a specific canonical variable gives the symmetric Poisson bracket form of the Hamiltonian equations,

$$\dot{q}_i = \{q_i, H\} \quad \text{and} \quad \dot{p}_i = \{p_i, H\} \qquad \forall i, j = 1, 2, \ldots, f \tag{2.85}$$

since the canonical variables feature vanishing partial derivatives with respect to time. Eq. (2.84) implies the important general conservation theorem of Hamiltonian dynamics: any phase space function $u = u(q, p)$ which does not *explicitly* depend on time is conserved or a constant of the motion if and only if its Poisson bracket with the Hamiltonian vanishes,

$$\{u, H\} = 0 \quad \Longleftrightarrow \quad u \text{ is constant of motion} \tag{2.86}$$

As a consequence of Eq. (2.82) with $w = H$ we mention that the Poisson bracket of two constants of motion u and v is itself a constant of motion. This statement is called *Poisson's theorem* and holds even when the constants of motion u and v do explicitly depend on time.

Finally we consider the angular momentum $\boldsymbol{l}$ for one particle in three dimensions as defined by Eq. (2.8). For Cartesian coordinates it can be directly shown that the components of angular momentum satisfy

$$\{l_i, l_j\} = \sum_{k=1}^{3} \varepsilon_{ijk} l_k \tag{2.87}$$

which is thus true for any canonical coordinate system. Since the Poisson bracket of two canonical momenta always has to vanish according to Eq. (2.79), it immediately follows that any two components of angular momentum cannot simultaneously be canonical momenta. However, $\boldsymbol{l}^2$ and *one* arbitrary component of angular momentum, e.g., l_z, may simultaneously be canonical momenta as

$$\{l^2, l_i\} = 0 \qquad \forall i = 1, 2, 3 \tag{2.88}$$

This result has important consequences for the transition to the quantum theory of systems with rotational symmetry (in our case: of atoms) as we shall see in chapters 6 and 9.

2.3.3
Canonical Transformations

Similarly to the ambiguity in the choice of the generalized coordinates in the Lagrangian formalism, the canonical variables are also not determined uniquely. The actual choice of canonical variables often depends on both the issue to be studied for a given physical system and personal preferences. In Hamiltonian mechanics there is, however, one important difference as compared to the Lagrangian formulation, since there are now $2f$ independent variables, which may be transformed into each other by suitable coordinate transformations. All transformations relating the original coordinates q and p to the new set of $2f$ coordinates, which shall be denoted as Q and P,

$$Q_i = Q_i(q, p, t) \quad \text{and} \quad P_i = P_i(q, p, t) \qquad \forall i = 1, \ldots, f \tag{2.89}$$

are called *canonical transformations* if they leave the form of the canonical equations (2.73) invariant. This defines a much larger class of transformations as compared to the coordinate transformations of the Lagrangian formulation, where only the set of f generalized coordinates may be transformed into another set of f generalized coordinates. In the Hamiltonian framework, however, coordinates and momenta may be transformed into each other, which may be used to exhibit the constants of motion. If (before or after a suitable canonical transformation) all constants of motion are momentum variables p_i, the Hamiltonian will not depend on the corresponding coordinate q_i due to the Hamiltonian equations (2.73). As in the Lagrangian framework such coordinates are called *cyclic coordinates*.

By no means all possible transformations of the form given in Eq. (2.89) are canonical. We close this section by mentioning an effective and simple criterion to test if a given transformation is canonical. Transformation equations that do not involve the time explicitly,

$$Q_i = Q_i(q, p) \quad \text{and} \quad P_i = P_i(q, p) \qquad \forall i = 1, \ldots, f \tag{2.90}$$

define a canonical transformation if and only if

$$\left\{Q_i, P_j\right\}_{q,p} = \delta_{ij} \quad \text{and} \quad \left\{Q_i, Q_j\right\}_{q,p} = \left\{P_i, P_j\right\}_{q,p} = 0 \qquad \forall i = 1, \ldots, f \quad (2.91)$$

i.e., if the new variables themselves satisfy the fundamental Poisson brackets of Eq. (2.79) with respect to the old variables.

2.4
Elementary Electrodynamics

Classical electrodynamics, i.e., Maxwell's unquantized theory for time-dependent electric and magnetic fields is inherently a covariant relativistic theory — in the sense of Einstein and Lorentz not Newton and Galilei — fitting perfectly well to the theory of special relativity as we shall understand in chapter 3. In this section only those basic aspects of elementary electrodynamics will thus be presented in their non-covariant form. These aspects are vital for the understanding of the advanced discussion in chapter 3.

2.4.1
Maxwell's Equations

Maxwell's equations, which were first presented in 1864 and published in 1865 [11], completely describe the classical behavior of electric and magnetic fields and — supported by the Lorentz force law — their interaction with charged particles and currents. In Gaussian units their differential form is given by

$$\text{div}\, \boldsymbol{E}(\boldsymbol{r}, t) = 4\pi\varrho(\boldsymbol{r}, t) \quad (2.92)$$

$$\text{curl}\, \boldsymbol{B}(\boldsymbol{r}, t) - \frac{1}{c}\frac{\partial \boldsymbol{E}(\boldsymbol{r}, t)}{\partial t} = \frac{4\pi}{c}\boldsymbol{j}(\boldsymbol{r}, t) \quad (2.93)$$

$$\text{curl}\, \boldsymbol{E}(\boldsymbol{r}, t) + \frac{1}{c}\frac{\partial \boldsymbol{B}(\boldsymbol{r}, t)}{\partial t} = 0 \quad (2.94)$$

$$\text{div}\, \boldsymbol{B}(\boldsymbol{r}, t) = 0 \quad (2.95)$$

The electric field $\boldsymbol{E}$ and the magnetic field $\boldsymbol{B}$ are 3-vectors depending in general on both space and time, and the constant c is the speed of light in vacuum. The charge density ϱ describes the distribution of charges in the system and may be discrete or continuous. For a system of N point particles with electric charges q_i at positions $\boldsymbol{r}_i$ it is given by

$$\varrho(\boldsymbol{r}, t) = \sum_{i=1}^{N} q_i\, \delta^{(3)}\left(\boldsymbol{r} - \boldsymbol{r}_i(t)\right) \quad (2.96)$$

The total charge Q_V of the system inside the volume V is given by the spatial integral

$$Q_V(t) \;=\; \int_V \mathrm{d}^3r\,\varrho(\boldsymbol{r},t) \;=\; \sum_{i=1}^{N} q_i \tag{2.97}$$

where the last equality holds for N point charges inside the volume V, of course. Similarly, the local current density $\boldsymbol{j}$ generated by this system of N (moving) point charges is the 3-vector given by

$$\boldsymbol{j}(\boldsymbol{r},t) \;=\; \sum_{i=1}^{N} q_i\,\boldsymbol{v}_i(t)\delta^{(3)}\left(\boldsymbol{r}-\boldsymbol{r}_i(t)\right) \;=\; \sum_{i=1}^{N} q_i\,\dot{\boldsymbol{r}}_i(t)\delta^{(3)}\left(\boldsymbol{r}-\boldsymbol{r}_i(t)\right) \tag{2.98}$$

and the total current flowing through an area A is given by

$$I_A(t) \;=\; \int_A \boldsymbol{j}(\boldsymbol{r},t)\cdot \mathrm{d}\boldsymbol{\sigma} \tag{2.99}$$

with $\mathrm{d}\boldsymbol{\sigma} = \hat{\boldsymbol{n}}\mathrm{d}\sigma$ being an oriented normal area element of the area A.

The set of four Maxwell equations may be interpreted as follows: Eq. (2.92) is called Gauss' law and identifies charges as the sources of the electric field. Similarly, Ampère's law represented by Eq. (2.93) states that both a current density and a time-varying electric field give rise to a magnetic vortex field. The second term on the left hand side of Eq. (2.93) is called Maxwell's displacement current. Faraday's law of induction in Eq. (2.94) states that a time-varying magnetic field causes an electric vortex field. Finally Eq. (2.95) guarantees the absence of magnetic charges or monopoles. The first two equations are called the inhomogeneous Maxwell equations due to the presence of the inhomogeneities ϱ and $\boldsymbol{j}$ representing the sources of electric and magnetic fields. Consequently, Eqs. (2.94) and (2.95) are called the homogeneous Maxwell equations.

As a direct consequence of the inhomogeneous Maxwell equations the continuity equation

$$\frac{\partial \varrho}{\partial t} + \operatorname{div}\boldsymbol{j} \;=\; 0 \tag{2.100}$$

is obtained which is the differential or local form of charge conservation. It locally identifies the change of the electric charge density as the source of a current density in the adjacent neighborhood. By spatial integration over the volume V and application of Gauss' theorem as given by Eq. (A.19), cf. appendix A.1.3, it can be cast in its integral or global form,

$$\frac{\mathrm{d}Q_V}{\mathrm{d}t} + \oint_{\partial V} \boldsymbol{j}\cdot\mathrm{d}\boldsymbol{\sigma} \;=\; \dot{Q}_V + I_{\partial V} \;=\; 0 \tag{2.101}$$

with $I_{\partial V}$ being the current through the closed surface ∂V.

The electromagnetic field given by $\boldsymbol{E}$ and $\boldsymbol{B}$ acts on *any* charged particle in the system. The electromagnetic force on particle i with charge q_i is given by

$$\boldsymbol{F}_{L,i} = \boldsymbol{F}_L(\boldsymbol{r}_i, t) = q_i\left(\boldsymbol{E}(\boldsymbol{r}_i, t) + \frac{1}{c}\dot{\boldsymbol{r}}_i \times \boldsymbol{B}(\boldsymbol{r}_i, t)\right) \tag{2.102}$$

where $\boldsymbol{r}_i$ is the position vector of the corresponding particle. This force is precisely of the most general form of a conservative force in the sense of Eq. (2.34), since the electric field may always be expressed as the gradient of a potential as we shall see in section 2.4.5. While the force due to the electric field depends only on the *position* of the particle, the force resulting from the magnetic field acts only on *moving* particles and is always perpendicular to both their velocity and the magnetic field (through the vector product). The force given by Eq. (2.102) is called the *Lorentz force*. A derivation of its detailed expression given by Eq. (2.102) will be presented in chapter 3.

2.4.2 Energy and Momentum of the Electromagnetic Field

Consider a system of N point particles with charges q_i interacting with the electromagnetic field. In general both the particles and the fields are carrying energy, momentum and angular momentum. Conservation theorems will thus turn out to hold only for the sum of the particle and field components. We now discuss how these quantities are related to the electric and magnetic fields. A more detailed account can be found in the book by Jackson [12].

2.4.2.1 Energy and Poynting's Theorem

The Lorentz force in Eq. (2.102) acting on the N particles will change the total energy E_{mat} of the matter inside the system by

$$\mathrm{d}E_{\text{mat}} \overset{(2.38)}{=} \sum_{i=1}^{N} \boldsymbol{F}_{L,i} \cdot \mathrm{d}\boldsymbol{r}_i \overset{(2.102)}{=} \sum_{i=1}^{N} q_i \dot{\boldsymbol{r}}_i \cdot \boldsymbol{E}(\boldsymbol{r}_i, t)\, \mathrm{d}t \tag{2.103}$$

That is, electromagnetic energy of the field can be converted into mechanical energy of the particles and vice versa. Rewriting this expression gives

$$\frac{\mathrm{d}E_{\text{mat}}}{\mathrm{d}t} \overset{(2.98)}{=} \int_V \mathrm{d}^3r\, \boldsymbol{j}(\boldsymbol{r}, t) \cdot \boldsymbol{E}(\boldsymbol{r}, t) = \frac{\mathrm{d}}{\mathrm{d}t}\int_V \mathrm{d}^3r\, u_{\text{mat}}(\boldsymbol{r}, t) \tag{2.104}$$

where we have introduced the energy density u_{mat} of the matter in the last step, and V might be any sufficiently large volume to comprise all particles and fields. The essential trick is now to express the quantity $\boldsymbol{j} \cdot \boldsymbol{E}$ by the electromagnetic fields only,

$$\begin{aligned} \boldsymbol{j}\cdot\boldsymbol{E} &\overset{(2.93)}{=} \frac{c}{4\pi}\left(\boldsymbol{E}\cdot\operatorname{curl}\boldsymbol{B} - \frac{1}{c}\boldsymbol{E}\cdot\frac{\partial\boldsymbol{E}}{\partial t}\right) \\ &\overset{(A.14)}{=} \frac{c}{4\pi}\left(-\operatorname{div}(\boldsymbol{E}\times\boldsymbol{B}) + \boldsymbol{B}\cdot\operatorname{curl}\boldsymbol{E} - \frac{1}{2c}\frac{\partial\boldsymbol{E}^2}{\partial t}\right) \\ &\overset{(2.94)}{=} -\frac{c}{4\pi}\operatorname{div}(\boldsymbol{E}\times\boldsymbol{B}) - \frac{1}{8\pi}\frac{\partial}{\partial t}(\boldsymbol{E}^2+\boldsymbol{B}^2) \\ &= -\operatorname{div}\boldsymbol{S} - \frac{\partial u_{\text{em}}}{\partial t} \end{aligned} \tag{2.105}$$

where we have introduced the *energy density* u_{em} of the electromagnetic field,

$$u_{\text{em}} = u_{\text{em}}(\boldsymbol{r},t) = \frac{1}{8\pi}(\boldsymbol{E}^2+\boldsymbol{B}^2) \tag{2.106}$$

and *Poynting's vector* $\boldsymbol{S}$ (i.e., the energy–current density of the field) given by

$$\boldsymbol{S} = \boldsymbol{S}(\boldsymbol{r},t) = \frac{c}{4\pi}(\boldsymbol{E}\times\boldsymbol{B}) \tag{2.107}$$

The important identity in Eq. (2.105) relates the energy density of the electromagnetic field with the currents flowing in the system and is sometimes denoted as *Poynting's theorem.* Inserting these results in Eq. (2.104) and taking the arbitrariness of the volume V into account, we find

$$\frac{\partial}{\partial t}\Big(u_{\text{mat}}(\boldsymbol{r},t) + u_{\text{em}}(\boldsymbol{r},t)\Big) = -\operatorname{div}\boldsymbol{S}(\boldsymbol{r},t) \tag{2.108}$$

Note that the total time derivative acting on the integral in Eq. (2.104) had to be replaced by a partial time derivative since the order of integration and differentiation has been interchanged. Subsequent spatial integration over the volume V and application of Gauss' theorem (cf. appendix A.1.3) yields

$$\frac{\mathrm{d}}{\mathrm{d}t}\Big(E_{\text{mat}} + E_{\text{em}}\Big) = -\oint_{\partial V}\boldsymbol{S}\cdot\mathrm{d}\boldsymbol{\sigma} \tag{2.109}$$

For sufficiently large volumes V the electromagnetic fields are completely embedded in this volume V and no fields are leaving this volume. Hence the right hand side of Eq. (2.109) vanishes and

$$E_{\text{mat}} + E_{\text{em}} \equiv E_{\text{tot}} = \text{const} \tag{2.110}$$

holds. In the absence of external forces the total energy of a system consisting of matter and electromagnetic fields is thus conserved.

2.4.2.2 **Momentum and Maxwell's Stress Tensor**

The investigation of the momentum transfer occurring in the system requires a similar reasoning. According to Eq. (2.2) the change of the total momentum of the particles is given by

$$\begin{aligned}\frac{\mathrm{d}\boldsymbol{P}_{\mathrm{mat}}}{\mathrm{d}t} &= \sum_{i=1}^{N} \dot{\boldsymbol{p}}_i \stackrel{(2.2)}{=} \sum_{i=1}^{N} \boldsymbol{F}_{\mathrm{L},i} \stackrel{(2.102)}{=} \sum_{i=1}^{N} q_i \left(\boldsymbol{E}(\boldsymbol{r}_i, t) + \frac{1}{c} \dot{\boldsymbol{r}}_i \times \boldsymbol{B}(\boldsymbol{r}_i, t) \right) \\ &\underset{(2.98)}{\stackrel{(2.96)}{=}} \int_V \mathrm{d}^3 r \left(\varrho(\boldsymbol{r}, t) \boldsymbol{E}(\boldsymbol{r}, t) + \frac{1}{c} \boldsymbol{j}(\boldsymbol{r}, t) \times \boldsymbol{B}(\boldsymbol{r}, t) \right) \\ &= \frac{\mathrm{d}}{\mathrm{d}t} \int_V \mathrm{d}^3 r \, g_{\mathrm{mat}}(\boldsymbol{r}, t) \end{aligned} \tag{2.111}$$

where the total *momentum density* g_{mat} of the particles has been introduced. Again, expressing the sources ϱ and $\boldsymbol{j}$ on the right hand side of this equation by the electromagnetic fields only, we arrive at

$$\begin{aligned}\varrho \boldsymbol{E} + \frac{1}{c} \boldsymbol{j} \times \boldsymbol{B} &= \frac{1}{4\pi} \Big(\boldsymbol{E} \, (\mathrm{div} \boldsymbol{E}) + \boldsymbol{B} \, (\mathrm{div} \boldsymbol{B}) - \boldsymbol{E} \times (\mathrm{curl} \, \boldsymbol{E}) \\ &\quad - \boldsymbol{B} \times (\mathrm{curl} \, \boldsymbol{B}) \Big) - \frac{\partial}{\partial t} \left(\frac{1}{c^2} \boldsymbol{S} \right) \end{aligned} \tag{2.112}$$

where we have heavily taken advantage of all four Maxwell equations (2.92)–(2.95), the definition of Poynting's vector Eq. (2.107), and the antisymmetry of cross products. With the *momentum density of the electromagnetic field* g_{em},

$$g_{\mathrm{em}}(\boldsymbol{r}, t) = \frac{1}{c^2} \boldsymbol{S} = \frac{1}{4\pi c} (\boldsymbol{E} \times \boldsymbol{B}) \tag{2.113}$$

and introducing *Maxwell's stress tensor* T_{ij} by

$$T_{ij} = \frac{1}{4\pi} \left[E_i E_j + B_i B_j - \tfrac{1}{2} \delta_{ij} (\boldsymbol{E}^2 + \boldsymbol{B}^2) \right] \tag{2.114}$$

we may combine Eqs. (2.111) and (2.112) to obtain

$$\frac{\partial}{\partial t} \left[g_{\mathrm{mat},i}(\boldsymbol{r}, t) + g_{\mathrm{em},i}(\boldsymbol{r}, t) \right] = \sum_{j=1}^{3} \partial_j T_{ij}(\boldsymbol{r}, t) \tag{2.115}$$

This is the local or differential form of conservation of total momentum. Again, spatial integration over a sufficiently large volume V and application of Gauss' theorem (A.19) yields the global conservation theorem for the total momentum of a closed system,

$$\frac{\mathrm{d}}{\mathrm{d}t} \left(P_{\mathrm{mat},i} + P_{\mathrm{em},i} \right) = - \sum_{j=1}^{3} \oint_{\partial V} T_{ij} n_j \mathrm{d}\sigma = 0 \tag{2.116}$$

where n_j is the jth component of a normal vector perpendicular to the surface element $\mathrm{d}\sigma$. As a consequence, $\boldsymbol{P}_{\mathrm{tot}} = \boldsymbol{P}_{\mathrm{mat}} + \boldsymbol{P}_{\mathrm{em}}$ is constant for such systems.

2.4.2.3 **Angular Momentum**

Also for the total angular momentum $\boldsymbol{L}_{\text{tot}} = \boldsymbol{L}_{\text{mat}} + \boldsymbol{L}_{\text{em}}$ a conservation theorem for closed systems may be found by similar considerations. The total angular momentum of the electromagnetic field, $\boldsymbol{L}_{\text{em}}$, is obtained by

$$\boldsymbol{L}_{\text{em}}(t) = \int \mathrm{d}^3 r\, \boldsymbol{l}_{\text{em}}(\boldsymbol{r}, t) \tag{2.117}$$

and the angular momentum density $\boldsymbol{l}_{\text{em}}$ of the electromagnetic field is given by

$$\boldsymbol{l}_{\text{em}} = \boldsymbol{l}_{\text{em}}(\boldsymbol{r}, t) = \boldsymbol{r} \times \boldsymbol{g}_{\text{em}} = \frac{1}{c^2} \boldsymbol{r} \times \boldsymbol{S} \tag{2.118}$$

The close analogy of this relation for fields to the relation of angular momentum and linear momentum for particles, $\boldsymbol{l}_{\text{mat}} = \boldsymbol{l} = \boldsymbol{r} \times \boldsymbol{p}$, is evident.

2.4.3
Plane Electromagnetic Waves in Vacuum

We will now derive the wave equations for electromagnetic fields in vacuum, i.e., in the absence of charges and currents within an infinite volume of space. Taking the curl of Eqs. (2.93) and (2.94), and making use of the vanishing divergences in vacuum and of Eq. (A.16), we obtain the wave equations

$$\Delta \boldsymbol{E}(\boldsymbol{r}, t) = \frac{1}{c^2} \frac{\partial^2 \boldsymbol{E}(\boldsymbol{r}, t)}{\partial t^2} \quad \text{and} \quad \Delta \boldsymbol{B}(\boldsymbol{r}, t) = \frac{1}{c^2} \frac{\partial^2 \boldsymbol{B}(\boldsymbol{r}, t)}{\partial t^2} \tag{2.119}$$

for the propagation of the electric and magnetic field in vacuum. Each Cartesian component of the electric and magnetic field thus satisfies a scalar wave equation with solutions

$$\boldsymbol{E}(\boldsymbol{r}, t) = Re\Big\{ E_0 \exp\big[\mathrm{i}(\boldsymbol{k} \cdot \boldsymbol{r} - \omega t)\big] \Big\}\, \boldsymbol{e}_1 \tag{2.120}$$

$$\boldsymbol{B}(\boldsymbol{r}, t) = Re\Big\{ B_0 \exp\big[\mathrm{i}(\boldsymbol{k} \cdot \boldsymbol{r} - \omega t)\big] \Big\}\, \boldsymbol{e}_2 \tag{2.121}$$

E_0 and B_0 are constant complex amplitudes, and $\boldsymbol{e}_1$ and $\boldsymbol{e}_2$ shall be *constant* unit vectors. As a consequence the electric field vector always points in the direction of $\boldsymbol{e}_1$, and such an electromagnetic wave is said to be *linearly polarized*. By superpositions of two such linearly polarized waves with different phases and amplitudes so-called *elliptically* polarized waves may be constructed. However, we do not need to further discuss this possibility here.

The *wave vector* $\boldsymbol{k}$ defines the direction of propagation of the waves and is related to the angular frequency ω by

$$\omega = \omega(k) = kc \qquad (k = |\boldsymbol{k}| > 0,\ \omega > 0) \tag{2.122}$$

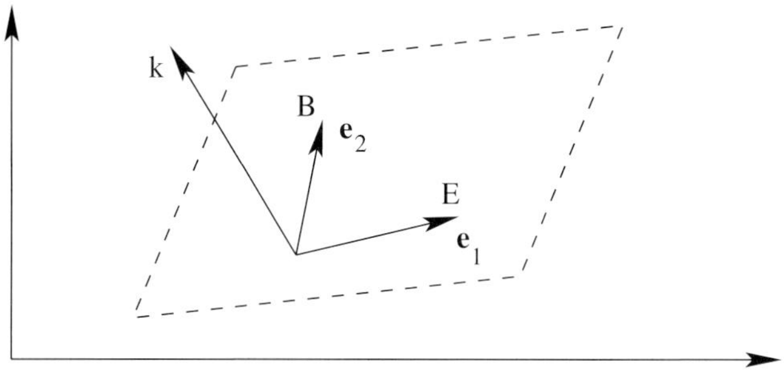

Figure 2.4 Wave vector k and the two perpendicular unit vectors e_1 and e_2 defining a linearly polarized electromagnetic wave.

with the speed of light c being the speed of propagation of the electromagnetic waves. As a consequence of the linear dispersion relation (2.122) electromagnetic waves in vacuum feature no dispersion, i.e., any wave packet will keep its shape as time goes by. The Maxwell equations (2.92) and (2.95) immediately imply that both the electric and the magnetic field are always perpendicular to the direction of propagation given by $\boldsymbol{k}$, $\boldsymbol{e}_1 \cdot \boldsymbol{k} = 0$ and $\boldsymbol{e}_2 \cdot \boldsymbol{k} = 0$. Electromagnetic waves are thus *transverse waves*. Insertion of the solutions given in Eqs. (2.120) and (2.121) into the Maxwell equation (2.94) yields the relation

$$\mathrm{i}\left[kE_0(\hat{\boldsymbol{k}} \times \boldsymbol{e}_1) - \frac{w}{c}B_0\boldsymbol{e}_2\right] \exp\left(\mathrm{i}[\boldsymbol{k} \cdot \boldsymbol{r} - \omega t]\right) = 0 \tag{2.123}$$

which immediately reveals two further constraints to be imposed on the solutions for the electric and magnetic fields,

$$\boldsymbol{e}_2 = \hat{\boldsymbol{k}} \times \boldsymbol{e}_1 \quad \text{and} \quad E_0 = B_0 \tag{2.124}$$

with $\hat{\boldsymbol{k}} = \boldsymbol{k}/k$ being the unit vector in the direction of $\boldsymbol{k}$. This situation is visualized in Figure 2.4.

2.4.4 Potentials and Gauge Symmetry

So far we have compiled some basic properties of electromagnetic fields. Both the electric and magnetic field are physical observables available by experiment. Since each of these fields consists of three Cartesian components — coupled to each other by a set of first-order differential equations — we might be tempted to assume that Maxwell's theory of electromagnetism features six independent dynamical fields. However, this perception would be completely wrong. As a direct consequence of the *homogeneous* Maxwell equations the electric and magnetic fields are completely determined by a set of only four fields, commonly denoted as the *scalar potential* $\phi = \phi(\boldsymbol{r}, t)$ and the *vector potential* $\boldsymbol{A} = \boldsymbol{A}(\boldsymbol{r}, t)$, according to

$$\boldsymbol{E} = -\operatorname{grad}\phi - \frac{1}{c}\frac{\partial \boldsymbol{A}}{\partial t} \quad \text{and} \quad \boldsymbol{B} = \operatorname{curl}\boldsymbol{A} \tag{2.125}$$

After introduction of these potentials the homogeneous Maxwell equations (2.94) and (2.95) are thus identically satisfied due to general properties of differential vector calculus (cf. appendix A.1). The dynamical behavior of the potentials is determined by the inhomogeneous Maxwell equations, which in terms of the potentials read

$$-\Delta\phi - \frac{1}{c}\frac{\partial}{\partial t}(\operatorname{div}\boldsymbol{A}) = 4\pi\varrho \tag{2.126}$$

$$-\Delta\boldsymbol{A} + \frac{1}{c^2}\frac{\partial^2 \boldsymbol{A}}{\partial t^2} + \operatorname{grad}\left(\frac{1}{c}\frac{\partial\phi}{\partial t} + \operatorname{div}\boldsymbol{A}\right) = \frac{4\pi}{c}\boldsymbol{j} \tag{2.127}$$

Equations (2.125)–(2.127) are equivalent in all respects to Maxwell's four original equations in terms of $\boldsymbol{E}$ and $\boldsymbol{B}$.

By inspection of Eq. (2.125) we find that the scalar and vector potentials defining the electric and magnetic fields are not unique but bear a certain degree of arbitrariness. The physical observables, namely the electric and magnetic fields, remain unchanged under so-called *gauge transformations* of the form

$$\phi \longrightarrow \phi' = \phi - \frac{1}{c}\frac{\partial\chi}{\partial t} \tag{2.128}$$

$$\boldsymbol{A} \longrightarrow \boldsymbol{A}' = \boldsymbol{A} + \operatorname{grad}\chi \tag{2.129}$$

where the *gauge function* $\chi = \chi(\boldsymbol{r}, t)$ is any sufficiently smooth real-valued function of space and time. The electric and magnetic fields are thus said to be *gauge invariant*. Since any observable quantity necessarily has to feature gauge invariance it is immediately obvious that the potentials ϕ and $\boldsymbol{A}$ cannot be physical observables. This is strictly true in classical theory but no longer in quantum theory, where the scalar and vector potential might have measurable effects. This is impressively demonstrated by the so-called Aharonov–Bohm effect [13,14], for example. A very good discussion of this effect may be found in Refs. [15] and [16].

Nevertheless, the introduction of the potentials ϕ and $\boldsymbol{A}$ reduces the number of fields, and they are to be considered as the dynamical variables of the theory. Due to the gauge symmetry given by Eqs. (2.128) and (2.129) it is obvious that Maxwell's theory of electromagnetism features redundant degrees of freedom which will seriously hamper its quantization in chapter 7. The potentials ϕ and $\boldsymbol{A}$ themselves are denoted as *gauge potentials*.

Equations (2.126) and (2.127) are the most general form of the inhomogeneous Maxwell's equations for the gauge potentials valid for any chosen gauge. They are complicated coupled second-order differential equations, but they may be simplified by a suitable choice of gauge. Depending on the physical situation and the question to be studied several different gauges may be advantageous. Some prominent choices that can frequently be met in the literature are

$$\text{Lorentz gauge:} \quad \frac{1}{c}\frac{\partial \phi}{\partial t} + \operatorname{div}\boldsymbol{A} = 0 \tag{2.130}$$

$$\text{Coulomb gauge:} \quad \operatorname{div}\boldsymbol{A} = 0 \tag{2.131}$$

$$\text{Temporal gauge:} \quad \phi = 0 \tag{2.132}$$

$$\text{Axial gauge:} \quad A_z = 0 \tag{2.133}$$

For all these choices of gauge it can rather simply be shown that they establish valid constraints on the gauge potentials, i.e., one can always find gauge potentials satisfying the gauge condition. We briefly discuss two gauges here, which will be needed later in this book.

2.4.4.1 **Lorentz Gauge**

In Lorentz gauge given in Eq. (2.130) the inhomogeneous Maxwell equations (2.126) and (2.127) will be decoupled and read

$$-\Delta\phi + \frac{1}{c^2}\frac{\partial^2\phi}{\partial t^2} = 4\pi\varrho \tag{2.134}$$

$$-\Delta\boldsymbol{A} + \frac{1}{c^2}\frac{\partial^2\boldsymbol{A}}{\partial t^2} = \frac{4\pi}{c}\boldsymbol{j} \tag{2.135}$$

They can be conveniently abbreviated as

$$\Box\phi = 4\pi\varrho \quad \text{and} \quad \Box\boldsymbol{A} = \frac{4\pi}{c}\boldsymbol{j} \tag{2.136}$$

where the four-dimensional Laplacian or d'Alembert operator (cf. appendix A.2) has been employed. These equations are also called the *field equations* in Lorentz gauge. Due to the Lorentz gauge condition, at most three of the four gauge potentials are independent of one another. Furthermore, the Lorentz gauge condition does not completely fix the potentials, since *restricted gauge transformations* of the form given in Eqs. (2.128) and (2.129) with

$$\Box\chi = \left(\frac{1}{c^2}\frac{\partial^2}{\partial t^2} - \Delta\right)\chi = 0 \tag{2.137}$$

obviously preserve the Lorentz condition for the potentials. All potentials related to each other by such restricted gauge transformations are said to belong to the Lorentz gauge. The Lorentz gauge is especially useful in the absence of sources since the field equations (2.136) then take the form of homogeneous wave equations.

2.4.4.2 **Coulomb Gauge**

In this book we will mostly employ the Coulomb gauge of Eq. (2.131), which is also called *transverse* or *radiation gauge*. With this choice of gauge Eq. (2.126) is reduced to Poisson's equation

$$\Delta\phi = -4\pi\varrho \tag{2.138}$$

and is thus immediately solved by Poisson's integral

$$\phi(\boldsymbol{r},t) = \int d^3r' \frac{\varrho(\boldsymbol{r}',t)}{|\boldsymbol{r}-\boldsymbol{r}'|} \tag{2.139}$$

i.e., the scalar potential ϕ is given by the *instantaneous* Coulomb potential and thus completely determined by the charge density ϱ. Since the gauge condition itself is a further constraint on the potentials, only two of the four gauge fields are independent dynamical variables, which will give rise to only two transverse photons in the quantized theory (cf. chapter 7).

2.4.4.3 **Retarded Potentials**

From the previous section and, specifically, from inspection of the Poisson integral in Eq. (2.139) it is clear that the scalar potential is transmitted instantaneously; retardation effects do not show up in Coulomb gauge for the scalar potential (however, this is not true for the vector potential). We will later utilize this fact in section 3.5 to derive the retardation-free scalar potential in a special frame of reference in which Coulomb gauge holds.

In Lorentz gauge, however, both scalar and vector potentials explicitly incorporate a retardation term, because in this gauge the Maxwell Eqs. (2.134) and (2.135) have the most general solutions

$$\phi(\boldsymbol{r},t) = \int d^3r' \frac{\varrho(\boldsymbol{r}',t-|\boldsymbol{r}-\boldsymbol{r}'|/c)}{|\boldsymbol{r}-\boldsymbol{r}'|} \tag{2.140}$$

$$\boldsymbol{A}(\boldsymbol{r},t) = \frac{1}{c}\int d^3r' \frac{\boldsymbol{j}(\boldsymbol{r}',t-|\boldsymbol{r}-\boldsymbol{r}'|/c)}{|\boldsymbol{r}-\boldsymbol{r}'|} \tag{2.141}$$

Consequently, we arrive at the important point that it is the time derivative of the potentials in Eqs. (2.134) and (2.135), $\partial^2\phi/\partial t^2$ and $\partial^2\boldsymbol{A}/\partial t^2$, which produces the retardation effect $t - |\boldsymbol{r}-\boldsymbol{r}'|/c$.

2.4.5
Survey of Electro– and Magnetostatics

2.4.5.1 Electrostatics

In the molecular sciences, one often deals with *static* situations where all charges are fixed in space, i.e., the charge density $\varrho = \varrho(\boldsymbol{r})$ is stationary and there are no currents occurring in the system, $\boldsymbol{j} = 0$. Then all magnetic fields will obviously vanish and $\boldsymbol{A} = 0$. The field equations are thus simplified to a large extent and read

$$\text{div}\,\boldsymbol{E}(\boldsymbol{r}) = 4\pi\varrho(\boldsymbol{r}) \quad \text{and} \quad \text{curl}\,\boldsymbol{E}(\boldsymbol{r}) = 0 \tag{2.142}$$

The electric field is thus completely determined by the scalar potential $\phi = \phi(\boldsymbol{r})$ only according to

$$\boldsymbol{E}(\boldsymbol{r}) = -\,\text{grad}\,\phi(\boldsymbol{r}) \tag{2.143}$$

and the potential itself has to satisfy Poisson's equation

$$\Delta\phi(\boldsymbol{r}) = -4\pi\varrho(\boldsymbol{r}) \tag{2.144}$$

Taking advantage of the Green's function of the Laplacian given in appendix A.1.3 we can immediately calculate the scalar potential and electric field due to a given charge distribution ϱ as

$$\phi(\boldsymbol{r}) = \int d^3r' \frac{\varrho(\boldsymbol{r}')}{|\boldsymbol{r}-\boldsymbol{r}'|} \quad \text{and} \quad \boldsymbol{E}(\boldsymbol{r}) = \int d^3r' \varrho(\boldsymbol{r}') \frac{\boldsymbol{r}-\boldsymbol{r}'}{|\boldsymbol{r}-\boldsymbol{r}'|^3} \tag{2.145}$$

The total potential energy U of a test charge distribution $\varrho = \varrho(\boldsymbol{r})$ in an *external* electric field $\boldsymbol{E}_{\text{ext}} = -\nabla\phi_{\text{ext}}$ caused by the external charge distribution ϱ_{ext} is

$$U = \int d^3r\, \varrho(\boldsymbol{r})\, \phi_{\text{ext}}(\boldsymbol{r}) \overset{(2.145)}{=} \int d^3r \int d^3r' \frac{\varrho(\boldsymbol{r})\varrho_{\text{ext}}(\boldsymbol{r}')}{|\boldsymbol{r}-\boldsymbol{r}'|} \tag{2.146}$$

where ϱ will *not* contribute to the external electric field. In the limit of a single point charge q the expression for the potential energy is reduced to $U(\boldsymbol{r}) = q\phi_{\text{ext}}(\boldsymbol{r})$ by virtue of Eq. (2.96).

We now investigate the total potential energy of a given charge distribution, i.e., the work which is necessary to bring given charges together starting from infinity. For n point charges q_i this energy is given by

$$U = \frac{1}{2}\sum_{\substack{i,j=1\\ i\neq j}}^{n} \frac{q_i q_j}{|\boldsymbol{r}_i - \boldsymbol{r}_j|} = \frac{1}{2}{\sum_{i,j=1}^{n}}' \frac{q_i q_j}{|\boldsymbol{r}_i - \boldsymbol{r}_j|} = \sum_{i<j}^{n} \frac{q_i q_j}{|\boldsymbol{r}_i - \boldsymbol{r}_j|} \tag{2.147}$$

where the self-energy ($i = j$) has to be removed in the summation, of course. The prefactor of $1/2$ guarantees that each Coulomb interaction is only considered once and not twice. Alternatively the summation may be restricted to $i < j$. For continuous charge distributions ϱ this potential energy thus reads

$$U = \frac{1}{2}\int d^3r \int d^3r' \,\frac{\varrho(\boldsymbol{r})\varrho(\boldsymbol{r}')}{|\boldsymbol{r}-\boldsymbol{r}'|} \stackrel{(2.145)}{=} \frac{1}{2}\int d^3r\, \varrho(\boldsymbol{r})\phi(\boldsymbol{r}) \tag{2.148}$$

Here ϕ is the potential generated by the charge distribution ϱ itself and not an external potential as before. This explains the factor of $1/2$ as compared to Eq. (2.146). Exploiting the field equation (2.144) and integrating by parts the energy density u_e of a static electric field (cf. Eq. (2.106)) is immediately recovered,

$$U = -\frac{1}{8\pi}\int d^3r\, \phi(\boldsymbol{r})\Delta\phi(\boldsymbol{r}) = \frac{1}{8\pi}\int d^3r\, \boldsymbol{E}(\boldsymbol{r})^2 = \int d^3r\, u_e(\boldsymbol{r}) \tag{2.149}$$

Since the force $\boldsymbol{F}$ with which a given electric field $\boldsymbol{E}$ is acting upon any charge q is given by $\boldsymbol{F} = q\boldsymbol{E}$, Eq. (2.145) yields the familiar electrostatic Coulomb force or Coulomb interaction between charged particles. That is, the Coulomb force with which charge q_1 is acting upon charge q_2 is given by

$$\boldsymbol{F}_{12} = -q_1 q_2 \frac{\boldsymbol{r}_1 - \boldsymbol{r}_2}{|\boldsymbol{r}_1 - \boldsymbol{r}_2|^3} = -q_1 q_2 \frac{\hat{\boldsymbol{r}}_{12}}{r_{12}^2} \tag{2.150}$$

with $\hat{\boldsymbol{r}}_{12}$ denoting the unit vector pointing from $\boldsymbol{r}_2$ to $\boldsymbol{r}_1$.

2.4.5.2 **Magnetostatics**

Phenomena with stationary current density $\boldsymbol{j} = \boldsymbol{j}(\boldsymbol{r})$ and vanishing charge density ϱ are the subject of magnetostatics. For this situation the field equations read

$$\text{curl}\, \boldsymbol{B}(\boldsymbol{r}) = \frac{4\pi}{c}\boldsymbol{j}(\boldsymbol{r}) \quad \text{and} \quad \text{div}\, \boldsymbol{B}(\boldsymbol{r}) = 0 \tag{2.151}$$

That is, the magnetic field $\boldsymbol{B}$ is still given by Eq. (2.125) without any modification, and the continuity equation, Eq. (2.100), takes the form

$$\text{div}\, \boldsymbol{j}(\boldsymbol{r}) = 0 \tag{2.152}$$

Due to Eq. (A.16) the vector potential $\boldsymbol{A}$ has thus to satisfy the condition

$$\Delta\boldsymbol{A}(\boldsymbol{r}) - \text{grad div}\boldsymbol{A}(\boldsymbol{r}) = -\frac{4\pi}{c}\boldsymbol{j}(\boldsymbol{r}) \tag{2.153}$$

which is reduced to a Poisson-type equation only by application of the Coulomb gauge given by Eq. (2.131). In magnetostatics Coulomb gauge is hence the most favorable gauge. With this choice of gauge the vector potential and magnetic field due to a given stationary current density $\boldsymbol{j}$ are given by

$$A(r) = \frac{1}{c}\int d^3r' \frac{j(r')}{|r-r'|} \quad \text{and} \quad B(r) = \frac{1}{c}\int d^3r' j(r') \times \frac{r-r'}{|r-r'|^3} \tag{2.154}$$

The expression for the magnetic field $\boldsymbol{B}(\boldsymbol{r})$ given above is known as the Biot–Savart law. Since the magnetic force acting upon a moving charged particle

$$F = \frac{q}{c}\dot{r} \times B \tag{2.155}$$

cannot be derived as the gradient of a potential, it does not make any sense to define the concept of potential energy in magnetostatics. Finally, for a constant *and* homogeneous magnetic field B, e.g., in the z-direction, $\boldsymbol{B} = B\boldsymbol{e}_z$, Eq. (2.125) may be inverted to yield the vector potential as

$$A = \frac{1}{2} B \times r \tag{2.156}$$

2.4.6
One Classical Particle Subject to Electromagnetic Fields

The equation of motion for a nonrelativistic charged particle with charge q and mass m subject to time-dependent electromagnetic fields will now be derived within the Lagrangian as well as the Hamiltonian formalism. The fields will be treated as *external* fields without any dynamic at this stage. They just establish the forces acting upon the particle and thus govern its motion, but they are not affected by this motion. They do not represent dynamical variables.

Since there are no constraints Cartesian coordinates can be chosen as generalized coordinates to describe the motion of the particle. The electromagnetic fields are completely specified by the scalar potential ϕ and the vector potential $\boldsymbol{A}$, which may both explicitly depend on space and time. At this stage we cannot *derive* the Lagrangian for this system from first principles. Clearly, according to Eq. (2.146) the electrostatic potential energy is given by $q\phi$, but there is no obvious way for the implementation of the vector potential into the Lagrangian formalism at the moment. Thus we just have to guess the interaction term and legitimize it *a posteriori* by analyzing if it yields the correct equation of motion. Later in chapter 3 we will see how the coupling of the particle's motion to the vector potential arises naturally and uniquely from the principle of minimal coupling. We then start with the Lagrangian

$$L(r,\dot{r},t) = T(\dot{r}) - U(r,\dot{r},t) = \frac{1}{2}m\dot{r}^2 - q\phi(r,t) + \frac{q}{c}\dot{r}\cdot A(r,t) \tag{2.157}$$

This is the most prominent and important example of a velocity-dependent potential U. The evaluation of the Euler–Lagrange equations (2.52) together with the relations between electromagnetic fields and potentials in Eq. (2.125) yield

$$m\ddot{x}_i = q\left[E_i + \frac{1}{c}(\dot{\boldsymbol{r}} \times \boldsymbol{B})_i\right] \overset{(2.54)}{=} G_i \tag{2.158}$$

as the equation of motion for the ith component of $\boldsymbol{r}$. We emphasize again the crucial importance of distinguishing correctly between partial and total time derivatives for the evaluation of the Euler–Lagrange equations. The right-hand side of Eq. (2.158) is exactly the generalized force G_i for velocity-dependent potentials and may be identified with the ith component of the familiar *Lorentz force* $\boldsymbol{F}_L$. Summarizing these results the equation of motion can be expressed in 3-vector notation as

$$m\ddot{\boldsymbol{r}} = \boldsymbol{F}_L = q\left(\boldsymbol{E} + \frac{1}{c}\dot{\boldsymbol{r}} \times \boldsymbol{B}\right) \tag{2.159}$$

It is important to realize that for any system including a vector potential canonical and linear momentum do not coincide. According to Eq. (2.53) the canonical momentum $\boldsymbol{p}$ conjugate to $\boldsymbol{r}$ is given by

$$\boldsymbol{p} = m\dot{\boldsymbol{r}} + \frac{q}{c}\boldsymbol{A} \tag{2.160}$$

whereas due to Eq. (2.6) the linear (or kinematical or mechanical) momentum reads

$$\boldsymbol{\pi} = m\dot{\boldsymbol{r}} = \boldsymbol{p} - \frac{q}{c}\boldsymbol{A} \tag{2.161}$$

Note that we have introduced the symbol $\boldsymbol{\pi}$ for linear momentum here in order to better distinguish it from canonical momentum. This notational rigor is only needed for the discussion of this section and will thus be dropped elsewhere in the book. In most cases it will become obvious from the context to which kind of momentum we are referring. Obviously, it is rather the linear than the canonical momentum which is gauge invariant. The canonical momentum $\boldsymbol{p}$ satisfies the fundamental Poisson brackets of Eq. (2.79), of course, whereas the components of linear momentum, interpreted as phase space functions $\pi_i = \pi_i(\boldsymbol{r}, \boldsymbol{p}, t)$, feature nonvanishing but gauge invariant Poisson brackets,

$$\{\pi_i, \pi_j\} = \frac{q}{c}\sum_{k=1}^{3} \varepsilon_{ijk} B_k \tag{2.162}$$

Taking advantage of the canonical momentum we will now derive the Hamiltonian describing this nonrelativistic particle within electromagnetic fields. After expression of all velocities $\dot{x}_i$ by canonical momenta p_i in the

Legendre transformation of Eq. (2.69) and basic algebraic manipulations one arrives at

$$H(\boldsymbol{r},\boldsymbol{p},t) = \frac{1}{2m}\left[\boldsymbol{p} - \frac{q}{c}\boldsymbol{A}(\boldsymbol{r},t)\right]^2 + q\phi(\boldsymbol{r},t) \tag{2.163}$$

The canonical equations thus read

$$\begin{aligned} \dot{p}_i &= -\frac{\partial H}{\partial x_i} = \frac{q}{mc}\sum_{j=1}^{3}\left(p_j - \frac{q}{c}A_j\right)\frac{\partial A_j}{\partial x_i} - q\frac{\partial \phi}{\partial x_i} \\ \dot{x}_i &= \frac{\partial H}{\partial p_i} = \frac{1}{m}\left(p_i - \frac{q}{c}A_i\right) \qquad (i=1,2,3) \end{aligned} \tag{2.164}$$

and are equivalent to the Euler–Lagrange equations, of course. We finally mention that gauge transformations of the electromagnetic potentials correspond to gauge transformations of the Lagrangian L of the type given in Eq. (2.62). For this purpose consider the most general electromagnetic gauge transformation given by Eqs. (2.128) and (2.129) applied to the Lagrangian L defined by Eq. (2.157). After this gauge transformation the Lagrangian reads

$$L'(\boldsymbol{r},\dot{\boldsymbol{r}},t) = L(\boldsymbol{r},\dot{\boldsymbol{r}},t) + \frac{\mathrm{d}}{\mathrm{d}t}\left[\frac{q}{c}\chi(\boldsymbol{r},t)\right] \tag{2.165}$$

i.e., the effect of an electromagnetic gauge transformation on the Lagrangian is to add the total time derivative of the gauge function χ scaled by q/c.

2.4.7
Interaction of Two Moving Charged Particles

Moving charges exert not only time-independent electrostatic forces but also velocity-dependent, i.e., $\dot{\boldsymbol{r}}$-dependent forces as already highlighted by the expression of the Lorentz force in Eq. (2.159). While we have considered the case of a single charged particle moving in (external) $\boldsymbol{E}$ and $\boldsymbol{B}$ fields we should now consider the case where both of these fields are produced by a second moving charge (a situation met when elementary particles like electrons interact). For this, we need to know the scalar and vector potentials produced by the second charge at the position and time of the first charge. While it is straightforward to give expressions for both fields in the rest frame of the second charge, it cannot easily be given for any other frame of reference (like a laboratory frame in which both charges are moving) due to the finite propagation of the electromagnetic fields (i.e., the speed of light is not infinite). In fact, we need first introduce a transformation that allows us to transform scalar and vector potentials from one inertial frame of reference to another one, namely the Lorentz transformation. Therefore, the discussion of the interaction energy of two moving charges is postponed to section 3.5.

Further Reading

R. P. Feynman, R. B. Leighton, M. Sands, [15, 17]. *The Feynman Lectures on Physics I & II.*

Compiled half a century ago these marvelous and timeless two volumes of the Feynman lectures on classical mechanics (Vol. I) and electrodynamics (Vol. II) are still invaluable sources of inspiration for both students and researchers with very different scientific backgrounds. The presentation is rather focused on physical ideas and concepts than on mathematical details and is thus ideally suited as a primer.

H. Goldstein, [18]. *Classical Mechanics.*

Goldstein's classical presentation covers all aspects of classical mechanics on an elementary level suited for undergraduates. Especially the treatment of Hamiltonian mechanics with its various relations to quantum theory is given much space.

L. D. Landau, E. M. Lifshitz, [19]. *Course of Theoretical Physics. Vol. I: Mechanics.*

Starting from the principle of least action and Galileo's principle of relativity this legendary book presents many aspects of classical mechanics in an both elementary and condensed form. Besides a thorough discussion of symmetry the focus is on many classical applications of the mechanics of point particles like scattering, small oscillations, motion of rigid bodies, the Kepler problem and many more. It is a good starting point for a first course in theoretical mechanics.

L. D. Landau, E. M. Lifshitz, [20]. *Course of Theoretical Physics. Vol. II: The Classical Theory of Fields.*

The second volume of the Landau–Lifshitz series on theoretical physics continues directly after volume I and covers the classical theory of fields. It starts with an introduction of Einstein's principle of relativity and a discussion of the special theory of relativity for mechanics. It follows a quite complete presentation of classical electrodynamics including radiation phenomena and scattering of waves of different energy. It concludes with an introduction of gravitational fields, the theory of general relativity and classical relativistic cosmology.

V. I. Arnold, [21]. *Mathematical Methods of Classical Mechanics.*

This book gives a quite formal and mathematically challenging presentation of classical mechanics. It may be hard to follow for the beginner, but very enlightening for a second course in mechanics. The symplectic structure of Hamiltonian mechanics is presented in detail and coordinate-free expressions employing differential forms are given. A very detailed appendix of more than 200 pages explains the mathematical foundations.

J. D. Jackson, [12]. *Classical Electrodynamics.*

Jackson's standard textbook discusses a large variety of topics of classical electrodynamics in a fashion that is both comprehensible for students and interesting for advanced scientists. Almost all material of the book is presented in non-covariant 3-vector notation.

O. Jahn, V. V. Sreedhar, [22]. *The Maximal Invariance Group of Newton's Equations for a Free Point Particle.*

The maximal invariance (symmetry) group for a free point particle in nonrelativistic mechanics is shown to be a 12-parameter group instead of the 10-parameter Galilei group. This elementary but by no means trivial discussion may be of interest for the advanced reader but goes beyond the scope of this book.

3
Concepts of Special Relativity

Einstein's theory of special relativity relying on a modified principle of relativity is presented and the Lorentz transformations are identified as the natural coordinate transformations of physics. This necessarily leads to a modification of our perception of space and time and to the concept of a four-dimensional unified space–time. Its basic kinematical and dynamical implications on classical mechanics are discussed. Maxwell's gauge theory of electrodynamics is presented in its natural covariant 4-vector form.

3.1
Einstein's Relativity Principle and Lorentz Transformations

In the last chapter the basic framework of classical nonrelativistic mechanics has been developed. This theory crucially relies on the Galilean principle of relativity (cf. section 2.1.2), which does not match experimental results for high velocities and thus has to be replaced by the more general relativity principle of Einstein. This will directly lead to classical relativistic mechanics and electrodynamics, where again the term *classical* is used to distinguish this theory from the corresponding relativistic *quantum* theory to be presented in the later chapters of this book.

It should be emphasized that the material in this chapter, as being a generalization of the nonrelativistic theory discussed in chapter 2, is organized in a similar fashion in order to clearly exhibit both the analogies and the differences from the nonrelativistic framework. It is thus absolutely mandatory that the reader has fully grasped the essentials of sections 2.1 and 2.4.

3.1.1
Deficiencies of Newtonian Mechanics

Classical Newtonian mechanics is based on the Galilei transformations given by Eq. (2.13), which permit velocities larger than the speed of light in vacuum c. In general, according to the Galilei transformations, the speed of light, i.e., the speed of propagation of electromagnetic waves, should depend on the rel-

Relativistic Quantum Chemistry. Markus Reiher and Alexander Wolf

ISBN: 978-3-527-31292-4

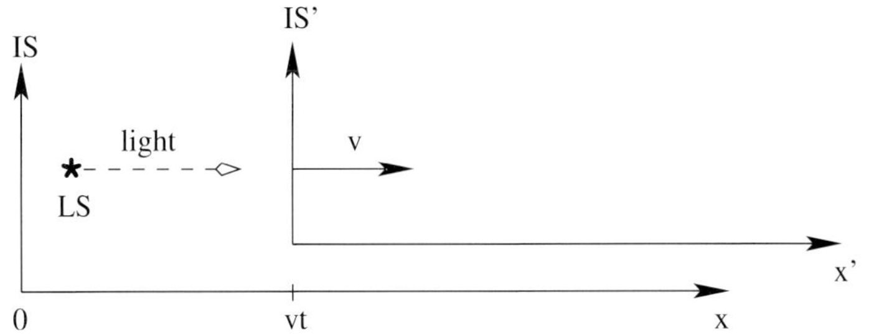

Figure 3.1 Schematic representation of a light source (LS) at rest in inertial frame IS emitting light in positive x-axis. For an observer in IS, the inertial frame IS$'$ is moving with velocity v in the same direction as the emitted electromagnetic wave. Given an observer in IS, who measures the speed of light c, according to the Galilei transformation given by Eq. (2.15), an observer in IS$'$ should measure a lower speed of light $c' = c - v$. However, this is not observed experimentally.

ative motion of the observer against the light source. Consider a light source at rest in inertial frame IS emitting light in the positive x-direction (cf. Figure 3.1). For an observer in IS, this light will propagate with velocity c. According to the Galilei transformation and Eq. (2.14), an observer in IS$'$, however, should measure the speed of light $c' = c - v < c$.

However, it has been verified in numerous experiments that all observers in all inertial frames, independently of their state of motion, will always measure the *same* speed of light,

$$c' = c \approx 299\,792\,\frac{\text{km}}{\text{s}} = \text{const.} \qquad \forall\ \text{IS}' \tag{3.1}$$

The first high-precision experiment of this kind was performed by Michelson and Morley in 1887 using their famous interferometer for velocities $v \approx 30$ km/s. This experiment has been repeated again and again by Michelson and Miller between 1902 and 1904 with much higher accuracy but the very same result: the speed of light is constant in all inertial frames of reference.

Another formal deficiency of Newtonian mechanics and thus the Galilei transformations is the existence of Maxwell's equations (2.92)–(2.95), which contain the speed of light c as a parameter. Maxwell's equations directly yield the wave equation (2.119), which states that the propagation of electromagnetic fields in vacuum occurs with speed c. Either Maxwell's equations are *not* valid in all inertial frames of reference or the Galilei transformations cannot be the correct coordinate transformations between inertial frames. Experimentally one finds that the Maxwell equations are valid for any inertial system.

There is an overwhelming plethora of further experimental evidence that proves the inadequacy of Newtonian mechanics and the Galilean transformations (at least for very high relative velocities). For example, the velocity of any particle or object has always been measured to be smaller than (massive particles) or equal to (massless particles) the velocity of light in vacuum. According to the Galilean transformation (2.13), however, velocities $v > c$ should be perfectly possible.

Furthermore, numerous modern technological achievements rely crucially on the precise description of space and time as given by the special theory of relativity as it will be presented in the rest of this chapter. For example, in order to achieve sufficient accuracy for the global positioning satellite (GPS) system, effects and predictions of special relativity have to be taken into account constructively. Also all modern accelerators employed in particle physics operate very successfully in the ultra-relativistic regime.

3.1.2
Relativity Principle of Einstein

For the reasons discussed above we have to abandon the Galilean principle of relativity and accept that Newton's laws cannot be fundamental laws of nature. We thus consider Maxwell's equations for electromagnetic fields to be valid in *all* inertial frames of reference and consequently obtain the *relativity principle of Einstein*:

1. All inertial frames are equivalent.
2. Maxwell's equations have the same form in all inertial frames.

The reader should compare this formulation with its nonrelativistic counterpart given in section 2.1.2. As a direct consequence of the covariance, i.e., invariance in form, of Maxwell's equations we obtain the desired result that the speed of light c is the *same* in all inertial frames of reference, cf. Eq (3.1). This immediately implies that the coordinate transformations relating the description of events in different inertial frames can no longer be the Galilei transformations but have to be replaced by more suitable coordinate transformations. The explicit form of these so-called Lorentz transformations will be derived in the next section.

As a prerequisite for the analysis of the implications of this new relativity principle we consider two events E_1 and E_2 connected by a light signal, e.g., the emission and subsequent absorption of a photon. An observer in IS will describe this process by the two events

$$E_1 : (t_1, \boldsymbol{r}_1) \quad \text{and} \quad E_2 : (t_2, \boldsymbol{r}_2) \qquad \text{(in IS)} \tag{3.2}$$

Since these two events are connected by a light signal propagating with the constant speed of light c, by application of the fundamental definition of velocity the observer will find the relation

$$\frac{|\boldsymbol{r}_2 - \boldsymbol{r}_1|}{t_2 - t_1} = c \quad \Longleftrightarrow \quad c^2(t_2 - t_1)^2 - (\boldsymbol{r}_2 - \boldsymbol{r}_1)^2 = 0 \tag{3.3}$$

This observation motivates the definition of the *four-dimensional distance* or *space-time interval* s_{12} between any two events by

$$s_{12}^2 \equiv c^2(t_2 - t_1)^2 - (\boldsymbol{r}_2 - \boldsymbol{r}_1)^2 = c^2(t_2 - t_1)^2 - r_{12}^2 \tag{3.4}$$

where we have employed the shorthand notation given by Eq. (2.19) in the last step. According to Eq. (3.3) the observer in IS will thus measure the squared space-time interval $s_{12}^2 = 0$ for two events connected by a light signal. Due to the constant speed of light c any other observer in any other inertial frame of reference IS′ will measure exactly the same four-dimensional distance $s_{12}'^2 = 0$. We have thus arrived at the very important result that the space-time interval between any two events *connected by a light signal* is invariant under Lorentz transformations, i.e., the same in all inertial frames of reference,

$$s_{12}^2 = s_{12}'^2 = 0 \qquad \forall\ \text{IS}' \qquad \text{(for light signal)} \tag{3.5}$$

However, if we assume homogeneity (i.e., translational invariance) of space and time and isotropy (i.e., rotational invariance) of space, the invariance of the space-time interval will also hold for *any* other two events, i.e., events connected by a light signal or not, i.e.,

$$s_{12}^2 = s_{12}'^2 \qquad \forall\ \text{IS}' \qquad \text{(for any 2 events)} \tag{3.6}$$

The formal (technical) proof of the implication leading from Eq. (3.5) to Eq. (3.6) is given in appendix C.1. Here we just mention that Eq. (3.6) may easily be tested experimentally by measuring the velocity v of a uniformly moving particle or object in different inertial frames of reference. For such a particle the space-time interval (in IS) is given by $s_{12}^2 = (c^2 - v^2) \cdot (t_2 - t_1)^2$, and experimentally one will always find the *same* interval value for all inertial frames of reference.

We note in passing that due to the definition of the space-time interval s_{12}^2 by Eq. (3.4) it is obvious that s_{12}^2 may have either positive or negative values or may even vanish as in the case of light rays. For $s_{12}^2 > 0$ the distance between the two events is denoted as *time-like* and for $s_{12}^2 < 0$ as *space-like*, whereas the distance between events connected via a light signal with $s_{12}^2 = 0$ is called *lightlike*.

The relativity principle of Einstein or, more precisely, the principle of constant speed of light led us to the definition of the four-dimensional distance s_{12}^2 given by Eq. (3.4), which we recognized as an invariant quantity under Lorentz transformations relating different inertial frames of reference. It is thus advisible to adopt a more suitable notation in order to reflect the structure of space and time as imposed by our theory. We will thus introduce the four-dimensional space-time (column) vector

$$x = (x^\mu) = \begin{pmatrix} x^0 \\ x^1 \\ x^2 \\ x^3 \end{pmatrix} = \begin{pmatrix} ct \\ x \\ y \\ z \end{pmatrix} = \begin{pmatrix} ct \\ x^i \end{pmatrix} = \begin{pmatrix} ct \\ \boldsymbol{r} \end{pmatrix} \tag{3.7}$$

comprising both time and space coordinates of an event E. In order to simplify our notation we will sometimes denote this 4-vector sloppily by the row vector $(ct, \boldsymbol{r})$. However, in a strict mathematical sense we will always refer to x as a column 4-vector. We have chosen Greek indices ($\mu = 0, \ldots, 3$) to label space-time coordinates, whereas $\mu = 0$ always refers to the time-like coordinate and Latin indices ($i = 1, 2, 3$) shall label the spatial components of this vector. For reasons which will become obvious in section 3.1.4 we have chosen *superscript* indices in order to denote each (contravariant) component x^μ ($\mu = 0, \ldots, 3$) of the space-time vector. If we further define the symmetric (4×4)-dimensional matrix g, which will be denoted as the *metric* or *metric tensor*, by

$$g = (g_{\mu\nu}) = \begin{pmatrix} 1 & 0 & 0 & 0 \\ 0 & -1 & 0 & 0 \\ 0 & 0 & -1 & 0 \\ 0 & 0 & 0 & -1 \end{pmatrix} = \text{diag}(1, -1, -1, -1) \tag{3.8}$$

we may express the squared four-dimensional distance s_{12}^2 between two events as

$$\begin{aligned} s_{12}^2 &= \sum_{\mu,\nu=0}^{3} g_{\mu\nu} \left(x_2^\mu - x_1^\mu\right) \left(x_2^\nu - x_1^\nu\right) \\ &= g_{\mu\nu}(x_2^\mu - x_1^\mu)(x_2^\nu - x_1^\nu) \;=\; c^2(t_2 - t_1)^2 - \boldsymbol{r}_{12}^2 \end{aligned} \tag{3.9}$$

where *Einstein's summation convention* has been employed in the second line of Eq. (3.9), i.e., all Greek indices occurring twice, once as subscript and once as superscript, are summed from 0 to 3, and all Latin indices (so-called spatial indices) occurring twice are summed from 1 to 3. This summation convention does not have any sophisticated meaning but is solely employed in order to simplify our notation. We will thus use it for the rest of this chapter unless explicitly stated otherwise. The flat four-dimensional space described by space-time vectors x as given by Eq. (3.7) and equipped with the metric g is denoted as the *Minkowski space*. Consequently, the metric g is sometimes also denoted as the *Minkowski tensor*.

Finally, we can now compactly express the squared four-dimensional distance ds^2 between two infinitesimal neighboring events, i.e., $x_2 = x_1 + dx$, as

$$ds^2 = \sum_{\mu,\nu=0}^{3} g_{\mu\nu} dx^\mu dx^\nu = g_{\mu\nu} dx^\mu dx^\nu = c^2 dt^2 - d\boldsymbol{r}^2 \tag{3.10}$$

This could equally well be expressed in closed form, i.e., using a coordinate-free formulation, as

$$ds^2 = dx^T g \, dx \tag{3.11}$$

3.1.3
Lorentz Transformations

3.1.3.1 Definition of General Lorentz Transformations

We are now in a position to determine those coordinate transformations between two inertial frames within the four-dimensional space-time which leave the space-time interval ds^2 invariant. The new coordinates x' (in IS$'$) have to be functions of the old coordinates x (in IS), i.e., $x' = x'(x)$. Due to the homogeneity of space and time, however, the relationship between the old and new coordinates has to be linear, i.e.,

$$x' = \Lambda x + a \qquad \Longleftrightarrow \qquad x'^\mu = \Lambda^\mu{}_\nu x^\nu + a^\mu \tag{3.12}$$

The entries of the (4×4)-matrix Λ and the 4-vector a have to be constant, i.e., *independent* of the coordinates x, since otherwise the transformation would be different at different positions in space-time. Furthermore, the entries have to be real-valued, since space-time coordinates cannot be complex numbers. The 4-vector a simply represents trivial temporal or spatial shifts of the origin of IS$'$ with reference to the origin of IS, such that the space-time coordinate differential is given by

$$dx' = \Lambda \, dx \qquad \Longleftrightarrow \qquad dx'^\mu = \Lambda^\mu{}_\nu \, dx^\nu \tag{3.13}$$

In order to avoid any misunderstanding we explicitly write down the matrix for Λ,

$$\Lambda = \left(\Lambda^\mu{}_\nu\right) = \begin{pmatrix} \Lambda^0{}_0 & \Lambda^0{}_1 & \Lambda^0{}_2 & \Lambda^0{}_3 \\ \Lambda^1{}_0 & \Lambda^1{}_1 & \Lambda^1{}_2 & \Lambda^1{}_3 \\ \Lambda^2{}_0 & \Lambda^2{}_1 & \Lambda^2{}_2 & \Lambda^2{}_3 \\ \Lambda^3{}_0 & \Lambda^3{}_1 & \Lambda^3{}_2 & \Lambda^3{}_3 \end{pmatrix} \tag{3.14}$$

and emphasize that the first (left and upper) index μ denotes the row and the second (right and lower) index ν denotes the column of the corresponding entry, respectively. Entries of coordinate transformation matrices Λ will always be labeled according to this convention. Consequently, the entries of the transposed matrix Λ^T are given as

$$(\Lambda^T)^{\mu}{}_{\nu} = \Lambda^{\nu}{}_{\mu} \tag{3.15}$$

The coordinate transformations from IS to IS′ given by Eq. (3.12) are called *Lorentz transformations* if they leave the space-time interval s_{12}^2 for *any* two events invariant. In appendix C.2 it is shown that Lorentz transformations of the form given by Eq. (3.12) are indeed the unique transformations within a flat four-dimensional space-time that achieve this goal. If we had required only the invariance of the four-dimensional space-time interval for light rays, i.e., only if $s_{12}^2 = 0$, we would have arrived at a more general set of *nonlinear* coordinate transformations. However, this 15-parameter group of transformations called the *conformal group* would not retain the homogeneity of space and time and had thus to be rejected.

Thus, we can now determine the defining condition for Λ by exploiting the invariance constraint for $\mathrm{d}s^2$ as given by Eq. (3.6),

$$\mathrm{d}s'^2 \overset{(3.11)}{=} \mathrm{d}x'^T g\,\mathrm{d}x' \overset{(3.13)}{=} \mathrm{d}x^T \Lambda^T g\,\Lambda\,\mathrm{d}x \overset{!}{=} \mathrm{d}s^2 \overset{(3.11)}{=} \mathrm{d}x^T g\,\mathrm{d}x \tag{3.16}$$

which immediately yields

$$\Lambda^T g\,\Lambda = g \qquad \Longleftrightarrow \qquad \Lambda^{\alpha}{}_{\mu}\, g_{\alpha\beta}\, \Lambda^{\beta}{}_{\nu} = g_{\mu\nu} \tag{3.17}$$

Eq. (3.17) is the fundamental and most important property of Lorentz transformations which completely specifies these transformations.

It cannot be overemphasized here that space and time in relativistic mechanics are relative quantities depending on the motion of the observer. Space and time coordinates are no longer independent of each other but may be transformed into each other by means of Eq. (3.12). This has to be compared to classical Newtonian mechanics, where space and time are *absolute* quantities which can always be clearly distinguished from each other.

3.1.3.2 **Classification of Lorentz Transformations**

Due to the symmetric form of the metric g Eq. (3.17) imposes only 10 constraints on the possible values of the elements of Λ, i.e., Λ contains in general 6 free parameters. Together with the 4 free parameters comprised by the 4-vector a it is obvious that the most general Lorentz transformation contains 10 free parameters. The set of all Lorentz transformations thus represents a non-Abelian 10-parameter group and is denoted as the *inhomogeneous Lorentz group* or *Poincaré group*. This should be compared to the 10-parameter Galilei group of nonrelativistic Newtonian mechanics we encountered in section 2.1.2.

In most cases of practical interest we will restrict ourselves to the case where $a = 0$. This defines the *homogeneous Lorentz transformations* which form a 6-parameter subgroup of the Poincaré group, the so-called *homogeneous Lorentz group*. It should be obvious from the comparison with the Galilei transformations given by Eq. (2.13) that each homogeneous Lorentz transformation might be characterized by 3 angles defining a (constant) spatial rotation of the coordinate axes and the 3 components of a velocity vector $\boldsymbol{v}$. A homogeneous Lorentz transformation without rotation, i.e., a Lorentz transformation which leaves the coordinate axes of IS′ parallel to those of IS, is called a *general Lorentz boost* $\Lambda(\boldsymbol{v})$ and depends only on the relative velocity $\boldsymbol{v}$ between IS and IS′.

For a further classification of the set of all Lorentz transformations we need some elementary properties of these transformations. From Eq. (3.17) and det $g = -1$ it is immediately clear that Lorentz transformations feature a determinant det $\Lambda = \pm 1$. And according to Eq. (3.17) with $\mu = \nu = 0$ it follows that $|\Lambda^0{}_0| \geq 1$. We can thus define the subgroup of *proper Lorentz transformations* by the condition

$$\Lambda^0{}_0 \geq 1 \quad \text{and} \quad \det\Lambda = +1 \qquad \text{(proper Lorentz transformations)} \tag{3.18}$$

Proper Lorentz transformations can always be continuously converted into the identical transformation given by $\Lambda = 1 = id$. *Improper Lorentz transformations* involve either space inversion ($\Lambda^0{}_0 \geq 1$, det $\Lambda = -1$) or time reversal ($\Lambda^0{}_0 \leq -1$, det $\Lambda = -1$) or both. These improper transformations do not play any role in quantum chemistry and will thus not be considered in this presentation.

Pure rotations of the coordinate axes of IS′ with reference to those of IS also belong to the homogeneous Lorentz transformations and are given by

$$\Lambda(R) = \begin{pmatrix} 1 & 0 \\ 0 & R \end{pmatrix} \tag{3.19}$$

where the general orthogonal (3×3)-matrix R is defined by Eq. (2.20). Any homogeneous Lorentz transformation may be expressed as the product of a simple rotation and a general Lorentz boost as

$$\Lambda = \Lambda(R)\,\Lambda(\boldsymbol{v}) \tag{3.20}$$

3.1.3.3 **Inverse Lorentz Transformation**

So far we have considered the Lorentz transformation Λ from IS to IS′ given by Eq. (3.12). Due to the group structure of Lorentz transformations the inverse transformation Λ^{-1} will always exist and mediate the transition from IS′ back to IS,

$$x = \Lambda^{-1}(x' - a) \qquad \Longleftrightarrow \qquad x^\mu = \left(\Lambda^{-1}\right)^\mu{}_\nu \left(x'^\nu - a^\nu\right) \tag{3.21}$$

The trivial constant space-time shift a does not affect the determination of the inverse transformation and will thus be neglected in the following. Anticipating the discussion in section 3.1.4, it is further convenient to define the fully contravariant components $g^{\mu\nu}$ of the metric to be identical to the fully covariant components $g_{\mu\nu}$ as already introduced in Eq. (3.8), $g^{\mu\nu} = g_{\mu\nu}$, and to note that the metric is both symmetric and its own inverse,

$$g^T = g \quad \Longleftrightarrow \quad g_{\mu\nu} = g_{\nu\mu}\,, \quad g^{\mu\nu} = g^{\nu\mu} \tag{3.22}$$

$$g\,g = \mathbf{1} = id \quad \Longleftrightarrow \quad g^{\mu\nu}\,g_{\nu\sigma} = g^{\mu}{}_{\sigma} = \delta^{\mu}{}_{\sigma} \tag{3.23}$$

The four-dimensional Kronecker symbol $\delta^{\mu}{}_{\nu}$ is a straightforward generalization of the corresponding three-dimensional symbol δ_{ij} as given by Eq. (2.24), where the indices have now been chosen as super- or subscripts in order to support Einstein's summation convention,

$$\delta^{\mu}{}_{\nu} = \begin{cases} 1 \ , & \mu = \nu \\ 0 \ , & \mu \neq \nu \end{cases} \tag{3.24}$$

After these preliminaries, multiplication of Eq. (3.17) with the metric g from the left immediately yields an equation for the inverse transformation,

$$\Lambda^{-1} = g\,\Lambda^T g \quad \Longleftrightarrow \quad (\Lambda^{-1})^{\mu}{}_{\nu} = g_{\nu\alpha}\,g^{\mu\beta}\,\Lambda^{\alpha}{}_{\beta} \tag{3.25}$$

where Eqs. (3.15) and (3.22) have been employed. Inspection of Eq. (3.25) suggests that, in order to simplify the notation for matrix equations, it might be advantageous to introduce the shorthand notation $\bar{\Lambda} = \Lambda^{-1}$ for the inverse transformation and denote its entries by the new symbol $\Lambda_{\nu}{}^{\mu}$,

$$\bar{\Lambda} = \Lambda^{-1} = \left(\bar{\Lambda}^{\mu}{}_{\nu}\right) \equiv \left(\Lambda_{\nu}{}^{\mu}\right) \quad \text{and} \quad \Lambda_{\nu}{}^{\mu} \equiv g_{\nu\alpha}\,g^{\mu\beta}\,\Lambda^{\alpha}{}_{\beta} \tag{3.26}$$

This allows us to compactly formulate matrix equations containing the inverse transformation like

$$\bar{\Lambda}^{\mu}{}_{\nu}\,\Lambda^{\nu}{}_{\sigma} \overset{(3.26)}{=} \Lambda_{\nu}{}^{\mu}\,\Lambda^{\nu}{}_{\sigma} \overset{(3.26)}{=} g_{\nu\alpha}\,g^{\mu\beta}\,\Lambda^{\alpha}{}_{\beta}\,\Lambda^{\nu}{}_{\sigma} \overset{(3.17)}{=} g^{\mu\beta}\,g_{\beta\sigma} \overset{(3.23)}{=} \delta^{\mu}{}_{\sigma} \tag{3.27}$$

For the sake of clarity we finally express Λ^{-1} explicitly by the elements of the original transformation Λ,

$$\Lambda^{-1} = \left((\Lambda^{-1})^{\mu}{}_{\nu}\right) = \begin{pmatrix} \Lambda^0{}_0 & -\Lambda^1{}_0 & -\Lambda^2{}_0 & -\Lambda^3{}_0 \\ -\Lambda^0{}_1 & \Lambda^1{}_1 & \Lambda^2{}_1 & \Lambda^3{}_1 \\ -\Lambda^0{}_2 & \Lambda^1{}_2 & \Lambda^2{}_2 & \Lambda^3{}_2 \\ -\Lambda^0{}_3 & \Lambda^1{}_3 & \Lambda^2{}_3 & \Lambda^3{}_3 \end{pmatrix} \tag{3.28}$$

Application of Eq. (3.26) yields the quite useful relations between the elements of any arbitrary Lorentz transformation Λ,

$$\Lambda^{\mu}{}_{\alpha}\,\Lambda_{\nu}{}^{\alpha} = \delta^{\mu}{}_{\nu} \quad \text{and} \quad \Lambda_{\mu}{}^{\alpha}\,\Lambda^{\mu}{}_{\beta} = \delta^{\alpha}{}_{\beta} \tag{3.29}$$

3.1.4
Scalars, Vectors, and Tensors in Minkowski Space

The Lorentz transformations have been identified as those coordinate transformations of the four-dimensional space-time (Minkowski space) that leave the four-dimensional (squared) distance s_{12}^2 between any two events E_1 and E_2 invariant, $s_{12}^2 = s_{12}'^2$. We emphasize that both events have to be described with reference to the same inertial system, which may be IS or IS′. As a consequence of Eq. (3.12) the mixing of different components of the space-time vector x under Lorentz transformations is only due to the homogeneous Lorentz transformations described by the matrix Λ. We will thus focus on homogeneous transformations only in order to define scalars, vectors, and tensors by their transformation properties under Lorentz transformations. In contrast to the three-dimensional discussion in section 2.1.2, however, we will have to distinguish two kinds of vectors.

3.1.4.1 Contra- and Covariant Components

In section 3.1.2 we introduced the *contravariant* space-time 4-vector x^μ, whose components will be denoted by superscript indices and are given by Eq. (3.7). It is essential to realize that the metric g might be employed in order to lower (or raise) indices of any vector (within one inertial system IS) according to

$$x_\mu = g_{\mu\nu}\, x^\nu \qquad \Longleftrightarrow \qquad x^\mu = g^{\mu\nu}\, x_\nu \tag{3.30}$$

The quantities x_μ labeled with subscript indices are denoted as the *covariant* components of the 4-vector x and are given by

$$(x_\mu) = \begin{pmatrix} x_0 \\ x_1 \\ x_2 \\ x_3 \end{pmatrix} = \begin{pmatrix} ct \\ -x \\ -y \\ -z \end{pmatrix} = \begin{pmatrix} ct \\ x_i \end{pmatrix} = \begin{pmatrix} ct \\ -\boldsymbol{r} \end{pmatrix} \tag{3.31}$$

The effect of lowering the index of (the components of) a vector from superscript to subscript is to change the sign of the spatial components. We may thus express the four-dimensional distance between two infinitesimal neighboring events as

$$\mathrm{d}s^2 \overset{(3.10)}{=} g_{\mu\nu}\, \mathrm{d}x^\mu\, \mathrm{d}x^\nu \overset{(3.30)}{=} \mathrm{d}x^\mu\, \mathrm{d}x_\mu = \mathrm{d}x_\mu\, \mathrm{d}x^\mu = c^2\mathrm{d}t^2 - \mathrm{d}\boldsymbol{r}^2 \tag{3.32}$$

For later convenience we introduce the four-dimensional scalar product between any two 4-vectors a and b by

$$a \cdot b = a^T g\, b = a^\mu b_\mu = a^0 b^0 - a^i\, b^i = a^0 b^0 - \boldsymbol{a} \cdot \boldsymbol{b} \tag{3.33}$$

As a consequence, the four-dimensional distance between two infinitesimal neighboring events may now also be expressed as

$$\mathrm{d}s^2 = \mathrm{d}x \cdot \mathrm{d}x = \mathrm{d}x^T g\, \mathrm{d}x = g_{\mu\nu}\, \mathrm{d}x^\mu\, \mathrm{d}x^\nu \tag{3.34}$$

Due to the diagonal structure of the metric g their fully contravariant and covariant components are identical,

$$(g^{\mu\nu}) = (g_{\mu\nu}) = \begin{pmatrix} 1 & 0 & 0 & 0 \\ 0 & -1 & 0 & 0 \\ 0 & 0 & -1 & 0 \\ 0 & 0 & 0 & -1 \end{pmatrix} \tag{3.35}$$

which has already been anticipated in section 3.1.3.

3.1.4.2 Transformation Properties of Scalars, Vectors, and Tensors

After the preliminaries presented above we can now precisely define vectors and tensors in Minkowski space by their transformation properties under Lorentz transformations. Each four-component quantity A, which features the same transformation property as the contravariant space-time vector x^μ as given by Eq. (3.12),

$$A'^\mu = \Lambda^\mu{}_\nu A^\nu \tag{3.36}$$

is called a *contravariant Lorentz 4-vector* or just a *vector*. The components of such a contravariant vector will be denoted by superscript indices A^μ. We can now investigate the transformation property of the covariant components x_μ of the space-time vector x,

$$x'_\mu \overset{(3.30)}{=} g_{\mu\nu} x'^\nu \overset{(3.36)}{=} g_{\mu\nu} \Lambda^\nu{}_\alpha x^\alpha \overset{(3.30)}{=} g_{\mu\nu} g^{\alpha\beta} \Lambda^\nu{}_\alpha x_\beta \overset{(3.26)}{=} \Lambda_\mu{}^\beta x_\beta \tag{3.37}$$

and find that they are transformed with the inverse transformation $\bar{\Lambda}$ rather than the original transformation Λ. We thus define *any* four-component quantity B, whose components feature the same transformation property as the covariant components x_μ of the space-time vector x,

$$B'_\mu = \Lambda_\mu{}^\nu B_\nu \tag{3.38}$$

as a *covariant Lorentz 4-vector* or just a *co-vector*.

In contrast to the three-dimensional situation of nonrelativistic mechanics there are now *two kinds* of vectors within the four-dimensional Minkowski space. Contravariant vectors transform according to Eq. (3.36) whereas covariant vectors transform acoording to Eq. (3.38) by the transition from IS to IS′. The reason for this crucial feature of Minkowski space is solely rooted in the structure of the metric g given by Eq. (3.8) which has been shown to be responsible for the central structure of Lorentz transformations as given by Eq. (3.17). As a consequence, the transposed Lorentz transformation Λ^T no longer represents the inverse transformation Λ^{-1}. As we have seen, the inverse Lorentz transformation is now more involved and given by Eq. (3.25).

Similarly to the nonrelativistic situation (cf. Eq. (2.28)) the components of the transformation matrix Λ may be expressed as derivatives of the new coordinates with respect to the old ones or vice versa (see also appendix A.2),

$$\Lambda^{\mu}{}_{\nu} = \frac{\partial x'^{\mu}}{\partial x^{\nu}} = \partial_{\nu} x'^{\mu} \quad \text{and} \quad \Lambda_{\mu}{}^{\nu} = \frac{\partial x^{\nu}}{\partial x'^{\mu}} = \partial'_{\mu} x^{\nu} \tag{3.39}$$

where we have employed a shorthand notation for the 4-gradient defined by

$$\partial_{\mu} \equiv \frac{\partial}{\partial x^{\mu}} = \left(\frac{1}{c}\frac{\partial}{\partial t}, \nabla \right) = \left(\frac{1}{c}\frac{\partial}{\partial t}, \frac{\partial}{\partial x}, \frac{\partial}{\partial y}, \frac{\partial}{\partial z} \right) \tag{3.40}$$

The 4-gradient has been written as a row vector above solely for our convenience; it still is to be interpreted mathematically as a column vector, of course. Being defined as the derivative with respect to the contravariant components x^{μ}, the 4-gradient ∂_{μ} is *naturally* a covariant vector since its transformation property under Lorentz transformations is given by

$$\partial'_{\mu} = \frac{\partial}{\partial x'^{\mu}} = \frac{\partial x^{\nu}}{\partial x'^{\mu}} \frac{\partial}{\partial x^{\nu}} \overset{(3.39)}{=} \Lambda_{\mu}{}^{\nu} \partial_{\nu} \tag{3.41}$$

where we have employed the chain rule to $x = x(x')$ at the second equality. Accordingly, the contravariant components ∂^{μ} of the 4-gradient transform like the contravariant components x^{μ} of the space-time vector under Lorentz transformations,

$$\partial'^{\mu} \overset{(3.30)}{=} g^{\mu\nu} \partial'_{\nu} \overset{(3.41)}{=} g^{\mu\nu} \Lambda_{\nu}{}^{\alpha} \partial_{\alpha} \overset{(3.30)}{=} g^{\mu\nu} g_{\alpha\beta} \Lambda_{\nu}{}^{\alpha} \partial^{\beta} \overset{(3.26)}{=} \Lambda^{\mu}{}_{\beta} \partial^{\beta} \tag{3.42}$$

Any n-index quantity T with 4^n components $T^{\mu_1\mu_2\ldots\mu_n}$ is called a *contravariant Lorentz tensor* of the nth rank if its components transform like the n-fold product of contravariant vectors under Lorentz transformations,

$$T'^{\mu_1\mu_2\ldots\mu_n} = \Lambda^{\mu_1}{}_{\nu_1} \Lambda^{\mu_2}{}_{\nu_2} \ldots \Lambda^{\mu_n}{}_{\nu_n} T^{\nu_1\nu_2\ldots\nu_n} \tag{3.43}$$

Similarly, the covariant components of any tensor may be obtained by lowering indices from super- to subscripts by application of the metric g,

$$T_{\mu_1\mu_2\ldots\mu_n} = g_{\mu_1\nu_1} g_{\mu_2\nu_2} \cdots g_{\mu_n\nu_n} T^{\nu_1\nu_2\ldots\nu_n} \tag{3.44}$$

Consequently, the transformation property of a *covariant Lorentz tensor* under Lorentz transformations is thus given as the one of an n-fold product of covariant vectors,

$$T'_{\mu_1\mu_2\ldots\mu_n} = \Lambda_{\mu_1}{}^{\nu_1} \Lambda_{\mu_2}{}^{\nu_2} \ldots \Lambda_{\mu_n}{}^{\nu_n} T_{\nu_1\nu_2\ldots\nu_n} \tag{3.45}$$

Tensor components comprising both contravariant and covariant indices like $T^{\mu}{}_{\nu}$ are denoted mixed tensors of the corresponding rank. Each index features its corresponding transformation property, e.g.,

$$T'^{\mu}{}_{\nu} = \Lambda^{\mu}{}_{\alpha} \Lambda_{\nu}{}^{\beta} T^{\alpha}{}_{\beta} \tag{3.46}$$

According to this definition, a tensor of the first rank is simply a vector.

A very important second-rank tensor is the metric g, whose contra- ($g^{\mu\nu}$) and covariant ($g_{\mu\nu}$) components coincide (cf. Eq. (3.35)) and are exactly *the same* in all frames of reference since

$$\begin{aligned} g'_{\mu\nu} &\overset{(3.45)}{=} \Lambda_\mu{}^\alpha \Lambda_\nu{}^\beta g_{\alpha\beta} \overset{(3.17)}{=} \Lambda_\mu{}^\alpha \Lambda_\nu{}^\beta \Lambda^\sigma{}_\alpha \Lambda^\tau{}_\beta g_{\sigma\tau} \\ &\overset{(3.29)}{=} \delta^\sigma{}_\mu \delta^\tau{}_\nu g_{\sigma\tau} \overset{(3.24)}{=} g_{\mu\nu} \end{aligned} \tag{3.47}$$

Though the metric g has been defined as constant by Eq. (3.8), we have just shown that it perfectly fits into the definition of a tensor and features the correct transformation property under Lorentz transformations. Another very useful fourth-rank Lorentz tensor is the totally antisymmetric Levi–Cività (pseudo-)tensor $\varepsilon^{\alpha\beta\gamma\delta}$, whose contravariant components are defined by

$$\varepsilon^{\alpha\beta\gamma\delta} = \left\{ \begin{array}{rcl} +1 & & (\alpha\beta\gamma\delta) \text{ is an even permutation of } (0123) \\ -1 & \text{if} & (\alpha\beta\gamma\delta) \text{ is an odd permutation of } (0123) \\ 0 & & \text{else} \end{array} \right\} \tag{3.48}$$

The Levi–Cività tensor is a natural generalization of the totally antisymmetric third-rank tensor ε_{ijk} as defined by Eq. (2.9). For Lorentz transformations Λ with $\det \Lambda = +1$ the Levi–Cività pseudo-tensor is indeed a tensor which has the same values in all frames of reference,

$$\varepsilon'^{\alpha\beta\gamma\delta} = \Lambda^\alpha{}_\mu \Lambda^\beta{}_\nu \Lambda^\gamma{}_\rho \Lambda^\delta{}_\sigma \varepsilon^{\mu\nu\rho\sigma} = \varepsilon^{\alpha\beta\gamma\delta} \tag{3.49}$$

Its fully covariant components are given by

$$\varepsilon_{\alpha\beta\gamma\delta} = g_{\alpha\mu} g_{\beta\nu} g_{\gamma\rho} g_{\delta\sigma} \varepsilon^{\mu\nu\rho\sigma} = -\varepsilon^{\alpha\beta\gamma\delta} \tag{3.50}$$

We emphasize again that any tensor of the second rank can always be expressed as a matrix, but that not every matrix is a tensor. Any tensor is uniquely defined within *one* given inertial system IS, and its components may be transformed to another coordinate system IS′. This transition to another coordinate system is described by Lorentz transformation matrices Λ, which are thus *not* tensors at all but mediate the change of coordinates. The matrices Λ are not defined with respect to one specific IS, but relate two inertial systems IS and IS′. Nevertheless, due to Eq. (3.26) the indices of components $\Lambda^\mu{}_\nu$ of Lorentz transformation matrices may be raised or lowered as if they were tensors.

Quantities without any indices like the mass m or the space-time interval ds^2, which are not only covariant but invariant under Lorentz transformations, are called *Lorentz scalars* or zero-rank tensors. They have exactly the same value in all inertial frames of reference. A very important scalar operator for both relativistic mechanics and electrodynamics is the *d'Alembert operator*

$$\Box \equiv \partial_\mu \partial^\mu = \frac{1}{c^2}\frac{\partial^2}{\partial t^2} - \Delta \tag{3.51}$$

We close this section by emphasizing again that due to the homogeneity and isotropy of space-time *all* fundamental laws of nature have to be covariant (i.e., invariant in form) under Lorentz transformations, i.e., they have to be tensor equations. If the equations changed form under coordinate transformations, some directions or positions in space-time would be singled out which would violate our assumption of homogeneity and isotropy. In other words, *any* equation which is *not* a tensor equation (or cannot be cast in tensor form) can in principle not be a fundamental equation of nature. This crucial role of tensor equations for *all* branches of physics is by far the most important reason for covering scalars, vectors, and tensors in such great depth in this section and at the beginning of a book on relativistic quantum chemistry. All technical details presented above like the shuffling of indices or the manipulation of equations are primarily needed for actual calculations and thus only of minor importance.

The discussion of four-dimensional Minkowski vectors and tensors has been presented in close analogy to the nonrelativistic discussion in section 2.1.2. The similarities and differences between these two frameworks are schematically compared to each other in Table 3.1.

Table 3.1 Comparison of three- and four-dimensional quantities relevant for nonrelativistic and relativistic classical mechanics, respectively. See text for further details.

3D	quantity	4D
$\boldsymbol{r}$	position vector	$x^\mu = (ct, \boldsymbol{r})$
$x'_i = \sum_{j=1}^{3} R_{ij} x_j$	transformations	$x'^\mu = \Lambda^\mu{}_\nu x^\nu, \ x'_\mu = \Lambda_\mu{}^\nu x_v$
$R^{-1} = R^T$	inverse transformation	$\Lambda^{-1} = g\,\Lambda^T g$
$R^T R = 1$	defining property	$\Lambda^T g \Lambda = g$
$\mathrm{d}\boldsymbol{r}^2 = \sum_{i,j=1}^{3} \delta_{ij} \mathrm{d}x_i \mathrm{d}x_j$	distance	$\mathrm{d}s^2 = g_{\mu\nu} \mathrm{d}x^\mu \mathrm{d}x^\nu = c^2 \mathrm{d}t^2 - \mathrm{d}\boldsymbol{r}^2$
$\delta_{ij} = \mathrm{diag}(1,1,1)$	metric	$g_{\mu\nu} = \mathrm{diag}(1,-1,-1,-1)$

3.2 Kinematical Effects in Special Relativity

3.2.1 Explicit Form of Special Lorentz Transformations

3.2.1.1 Lorentz Boost in One Direction

So far, only general properties of Lorentz transformations have been investigated but no explicit expression for the transformation matrix Λ has yet been given. We are now going to derive the transformation matrix Λ for a Lorentz boost in x-direction in a very clear and elementary fashion. For $t = t' = 0$ the two inertial frames IS and IS′ shall coincide, and the constant motion of IS′ relative to IS shall be described by the velocity vector $\boldsymbol{v} = v\boldsymbol{e}_x$, cf. Figure 3.2. Since the y- and z-directions are not affected by this transformation, we explicitly write down the transformation given by Eq. (3.12) (for $a = 0$) for the relevant subspace

$$\begin{pmatrix} ct' \\ x' \end{pmatrix} = \begin{pmatrix} \Lambda^0{}_0 & \Lambda^0{}_1 \\ \Lambda^1{}_0 & \Lambda^1{}_1 \end{pmatrix} \begin{pmatrix} ct \\ x \end{pmatrix}, \qquad y' = y\,, \quad z' = z \tag{3.52}$$

In order to derive the explicit form of Λ we consider two light signals, which are emitted at $t = t' = 0$ from the origin ($x = x' = 0$), one to the left (i.e., in negative x-direction) and one to the right (i.e., in positive x-direction). For each time $t_1 > 0$ the first light signal to the left defines an event E_1 whose coordinates in IS and IS′ are — due to the constant speed of light — given by

$$\mathrm{IS} \;:\; x_1 + ct_1 \stackrel{!}{=} 0 \tag{3.53}$$

$$\mathrm{IS}' \;:\; x_1' + ct_1' \stackrel{(3.52)}{=} (\Lambda^1{}_1 + \Lambda^0{}_1)x_1 + (\Lambda^0{}_0 + \Lambda^1{}_0)ct_1 \stackrel{!}{=} 0 \tag{3.54}$$

Multiplying Eq. (3.53) by $(\Lambda^0{}_0 + \Lambda^1{}_0)$ and subtracting Eq. (3.54) immediately implies the relation

$$\Lambda^0{}_0 + \Lambda^1{}_0 - \Lambda^0{}_1 - \Lambda^1{}_1 = 0 \tag{3.55}$$

Analogously, for any time $t_2 > 0$ the second light signal to the right defines an event E_2 whose coordinates in IS and IS′ are given by

$$\mathrm{IS} \;:\; x_2 - ct_2 \stackrel{!}{=} 0 \tag{3.56}$$

$$\mathrm{IS}' \;:\; x_2' - ct_2' \stackrel{(3.52)}{=} (\Lambda^1{}_1 - \Lambda^0{}_1)x_2 - (\Lambda^0{}_0 - \Lambda^1{}_0)ct_2 \stackrel{!}{=} 0 \tag{3.57}$$

Multiplication of Eq. (3.56) by $(\Lambda^0{}_0 - \Lambda^1{}_0)$ and subtracting Eq. (3.57) yields

$$\Lambda^0{}_0 - \Lambda^1{}_0 + \Lambda^0{}_1 - \Lambda^1{}_1 = 0 \tag{3.58}$$

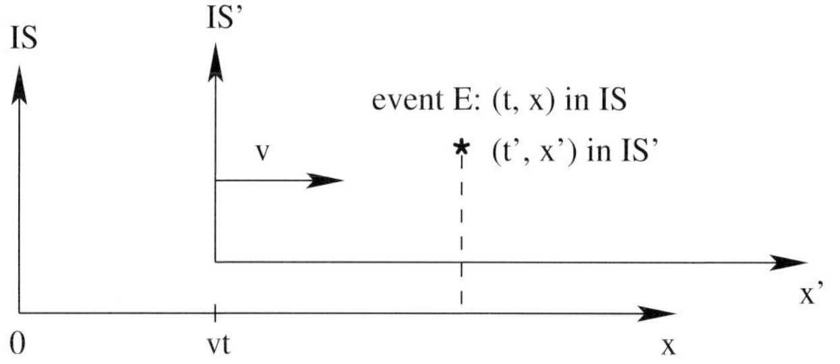

Figure 3.2 Illustration of a Lorentz boost in x-direction. For $t = t' = 0$ the two inertial frames IS and IS$'$ coincide, and for later times the coordinates of the event E with respect to IS and IS$'$ are related to each other according to Eq. (3.67).

Addition of Eqs. (3.55) and (3.58) yields $\Lambda^0{}_0 = \Lambda^1{}_1$, and subtraction of Eq. (3.55) from Eq. (3.58) gives $\Lambda^1{}_0 = \Lambda^0{}_1$, such that only two parameters of Λ remain to be determined.

We now consider the origin (O) of IS$'$. For any time t it is given by $x'_O = 0$ and $x_O = vt$. But due to the Lorentz transformation given by Eq. (3.52) it must also be true that

$$x'_O = \Lambda^1{}_0\, ct + \Lambda^1{}_1\, x_O = \Lambda^1{}_0\, ct + \Lambda^0{}_0\, vt = 0 \tag{3.59}$$

We thus find $\Lambda^1{}_0 = -(v/c)\Lambda^0{}_0$ and may write down the desired Lorentz transformation as

$$\begin{pmatrix} ct' \\ x' \end{pmatrix} = \begin{pmatrix} \Lambda^0{}_0 & -\frac{v}{c}\Lambda^0{}_0 \\ -\frac{v}{c}\Lambda^0{}_0 & \Lambda^0{}_0 \end{pmatrix} \begin{pmatrix} ct \\ x \end{pmatrix}, \qquad y' = y\,, \quad z' = z \tag{3.60}$$

containing only one yet undetermined parameter $\Lambda^0{}_0$.

So far we have only exploited the principle of constant speed of light in all inertial frames, but not the (first part of the) relativity principle itself, cf. section 3.1.2. Due to the relativity principle the Lorentz back transformation from IS$'$ to IS must have the same form as given by Eq. (3.60) with v replaced by $-v$,

$$\begin{pmatrix} ct \\ x \end{pmatrix} = \begin{pmatrix} \Lambda^0{}_0 & +\frac{v}{c}\Lambda^0{}_0 \\ +\frac{v}{c}\Lambda^0{}_0 & \Lambda^0{}_0 \end{pmatrix} \begin{pmatrix} ct' \\ x' \end{pmatrix}, \qquad y = y'\,, \quad z = z' \tag{3.61}$$

Considering the Lorentz back and forth transformations for the x-coordinate, for example, we find

$$\begin{aligned} x &\overset{(3.61)}{=} \Lambda^0{}_0(x' + vt') \overset{(3.60)}{=} \Lambda^0{}_0\Big[\Lambda^0{}_0(x - vt) + v\Lambda^0{}_0\Big(t - \frac{v}{c^2}x\Big)\Big] \\ &= (\Lambda^0{}_0)^2\left(1 - \frac{v^2}{c^2}\right) x \end{aligned} \tag{3.62}$$

This determines the last free parameter $\Lambda^0{}_0$, for which it is convenient to introduce the new symbol γ,

$$\Lambda^0{}_0 = \frac{1}{\sqrt{1-\frac{v^2}{c^2}}} \equiv \gamma = \gamma(v) \tag{3.63}$$

Note that we have restricted ourselves to proper Lorentz transformations with $\Lambda^0{}_0 \geq 1$, cf. Eq. (3.18). Application of the Lorentz back and forth transformations to the t-coordinate yields exactly the same result, of course. The *Lorentz factor* γ depends only on the relative velocity v between IS and IS$'$, is strictly greater than 1, and tends to unity and features a smooth and vanishing derivative with respect to v for $v \to 0$,

$$\gamma(v) \geq 1\,,\ \forall v \quad , \quad \gamma(v=0) = 1 \quad , \quad \frac{\mathrm{d}\gamma(v)}{\mathrm{d}v} \longrightarrow 0 \ \text{ for } \ v \overset{v>0}{\longrightarrow} 0 \tag{3.64}$$

For small velocities $v \ll c$ the Lorentz factor might be approximated by the first terms of its series expansion,

$$\gamma(v) \approx 1 + \frac{1}{2}\frac{v^2}{c^2} + \frac{3}{8}\frac{v^4}{c^4} + \cdots \tag{3.65}$$

which is often conveniently employed in special relativity in order to recover the nonrelativistic limit. Furthermore, for $v \to c$ the Lorentz factor $\gamma(v)$ tends to infinity and is singular, i.e., not defined for $v = c$. We may now explicitly write down the Lorentz transformation for a boost in x-direction as given by Eq. (3.52) as

$$\begin{pmatrix} ct' \\ x' \end{pmatrix} = \begin{pmatrix} \gamma & -\frac{v}{c}\gamma \\ -\frac{v}{c}\gamma & \gamma \end{pmatrix} \begin{pmatrix} ct \\ x \end{pmatrix} , \qquad y' = y\,, \quad z' = z \tag{3.66}$$

which may also be expressed in the more familiar form

$$x' = \gamma(x - vt)\,, \quad y' = y\,, \quad z' = z\,, \quad t' = \gamma\left(t - \frac{v}{c^2}x\right) \tag{3.67}$$

For $v = 0$ Eq. (3.67) for a Lorentz boost in x-direction reduces to the Galilei boost as given by Eq. (2.15). But for nonvanishing velocities v the relativistic transformation is more involved and mixes space- and time-coordinates.

It is sometimes convenient to have another representation of the Lorentz boost in x-direction at hand. This will, by the way, yield yet another (more compact) derivation of the transformation given by Eq. (3.67). As a direct consequence of the fundamental relation for Lorentz transformations as given by Eq. (3.17) we find for the relevant t-x-subspace the following three relations,

$$\left(\Lambda^0{}_0\right)^2 - \left(\Lambda^1{}_0\right)^2 = +1 \tag{3.68}$$

$$\Lambda^0{}_0\,\Lambda^0{}_1 - \Lambda^1{}_0\,\Lambda^1{}_1 = 0 \tag{3.69}$$

$$\left(\Lambda^0{}_1\right)^2 - \left(\Lambda^1{}_1\right)^2 = -1 \tag{3.70}$$

These equations establish a set of constraints on the desired Lorentz transformation. We find by inspection (cf. also appendix C.3) that these equations are satisfied by the choice

$$\Lambda^0{}_0 = \Lambda^1{}_1 = \cosh\psi\,, \qquad \Lambda^1{}_0 = \Lambda^0{}_1 = -\sinh\psi \tag{3.71}$$

for any $\psi \in \mathbb{R}$. The possible solution with $\Lambda^0{}_0 = -\cosh\psi$ has been disregarded, since for $v \to 0$ we want to recover the identical transformation with $\Lambda^0{}_0 = +1$. Consideration of the origin of IS′ again, as given by Eq. (3.59), yields the relation

$$\psi = \operatorname{artanh}\frac{v}{c} \tag{3.72}$$

The quantity ψ is denoted as the *rapidity* and is related to the Lorentz factor γ according to (cf. appendix C.3)

$$\gamma = \cosh\psi = \frac{1}{\sqrt{1-\frac{v^2}{c^2}}} \tag{3.73}$$

such that we may write Eq. (3.66) equivalently as

$$\begin{pmatrix} ct' \\ x' \end{pmatrix} = \begin{pmatrix} \cosh\psi & -\sinh\psi \\ -\sinh\psi & \cosh\psi \end{pmatrix} \begin{pmatrix} ct \\ x \end{pmatrix}, \quad y' = y\,, \quad z' = z \tag{3.74}$$

The form of the Lorentz transformation for a boost in x-direction as given by Eqs. (3.63) and (3.67) suggests that the relative velocity v between IS and IS′ is always smaller than the speed of light c. This is indeed true and will be shown to follow from the relativistic equation of motion, i.e., the relativistic dynamics, to be discussed in section 3.3.

3.2.1.2 **General Lorentz Boost**

We now consider a general Lorentz boost from the inertial system IS to the inertial system IS′, which moves with constant velocity $\boldsymbol{v} = \sum v_i \boldsymbol{e}_i$ relative to IS. Note that the velocity $\boldsymbol{v}$ is a 3-vector and thus its components v_i are labeled with subscript indices, e.g., $v_1 = v_x$. This must not be confused with the covariant components of a 4-vector. The velocity vector $\boldsymbol{v}$ may point in any direction, but the coordinate axes of IS and IS′ must be parallel to each other, i.e., there is no constant rotation between the axes of the two systems. For a particle at rest in IS (for example at the origin $\boldsymbol{r} = 0$) we have $d\boldsymbol{r} = 0$ and can thus immediately write down the Lorentz transformation given by Eq. (3.13) as

$$dx'^0 = c\,dt' = \Lambda^0{}_0\, c\,dt\,, \qquad dx'^i = \Lambda^i{}_0\, c\,dt\,, \qquad i = 1,2,3 \tag{3.75}$$

Since this particle moves with velocity $-\boldsymbol{v}$ for an observer in IS$'$, it holds that

$$\frac{dx'^i}{dt'} = -v_i \overset{(3.75)}{=} \frac{c\Lambda^i{}_0}{\Lambda^0{}_0} \quad \Longrightarrow \quad \Lambda^i{}_0 = -\frac{v_i}{c}\Lambda^0{}_0 \tag{3.76}$$

Similarly, consideration of a particle at rest in IS$'$, which consequently moves in IS with velocity $\boldsymbol{v}$, we find the same relation, i.e., $\Lambda^0{}_i = \Lambda^i{}_0$. Exploiting the fundamental equation for Lorentz transformations as given by Eq. (3.17) for $\mu = \nu = 0$, we immediately find a further relation between $\Lambda^0{}_0$ and $\Lambda^i{}_0$,

$$\left(\Lambda^0{}_0\right)^2 - \sum_{i=1}^{3}\left(\Lambda^i{}_0\right)^2 = 1 \tag{3.77}$$

Combining Eqs. (3.76) and (3.77) yields the Lorentz factor $\gamma = \gamma(\boldsymbol{v})$ to read

$$\Lambda^0{}_0 = \gamma = \gamma(\boldsymbol{v}) = \frac{1}{\sqrt{1-\frac{\boldsymbol{v}^2}{c^2}}} \tag{3.78}$$

which again is a scalar under Lorentz transformations.

In order to finally determine the desired transformation matrix $\Lambda(\boldsymbol{v})$ for a general Lorentz boost, we note that we can always first rotate IS$'$ to IS$''$ by application of $\Lambda(R)$ as given by Eq. (3.19) such that the new coordinate axis x'' is parallel to $\boldsymbol{v}$, then apply the familiar Lorentz boost $\Lambda(v)$ in x''-direction as given by Eq. (3.66), and finally rotate the new system IS$'''$ back to IS$'$ by the inverse rotation $\Lambda(R^T)$,

$$\Lambda(\boldsymbol{v}) = \Lambda(R^T)\,\Lambda(v)\,\Lambda(R) \tag{3.79}$$

The effect of the rotations as described by $\Lambda(R)$ and $\Lambda(R^T)$ on the Lorentz boost matrix $\Lambda(v)$ is as follows: the element $\Lambda^0{}_0$ remains unaffected, the elements $\Lambda^i{}_0$ and $\Lambda^0{}_i$ transform like a 3-vector, and the elements $\Lambda^i{}_j$ transform like a 3-tensor of second rank. Since the transformation matrix $\Lambda(\boldsymbol{v})$ must reduce to the familiar Lorentz boost in x-direction as given by Eq. (3.66) for $\boldsymbol{v} = v\boldsymbol{e}_1$, we thus must have

$$\Lambda^i{}_0 = \Lambda^0{}_i = -\gamma\frac{v_i}{c} \tag{3.80}$$

which we have already found above by direct calculation. Since $\delta_{ij} + v_i v_j(\gamma - 1)/v^2$ is indeed a 3-tensor of second rank with the correct limit for $\boldsymbol{v} = v\boldsymbol{e}_1$, the final transformation matrix for a general Lorentz boost is given by

$$\Lambda(\boldsymbol{v}) = \begin{pmatrix} \gamma & -\gamma v_1/c & -\gamma v_2/c & -\gamma v_3/c \\ -\gamma v_1/c & & & \\ -\gamma v_2/c & & \delta_{ij} + \dfrac{v_i v_j(\gamma-1)}{v^2} & \\ -\gamma v_3/c & & & \end{pmatrix} \tag{3.81}$$

If one feels uncomfortable with our vector- and tensor-reasoning above, one is strongly recommended to verify that Eq. (3.81) represents the correct transformation matrix by explicit calculation of Eq. (3.17).

The matrix for a general Lorentz boost as given by Eq. (3.81) is symmetric, $\Lambda(\boldsymbol{v}) = [\Lambda(\boldsymbol{v})]^T$. Its inverse

$$\left[\Lambda(\boldsymbol{v})\right]^{-1} = \Lambda(-\boldsymbol{v}) \tag{3.82}$$

is thus easily obtained from $\Lambda(\boldsymbol{v})$ just by switching the sign of the $\Lambda^i{}_0$ and $\Lambda^0{}_i$ elements. However, a general Lorentz transformation is *not* symmetric, since rotations $\Lambda(R)$ are not symmetric and any Lorentz transformation might be expressed as the product of a general Lorentz boost and a rotation, cf. Eq. (3.20).

In this section we have presented several approaches to looking at the Lorentz transformations for boosts. The reader is strongly encouraged to play around with all of them a little bit in order to convince himself or herself of the consistency of the above equations and to develop some familiarity with these crucial transformations.

3.2.2 Length Contraction, Time Dilation, and Proper Time

3.2.2.1 Length Contraction

We consider again the canonical situation of two inertial frames IS and IS′ related by a Lorentz boost in x-direction. At time $t = t' = 0$ the origins of the two frames shall coincide, cf. Figure 3.3. Now a stick (or any other object) of length l_0 at rest in IS′ is located along the x-axis such that its beginning is in the origin at $x'_1 = 0$. Its end is thus at $x'_2 = l_0$, and clearly, an observer in IS′ will measure the length of this stick to be

$$x'_2 - x'_1 = l_0 \qquad \text{(length in IS}'\text{)} \tag{3.83}$$

This length of an object at rest is denoted as its *proper length* l_0.

In order to measure the length of this object in IS, one has to measure the positions of its beginning x_1 and its end x_2 at the *same* time t. Without loss of generality we might always choose $t = 0$. We thus have to consider the two events

$$\text{E}_1 : \quad (t_1 = 0,\, x_1) \ \text{in IS} \quad \longrightarrow \quad (t'_1,\, x'_1 = 0) \ \text{in IS}' \tag{3.84}$$

$$\text{E}_2 : \quad (t_2 = 0,\, x_2) \ \text{in IS} \quad \longrightarrow \quad (t'_2,\, x'_2 = l_0) \ \text{in IS}' \tag{3.85}$$

The coordinates of any of these two events in IS and IS′ are related via the Lorentz boost given by Eq. (3.67). Taking advantage of $t_1 = t_2 = 0$ directly

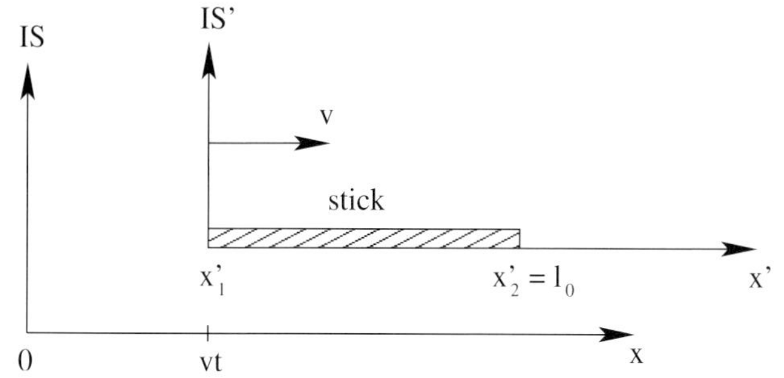

Figure 3.3 Illustration of the situation for the length contraction of a moving body. The stick with proper length l_0 is at rest in IS$'$, whereas it is moving with velocity v for an observer in IS. For $t = t' = 0$ the two inertial frames IS and IS$'$ coincide, and for later times the coordinates of events E with respect to IS and IS$'$ are related to each other according to Eq. (3.67).

yields $x_1 = 0$ and $x_2 = l_0/\gamma$. Since the length l of the stick in IS is given as the distance between its beginning and end, $l = x_2 - x_1$, one finds for the length in IS

$$l = \frac{l_0}{\gamma} = l_0\sqrt{1 - \frac{v^2}{c^2}} \qquad \text{(length in IS)} \tag{3.86}$$

The length of moving bodies is thus always smaller than their proper length, or more precisely, the length of an object to be measured by an observer is the smaller the faster the observer and the body are moving relative to each other. This phenomenon is called *length contraction* or *Fitzgerald–Lorentz contraction* of moving bodies.

The derivation above was focused on a Lorentz boost in the direction of the stick, i.e., the relative motion of the two frames IS and IS$'$, described by the velocity vector v, and the stick were parallel. Due to the Lorentz boost given by Eq. (3.67) no length contraction will occur in directions perpendicular to v. This immediately yields the general expression for the length contraction,

$$l_{\|} = l_{0,\|}\sqrt{1 - \frac{v^2}{c^2}} \qquad \text{and} \qquad l_{\perp} = l_{0,\perp} \tag{3.87}$$

3.2.2.2 Time Dilation

Similar to the effect on lengths, moving clocks show different times as compared to those at rest. In order to exhibit this peculiarity of special relativity, we again consider the situation of a Lorentz boost as illustrated in Figure 3.2. Now a clock is placed at the origin of IS$'$ which shall be at rest in IS$'$. With this clock we may measure the time between any two events E_1 and E_2 in IS$'$.

Now 2 observers are placed on the x-axis of IS at positions x_1 and x_2. Without loss of generality we may always assume $x_1 = 0$. We define the two events

E_1 and E_2 by the passing of the IS′-clock at the two observers at positions x_1 and x_2, which are at rest in IS. We thus have to consider the two events

$$E_1: \quad (t_1 = 0, x_1 = 0) \text{ in IS} \quad \longrightarrow \quad (t_1', x_1' = 0) \text{ in IS}' \tag{3.88}$$

$$E_2: \quad (t_2, x_2 = vt_2) \text{ in IS} \quad \longrightarrow \quad (t_2', x_2' = 0) \text{ in IS}' \tag{3.89}$$

For an observer in IS′ the time difference between these two events is trivially given by

$$t_2' - t_1' \overset{(3.67)}{=} t_2' \equiv \tau \qquad \text{(time difference in IS}') \tag{3.90}$$

and is called the *proper time* of the clock. Since the coordinates of any of these two events in IS and IS′ are again related via Eq. (3.67) one immediately finds $t_1' = 0$ and $t_2' = t_2/\gamma$. With $t \equiv t_2 - t_1 = t_2$ denoting the time difference in IS, the relation between the time differences (between the two events) in IS and IS′ is given by

$$t = \gamma\tau = \frac{\tau}{\sqrt{1 - \frac{v^2}{c^2}}} \qquad \text{(time difference in IS)} \tag{3.91}$$

The time difference t in IS is thus always larger than the time difference τ in IS′, where the clock is at rest, i.e., a moving clock loses time as compared to a clock at rest. This phenomenon is called relativistic *time dilation* and might briefly be summarized by the statement that "proper time is always the shortest time between two events".

3.2.2.3 **Proper Time**

We close this section by considering the more general case of a time-dependent movement of the clock. With reference to the inertial frame IS the movement of the clock is now described by the velocity $v(t)$. In this case no global inertial frame IS′ might be found where the clock is at rest. However, for each moment t we might consider the inertial frame IS′ which is moving with the instantaneous velocity $v(t)$ relative to IS. In IS′ the clock is thus at rest at time t. According to Eq. (3.91) the infinitesimal time intervals $d\tau$ in IS′ and dt in IS are thus related by

$$d\tau = \frac{dt}{\gamma(v)} = dt\sqrt{1 - \frac{v(t)^2}{c^2}} \tag{3.92}$$

The proper time between two events E_1 at time t_1 and E_2 at time t_2 (in IS) as shown by the clock in IS′ is thus given by the integral

$$\tau = \int_{\tau_1}^{\tau_2} d\tau = \int_{t_1}^{t_2} dt\sqrt{1 - \frac{v(t)^2}{c^2}} \tag{3.93}$$

The most important property of the proper time τ is its invariance under Lorentz transformations, i.e., τ is a Lorentz scalar and thus independent of the inertial system. Being defined as the time difference in the inertial system where the clock is at rest, this statement is obviously true. However, in order to formally prove this statement we consider the infinitesimal distance $\mathrm{d}s$ for the clock. According to Eq. (3.10) it is given by

$$\mathrm{d}s = \sqrt{c^2\mathrm{d}t^2 - \mathrm{d}\boldsymbol{r}^2} = c\sqrt{1 - \frac{v(t)^2}{c^2}}\,\mathrm{d}t \overset{(3.92)}{=} c\,\mathrm{d}\tau \tag{3.94}$$

Since $\mathrm{d}s$ is a scalar, also $\mathrm{d}\tau$ is a scalar and thus invariant under Lorentz transformations.

Numerous high-precision experiments have been performed in order to test the counter-intuitive relativistic effects presented above. We emphasize that not a single piece of experimental evidence so far could falsify any of these effects or even cause the slightest doubt about their correctness.

3.2.3 Addition of Velocities

We may encounter situations where one particle or system is moving (with time-dependent velocity $\boldsymbol{v}(t)$) relative to an inertial system IS′, which itself is moving relative to another inertial system IS. Let $\boldsymbol{v}_1$ denote the velocity of IS′ relative to IS. At each time t we may consider the inertial system IS″ where the particle is at rest, i.e., IS″ is moving with velocity $\boldsymbol{v}_2 = \boldsymbol{v}(t)$ relative to IS′. This general situation is illustrated in Figure 3.4. In the following we will investigate the velocity $\boldsymbol{V} = \boldsymbol{V}(\boldsymbol{v}_1, \boldsymbol{v}_2)$ with which IS″ is moving relative to IS. Due to the group structure of Lorentz transformations, the Lorentz transformation from IS to IS″ may always be obtained by successive application of the transformation from IS to IS′ followed by the transformation from IS′ to IS″. Consequently, the transformation matrix $\Lambda(\boldsymbol{V})$ for the boost from IS to IS″ is simply given by the matrix product of the boost matrices for the transformations from IS to IS′, $\Lambda(\boldsymbol{v}_1)$, and IS′ to IS″, $\Lambda(\boldsymbol{v}_2)$,

$$\Lambda(\boldsymbol{V}) = \Lambda(\boldsymbol{v}_2)\,\Lambda(\boldsymbol{v}_1) \tag{3.95}$$

3.2.3.1 Parallel Velocities

If the two velocities $\boldsymbol{v}_1$ and $\boldsymbol{v}_2$ are parallel to each other, the situation is simplified to a large extent. We can always rotate all three inertial systems simultaneously such that $\boldsymbol{v}_1$ and $\boldsymbol{v}_2$ are parallel to the x-axes of all three frames of reference. Consequently, all three transformation matrices occurring in Eq. (3.95) are of the simple form given by Eq. (3.66) or Eq. (3.74), respectively. Eq. (3.95)

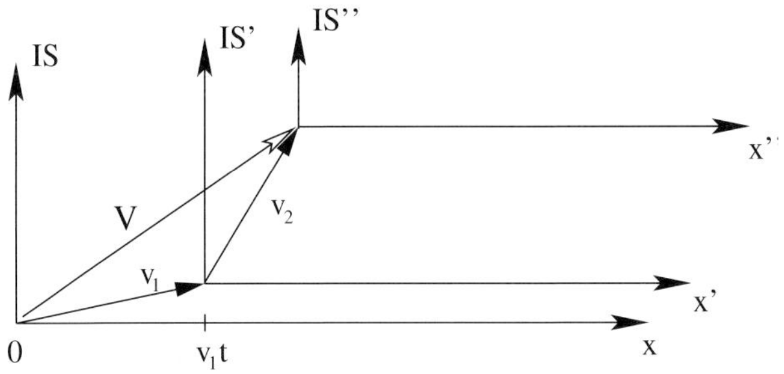

Figure 3.4 Illustration of three inertial systems moving relative to each other. IS′ is moving with velocity v_1 relative to IS, whereas IS″ is moving with velocity v_2 relative to IS′. The velocities v_1 and v_2 will in general not be parallel to each other. Without loss of generality, however, the axes of the inertial systems may be assumed to be parallel to each other.

reduces thus to

$$\begin{aligned}\Lambda(\boldsymbol{V}) &\overset{(3.74)}{=} \begin{pmatrix} \cosh\psi_2 & -\sinh\psi_2 \\ -\sinh\psi_2 & \cosh\psi_2 \end{pmatrix} \cdot \begin{pmatrix} \cosh\psi_1 & -\sinh\psi_1 \\ -\sinh\psi_1 & \cosh\psi_1 \end{pmatrix} \\ &\overset{(C.34)}{=} \begin{pmatrix} \cosh(\psi_1+\psi_2) & -\sinh(\psi_1+\psi_2) \\ -\sinh(\psi_1+\psi_2) & \cosh(\psi_1+\psi_2) \end{pmatrix} \\ &= \begin{pmatrix} \cosh\psi & -\sinh\psi \\ -\sinh\psi & \cosh\psi \end{pmatrix}\end{aligned} \tag{3.96}$$

where, according to Eq. (3.72), the rapidities are given by

$$\psi_1 = \operatorname{artanh}\frac{v_1}{c}, \qquad \psi_2 = \operatorname{artanh}\frac{v_2}{c}, \qquad \psi = \operatorname{artanh}\frac{V}{c} \tag{3.97}$$

For parallel velocities the rapidities are thus simply additive, $\psi = \psi_1 + \psi_2$. Furthermore, according to Eq. (3.96) the Lorentz boosts from IS to IS′ and IS′ to IS″ commute,

$$\Lambda(\boldsymbol{v}_1)\,\Lambda(\boldsymbol{v}_2) = \Lambda(\boldsymbol{v}_2)\,\Lambda(\boldsymbol{v}_1) \qquad (\text{for } \boldsymbol{v}_1 \parallel \boldsymbol{v}_2) \tag{3.98}$$

This may be compared with the successive application of three-dimensional rotations, which also only commute if they are applied about the same axis of rotation. In order to derive an explicit expression for the velocity $\boldsymbol{V} = \boldsymbol{V}(\boldsymbol{v}_1, \boldsymbol{v}_2)$, we consider

$$\begin{aligned}\frac{V}{c} &\overset{(3.97)}{=} \tanh\psi \overset{(3.96)}{=} \tanh(\psi_1+\psi_2) \\ &\overset{(C.36)}{=} \frac{\tanh\psi_1 + \tanh\psi_2}{1+\tanh\psi_1\tanh\psi_2} \overset{(3.97)}{=} \frac{1}{c}\,\frac{v_1+v_2}{1+v_1v_2/c^2}\end{aligned} \tag{3.99}$$

For parallel velocities $v_1 \parallel v_2$ the resulting velocity V with which IS″ is moving relative to IS is thus given by

$$V = \frac{v_1 + v_2}{1 + v_1 v_2 / c^2} \qquad (\text{for } \boldsymbol{v}_1 \parallel \boldsymbol{v}_2) \tag{3.100}$$

It is instructive to consider the limits of Eq. (3.100), which is denoted as the addition theorem for (parallel) relativistic velocities. For nonrelativistic velocities $v_1 \ll c$ and $v_2 \ll c$ Eq. (3.100) simply reduces to the familiar nonrelativistic Galilean addition of velocities, $V = v_1 + v_2$, cf. Eq. (2.14). In the ultra-relativistic limit, however, where at least one of the velocities v_1 or v_2 approaches the speed of light c (from below), the resulting velocity V also tends to c. Furthermore, if at least one of the velocities v_1 or v_2 equals the speed of light, the resulting velocity V will also equal the speed of light; this is the case for photons, for example. Once again, this is a further precursor of the impossibility of velocities which are larger than the speed of light, cf. section 3.3.

3.2.3.2 **General Velocities**

For general, i.e., non-parallel velocities $\boldsymbol{v}_1$ and $\boldsymbol{v}_2$ we cannot simplify the problem by application of a suitable rotation of the coordinate axes. Now all three Lorentz boost matrices occurring in Eq. (3.95) are of the most general form given by Eq. (3.81) with a yet undetermined resulting velocity $\boldsymbol{V}$. We thus have to evaluate Eq. (3.95) directly. In order to achieve this task, it is convenient to introduce the abbreviations

$$\gamma_1 = \gamma(\boldsymbol{v}_1), \qquad \gamma_2 = \gamma(\boldsymbol{v}_2), \qquad \gamma = \gamma(\boldsymbol{V}) \tag{3.101}$$

where the Lorentz factors $\gamma(\boldsymbol{v})$ are defined according to Eq. (3.78). We can now calculate the 00-element of $\Lambda(\boldsymbol{V})$,

$$\begin{aligned}
\gamma &= [\Lambda(\boldsymbol{V})]^0{}_0 = \left[\Lambda(\boldsymbol{v}_2)\right]^0{}_\mu \left[\Lambda(\boldsymbol{v}_1)\right]^\mu{}_0 \\
&= \left[\Lambda(\boldsymbol{v}_2)\right]^0{}_0 \left[\Lambda(\boldsymbol{v}_1)\right]^0{}_0 + \sum_{i=1}^{3} \left[\Lambda(\boldsymbol{v}_2)\right]^0{}_i \left[\Lambda(\boldsymbol{v}_1)\right]^i{}_0 \\
&= \gamma_1\gamma_2 + \sum_{i=1}^{3} \gamma_1\gamma_2 \frac{v_{1,i}\, v_{2,i}}{c^2} = \gamma_1\gamma_2 \left(1 + \frac{\boldsymbol{v}_1 \cdot \boldsymbol{v}_2}{c^2}\right)
\end{aligned} \tag{3.102}$$

Similarly the $0j$-element of $\Lambda(\boldsymbol{V})$ may be obtained to read

$$\begin{aligned}
-\gamma\frac{V_j}{c} &= [\Lambda(\boldsymbol{V})]^0{}_j = \left[\Lambda(\boldsymbol{v}_2)\right]^0{}_\mu \left[\Lambda(\boldsymbol{v}_1)\right]^\mu{}_j \\
&= \left[\Lambda(\boldsymbol{v}_2)\right]^0{}_0 \left[\Lambda(\boldsymbol{v}_1)\right]^0{}_j + \sum_{i=1}^{3} \left[\Lambda(\boldsymbol{v}_2)\right]^0{}_i \left[\Lambda(\boldsymbol{v}_1)\right]^i{}_j
\end{aligned}$$

$$= -\gamma_1\gamma_2\frac{v_{1,j}}{c} - \sum_{i=1}^{3}\gamma_2\frac{v_{2,i}}{c}\left(\delta_{ij} + (\gamma_1 - 1)\frac{v_{1,i}\,v_{1,j}}{v_1^2}\right) \tag{3.103}$$

We directly find $\boldsymbol{V} = \sum_{j=1}^{3} V_j \boldsymbol{e}_j$ from Eq. (3.103) to read

$$\boldsymbol{V} = \frac{\gamma_1\gamma_2}{\gamma}\left(\boldsymbol{v}_1 + \frac{\boldsymbol{v}_2}{\gamma_1} + \frac{\gamma_1 - 1}{\gamma_1}\frac{\boldsymbol{v}_1\cdot\boldsymbol{v}_2}{v_1^2}\boldsymbol{v}_1\right) \tag{3.104}$$

In order to express $\boldsymbol{V}$ solely by $\boldsymbol{v}_1$ and $\boldsymbol{v}_2$ it is convenient to eliminate the Lorentz factor γ in Eq. (3.104) by employing the explicit expression for γ as given by Eq. (3.102). We thus arrive at

$$\boldsymbol{V} = \frac{\boldsymbol{v}_1 + \boldsymbol{v}_2 + \left(\boldsymbol{v}_2 - \frac{\boldsymbol{v}_1\cdot\boldsymbol{v}_2}{v_1^2}\boldsymbol{v}_1\right)\left(\sqrt{1 - \frac{v_1^2}{c^2}} - 1\right)}{1 + \frac{\boldsymbol{v}_1\cdot\boldsymbol{v}_2}{c^2}} \tag{3.105}$$

Eq. (3.105) constitutes the addition theorem for general relativistic velocities. For parallel velocities $\boldsymbol{v}_1 \parallel \boldsymbol{v}_2$ it reduces to Eq. (3.100). Note that for general, i.e., non-parallel velocities $\boldsymbol{v}_1$ and $\boldsymbol{v}_2$ the result is *not* symmetric in $\boldsymbol{v}_1$ and $\boldsymbol{v}_2$. It thus makes a difference whether we first transform with $\Lambda(\boldsymbol{v}_1)$ and then with $\Lambda(\boldsymbol{v}_2)$ or vice versa. This is most dramatically seen for orthogonal velocities $\boldsymbol{v}_1 \perp \boldsymbol{v}_2$, for which the resulting velocity $\boldsymbol{V}$ is given by

$$\boldsymbol{V} = \boldsymbol{v}_1 + \boldsymbol{v}_2\sqrt{1 - \frac{v_1^2}{c^2}} \qquad (\text{for } \boldsymbol{v}_1 \perp \boldsymbol{v}_2) \tag{3.106}$$

General Lorentz boosts $\Lambda(\boldsymbol{v})$ are thus in general *not* commutative, whereas boosts for parallel velocities commute, cf. Eq. (3.98).

3.3 Relativistic Dynamics

So far we have only considered kinematical effects of special relativity, but no forces or energies have been discussed yet. We will close this gap in this section with the discussion of dynamical aspects of special relativity. This will reveal some of the most remarkable and famous peculiarities of this theory, which have a striking influence on relativistic quantum chemistry.

3.3.1 Elementary Relativistic Dynamics

3.3.1.1 Trajectories and Relativistic Velocity

In nonrelativistic mechanics the trajectory of a particle (as given by a sequence of events, cf. section 2.1.2) is described by $\boldsymbol{r} = \boldsymbol{r}(t)$, whereas the time variable t simply parameterizes the propagation of the particle. The reader should recall here the absolute meaning of time t in nonrelativistic physics. In relativistic theories, however, ordinary time t no longer features any absolute meaning but depends on the chosen frame of reference. Furthermore, it may often be advantageous to treat space and time on an equal footing and thus employ another parameter for the parameterization of the trajectory. The natural substitution for ordinary time t is proper time τ, i.e., we may describe the trajectory by $x^\mu = x^\mu(\tau)$. The relationship between t and τ is given by Eq. (3.92), which can be expressed as $\mathrm{d}t = \gamma(\boldsymbol{v})\mathrm{d}\tau$. Since the trajectory $\boldsymbol{r}(t)$ uniquely determines $\boldsymbol{v}(t) = \dot{\boldsymbol{r}}(t)$ and thus $t = t(\tau)$, also $x^0 = ct(\tau) = x^0(\tau)$ and $x^i = x^i(t(\tau)) = x^i(\tau)$ are fixed. The four functions $x^\mu(\tau)$ thus contain exactly the same information as the three functions $x^i(t)$,

$$x^i = x^i(t) \quad \xleftrightarrow{t=t(\tau)} \quad x^\mu = x^\mu(\tau) \tag{3.107}$$

It is thus natural to generalize the nonrelativistic three-dimensional definition of velocity $\boldsymbol{v} = \mathrm{d}\boldsymbol{r}/\mathrm{d}t$ by the so-called *4-velocity* u^μ given by

$$u^\mu = \frac{\mathrm{d}x^\mu}{\mathrm{d}\tau} \overset{(3.92)}{=} \gamma(c, \boldsymbol{v}) \tag{3.108}$$

where the explicit form of the 4-vector x^μ as given by Eq. (3.7) has been employed in the last step. Note that the right hand side of this equation has conveniently been denoted as a row vector, whereas it still has to be interpreted as a column vector, of course. Since proper time τ is a Lorentz scalar and space-time x^μ is a Lorentz vector, the 4-velocity u^μ is also a Lorentz 4-vector. It features the same four-dimensional length in all frames of reference,

$$u \cdot u = u^T g\, u = u^\mu u_\mu = \gamma^2 \left(c^2 - v^2\right) = c^2 \tag{3.109}$$

3.3.1.2 Relativistic Momentum and Energy

Similarly to the four-dimensional velocity u^μ also the *4-momentum* p^μ is defined in analogy to the nonrelativistic set-up, i.e.,

$$p^\mu = m u^\mu = m \frac{\mathrm{d}x^\mu}{\mathrm{d}\tau} = \gamma m(c, \boldsymbol{v}) = \left(\frac{E}{c}, \boldsymbol{p}\right) \tag{3.110}$$

Obviously, the relativistic momentum p^μ is also a Lorentz 4-vector, where we have introduced the relativistic (kinematical) 3-momentum

$$\boldsymbol{p} = \gamma m \boldsymbol{v} = \frac{m\boldsymbol{v}}{\sqrt{1 - \frac{\boldsymbol{v}^2}{c^2}}} \tag{3.111}$$

which differs from the nonrelativistic kinematical momentum defined by Eq. (2.6) by the factor γ, and the relativistic energy is given by

$$E = \gamma mc^2 = \frac{mc^2}{\sqrt{1 - \frac{\boldsymbol{v}^2}{c^2}}} \tag{3.112}$$

It should be emphasized that the relativistic energy E does *not* equal mc^2, since the famous expression $E = m(v)c^2$ assumes a velocity-dependent mass $m(v) = \gamma m$. However, we will consequently *not* adopt any concept of velocity-dependent mass (or relativistic mass) throughout this presentation. Mass m will always denote the rest mass, which is independent of the chosen frame of reference and thus a fundamental property of the particle.

The deeper reason for calling E and $\boldsymbol{p}$ the relativistic energy and momentum, respectively, is that they are *conserved* quantities in the following sense. If an observer in one specific inertial frame of reference IS sees that E and/or $\boldsymbol{p}$ are conserved throughout a reaction or process, so does any other observer in another frame of reference IS′ related to IS via a Lorentz transformation. This unique feature of energy and momentum directly follows from the 4-vector property of p^μ, since the change $\Delta p'^\mu$ in energy or momentum in IS′ is related to those in IS by

$$\Delta p'^\mu = \Delta \Lambda^\mu{}_\nu p^\nu = \Lambda^\mu{}_\nu \Delta p^\nu \tag{3.113}$$

The last equality holds since the Lorentz transformation Λ only depends on the relative motion between IS and IS′ and not on specific features of the system under investigation; neither does it feature an explicit space-time dependence. If the change in energy and/or momentum vanishes in IS, it will thus also vanish in IS′. We mention without proof that energy E and momentum $\boldsymbol{p}$ are the *only* functions of velocity whose conservation is Lorentz invariant. Furthermore, if momentum $\boldsymbol{p}$ is conserved, so is the energy E. This is most easily seen by considering two inertial frames IS and IS′ with

$$\Delta \boldsymbol{p} = 0 \quad (\text{in IS}) \qquad \text{and} \qquad \Delta \boldsymbol{p}' = 0 \quad (\text{in IS}') \tag{3.114}$$

Since momentum is a 4-vector, the momentum changes in IS and IS′ are related to each other via a Lorentz transformation, $\Delta p'^i = \Lambda^i{}_\alpha \Delta p^\alpha$ $(i = 1, 2, 3)$. Applying the assumed momentum conservation as expressed by Eq. (3.114) to this relation we find $0 = \Lambda^i{}_0 \, \Delta p^0$, and since $\Lambda^i{}_0$ does not vanish in general we must have $\Delta p^0 = 0$, i.e., the energy $E = cp^0$ is conserved.

From the discussion so far it should be obvious that in relativistic theories a clear distinction between energy and momentum is no longer possible. Energy and momentum represent the four components of a 4-vector and thus transform into each other under Lorentz transformations. This peculiarity of special relativity is often reflected by denoting the 4-vector p^μ as the *energy–momentum* vector in order to emphasize the unity of energy and momentum in relativistic theories. This may be compared to the space-time vector x^μ which combines both space and time coordinates in a single 4-vector.

As a direct consequence of the expression for the relativistic energy given by Eq. (3.112) one immediately finds that the energy for *massive* particles ($m > 0$) diverges to infinity for $v \rightarrow c$. It is thus not possible for any *massive* particle to move with a velocity v equal to or greater than the speed of light c, since this would require infinite amounts of energy. This also excludes the possibility of inertial frames of reference moving with the speed of light c or even faster relative to each other. This perfectly corresponds to our explicit expressions for Lorentz transformations which become singular for $v \rightarrow c$, cf. section 3.2.1. For *massless* particles ($m = 0$) with nonvanishing energy E, however, Eq. (3.112) implies that they must necessarily move with the speed of light c, since otherwise the energy would constantly vanish. This is then true for *any* inertial frame of reference, since due to Eqs. (3.100) or (3.105) a particle velocity $v = c$ is not modified under Lorentz transformations. Nature features indeed such massless particles propagating with the speed of light in any IS. The most famous and important example of this are photons, i.e., electromagnetic waves.

For later convenience we briefly mention the four-dimensional scalar product between energy–momentum and space-time explicitly, which is given by

$$p \cdot x = p^\mu x_\mu = Et - \boldsymbol{p} \cdot \boldsymbol{r} \tag{3.115}$$

3.3.1.3 **Energy–Momentum Relation**

As a direct consequence of Eq. (3.109), the four-dimensional length of the relativistic momentum has the same value in all frames of reference,

$$p \cdot p = p^T g \, p = p^\mu p_\mu \overset{(3.110)}{=} m^2 u^\mu u_\mu \overset{(3.109)}{=} m^2 c^2 \tag{3.116}$$

Together with Eq. (3.110) this immediately yields the famous *energy–momentum relation*

$$E^2 = c^2 \boldsymbol{p}^2 + m^2 c^4 \tag{3.117}$$

which is also sometimes denoted as the *energy–momentum dispersion relation*. It is exactly this relation which will later give rise to the existence of anti-particles with negative energy, since Eq. (3.117) obviously features two roots, a positive-energy solution (related to normal 'particles') and a negative-energy solution (somehow related to 'anti-particles '),

$$E = \begin{cases} +\sqrt{c^2\boldsymbol{p}^2 + m^2c^4} & = & +\gamma mc^2 \quad \text{for} \quad E > 0 \\ -\sqrt{c^2\boldsymbol{p}^2 + m^2c^4} & = & -\gamma mc^2 \quad \text{for} \quad E < 0 \end{cases} \tag{3.118}$$

The existence of negative-energy solutions is thus an intrinsic feature of *every* relativistic theory, and we will have to find suitable means and interpretations to cope with this peculiarity (cf. chapter 5).

The energy of a free particle as given by Eq. (3.112) may always be written as the sum of two terms,

$$E = \gamma mc^2 = \underbrace{mc^2}_{E_0} + \underbrace{(\gamma - 1)mc^2}_{E_{\text{kin}}} \tag{3.119}$$

where the first term E_0 denotes the rest energy of the particle and the second term E_{kin} denotes its relativistic kinetic energy, i.e., the energy required to accelerate the particle from rest to velocity $\boldsymbol{v}$. For low velocities $\boldsymbol{v}$ and thus low momenta $\boldsymbol{p}$, i.e., $|\boldsymbol{v}| \ll c$ and thus $|\boldsymbol{p}| \ll mc$, respectively, we may approximate the energy expression as given by Eq. (3.119) by the leading order terms of its series expansion,

$$E \overset{(3.65)}{=} mc^2 + \frac{1}{2}mv^2 + \frac{3}{8}\frac{mv^4}{c^2} + \dots \qquad (\text{for } |\boldsymbol{v}| \ll c) \tag{3.120}$$

The second term on the right hand side represents the familiar nonrelativistic kinetic energy and the third term is the first relativistic correction; higher-order corrections have been suppressed here. We could have equally well expanded the energy–momentum relation as given by Eq. (3.117) to arrive at

$$E = mc^2 + \frac{\boldsymbol{p}^2}{2m} - \frac{1}{8}\frac{p^4}{m^3c^2} + \dots \qquad (\text{for } |\boldsymbol{p}| \ll mc) \tag{3.121}$$

We emphasize that the prefactor of the first relativistic correction term differs from the corresponding prefactor in Eq. (3.120) since the relativistic momentum $\boldsymbol{p}$ itself contains the Lorentz factor γ, which has been expanded in Eq. (3.120) but *not* in Eq. (3.121).

For ultra-relativistic particles with $|\boldsymbol{p}| \gg mc$ the rest energy contribution to the energy E is negligible and we thus find $E \approx c|\boldsymbol{p}|$. The energy–momentum relation (3.117) also holds for massless particles with $m = 0$ and reads $E = c|\boldsymbol{p}|$ (for positive-energy solutions).

3.3.2 Equation of Motion

In order to derive the relativistic analog of the Newtonian equation of motion, we again consider the motion of a single particle with velocity $\boldsymbol{v}$ with reference to the inertial frame IS. The instantaneous rest frame of the particle at time t will be denoted as IS$'$. Since the particle is at rest in IS$'$ at time t (and will thus move only with small velocities $v \ll c$ at times $t \pm \mathrm{d}t$) we *assume* Newton's classical equation of motion as given by Eq. (2.2) to hold *exactly* in IS$'$,

$$\boldsymbol{F} = \frac{\mathrm{d}\boldsymbol{p}'_N}{\mathrm{d}t'} = m\frac{\mathrm{d}\boldsymbol{v}'}{\mathrm{d}t'} = m\frac{\mathrm{d}^2\boldsymbol{r}'}{\mathrm{d}t'^2} \tag{3.122}$$

Here $\boldsymbol{F}$ denotes the familiar nonrelativistic Newtonian force, which is independent of the chosen frame of reference (cf. section 2.1.2) and does thus not bear a prime. The classical nonrelativistic Newtonian momentum is denoted as $\boldsymbol{p}'_N = m\boldsymbol{v}'$, and m is again the rest mass of the particle. Newton's equation of motion is well tested and established for small velocities and we do thus not have any reason to question its validity in IS$'$. It cannot be overemphasized, however, that Eq. (3.122) is only valid in IS$'$ but not in other inertial frames of reference. In order to find the correct equation of motion for general inertial frames of reference we will establish a covariant 4-vector equation which reduces to Eq. (3.122) if the particle is at rest.

3.3.2.1 Minkowski Force

According to our discussion of the 4-velocity in the last section the natural covariant generalization of Eq. (3.122) is given by

$$f^\mu = \frac{\mathrm{d}p^\mu}{\mathrm{d}\tau} = m\frac{\mathrm{d}u^\mu}{\mathrm{d}\tau} = m\frac{\mathrm{d}^2x^\mu}{\mathrm{d}\tau^2} \tag{3.123}$$

where the 4-vector f^μ on the left hand side of this equation is denoted as the *Minkowski force* and is the only yet undetermined quantity. Eq. (3.123) is the relativistic equation of motion — the relativistic pendant to Newton's second axiom given by Eq. (2.2) — and as a covariant 4-vector equation it will hold in *any* inertial system. We will now determine the Minkowski force by the requirement that Eq. (3.123) reduces to Eq. (3.122) in IS$'$. We thus have to express the 4-acceleration $\mathrm{d}u^\mu/\mathrm{d}\tau$ by the three-dimensional quantities $\boldsymbol{v}$ and $\dot{\boldsymbol{v}} = \mathrm{d}\boldsymbol{v}/\mathrm{d}t$. We first note that

$$\dot{\gamma}(t) = \frac{\mathrm{d}\gamma(t)}{\mathrm{d}t} = \frac{\mathrm{d}}{\mathrm{d}t}\left[\left(1 - \frac{v^2(t)}{c^2}\right)^{-1/2}\right] = \gamma^3\,\frac{\dot{\boldsymbol{v}}\cdot\boldsymbol{v}}{c^2} \tag{3.124}$$

and thus arrive at

$$\frac{\mathrm{d}u^\mu}{\mathrm{d}\tau} \overset{(3.92)}{=} \gamma\frac{\mathrm{d}}{\mathrm{d}t}(\gamma c, \gamma\boldsymbol{v}) \overset{(3.124)}{=} \gamma^4\frac{\dot{\boldsymbol{v}}\cdot\boldsymbol{v}}{c^2}\,(c, \boldsymbol{v}) + (0, \gamma^2\dot{\boldsymbol{v}}) \tag{3.125}$$

Eq. (3.125) holds for *any* inertial frame of reference, and thus it is also true in IS$'$ where $\boldsymbol{v}' = 0$, which immediately yields

$$\frac{\mathrm{d}u'^{\mu}}{\mathrm{d}\tau} = (0, \dot{\boldsymbol{v}}') = \left(0, \frac{\mathrm{d}\boldsymbol{v}'}{\mathrm{d}t'}\right) \tag{3.126}$$

Insertion of this result in the equation of motion (3.123) yields the Minkowski force in IS$'$,

$$f' = (f'^{\mu}) = m\left(\frac{\mathrm{d}u'^{\mu}}{\mathrm{d}\tau}\right) = m\left(0, \frac{\mathrm{d}\boldsymbol{v}'}{\mathrm{d}t'}\right) \overset{(3.122)}{=} (0, \boldsymbol{F}) \tag{3.127}$$

The Minkowski force f in any other inertial frame IS, which moves with velocity $-\boldsymbol{v}$ relative to IS$'$, might thus be obtained via a Lorentz transformation,

$$f = \Lambda(-\boldsymbol{v})f' \qquad \longleftrightarrow \qquad f^{\mu} = \left[\Lambda(-\boldsymbol{v})\right]^{\mu}{}_{\nu} f'^{\nu} \tag{3.128}$$

For the simple case of a Lorentz boost in x-direction with $\boldsymbol{v} = v\boldsymbol{e}_1$ the relation between the Newtonian force $\boldsymbol{F} = (F_1, F_2, F_3)$ and the Minkowski force f^{μ} is by virtue of Eq. (3.66) given by

$$f = (f^{\mu}) = \Lambda(-v)f' \overset{(3.127)}{=} \left(\gamma\frac{v}{c}F_1, \gamma F_1, F_2, F_3\right) \tag{3.129}$$

For the general case of an arbitrary Lorentz boost $\Lambda(\boldsymbol{v})$ we note that the inverse transformation $\Lambda(-\boldsymbol{v})$ is easily obtained from Eq. (3.81) by switching the signs of the velocity components v_i (cf. Eq. (3.82)). The Minkowski force f in IS is thus given by

$$\begin{aligned} f = (f^{\mu}) &\overset{(3.128)}{=} \left(\gamma\sum_{i=1}^{3}\frac{v_i F_i}{c}, \left\{\sum_{j=1}^{3}\left[\delta_{ij} + \frac{v_i v_j(\gamma-1)}{v^2}\right]\cdot F_j\right\}_{i=1,2,3}\right) \\ &= \left(\gamma\frac{\boldsymbol{v}\cdot\boldsymbol{F}}{c}, \left[F_i + (\gamma-1)\frac{v_i(\boldsymbol{v}\cdot\boldsymbol{F})}{v^2}\right]_{i=1,2,3}\right) \\ &= \left(\gamma\frac{\boldsymbol{v}\cdot\boldsymbol{F}}{c}, \boldsymbol{F} + (\gamma-1)\frac{\boldsymbol{v}(\boldsymbol{v}\cdot\boldsymbol{F})}{v^2}\right) \end{aligned} \tag{3.130}$$

3.3.2.2 **Lorentz Force**

As the most important example of a Minkowski force in special relativity we will now explicitly consider the Lorentz force on a charged particle, which has already been encountered in section 2.4. In IS$'$, where the particle is at rest, it experiences a force

$$\boldsymbol{F} = q\boldsymbol{E}' = q(\boldsymbol{E}'_{\parallel} + \boldsymbol{E}'_{\perp}) \tag{3.131}$$

where the electric field $\boldsymbol{E}'$ has been artificially decomposed into its parallel and orthogonal components, $\boldsymbol{E}'_{\parallel}$ and $\boldsymbol{E}'_{\perp}$, respectively, relative to the velocity

v, with which IS′ is moving relative to IS. In order to derive the Minkowski force f in IS, we have to anticipate the relation between electromagnetic fields in IS and IS′, which will not be derived until section 3.4. Here we just mention this relation, cf. Eq. (3.183),

$$\boldsymbol{E}'_{\parallel} = \boldsymbol{E}_{\parallel} \quad \text{and} \quad \boldsymbol{E}'_{\perp} = \gamma \left(\boldsymbol{E}_{\perp} + \frac{1}{c} \boldsymbol{v} \times \boldsymbol{B} \right) \tag{3.132}$$

Now the electric field $\boldsymbol{E}'$ in IS′ may be expressed by the components of the electric and magnetic fields in IS,

$$\begin{aligned} \boldsymbol{E}' &= \boldsymbol{E}'_{\parallel} + \boldsymbol{E}'_{\perp} = \boldsymbol{E}_{\parallel} + \gamma \left(\boldsymbol{E}_{\perp} + \frac{1}{c} \boldsymbol{v} \times \boldsymbol{B} \right) \\ &= \underbrace{\boldsymbol{E}_{\parallel} + \boldsymbol{E}_{\perp}}_{\boldsymbol{E}} - (1-\gamma) \boldsymbol{E}_{\perp} + \gamma \frac{1}{c} \boldsymbol{v} \times \boldsymbol{B} \end{aligned} \tag{3.133}$$

Note that we have artificially extracted the electric field $\boldsymbol{E}$ in the last equality of Eq. (3.133), which represents an exact relation. Inserting this expression for the electric field $\boldsymbol{E}'$ into the relation for the Minkowski force as given by Eq. (3.130) directly yields the four-dimensional Lorentz force f in IS,

$$\begin{aligned} f = (f^{\mu}) &= \left(\gamma \frac{q}{c} \boldsymbol{v} \cdot \boldsymbol{E}' \,,\, q\boldsymbol{E}' + (\gamma - 1) \frac{q\boldsymbol{v}\,(\boldsymbol{v} \cdot \boldsymbol{E}')}{v^2} \right) \\ &\overset{(3.133)}{=} \left(\gamma \frac{q}{c} \boldsymbol{v} \cdot \boldsymbol{E} \,,\, q \left[\boldsymbol{E} - (1-\gamma)\boldsymbol{E}_{\perp} + \gamma \frac{1}{c} \boldsymbol{v} \times \boldsymbol{B} \right.\right. \\ &\qquad \left.\left. + (\gamma - 1) \underbrace{\frac{\boldsymbol{v}\,(\boldsymbol{v} \cdot \boldsymbol{E})}{v^2}}_{\boldsymbol{E}_{\parallel}} \right] \right) \\ &= \gamma q \left(\frac{\boldsymbol{v} \cdot \boldsymbol{E}}{c} \,,\, \boldsymbol{E} + \frac{1}{c} \boldsymbol{v} \times \boldsymbol{B} \right) \end{aligned} \tag{3.134}$$

where we have repeatedly taken advantage of the fact that only the first term of $\boldsymbol{E}'$ as given by Eq. (3.133) features a nonvanishing scalar product with the velocity $\boldsymbol{v}$, $\boldsymbol{v} \cdot \boldsymbol{E}' = \boldsymbol{v} \cdot \boldsymbol{E}$, due to the definition of the parallel and perpendicular components of the electric field relative to the velocity $\boldsymbol{v}$. Eq. (3.134) represents the correct four-dimensional relativistic force in IS.

Insertion of the force f into the equation of motion (3.123) yields the relativistic equations of motion for a charged particle subject to an external electromagnetic field. For $\mu = 0$ it reads

$$\frac{\mathrm{d}}{\mathrm{d}t}(\gamma m c^2) = \frac{\mathrm{d}E}{\mathrm{d}t} = q\boldsymbol{v} \cdot \boldsymbol{E} \tag{3.135}$$

and for $\mu = i$ it is given by

$$\frac{\mathrm{d}}{\mathrm{d}t}(\gamma m \boldsymbol{v}) = \frac{\mathrm{d}\boldsymbol{p}}{\mathrm{d}t} = q \left(\boldsymbol{E} + \frac{1}{c} \boldsymbol{v} \times \boldsymbol{B} \right) = \boldsymbol{F}_L \tag{3.136}$$

We close this section by mentioning that the familiar Lorentz force F_L as given by Eq. (3.136) is the correct nonrelativistic, i.e., low-velocity approximation to the Newtonian force F since,

$$\begin{aligned} F &\overset{(3.131)}{=} qE' \overset{(3.133)}{=} q\left(E-(1-\gamma)E_{\perp}+\gamma\frac{1}{c}v\times B\right) \\ &\approx q\left(E+\frac{1}{c}v\times B\right)+\mathcal{O}\left(\frac{v^2}{c^2}\right) = F_L+\mathcal{O}\left(\frac{v^2}{c^2}\right) \end{aligned} \tag{3.137}$$

The explicit form of the Lorentz force F_L as derived in this section has been shown to be an immediate consequence of the transformation properties of the electric and magnetic fields under Lorentz transformations as given by Eq. (3.132).

3.3.3
Lagrangian and Hamiltonian Formulation

So far we have only considered elementary relativistic mechanics based on the equation of motion given by Eq. (3.123). Similarly to the nonrelativistic discussion in chapter 2 we will now derive the Lagrangian and Hamiltonian formulation of relativistic mechanics.

3.3.3.1 **Relativistic Free Particle**

According to Eq. (3.107) the trajectory of a relativistic particle may equally well be described by the functions $r(t)$ or $x(\tau)$. Consequently, also the Lagrangian might be expressed in terms of either $r(t)$ or $x(\tau)$ and its corresponding derivatives with respect to t or τ, respectively. We thus have to consider the Lagrangians

$$L_1 = L_1(r, v, t) \qquad \text{and} \qquad L_2 = L_2(x, u) \tag{3.138}$$

Note that L_2 does not explicitly depend on proper time τ, since according to Eq. (3.92) τ is uniquely determined by the space-time vector x and the 4-velocity u. For a better comparison between the three-dimensional formulation (L_1) and the explicitly covariant formulation (L_2) we have employed the velocity $v = \dot{r}$ (instead of $\dot{r}$ itself) in Eq. (3.138). Both Lagrangians L_1 and L_2 do not represent physical observables and are thus not uniquely determined. According to the Hamiltonian principle of least action given by Eq. (2.47), $\delta S = 0$, they only have to yield the same equation of motion. This is in particular guaranteed if even the actions themselves are identical, i.e.,

$$S = \int_{t_1}^{t_2} dt\, L_1(r, v, t) \overset{!}{=} \int_{\tau_1}^{\tau_2} d\tau\, L_2(x, u) \tag{3.139}$$

We first consider the three-dimensional formulation employing L_1. For a free particle we require the usual space-time symmetries and conservation laws to be valid (cf. section 2.2.3): due to spatial translational invariance (i.e., momentum conservation) L_1 must not explicitly depend on $\boldsymbol{r}$, due to temporal translational invariance (i.e., energy conservation) it must not explicitly depend on t, and due to rotational invariance it must not depend on the direction of velocity, but only on its magnitude $v_1 = |\boldsymbol{v}_1|$, i.e., $L_1 = L_1(v) = f_1(v^2)$; here f_1 denotes a yet undetermined function. According to Eq. (2.52) the Euler–Lagrange equations for L_1 thus read

$$\frac{\mathrm{d}}{\mathrm{d}t}\frac{\partial L_1}{\partial v_i} = 0\,, \qquad i = 1,2,3 \tag{3.140}$$

These three equations have to equal the spatial components of the elementary equation of motion as given by Eq. (3.123), which for a free particle with vanishing Minkowski force $f^\mu = 0$ read

$$\frac{\mathrm{d}}{\mathrm{d}t}\Big(\gamma(v(t))\, m\, v_i(t)\Big) = \frac{\mathrm{d}}{\mathrm{d}t}\left(\frac{m\, v_i(t)}{\sqrt{1 - v^2(t)/c^2}}\right) = 0\,, \qquad i = 1,2,3 \tag{3.141}$$

Comparison of the last two equations immediately yields

$$L_1(v) = -C\sqrt{1 - \frac{v^2}{c^2}} = -mc^2\sqrt{1 - \frac{v^2}{c^2}} \tag{3.142}$$

where the constant $C = mc^2$ has arbitrarily been fixed in order to recover the familiar nonrelativistic result (up to the negative rest energy as an additive constant) for velocities $v \ll c$. Note that in relativistic theories the Lagrangian is obviously no longer given as the difference between kinetic and potential energy.

Now we repeat our considerations for the explicitly covariant Lagrangian L_2. Due to the homogeneity of space-time, i.e., translational invariance in space and time L_2 must not depend on x, and due to rotational invariance it must not depend on the 4-velocity u directly, but only on its four-dimensional length $u^\mu u_\mu$, i.e., $L_2 = L_2(u) = f_2(u^\mu u_\mu)$. Analogously to the nonrelativistic situation discussed in section 2.2.2 we can derive the Euler–Lagrange equations by the Hamiltonian principle $\delta S = 0$ for any arbitrary four-dimensional variation δx with vanishing variation at the boundaries of the path,

$$\begin{aligned}\delta S &= S[x + \delta x] - S[x] = \int_{\tau_1}^{\tau_2} \mathrm{d}\tau \left\{\frac{\partial L_2}{\partial x^\mu}\delta x^\mu + \frac{\partial L_2}{\partial u^\mu}\delta u^\mu\right\} \\ &= \int_{\tau_1}^{\tau_2} \mathrm{d}\tau \left\{\left(\frac{\partial L_2}{\partial x^\mu} - \frac{\mathrm{d}}{\mathrm{d}\tau}\frac{\partial L_2}{\partial u^\mu}\right)\delta x^\mu\right\} \stackrel{!}{=} 0\,, \quad \forall \delta x^\mu\end{aligned} \tag{3.143}$$

where we have employed integration by parts and Eq. (3.108). The covariant four-dimensional Euler–Lagrange equations thus read

$$\frac{\partial L_2}{\partial x^\mu} - \frac{\mathrm{d}}{\mathrm{d}\tau}\frac{\partial L_2}{\partial u^\mu} = 0\,, \quad \forall \mu = 0,1,2,3 \tag{3.144}$$

For a free particle this is reduced to

$$\frac{\mathrm{d}}{\mathrm{d}\tau}\frac{\partial L_2}{\partial u^\mu} = \frac{\mathrm{d}}{\mathrm{d}\tau}\left[2u_\mu f_2'(u^\mu u_\mu)\right] = 2f_2'(c^2)\frac{\mathrm{d}u_\mu}{\mathrm{d}\tau} = 0 \tag{3.145}$$

where f_2' denotes the derivative with respect to the argument (i.e., $u^\mu u_\mu$ here). The equation of motion for a free particle is thus given by

$$\frac{\mathrm{d}u_\mu}{\mathrm{d}\tau} = 0 \tag{3.146}$$

which is equivalent to Eq. (3.141) and does not depend at all on the function f_2, i.e., the specific functional form of the Lagrangian. In order to fix the form of the Lagrangian L_2 we require its action to be the same as that for L_1,

$$\begin{aligned} L_1(v)\,\mathrm{d}t &= -mc^2\sqrt{1-\frac{v^2}{c^2}}\,\mathrm{d}t = -mc\sqrt{c^2-\left(\frac{\mathrm{d}\boldsymbol{r}}{\mathrm{d}t}\right)^2}\,\mathrm{d}t \\ &\overset{(3.92)}{=} -mc\sqrt{\left(\frac{\mathrm{d}(ct)}{\mathrm{d}\tau}\right)^2+\left(\frac{\mathrm{d}\boldsymbol{r}}{\mathrm{d}\tau}\right)^2}\,\mathrm{d}\tau \\ &\overset{(3.108)}{=} -mc\sqrt{u^\mu u_\mu}\,\mathrm{d}\tau = L_2(u)\,\mathrm{d}\tau \end{aligned} \tag{3.147}$$

where one should notice that $v = v(t(\tau))$. It is crucial to note that we must not insert the identity given by Eq. (3.109) into the Lagrangian itself, since it is the functional form of the Lagrangian which is important and not its numerical value. The covariant Lagrangian thus reads

$$L_2(u) = -mc\sqrt{u^\mu u_\mu} \tag{3.148}$$

3.3.3.2 Relativistic Particle Subject to External Electromagnetic Fields

We now consider one relativistic particle with charge q subject to an external, i.e., nondynamical electromagnetic field. As already discussed in section 2.4, the electromagnetic field may always be completely described by the scalar potential ϕ and the vector potential $\boldsymbol{A}$, cf. Eq. (2.125), whose four components will now be combined to the contravariant Lorentz 4-vector $A^\mu = (\phi, \boldsymbol{A})$ also denoted as the *4-potential*. In order to define the Lagrangian describing the motion of the particle subject to the external electromagnetic field we have to specify a suitable interaction term and add it to the Lagrangian for the free particle as given by Eq. (3.148). Since the electromagnetic force is linear in the field strengths $\boldsymbol{E}$ and $\boldsymbol{B}$, the Lagrangian has also to be a linear function of the 4-potential A^μ. Since the Lagrangian is a Lorentz scalar, the yet undetermined interaction term has to be a Lorentz scalar, too. Since we already know that the Lorentz force $\boldsymbol{F}_L$ has a velocity-dependent component, cf. Eq. (3.136), the easiest choice for the interaction term will be proportional to $A_\mu u^\mu$, which motivates the Lagrangian

$$L_2(x,u) = -mc\sqrt{u^\mu u_\mu} - \frac{q}{c}A_\mu(x)u^\mu \tag{3.149}$$

The prefactor of the interaction term has been suitably chosen in order to yield the correct equation of motion. We emphasize that the 4-potential $A^\mu = A^\mu(x(\tau))$ represents an external field only and thus the Lagrangian does not *dynamically* depend on A^μ. Straightforward evaluation of the two terms of the Euler–Lagrange equations as given by Eq. (3.144) yields

$$\frac{\partial L_2}{\partial x^\mu} = -\frac{q}{c}\frac{\partial A_\nu}{\partial x^\mu}u^\nu \tag{3.150}$$

$$\frac{\mathrm{d}}{\mathrm{d}\tau}\frac{\partial L_2}{\partial u^\mu} = \frac{\mathrm{d}}{\mathrm{d}\tau}\left(-mu_\mu - \frac{q}{c}A_\mu\right) = -m\frac{\mathrm{d}u_\mu}{\mathrm{d}\tau} - \frac{q}{c}\frac{\partial A_\mu}{\partial x^\nu}u^\nu \tag{3.151}$$

which can be suitably rearranged to read

$$m\frac{\mathrm{d}u_\mu}{\mathrm{d}\tau} = \frac{q}{c}\left(\frac{\partial A_\nu}{\partial x^\mu} - \frac{\partial A_\mu}{\partial x^\nu}\right)u^\nu = \frac{q}{c}\underbrace{(\partial_\mu A_\nu - \partial_\nu A_\mu)}_{F_{\mu\nu}}u^\nu = \frac{q}{c}F_{\mu\nu}u^\nu \tag{3.152}$$

where we have introduced the field tensor $F_{\mu\nu}$ in the last step. After some tedious algebra and exploitation of Eq. (2.125), the 0-component of this equation of motion is recovered to be given by Eq. (3.135), whereas the 3 spatial components of Eq. (3.152) represent just the familar Lorentz force law already given by Eq. (3.136). In section 3.4 we will intensively come back to this setting and find easier ways of evaluating the equation of motion.

In analogy to our argumentation for the free particle above, we require the three-dimensional formulation of the Lagrangian $L_1 = L_1(\boldsymbol{r},\boldsymbol{v},t)$ to yield exactly the same action S as the explicitly covariant Lagrangian $L_2 = L_2(x,u)$, i.e., the interaction term has to be rewritten as

$$-\frac{q}{c}A_\mu u^\mu \,\mathrm{d}\tau = \left(-q\phi + \frac{q}{c}\boldsymbol{A}\cdot\boldsymbol{v}\right)\mathrm{d}t \tag{3.153}$$

which recovers the familiar interaction terms already encountered in section 2.4.6, cf. Eq. (2.157). The corresponding Lagrangian is thus given by

$$L_1(\boldsymbol{r},\boldsymbol{v},t) = -mc^2\sqrt{1-\frac{v^2}{c^2}} - q\phi(\boldsymbol{r},t) + \frac{q}{c}\boldsymbol{v}\cdot\boldsymbol{A}(\boldsymbol{r},t) \tag{3.154}$$

The corresponding generalized momentum is given by

$$\boldsymbol{p} = \left(\frac{\partial L_1}{\partial v_i}\right)_{i=1,2,3} = \gamma m\boldsymbol{v} + \frac{q}{c}\boldsymbol{A} \tag{3.155}$$

which yields for the velocity expressed by the momentum, $v = v(r, p, t)$,

$$v = \frac{p - \frac{q}{c}A}{\sqrt{m^2 + \frac{1}{c^2}\left(p - \frac{q}{c}A\right)^2}} \tag{3.156}$$

This finally gives rise to the relativistic Hamiltonian for a particle subject to an external electromagnetic field,

$$\begin{aligned} H_1(r, p, t) &= p \cdot v(r, p, t) - L_1(r, v(r, p, t), t) \\ &= c\sqrt{m^2c^2 + \left(p - \frac{q}{c}A(r, t)\right)^2} + q\phi(r, t) \end{aligned} \tag{3.157}$$

3.4 Covariant Electrodynamics

In section 2.4 a brief summary of Maxwell's theory of electrodynamics has been presented in its classical, three-dimensional form. Since electrodynamics intrinsically is a relativistic gauge field theory, the structure and symmetry properties of this theory become much more apparent in its natural, explicitly covariant 4-vector formulation. We will thus now reformulate the most central apsects of electrodynamics employing 4-vectors and 4-tensors and finally *derive* the Maxwell equations as the equations of motion for the gauge field.

3.4.1 Ingredients

3.4.1.1 Charge–Current Density

We begin by introducing the basic ingredients of electrodynamics in covariant notation. The *charge–current density*

$$j(x) = (j^\mu(x)) = (c\varrho(x), j(x)) = (c\varrho(r, t), j(r, t)) \tag{3.158}$$

combines the charge density ϱ as defined by Eq. (2.96) and the current density j of Eq. (2.98) to give a four-component quantity. Taking advantage of the definition of the 4-gradient ∂_μ as defined by Eq. (3.40) we can immediately re-express the continuity equation given by Eq. (2.100) in apparently covariant form,

$$\partial_\mu j^\mu = \frac{\partial\varrho}{\partial t} + \operatorname{div} j = 0 \tag{3.159}$$

The continuity equation is a direct consequence of the inhomogeneous Maxwell equations, which are valid in *all* inertial frames of reference. As a consequence, the continuity equation is also valid in *all* inertial frames of reference, and its left hand side $\partial_\mu j^\mu$ thus represents a Lorentz scalar. Since the 4-gradient is a Lorentz vector, the charge–current density j^μ has also to be a Lorentz 4-vector, i.e., its transformation property under Lorentz transformations is given by

$$j'^\mu = \Lambda^\mu{}_\nu j^\nu \tag{3.160}$$

We mention without proof that the total charge Q_V of a system inside the volume V, cf. Eq. (2.97),

$$Q_V(t) = \int_V d^3r\, \varrho(\boldsymbol{r},t) = \frac{1}{c}\int_V d^3r\, j^0(x) \tag{3.161}$$

can also formally be shown to be a Lorentz scalar, i.e., the charge of a particle is independent of its motion and velocity.

3.4.1.2 **Gauge Field**

The six components of the electric and magnetic fields, $\boldsymbol{E}$ and $\boldsymbol{B}$, respectively, do not represent independent variables, cf. section 2.4. As a direct consequence of the homogeneous Maxwell equations the physically observable fields $\boldsymbol{E}$ and $\boldsymbol{B}$ are uniquely determined by the scalar potential ϕ and the vector potential $\boldsymbol{A}$. We now combine the four components of these potentials to the *gauge field 4-vector* $A^\mu = A^\mu(x)$ defined by

$$A(x) = \left(A^\mu(x)\right) = (\phi(x), \boldsymbol{A}(x)) = (\phi(\boldsymbol{r},t), \boldsymbol{A}(\boldsymbol{r},t)) \tag{3.162}$$

So far we have just defined another four-component quantity A^μ, but by now it is not clear whether it properly transforms under Lorentz transformations in order to justify the phrase *4-vector*. In order to prove the transformation property of the gauge field, we re-express the inhomogeneous Maxwell equations in Lorentz gauge as given by Eq. (2.136) in explicitly covariant form by employment of the charge–current density j^μ and the gauge field A^μ,

$$\Box A^\mu = \frac{4\pi}{c} j^\mu \tag{3.163}$$

Since the d'Alembert operator $\Box$ is a Lorentz scalar, cf. Eq. (3.51), and the charge–current density j^μ has been shown to be a Lorentz 4-vector, it is immediately obvious that the gauge field A^μ also represents a Lorentz 4-vector and transforms according to Eq. (3.36) under Lorentz transformations.

In section 2.4 we have seen that the electric and magnetic fields are invariant under gauge transformations of the form given by Eqs. (2.128) and (2.129). Employing the covariant notation developed so far we can compactly recast these gauge transformations into the form

$$A^\mu \longrightarrow A'^\mu = A^\mu - \partial^\mu \chi \tag{3.164}$$

where $\chi = \chi(x)$ again represents a sufficiently smooth gauge function. We note in passing that due to A^μ being a 4-vector, the Lorentz gauge condition $\partial_\mu A^\mu = 0$, cf. Eq. (2.130), is a Lorentz scalar, i.e., if it holds in one specific frame of reference it will also hold in any other inertial frame.

3.4.1.3 **Field Strength Tensor**

We will now explore how the electric and magnetic fields fit into the explicitly covariant framework developed so far. Since the fields $\boldsymbol{E}$ and $\boldsymbol{B}$ are physical observables and thus gauge invariant quantities, we have to construct a gauge invariant quantity based on the gauge field A^μ. The simplest choice is the so-called *field strength tensor* $F = \left(F^{\mu\nu}\right)$ defined by

$$F^{\mu\nu} = \partial^\mu A^\nu - \partial^\nu A^\mu \overset{(2.125)}{=} \begin{pmatrix} 0 & -E_x & -E_y & -E_z \\ E_x & 0 & -B_z & B_y \\ E_y & B_z & 0 & -B_x \\ E_z & -B_y & B_x & 0 \end{pmatrix} \tag{3.165}$$

which is – due to A^μ and ∂^ν being 4-vectors – an antisymmetric Lorentz tensor of rank 2. Even if we did not explicitly write down the 16 gauge invariant components of this tensor, its gauge invariance could immediately be shown by application of a gauge transformation as given by Eq. (3.164),

$$\begin{aligned} F^{\mu\nu} \longrightarrow F'^{\mu\nu} &= \partial^\mu (A^\nu - \partial^\nu \chi) - \partial^\nu (A^\mu - \partial^\mu \chi) \\ &= \partial^\mu A^\nu - \partial^\nu A^\mu = F^{\mu\nu} \end{aligned} \tag{3.166}$$

As for every contravariant Lorentz tensor its indices might be lowered by application of the metric g,

$$F_{\mu\nu} = g_{\mu\alpha}\, g_{\nu\beta}\, F^{\alpha\beta} \overset{(3.165)}{=} \begin{pmatrix} 0 & E_x & E_y & E_z \\ -E_x & 0 & -B_z & B_y \\ -E_y & B_z & 0 & -B_x \\ -E_z & -B_y & B_x & 0 \end{pmatrix} \tag{3.167}$$

i.e., the covariant components $F_{\mu\nu}$ of the field strength tensor are obtained by simply changing the sign of the components of the electric field. It is often convenient to express the physically observable fields by the components of the field strength tensor,

$$E_i = -F^{0i} = F_{0i} \qquad \text{and} \qquad B_i = -\frac{1}{2}\varepsilon_{ijk} F^{jk} \tag{3.168}$$

where it has to be emphasized again that the electric and magnetic fields do not represent 4-vectors and thus feature no distinction between contravariant and covariant components, e.g., $E_1 = E_x$, etc.

We will now try to formulate the Maxwell equations in terms of the gauge invariant field strength tensor $F^{\mu\nu}$. By direct inspection of Gauss' and Ampère's laws as given by Eqs. (2.92) and (2.93), respectively, and by exploiting the definition of the charge–current density and the field strength tensor as given by Eqs. (3.158) and (3.165), respectively, it is immediately obvious that the inhomogeneous Maxwell equations may be expressed as

$$\partial_\mu F^{\mu\nu} = \frac{4\pi}{c}\, j^\nu \tag{3.169}$$

This equation is independent of the chosen gauge. For $\nu = 0$ Gauss' law is recovered and for $\nu = i$ Eq. (3.169) reduces to Ampère's law. Expressed in terms of the gauge potential A^μ, the homogeneous Maxwell equations as given by Eqs. (2.94) and (2.95) are automatically satisfied by the mere existence of the gauge field itself, cf. section 2.4. In order to conveniently express the homogeneous Maxwell equations in terms of the field strength, we introduce the *dual field strength tensor* $\widetilde{F}^{\mu\nu}$ defined by

$$\widetilde{F}^{\mu\nu} = \frac{1}{2}\,\varepsilon^{\mu\nu\sigma\tau}\,F_{\sigma\tau} = \varepsilon^{\mu\nu\sigma\tau}\,\partial_\sigma A_\tau = \begin{pmatrix} 0 & -B_x & -B_y & -B_z \\ B_x & 0 & E_z & -E_y \\ B_y & -E_z & 0 & E_x \\ B_z & E_y & -E_x & 0 \end{pmatrix} \tag{3.170}$$

The four-dimensional totally antisymmetric (pseudo-)tensor $\varepsilon^{\mu\nu\sigma\tau}$ has been introduced earlier, cf. Eq. (3.48). The dual field strength tensor $\widetilde{F}$ is obtained from the field strength tensor F by a so-called *duality transformation* given by

$$\boldsymbol{E} \longrightarrow \boldsymbol{B} \quad \text{and} \quad \boldsymbol{B} \longrightarrow -\boldsymbol{E} \quad \text{(duality transformation)} \tag{3.171}$$

In terms of the dual field strength tensor the homogeneous Maxwell equations simply read

$$\partial_\mu \widetilde{F}^{\mu\nu} = 0 \tag{3.172}$$

which follows directly from the definition of the dual field strength tensor $\widetilde{F}^{\mu\nu}$ given by Eq. (3.170). The homogeneous Maxwell equations expressed in this compact form represent the so-called *Bianchi identities* of the gauge theory of electrodynamics. Their deeper relation to the inhomogeneous Maxwell equations is as follows. In the absence of sources for the electromagnetic field, i.e., $\varrho = \boldsymbol{j} = 0$, the homogeneous Maxwell equations are simply obtained from the inhomogeneous Maxwell equations by a duality transformation (DT) of the form given by Eq. (3.171),

$$\partial_\mu F^{\mu\nu} = 0 \quad \overset{\text{DT}}{\longrightarrow} \quad \partial_\mu \widetilde{F}^{\mu\nu} = 0 \tag{3.173}$$

For $\nu = 0$ the homogeneous Maxwell equation (3.172) states the absence of magnetic monopoles, cf. Eq. (2.95) and for $\nu = i$ this equation represents Faraday's law of induction.

We close this section on the field strength tensor by a discussion of two important invariants of the electromagnetic field, which are both Lorentz and gauge invariant. They can be evaluated by direct calculation and read,

$$F^{\mu\nu} F_{\mu\nu} = 2\left(\boldsymbol{B}^2 - \boldsymbol{E}^2\right) \tag{3.174}$$

$$F^{\mu\nu} \widetilde{F}_{\mu\nu} = -4\left(\boldsymbol{E} \cdot \boldsymbol{B}\right) \tag{3.175}$$

Eq. (3.174) expresses the fact that only the weaker, i.e., smaller of the two fields $\boldsymbol{E}$ and $\boldsymbol{B}$ may be transformed away by a suitable Lorentz transformation, but not both fields. Eq. (3.175) states that the electric or magnetic field may only be transformed away by a suitable Lorentz transformation if the fields are orthogonal to each other. If the electric and magnetic fields are orthogonal to each other in one specific IS, then they are orthogonal to each other in any frame of reference.

3.4.2
Transformation of Electromagnetic Fields

We are now in a position to discuss the transformation behavior of electric and magnetic fields under a change of the coordinate system, i.e., their transformation properties under Lorentz transformations. Since the electric and magnetic fields do *not* represent Lorentz 4-vectors, we have to analyze the transformation property of the field strength tensor $F^{\mu\nu}$ instead.

We thus consider the inertial frame IS′ moving with velocity $\boldsymbol{v}$ relative to the inertial frame IS. Without loss of generality and in order to simplify our discussion we restrict ourselves to the case where this motion is parallel to the x-axis, i.e., $\boldsymbol{v} = v\boldsymbol{e}_x$. At $t = t' = 0$ the two inertial frames must coincide, cf. Figure 3.2. The components of the electric and magnetic fields are given by the field strength tensor $F^{\mu\nu}$ according to Eq. (3.168). Since the transformation property of the second rank tensor $F^{\mu\nu}$ is given by

$$F'^{\mu\nu} = \Lambda^{\mu}{}_{\sigma} \Lambda^{\nu}{}_{\tau} F^{\sigma\tau} \qquad \Longleftrightarrow \qquad F' = \Lambda F \Lambda^T \tag{3.176}$$

we find by direct application of the special Lorentz boost in x-direction as given by Eq. (3.66) the transformation behavior of the components of the electric and magnetic fields. For example,

$$\begin{aligned} E'_1 &= E'_x = -F'^{01} = -\Lambda^{0}{}_{\alpha} \Lambda^{1}{}_{\beta} F^{\alpha\beta} = -\Lambda^{0}{}_{1} \Lambda^{1}{}_{0} F^{10} - \Lambda^{0}{}_{0} \Lambda^{1}{}_{1} F^{01} \\ &= -\frac{v^2}{c^2}\gamma^2 E_x + \gamma^2 E_x = \gamma^2 E_x\left(1 - \frac{v^2}{c^2}\right) = E_x = E_1 \end{aligned} \tag{3.177}$$

Similar calculations yield for the other components of the electric field

$$E'_2 = E'_y = -F'^{02} = \gamma\Big(E_y - \frac{v}{c}B_z\Big) \tag{3.178}$$

$$E'_3 = E'_z = -F'^{03} = \gamma\Big(E_z + \frac{v}{c}B_y\Big) \tag{3.179}$$

The transformation behavior for the components of the magnetic field could similarly be derived by exploiting Eq. (3.168). A more elegant and direct approach, however, is based on the duality transformation given by Eq. (3.171), which immediately yields

$$B'_1 = B'_x = B_x \tag{3.180}$$

$$B'_2 = B'_y = \gamma\Big(B_y + \frac{v}{c}E_z\Big) \tag{3.181}$$

$$B'_3 = B'_z = \gamma\Big(B_z - \frac{v}{c}E_y\Big) \tag{3.182}$$

Since our constraint of a relative motion in x-direction (i.e., $\boldsymbol{v} = v\boldsymbol{e}_x$) was arbitrary and only imposed in order to simplify the calculations above, we may find the transformation properties of electric and magnetic fields for the general motion of IS$'$ relative to IS as described by the velocity $\boldsymbol{v}$ to read

$$\boldsymbol{E}'_{\parallel} = \boldsymbol{E}_{\parallel} \quad \text{and} \quad \boldsymbol{E}'_{\perp} = \gamma\left(\boldsymbol{E}_{\perp} + \frac{1}{c}\boldsymbol{v}\times\boldsymbol{B}\right) \tag{3.183}$$

$$\boldsymbol{B}'_{\parallel} = \boldsymbol{B}_{\parallel} \quad \text{and} \quad \boldsymbol{B}'_{\perp} = \gamma\left(\boldsymbol{B}_{\perp} - \frac{1}{c}\boldsymbol{v}\times\boldsymbol{E}\right) \tag{3.184}$$

where we have splitted the fields in a component parallel to the velocity $\boldsymbol{v}$ and a component perpendicular to $\boldsymbol{v}$. The equation $\boldsymbol{E} = \boldsymbol{E}_{\parallel} + \boldsymbol{E}_{\perp}$ obviously holds. The central result derived above is that electric and magnetic fields are transformed into each other under Lorentz transformations. This is a completely natural result, since a moving charge may represent a current – and thus a source for the magnetic field – in one inertial system (in IS), whereas it is at rest and thus does not give rise to a magnetic field in another inertial system (in IS$'$).

3.4.3
Lagrangian Formulation and Equations of Motion

3.4.3.1 Lagrangian for the Electrodynamic Field

We will now construct the Lagrangian for electrodynamics, where the gauge field A^{μ} represents the dynamical variable. The six components of the electric and magnetic field cannot be chosen independently of each other, since they are uniquely determined by the four components of the gauge field, cf. Eqs. (2.125) or (3.165) and (3.168). They thus do not represent suitable generalized coordinates, cf. section 2.2.1, and have to be replaced by the gauge field for the construction of the Lagrangian density of electrodynamics. We recall here that we are dealing with a field theory, where the dynamical variables are no longer given by discrete particle coordinates but by continuous real-valued functions of space and time, $A^\mu = A^\mu(x)$. As a consequence the dynamic behavior of the system is completely described by a suitably chosen Lagrangian density $\mathcal{L}$, whose spatial integration over the available space yields the traditional Lagrangian, cf. section 2.2.2.

Since the Maxwell equations are linear in the fields – both the gauge fields ϕ and $\boldsymbol{A}$ and the physically observable fields $\boldsymbol{E}$ and $\boldsymbol{B}$ – the Lagrangian can only be quadratic in the fields or, equivalently, the field strength tensor $F^{\mu\nu}$. The kinetic contribution of the gauge field to the Lagrangian, or more precisely, the action $S = \frac{1}{c}\int d^4x\mathcal{L}$, must thus represent a Lorentz and gauge invariant term containing the quadratic field strength tensor. The simplest choice subject to these constraints is given by

$$\mathcal{L}_{\text{rad}}(A, \partial A) = -\frac{1}{16\pi} F^{\mu\nu}(x) F_{\mu\nu}(x) \tag{3.185}$$

where the prefactor has been suitably chosen in order to yield the correct equation of motion. This Lagrangian explicitly depends only on the dynamical variables $A = (A^\mu)$ and their derivatives $\partial A = (\partial_\mu A^\nu)$, and only implicitly on the space-time coordinates $x = (x^\mu)$. It has been given the subscript "rad" to express its relation to the radiation field of electrodynamics, cf. chapter 7. Its Lorentz invariance is obvious and its gauge invariance has been proven by Eq. (3.166).

We now have to construct the interaction term between the dynamical variables, i.e., the gauge field A^μ and the sources of the electrodynamical field, i.e., charged particles giving rise to a charge–current density j^μ, cf. Eq. (3.158). Again, the contribution of this interaction term to the action S has to be Lorentz and gauge invariant, and the simplest choice is thus given by

$$\mathcal{L}_{\text{int}}(A, \partial A, x, u) = -\frac{1}{c} A_\mu(x) j^\mu(x) \tag{3.186}$$

This interaction Lagrangian density may depend explicitly on the space-time coordinates x and the 4-velocity u via the charge–current density j^μ. However, as far as only the equation of motion for the electrodynamic field is concerned they do not represent dynamical variables. Lorentz invariance of this interaction term is obvious, and gauge invariance of the corresponding action is a direct consequence of the continuity equation for the charge–current density

j^μ, cf. Eq. (3.159),

$$
\begin{aligned}
S_{\text{int}} \longrightarrow S'_{\text{int}} &= \frac{1}{c}\int d^4x \mathcal{L}_{\text{int}}(A', \partial A', x, u) \stackrel{(3.186)}{=} -\frac{1}{c^2}\int d^4x\, A'_\mu j^\mu \\
&\stackrel{(3.164)}{=} -\frac{1}{c^2}\int d^4x A_\mu j^\mu + \frac{1}{c^2}\int d^4x (\partial_\mu \chi) j^\mu \\
&\stackrel{(p.I.)}{=} S_{\text{int}} - \frac{1}{c^2}\int d^4x (\partial_\mu j^\mu)\chi \stackrel{(3.159)}{=} S_{\text{int}}
\end{aligned}
\tag{3.187}
$$

The Lagrangian density $\mathcal{L}_{\text{em}}$ for the electromagnetic field is thus given as the sum of the kinetic term $\mathcal{L}_{\text{rad}}$ and the interaction term $\mathcal{L}_{\text{int}}$,

$$
\mathcal{L}_{\text{em}}(A, \partial A, x, u) = \mathcal{L}_{\text{rad}} + \mathcal{L}_{\text{int}} = -\frac{1}{16\pi} F^{\mu\nu} F_{\mu\nu} - \frac{1}{c} A_\mu j^\mu \tag{3.188}
$$

3.4.3.2 **Minimal Coupling**

We will now examine one of the most central features of the interaction between the electromagnetic field and charged matter as described by the interaction Lagrangian density $\mathcal{L}_{\text{int}}$ given by Eq. (3.186). As has been explicitly shown above, this form for $\mathcal{L}_{\text{int}}$ is the simplest choice compatible with all constraints, i.e., Lorentz and gauge invariance and containing both the gauge field A^μ and the charge–current density (i.e., matter) j^μ that couples the electrodynamical field to charged matter. It is thus often denoted as the *minimal coupling* term or the term arising from minimal coupling of the gauge field to matter. It cannot be overemphasized that it is exactly the minimal coupling procedure which ensures gauge invariance of the corresponding action S_{int}, cf. Eq. (3.187). This relation is crucial for every gauge theory and is extensively taken advantage of in all gauge theories of modern physics, e.g., for electroweak and strong interactions.

We will now demonstrate how the interaction Lagrangian density $\mathcal{L}_{\text{int}}$ and its corresponding Lagrangian L_{int} or action S_{int} are related to the well-known and more familiar interaction terms for a single charged particle subject to an external electromagnetic field as given by Eq. (2.157) or (3.149). For this we rewrite the charge–current density j^μ for one moving particle with charge q, whose motion in IS is described by the trajectory $\boldsymbol{r}_1(t)$,

$$
\begin{aligned}
j = (j^\mu) &\stackrel{(3.158)}{=} (c\varrho, \boldsymbol{j}) \stackrel{(2.96),(2.98)}{=} q\,\delta^{(3)}(\boldsymbol{r} - \boldsymbol{r}_1(t))\,(c, \boldsymbol{v}) \\
&= c\,q\,\delta^{(3)}(\boldsymbol{r} - \boldsymbol{r}_1(t))\,\underbrace{\gamma(c, \boldsymbol{v})}_{u}\;\underbrace{\frac{1}{c\gamma}}_{d\tau/dx^0} \\
&= c\,q \int d\tau \delta^{(4)}(x - x_1(\tau))\, u(\tau)
\end{aligned}
\tag{3.189}
$$

Inserting this expression for j^μ into the interaction term given by Eq. (3.186) yields the action

$$\begin{aligned} S_{\text{int}} &= -\frac{1}{c^2}\int \mathrm{d}^4x\, A_\mu(x) j^\mu(x) \\ &\stackrel{(3.189)}{=} -\frac{1}{c^2}\int \mathrm{d}^4x\, A_\mu(x)\, c\, q \int \mathrm{d}\tau\, \delta^{(4)}(x - x_1(\tau))\, u^\mu(\tau) \\ &= -\frac{q}{c}\int \mathrm{d}\tau\, A_\mu(x_1(\tau))\, u^\mu(\tau) \end{aligned} \tag{3.190}$$

i.e., we have recovered the interaction term of the familiar Lagrangian L_2 given by Eq. (3.149). As has explicitly been shown in section 3.3.3 the expression for the action can be recast into the well-known three-dimensional form

$$S_{\text{int}} = \int \mathrm{d}t \underbrace{\left\{ -q\phi(\boldsymbol{r}_1, t) + \frac{q}{c}\boldsymbol{v}_1 \cdot \boldsymbol{A}(\boldsymbol{r}_1, t) \right\}}_{L_{1,\text{int}}} \tag{3.191}$$

where the motion of the charged particle is here described by $\boldsymbol{r}_1(t)$ and the interaction term of the three-dimensional Lagrangian L_1 defined by Eq. (3.154) has been employed. It is exactly this Lagrangian L_1 which gives rise to the minimal coupling expression for the conjugate generalized momentum $\boldsymbol{p}_1 = \gamma m \boldsymbol{v}_1 + \frac{q}{c}\boldsymbol{A}$.

We conclude this discussion by emphasizing again that the familiar forms of the three- and four-dimensional interaction Lagrangians are a direct consequence of the minimal coupling procedure given by Eq. (3.186) ensuring gauge invariance of the corresponding action.

3.4.3.3 **Euler–Lagrange Equations**

We will now finally *derive* Maxwell's equations as the Euler-Lagrange equations, i.e., the equations of motion for the gauge field A^μ being described by the Lagrangian density $\mathcal{L}_{\text{em}}$ defined by Eq. (3.188). In complete analogy to classical mechanics presented in chapter 2 the equations of motion are derived from the Lagrangian density by the Hamiltonian principle of least action. We thus will vary the dynamical degrees of freedom, i.e., the four components of the gauge field A^μ,

$$A \longrightarrow A + \delta A \quad \Longleftrightarrow \quad A^\mu \longrightarrow A^\mu + \delta A^\mu \tag{3.192}$$

where the variation at the boundaries of the available space-time will vanish, and determine the specific field configuration which guarantees stationary action. This particular field configuration will be a solution of the equations of motion to be derived. In analogy to the discussion in section 2.2.2, the variation of the action based on the Lagrangian density $\mathcal{L}_{\mathrm{em}}(A, \partial A, x, u)$ with respect to the gauge field A^μ is found to read

$$\begin{aligned}
\delta S[A(x)] &= S[A+\delta A] - S[A] \\
&= \frac{1}{c}\int \mathrm{d}^4x \left(\frac{\partial \mathcal{L}_{\mathrm{em}}}{\partial A^\mu}\delta A^\mu + \frac{\partial \mathcal{L}_{\mathrm{em}}}{\partial(\partial^\nu A^\mu)} \underbrace{\delta(\partial^\nu A^\mu)}_{\partial^\nu(\delta A^\mu)} \right) \\
&\overset{(p.I.)}{=} \frac{1}{c}\int \mathrm{d}^4x \left(\frac{\partial \mathcal{L}_{\mathrm{em}}}{\partial A^\mu} - \partial^\nu \frac{\partial \mathcal{L}_{\mathrm{em}}}{\partial(\partial^\nu A^\mu)} \right) \delta A^\mu \overset{!}{=} 0 \quad \forall \delta A^\mu
\end{aligned} \tag{3.193}$$

Since the variation of the action has to vanish for all variations of the gauge field this immediately yields the Euler–Lagrange equations for the gauge field A^μ,

$$\frac{\partial}{\partial x_\nu} \frac{\partial \mathcal{L}_{\mathrm{em}}}{\partial(\partial^\nu A^\mu)} = \frac{\partial \mathcal{L}_{\mathrm{em}}}{\partial A^\mu} \tag{3.194}$$

Insertion of the explicit form of $\mathcal{L}_{\mathrm{em}}$ as given by Eq. (3.188) into this Euler–Lagrange equation and directly evaluating the corresponding derivatives yields the equations of motion for the gauge field A^μ,

$$\partial_\mu F^{\mu\nu} = \frac{4\pi}{c} j^\nu \tag{3.195}$$

Eq. (3.195) just represents the inhomogeneous Maxwell equations in covariant form, cf. Eq. (3.169). We have thus *derived* the inhomogeneous Maxwell equations as the natural equations of motion for the gauge potential A^μ. The sources, as described by the charge–current density j^μ, are considered as external variables which do not represent dynamical degrees of freedom, i.e., only the action (or effect) of the sources on the gauge fields A^μ is taken into account.

In general, however, also the electromagnetic fields, as described by the gauge potentials A^μ will act upon the charged particles (i.e., the matter) within the system under consideration. In order to investigate this coupled interaction between electromagnetic fields and particles we have to consider the total action S_{tot} of the system, where both the fields A^μ and the particle coordinates x^μ represent dynamical degrees of freedom. In order to simplify the notation we will again restrict ourselves to the case of only a single charged particle with mass m and charge q described by the trajectories $x_1(\tau)$ or $\boldsymbol{r}_1(t)$, respectively. The total action is then given by

$$\begin{aligned}
S_{\mathrm{tot}} &= S_{\mathrm{tot}}[A, x] = S_{\mathrm{rad}} + S_{\mathrm{mat}} + S_{\mathrm{int}} \\
&= \frac{1}{c}\int \mathrm{d}^4x\, \mathcal{L}_{\mathrm{rad}}(A, \partial A) + \int \mathrm{d}t L_1(\boldsymbol{r}_1, \boldsymbol{v}_1, t) \\
&= -\frac{1}{16\pi c}\int \mathrm{d}^4x\, F^{\mu\nu}F_{\mu\nu} + \int \mathrm{d}\tau\, L_2(x_1, u_1)
\end{aligned} \tag{3.196}$$

where the three- or four-dimensional Lagrangians L_1 and L_2, respectively, which describe the dynamics of the particle, are given by

$$L_1(\boldsymbol{r}_1, \boldsymbol{v}_1, t) \stackrel{(3.154)}{=} \underbrace{-mc^2\sqrt{1-\frac{v_1^2}{c^2}}}_{L_{1,\mathrm{mat}}} \underbrace{- q\phi(\boldsymbol{r}_1, t) + \frac{q}{c}\boldsymbol{v}_1 \cdot \boldsymbol{A}(\boldsymbol{r}_1, t)}_{L_{1,\mathrm{int}}} \tag{3.197}$$

and

$$L_2(x_1, u_1) \stackrel{(3.149)}{=} \underbrace{-mc\sqrt{u_1^\mu u_{1,\mu}}}_{L_{2,\mathrm{mat}}} \underbrace{- \frac{q}{c} A_\mu(x_1(\tau))\, u_1^\mu}_{L_{2,\mathrm{int}}} \tag{3.198}$$

Application of a variation of the four-dimensional path $x_1(\tau)$ of the particle to the Lagrangian L_2 immediately yields the covariant Euler–Lagrange equations for the charged particle,

$$m\frac{du_{1,\mu}}{d\tau} = \frac{q}{c}F_{\mu\nu}u_1^\nu \tag{3.199}$$

which have already been discussed in section 3.3.3, cf. Eq. (3.152).

It cannot be overemphasized that by specification of the action S_{tot} the dynamics of both the electrodynamic and the material degrees of freedom, i.e., the gauge field A^μ and the particles x^μ, is completely determined. The equations of motion for the gauge field are given by Eq. (3.195) and are recovered to be the inhomogeneous Maxwell equations. The equations of motion for the particles are given by Eq. (3.199) and comprise the familiar Lorentz force law. The time evolution of the gauge fields and the particles as given by the two sets of equations of motion are non-trivially coupled via the interaction term S_{int}.

3.5
Interaction of Two Moving Charged Particles

We have encountered the interaction of two moving charged particles already in section 2.4.7. This type of interaction is pivotal for the description of the elementary particles in molecular science. However, in section 2.4.7 we claimed that it is possible to correctly describe their interaction only within the framework of a relativistically correct electrodynamical theory. After what has been said about Lorentz transformations, we are now in a position to tackle this issue. Of course, since the Lorentz transformation is defined for uniformly moving particles only, we again have to take advantage of the fact that for the case of *accelerated* motion for every instant of time (or for a sufficiently small time

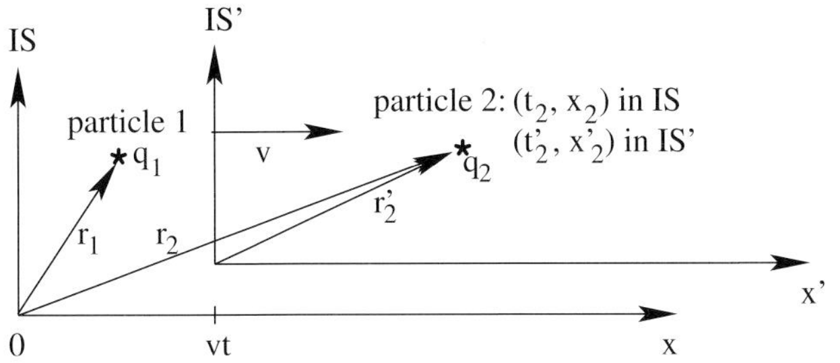

Figure 3.5 Two electromagnetically interacting moving charges in an inertial frame of reference IS. The frame IS′ is chosen to move with the same speed as particle 2, $v = \dot{r}_2$. For the sake of simplicity a Lorentz boost in x-direction as in section 3.2.1.1 is depicted, a restriction solely made for the readability of the figure but not imposed onto the derivations in the text.

interval) the unaccelerated picture can be adopted. We will later see that the interaction energy can be expressed in terms of the particle velocities, which may, of course, change with time. We shall provide a detailed presentation here since the matter is hardly discussed exhaustively in other sources despite of its fundamental significance for chemistry, the science of interacting electromagnetic charges. The presentation in Jackson's classic textbook is likely to be the most elaborate one but it is also very compact [12]. Darwin's original account is very compact [23] but has been worked out in the appendix of Szasz' book [24]. The latter presentations are staged within only one system of inertia, in which the electromagnetic interaction between moving charges is delayed by the speed of light and the speed of the charges. However, a more satisfactory account that fully meets the ideas of the theory of special relativity is required.

The situation of two interacting charges q_1 and q_2 is depicted in Figure 3.5. We choose the most general inertial frame of reference IS in which both particles — at positions $\boldsymbol{r}_1(t)$ and $\boldsymbol{r}_2(t)$ — are moving. In order to derive an expression for the interaction energy, we start by considering the scalar and vector potentials, ϕ_2 and $\boldsymbol{A}_2$, respectively, produced by particle 2 and exerted on particle 1. It is important to understand that we are looking for these quantities at the position of particle 1, namely at $\boldsymbol{r}_1$. According to Eq. (2.157), the potential energy of particle 1 in the field of particle 2 at time t in IS then reads

$$V_{12}(t) = q_1\phi_2(\boldsymbol{r}_1, t) - \frac{q_1}{c}\,\dot{\boldsymbol{r}}_1(t)\cdot \boldsymbol{A}_2(\boldsymbol{r}_1, t) \tag{3.200}$$

3.5.1
Scalar and Vector Potentials of a Charge at Rest

In order to derive expressions for the scalar and vector potentials produced by particle 2 we choose a frame of reference IS′ which is most suitable for this purpose. This is the case when IS′ is moving with the same speed, v, in the same direction as particle 2,

$$\boldsymbol{v} = \dot{\boldsymbol{r}}_2 \quad \text{and accordingly} \quad v = |\dot{\boldsymbol{r}}_2| \tag{3.201}$$

so that particle 2 is at rest in IS′. It must be noted that choosing the x-direction as the direction in which particle 2 moves will not formally simplify the derivation — by contrast, the derivation would become inconsistent as we shall see. This is due to the fact that we later have to transform the three-dimensional distances back to IS by a suitable Lorentz transformation (see the scalar potential below).

Since particle 2 is observed to be at rest in IS′, the vector potential in this frame of reference vanishes

$$\boldsymbol{A}'_2(\boldsymbol{r}', t') = 0 \quad \forall \boldsymbol{r}' \tag{3.202}$$

since a charge at rest does not produce magnetic fields (an aspect fully covered by Maxwell's theory). This is the essential step in the derivation since no assumption on what the vector potential looks like needs to be made. Only the scalar potential, i.e., the standard Coulomb potential of a point charge, contributes to the interaction energy. This also implies that div$\boldsymbol{A}$=0 and thus Coulomb gauge applies (see section 2.4.4.2). The scalar potential at any position $\boldsymbol{r}'$ in IS′ is given by

$$\phi'_2(\boldsymbol{r}', t') = \frac{q_2}{|\boldsymbol{r}' - \boldsymbol{r}'_2|} \tag{3.203}$$

The first charge q_1 experiences a stationary potential $\phi'_2(\boldsymbol{r}', t') = \phi'_2(\boldsymbol{r}')$ of the resting charge q_2 so that the electromagnetic interaction energy between charge q_1 at position $\boldsymbol{r}'_1(t'_1)$ and charge q_2 at position $\boldsymbol{r}'_2$ in IS′ reads

$$V'_{12}(t'_1) = q_1 \phi'_2(\boldsymbol{r}'_1, t'_1) = \frac{q_1 q_2}{|\boldsymbol{r}'_1(t'_1) - \boldsymbol{r}'_2|} \tag{3.204}$$

in analogy to Eq. (3.200) and taking into account Eq. (3.202). The explicit inclusion of time t'_1 solely denotes that this time fixes the position $\boldsymbol{r}'_1$ where the potential is probed. By construction, the position of the second charge remains fixed, $\boldsymbol{r}'_2 = \boldsymbol{r}'_2(t'_2) = \boldsymbol{r}'_2(t'_1)$ in IS′. Hence, retardation effects do not show up in the expression for the interaction energy.

A note on retardation in IS′ is advisable. The transmission of the electrostatic interaction takes time, i.e., it is retarded. But since, according to Eq. (3.204), the interaction energy does only depend on the position $\boldsymbol{r}'_1$ at which the interaction energy is probed, the question arises where retardation is hidden. Note that t'_1 simply determines the position $\boldsymbol{r}'_1$ at which the scalar potential of charge 2 is probed. With this in mind, we may also add that for a point charge at rest we can directly reduce the Poisson integral — cf. Eq. (2.139) in Coulomb gauge — to the expression given above since the charge density distribution is not affected by retardation effects in IS′. The Poisson integral can then be evaluated directly by virtue of a Dirac δ-distribution to the expression given by Eq. (3.204). It is then not necessary to consider the more general case of a moving extended charge distribution first.

Thus, the particular choice of IS′ has the advantage that the interaction energy is not affected by retardation effects. The reason for this is simply that the field experienced by charge 1 at time t'_1 when this charge is at position $\boldsymbol{r}'_1$ has been transmitted by charge 2 at time t'_2. The retardation time t'_{12} can be calculated from the expression

$$t'_{12} = \frac{\boldsymbol{r}'_1(t'_1) - \boldsymbol{r}'_2(t'_2)}{c} = \frac{\boldsymbol{r}'_1(t'_1) - \boldsymbol{r}'_2(t'_1)}{c} \tag{3.205}$$

since charge 2 is at rest. As a consequence, the distance $|\boldsymbol{r}'_1(t'_1) - \boldsymbol{r}'_2|$ relevant to the Coulombic interaction energy does not need to be adjusted for retardation effects in IS′. Retardation effects on the interaction energy will show up, however, in IS. They are automatically produced by a proper Lorentz transformation from IS′ back to IS. We should note that this route is usually not taken when the interaction energy is derived. Instead, Darwin and therefore also Szasz consider the retardation directly in IS without invoking IS′. We will come back to this procedure later when the general derivation has been finished.

3.5.2
Retardation from Lorentz Transformation

To obtain the interaction energy in IS we Lorentz transform the coordinates and the 4-vector $(\phi'_2, \boldsymbol{A}'_2)$. With the Lorentz transformation $\Lambda(\boldsymbol{v})$ — as given by Eq. (3.81) — of the coordinates from IS to IS′ we obtain the space-like coordinate $\boldsymbol{r}'$ from the space-time coordinate $(ct, \boldsymbol{r})$ in IS as

$$\boldsymbol{r}' = \boldsymbol{r} + \boldsymbol{v}\left[(\gamma - 1)\frac{\boldsymbol{v}\cdot\boldsymbol{r}}{v^2} - \gamma t\right] \tag{3.206}$$

For Eq. (3.203) we measure the distance between $\boldsymbol{r}'$ and $\boldsymbol{r}'_2$ in IS′ at time t'. In terms of coordinates of the original frame of reference IS the distance vector $\boldsymbol{r}' - \boldsymbol{r}'_2$ can be calculated from Eq. (3.206) employed for $\boldsymbol{r}'$ and for $\boldsymbol{r}'_2$ so that we obtain

$$\boldsymbol{r}' - \boldsymbol{r}'_2 = \boldsymbol{r} - \boldsymbol{r}_2 + \boldsymbol{v}\left[(\gamma - 1)\frac{\boldsymbol{v}\cdot(\boldsymbol{r} - \boldsymbol{r}_2)}{v^2} - \gamma(t - t_2)\right] \tag{3.207}$$

Consequently, a distance measurement which yields $|\boldsymbol{r}' - \boldsymbol{r}'_2|$ at time t' in IS′ requires in IS the space *and* time coordinates $(ct, \boldsymbol{r})$ of the point of interest as well as those of charge 2, i.e., $(ct_2, \boldsymbol{r}_2)$.

If we choose $\boldsymbol{r} = \boldsymbol{r}_1$, we can evaluate the scalar potential at the position of the first particle. Substituting the distance vector of Eq. (3.207) into Eq. (3.203) we finally get

$$\phi_2'(\boldsymbol{r}_1', t_1') = \frac{q_2}{\left| \boldsymbol{r}_1 - \boldsymbol{r}_2 + \boldsymbol{v} \left[(\gamma - 1) \frac{\boldsymbol{v} \cdot (\boldsymbol{r}_1 - \boldsymbol{r}_2)}{v^2} - \gamma (t_1 - t_2) \right] \right|} \tag{3.208}$$

Hence, the positions of both particles at $\boldsymbol{r}_1'$ and $\boldsymbol{r}_2'$ in IS′ connect two events in IS at different, but connected times t_1 and t_2. While, by construction, we may choose t_1 arbitrarily, t_2 is completely determined by the time $t_{12} = t_1 - t_2$ needed to transmit the interaction from charge 2 to charge 1 with the speed of light. The scalar potential thus contains a retardation term, which arises here naturally through the Lorentz transformation. Hence, the correct relativistic treatment of the interaction problem automatically introduces the retardation effect.

3.5.3
General Expression for the Interaction Energy

In order to calculate the interaction energy of particles 1 and 2 we need to transform these potentials to the original frame of reference IS by the inverse Lorentz boost $\Lambda(\boldsymbol{v})^{-1}$ as given by Eqs. (3.81) and (3.82),

$$\phi_2(\boldsymbol{r}_1, t_1) = \gamma \phi_2' + \gamma \frac{\dot{\boldsymbol{r}}_2}{c} \boldsymbol{A}_2' \overset{(3.202)}{=} \gamma \phi_2'(\boldsymbol{r}_1', t_1') \tag{3.209}$$

$$\begin{aligned} \boldsymbol{A}_2(\boldsymbol{r}_1, t_1) &= \gamma \frac{\dot{\boldsymbol{r}}_2}{c} \phi_2' + \sum_{i=1}^{3} \left(\sum_{j=1}^{3} \left[\delta_{ij} + \frac{v_i v_j (\gamma - 1)}{v^2} \right] A_{2,j}' \right) \boldsymbol{e}_i \\ &\overset{(3.202)}{=} \gamma \frac{\dot{\boldsymbol{r}}_2}{c} \phi_2'(\boldsymbol{r}_1', t_1') \end{aligned} \tag{3.210}$$

and obtain the interaction energy at time t_1 according to Eq. (3.200) as

$$V_{12}(t_1) = \gamma(\dot{\boldsymbol{r}}_2)\, q_1 \phi_2'(\boldsymbol{r}_1', t_1', \gamma(\dot{\boldsymbol{r}}_2)) \left(1 - \frac{\dot{\boldsymbol{r}}_1 \cdot \dot{\boldsymbol{r}}_2}{c^2} \right) \tag{3.211}$$

in which $\gamma(\dot{\boldsymbol{r}}_2) = 1 \Big/ \sqrt{1 - \dot{\boldsymbol{r}}_2^2 / c^2}$ and where we explicitly noted the dependence of ϕ_2' on $\gamma(\dot{\boldsymbol{r}}_2)$. Accordingly, we could have started from particle 1 and would have obtained the same interaction energy but with exchanged particle indices,

$$V_{21}(t_2) = \gamma(\dot{\boldsymbol{r}}_1)\, q_2 \phi_1'(\boldsymbol{r}_2', t_2', \gamma(\dot{\boldsymbol{r}}_1)) \left(1 - \frac{\dot{\boldsymbol{r}}_2 \cdot \dot{\boldsymbol{r}}_1}{c^2} \right) \tag{3.212}$$

The interaction energy $V_{12}(t_1)$ is very involved because of the lengthy expression of ϕ_2' defined in Eq. (3.208), which contains the velocity of charge 2 as well as two different times t_1 and t_2 in the denominator (which would create insurmountable difficulties upon a later transition to quantum mechanics by relying on the correspondence principle). Therefore, it is desirable to calculate the interaction energy solely from quantities defined at t_1 (and not at t_2).

3.5.4
Interaction Energy at One Instant of Time

It can be anticipated that it will also be convenient to derive an approximation to $V_{12}(t)$ for velocities that are small compared to the speed of light c. Apparently, all square root expressions in Eq. (3.211) should be expanded in powers of $1/c$. For the sake of clarity we define the difference vector $\boldsymbol{r}_{12}$ as well as its length r_{12}

$$\boldsymbol{r}_{12}(t_1, t_2) = \boldsymbol{r}_1(t_1) - \boldsymbol{r}_2(t_2) \quad \text{and} \quad r_{12}(t_1, t_2) = |\boldsymbol{r}_{12}| = \sqrt{\boldsymbol{r}_{12}^2} \tag{3.213}$$

where we have explicitly denoted that up to this point the position of charge q_1 is given at t_1, while that of q_2 is to be taken at t_2. The electromagnetic wave transmitting the interaction from charge 2 to charge 1 leaves $\boldsymbol{r}_2$ at t_2 and arrives then at $\boldsymbol{r}_1$ after a time delay of t_{12},

$$t_{12} = t_1 - t_2 = \frac{r_{12}(t_1, t_2)}{c} \tag{3.214}$$

For the scalar coefficient in brackets in the denominator of Eq. (3.208) we define the abbreviation $a(\dot{\boldsymbol{r}}_2)$,

$$a(\dot{\boldsymbol{r}}_2) \equiv [\gamma(\dot{r}_2) - 1] \frac{\dot{\boldsymbol{r}}_2 \cdot \boldsymbol{r}_{12}}{\dot{\boldsymbol{r}}_2^2} - \gamma(\dot{r}_2) \frac{r_{12}}{c} \tag{3.215}$$

where we explicitly replaced the relative speed v of IS′ against IS by $\dot{r}_2$. With these definitions, the scalar potential ϕ_2' can be written in a compact form as

$$\phi_2'(\boldsymbol{r}_1', t_1', \dot{\boldsymbol{r}}_2) = \frac{q_2}{|\boldsymbol{r}_{12} + a(\dot{\boldsymbol{r}}_2)\dot{\boldsymbol{r}}_2|} = \frac{q_2}{\sqrt{r_{12}^2 + 2a(\dot{\boldsymbol{r}}_2)\boldsymbol{r}_{12} \cdot \dot{\boldsymbol{r}}_2 + a^2(\dot{\boldsymbol{r}}_2)\dot{\boldsymbol{r}}_2^2}} \tag{3.216}$$

However, for later convenience we rewrite the denominator as

$$\phi_2'(\boldsymbol{r}_1', t_1', \dot{\boldsymbol{r}}_2) = \frac{q_2}{r_{12}\sqrt{1 + 2a(\dot{\boldsymbol{r}}_2)\frac{\boldsymbol{r}_{12} \cdot \dot{\boldsymbol{r}}_2}{r_{12}^2} + a^2(\dot{\boldsymbol{r}}_2)\frac{\dot{\boldsymbol{r}}_2^2}{r_{12}^2}}} \tag{3.217}$$

3.5.4.1 **Taylor Expansion of Potential and Energy**

In order to simplify this expression we first need to expand the factor $\gamma(\dot{\boldsymbol{r}}_2)$ in terms of c^{-1},

$$\gamma(\dot{\boldsymbol{r}}_2) = \frac{1}{\sqrt{1-\frac{\dot{\boldsymbol{r}}_2^2}{c^2}}} \approx 1 + \frac{1}{2}\frac{\dot{\boldsymbol{r}}_2^2}{c^2} + O(c^{-4}) \tag{3.218}$$

according to

$$(1 \pm x)^{-1/2} = 1 \mp \frac{1}{2}x + \frac{3}{8}x^2 + O(x^3) \tag{3.219}$$

Naturally, an expansion in terms of the reciprocal speed of light will finally result in a leading-order nonrelativistic description (with instantaneous transmission) of the interaction between two charges together with low-order relativistic corrections.

Now, $a(\dot{\boldsymbol{r}}_2)$ can be approximated by

$$a(\dot{\boldsymbol{r}}_2) \approx -\frac{r_{12}}{c} + \frac{1}{2}\frac{\dot{\boldsymbol{r}}_2 \cdot \boldsymbol{r}_{12}}{c^2} + O(c^{-3}) \tag{3.220}$$

so that we obtain for ϕ_2',

$$\phi_2' \approx \frac{q_2}{r_{12}\sqrt{1 + 2\left[-\frac{r_{12}}{c} + \frac{1}{2}\frac{\dot{\boldsymbol{r}}_2 \cdot \boldsymbol{r}_{12}}{c^2}\right]\frac{\boldsymbol{r}_{12} \cdot \dot{\boldsymbol{r}}_2}{r_{12}^2} + \left[\frac{r_{12}^2}{c^2}\right]\frac{\dot{\boldsymbol{r}}_2^2}{r_{12}^2} + O(c^{-3})}} \tag{3.221}$$

which can be rearranged to read

$$\begin{aligned} \phi_2' &\approx \frac{q_2}{r_{12}}\left[1 + \frac{\boldsymbol{r}_{12} \cdot \dot{\boldsymbol{r}}_2}{r_{12}c} - \frac{1}{2}\frac{(\boldsymbol{r}_{12} \cdot \dot{\boldsymbol{r}}_2)^2}{r_{12}^2c^2} - \frac{1}{2}\frac{\dot{\boldsymbol{r}}_2^2}{c^2} + \frac{3}{2}\frac{(\boldsymbol{r}_{12} \cdot \dot{\boldsymbol{r}}_2)^2}{r_{12}^2c^2}\right] + O(c^{-3}) \\ &= \frac{q_2}{r_{12}}\left[1 + \frac{\boldsymbol{r}_{12} \cdot \dot{\boldsymbol{r}}_2}{r_{12}c} + \frac{(\boldsymbol{r}_{12} \cdot \dot{\boldsymbol{r}}_2)^2}{r_{12}^2c^2} - \frac{1}{2}\frac{\dot{\boldsymbol{r}}_2^2}{c^2}\right] + O(c^{-3}) \end{aligned} \tag{3.222}$$

after expansion of the square root in the denominator according to Eq. (3.219) (note that the quadratic term of Eq. (3.219) also produces one nonnegligible term of order $O(c^{-2})$ explicitly written out in the first line of the last equation). Insertion of ϕ_2' into Eq. (3.211) yields the interaction energy V_{12} up to order $O(c^{-3})$,

$$\begin{aligned} V_{12}(t_1) &\approx \left[1 + \frac{1}{2}\frac{\dot{\boldsymbol{r}}_2^2}{c^2}\right] \times q_1 \times \frac{q_2}{r_{12}}\left[1 + \frac{\boldsymbol{r}_{12} \cdot \dot{\boldsymbol{r}}_2}{r_{12}c} + \frac{(\boldsymbol{r}_{12} \cdot \dot{\boldsymbol{r}}_2)^2}{r_{12}^2c^2} - \frac{1}{2}\frac{\dot{\boldsymbol{r}}_2^2}{c^2}\right] \\ &\quad \times \left[1 - \frac{\dot{\boldsymbol{r}}_1 \cdot \dot{\boldsymbol{r}}_2}{c^2}\right] + O(c^{-3}) \\ &= \frac{q_1q_2}{r_{12}}\left[1 + \frac{\boldsymbol{r}_{12} \cdot \dot{\boldsymbol{r}}_2}{r_{12}c} + \frac{(\boldsymbol{r}_{12} \cdot \dot{\boldsymbol{r}}_2)^2}{r_{12}^2c^2} - \frac{\dot{\boldsymbol{r}}_1 \cdot \dot{\boldsymbol{r}}_2}{c^2}\right] + O(c^{-3}) \end{aligned} \tag{3.223}$$

Hidden in the above expression is the fact that all quantities that carry the index of the second charge 2 are to be taken at time t_2 although the interaction energy is sought at time t_1.

3.5.4.2 Variables of Charge Two at Time of Charge One

After having introduced expansions in powers of the speed of light, we may proceed further and rewrite the interaction energy expressed by quantities at time t_1 only. For this purpose, we introduce the Taylor series expansions around time t_1 of the position and velocity of the second charge,

$$\boldsymbol{r}_2(t_2) = \boldsymbol{r}_2(t_1 - t_{12}) = \boldsymbol{r}_2(t_1) - \dot{\boldsymbol{r}}_2(t_1)t_{12} + \frac{1}{2}\ddot{\boldsymbol{r}}_2(t_1)t_{12}^2 + O(t_{12}^3) \tag{3.224}$$

$$\dot{\boldsymbol{r}}_2(t_2) = \dot{\boldsymbol{r}}_2(t_1 - t_{12}) = \dot{\boldsymbol{r}}_2(t_1) - \ddot{\boldsymbol{r}}_2(t_1)t_{12} + \frac{1}{2}\dddot{\boldsymbol{r}}_2(t_1)t_{12}^2 + O(t_{12}^3) \tag{3.225}$$

from which we first obtain an expression of the distance vector,

$$\begin{aligned} \boldsymbol{r}_{12}(t_1,t_2) &= \boldsymbol{r}_1(t_1) - \boldsymbol{r}_2(t_2) \\ &\overset{(3.224)}{=} \underbrace{\boldsymbol{r}_1(t_1) - \boldsymbol{r}_2(t_1)}_{\equiv \boldsymbol{r}_{12}(t_1)} + \dot{\boldsymbol{r}}_2(t_1)t_{12} - \frac{1}{2}\ddot{\boldsymbol{r}}_2(t_1)t_{12}^2 + O(t_{12}^3) \end{aligned} \tag{3.226}$$

where we introduced the distance at time t_1, $\boldsymbol{r}_{12}(t_1)$. From Eq. (3.226) we readily obtain the square of the distance

$$\begin{aligned} r_{12}^2(t_1,t_2) &= r_{12}^2(t_1) + 2\boldsymbol{r}_{12}(t_1)\cdot\dot{\boldsymbol{r}}_2(t_1)t_{12} \\ &\quad + \left(\dot{r}_2^2(t_1) - \boldsymbol{r}_{12}(t_1)\cdot\ddot{\boldsymbol{r}}_2(t_1)\right)t_{12}^2 + O(t_{12}^3) \end{aligned} \tag{3.227}$$

and the inverse of the distance

$$\begin{aligned} \frac{1}{r_{12}(t_1,t_2)} &= \frac{1}{\sqrt{r_{12}^2(t_1,t_2)}} = \frac{1}{r_{12}(t_1)}\left[1 - \frac{\boldsymbol{r}_{12}(t_1)\cdot\dot{\boldsymbol{r}}_2(t_1)}{r_{12}^2(t_1)}t_{12}\right. \\ &\quad - \frac{1}{2}\frac{\dot{r}_2^2(t_1) - \boldsymbol{r}_{12}(t_1)\cdot\ddot{\boldsymbol{r}}_2(t_1)}{r_{12}^2(t_1)}t_{12}^2 \\ &\quad \left. + \frac{3}{8}\left(2\frac{\boldsymbol{r}_{12}(t_1)\cdot\dot{\boldsymbol{r}}_2(t_1)}{r_{12}^2(t_1)}t_{12}\right)^2\right] + O(t_{12}^3) \end{aligned} \tag{3.228}$$

by utilizing Eq. (3.219) again. For the square of the inverse distance we then simply obtain

$$\frac{1}{r_{12}^2(t_1,t_2)} = \frac{1}{r_{12}^2(t_1)} - 2\frac{\boldsymbol{r}_{12}(t_1)\cdot\dot{\boldsymbol{r}}_2(t_1)}{r_{12}^4(t_1)}t_{12} + O(t_{12}^2) \tag{3.229}$$

where we have explicitly included only terms up to order $O(t_{12})$ since higher-order terms will later not be required because of the $1/c^2$ prefactor in Eq. (3.223).

3.5.4.3 Final Expansion of the Interaction Energy

If we now substitute in the expression of the interaction energy given by Eq. (3.223) all quantities that depend explicitly on t_2 by their Taylor expansions,

$$
\begin{aligned}
V_{12}(t_1) &= q_1 q_2 \underbrace{\left[\frac{1}{r_{12}} - \frac{\boldsymbol{r}_{12}\cdot\dot{\boldsymbol{r}}_2}{r_{12}^3}t_{12} - \frac{1}{2}\frac{\dot{\boldsymbol{r}}_2^2 - \boldsymbol{r}_{12}\cdot\ddot{\boldsymbol{r}}_2}{r_{12}^3}t_{12}^2 + \frac{3}{2}\frac{(\boldsymbol{r}_{12}\cdot\dot{\boldsymbol{r}}_2)^2}{r_{12}^5}t_{12}^2\right]}_{1/r_{12}(t_1,t_2)\ \text{Eq. (3.228)}} \\
&\quad \times \left\{1 + \frac{1}{c}\underbrace{\left[\frac{1}{r_{12}} - \frac{\boldsymbol{r}_{12}\cdot\dot{\boldsymbol{r}}_2}{r_{12}^3}t_{12}\right]}_{1/r_{12}(t_1,t_2)\ \text{Eq. (3.228)}} \times \left[\underbrace{(\boldsymbol{r}_{12} + \dot{\boldsymbol{r}}_2 t_{12})}_{\boldsymbol{r}_{12}(t_1,t_2)\ \text{Eq. (3.226)}} \cdot \underbrace{(\dot{\boldsymbol{r}}_2 - \ddot{\boldsymbol{r}}_2 t_{12})}_{\dot{\boldsymbol{r}}_2(t_2)\ \text{Eq. (3.225)}}\right]\right. \\
&\quad \left. + \frac{1}{c^2}\underbrace{\left[\frac{1}{r_{12}^2}\right]}_{\substack{1/r_{12}^2(t_1,t_2)\\ \text{Eq. (3.229)}}} \times \underbrace{(\boldsymbol{r}_{12}\cdot\dot{\boldsymbol{r}}_2)^2}_{\substack{\text{only lowest-order}\\ \text{terms survive}}} - \frac{1}{c^2}\dot{\boldsymbol{r}}_1 \cdot \underbrace{\dot{\boldsymbol{r}}_2}_{\substack{\dot{\boldsymbol{r}}_2(t_2)\\ \text{Eq. (3.225)}}}\right\} + O(c^{-3}) \qquad (3.230)
\end{aligned}
$$

(all terms on the right hand side — except t_{12} — depend only on t_1 and this dependence has therefore been dropped). We have all expansions already adjusted to the order really needed for an overall order of $O(c^{-2})$, where it is to be noted that t_{12} is of order $O(1/c)$ according to Eq. (3.214). Carrying out the multiplications in the above equation and retaining only all terms up to order $\mathcal{O}(c^{-2})$ we obtain for the term in curly brackets

$$
\begin{aligned}
\left\{\cdots\right\} &= \left\{1 + \frac{\boldsymbol{r}_{12}\cdot\dot{\boldsymbol{r}}_2}{c r_{12}} + \underbrace{\frac{\dot{\boldsymbol{r}}_2^2 - \boldsymbol{r}_{12}\cdot\ddot{\boldsymbol{r}}_2}{r_{12}}\frac{t_{12}}{c}}_{=\frac{\dot{\boldsymbol{r}}_2^2 - \boldsymbol{r}_{12}\cdot\ddot{\boldsymbol{r}}_2}{c^2} + O(c^{-3})}\right. \\
&\quad \left. \underbrace{-\frac{(\boldsymbol{r}_{12}\cdot\dot{\boldsymbol{r}}_2)^2}{c r_{12}^3}t_{12} + \frac{(\boldsymbol{r}_{12}\cdot\dot{\boldsymbol{r}}_2)^2}{c r_{12}^2}}_{=0+O(c^{-3})} - \frac{\dot{\boldsymbol{r}}_1\cdot\dot{\boldsymbol{r}}_2}{c^2} + O(c^{-3})\right\} \\
&= \left\{1 + \frac{\boldsymbol{r}_{12}\cdot\dot{\boldsymbol{r}}_2}{r_{12}}c^{-1} + \left[\dot{\boldsymbol{r}}_2^2 - \boldsymbol{r}_{12}\cdot\ddot{\boldsymbol{r}}_2 - \dot{\boldsymbol{r}}_1\cdot\dot{\boldsymbol{r}}_2\right]c^{-2} + O(c^{-3})\right\} \qquad (3.231)
\end{aligned}
$$

again with t_{12} being of order $O(1/c)$. If we insert this expression into Eq. (3.223) and factor out we get

$$
\begin{aligned}
V_{12}(t_1) &= q_1 q_2 \left[\frac{1}{r_{12}} \left\{ 1 + \frac{\boldsymbol{r}_{12}\cdot\dot{\boldsymbol{r}}_2}{c r_{12}} + \frac{\dot{\boldsymbol{r}}_2^2 - \boldsymbol{r}_{12}\cdot\ddot{\boldsymbol{r}}_2 - \dot{\boldsymbol{r}}_1\cdot\dot{\boldsymbol{r}}_2}{c^2} \right\} \right. \\
&\quad - \frac{\boldsymbol{r}_{12}\cdot\dot{\boldsymbol{r}}_2}{r_{12}^3} t_{12} \left\{ 1 + \frac{\boldsymbol{r}_{12}\cdot\dot{\boldsymbol{r}}_2}{c r_{12}} \right\} - \frac{1}{2}\frac{\dot{\boldsymbol{r}}_2^2 - \boldsymbol{r}_{12}\cdot\ddot{\boldsymbol{r}}_2}{r_{12}^3} t_{12}^2 \\
&\quad \left. + \frac{3}{2}\frac{(\boldsymbol{r}_{12}\cdot\dot{\boldsymbol{r}}_2)^2}{r_{12}^5} t_{12}^2 + O(c^{-3}) \right]
\end{aligned}
\tag{3.232}
$$

3.5.4.4 **Expansion of the Retardation Time**

Now, only the term linear in t_{12} requires special attention in order to maintain a consistent order of $O(c^{-2})$. For this purpose, we need to study the Taylor expansion of t_{12} in Eq. (3.214), for which we first need to expand the distance $r_{12}(t_1, t_2)$ as the square root of Eq. (3.227), $\sqrt{\boldsymbol{r}_{12}^2(t_1, t_2)}$,

$$
\begin{aligned}
r_{12}(t_1, t_2) &= r_{12}\sqrt{1 + 2\frac{\boldsymbol{r}_{12}\cdot\dot{\boldsymbol{r}}_2}{r_{12}^2} t_{12} + \frac{\dot{\boldsymbol{r}}_2^2 - \boldsymbol{r}_{12}\cdot\ddot{\boldsymbol{r}}_2}{r_{12}^2} t_{12}^2 + O(t_{12}^3)} \\
&= r_{12} + \frac{\boldsymbol{r}_{12}\cdot\dot{\boldsymbol{r}}_2}{r_{12}} t_{12} + \frac{1}{2}\frac{\dot{\boldsymbol{r}}_2^2 - \boldsymbol{r}_{12}\cdot\ddot{\boldsymbol{r}}_2}{r_{12}} t_{12}^2 \\
&\quad - \frac{r_{12}}{8}\left[2\frac{\boldsymbol{r}_{12}\cdot\dot{\boldsymbol{r}}_2}{r_{12}^2} t_{12} \right]^2 + O(t_{12}^3)
\end{aligned}
\tag{3.233}
$$

so that we can write for t_{12}

$$
t_{12} = \frac{r_{12}(t_1, t_2)}{c} = \frac{r_{12}}{c}\left(1 + \frac{\boldsymbol{r}_{12}\cdot\dot{\boldsymbol{r}}_2}{r_{12}^2} t_{12} \right) + O(c^{-3}) \tag{3.234}
$$

Unfortunately, we encounter a new t_{12} on the right hand side. But since this term will finally be of order $O(c^{-2})$, we may approximate $t_{12} \approx r_{12}(t_1)/c$ and arrive at the final low-order expansion for t_{12},

$$
t_{12} = \frac{r_{12}}{c} + \frac{\boldsymbol{r}_{12}\cdot\dot{\boldsymbol{r}}_2}{c^2} + O(c^{-3}) \tag{3.235}
$$

(of course, this expansion is also obtained if Eq. (3.234) is simply rearranged to solve for t_{12} and the resulting $1/[1 - (\boldsymbol{r}_{12}\cdot\dot{\boldsymbol{r}}_2)/(c r_{12})]$ is expanded up to order $O(c^{-3})$.

3.5.4.5 **General Darwin Interaction Energy**

Remember that all terms whose time dependence has not explicitly been noted are evaluated at time t_1. Eq. (3.235) finally allows us to write the interaction energy without any reference to the retardation time t_{12} and to assign all terms a unique order in $1/c$,

$$V_{12}(t_1) = q_1 q_2 \left[\frac{1}{r_{12}} + \frac{1}{2} \frac{\dot{r}_2^2 - \boldsymbol{r}_{12} \cdot \ddot{\boldsymbol{r}}_2 - 2\dot{\boldsymbol{r}}_1 \cdot \dot{\boldsymbol{r}}_2}{r_{12} c^2} - \frac{1}{2} \frac{(\boldsymbol{r}_{12} \cdot \dot{\boldsymbol{r}}_2)^2}{r_{12}^3 c^2} \right] + O(c^{-3}) \quad (3.236)$$

This is the interaction energy of two moving particles derived by Darwin [23]. Darwin, however, started his derivation of the interaction energy of two moving particles from an expression for the scalar and vector potentials of a single moving charge

$$\phi_2 = \gamma \phi_2' = \frac{q_2}{r_{12} - \frac{\boldsymbol{r}_{12} \cdot \dot{\boldsymbol{r}}_2}{c}} \quad (3.237)$$

$$\boldsymbol{A}_2 = \frac{\dot{\boldsymbol{r}}_2}{c} \phi_2 = \frac{\dot{\boldsymbol{r}}_2 q_2}{c r_{12} - \boldsymbol{r}_{12} \cdot \dot{\boldsymbol{r}}_2} \quad (3.238)$$

which are a form of the (retarded) Liénard–Wiechert potentials of a single moving charge (note again that index 1 refers also to t_1 while all quantities that carry the index of charge 2 are to be taken at time t_2 in IS). Such retarded potentials have been derived by Liénard, by Wiechert and also by others around 1900 [25] and they are, of course, related to the most general expression given in Eq. (3.208). In his famous lectures [15], Feynman derives the Liénard–Wiechert potential based on geometrical considerations and *derives* in this manner the Lorentz transformation — though a special form of a Lorentz boost in x-direction is obtained with $y = y'$ and $z = z'$ so that only the time and x coordinates mix. This special form of the Lorentz transformation is often employed as it is seemingly simpler to use; we started in this section, however, from the most general form of the Lorentz transformation in this section as we had to deal with a general three-dimensional distance between two particles (calculated from *all three* spatial coordinates) in IS′ for which a transformation that treats the spatial coordinates differently would not be appropriate for further manipulation (see above).

In order to arrive at Darwin's approximation to the interaction energy for any given instant of time from Eqs. (3.237) and (3.238) we recall the expansions for $\boldsymbol{r}_{12}$, $\dot{\boldsymbol{r}}_2$, and r_{12}, i.e., Eqs. (3.226), (3.225) and (3.233) above, to obtain for the denominator of Eq. (3.237)

$$\begin{aligned} r_{12} - \frac{\boldsymbol{r}_{12} \cdot \dot{\boldsymbol{r}}_2}{c} &= r_{12} + \frac{\boldsymbol{r}_{12} \cdot \dot{\boldsymbol{r}}_2}{r_{12}} t_{12} + \frac{1}{2} \frac{\dot{r}_2^2 - \boldsymbol{r}_{12} \cdot \ddot{\boldsymbol{r}}_2}{r_{12}} t_{12}^2 - \frac{1}{2} \frac{(\boldsymbol{r}_{12} \cdot \dot{\boldsymbol{r}}_2)^2}{r_{12}^3} t_{12}^2 \\ &\quad - \frac{1}{c} \left[\boldsymbol{r}_{12} \cdot \dot{\boldsymbol{r}}_2 + \left(\dot{r}_2^2 - \boldsymbol{r}_{12} \cdot \ddot{\boldsymbol{r}}_2 \right) t_{12} \right] + O(c^{-3}) \\ &= r_{12} + \frac{1}{2} \frac{(\boldsymbol{r}_{12} \cdot \dot{\boldsymbol{r}}_2)^2}{r_{12} c^2} - \frac{1}{2} \frac{\dot{r}_2^2 - \boldsymbol{r}_{12} \cdot \ddot{\boldsymbol{r}}_2}{c^2} + O(c^{-3}) \end{aligned} \quad (3.239)$$

This expression can now be inserted in Eq. (3.237)

$$\phi_2 = \frac{q_2}{r_{12}\left[1 - \frac{1}{2c^2}\left(\dot{r}_2^2 - \boldsymbol{r}_{12}\cdot\ddot{\boldsymbol{r}}_2 - \frac{(\boldsymbol{r}_{12}\cdot\dot{\boldsymbol{r}}_2)^2}{r_{12}^2}\right)\right] + O(c^{-3})} \tag{3.240}$$

and Darwin's approximation to ϕ_2 is obtained by recalling that $(1-x)^{-1} = 1 + x + O(x^2)$,

$$\phi_2 = \frac{1}{r_{12}}\left[1 + \frac{1}{2c^2}\left(\dot{r}_2^2 - \boldsymbol{r}_{12}\cdot\ddot{\boldsymbol{r}}_2 - \frac{(\boldsymbol{r}_{12}\cdot\dot{\boldsymbol{r}}_2)^2}{r_{12}^2}\right)\right] + O(c^{-3}) \tag{3.241}$$

Eq. (3.238) already contains a prefactor of $1/c$, and thus the vector potential is easily obtained to the same order as

$$\boldsymbol{A}_2 = \frac{\dot{\boldsymbol{r}}_2 q_2}{c r_{12}} + O(c^{-3}) \tag{3.242}$$

so that, according to Eq. (3.200), the total interaction energy reads

$$V_{12} = \frac{q_1 q_2}{r_{12}}\left[1 + \frac{1}{2c^2}\left(\dot{r}_2^2 - \boldsymbol{r}_{12}\cdot\ddot{\boldsymbol{r}}_2 - \frac{(\boldsymbol{r}_{12}\cdot\dot{\boldsymbol{r}}_2)^2}{r_{12}^2}\right)\right] - \frac{q_1 q_2}{c^2}\frac{\dot{\boldsymbol{r}}_1\cdot\dot{\boldsymbol{r}}_2}{r_{12}} + O(c^{-3}) \tag{3.243}$$

which is identical to Eq. (3.236) already given above.

3.5.5 Symmetrized Darwin Interaction Energy

Eq. (3.236) and, thus, also Eq. (3.243) clearly demonstrate that the approximate expression for the interaction energy which contains only particle positions, velocities and accelerations to be taken at the same time, is *not* symmetric with respect to the particle labels. This is already evident by inspection of the velocities, which enter the interaction energy unsymmetrically as the velocity of charge 2 is distinguished. Darwin in his seminal paper, however, was looking for an approximate interaction energy which is symmetric in the particle labels. In order to arrive at such an expression, he utilized the freedom of adding a total time derivative of a function to the Langrangian and thus to the potential energy as this does not change the Euler–Lagrange equations (as explained in detail in section 2.2.3.1). For this gauge transformation of the Lagrangian, Darwin suggested to apply as the function $F(q,t)$ in Eq. (2.62) the following choice

$$F(\dot{\boldsymbol{r}}_2, \boldsymbol{r}_{12}, r_{12}) = \frac{q_1 q_2}{2c^2}\frac{\boldsymbol{r}_{12}\cdot\dot{\boldsymbol{r}}_2}{r_{12}} \tag{3.244}$$

apart from an additional $m_2c^2\gamma(\dot{\boldsymbol{r}}_1)$ term that was included to symmetrize the kinetic energy, but which we neglect here as we deal with the potential energy only. The total time derivative of $F(\dot{\boldsymbol{r}}_2, \boldsymbol{r}_{12}, r_{12})$,

$$\frac{\mathrm{d}}{\mathrm{d}t}F(\dot{\boldsymbol{r}}_2, \boldsymbol{r}_{12}, r_{12}) = \frac{q_1q_2}{2c^2}\frac{\mathrm{d}}{\mathrm{d}t}\left(\frac{\boldsymbol{r}_{12}\cdot\dot{\boldsymbol{r}}_2}{r_{12}}\right) \tag{3.245}$$

is — according to Eq. (2.62) — to be added to the potential energy. In order to arrive at the explicit total time derivative we note that, first,

$$\begin{aligned} \frac{\mathrm{d}}{\mathrm{d}t}\left[\boldsymbol{r}_{12}\cdot\dot{\boldsymbol{r}}_2\right] = \frac{\mathrm{d}}{\mathrm{d}t}\left[(\boldsymbol{r}_1-\boldsymbol{r}_2)\cdot\dot{\boldsymbol{r}}_2\right] &= \boldsymbol{r}_1\cdot\ddot{\boldsymbol{r}}_2 + \dot{\boldsymbol{r}}_1\cdot\dot{\boldsymbol{r}}_2 - \dot{\boldsymbol{r}}_2^2 - \boldsymbol{r}_2\cdot\ddot{\boldsymbol{r}}_2 \\ &= \boldsymbol{r}_{12}\cdot\ddot{\boldsymbol{r}}_2 + \dot{\boldsymbol{r}}_1\cdot\dot{\boldsymbol{r}}_2 - \dot{\boldsymbol{r}}_2^2 \end{aligned} \tag{3.246}$$

and, second,

$$\begin{aligned} \frac{\mathrm{d}}{\mathrm{d}t}r_{12} = \frac{\mathrm{d}}{\mathrm{d}t}\sqrt{(\boldsymbol{r}_1-\boldsymbol{r}_2)^2} &= \frac{1}{2\sqrt{(\boldsymbol{r}_1-\boldsymbol{r}_2)^2}}2\left(\boldsymbol{r}_1-\boldsymbol{r}_2\right)\left(\dot{\boldsymbol{r}}_1-\dot{\boldsymbol{r}}_2\right) \\ &= \frac{1}{r_{12}}\left[\boldsymbol{r}_{12}\cdot(\dot{\boldsymbol{r}}_1-\dot{\boldsymbol{r}}_2)\right] \end{aligned} \tag{3.247}$$

so that the quotient rule finally yields

$$\begin{aligned} \frac{\mathrm{d}}{\mathrm{d}t}\left(\frac{\boldsymbol{r}_{12}\cdot\dot{\boldsymbol{r}}_2}{r_{12}}\right) &= \frac{1}{r_{12}^2}\left[\left(\boldsymbol{r}_{12}\cdot\ddot{\boldsymbol{r}}_2 + \dot{\boldsymbol{r}}_1\cdot\dot{\boldsymbol{r}}_2 - \dot{\boldsymbol{r}}_2^2\right)r_{12} - \frac{\left[\boldsymbol{r}_{12}\cdot(\dot{\boldsymbol{r}}_1-\dot{\boldsymbol{r}}_2)\right](\boldsymbol{r}_{12}\cdot\dot{\boldsymbol{r}}_2)}{r_{12}}\right] \\ &= \frac{\boldsymbol{r}_{12}\cdot\ddot{\boldsymbol{r}}_2 + \dot{\boldsymbol{r}}_1\cdot\dot{\boldsymbol{r}}_2 - \dot{\boldsymbol{r}}_2^2}{r_{12}} + \frac{(\boldsymbol{r}_{12}\cdot\dot{\boldsymbol{r}}_2)^2}{r_{12}^3} - \frac{(\boldsymbol{r}_{12}\cdot\dot{\boldsymbol{r}}_1)(\boldsymbol{r}_{12}\cdot\dot{\boldsymbol{r}}_2)}{r_{12}^3} \end{aligned} \tag{3.248}$$

The 'gauge transformed' potential energy $\tilde{V}_{12}$ then reads

$$\tilde{V}_{12} = V_{12} + \frac{\mathrm{d}}{\mathrm{d}t}F = \frac{q_1q_2}{r_{12}} - \frac{q_1q_2}{2c^2}\left[\frac{\dot{\boldsymbol{r}}_1\cdot\dot{\boldsymbol{r}}_2}{r_{12}} + \frac{(\boldsymbol{r}_{12}\cdot\dot{\boldsymbol{r}}_1)(\boldsymbol{r}_{12}\cdot\dot{\boldsymbol{r}}_2)}{r_{12}^3}\right] + O(c^{-3}) \tag{3.249}$$

by adding Eq. (3.236) and (3.248). Hence, we recover the instantaneous Coulomb interaction as the first term and then a correction of the order $O(c^{-2})$, which is symmetric in the particle coordinates by construction.

Finally, we note that we may express this interaction energy equally well in terms of the momenta $\boldsymbol{p}_i = m_i\dot{\boldsymbol{r}}_i$,

$$\tilde{V}_{12} = \frac{q_1q_2}{r_{12}} - \frac{q_1q_2}{2c^2m_1m_2}\left[\frac{\boldsymbol{p}_1\cdot\boldsymbol{p}_2}{r_{12}} + \frac{(\boldsymbol{r}_{12}\cdot\boldsymbol{p}_1)(\boldsymbol{r}_{12}\cdot\boldsymbol{p}_2)}{r_{12}^3}\right] + O(c^{-3}) \tag{3.250}$$

which appears to be trivial at this stage, but which can create confusion in quantum theory (see section 8.1).

Finally, we emphasize that the last equation results from the specific choice of gauge function for the total time derivative added to the Lagrangian. Hence, *the choice of the somewhat arbitrary gauge function*, which solely fulfills the boundary condition to finally eliminate all terms that are not symmetric in the particle indices, *determines the final expression for the interaction energy*. In addition, one could equally well apply the gauge tranformation to the scalar and vector potentials as in Eqs. (2.128) and (2.129). According to Eq. (2.165), it can be accomplished to produce the same interaction energy. A derivation of the Darwin Lagrangian that employs the Coulomb gauge for the electromagnetic potentials can be found in Jackson's textbook [12]. There, the Darwin potential energy is obtained via the Poisson equation which yields the *instantaneous* scalar potential in Coulomb gauge. Then, the transverse vector potential of a moving point charge for velocities which are small as compared to the speed of light c is derived from the transverse part of the current density, which stems from a decomposition of the total current density into a transverse, $\boldsymbol{j}_t$, and longitudinal, $\boldsymbol{j}_l$, component so that $\boldsymbol{j} = \boldsymbol{j}_t + \boldsymbol{j}_l$ with $\nabla \cdot \boldsymbol{j}_t = 0$ and $\nabla \times \boldsymbol{j}_l = 0$.

Harriman [26] gives a derivation which starts from the retarded potentials, Eqs. (2.140) and (2.141), for which he assumes Taylor expansions of the charge and current density of the integrand in terms of the retardation time $|\boldsymbol{r} - \boldsymbol{r}'|/c$,

$$\rho(\boldsymbol{r}', t') = \rho(\boldsymbol{r}', t) - \dot{\rho}(\boldsymbol{r}', t)\frac{|\boldsymbol{r} - \boldsymbol{r}'|}{c} + \frac{1}{2}\ddot{\rho}(\boldsymbol{r}', t)\frac{|\boldsymbol{r} - \boldsymbol{r}'|^2}{c^2} + \cdots \quad (3.251)$$

$$\boldsymbol{j}(\boldsymbol{r}', t') = \boldsymbol{j}(\boldsymbol{r}', t) - \dot{\boldsymbol{j}}(\boldsymbol{r}', t)\frac{|\boldsymbol{r} - \boldsymbol{r}'|}{c} + \frac{1}{2}\ddot{\boldsymbol{j}}(\boldsymbol{r}', t)\frac{|\boldsymbol{r} - \boldsymbol{r}'|^2}{c^2} + \cdots \quad (3.252)$$

where the derivatives are to be taken with respect to the retardation time at ($\boldsymbol{r}'$,t) and have therefore been denoted by a dot. To evaluate these expressions it is then necessary to find an expression for the densities. They are for point particle 2 given by

$$\rho(\boldsymbol{r}', t) = q_2\delta^{(3)}(\boldsymbol{r}' - \boldsymbol{r}_2) \quad \text{and} \quad \boldsymbol{j}(\boldsymbol{r}', t) = q_2\delta^{(3)}(\boldsymbol{r}' - \boldsymbol{r}_2)\dot{\boldsymbol{r}}_2 \quad (3.253)$$

and finally yield — again, of course, after a gauge transformation to obtain the symmetrized form — the electromagnetic potentials and then the Darwin potential energy.

Further Reading

S. Weinberg, [27]. *Gravitation and Cosmology*.

This wonderful presentation of Einstein's general theory of relativity and its application to cosmology gives an almost complete introduction to special relativity at the beginning of the book. The reader will also find an interesting outline of the historical development of this epochal theory.

D. Bohm, [28]. *The Special Theory of Relativity.*

This book by Bohm gives a clear and readable introduction to the concepts and important applications of the special theory of relativity. It is more text based than many modern presentations of special relativity and puts great emphasis on explaining the basic ideas and assumptions rather than developing the advanced mathematical framework based on tensors and the Ricci calculus. It can highly be recommended as an introduction for the beginner.

W. Pauli, [29]. *Relativitätstheorie.*

At the age of 21 Wolfgang Pauli published his famous summary of the general theory of relativity, which is frequently read and used even today. Though technically quite demanding, this ingenious presentation is still an invaluable source of knowledge and inspiration for both the experienced expert and the beginner with only little previous exposure to this field.

D. E. Mook, T. Vargish, [30]. *Inside Relativity.*

This is an unconventional yet highly readable presentation of both the special and general theory of relativity. The authors avoid any complicated mathematical formulae and present the principles and ideas underlying the whole theory purely text based. It can be recommended for beginners and also for experienced readers who will find the many biographical and historical anecdotes amusing.

S. Parrott, [31]. *Relativistic Electrodynamics and Differential Geometry.*

Parrott's highly mathematical presentation should certainly not be employed for a first contact with this subject. However, for the experienced reader it represents a wonderful compilation and discussion of the essential and highly non-trivial geometrical asects and implications of special relativity. The complete theory is formulated in terms of differential forms, i.e., in coordinate-free external calculus, which reveals a completely new perspective on relativity gaining more and more importance for the understanding of modern physics.

F. Gross, [32]. *Relativistic Quantum Mechanics and Field Theory.*

Like many other books on relativistic quantum field theory, the book by Gross also contains a thorough introduction to special relativity. All relevant aspects of special relativity are developed in a concise notation which is applied to quantum mechanics and quantum field theory in later parts of the book.

4
Basics of Quantum Mechanics

The study of matter at a molecular and sub-molecular level requires a quantum mechanical framework. In this chapter, we provide an overview of the essential concepts and notation of quantum theory needed to understand the development of relativistic many-electron quantum mechanics. This introductory overview starts from basic axioms with a focus on the nonrelativistic theory. Gauge and Lorentz invariance properties as introduced in the classical theory are then discussed for the basic quantum mechanical equations in the following two chapters.

The theory of molecular science has to be a custom-made theory of matter on a molecular scale and at energies accessible by thermal motion or photoexcitation. Specifically, we may choose as elementary particles, which compose molecular aggregates, electrons and atomic nuclei to be treated by this theory. It will thus not be necessary to explicitly consider protons or neutrons or even more fundamental particles like quarks. By choosing electrons and atomic nuclei as *the* elementary particles of relativistic quantum chemistry, in this book we will elevate the rather general and fundamental quantum mechanical many-electron theory of quantum electrodynamics to a level that allows us to perform actual calculations and simple qualitative considerations of specific systems. Despite these restrictions, a theory for processes at a molecular and sub-molecular scale still has to be quantum mechanical in nature.

In this section, we shall provide a discussion of elements of quantum mechanics, which have been selected mainly for the purpose of serving the development of relativistic quantum chemistry in later chapters. Unfortunately, in view of the complexity of the subject and the limited space in this book, this conceptual discussion of quantum theory cannot be discussed in satisfactory depth. Also, some aspects will be fully developed only in later chapters.

In contrast to Newtonian mechanics, we do not have an intuitive access to quantum mechanics. In the case of classical mechanics, Lindsay and Margenau [33] coined the term *principle of elementary abstraction* meaning elementary abstraction from macroscopic observations like the definition of the velocity as the differential quotient of a length per (infinitesimally) small time interval. Therefore, we proceed to formulate a few non-intuitive postulates which serve

Relativistic Quantum Chemistry. Markus Reiher and Alexander Wolf

ISBN: 978-3-527-31292-4

as a basis for this theory, though we have no rationale for the introduction of the specific form of these postulates except that they lead to a formalism which allows us to calculate numbers for observables that turn out to be consistent with experimental observations.

Several possibilities of formulating these postulates exist. The collection and presentation of postulates which we consider useful for the development of a relativistic many-electron theory are given in the following sections. Also, we avoid any discussion of the wave-particle dualism of matter since we are interested in the presentation of a formalism that eventually allows us to describe and predict matter on a molecular scale. It is thus most convenient to think of electrons and atomic nuclei simply as particles rather than as waves. The following sections introduce five basic axioms of quantum theory.

4.1 The Quantum Mechanical State

It is convenient to define a *state* of a system in quantum mechanics. All (physical) information that can be known about a quantum mechanical system is contained in a quantum mechanical state function, Ψ_n, which is also called a *wave function* mainly for historical reasons. In order to be able to distinguish different states of a system we introduce the subscript n to label these different states. The term 'quantum mechanical system' will denote an elementary particle or a collection of elementary particles. In chemistry, it is a collection of electrons and atomic nuclei constituting an atom, a molecule or an assembly of atoms and molecules.

4.1.1 Bracket Notation

The n-th quantum mechanical state Ψ_n can be described independently of its representation by employing Dirac's *bra-ket notation,* $|\Psi_n\rangle = |n\rangle$. This notation has the advantage of providing a universal notation so that there is no primary need to refer to a certain set of dynamical variables (e.g., spatial or momentum variables) or to a certain mathematical representation of the state as a function or a vector. The *ket* $|n\rangle$ refers to Ψ_n and the *bra* $\langle n|$ to its complex conjugate $\Psi_n^\star$ (this is to be replaced by $\Psi_n^\dagger = \Psi_n^{\star,T}$ if the state Ψ_n contains more than one component and is thus a vector quantity; $\dagger$ is pronounced 'dagger'). More precisely, the bra vector $\langle n|$ is not an element of the same vector space but of the dual vector space, which is the space of all linear mappings

$$\begin{aligned} \langle n| : \qquad & V \rightarrow \mathbb{R}, \\ & |m\rangle \mapsto \langle n|m\rangle \end{aligned} \tag{4.1}$$

where $\langle n|m\rangle$ denotes the scalar product on the vector space V — the space of square-integrable functions (cf. section 4.1.4). This scalar product is defined as the integral over all dynamical variables denoted τ,

$$\langle n|m\rangle = \int_{-\infty}^{+\infty} d\tau \;\Psi_n^\star(\tau)\Psi_m(\tau) \tag{4.2}$$

4.1.2
Expansion in a Complete Basis Set

A state can be expressed in terms of known basis functions $\Phi_k(\tau)$,

$$\Psi_n(\tau) = \sum_{k=1}^{K} C_{nk}\Phi_k(\tau) \tag{4.3}$$

with yet unknown expansion coefficients. The basis functions may always be chosen to constitute an orthonormal set,

$$\int_{-\infty}^{+\infty} d\tau \;\Phi_k^\star(\tau)\Phi_l(\tau) = \delta_{kl} \tag{4.4}$$

Then, the bracket in this basis set expansion reduces to

$$\langle n|m\rangle = \sum_{k,l=1}^{K} C_{nk}^\star C_{ml} \int_{-\infty}^{+\infty} d\tau \;\Phi_k^\star(\tau)\Phi_l(\tau) \overset{(4.4)}{=} \sum_{k=1}^{K} C_{nk}^\star C_{mk} \equiv \boldsymbol{C}_n^\dagger \boldsymbol{C}_m \tag{4.5}$$

where $\boldsymbol{C}_m$ denotes a vector of all expansion coefficients C_{mk} of state $|m\rangle$. The number of basis functions K is in general infinite if the representation is to be exact. If K=∞ the basis is then said to be *complete*. In this case, we find

$$\sum_{k=1}^{\infty} |\Phi_k\rangle\langle\Phi_k| = \mathbf{1} = id \tag{4.6}$$

which is also known as a resolution of the identity. If, however, K is finite, the basis is said to be *finite*.

4.1.3
Born Interpretation

The quantum mechanical state $|n\rangle$ has no direct physical interpretation, but its absolute square, $|\Psi_n|^2$=$\Psi_n^\star\Psi_n$, can be interpreted as a probability density distribution. This so-called *Born interpretation* implies for a single particle that the wave function has to be normalized, i.e., integration over all dynamical variables of a system must yield unity,

$$\langle n|n\rangle \overset{!}{=} 1 \tag{4.7}$$

because the probability of finding a single particle somewhere in space must be one at any time. This concept can be generalized for a system containing an arbitrary number of particles. Given N elementary particles with positions $\{\boldsymbol{r}_i\}$, we may consider a state function $\Psi_n(\boldsymbol{r}_1, \boldsymbol{r}_2, \ldots, \boldsymbol{r}_N)$ describing the n-th state. This state function will still be normalized to one according to Eq. (4.7), although it describes N particles. Eq. (4.7) would read in this case

$$\langle n|n\rangle = \int_{-\infty}^{+\infty} \mathrm{d}^3 r_1 \cdots \int_{-\infty}^{+\infty} \mathrm{d}^3 r_N \, \Psi_n^\star(\boldsymbol{r}_1, \boldsymbol{r}_2, \ldots, \boldsymbol{r}_N) \Psi_n(\boldsymbol{r}_1, \boldsymbol{r}_2, \ldots, \boldsymbol{r}_N) \stackrel{!}{=} 1 \quad (4.8)$$

Accordingly, the integrand of Eq. (4.8) $|\Psi_n(\boldsymbol{r}_1, \boldsymbol{r}_2, \ldots, \boldsymbol{r}_N)|^2 \mathrm{d}^3 r_1 \ldots \mathrm{d}^3 r_N$ is the probability of finding particle 1 in volume element $\mathrm{d}^3 r_1$, particle 2 in volume element $\mathrm{d}^3 r_2$ and so forth.

We define the particle density distribution of the quantum mechanical state Ψ_n as

$$\rho_n(\boldsymbol{r}) \equiv N \int_{-\infty}^{+\infty} \mathrm{d}^3 r_2 \cdots \int_{-\infty}^{+\infty} \mathrm{d}^3 r_N \, \Psi_n^\star(\boldsymbol{r}, \boldsymbol{r}_2, \ldots, \boldsymbol{r}_N) \Psi_n(\boldsymbol{r}, \boldsymbol{r}_2, \ldots, \boldsymbol{r}_N) \quad (4.9)$$

so that $[\rho_n(\boldsymbol{r})/N]\, \mathrm{d}^3 r$ is the probability of finding *any* particle in the volume element $\mathrm{d}^3 r$. For the sake of brevity, we usually drop the state index n and write simply $\rho(\boldsymbol{r})$ for the particle density distribution as it will be clear from the context whether this is a ground state density or the density of a state higher in energy.

After integration of the density of any state over all space the number of particles is recovered,

$$\int_{-\infty}^{+\infty} \mathrm{d}^3 r \, \rho(\boldsymbol{r}) = N \quad (4.10)$$

In the case of purely electronic systems, the particle density is also called *electron density*. Trivially related to the electron density is the *charge density*

$$\rho_c(\boldsymbol{r}) = q_e \rho(\boldsymbol{r}) = -e\rho(\boldsymbol{r}) \quad (4.11)$$

which is simply the charge-weighted particle density. Consequently, $-e\rho(\boldsymbol{r})\mathrm{d}^3 r$ represents the (electronic) charge in the infinitesimally small volume element $\mathrm{d}^3 r$. For the probability in a finite volume, this needs to be integrated.

4.1.4
State Vectors in Hilbert Space

Mathematically, any physical state of a system is represented by a vector $|\Psi_n\rangle$ in a corresponding Hilbert space $\mathcal{H}$. In general, a Hilbert space is a linear space (i.e., vector space) of infinite dimension, where the vectors are square-integrable complex-valued functions of space and time. The integrability constraint,

$$\int_{-\infty}^{+\infty} d^3r\, \Psi_n^\star(\boldsymbol{r})\Psi_m(\boldsymbol{r}) < \infty \tag{4.12}$$

has to be imposed in order to guarantee normalizability as demanded by Eq. (4.8). Such a Hilbert space is associated with any physical system, we may have one $\mathcal{H}_1$ for particle 1 and another one $\mathcal{H}_2$ for particle 2. The Hilbert space $\mathcal{H}_{\rm tot}$ for a system comprising both particles can be constructed from the single-particle Hilbert spaces via a direct product

$$\mathcal{H}_{\rm tot} = \mathcal{H}_1 \otimes \mathcal{H}_2 \tag{4.13}$$

This concept of direct products — also called Kronecker or tensor products — will be elaborated in detail in section 8.4.

Formally, any physical observable is then a linear, hermitean operator $\hat{O}$,

$$\hat{O}^\dagger = \hat{O} \tag{4.14}$$

which is also called a self-adjoint operator. $\hat{O}$ acts on functions $|n\rangle$ of the Hilbert space,

$$\begin{aligned} \hat{O}: \quad & \mathcal{H} \to \mathcal{H} \\ & |n\rangle \mapsto \hat{O}|n\rangle \end{aligned} \tag{4.15}$$

Any statement concerning the physical properties of the system has thus to be expressed in terms of mathematical operations defined on this Hilbert space.

Upon our transition from classical mechanics to quantum mechanics, we distinguish operators from their classical analogs by a hat. This is convenient here but will make all further derivations quite clumsy. Therefore, the hat will be omitted as a designator for an operator later in this and then in all following chapters. It should be obvious from the context which quantity is an operator and which is not.

So far we have solely postulated the existence of a quantum mechanical state from which physical information may be extracted. How this 'extraction' can be achieved will be the subject of the postulates of section 4.3. In the next section we first introduce the equation of motion for quantum mechanical states that governs their time evolution.

4.2 The Equation of Motion

The evolution of a quantum mechanical state in time is described by the following equation of motion:

$$i\hbar \frac{\partial}{\partial t} \Psi_n(\boldsymbol{r}, t) = \hat{H} \Psi_n(\boldsymbol{r}, t) \tag{4.16}$$

We have now introduced an explicit dependence of Ψ_n on time t as well as the *Hamiltonian operator* $\hat{H}$ that operates on Ψ_n. This operator will later be shown to represent the energy observable, since it is related to the Hamiltonian function in classical physics known from chapters 2 and 3. The equation of motion and thus the explicit form of the Hamiltonian operator $\hat{H}$ cannot be rigorously deduced. The most prominent rationale for $\hat{H}$ is the correspondence principle which relates $\hat{H}$ to the Hamiltonian function once the explicit form of the equation of motion is found (see section 4.3.5).

4.2.1 Restrictions on the Fundamental Quantum Mechanical Equation

It is the fundamental Eq. (4.16), which has to obey the principle of Einstein's theory of special relativity, namely of being invariant in form in different inertial frames of reference. Hence, the choices for the Hamiltonian operator $\hat{H}$ are further limited by the requirement of form invariance of the whole equation under Lorentz transformations which will be discussed in detail in chapter 5.

Apart from Lorentz covariance the quantum mechanical state equation must obey certain mathematical criteria: (i) it must be homogeneous in order to fulfill Eq. (4.7) for all times, and (ii) it must be a linear equation so that linear combinations of solutions are also solutions. The latter requirement is often denoted as the *superposition principle*, which is required for the description of interference phenomena. However, it is equally well justified to regard these requirements as the consequences of the equation of motion in accordance with experiment if the equation of motion and the form of the Hamiltonian operator are postulated.

4.2.2 Time Evolution and Probabilistic Character

Since the equation of motion, Eq. (4.16), contains a first derivative in time, the quantum state $\Psi(\boldsymbol{r}, t)$ at time t is solely determined by the initial state $\Psi(\boldsymbol{r}, 0)$, which, mathematically speaking, fixes the only integration constant. However, this fact demonstrates that the time evolution of a quantum mechanical system as described by Eq. (4.16) is a purely deterministic process, because a quantum mechanical state evolves uniquely according to the linear differential equation of motion at time t out of a state at an earlier time. The probabilistic character of quantum mechanics solely arises from the measurement process for which no satisfactory and consistent mathematical description is available (Born's interpretation aims at coping with this problem; see also section 4.3).

4.2.3 Stationary States

If the Hamiltonian operator $\hat{H}$ does *not* change with time, spatial and time coordinates can be separated in the ansatz for the state

$$\Psi_n(\boldsymbol{r},t) = \psi_n(t)\Psi_n(\boldsymbol{r}) \tag{4.17}$$

where we use the same symbol for the complete wave function and for the part that does no longer depend on time, $\Psi_n(\boldsymbol{r})$. This time-independent state is therefore called a *stationary state*. Inserting the ansatz into Eq. (4.16) yields

$$\Psi_n(\boldsymbol{r})\mathrm{i}\hbar\frac{\partial\psi_n(t)}{\partial t} = \psi_n(t)\hat{H}\Psi_n(\boldsymbol{r}) \tag{4.18}$$

Now, this equation can be rearranged in such a way that the left hand side solely depends on time, while the right-hand side then solely depends on position coordinates, which implies that both sides of the equation must be a constant independent of space and time. If we call this separation constant E_n, we may write

$$\frac{1}{\psi_n(t)}\mathrm{i}\hbar\frac{\partial\psi_n(t)}{\partial t} = \frac{1}{\Psi_n(\boldsymbol{r})}\hat{H}\Psi_n(\boldsymbol{r}) \equiv E_n \tag{4.19}$$

This equation can be split into two,

$$\mathrm{i}\hbar\frac{\partial\psi_n(t)}{\partial t} = E_n\psi_n(t) \tag{4.20}$$

$$\hat{H}\Psi_n(\boldsymbol{r}) = E_n\Psi_n(\boldsymbol{r}) \tag{4.21}$$

The separation constant E_n thus becomes an energy eigenvalue of the system since Eq. (4.21) is nothing but the eigenvalue equation of the Hamiltonian $\hat{H}$, which, as mentioned before, represents the energy observable. Eq. (4.20) can be integrated at once to yield

$$\psi_n(t) = \exp\left(-\frac{\mathrm{i}}{\hbar}E_n t\right) \tag{4.22}$$

while the solution of Eq. (4.21) depends on the explicit representation of $\hat{H}$, which is dependent on the particular system under consideration.

4.3
Observables

In section 4.1 it was stated that all physical information is contained in the quantum mechanical state Ψ, but the question is how this information can be extracted. Any observable of a physical system is described by an hermitean operator acting on the corresponding Hilbert space. We further postulate that any experimentally measured value o_k of a physical observable O must be identical to one of the eigenvalues of the corresponding operator $\hat{O}$,

$$\hat{O}\Phi_k = o_k \Phi_k \tag{4.23}$$

All eigenstates $\{\Phi_k\}$ of the hermitean operator form a complete orthonormal set, $\langle\Phi_k|\Phi_l\rangle = \delta_{kl}$, so that the actual state Ψ of the system can always be expressed as a superposition of these eigenstates according to Eq. (4.3). Once a measurement is performed on the system this superposition collapses to one of the eigenstates and the measured value is the corresponding eigenvalue. The probability of measuring some eigenvalue o_k is determined by the overlap of the corresponding eigenstate Φ_k and the actual state Ψ, $\langle\Phi_k|\Psi\rangle$.

4.3.1
Expectation Values

We now define the mean result for a measured observable. But since a measurement forces any system into an eigenstate we cannot take such a mean value from a sequence of measurements on the same system but need to introduce a work around. For the definition of an average, we need to have an ensemble of identically prepared systems each described by the state vector Ψ and subjected to the measurement process. Then, we may define the average result of these measurements as

$$o = \langle\hat{O}\rangle_\Psi = \frac{\langle\Psi|\hat{O}|\Psi\rangle}{\langle\Psi|\Psi\rangle} \tag{4.24}$$

which is called the *expectation value*. The average is over all dynamical variables on which Ψ depends. For a normalized state function $\Psi(\{\boldsymbol{r}_i\})$ the denominator of Eq. (4.24) is simply one and we obtain

$$o = \langle\Psi|\hat{O}|\Psi\rangle = \int_{-\infty}^{\infty} d^3r_1 \cdots \int_{-\infty}^{\infty} d^3r_N \Psi(\{\boldsymbol{r}_i\})^\dagger \hat{O}(\{\boldsymbol{r}_i\}) \Psi(\{\boldsymbol{r}_i\}) \tag{4.25}$$

where we assume that Ψ describes N particles with positions $\{\boldsymbol{r}_i\}$ as the only dynamical variables, so that the integration can be made explicit. Moreover, we have generalized the complex conjugate $\Psi^\star$ to become $\Psi^\dagger$, which means transposition and complex conjugation of Ψ in cases in which Ψ is of vectorial form.

The expectation value is the mean value of a large number of measurements of the observable. If the state Ψ is given by the superposition of Eq. (4.3), the expectation value becomes

$$o = \sum_{kl} C_k^\star C_l \langle \Phi_k | \hat{O} | \Phi_l \rangle \stackrel{(4.23)}{=} \sum_k |C_k|^2 \, o_k \tag{4.26}$$

where we have dropped the state label n for the sake of simplicity. Hence, the expectation value must not be confused with the most probable eigenvalue, which is determined by the largest squared weight $|C_k|^2$ in this expansion. The expectation value coincides with an eigenvalue o_k if the system has been prepared in the eigenstate Φ_k of the operator corresponding to the observable.

4.3.2 Hermitean Operators

Given a general operator $\hat{A}$, its adjoint operator $\hat{A}^\dagger$ is defined by

$$\langle \Psi_n | \hat{A} \Psi_m \rangle = \langle \hat{A}^\dagger \Psi_n | \Psi_m \rangle \tag{4.27}$$

which reads in the case of an hermitean operator $\hat{O} = \hat{O}^\dagger$ as

$$\langle \Psi_n | \hat{O} \Psi_m \rangle = \langle \hat{O} \Psi_n | \Psi_m \rangle \equiv \langle \Psi_n | \hat{O} | \Psi_m \rangle \tag{4.28}$$

i.e., for an hermitean operator it does not make any difference whether the operator acts to the left or to the right within a scalar product. This is reflected by the symmetric formulation of the Dirac bracket in Eq. (4.28) where the operator could be separated from the states by two vertical dashes, a notation that has already been anticipated in the notation for the expectation value in Eq. (4.24). We note without proof that all hermitean operators feature strictly real eigenvalue spectra, so that $o_n \in \mathbb{R}$, which is a mandatory property for observables as they cannot be complex numbers.

Unfortunately, there exists no recipe to find the explicit form of the operator $\hat{O}$ to be associated with the observable O. Hence, the main question is how to find the proper operator representation for a given observable or, to put it the other way around, how to identify those hermitean operators that correspond to observable quantities. Typical observables are energy, canonical momentum, and angular momentum.

4.3.3 Unitary Transformations

In order to investigate the uniqueness of a quantum mechanical state function Ψ, we study *unitary transformations*. The operator U on a Hilbert space $\mathcal{H}$ is called *unitary*, if

$$UU^\dagger = U^\dagger U = U^{\star,T} U = \mathbf{1} \tag{4.29}$$

and, hence, its inverse U^{-1} is equal to its adjoint $U^{\dagger}$. This is equivalent to the following properties for a unitary operator U:

1) U is invertible, i.e., U^{-1} exists,
2) U preserves all scalar products, i.e.,
$\langle U\Psi_n|U\Psi_m\rangle = \langle\Psi_n|\Psi_m\rangle \ \forall \ \Psi_n, \Psi_m \in \mathcal{H}$.

A unitary operator leaves the lengths of vectors and angles between them unchanged. The transformations mediated through the action of such an operator are therefore analogous to those of rotations in three-dimensional Euclidean space.

If U is unitary and $\hat{O}$ is an hermitean operator, then $U\hat{O}U^{\dagger}$ is also hermitean since $(U\hat{O}U^{\dagger})^{\dagger} = (U^{\dagger})^{\dagger}\hat{O}^{\dagger}U^{\dagger} = U\hat{O}U^{\dagger}$. Moreover, since

$$\hat{O}(\underbrace{U^{\dagger}U}_{1})\Psi = a\Psi \quad \stackrel{U\cdot}{\Longrightarrow} \quad U\hat{O}U^{\dagger}U\Psi = aU\Psi \tag{4.30}$$

$\hat{O}$ and $U\hat{O}U^{\dagger}$ possess the same set of eigenvalues,

$$\hat{O}\Psi = a\Psi \quad \text{and} \quad (U\hat{O}U^{\dagger})(U\Psi) = a(U\Psi) \tag{4.31}$$

Also, expectation values over any operator $\hat{O}$ are invariant under unitary transformations,

$$\langle\hat{O}\rangle = \langle\Psi|\hat{O}|\Psi\rangle = \langle\Psi|\underbrace{U^{\dagger}U}_{1}\hat{O}\underbrace{U^{\dagger}U}_{1}|\Psi\rangle = \langle U\Psi|U\hat{O}U^{\dagger}|U\Psi\rangle \tag{4.32}$$

Consequently, any quantum mechanical state Ψ is only defined up to an arbitrary unitary transformation: a physical state may thus be equivalently described by Ψ or by $(U\Psi)$ and the integral (also called matrix element) containing an operator may be calculated as $\langle\Psi|\hat{O}|\Psi\rangle$ or equivalently as $\langle U\Psi|(U\hat{O}U^{\dagger})|U\Psi\rangle$. This is important for the calculation of observables and expectation values of observables (as discussed here and also refered to in later chapters).

4.3.4
Heisenberg Equation of Motion

The equation of motion, Eq. (4.16), defines the so-called *Schrödinger picture*, where the state Ψ is propagated in time while the operators (observables) are stationary, e.g., $\hat{H} \rightarrow \hat{H}^{(S)}$ where the superscript '(S)' now explicitly denotes the Schrödinger picture. This holds for all 'ordinary' observables like energy, momentum, and position rather than for a system interacting with a time-dependent external potential (like an electromagnetic light wave). For the sake of brevity we skip the state index n. It is now possible to formulate a different dynamics called the *Heisenberg picture* which rests on stationary states but time-propagated operators. To deduce the Heisenberg picture, we introduce the unitary transformation

$$U(t) = \exp\left(-\frac{\mathrm{i}}{\hbar}\hat{H}^{(S)}t\right) \quad \Rightarrow \quad U^{\dagger}(t) = \exp\left(\frac{\mathrm{i}}{\hbar}\hat{H}^{(S)}t\right) \tag{4.33}$$

since $\hat{H}^{(S),\dagger} = \hat{H}^{(S)}$. This unitary transformation allows us to describe the propagation of any state $\Psi(t)$ in time

$$\Psi(t) = U(t)\,\Psi(t_0) \tag{4.34}$$

where $\Psi(t_0)$=$\Psi(t$=$0)$ is *time-independent*. The ansatz for $\Psi(t)$ obviously fulfills the equation of motion given in Eq. (4.16), which may easily be verified by insertion. According to what has been said in the preceding section about unitary transformations, we have two sets of operators,

$$\hat{O}^{(H)} = U^{\dagger}(t)\,\hat{O}^{(S)}\,U(t) \tag{4.35}$$

where the superscript '(H)' denotes the Heisenberg picture, respectively. That is, the unitary transformation $U(t)$ allows us to switch from an operator in the Schrödinger picture, $\hat{O}^{(S)}$, to the corresponding one in the Heisenberg picture. Of course, expectation values are the same in both pictures as the physical assertions must remain unchanged,

$$\begin{aligned} o &= \langle\Psi(t)|\hat{O}^{(S)}|\Psi(t)\rangle = \langle U(t)\Psi(t_0)|\hat{O}^{(S)}|U(t)\Psi(t_0)\rangle \\ &= \langle\Psi(t_0)|U^{\dagger}(t)\hat{O}^{(S)}U(t)|\Psi(t_0)\rangle \overset{(4.35)}{=} \langle\Psi(t_0)|\hat{O}^{(H)}|\Psi(t_0)\rangle \end{aligned} \tag{4.36}$$

The time dependence vanishes in the expectation value because the time-dependent part of the state, Eq. (4.22), is multiplied by its complex conjugate to yield unity, $\exp(+\mathrm{i}Et/\hbar)\exp(-\mathrm{i}Et/\hbar) = 1$. Since the Hamiltonian $\hat{H}^{(S)}$ commutes with $U(t)$

$$\hat{H}^{(S)}U(t) = U(t)\hat{H}^{(S)} \tag{4.37}$$

the Hamiltonian turns out to be the same in both pictures,

$$\hat{H}^{(H)} = U^{\dagger}(t)\hat{H}^{(S)}U(t) = \hat{H}^{(S)}U^{\dagger}(t)U(t) \overset{(4.29)}{=} \hat{H}^{(S)} \equiv \hat{H} \tag{4.38}$$

If we now differentiate Eq. (4.35), we obtain

$$\frac{\mathrm{d}\hat{O}^{(H)}}{\mathrm{d}t} = \frac{\mathrm{d}U^{\dagger}(t)}{\mathrm{d}t}\,\hat{O}^{(S)}\,U(t) + U^{\dagger}(t)\,\hat{O}^{(S)}\,\frac{\mathrm{d}U(t)}{\mathrm{d}t} \tag{4.39}$$

Because $\hat{O}^{(S)}$ does not depend on time and its time derivative vanishes. The total time derivative of $U(t)$ follows by differentiation of Eq. (4.33),

$$\frac{\mathrm{d}U(t)}{\mathrm{d}t} = -\frac{\mathrm{i}}{\hbar}\hat{H}\,U(t) \tag{4.40}$$

where we used the fact that the Hamiltonian, like all other operators in the Schrödinger picture, does *not* depend on time. Not even the position operators in the Schrödinger Hamiltonian can be considered to be functions of time because this would imply some sort of trajectory, which is a classical concept that is *not* valid in quantum mechanics. With Eq. (4.40) and its complex conjugate,

$$\frac{dU^\dagger(t)}{dt} = \frac{i}{\hbar} U^\dagger(t)\hat{H}^\dagger = \frac{i}{\hbar} U^\dagger(t)\hat{H} \tag{4.41}$$

we can then write for the total time derivative of $\hat{O}^{(H)}$ in Eq. (4.39)

$$\frac{d\hat{O}^{(H)}}{dt} = \frac{i}{\hbar} U^\dagger(t)\hat{H}\,\hat{O}^{(S)}\,U(t) - U^\dagger(t)\,\hat{O}^{(S)}\,\frac{i}{\hbar}\hat{H}U(t) \tag{4.42}$$

In order to introduce the Heisenberg-picture observable also on the right hand side we insert unities through Eq. (4.29) twice to obtain

$$\begin{aligned}\frac{d\hat{O}^{(H)}}{dt} &= \frac{i}{\hbar} U^\dagger(t)\hat{H}\,U(t)U^\dagger(t)\,\hat{O}^{(S)}\,U(t) - U^\dagger(t)\,\hat{O}^{(S)}\,U(t)U^\dagger(t)\,\frac{i}{\hbar}\hat{H}U(t) \\ &\overset{(4.38)}{=} \frac{i}{\hbar}\hat{H}\,\underbrace{U^\dagger(t)\,\hat{O}^{(S)}\,U(t)}_{\hat{O}^{(H)}} - \underbrace{U^\dagger(t)\,\hat{O}^{(S)}\,U(t)}_{\hat{O}^{(H)}}\,\frac{i}{\hbar}\hat{H}\end{aligned} \tag{4.43}$$

so that we finally obtain Heisenberg's equation of motion

$$\frac{d\hat{O}^{(H)}}{dt} = \frac{i}{\hbar}\left[\hat{H}, \hat{O}^{(H)}\right] = \frac{1}{i\hbar}\left[\hat{O}^{(H)}, \hat{H}\right] \tag{4.44}$$

where now the operator $\hat{O}^{(H)}$ is propagated in time instead of the state function $\Psi(t_0)$, which remains stationary.

Eq. (4.44) allows us to make the connection to classical physics. Comparing this equation with the classical equation of motion as written in Eq. (2.84) we note that they are of similar form. They are true analogs if we consider the commutator $\left[\hat{O}^{(H)}, \hat{H}\right]/(i\hbar)$ to be the quantum mechanical analog of the Poisson bracket $\{O, H\}$ for the same observable $O = O(p, q)$ (and the classical Hamiltonian function H) that does not explicitly depend on time. The commutator between two operators is, in general, defined as

$$[A, B] = AB - BA \tag{4.45}$$

In addition, we note that if $\hat{O}^{(H)}$ commutes with the Hamiltonian operator $\hat{H}$, the time derivative of $\hat{O}^{(H)}$ will be zero and, hence, a constant of the motion — in close analogy to the classical case given in Eq. (2.86).

4.3.5
Hamiltonian in Nonrelativistic Quantum Theory

The last section has already indicated some analogies between quantum mechanics and classical mechanics. In accordance with (nonrelativistic) Hamiltonian mechanics, Schrödinger quantum mechanics may be devised through the correspondence principle. The correspondence principle is no universal rule and only of limited applicability (and, of course, this recipe will fail for all observables that have no classical analog — like spin, see below). It solely expresses the desire to have some sort of connection to the classical world, which one wants to think of as a limiting case of quantum mechanics for large quantum numbers n or vanishing $\hbar$.

The correspondence principle states that the Schrödinger Hamiltonian may be derived through a substitution of $\boldsymbol{p}$ and $\boldsymbol{r}$ in the total energy function H of Hamiltonian mechanics by operator expressions $\hat{\boldsymbol{p}}$ and $\hat{\boldsymbol{r}}$. Thus, the form of $\hat{H}$ is known in terms of $\hat{\boldsymbol{p}}$ and $\hat{\boldsymbol{r}}$,

$$\underbrace{H = \frac{\boldsymbol{p}^2}{2m} + V(\boldsymbol{r})}_{\text{(classical) Hamiltonian mechanics}} \qquad \longrightarrow \qquad \underbrace{\hat{H} = \frac{\hat{\boldsymbol{p}}^2}{2m} + \hat{V}(\hat{\boldsymbol{r}})}_{\text{Schrödinger quantum mechanics}} \tag{4.46}$$

for a *single* particle of mass m in an external scalar potential V without vector potentials (see section 5.4.3 for the derivation of the Schrödinger Hamiltonian in the presence of vector potentials). Therefore, the name Schrödinger quantum mechanics has become a synonym for *non*relativistic quantum mechanics as it is related to the nonrelativistic total energy H of chapter 2. The generalization to the relativistic regime is not straightforward and therefore done in a separate chapter, namely in chapter 5.

The interaction energy operator $\hat{V}(\hat{\boldsymbol{r}})$ solely depends on the position operator $\hat{\boldsymbol{r}}$, and thus its explicit form can only be given once $\hat{\boldsymbol{r}}$ as well as the type of interaction is defined. Also, the explicit form of the kinetic energy operator,

$$\hat{T} = \frac{\hat{\boldsymbol{p}}^2}{2m} \tag{4.47}$$

is not known yet. We may add, however, that this single-particle case can be easily generalized to N particles,

$$\hat{H} = \sum_{i=1}^{N} \frac{\hat{\boldsymbol{p}}_i^2}{2m_i} + V(\hat{\boldsymbol{r}}_1, \hat{\boldsymbol{r}}_1, \ldots, \hat{\boldsymbol{r}}_N) \tag{4.48}$$

considering Eqs. (2.75) and (2.147), and we come back to this issue in section 8.2.

At this stage, we need to postulate the momentum and position operators to make the Hamiltonian operator explicit. Following Schrödinger, we choose

$$\hat{p} \equiv -\mathrm{i}\hbar\nabla = \frac{\hbar}{\mathrm{i}}\nabla \tag{4.49}$$

$$\hat{r} \equiv r \tag{4.50}$$

because this yields energy eigenvalues as solutions of Eq. (4.21) whose differences are in good agreement with transition energies measured in spectroscopy (for instance, all spectroscopic series of hydrogen atomic spectroscopy can be explained with remarkable accuracy). The momentum operator thus turns out to be a differential operator, while the position operator is a multiplicative operator. Here, it is important to understand that this choice *must be postulated*; it cannot be rigorously derived but constitutes a postulate of quantum mechanics justified by the success in explaining spectroscopic measurements. We come back to this choice in section 4.3.6 in order to better understand the relationship between both operators. The above choice of operators yields for the squared momentum operator

$$\hat{p}^2 = \hat{p}\hat{p} = (-\mathrm{i}\hbar\nabla)(-\mathrm{i}\hbar\nabla) = -\hbar^2\nabla^2 = -\hbar^2\Delta \tag{4.51}$$

where we introduced the scalar operator Δ

$$\Delta = \frac{\partial^2}{\partial x^2} + \frac{\partial^2}{\partial y^2} + \frac{\partial^2}{\partial z^2} \tag{4.52}$$

which is the *Laplacian operator* already encountered in chapter 2. We may add that the (nonrelativistic) equation of motion, Eq. (4.16), for a single particle,

$$\mathrm{i}\hbar\frac{\partial}{\partial t}\Psi_n(\boldsymbol{r},t) = \left[-\frac{\hbar^2}{2m}\Delta + \hat{V}(\hat{\boldsymbol{r}})\right]\Psi_n(\boldsymbol{r},t) \tag{4.53}$$

then contains a first derivative with respect to time and second derivatives with respect to the spatial coordinates. The space-time coordinates are thus not treated in the same way and invariance in form under a Lorentz coordinate transformation does not hold. This is expected since we started our analogy from the *nonrelativistic* Hamiltonian function of classical physics.

We require any physical observable to be hermitean and should therefore demonstrate this property for the Hamiltonian just 'deduced'. The hermiticity of the scalar potential is easily seen

$$\langle\Psi|\hat{V}\Psi\rangle^\star = \langle\hat{V}\Psi|\Psi\rangle \tag{4.54}$$

because of the fact that it is a real and multiplicative operator. For the kinetic energy part in Eq. (4.48) it is sufficient to show that the momentum operator is hermitean, which we see after integration by parts for all its components

$$\langle \boldsymbol{p}\Psi|\Psi\rangle^\star = \langle\Psi|(-\mathrm{i}\hbar)\nabla\Psi\rangle^\star = -\langle\nabla\Psi|(-\mathrm{i}\hbar)\Psi\rangle^\star = -\langle\mathrm{i}\hbar\nabla\Psi|\Psi\rangle = \langle\boldsymbol{p}\Psi|\Psi\rangle \tag{4.55}$$

and noting that the function $(\Psi^\star\Psi)$ vanishes at the boundaries $+\infty$ and $-\infty$ because of Eq. (4.7).

4.3.6
Commutation Relations for Position and Momentum Operators

Measurement of observables in different directions, i and j, corresponds to a sequential application of operators for these two directions. If their results are to be independent of what is measured first, the commutator of the product of both operators must vanish. For the conjugate position and momentum operators we find

$$\left[x_i, p_j\right] = 0 \quad \forall\, i \neq j \quad \text{and} \quad \left[x_i, x_j\right] = \left[p_i, p_j\right] = 0 \quad \forall\, i, j \in \{x, y, z\} \tag{4.56}$$

whereas conjugate operators in the same direction i do *not* commute

$$[x_i, p_i] = \mathrm{i}\hbar \qquad \forall\, i \in \{x, y, z\} \tag{4.57}$$

Hence, if we measure momentum and position in the same direction, the result depends on what has been measured first. These conditions on the commutators are required in order to fulfill the Heisenberg uncertainty relation. The founders of quantum mechanics noted that the only guiding principle for the new quantum theory must be the requirement that results of observations must be reproduced by the theory even if this then collides with classical concepts. The uncertainty relation may be deduced after a couple of steps have been taken starting with the definition of the dispersion of a measurement as the square of the deviation of the actual measurement (expressed by the operator) and the expectation value. In this derivation, which can be found, for instance, in Ref. [16], the nonvanishing commutator of conjugate variables plays a decisive role.

Note another analogy to classical physics, namely that the commutator relations for conjugate variables resemble the Poisson brackets for these variables introduced in section 2.3.2 in the framework of classical mechanics.

Eq. (4.57) poses a constraint on the choice of momentum and position operators. From the mathematical point of view, two ansätze fulfill the commutator of Eq. (4.57), first,

$$\begin{aligned} \hat{\boldsymbol{p}} &= -\mathrm{i}\hbar\nabla_{\boldsymbol{r}} + F(\hat{\boldsymbol{r}}) & & & \hat{\boldsymbol{p}} &= -\mathrm{i}\hbar\nabla_{\boldsymbol{r}} \\ & & &\text{or} & & \\ \hat{\boldsymbol{r}} &= \boldsymbol{r} & & & \hat{\boldsymbol{r}} &= \boldsymbol{r} + G(\hat{\boldsymbol{p}}) \end{aligned} \tag{4.58}$$

where $\hat{\boldsymbol{r}}$ is a multiplicative operator so that $\hat{\boldsymbol{p}}$ must contain the derivative with respect to this position coordinate, and, second,

$$\begin{aligned} \hat{\boldsymbol{p}}' &= \boldsymbol{p} & & & \hat{\boldsymbol{p}}' &= \boldsymbol{p} + F'(\hat{\boldsymbol{r}}') \\ & & &\text{or} & & \\ \hat{\boldsymbol{r}}' &= \mathrm{i}\hbar\nabla_{\boldsymbol{p}} + G'(\hat{\boldsymbol{p}}') & & & \hat{\boldsymbol{r}}' &= \mathrm{i}\hbar\nabla_{\boldsymbol{p}} \end{aligned} \tag{4.59}$$

where now $\hat{\boldsymbol{p}}$ is a multiplicative operator and hence $\hat{\boldsymbol{r}}$ must contain the derivative with respect to this momentum coordinate. The operator $\nabla_{\boldsymbol{r}}$ denotes the standard gradient vector, while $\nabla_{\boldsymbol{p}}$ is defined as

$$\nabla_{\boldsymbol{p}} = \left(\frac{\partial}{\partial p_x}, \frac{\partial}{\partial p_y}, \frac{\partial}{\partial p_z}\right)^T \tag{4.60}$$

Hence, one of the operators must be a differential operator while the other must be a simple multiplicative operator of the same variable. The first choice is called the *position-space representation*, while the second is called the *momentum-space representation*. Of course, one may add constants to these definitions but they are chosen to be zero since they would represent arbitrary shifts. Further, we must require that all arbitrary functions of position and momentum vanish,

$$F(\hat{\boldsymbol{r}}) = G(\hat{\boldsymbol{p}}) = F'(\hat{\boldsymbol{r}}) = G'(\hat{\boldsymbol{p}}) = 0 \tag{4.61}$$

in order to arrive at valid operators consistent with the correspondence principle as discussed in the last section.

Note that the correspondence principle of Eq. (4.46) is less obvious if we had chosen to write the operators in momentum space, where V takes a complicated form owing to the Fourier transformation that connects both representations (see section 6.10 for an example).

4.3.7
The Schrödinger Velocity Operator

As in classical physics, the conjugate quantity to the position operator is — according to the last section — the momentum operator. For this reason, velocity operators play a minor role in quantum mechanics. However, Heisenberg's equation of motion, Eq. (4.44), can be utilized to define the velocity operator $\dot{\hat{\boldsymbol{r}}}$,

$$\hat{\boldsymbol{v}} \equiv \dot{\hat{\boldsymbol{r}}} = \frac{\mathrm{d}\hat{\boldsymbol{r}}}{\mathrm{d}t} = \frac{\mathrm{i}}{\hbar}[\hat{H}, \hat{\boldsymbol{r}}] \tag{4.62}$$

Of course, the explicit form of the velocity operator depends on $\hat{H}$. For the Schrödinger Hamiltonian, Eq. (4.46), we obtain the Schrödinger velocity operator

$$\dot{\hat{\boldsymbol{r}}} = \frac{\mathrm{i}}{\hbar}\left[\frac{\hat{\boldsymbol{p}}^2}{2m} + \hat{V}(\boldsymbol{r}), \hat{\boldsymbol{r}}\right] = \frac{\mathrm{i}}{\hbar}\left[\frac{\hat{\boldsymbol{p}}^2}{2m}, \hat{\boldsymbol{r}}\right] = \frac{\mathrm{i}}{2m\hbar}\left(\hat{\boldsymbol{p}}[\hat{\boldsymbol{p}}, \hat{\boldsymbol{r}}] + [\hat{\boldsymbol{p}}, \hat{\boldsymbol{r}}]\hat{\boldsymbol{p}}\right) \stackrel{(4.57)}{=} \frac{\hat{\boldsymbol{p}}}{m} \tag{4.63}$$

simply as the canonical momentum operator divided by the mass, in close analogy to classical mechanics. Moreover, since position and momentum operators of different particles always commute, the equation above also holds for the velocity operator of any particle in a many-particle system described by $\hat{H}$: it is always related to the momentum operator of the particle under consideration.

4.3.8
Ehrenfest and Hellmann–Feynman Theorems

Connected to our derivation of the Heisenberg equation of motion is Ehrenfest's theorem for the time evolution of expectation values. We leave aside all explicit dependences of the state function and of the operator on dynamical variables and thereby make no reference to the particular choice of picture. If we consider the expectation value of an observable O for a normalized state Ψ, Eq. (4.25), then its time derivative is given by

$$\begin{aligned}\frac{\mathrm{d}}{\mathrm{d}t}\langle\hat{O}\rangle_\Psi &= \int_{-\infty}^{\infty}\mathrm{d}^3r_1\cdots\int_{-\infty}^{\infty}\mathrm{d}^3r_N\left[\frac{\partial\Psi^\dagger}{\partial t}\hat{O}\Psi+\Psi^\dagger\frac{\partial\hat{O}}{\partial t}\Psi+\Psi^\dagger\hat{O}\frac{\partial\Psi}{\partial t}\right]\\ &\overset{(4.16)}{=} \frac{\mathrm{i}}{\hbar}\left\langle[\hat{H},\hat{O}]\right\rangle_\Psi+\left\langle\frac{\partial\hat{O}}{\partial t}\right\rangle_\Psi\end{aligned} \tag{4.64}$$

where the total time derivative has been replaced on the right hand side by the partial time derivative since all quantities in the integrand depend at most on position and time (in position-space representation), and the positions do *not* depend on time as there are no trajectories in quantum mechanics. In this equation we have explicitly emphasized that it holds for a system comprising an arbitrary number of particles N. It is not restricted to a single particle. Again, as for the Heisenberg equation of motion, we note the analogy with the classical equation of motion as written in Eq. (2.84).

If Ψ is an eigenfunction of $\hat{H}$ or if $\hat{O}$ commutes with $\hat{H}$, then the first term vanishes and the derivative of the expectation value is equal to the derivative of the operator,

$$\frac{\mathrm{d}\langle\hat{O}\rangle_\Psi}{\mathrm{d}t} \overset{(4.21)}{=} \left\langle\frac{\partial\hat{O}}{\partial t}\right\rangle_\Psi \tag{4.65}$$

For the special choice that $\hat{O}$ is the Hamiltonian $\hat{H}$, we have

$$\frac{\mathrm{d}\langle\hat{H}\rangle_\Psi}{\mathrm{d}t} \overset{(4.16)}{=} \left\langle\frac{\partial\hat{H}}{\partial t}\right\rangle_\Psi \tag{4.66}$$

which is always fulfilled if Ψ is a true quantum mechanical state function as defined by Eq. (4.16). This is a special form of the Hellmann–Feynman theorem [34–36], which can be generalized for first derivatives of any parameter

λ, on which the wave function may depend, if we start again from Eq. (4.64)

$$\begin{aligned}\frac{\mathrm{d}}{\mathrm{d}\lambda}\langle \hat{H}\rangle_\Psi &= \int_{-\infty}^{\infty}\mathrm{d}^3r_1\cdots\int_{-\infty}^{\infty}\mathrm{d}^3r_N\left[\frac{\mathrm{d}\Psi^\dagger}{\mathrm{d}\lambda}\hat{H}\Psi+\Psi^\dagger\frac{\mathrm{d}\hat{H}}{\mathrm{d}\lambda}\Psi+\Psi^\dagger\hat{H}\frac{\mathrm{d}\Psi}{\mathrm{d}\lambda}\right]\\ &\overset{(4.21)}{=} \left\langle\frac{\mathrm{d}\hat{H}}{\mathrm{d}\lambda}\right\rangle_\Psi \end{aligned} \tag{4.67}$$

where we imposed that Ψ must be an eigenfunction of $\hat{H}$. Observe that we now did not switch to partial derivatives with respect to λ on the right hand side because λ may be a most general variable whose variation might also affect quantities that depend implicitly on it (see, for example, the nuclear positions in section 10.5 later in this book).

Finally, we consider in this section the force law in quantum mechanics (Ehrenfest, 1927). From Heisenberg's equation of motion employing the Schrödinger Hamiltonian we have

$$\frac{\mathrm{d}\hat{\boldsymbol{p}}}{\mathrm{d}t}=\frac{1}{\mathrm{i}\hbar}\left[\hat{\boldsymbol{p}},\hat{H}\right]\overset{(4.46)}{=}\frac{1}{\mathrm{i}\hbar}\left[\hat{\boldsymbol{p}},\hat{V}(\hat{\boldsymbol{r}})\right]\overset{(4.57)}{=}-\nabla\hat{V}(\hat{\boldsymbol{r}}) \tag{4.68}$$

Differentiating the already established Schrödinger velocity operator from the last section with respect to time, we obtain

$$\ddot{\hat{\boldsymbol{r}}}=\frac{\mathrm{d}\dot{\hat{\boldsymbol{r}}}}{\mathrm{d}t}\overset{(4.63)}{=}\frac{1}{m}\frac{\mathrm{d}\hat{\boldsymbol{p}}}{\mathrm{d}t} \tag{4.69}$$

so that Eq. (4.68) can be written as

$$m\ddot{\hat{\boldsymbol{r}}}=-\nabla\hat{V}(\hat{\boldsymbol{r}}) \tag{4.70}$$

which is the quantum analog of Newton's second law, Eqs. (2.2) and (2.37). Taking the expectation value of both sides of this equation with respect to a Heisenberg state, which does not change with time, we obtain

$$m\left\langle\ddot{\hat{\boldsymbol{r}}}\right\rangle_{\Psi(t_0)}\overset{(4.65)}{=}m\frac{\mathrm{d}^2}{\mathrm{d}t^2}\langle\boldsymbol{r}\rangle_{\Psi(t_0)}=-\left\langle\nabla\hat{V}(\hat{\boldsymbol{r}})\right\rangle_{\Psi(t_0)} \tag{4.71}$$

4.3.9
Current Density and Continuity Equation

We have already encountered the particle density distribution in the context of the Born interpretation in section 4.1. After what has been said about observables, it should be possible to assign an operator to this observable also. We can deduce the explicit form of the operator $\hat{\rho}_{\boldsymbol{r}}$, which represents the particle density at a given position $\boldsymbol{r}$ in space, by relating its expectation value for an N-particle system,

$$\langle\hat{\rho}_{\boldsymbol{r}}\rangle\equiv\int_{-\infty}^{\infty}\mathrm{d}^3r_1\int_{-\infty}^{\infty}\mathrm{d}^3r_2\cdots\int_{-\infty}^{\infty}\mathrm{d}^3r_N\,\Psi^\dagger(\boldsymbol{r}_1,\boldsymbol{r}_2,\ldots,\boldsymbol{r}_N,t)\,\hat{\rho}_{\boldsymbol{r}}\,\Psi(\boldsymbol{r}_1,\boldsymbol{r}_2,\ldots,\boldsymbol{r}_N,t)$$

(4.72)

to the Born interpretation $\rho(\boldsymbol{r},t) = \langle \hat{\rho}_{\boldsymbol{r}} \rangle$ so that we must fulfill

$$\langle \hat{\rho}_{\boldsymbol{r}} \rangle = N \int_{-\infty}^{\infty} \mathrm{d}^3 r_2 \cdots \int_{-\infty}^{\infty} \mathrm{d}^3 r_N \, \Psi^\dagger(\boldsymbol{r},\boldsymbol{r}_2,\ldots,\boldsymbol{r}_N,t)\Psi(\boldsymbol{r},\boldsymbol{r}_2,\ldots,\boldsymbol{r}_N,t) \tag{4.73}$$

In order to fulfill Eq. (4.73), the operator $\hat{\rho}_{\boldsymbol{r}}$ can be explicitly and consistently written in position-space representation as

$$\hat{\rho}_{\boldsymbol{r}} = \sum_{i=1}^{N} \delta^{(3)}(\boldsymbol{r}_i - \boldsymbol{r}) \tag{4.74}$$

where $\boldsymbol{r}$ is a general position coordinate as before, while $\boldsymbol{r}_i$ denotes the coordinate of the ith particle.

For the time evolution of the density, we utilize Ehrenfest's theorem

$$\frac{\mathrm{d}\rho(\boldsymbol{r},t)}{\mathrm{d}t} = \frac{\mathrm{d}}{\mathrm{d}t}\langle \hat{\rho}_{\boldsymbol{r}} \rangle \overset{(4.64)}{=} \frac{\mathrm{i}}{\hbar}\left\langle [\hat{H},\hat{\rho}_{\boldsymbol{r}}] \right\rangle + \left\langle \frac{\partial \hat{\rho}_{\boldsymbol{r}}}{\partial t} \right\rangle \tag{4.75}$$

where the partial derivative on the right hand side of Eq. (4.75) vanishes because the density operator does not explicitly depend on time as there are no (classical) trajectories, $\boldsymbol{r} = \boldsymbol{r}(t)$, in quantum mechanics. Accordingly, we have

$$\frac{\mathrm{d}\rho(\boldsymbol{r},t)}{\mathrm{d}t} = \frac{\partial \rho(\boldsymbol{r},t)}{\partial t} \tag{4.76}$$

For the Hamiltonian, we choose again the Schrödinger Hamiltonian, but now for an N-particle system of Eq. (4.48), in which all kinetic energy operators of the individual particles are simply summed as in classical physics. All constant and potential energy operators in $V(\boldsymbol{r}_1,\boldsymbol{r}_1,\ldots,\boldsymbol{r}_N)$ in the Hamiltonian commute with the density operator $\hat{\rho}_{\boldsymbol{r}}$ and thus cancel. In nonrelativistic quantum mechanics the time propagation of the density then simplifies to

$$\begin{aligned} \frac{\partial}{\partial t}\rho(\boldsymbol{r},t) &= \frac{\mathrm{i}}{\hbar}\left\langle \left[\sum_{i=1}^{N} \frac{\boldsymbol{p}_i^2}{2m_i}, \hat{\rho}_{\boldsymbol{r}} \right] \right\rangle \overset{(4.74)}{=} \left\langle \left[\sum_{i=1}^{N} \frac{\boldsymbol{p}_i^2}{2m_i}, \sum_{j=1}^{N} \delta^{(3)}(\boldsymbol{r}_j - \boldsymbol{r}) \right] \right\rangle \\ &= \frac{\mathrm{i}}{\hbar}\left\langle \sum_{i=1}^{N} \left[\frac{\boldsymbol{p}_i^2}{2m_i}, \delta^{(3)}(\boldsymbol{r}_i - \boldsymbol{r}) \right] \right\rangle \end{aligned} \tag{4.77}$$

If we assume that all particles are physically indistinguishable, which is a natural assumption considering the fact that we are genuinely interested in particle distributions of the same type of particles from which other cases can be deduced, we may write

$$\frac{\partial \rho(\boldsymbol{r},t)}{\partial t} = \frac{\mathrm{i}}{\hbar} N \left\langle \left[\frac{\boldsymbol{p}_1^2}{2m_1}, \delta^{(3)}(\boldsymbol{r}_1 - \boldsymbol{r}) \right] \right\rangle \tag{4.78}$$

This equation can be re-cast to finally yield the (nonrelativistic) *current density* of a system of N particles all with the same mass $m \equiv m_i$. For this we resolve $\boldsymbol{p}_1 = -i\hbar\nabla_1$ to obtain the *continuity equation* — recall section 3.4 — for an N-particle system,

$$\begin{aligned}\frac{\partial\rho(\boldsymbol{r},t)}{\partial t} &= N\frac{\hbar^2}{2m\,\mathrm{i}}\left\langle\Psi\middle|\nabla_1^2\delta^{(3)}(\boldsymbol{r}_1-\boldsymbol{r})-\delta^{(3)}(\boldsymbol{r}_1-\boldsymbol{r})\nabla_1^2\middle|\Psi\right\rangle\\ &= N\frac{\hbar^2}{2m\,\mathrm{i}}\left\{\left\langle\nabla_1^2\Psi\middle|\delta^{(3)}(\boldsymbol{r}_1-\boldsymbol{r})\middle|\Psi\right\rangle-\left\langle\Psi\middle|\delta^{(3)}(\boldsymbol{r}_1-\boldsymbol{r})\middle|\nabla_1^2\Psi\right\rangle\right\}\end{aligned} \tag{4.79}$$

where we exploited the hermiticity of the Hamiltonian. In the most simple case of a single particle, where $\Psi(\boldsymbol{r}_1,\ldots,\boldsymbol{r}_N,t)\rightarrow\Psi(\boldsymbol{r},t)$, Eq. (4.79) reduces to

$$\begin{aligned}\frac{\partial}{\partial t}\left(\Psi^\star\Psi\right) &= \frac{\hbar^2}{2m\,\mathrm{i}}\left\{(\nabla^2\Psi)^\star\Psi-\Psi^\star\nabla^2\Psi\right\}\\ &= -\nabla\cdot\underbrace{\frac{\hbar^2}{2m\,\mathrm{i}}\left\{\Psi^\star\nabla\Psi-(\nabla\Psi)^\star\Psi\right\}}_{\equiv\boldsymbol{j}}\end{aligned} \tag{4.80}$$

where we introduced the nonrelativistic current density $\boldsymbol{j}$ for the time-dependent state $\Psi(\boldsymbol{r},t)$. The last derivation step may be most easily verified in reversed direction. The continuity equation can also be derived by starting directly from the quantum mechanical equation of motion, Eq. (4.16), which can be combined with its complex-conjugate after multiplication of both equations with $\Psi^\star$ and Ψ, respectively.

Hence, the continuity equation for an N-particle system expresses the natural observation that a change of density at a given position with time must result in a current from or to the environment of that position. The equation can be instrumentalized to *define* the density distribution, and we will rely on this option in the relativistic framework in chapters 5, 8 and 12.

4.4 Angular Momentum and Rotations

To describe rotational motions in quantum mechanics, it is convenient to switch from Cartesian to polar coordinates. In *classical* mechanics we may write the square of the angular momentum of a single particle as

$$\boldsymbol{l}^2 = (\boldsymbol{r}\times\boldsymbol{p})^2 = \boldsymbol{r}^2\boldsymbol{p}^2-(\boldsymbol{r}\cdot\boldsymbol{p})^2 = r^2p^2-(\boldsymbol{r}\cdot\boldsymbol{p})^2 \tag{4.81}$$

so that the square of the momentum becomes

$$\boldsymbol{p}^2 = p_r^2+\frac{\boldsymbol{l}^2}{r^2} \tag{4.82}$$

The last equation introduces the square of the radial momentum p_r^2, which is explicitly given by

$$p_r^2 = \frac{(\boldsymbol{r} \cdot \boldsymbol{p})^2}{r^2} = \left(\frac{\boldsymbol{r} \cdot \boldsymbol{p}}{r}\right) \tag{4.83}$$

Accordingly, we may write the absolute value of a momentum along a radial coordinate,

$$p_r \equiv \frac{\boldsymbol{r}}{r} \cdot \boldsymbol{p} = \hat{\boldsymbol{r}} \cdot \boldsymbol{p} \tag{4.84}$$

Note that the hat in $\hat{\boldsymbol{r}}$ denotes the *unit* vector in $\boldsymbol{r}$ direction and that we have omitted the 'transpose' sign in the scalar product for the sake of brevity here and in all following scalar products. The vectorial quantity $\boldsymbol{p}_r$ of length p_r and direction $\hat{\boldsymbol{r}}$ cannot be obtained from the study of a scalar quantity like the energy but requires consideration of relations for momenta. With these definitions, the classical kinetic energy for a single particle can then be written as

$$T = \frac{\boldsymbol{p}^2}{2m} = \frac{p_r^2}{2m} + \frac{\boldsymbol{l}^2}{2mr^2} \tag{4.85}$$

which for fixed radial coordinate r simplifies to the pure rotational energy expression

$$T_{\text{rot}} = \frac{\boldsymbol{l}^2}{2mr^2} = \frac{\boldsymbol{l}^2}{2I} \tag{4.86}$$

with $I \equiv mr^2$ being the moment of inertia of the point-like particle under consideration.

4.4.1 Orbital Angular Momentum

According to the correspondence principle, the angular momentum operator $\hat{\boldsymbol{l}}$ in quantum mechanics is formally defined in close analogy to classical mechanics,

$$\hat{\boldsymbol{l}} \equiv \hat{\boldsymbol{r}} \times \hat{\boldsymbol{p}} = \begin{pmatrix} \hat{y}\hat{p}_z - \hat{z}\hat{p}_y \\ \hat{z}\hat{p}_x - \hat{x}\hat{p}_z \\ \hat{x}\hat{p}_y - \hat{y}\hat{p}_x \end{pmatrix} \equiv \begin{pmatrix} \hat{l}_x \\ \hat{l}_y \\ \hat{l}_z \end{pmatrix} \tag{4.87}$$

from which the squared angular momentum can be directly calculated. The result, however, now contains an additional term when compared to Eq. (4.81),

$$\hat{\boldsymbol{l}}^2 = (\hat{\boldsymbol{r}} \times \hat{\boldsymbol{p}})^2 = \hat{\boldsymbol{r}}^2\hat{\boldsymbol{p}}^2 - (\hat{\boldsymbol{r}} \cdot \hat{\boldsymbol{p}})^2 + i\hbar\hat{\boldsymbol{r}} \cdot \hat{\boldsymbol{p}} \tag{4.88}$$

which is solely due to the non-commutation of position and momentum operator (cf. section 4.3.6). Note that the dimension of the squared angular momentum is the dimension of an action and, hence, the unit of Planck's quantum of action h. Accordingly, every term in Eq. (4.88) contains a squared Planck's constant, $h^2 = 4\pi^2\hbar^2$, originating from the canonical momentum operator.

It is now appropriate to explicitly introduce polar coordinates:

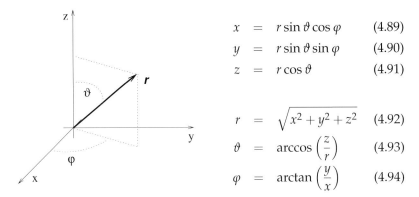

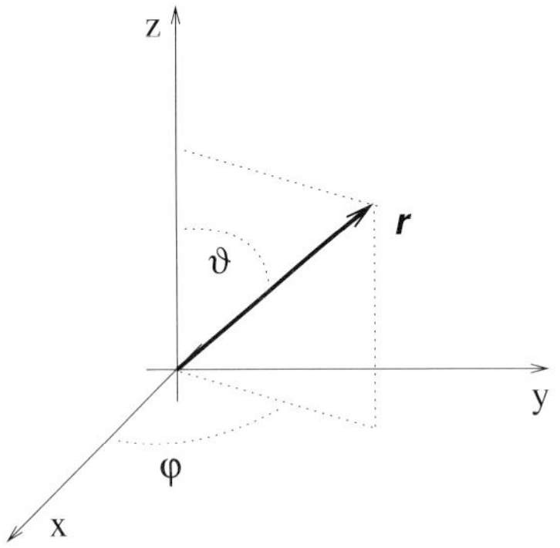

$$x = r\sin\vartheta\cos\varphi \tag{4.89}$$

$$y = r\sin\vartheta\sin\varphi \tag{4.90}$$

$$z = r\cos\vartheta \tag{4.91}$$

$$r = \sqrt{x^2+y^2+z^2} \tag{4.92}$$

$$\vartheta = \arccos\left(\frac{z}{r}\right) \tag{4.93}$$

$$\varphi = \arctan\left(\frac{y}{x}\right) \tag{4.94}$$

The projection of the momentum operator onto the position operator is given by

$$\hat{\boldsymbol{r}}\cdot\hat{\boldsymbol{p}} = -\mathrm{i}\hbar(\boldsymbol{r}\cdot\nabla) = -\mathrm{i}\hbar\left(x\frac{\partial}{\partial x}+y\frac{\partial}{\partial y}+z\frac{\partial}{\partial z}\right) \tag{4.95}$$

and can be re-written in spherical coordinates using the chain rule

$$\frac{\partial}{\partial u} = \frac{\partial r}{\partial u}\frac{\partial}{\partial r}+\frac{\partial\vartheta}{\partial u}\frac{\partial}{\partial\vartheta}+\frac{\partial\varphi}{\partial u}\frac{\partial}{\partial\varphi}, \quad u\in\{x,y,z\} \tag{4.96}$$

The explicit evaluation of the scalar product then yields

$$\hat{\boldsymbol{r}}\cdot\hat{\boldsymbol{p}} = -\mathrm{i}\hbar r\frac{\partial}{\partial r} \tag{4.97}$$

Therefore, we may now write Eq. (4.88) as

$$\hat{\boldsymbol{l}}^2 = \hat{\boldsymbol{r}}^2\hat{\boldsymbol{p}}^2 - \left(-\mathrm{i}\hbar r\frac{\partial}{\partial r}\right)^2 + \mathrm{i}\hbar\left(-\mathrm{i}\hbar r\frac{\partial}{\partial r}\right) = \hat{\boldsymbol{r}}^2\hat{\boldsymbol{p}}^2 + \hbar^2\left[\left(r\frac{\partial}{\partial r}\right)^2 + r\frac{\partial}{\partial r}\right] \tag{4.98}$$

from which we can obtain an expression of the squared canonical momentum operator,

$$\hat{\boldsymbol{p}}^2 = \frac{1}{r^2}\hat{\boldsymbol{l}}^2 - \frac{\hbar^2}{r^2}\left[\left(r\frac{\partial}{\partial r}\right)^2 + r\frac{\partial}{\partial r}\right]. \tag{4.99}$$

We note that

$$\left[\left(r\frac{\partial}{\partial r}\right)^2 + r\frac{\partial}{\partial r}\right] = \frac{\partial^2}{\partial r^2} + \frac{2}{r}\frac{\partial}{\partial r} = \left(\frac{1}{r}\frac{\partial}{\partial r}r\right)^2 \tag{4.100}$$

and therefore define the radial momentum operator according to

$$\hat{p}_r = -\mathrm{i}\hbar\frac{1}{r}\frac{\partial}{\partial r}r = -\mathrm{i}\hbar\left(\frac{\partial}{\partial r} + \frac{1}{r}\right) \tag{4.101}$$

At first sight, this result is surprising as the radial momentum operator is not simply the derivative with respect to the radial coordinate r but contains an additive term $\propto 1/r$. Nevertheless, the definition of the radial momentum operator $\hat{p}_r$ as given above still fulfills the standard commutation relation with the conjugated radial position operator,

$$[\hat{r}, \hat{p}_r] = \mathrm{i}\hbar \tag{4.102}$$

because the additional term $\propto 1/r$ of Eq. (4.101) commutes with r as already noted in Eq. (4.58).

With the definition of a radial momentum in Eq. (4.101) we can now write the kinetic energy operator in polar coordinates as

$$\hat{T} = \frac{\hat{\boldsymbol{p}}^2}{2m} = \frac{1}{2m}\left[\frac{1}{r^2}\hat{\boldsymbol{l}}^2 - \frac{\hbar^2}{r^2}\left(\frac{1}{r}\frac{\partial}{\partial r}r\right)^2\right] = \frac{\hat{\boldsymbol{l}}^2}{2mr^2} + \frac{\hat{p}_r^2}{2m} \tag{4.103}$$

Thus, in agreement with the correspondence principle, the operator for the rotational energy of a single particle with fixed distance r from the rotational axis (rigid rotor) reads

$$\hat{T}_{\text{rot}} = \frac{\hat{\boldsymbol{l}}^2}{2mr^2} \tag{4.104}$$

While our reasoning has relied on the correspondence principle so far, we should note that all steps can be proved by coordinate transformation of the kinetic energy operator from Cartesian to polar coordinates. That is, the validity of Eq. (4.103) can also be shown by transforming the Laplacian Δ in Cartesian coordinates to spherical polar coordinates. It then turns out that the components of the angular momentum operator read,

$$l_x = \frac{\hbar}{\mathrm{i}}\left(-\sin\varphi\frac{\partial}{\partial\vartheta} - \cos\varphi\cot\vartheta\frac{\partial}{\partial\varphi}\right) \tag{4.105}$$

$$l_y = \frac{\hbar}{\mathrm{i}}\left(\cos\varphi\frac{\partial}{\partial\vartheta} - \sin\varphi\cot\vartheta\frac{\partial}{\partial\varphi}\right) \tag{4.106}$$

$$l_z = \frac{\hbar}{\mathrm{i}}\frac{\partial}{\partial\varphi} \tag{4.107}$$

which then yields the squared angular momentum operator in polar coordinates,

$$\hat{l}^2 = l_x^2 + l_y^2 + l_z^2 = -\hbar^2 \left[\frac{1}{\sin\vartheta} \frac{\partial}{\partial\vartheta} \left(\sin\vartheta \frac{\partial}{\partial\vartheta} \right) + \frac{1}{\sin^2\vartheta} \frac{\partial^2}{\partial\varphi^2} \right] \tag{4.108}$$

the so-called Legendre operator. In order to stress that these operator equations are of the same form as the corresponding equations in classical mechanics, the operators have all been identified by a hat. In the following, however, we skip these hats for the sake of brevity. The components of the angular momentum operator fulfill the following commutation relations

$$[l_x, l_y] = i\hbar l_z\,, \quad [l_y, l_z] = i\hbar l_x\,, \quad [l_z, l_x] = i\hbar l_y \tag{4.109}$$

$$[l^2, l_i] = 0\,, \quad \text{for} \quad i \in \{x, y, z\} \tag{4.110}$$

which can be proved by insertion of the components defined by Eqs. (4.105)–(4.107). Note that the first set of three commutators can be conveniently written with the Levi–Cività symbol as

$$[l_i, l_j] = i\hbar \sum_{k=1}^{3} \epsilon_{ijk}\, l_k\,, \quad \text{for} \quad i, j, k \in \{x, y, z\} \tag{4.111}$$

Since the (scalar) Legendre operator enters the quantum mechanical expression for the rotational energy, Eqs. (4.103) and (4.104), its eigenfunctions are also the eigenfunctions of $\hat{T}_{rot}$. The eigenfunctions of the Legendre operator $\hat{l}^2$

$$l^2 Y_{lm}(\vartheta, \varphi) = l(l+1)\,\hbar^2\, Y_{lm}(\vartheta, \varphi) \tag{4.112}$$

are the spherical harmonics $Y_{lm}(\vartheta, \varphi)$. Pictures of these functions can be found in almost any textbook on quantum mechanics or quantum chemistry. The spherical harmonics are simultaneous eigenfunctions of the l_z operator,

$$l_z Y_{lm}(\vartheta, \varphi) = m\,\hbar\, Y_{lm}(\vartheta, \varphi) \tag{4.113}$$

owing to the commutation relations given above. The *orbital angular momentum quantum number* l can take the values $0, 1, 2, 3, \ldots$ (also know as *azimuthal* quantum number) and the *magnetic quantum number* m must be in the interval $[-l, -l+1, \ldots, l]$ (also known as orientational quantum number). The eigenfunctions can be efficiently constructed through the definition of ladder operators, which is standard in nonrelativistic quantum mechanics and therefore omitted here. The general expression for the spherical harmonics reads [37]

$$Y_{lm}(\vartheta, \varphi) = (-1)^m \sqrt{\frac{2l+1}{4\pi} \frac{(l-m)!}{(l+m)!}}\, P_{lm}(\cos\vartheta) e^{im\varphi} \tag{4.114}$$

in polar coordinates, where we introduced the associated Legendre polynomials,

$$P_{lm}(\xi) = (1-\xi^2)^{\frac{m}{2}} \frac{\mathrm{d}^m}{\mathrm{d}\xi^m} P_l(\xi) \tag{4.115}$$

with the Legendre polynomials given by

$$P_l(\xi) = \frac{1}{2^l l!} \frac{\mathrm{d}^l}{\mathrm{d}\xi^l} (\xi^2 - 1)^l \tag{4.116}$$

In our case, these polynomials read

$$P_l(\cos\vartheta) = \frac{1}{2^l l!} \frac{\mathrm{d}^l (\cos^2\vartheta - 1)^l}{\mathrm{d}(\cos\vartheta)^l} = \frac{(-1)^l}{2^l l!} \frac{\mathrm{d}^l \sin^{2l}\vartheta}{\mathrm{d}(\cos\vartheta)^l} \tag{4.117}$$

and hence

$$P_{lm}(\cos\vartheta) = \frac{(-1)^l}{2^l l!} \sin^m\vartheta \frac{\mathrm{d}^{l+m} \sin^{2l}\vartheta}{\mathrm{d}\cos^{l+m}\vartheta} \tag{4.118}$$

which can be written explicitly for the lowest four Legendre polynomials

$$P_0 = 1,\ P_1 = \cos\vartheta,\ P_2 = \frac{1}{2}(3\cos^2\vartheta - 1),\ P_3 = \frac{1}{2}(5\cos^3\vartheta - 3\cos\vartheta) \tag{4.119}$$

The spherical harmonics Y_{lm} are defined with a specific phase factor that may be different in different presentations. Here, we followed Edmonds [37], which is the convention used by Condon and Shortley [38]. These Y_{lm} differ from those of Bethe [39] by a factor of $(-1)^m$. Compared to Schiff [40] the Y_{lm} are equal for negative values of m, while for positive values they differ by the factor $(-1)^m$.

The following list provides some explicit formulae of the (normalized) spherical harmonics for the small angular momentum quantum numbers,

$$Y_{00} = \sqrt{\frac{1}{4\pi}},\quad Y_{10} = \sqrt{\frac{3}{4\pi}}\cos\vartheta,\quad Y_{20} = \sqrt{\frac{5}{16\pi}}(3\cos^2\vartheta - 1) \tag{4.120}$$

$$Y_{1(\mp1)} = \pm\sqrt{\frac{3}{8\pi}}\sin\vartheta e^{\mp i\varphi},\quad Y_{2(\mp1)} = \pm\sqrt{\frac{15}{8\pi}}\sin\vartheta\cos\vartheta e^{\mp i\varphi} \tag{4.121}$$

$$Y_{2(\mp2)} = \sqrt{\frac{15}{32\pi}}\sin^2\vartheta e^{\mp 2i\varphi} \tag{4.122}$$

The spherical harmonics fulfill the completeness condition,

$$\sum_{l=0}^{\infty}\sum_{m=-l}^{l} Y_{lm}(\vartheta,\varphi) Y^{\star}_{lm}(\vartheta',\varphi') = \frac{1}{\sin\vartheta}\delta(\vartheta-\vartheta')\delta(\varphi-\varphi') \tag{4.123}$$

and are orthogonal,

$$\langle Y_{lm}|Y_{l'm'}\rangle = \int_0^\pi \sin\vartheta d\vartheta \int_0^{2\pi} d\varphi Y^\star_{lm}(\vartheta,\varphi) Y_{l'm'}(\vartheta,\varphi) = \delta_{ll'}\delta_{mm'} \tag{4.124}$$

they satisfy an addition theorem

$$\sum_{m=-l}^{l} Y_{lm}(\vartheta,\varphi) Y^\star_{lm}(\vartheta',\varphi') = \frac{2l+1}{4\pi} P_l(\cos\Theta) \tag{4.125}$$

with Θ being the angle between the vectors (r,ϑ,φ) and (r',ϑ',φ') and feature the following symmetry

$$Y_{l(-m)} = (-1)^m Y^\star_{lm} \tag{4.126}$$

Inversion of coordinates, $\boldsymbol{r} \rightarrow -\boldsymbol{r}$, can be formally expressed by the parity operator P. In polar coordinates, inversion of coordinates reduces to changes of the angles, $r \rightarrow r$, $\varphi \rightarrow (\varphi+\pi)_{\text{mod}2\pi}$ and $\vartheta \rightarrow \pi - \vartheta$. For the spherical harmonics under parity transformation we find

$$PY_{lm}(\vartheta,\varphi) = Y_{lm}(\pi-\vartheta,(\varphi+\pi)_{\text{mod}2\pi}) = (-1)^l Y_{lm}(\vartheta,\varphi) \tag{4.127}$$

Hence, for even l spherical harmonics have even parity, while odd l lead to odd parity and, hence, sign inversion under point reflection at the origin of the coordinate system.

The spherical harmonics are often split into real and imaginary parts, which is straightforward via Euler's equation employed for the exponential $\exp(\mathrm{i}m\varphi)$. In quantum chemistry, the *Cartesian* spherical harmonics in particular become important. By noting that

$$e^{\mathrm{i}\varphi} = \frac{x+\mathrm{i}y}{\sqrt{x^2+y^2}} \tag{4.128}$$

and

$$\vartheta = \arcsin\left(\sqrt{\frac{x^2+y^2}{x^2+y^2+z^2}}\right) = \arccos\left(\frac{z}{\sqrt{x^2+y^2+z^2}}\right) \tag{4.129}$$

we may obtain for $l=1$,

$$Y_{10} = \frac{1}{2}\sqrt{\frac{3}{\pi}}\frac{z}{\sqrt{x^2+y^2+z^2}} \quad \text{and} \quad Y_{1(\pm1)} = \frac{1}{2}\sqrt{\frac{3}{2\pi}}\frac{x\pm\mathrm{i}y}{\sqrt{x^2+y^2+z^2}} \tag{4.130}$$

4.4.2
Coupling of Angular Momenta

If a system consists of particles that all feature an angular momentum $\boldsymbol{j}_i$, it is desirable to assign a total angular momentum $\boldsymbol{J}$ to the whole system. Since this section considers general angular momenta, we have changed the notation and used instead of the symbol for the orbital angular momentum $\boldsymbol{l}$ the symbol $\boldsymbol{j}$. The simplest such system is a system of two particles with two angular momenta $\boldsymbol{j}_1$ and $\boldsymbol{j}_2$, which we shall consider here (see the monographs in Refs. [37, 41] for detailed accounts that also introduce a general tensor formalism to deal with angular momentum eigenfunctions).

In quantum mechanics, the total angular momentum operator $\boldsymbol{J}$ is simply the sum of the two individual angular-momentum vector operators

$$\boldsymbol{J} = \boldsymbol{j}_1 + \boldsymbol{j}_2 \tag{4.131}$$

in close analogy to classical physics. For the total angular momentum, the typical angular-momentum eigenvalue equations encountered in the last section hold

$$\boldsymbol{J}^2 \,|JM_J\rangle = J(J+1)\hbar^2 \,|JM_J\rangle \tag{4.132}$$

$$J_z \,|JM_J\rangle = M_J\hbar \,|JM_J\rangle \tag{4.133}$$

where Dirac's ket symbol denotes the yet unknown eigenfunctions that are distinguished by their eigenvalues, i.e., by the *total* angular-momentum quantum numbers J and M_J. The properties already discussed for a single orbital angular momentum transfer to the case of total orbital angular momentum as can be shown by making all formal equations given in this section explicit, i.e., by recalling the definitions of Eqs. (4.105)–(4.107) and starting to insert them into Eq. (4.131).

Since the uncoupled states $|j_1 m_{j(1)}\rangle$ and $|j_2 m_{j(2)}\rangle$ each form a complete basis, the (direct) product basis built from them is a complete basis for the Hilbert space representation of the total eigenstate $|JM_J\rangle$ in accord with Eq. (4.13). However, in this product basis $\boldsymbol{J}^2$ is not diagonal. In order to arrive at a coupled representation we use a proper linear combination of the uncoupled product states,

$$|JM_J\rangle = \sum_{m_{j(1)}, m_{j(2)}} |j_1 m_{j(1)}\rangle |j_2 m_{j(2)}\rangle \,\langle j_1 m_{j(1)} j_2 m_{j(2)} | JM_J\rangle \tag{4.134}$$

The projection coefficients $\langle j_1 m_{j(1)} j_2 m_{j(2)} | JM_J\rangle$ are called Clebsch–Gordan vector coupling coefficients. These coefficients, which are required to produce an eigenfunction from an expansion into the uncoupled product basis, are defined for $J = j_1 + j_2$ as [42]

$$\langle j_1 m_{j(1)} j_2 m_{j(2)} | JM_J\rangle = \sqrt{\frac{(2j_1)!(2j_2)!}{(2J)!}} \times \sqrt{\frac{(J+M_J)!(J-M_J)!}{(j_1+m_{j(1)})!(j_1-m_{j(1)})!(j_2+m_{j(2)})!(j_2-m_{j(2)})!}} \tag{4.135}$$

Related to the Clebsch–Gordan vector coupling coefficients are Wigner's 3j-symbols

$$\begin{pmatrix} j_1 & j_2 & J \\ m_{j(1)} & m_{j(2)} & -M_J \end{pmatrix} = \frac{(-1)^{j_1-j_2+M_J}}{\sqrt{2J+1}} \langle j_1 m_{j(1)} j_2 m_{j(2)} | J M_J \rangle \tag{4.136}$$

which allow one to express symmetry properties of vector coupling coefficients in a most convenient way [37]. Although we do not give any of the important properties of Clebsch–Gordan coefficients and Wigner 3j-symbols, for which we refer to the above-quoted references, we may emphasize their most important property. This is the fact that they can be calculated from the quantum numbers only!

4.4.3 Spin

In 1922 Stern and Gerlach demonstrated in a famous experiment that a beam of silver atoms is split in an inhomogeneous magnetic field into two parts. This points to a magnetic dipole, originating from an angular momentum, that is interacting with the magnetic field. One may argue that silver atoms containing 47 electrons are already quite complex systems to consider the coupling of all individual angular momenta of these particles, but the experiment can also be carried out with a beam of hydrogen atoms [43]. According to Schrödinger's nonrelativistic quantum mechanics, however, the hydrogen atom, which can be treated analytically as we shall see in chapter 6, does not possess an (orbital) angular momentum in its ground state and would not give rise to beam splitting within a magnetic field. But, apparently, we are observing an angular momentum since the hydrogen-atom beam splits in a Stern–Gerlach experiment.

The effect has also been observed spectroscopically as a splitting of lines in the presence of a static magnetic field, for instance, in the Lyman series of hydrogen. In general, the spectroscopic effect is called the Zeeman effect and is dominated by orbital angular momentum, but not all features in a line spectrum affected by a static magnetic field can be explained by orbital angular momentum alone (anomalous Zeeman effect, Paschen–Back effect). This *duplexity problem* was resolved by the rather phenomenological introduction of an additional intrinsic angular momentum called *spin* [44], which has no classical analog. For properties of this kind, the correspondence principle becomes obviously meaningless. The analysis of this spin property showed the typical behavior of quantum angular momentum. In particular, this observable can be characterized by two eigenvalue equations

$$\mathbf{s}^2 \rho_{sm_s} = s(s+1)\hbar^2 \rho_{sm_s} \tag{4.137}$$

$$s_z \rho_{sm_s} = m_s \hbar \rho_{sm_s} \quad \text{with} \quad m_s \in [-s, -s+1, \ldots, s] \tag{4.138}$$

which are of the same type as the angular-momentum eigenvalue equation. For this reason, the new spin property is obviously an *internal* angular momentum of an elementary particle. The quantum number s may adopt positive integer or half-integer values or can be zero.

The explicit form of the spin eigenfunction ρ_{sm_s} depends on the system under consideration. Since we will mostly consider ensembles of electrons in this account, the spin of a single electron is most important to us. Experimentally, it is found that an electron is to be described by a spin quantum number $s = 1/2$. The eigenvalue equations may be written in vector form with the electron spin operator

$$\boldsymbol{s} = \frac{\hbar}{2}\boldsymbol{\sigma} \tag{4.139}$$

where $\boldsymbol{\sigma}$ is a vector consisting of the three Pauli spin matrices

$$\sigma_x = \begin{pmatrix} 0 & 1 \\ 1 & 0 \end{pmatrix}, \quad \sigma_y = \begin{pmatrix} 0 & -\mathrm{i} \\ \mathrm{i} & 0 \end{pmatrix}, \quad \sigma_z = \begin{pmatrix} 1 & 0 \\ 0 & -1 \end{pmatrix} \tag{4.140}$$

Consequently, the squared spin operator reads

$$\boldsymbol{s}^2 = \frac{\hbar^2}{4}\begin{pmatrix} 3 & 0 \\ 0 & 3 \end{pmatrix} = \frac{1}{2}\left(\frac{1}{2}+1\right)\hbar^2 \mathbf{1}_2 \tag{4.141}$$

which possesses two different eigenvectors

$$\rho_{\frac{1}{2}\frac{1}{2}} = \begin{pmatrix} 1 \\ 0 \end{pmatrix} \quad \text{and} \quad \rho_{\frac{1}{2}(-\frac{1}{2})} = \begin{pmatrix} 0 \\ 1 \end{pmatrix} \tag{4.142}$$

with magnetic spin quantum numbers $m_s = 1/2$ and $m_s = -1/2$ as can be easily verified by insertion into Eqs. (4.137) and (4.138). The z-component of the spin operator for spin-1/2 particles reads

$$s_z = \frac{\hbar}{2}\sigma_z = \frac{\hbar}{2}\begin{pmatrix} 1 & 0 \\ 0 & -1 \end{pmatrix} \tag{4.143}$$

According to what has been said in section 4.3 with respect to the quantum mechanical measurement process we note that each individual measurement of the spin's z-component of a particle, which has always been prepared in the same way, yields either of the eigenvalues, $1/2$ or $-1/2$, respectively. The expectation value of very many such measurements on equally prepared particles is, however, zero.

Like any angular momentum, spin angular momentum can be coupled to another spin angular momentum along the lines sketched in the last section. Two single spin operators couple to a total spin operator $\boldsymbol{S}$ via

$$\boldsymbol{S} = \boldsymbol{s}_1 + \boldsymbol{s}_2 \tag{4.144}$$

If all spins of a collection of particles, say electrons, are coupled to one total spin angular momentum $\boldsymbol{S}$, the corresponding total spin eigenfunction ρ_{SM_S} is constructed.

In addition, all orbital angular momenta of these electrons (in the spherical symmetry of an atom) can be coupled to a total $\boldsymbol{L}$ with corresponding eigenfunction $|LM_L\rangle$. This coupling scheme is called Russell–Saunders or LS-coupling. The resulting product functions turn out to be suitable for the angle- and spin-dependent parts of the nonrelativistic eigenfunctions of many-electron atoms. In particular, in Schrödinger's theory the Hamiltonian for any atom commutes with the squared total angular momentum operator as well as with the squared total spin operator (compare chapter 6). Therefore, the eigenstates of atoms can be classified according to the irreducible representations of the group of rotations, i.e., by the eigenvalue labels L and S. This has led to an important notation in atomic spectroscopy, which assigns a so-called *term symbol* $^{(2S+1)}L$ to a total atomic wave function (or to a set of degenerate wave functions, i.e., to a set of wave functions featuring the same eigenvalue) in order to denote its symmetry properties. Of course, the spherical symmetry only holds for atoms, and such LS-term symbols can only be assigned to them. However, for linear molecules the wave function is an eigenfunction of the z-component of the total angular momentum, which is thus still a good quantum number and the ΛS-coupling scheme can be used [45, p. 216]. The notation ΛS refers to the eigenvalues of the z-component of the total angular momentum and of the total spin operator, respectively. For general molecules, however, no orbital angular momentum symmetry survives, and only point group symmetry may be used to classify wave functions according to the irreducible representations of the point group valid for the frame of nuclear coordinates. Nevertheless, total spin will always be a good quantum number for molecules in nonrelativistic quantum mechanics. And, hence, all (nonrelativistic) quantum states can be classified according to the spin multiplicity $2S+1$, which denotes the degeneracy of a term with respect to spin. Recall that $M_S = -S, -S+1, \ldots, S$ spin functions arise for a given total spin quantum number S — none of them being distinguished in the nonrelativistic Hamiltonian, which is not sensitive with respect to spin, so that they all yield the same total energy eigenvalue. The degeneracy with respect to spin is therefore $2S+1$. We call states with $2S+1$=1 *singlet* states (S=0), with $2S+1$=2 *doublet* states (S=1/2), with $2S+1$=3 *triplet* states (S=1), with $2S+1$=4 *quartet* states (S=3/2), with $2S+1$=5 *quintet* states (S=2), and so forth, in accordance with the number of lines observable in spectroscopic experiments with external fields applied.

4.4.4 Coupling of Orbital and Spin Angular Momenta

As an angular momentum, spin can couple with the orbital angular momentum to yield a *total angular momentum for a single particle.* The corresponding total angular momentum operator $\boldsymbol{j}$ for a single particle is simply defined as the vectorial sum of all angular momenta present, that is the sum of the orbital angular momentum operator $\boldsymbol{l}$ and the spin momentum operator $\boldsymbol{s}$,

$$\boldsymbol{j} = \boldsymbol{l} \otimes \mathbf{1}_2 + \mathbf{1}_1 \otimes \boldsymbol{s} \quad \rightarrow \quad \boldsymbol{l}\mathbf{1}_2 + \boldsymbol{s} \tag{4.145}$$

with

$$\boldsymbol{l}\mathbf{1}_2 \equiv \boldsymbol{l} \otimes \mathbf{1}_2 = (l_x\mathbf{1}_2, l_y\mathbf{1}_2, l_z\mathbf{1}_2) \tag{4.146}$$

As already discussed in section 4.4.2 for the coupling of two general angular momenta, we seek a recipe to generate the eigenstates of the total angular momentum from those of the individual angular momenta, i.e., from the uncoupled product representation of orbital angular momentum and spin eigenfunctions.

The eigenvalue equation for the squared two-component total angular momentum $\boldsymbol{j}^2$ is of the same form as those already discussed for orbital and for spin angular momentum. They can be written as

$$\boldsymbol{j}^2 \chi_{jm_j} = j(j+1)\,\hbar^2\,\chi_{jm_j} \tag{4.147}$$

The eigenstates χ_{jm_j} also fulfill the equation for one component of $\boldsymbol{j}$,

$$j_z \chi_{jm_j} = m_j\,\hbar\,\chi_{jm_j} \tag{4.148}$$

as the subscript m_j already indicated. The χ_{jm_j} are called spherical spinors or Pauli spinors and we construct them from the uncoupled product states $\phi_{lm_l} = Y_{lm_l}\rho_{\frac{1}{2}m_s}$, of which the two possible combinations can be explicitly written as

$$\phi_{l(m_j-\frac{1}{2})} = Y_{l(m_j-\frac{1}{2})}\begin{pmatrix}1\\0\end{pmatrix} \quad \text{and} \quad \phi_{l(m_j+\frac{1}{2})} = Y_{l(m_j+\frac{1}{2})}\begin{pmatrix}0\\1\end{pmatrix} \tag{4.149}$$

The subscripts of these product states already indicate that two different m_l components need to be combined with the spin eigenfunctions. The reason for this is that in order to yield the coupled states with given m_j, we must be able to fulfill

$$m_j = m_l + m_s \quad \text{originating from} \quad j_z = l_z + s_z \tag{4.150}$$

so that

$$m_l = m_j - m_s = m_j \mp \frac{1}{2} \tag{4.151}$$

resulting in two different spherical harmonics Y_{lm_l} and $Y_{lm_l'}$ to enter the product basis so that $m_l = m_j - 1/2$ and $m_l' = m_j + 1/2$. From the product basis of Eq. (4.149) a Clebsch–Gordan summation yields the coupled eigenstates

$$\chi_{jm_j}^{(\pm)} = \sum_{m_s} Y_{(j\mp\frac{1}{2})(m_j-m_s)} \rho_{\frac{1}{2}m_s} \langle (j\mp\tfrac{1}{2})\,(m_j-m_s)\,\tfrac{1}{2}\,m_s | j\,m_j\rangle \tag{4.152}$$

$$= \begin{pmatrix} Y_{(j\mp\frac{1}{2})(m_j-\frac{1}{2})} & \langle (j\mp\frac{1}{2})\,(m_j-\frac{1}{2})\,\frac{1}{2}\,\frac{1}{2}|j\,m_j\rangle \\ Y_{(j\mp\frac{1}{2})(m_j+\frac{1}{2})} & \langle (j\mp\frac{1}{2})\,(m_j+\frac{1}{2})\,\frac{1}{2}\,(-\frac{1}{2})|j\,m_j\rangle \end{pmatrix} \tag{4.153}$$

The vector coupling coefficients $\langle l\,m_l\,s\,m_s|j\,m_j\rangle$ are the Clebsch–Gordan coefficients from section 4.4.2. For $j = l + \frac{1}{2}$ we explicitly obtain by resolving the Clebsch–Gordan coefficients

$$\chi_{jm_j}^{(+)} = \sqrt{\frac{l+m_j+1/2}{2l+1}}\phi_{l(m_j-\frac{1}{2})} + \sqrt{\frac{l-m_j+1/2}{2l+1}}\phi_{l(m_j+\frac{1}{2})}$$

$$= \begin{pmatrix} \sqrt{\frac{l+m_j+1/2}{2l+1}}Y_{l(m_j-\frac{1}{2})} \\ \sqrt{\frac{l-m_j+1/2}{2l+1}}Y_{l(m_j+\frac{1}{2})} \end{pmatrix} \tag{4.154}$$

and, accordingly, for $j = l - \frac{1}{2}$ and $l > 0$,

$$\chi_{jm_j}^{(-)} = -\sqrt{\frac{l-m_j+1/2}{2l+1}}\phi_{l(m_j-\frac{1}{2})} + \sqrt{\frac{l+m_j+1/2}{2l+1}}\phi_{l(m_j+\frac{1}{2})}$$

$$= \begin{pmatrix} -\sqrt{\frac{l-m_j+1/2}{2l+1}}Y_{l(m_j-\frac{1}{2})} \\ \sqrt{\frac{l+m_j+1/2}{2l+1}}Y_{l(m_j+\frac{1}{2})} \end{pmatrix} \tag{4.155}$$

At this point, we should make a comment on the sign choice in $\chi_{jm_j}^{(-)}$. Other authors [46] multiply $\chi_{jm_j}^{(-)}$ by (-1), which — as we may anticipate here — would affect Eqs. (4.170) and (4.171) and as a consequence also Eq. (4.174) below as well as the κ quantum number in chapter 6. The presentation in chapter 6 is, however, consistent with the present sign convention. The choice of phases in these functions is also crucial for the evaluation of matrix elements as required later in chapter 9. This issue must always be carefully considered before employing analytic expressions in actual calculations (see also Grant's notes on this issue [47, p. 134]).

Recalling the definitions of the angular momenta, Eqs. (4.87) and (4.139), we may explicitly show that the following set of eigenvalue equations hold for the coupled spherical spinors of Eqs. (4.154) and (4.155)

$$j^2 \chi_{jm_j}^{(\pm)} = j(j+1)\,\hbar^2\,\chi_{jm_j}^{(\pm)}, \quad j = \frac{1}{2}, \frac{3}{2}, \dots \tag{4.156}$$

$$l^2 \chi_{jm_j}^{(\pm)} = l(l+1)\,\hbar^2\,\chi_{jm_j}^{(\pm)}, \quad l = j \mp \frac{1}{2} \tag{4.157}$$

$$s^2 \chi_{jm_j}^{(\pm)} = s(s+1)\,\hbar^2\,\chi_{jm_j}^{(\pm)}, \quad s = \frac{1}{2} \tag{4.158}$$

$$j_z \chi_{jm_j}^{(\pm)} = m_j\,\hbar\,\chi_{jm_j}^{(\pm)}, \qquad m_j = -j, -j+1, \dots, j \tag{4.159}$$

Hence, the sign in the subscript of $\chi_{jm_j}^{(\pm)}$ only affects the eigenvalue of the equation for $\boldsymbol{l}^2$. For later convenience in chapter 6, we investigate the Pauli spinors of Eqs. (4.154) and (4.155) further (cf. chapter 6). In particular, the operator $(\boldsymbol{\sigma}\cdot\boldsymbol{r})/r = (\boldsymbol{\sigma}\cdot\hat{\boldsymbol{r}})$ will become important. This is a scalar operator, which therefore does not change the total angular momentum. This (2×2)-matrix $\boldsymbol{\sigma}\cdot\hat{\boldsymbol{r}}$ can be written explicitly as

$$\boldsymbol{\sigma}\cdot\hat{\boldsymbol{r}} = \left(\frac{\boldsymbol{\sigma}\cdot\boldsymbol{r}}{r}\right) = \frac{1}{r}(\sigma_x x + \sigma_y y + \sigma_z z) = \frac{1}{r}\begin{pmatrix} z & x - \mathrm{i}y \\ x + \mathrm{i}y & -z \end{pmatrix} \tag{4.160}$$

where we simply used the transformation to spherical coordinates of Eqs. (4.89)–(4.91) and Euler's formula $\exp(\pm\mathrm{i}\varphi) = \cos\varphi \pm \mathrm{i}\sin\varphi$ to rewrite the operator in terms of the new coordinates

$$\boldsymbol{\sigma}\cdot\hat{\boldsymbol{r}} = \begin{pmatrix} \cos\vartheta & \sin\vartheta e^{-\mathrm{i}\varphi} \\ \sin\vartheta e^{+\mathrm{i}\varphi} & -\cos\vartheta \end{pmatrix} \tag{4.161}$$

The fact that this operator does not change the total angular momentum can also be shown by evaluating the commutator, which turns out to vanish

$$\left[(\boldsymbol{\sigma}\cdot\hat{\boldsymbol{r}}), \boldsymbol{j}^2\right] = 0 \tag{4.162}$$

Hence, $\boldsymbol{j}^2$ and $(\boldsymbol{\sigma}\cdot\hat{\boldsymbol{r}})$ possess the same eigenfunctions, namely $\chi_{jm_j}^{(\pm)}$. Therefore, if $(\boldsymbol{\sigma}\cdot\hat{\boldsymbol{r}})$ operates on one of the eigenstates it *must not* change the total angular momentum quantum number j and can thus only switch between the two spherical spinors that are possible for a given j,

$$\left(\frac{\boldsymbol{\sigma}\cdot\boldsymbol{r}}{r}\right)\chi_{jm_j}^{(\xi)} = \alpha\chi_{jm_j}^{(\zeta)} \quad \text{with} \quad \xi,\zeta \in \{+,-\} \tag{4.163}$$

In order to elucidate whether $(\boldsymbol{\sigma}\cdot\hat{\boldsymbol{r}})$ leaves the spherical spinor unchanged or whether it switches from the $(+)$ to the $(-)$ spinor of the pair (or vice versa), we study the parity properties of the spherical spinors. By operating with the parity operator P on the Pauli spinors we find

$$P\chi_{jm_j}^{(\pm)}(\boldsymbol{r}) = \chi_{jm_j}^{(\pm)}(-\boldsymbol{r}) = (-1)^l\chi_{jm_j}^{(\pm)}(\boldsymbol{r}) \tag{4.164}$$

because of the parity of the spherical harmonics, Eq. (4.127). Because $l = l^{(+)} = j - 1/2$ in the case of $\chi_{jm_j}^{(+)}$ and $l = l^{(-)} = j + 1/2$ in the case of $\chi_{jm_j}^{(-)}$, the parity of a pair of Pauli spinors for given (j, m_j) is not the same. If $l^{(+)}$ is even, $\chi_{jm_j}^{(+)}$ will possess even parity while $l^{(-)}$ is then necessarily odd (as $\Delta l = 1$), and hence $\chi_{jm_j}^{(-)}$ is of odd parity (or vice versa). Since the operator $(\boldsymbol{\sigma} \cdot \hat{\boldsymbol{r}})$ is obviously of odd parity,

$$P(\boldsymbol{\sigma} \cdot \hat{\boldsymbol{r}}) = [\boldsymbol{\sigma} \cdot (-\hat{\boldsymbol{r}})] = -(\boldsymbol{\sigma} \cdot \hat{\boldsymbol{r}}) \tag{4.165}$$

it changes the parity of the spherical spinor in Eq. (4.163), and therefore we have

$$\left(\frac{\boldsymbol{\sigma} \cdot \boldsymbol{r}}{r}\right) \chi_{jm_j}^{(\pm)} = \alpha \chi_{jm_j}^{(\mp)} \tag{4.166}$$

Now, we need to determine the yet unknown prefactor α. For this purpose, we study the squared operator $(\boldsymbol{\sigma} \cdot \hat{\boldsymbol{r}})^2$. Recalling Dirac's relation for its evaluation (see appendix D.2), we find

$$(\boldsymbol{\sigma} \cdot \boldsymbol{r})(\boldsymbol{\sigma} \cdot \boldsymbol{r}) = |\boldsymbol{r}|^2 \mathbf{1}_2 = r^2 \mathbf{1}_2 \quad \Rightarrow \quad \left(\frac{\boldsymbol{\sigma} \cdot \boldsymbol{r}}{r}\right)\left(\frac{\boldsymbol{\sigma} \cdot \boldsymbol{r}}{r}\right) = \mathbf{1}_2 \tag{4.167}$$

since the vector product of $\boldsymbol{r}$ with itself is zero, $\boldsymbol{r} \times \boldsymbol{r} = 0$, while the corresponding scalar product yields the squared length r^2. The corresponding eigenvalue equation,

$$\left(\frac{\boldsymbol{\sigma} \cdot \boldsymbol{r}}{r}\right)^2 \chi_{jm_j}^{(\pm)} = \alpha^2 \chi_{jm_j}^{(\pm)} \stackrel{(4.167)}{=} \chi_{jm_j}^{(\pm)} \tag{4.168}$$

thus yields α^2=1. As a last step, we now need to decide whether $\alpha = 1$ or $\alpha = -1$ in Eq. (4.168). In order to accomplish this, we study a special choice of coordinates. Obviously, everything that has been said so far is valid for *any* choice of coordinates (r, φ, ϑ). We may therefore select a special set of coordinates that allows us to determine the sign of α. We choose $\vartheta = 0$, so that the Cartesian coordinates then become $x = y = 0$ and $z = r$ and therefore Eqs. (4.160) and (4.161) become

$$\left(\frac{\boldsymbol{\sigma} \cdot \boldsymbol{r}}{r}\right) \stackrel{\vartheta=0}{=} \sigma_z = \begin{pmatrix} 1 & 0 \\ 0 & -1 \end{pmatrix} \tag{4.169}$$

For $\vartheta = 0$, Eq. (4.168) can be simplified because only those spherical harmonics survive whose magnetic quantum number is zero, i.e., all other spherical harmonics include a vanishing sine function, $\sin \vartheta = 0$. Nonvanishing spherical harmonics occur if $m_j = 1/2$ or if $m_j = -1/2$. For the first case, $m_j = 1/2$, we deduce from Eq. (4.168) that

$$\left(\frac{\sigma \cdot r}{r}\right) \chi^{(+)}_{j(1/2)} \overset{\vartheta=0}{=} \begin{pmatrix} 1 & 0 \\ 0 & -1 \end{pmatrix} \begin{pmatrix} \sqrt{\dfrac{l^{(+)}+1}{2l^{(+)}+1}} Y_{l^{(+)}0}(\varphi,0) \\ 0 \end{pmatrix}$$

$$= - \begin{pmatrix} -\sqrt{\dfrac{l^{(-)}}{2l^{(+)}+1}} \sqrt{\dfrac{2l^{(+)}+1}{2l^{(-)}+1}} Y_{l^{(-)}0}(\varphi,0) \\ 0 \end{pmatrix} = -\chi^{(-)}_{j(1/2)}(\varphi,0) \quad (4.170)$$

and for the second choice $m_j = -1/2$, Eq. (4.168) yields

$$\left(\frac{\sigma \cdot r}{r}\right) \chi^{(+)}_{j(-1/2)} \overset{\vartheta=0}{=} \begin{pmatrix} 1 & 0 \\ 0 & -1 \end{pmatrix} \begin{pmatrix} 0 \\ \sqrt{\dfrac{l^{(+)}+1}{2l^{(+)}+1}} Y_{l^{(+)}0}(\varphi,0) \end{pmatrix}$$

$$= - \begin{pmatrix} 0 \\ \sqrt{\dfrac{l^{(-)}}{2l^{(+)}+1}} \sqrt{\dfrac{2l^{(+)}+1}{2l^{(-)}+1}} Y_{l^{(-)}0}(\varphi,0) \end{pmatrix} = -\chi^{(-)}_{j(-1/2)}(\varphi,0) \quad (4.171)$$

In order to better understand these equations, we note again that the l quantum number is different for a given j; in particular, we have $l^{(+)} = l^{(-)} - 1$. Therefore, the spherical harmonics are different in $\chi^{(-)}_{jm_j}(\varphi,0)$ and $\chi^{(+)}_{jm_j}(\varphi,0)$ for any given (j, m_j). However, for $\vartheta = 0$ all spherical harmonics with $m = 0$ take the particularly simple form

$$Y_{l0}(\varphi,0) = \sqrt{\frac{1}{4\pi}} \sqrt{2l+1} \quad (4.172)$$

so that the relation between two such functions is

$$Y_{l0}(\varphi,0) = \sqrt{\frac{2l+1}{2l'+1}} Y_{l'0}(\varphi,0) \quad (4.173)$$

As a final result we understand from Eqs. (4.170) and (4.171) that α carries a *negative* sign and we finally obtain

$$(\sigma \cdot \hat{r}) \chi^{(\pm)}_{jm_j} = -\chi^{(\mp)}_{jm_j} \quad (4.174)$$

We may note that the action of $(\sigma \cdot \hat{r})$ on a Pauli spinor can also be evaluated explicitly using the representation of the operator of Eq. (4.161) (for the necessary details see the appendix in the book by Bethe and Salpeter [48, p. 346]).

Finally, we should note that eigenfunctions of $\boldsymbol{j}$ (or of a total angular momentum $\boldsymbol{J}$, respectively) become an integral part of the eigenfunctions of atoms in the *relativistic* theory (jj-coupling, see also chapters 6 and 9). As discussed in the framework of the LS-coupling scheme of the nonrelativistic theory in section 4.4.3, we note that term symbols can also be assigned to the quantum states of atoms in the jj coupling scheme, but usually a combined symbol is given as $^{(2S+1)}L_J$. Again, for the special case of *linear* molecules, J_z is a good quantum number, giving rise to the $\omega\omega$-coupling scheme [45, p. 337], where ω denotes j_z and the resultant J_z is called Ω accordingly.

4.5 Pauli Antisymmetry Principle

The last axiom of quantum mechanics refers to symmetry properties of many-particle wave functions. This so-called Pauli principle postulates a certain symmetry with respect to exchange of the dynamical variables in the many-particle wave function depending on the spin of the set of particles under consideration. It is the only axiom which cannot be rationalized in any way by comparison with classical mechanics. Instead, it follows from a deeper and more general theory, namely from quantum field theory where it is also known as the *spin statistics theorem.*

Explicitly, the Pauli principle states that the wave function of an ensemble of particles with half-valued spin, which obey Fermi–Dirac quantum statistics and are therefore called *fermions*, must change its sign upon exchange of any two coordinates in the wave function. A fermionic many-particle wave function is then said to be *antisymmetric* under pair exchange of coordinates. By contrast, the wave function of a set of particles with integer spin, which obey Bose–Einstein quantum statistics and are therefore called *bosons*, does *not* change its sign upon pair permutation of any two coordinates in the function.

What has already been said about space and time coordinates in the preceding chapters suggests the obvious question for which coordinates the Pauli principle is valid. Do we need to apply the pair permutation to only spatial coordinates or to space-time coordinates? The permutation is to be applied to the spatial coordinates only since in quantum field theory the commutators are understood as equal-time commutation relations. Moreover, in nonrelativistic quantum mechanics this problem does not show up and we will later refer to space–spin coordinates that need to be exchanged for pair permutation. The situation will become more clear in section 8.6.5 once we have introduced the theoretical tools and background needed.

Further Reading

Many excellent and well-known classic texts on quantum mechanics are available such as the books by Feynman (lectures, volume III) [49], Messiah [50], Davydov [51], Landau and Lifshitz (volume III) [52], Merzbacher [53], Cohen-Tannoudji and co-authors [54], Schiff [40], Dirac [55], Bohm [56] and many others; we do not intend to provide a complete list here. Instead, we would like to draw the reader's attention to the following short list, which contains modern text books, conceptual discussions, and interesting modern developments in quantum theory:

C. J. Isham, [57]. *Lectures on Quantum Theory — Mathematical and Structural Foundations.*

Isham's compact text provides a conceptual introduction to quantum theory with remarks on its relativistic generalization skipping the explicit discussion of (standard) analytically solvable examples. Instead, Isham focuses on a discussion of the formal structure of the theory.

J. J. Sakurai, [16]. *Modern Quantum Mechanics.*

Sakurai presents a modern, very comprehensive, axiomatic introduction to quantum theory, which is concisely written. His presentation is a true jewel; in fact, it is also the only 'technical' textbook on quantum mechanics in our short list here.

M. Weissbluth, [58]. *Atoms and Molecules.*

Weissbluth's textbook provides an introduction to coupling of angular momenta in quantum mechanics and contains a survey of useful relations of Clebsch–Gordan coefficients as well as of Wigner $3j$-symbols. This book can be considered elementary and a useful first step to the more involved presentations of the theory of angular momentum.

A. R. Edmonds, [37]. *Angular Momentum in Quantum Mechanics.*

Edmonds' classic text on the theory of angular momentum is recommended for further studies on this subject. Our presentation was very brief in many respects. For instance, we have neither introduced ladder operators to explicitly construct the eigenfunctions of angular momentum eigenvalue equations nor have we deduced the expression for total angular momentum eigenstates in the coupled product basis. Edmonds' book fills all these gaps.

J. v. Neumann, [59]. *Mathematical Foundations of Quantum Mechanics.*

Von Neumann's early account of the mathematical structure and the basic principles of quantum mechanics can be warmly recommended. Many of the issues that have been only touched on in this chapter are discussed in depth by von Neumann.

Y. Aharonov, D. Rohrlich, [60]. *Quantum Paradoxes — Quantum Theory for the Perplexed.*

The current status of quantum theory including the most recent developments of the last ten to twenty years is sketched in this book. It nicely complements the books by Isham and Omnès.

R. Omnès, [61]. *The Interpretation of Quantum Mechanics.*

This book provides an up-to-date overview on (nonrelativistic) quantum theory from the point of view of potential misunderstandings and misconceptions which have led to various criticisms and "extensions" of quantum theory during the past decades. It does not, however, contain any reference to a relativistic formulation of quantum chemistry but focuses on the relation between classical physics and quantum physics. A more readable and quite instructive introduction to quantum mechanics from Omnès' point of view can be found in another book by the same author [62].

F. Strocchi, [63]. *An Introduction to the Mathematical Structure of Quantum Mechanics.*

This book is a clear, compact, and readable modern introduction into the mathematical structure of theory in physics (mostly quantum mechanics but with many connections to classical mechanics).

D. Giulini, E. Joos, C. Kiefer, J. Kupsch, I.-O. Stamatescu, H. D. Zeh, [64]. *Decoherence and the Appearance of a Classical World in Quantum Theory.*

In this unique book, many questions that arise beyond the standard streamlined presentation of quantum theory are addressed. The reader finds insightful essays on the emergence of classical physics from quantum physics and the decoherence mechanism, the measurement problem and the collapse of the wave function, and many other related subjects.

Part II

Dirac's Theory of the Electron

5 Relativistic Theory of the Electron

Having introduced the principles of special relativity in classical mechanics and electrodynamics as well as the foundations of quantum theory, we now discuss their unification in the relativistic, quantum mechanical description of the motion of a free electron. One might start right away with an appropriate ansatz for the basic equation of motion with arbitrary parameters to be chosen to fulfill boundary conditions posed by special relativity, which would lead us to the Dirac equation in standard notation. However, we proceed stepwise and derive the Klein–Gordon equation first so that the subsequent steps leading to Dirac's equation for a freely moving electron can be better understood.

5.1 Correspondence Principle and Klein–Gordon Equation

One postulate that has not explicitly been formulated as a basic axiom of quantum mechanics in the last chapter, because this postulate is valid for any physical theory, is that the equations of quantum mechanics have to be valid and invariant in form in all intertial reference frames. In this chapter, we take the first step toward a relativistic electronic structure theory and start to derive the basic quantum mechanical equation of motion for a single, freely moving electron, which shall obey the principles of relativity outlined in chapter 3. We are looking for a Hamiltonian which keeps Eq. (4.16) invariant in form under Lorentz transformations.

5.1.1 Classical Energy Expression and First Hints from the Correspondence Principle

We first consider the option to set up a quantum mechanical equation of motion which obeys the correspondence principle. If we apply the correspondence principle to the classical *nonrelativistic* kinetic energy expression $E = p^2/(2m)$ we arrive at the time-dependent Schrödinger equation, in which first derivatives with respect to time but second derivatives with respect to all spatial coordinates occur. Since time and spatial coordinates thus do not

Relativistic Quantum Chemistry. Markus Reiher and Alexander Wolf

ISBN: 978-3-527-31292-4

occur in the same form as it is required for covariance under Lorentz transformations, which mix time-like and space-like coordinates, it is apparent that the Schrödinger equation cannot fulfill the requirements of special relativity. It is not Lorentz covariant. Hence, we start out from the classical expression for the *relativistic* energy of the freely moving particle $E = \sqrt{\boldsymbol{p}^2c^2 + m_e^2c^4}$ and apply the substitutions,

$$E \longrightarrow i\hbar\frac{\partial}{\partial t} \tag{5.1}$$

$$\boldsymbol{p} \longrightarrow -i\hbar\nabla \tag{5.2}$$

according to the correspondence principle to obtain the final result as

$$E = \sqrt{\boldsymbol{p}^2c^2 + m_e^2c^4} \tag{5.3}$$

$$\downarrow \qquad\qquad \downarrow$$

$$i\hbar\frac{\partial}{\partial t}\,\Psi = \sqrt{-\hbar^2c^2\nabla^2 + m_e^2c^4}\;\Psi \tag{5.4}$$

The energy operator of Eq. (5.4) is known as the *square-root operator*. It is apparent that the square root of the spatial differentiation, ∇^2, on the right hand side of the resulting equation would be difficult to evaluate in position space. An expansion of the square root would lead to infinitely high derivatives with respect to the spatial coordinates so that time and spatial coordinates would be treated differently, again.

In order to circumvent the problem with the square root, we may use the squared energy expression to derive a quantum mechanical equation of motion,

$$E^2 = \boldsymbol{p}^2c^2 + m_e^2c^4 \tag{5.5}$$

$$\downarrow \qquad\qquad \downarrow$$

$$-\hbar^2\frac{\partial^2}{\partial t^2}\,\Psi = \left(-\hbar^2c^2\nabla^2 + m_e^2c^4\right)\,\Psi \tag{5.6}$$

The resulting Eq. (5.6) is a second-order differential equation for a freely moving particle, which treats time and spatial coordinates on the same footing. Eq. (5.6) is called the Klein–Gordon equation and was derived by W. Gordon in 1926, by O. Klein in 1927 and by others.

The Klein–Gordon equation for a freely moving particle can also be written in a compact, explicitly covariant form,

$$\left[\partial_\mu\partial^\mu + \left(\frac{m_ec}{\hbar}\right)^2\right]\Psi = \left[\Box + \left(\frac{m_ec}{\hbar}\right)^2\right]\Psi = 0 \tag{5.7}$$

where the covariant vector $\partial_\mu = \partial/\partial x^\mu$ with $x^\mu = (ct, \boldsymbol{x}) = (ct, x_1, x_2, x_3)$ was employed (see section 3.1). The covariance of the d'Alembertian $\Box = \partial_\mu\partial^\mu$

with respect to Lorentz transformation has already been shown in chapter 3 so that the covariance of the whole equation under such transformations is evident.

It is important to note that the Klein–Gordon Eq. (5.6) does not contain a reference to the quantum mechanical spin of a particle and can thus not explain its experimentally observed occurrence.

5.1.2
Solutions of the Klein–Gordon Equation

For a deeper understanding of the Klein–Gordon equation for a particle in field-free space we investigate its solutions. The Klein–Gordon eigenfunction in this case is given by the plane wave

$$\Psi(\boldsymbol{r},t) = \exp\left[\frac{\mathrm{i}}{\hbar}(Et - \boldsymbol{p}\cdot\boldsymbol{r})\right] \tag{5.8}$$

as we may verify easily by insertion of this function into the left hand side of Eq. (5.6),

$$-\hbar^2 \frac{\partial^2}{\partial t^2} \exp\left[\frac{\mathrm{i}}{\hbar}(Et - \boldsymbol{p}\cdot\boldsymbol{r})\right] = E^2 \exp\left[\frac{\mathrm{i}}{\hbar}(Et - \boldsymbol{p}\cdot\boldsymbol{r})\right] \tag{5.9}$$

and into its right hand side,

$$(-\hbar^2 c^2 \nabla^2 + m_e^2 c^4) \exp\left[\frac{\mathrm{i}}{\hbar}(Et - \boldsymbol{p}\cdot\boldsymbol{r})\right] = (c^2 \boldsymbol{p}^2 + m_e^2 c^4) \exp\left[\frac{\mathrm{i}}{\hbar}(Et - \boldsymbol{p}\cdot\boldsymbol{r})\right] \tag{5.10}$$

respectively. Comparison of coefficients in Eqs. (5.9) and (5.10) yields the eigenvalue

$$E^2 = c^2 \boldsymbol{p}^2 + m_e^2 c^4 \tag{5.11}$$

This corresponds then to the following energy values of the freely moving particle

$$E = \pm\sqrt{c^2 \boldsymbol{p}^2 + m_e^2 c^4} \tag{5.12}$$

which might represent the expected positive energies, but which also shows negative-energy values not straightforward to interpret in the present context. Hence, it is the quadratic form in Eq. (5.6) that leads to negative-energy eigenvalues for freely moving particles. Since freely moving particles possess a finite positive kinetic energy (or none if they are at rest), the negative energies are particularly difficult to explain. However, negative-energy states will later also show up in the context of the Dirac equation — first in section 5.3. Their significance will be discussed in section 5.5.

We may anticipate that the Klein–Gordon equation is not able to describe an electron for reasons given in the next section. However, in a properly quantized field-theoretical form (compare chapter 7) it describes neutral mesons of spin 0. Mesons are strongly interacting bosons, i.e., they are particles subject to the strong force (also called hadrons) with an integer spin subject to Bose–Einstein statistics (an example for a *fermionic* hadron would be the proton having spin 1/2).

5.1.3
The Klein–Gordon Density Distribution

The derivation so far has been based on the *ad hoc* hypothesis that the correspondence principle of chapter 4 is also beneficial in the relativistic regime. Despite the fact that the resulting equation is Lorentz covariant, the correspondence principle produces the yet unexplained negative-energy solutions, and more peculiarities are ahead. Hence, it is advisable to further study properties of the Klein–Gordon equation. An important observable to be analyzed is the density distribution, which is most appropriately derived from a continuity equation (compare sections 3.4 and 4.3.9). In order to derive a continuity equation from the Klein–Gordon equation, whose eigenfunction might play a role as a probability amplitude as in nonrelativistic Schrödinger quantum mechanics, we multiply Eq. (5.7) by $\Psi^\star$,

$$\Psi^\star \left[\partial_\mu \partial^\mu + \left(\frac{m_e c}{\hbar}\right)^2\right] \Psi = 0 \tag{5.13}$$

and consider also its complex conjugate,

$$\Psi \left[\partial_\mu \partial^\mu + \left(\frac{m_e c}{\hbar}\right)^2\right] \Psi^\star = 0 \tag{5.14}$$

Subtraction of Eq. (5.14) from Eq. (5.13) yields

$$\Psi^\star \partial_\mu \partial^\mu \Psi - \Psi \partial_\mu \partial^\mu \Psi^\star = 0 \tag{5.15}$$

which can be rewritten as

$$\partial_\mu \left[\Psi^\star \partial^\mu \Psi - \Psi \partial^\mu \Psi^\star\right] = 0 \tag{5.16}$$

because the product rule yields term by term

$$\partial_\mu \Psi^\star \partial^\mu \Psi = \frac{\partial}{\partial x^\mu}\left(\Psi^\star \frac{\partial \Psi}{\partial x_\mu}\right) = \Psi^\star \frac{\partial^2 \Psi}{\partial x^\mu \partial x_\mu} + \left(\frac{\partial \Psi^\star}{\partial x^\mu}\right)\left(\frac{\partial \Psi}{\partial x_\mu}\right) \tag{5.17}$$

$$\partial_\mu \Psi \partial^\mu \Psi^\star = \frac{\partial}{\partial x^\mu}\left(\Psi \frac{\partial \Psi^\star}{\partial x_\mu}\right) = \Psi \frac{\partial^2 \Psi^\star}{\partial x^\mu \partial x_\mu} + \left(\frac{\partial \Psi}{\partial x^\mu}\right)\left(\frac{\partial \Psi^\star}{\partial x_\mu}\right) \tag{5.18}$$

so that the last terms on the right hand sides of both Eqs. (5.17) and (5.18) cancel upon subtraction as demanded by Eq. (5.16).

In order to understand the physical meaning of Eq. (5.16), we recall the definition of the nonrelativistic current density of Eq. (4.80),

$$j = \frac{\hbar}{2m_e \mathrm{i}} \left[\Psi^\star \nabla \Psi - \Psi \nabla \Psi^\star\right] \tag{5.19}$$

and write Eq. (5.16) after multiplication with $\hbar/(2m_e \mathrm{i})$ as

$$\frac{\partial}{\partial t} \frac{\mathrm{i}\hbar}{2m_e c^2} \left[\Psi^\star \frac{\partial}{\partial t}\Psi - \Psi \frac{\partial}{\partial t}\Psi^\star\right] + \mathrm{div} j = 0 \tag{5.20}$$

This equation is the continuity equation,

$$\partial_\mu j^\mu = \dot{\rho} + \mathrm{div} j = 0 \tag{5.21}$$

with the following zeroth component of the 4-current $j^\mu = (c\rho, j)$,

$$\rho \equiv \frac{\mathrm{i}\hbar}{2m_e c^2} \left[\Psi^\star \frac{\partial}{\partial t}\Psi - \Psi \frac{\partial}{\partial t}\Psi^\star\right] \tag{5.22}$$

that defines the Klein–Gordon density distribution. Since the Klein–Gordon equation is a second-order differential equation in t, two integration constants arise which allow one to choose the initial values of Ψ and $\partial\Psi/\partial t$ independently of one another in such a way that ρ may take positive or negative values. Consequently, ρ is not positive definite and, hence, does not represent a probability density distribution, which must take strictly positive values for any coordinates t and x_k. Already for this reason, we shall reject the Klein–Gordon equation as a fundamental quantum mechanical equation, but we should note that a re-interpretation introduced by Feynman makes it a suitable equation for spinless particles [65].

Because of the unphysical feature of the Klein–Gordon density and the fact that spin does not emerge naturally (but would have to be included *a posteriori* as in the nonrelativistic framework) we are not able to deduce a fundamental relativistic quantum mechanical equation of motion for a freely moving electron. However, we may wonder which results of this section may be of importance for the derivation of such an equation of motion for the electron. Certainly, we would like to recover the plane wave solutions of Eq. (5.8) for the freely moving particle, but in order to introduce only a single integration constant (or the choice of a single initial value) for a positive definite density distribution we need to focus on *first-order differential equations in time*. These must also be *first-order differential equations in space* for the sake of Lorentz covariance.

5.2 Derivation of the Dirac Equation for a Freely Moving Electron

In 1928, Dirac proposed a new quantum mechanical equation for the electron [66, 67], which solved two problems at once, namely the Lorentz-covariance requirement and the duplexity of atomic states, which was accounted for by Goudsmit and Uhlenbeck's phenomenological introduction of spin. In fact, he showed how the dynamic spin variable is connected to Lorentz covariance — a connection that will become clear in the following. To derive this fundamental quantum mechanical equation for the electron, which features relativistic covariance, we set out with a basic ansatz for this equation based on the results of the preceding section,

$$\mathrm{i}\hbar\frac{\partial}{\partial t}\Psi = \left[\frac{\hbar c}{\mathrm{i}}\alpha^k\partial_k + \beta m_e c^2\right]\Psi \equiv H^D\Psi \tag{5.23}$$

where we had to introduce four parameters to be determined in order to avoid the flaws of the Klein–Gordon equation. Note that the components of the vector $\boldsymbol{\alpha}$,

$$\boldsymbol{\alpha} = (\alpha^i) = (\alpha^1, \alpha^2, \alpha^3) \tag{5.24}$$

are labeled with superscripts in order to employ Einstein's summation convention. From the point of view of relativity, we could have equally well chosen subscripts to denote the components, because the α^i do *not* represent components of a contravariant vector and, hence, do *not* transform according to Eq. (3.12).

5.2.1 Relation to the Klein–Gordon Equation

The Klein–Gordon equation simply implements the relativistic energy–momentum relation $E^2 = \boldsymbol{p}^2c^2 + m_e^2c^4$ for a freely moving particle. Accordingly, the solutions of the field-free Klein–Gordon equation are plane waves with an energy eigenvalue $E = \pm\sqrt{\boldsymbol{p}^2c^2 + m_e^2c^4}$, which must be reproduced by the solutions of the Dirac equation. In order to establish a relationship between Dirac and Klein–Gordon equations we apply the operator identity of Eq. (5.23), $\mathrm{i}\hbar\partial/\partial t = \hbar c/\mathrm{i}\alpha^k\partial_k + \beta m_e c^2$, to the left and right hand sides, respectively, of the Dirac Eq. (5.23)

$$\begin{aligned} -\hbar^2\frac{\partial^2}{\partial t^2}\Psi &= \left[\frac{\hbar c}{\mathrm{i}}\alpha^k\partial_k + \beta m_e c^2\right]\left[\frac{\hbar c}{\mathrm{i}}\alpha^k\partial_k + \beta m_e c^2\right]\Psi \\ &= -\frac{\hbar^2c^2}{2}\sum_{i,j=1}^{3}\left(\alpha^i\alpha^j + \alpha^j\alpha^i\right)\partial_i\partial_j\Psi \end{aligned} \tag{5.25}$$

$$+\frac{\hbar m_e c^3}{\mathrm{i}}\sum_{i=1}^{3}\left(\alpha^i\beta+\beta\alpha^i\right)\partial_i\Psi+\beta^2 m_e^2 c^4\Psi \tag{5.26}$$

where we introduced the anticommutator $\alpha^i\alpha^j + \alpha^j\alpha^i$ instead of the single product $\alpha^i\alpha^j$ and took care of the resulting double counting by the prefactor of $1/2$. We can now determine the unkown parameters by direct comparison with the Klein–Gordon equation [Eq. (5.6)] — the correspondence principle is thus implicitly encoded also in the Dirac equation. Comparison of coefficients in Eqs. (5.6) and (5.26) yields

$$\alpha^i\alpha^j+\alpha^j\alpha^i = 2\delta^{ij} \implies (\alpha^i)^2=1 \tag{5.27}$$

$$\alpha^i\beta+\beta\alpha^i = 0 \tag{5.28}$$

$$\beta^2 = 1 \implies \beta=\beta^{-1} \tag{5.29}$$

Thus, the α^k and β parameters must fulfill anticommutation relations

$$\{\alpha^i,\alpha^j\}=2\delta^{ij} \quad \text{and} \quad \{\alpha^i,\beta\}=0 \tag{5.30}$$

where the anticommutation operator is, in general, defined as

$$\{A,B\}=AB+BA \tag{5.31}$$

5.2.2 Explicit Expressions for the Dirac Parameters

From Eqs. (5.27)–(5.29) we may now determine explicit expressions for the four parameter α^1, α^2, α^3, and β. From $(\alpha^i)^2=\beta^2=1$ [see Eqs. (5.27) and (5.29)] it follows immediately that the parameters possess eigenvalues of either $+1$ or -1. If we now rewrite Eq. (5.28) as

$$\alpha^i=-\beta^{-1}\alpha^i\beta \overset{(5.29)}{=} -\beta\alpha^i\beta \tag{5.32}$$

we can utilize properties of the trace,

$$Tr(\alpha^i)=-Tr(\beta\alpha^i\beta)=-Tr(\alpha^i\beta^2)\overset{(5.29)}{=}-Tr(\alpha^i) \tag{5.33}$$

to understand that the trace can only be zero, $Tr(\alpha^i)=0$, and that the number of positive and negative eigenvalues of all α^i parameters must therefore be equal. Similarly, we obtain

$$\beta = -(\alpha^i)^{-1}\beta\alpha^i=-\alpha^i\beta\alpha^i \tag{5.34}$$

$$\implies Tr(\beta) = -Tr(\alpha^i\beta\alpha^i)=-Tr(\beta(\alpha^i)^2)=-Tr(\beta)=0 \tag{5.35}$$

Hence, β also features an equal number of positive and negative eigenvalues and thus an even dimension. As an important result we note that the parameters α^1, α^2, α^3, and β cannot simply be numbers but are of matrix form with *even* dimension.

The smallest possible dimension for the parameters would be two. The only possible 2×2 matrices *which are linearly independent* and feature the required properties (i.e., one eigenvalue which is +1, one which is −1 and a zero trace) are:

$$\begin{pmatrix} 0 & 1 \\ 1 & 0 \end{pmatrix} = \sigma_1 \, , \quad \begin{pmatrix} 0 & -\mathrm{i} \\ \mathrm{i} & 0 \end{pmatrix} = \sigma_2 \, , \quad \begin{pmatrix} 1 & 0 \\ 0 & -1 \end{pmatrix} = \sigma_3 \tag{5.36}$$

which are the three Pauli spin matrices $\boldsymbol{\sigma} = (\sigma_x, \sigma_y, \sigma_z)$. However, these are only three parameters, but our ansatz requires four parameters. Thus, the parameters cannot be two dimensional. The next even dimension is four and we find the following set of 4 × 4 matrices,

$$\alpha^i = \begin{pmatrix} 0 & \sigma_i \\ \sigma_i & 0 \end{pmatrix} \quad \text{and} \quad \beta = \begin{pmatrix} \mathbf{1}_2 & 0 \\ 0 & -\mathbf{1}_2 \end{pmatrix} \tag{5.37}$$

which possess all required properties given in Eqs. (5.27)–(5.29) as one may easily verify. Of course, this particular choice of the 4 × 4 matrices is neither unique nor is the dimension 4 the only dimension, which allows us to fulfill the required properties, i.e., higher even dimensions are also possible. How one may find other representations of these four parameters will be discussed in section 5.2.4.3. The particular choice of Eqs. (5.37) is called the *standard representation* of the Dirac matrices α^k and β.

Other representations are possible, and we give as a further example the Weyl representation

$$\alpha^i_{\text{Weyl}} = \begin{pmatrix} \sigma_i & 0 \\ 0 & -\sigma_i \end{pmatrix} \quad \text{and} \quad \beta_{\text{Weyl}} = \begin{pmatrix} 0 & \mathbf{1}_2 \\ \mathbf{1}_2 & 0 \end{pmatrix} \tag{5.38}$$

which has the effect that $c(\boldsymbol{\sigma} \cdot \boldsymbol{p})$ and $m_e c^2$ then change places in the Dirac matrix equation compared to the standard representation, i.e., $c(\boldsymbol{\sigma} \cdot \boldsymbol{p})$ is then found on the block-diagonal and $m_e c^2$ on the off-diagonal position in the matrix operator.

Because the Pauli matrices enter the Dirac matrices we already note the connection to the spin operator through Eq. (4.139). For the physical interpretation of the $\boldsymbol{\alpha}$ parameters we may quote Dirac [67]: *The α's are new dynamical variables which it is necessary to introduce in order to satisfy the conditions of the problem. They may be regarded as describing some internal motions of the electron, which for most purposes may be taken to be the spin of the electron postulated in previous theories.* However, they are also connected to the velocity operator to be shown in section 5.3.3, which was also known to Dirac as is clear from a footnote in a paper by Breit [68]. Since this matter is not that straightforward to interpret, Breit later devoted a whole new paper [69] to this issue.

Since the dimension of Ψ must necessarily be the same dimension as the one of the Dirac matrices α_k, β, we understand that it is, in general, an n-component vector of functions if the dimension of the Dirac matrices is n.

In the standard representation, the quantum mechanical state Ψ is a vector of four functions, called a 4-*spinor*.

5.2.3 Continuity Equation and Definition of the 4-Current

In order to derive the continuity equation corresponding to Eq. (5.23) and in turn to obtain an expression for the probability density distribution, we proceed as in section 5.1.3 and multiply Eq. (5.23) by the complex conjugate of Ψ. This adjoint row vector of Ψ reads $\Psi^\dagger = (\Psi_1^\star, \Psi_2^\star, ..., \Psi_n^\star)$ with $n = 4$ (but the derivation in this section is generally valid for other choices of n). We then obtain from Eq. (5.23),

$$i\hbar\Psi^\dagger \frac{\partial}{\partial t}\Psi = \frac{\hbar c}{i}\Psi^\dagger \alpha^k \partial_k \Psi + m_e c^2 \Psi^\dagger \beta \Psi \tag{5.39}$$

whose complex conjugate reads

$$-i\hbar \left(\frac{\partial}{\partial t}\Psi^\dagger\right)\Psi = -\frac{\hbar c}{i}(\partial_k \Psi^\dagger)\alpha^{k,\dagger}\Psi + m_e c^2 \Psi^\dagger \beta^\dagger \Psi. \tag{5.40}$$

As in section 5.1.3 we subtract Eqs. (5.40) from (5.39), divide by $(i\hbar)$ and obtain

$$\frac{\partial}{\partial t}(\Psi^\dagger\Psi) = -c\left[\Psi^\dagger\alpha^k\partial_k\Psi + (\partial_k\Psi^\dagger)\alpha^{k,\dagger}\Psi\right] + \frac{m_e c^2}{i\hbar}\left[\Psi^\dagger\beta\Psi - \Psi^\dagger\beta^\dagger\Psi\right] \tag{5.41}$$

The above equation can be simplified by noting that the parameters α^k, β are hermitean,

$$\alpha^{k,\dagger} = \alpha^k \quad \text{and} \quad \beta^\dagger = \beta \tag{5.42}$$

i.e., they are identical to the transposed conjugate-complex matrices, which can be verified for the standard choice presented in the preceding section. However, the true physical reason for the hermiticity of these matrices is the fact that the Dirac Hamiltonian H^D in Eq. (5.23) must be hermitean in order to obtain real eigenvalues. Consequently, the parameters α^k, β must be hermitean. Employing the properties in Eq. (5.42), Eq. (5.41) simplifies to

$$\frac{\partial}{\partial t}(\Psi^\dagger\Psi) = -c\left[\Psi^\dagger\alpha^k\partial_k\Psi + (\partial_k\Psi^\dagger)\alpha^k\Psi\right] \tag{5.43}$$

which yields the continuity equation [70],

$$\dot{\rho} = -\text{div}\boldsymbol{j} \tag{5.44}$$

if we define the Dirac density distribution of the single particle as

$$\rho \equiv \Psi^\dagger\Psi \tag{5.45}$$

and the Dirac current density as

$$\boldsymbol{j} \equiv c\Psi^{\dagger}\boldsymbol{\alpha}\Psi, \quad \text{i.e.,} \quad j^{k} \equiv c\Psi^{\dagger}\alpha^{k}\Psi \tag{5.46}$$

Both can be joined in an expression for the 4-current,

$$j^{\mu} = (j^{0}, j^{k}), \quad \text{with} \quad j^{0} \equiv c\rho \tag{5.47}$$

so that the continuity equation takes the particularly short form

$$\partial_{\mu} j^{\mu} = 0, \qquad \text{or written explicitly} \qquad \frac{1}{c}\frac{\partial}{\partial t} j^{0} + \frac{\partial}{\partial x^{k}} j^{k} = 0 \tag{5.48}$$

We should emphasize that the densities introduced above are pure particle densities. In order to have the charge density from the electron density of Eq. (5.45), one must multiply ρ by $q_e = -e$.

5.2.4 Lorentz Covariance of the Field-Free Dirac Equation

5.2.4.1 Covariant Form

The most important requirement for truly fundamental physical euqations is their invariance in form under Lorentz transformations (principle of relativity). To investigate the behavior of the Dirac equation in Eq. (5.23) under Lorentz transformations, we rewrite it as

$$-\mathrm{i}\hbar\beta\partial_{0}\Psi - \mathrm{i}\hbar\beta\partial_{i}\alpha^{i}\Psi + m_{e}c\Psi = 0 \tag{5.49}$$

and define a set of four new Dirac matrices $\gamma^{\mu} = (\gamma^{0}, \gamma^{i})$,

$$\gamma^{0} \equiv \beta = \begin{pmatrix} \mathbf{1}_2 & 0 \\ 0 & -\mathbf{1}_2 \end{pmatrix} \quad \text{and} \quad \gamma^{i} \equiv \beta\alpha^{i} = \begin{pmatrix} 0 & \sigma_i \\ -\sigma_i & 0 \end{pmatrix} \tag{5.50}$$

which possess the following properties

$$(\gamma^{0})^{2} = (\gamma^{i})^{2} = \mathbf{1}_4 \quad \text{and} \quad (\gamma^{0})^{\dagger} = \gamma^{0} \quad \text{while} \quad (\gamma^{i})^{\dagger} = -\gamma^{i}. \tag{5.51}$$

and fulfill anticommutation relations

$$\{\gamma^{\mu}, \gamma^{\nu}\} = \begin{cases} 0 & \text{if } \mu \neq \nu \\ 2 & \text{if } \mu = \nu \end{cases}, \quad \forall\, \mu, \nu \in \{0, 1, 2, 3\} \tag{5.52}$$

These properties can be easily verified by using Eqs. (5.27)–(5.29) and the definition of the γ^{μ} matrices in Eq. (5.50). After division by $\hbar$, the Dirac Eq. (5.49) may now be expressed in terms of the γ matrices as

$$\left[-\mathrm{i}\gamma^{\mu}\partial_{\mu} + \frac{m_e c}{\hbar}\right]\Psi = 0 \tag{5.53}$$

Here, it is important to understand that $\gamma^\mu \partial_\mu$ is a four-component object consisting of (4×4)-matrices rather than a Lorentz scalar, since γ^μ is not a Lorentz 4-vector. Moreover, though the equation already seems to be in covariant form, this still needs to be shown because we do not yet know how Ψ transforms under Lorentz transformations. In the following, we must determine the transformation properties of Ψ so that Eq. (5.53) is covariant under Lorentz transformations, which is a mandatory constraint for any true law of nature.

5.2.4.2 Lorentz Transformation of the Dirac Spinor

We consider two inertial frames of reference IS and IS′ related through a Lorentz transformation (recall section 3.1.3)

$$x' = \Lambda x + a \quad \text{and} \quad x = \Lambda^{-1}(x' - a) \tag{5.54}$$

of the space-time 4-vectors x^μ and x'^μ, where Λ is a Lorentz transformation matrix as given by Eq. (3.14) and a is a constant space-time shift. Because of the fact that the Dirac equation (5.23) [or Eq. (5.49), resp.] is linear in IS and must also be linear in IS′, Ψ and Ψ' must be related through the linear transformation f_Λ otherwise the transformation would create nonlinear components of the state in the new coordinate system. The quantum mechanical state Ψ' in IS′ can then be expressed by the components of Ψ defined in IS,

$$\Psi'(x') = f_\Lambda[\Psi(x)] \overset{(5.54)}{=} f_\Lambda[\Psi(\Lambda^{-1}(x' - a))] \tag{5.55}$$

and

$$\Psi(x) = f_\Lambda^{-1}\Psi'(x') \tag{5.56}$$

The (4×4)-matrix operator f_Λ acts on Dirac 4-spinors and is to be determined. Owing to the operator f_Λ the Lorentz transformation, which relates the coordinates of both inertial frames of reference IS and IS′, mixes the components Ψ_i of Ψ. Eq. (5.55) can be written for each component of the new state vector Ψ' as

$$\Psi'_\mu = \sum_{\nu=1}^{4} f_{\Lambda,\mu\nu}\Psi_\nu \tag{5.57}$$

To demonstrate the Lorentz covariance of Eq. (5.53) we introduce the coordinate tranformation step by step and start in analogy to Eq. (3.41) with the four-component differential operator ∂_μ,

$$\partial_\mu = \frac{\partial}{\partial x^\mu} = \frac{\partial x'^\nu}{\partial x^\mu}\frac{\partial}{\partial x'^\nu} \overset{(3.39)}{=} \Lambda^\nu{}_\mu \partial'_\nu \tag{5.58}$$

which yields, together with Eq. (5.55),

$$\left[-i\gamma^\mu \Lambda^\nu{}_\mu \partial'_\nu + \frac{m_e c}{\hbar}\right] f_\Lambda^{-1}\Psi' = 0 \tag{5.59}$$

After multiplication with the linear transformation matrix f_Λ from the left this then reads

$$\left[-\mathrm{i}\, f_\Lambda \gamma^\mu \Lambda^\nu{}_\mu \partial'_\nu f_\Lambda^{-1} + f_\Lambda \frac{m_e c}{\hbar} f_\Lambda^{-1}\right] \Psi' = 0 \tag{5.60}$$

which becomes

$$\left[-\mathrm{i}\left(f_\Lambda \gamma^\mu \Lambda^\nu{}_\mu f_\Lambda^{-1}\right) \partial'_\nu + \frac{m_e c}{\hbar}\right] \Psi' = 0 \tag{5.61}$$

For the next step it is important to note that the derivation of the explicit form of the entries in γ^μ, i.e., the α^i and β matrices presented in section 5.2.2, would yield the same result in any frame of reference. Consequently, the γ^μ matrices must not change under Lorentz transformations. If we now require that

$$\Lambda^\nu{}_\mu \gamma^\mu = f_\Lambda^{-1} \gamma^\nu f_\Lambda \tag{5.62}$$

which imposes a constraint on the yet unknown linear transformation f_Λ we obtain for the term in parentheses in Eq. (5.61)

$$f_\Lambda \gamma^\mu \Lambda^\nu{}_\mu f_\Lambda^{-1} = f_\Lambda \Lambda^\nu{}_\mu \gamma^\mu f_\Lambda^{-1} \overset{(5.62)}{=} f_\Lambda \left(f_\Lambda^{-1} \gamma^\nu f_\Lambda\right) f_\Lambda^{-1} = \gamma^\nu \tag{5.63}$$

and for the Dirac equation in IS′

$$\left[-\mathrm{i}\,\gamma^\nu \partial'_\nu + \frac{m_e c}{\hbar}\right] \Psi' = 0 \tag{5.64}$$

Thus it has been shown that the Dirac equation for a freely moving electron obviously fulfills the principle of relativity and is Lorentz covariant.

5.2.4.3 **Higher Level of Abstraction and Clifford Algebra**

For the sake of completeness, we should note that Feynman introduced a compact notation for the Dirac equation by defining the scalar product of the four-component vector of all γ matrices with any 4-vector $(a^\mu) = (a^0, \boldsymbol{a})$ by a slash through the symbol for this vector,

$$\not{a} = \gamma \cdot a = \gamma^\mu a_\mu = \gamma_\mu a^\mu = \gamma^0 a^0 - \boldsymbol{\gamma} \cdot \boldsymbol{a} \tag{5.65}$$

It is common practice to obtain the seemingly covariant components of the γ matrices from the seemingly contravariant ones by multiplication with the metric tensor of Eq. (3.8) as

$$\gamma_\nu = g_{\nu\mu} \gamma^\mu \tag{5.66}$$

We stress again that γ^μ is not a Lorentz 4-vector but a Lorentz scalar since it has exactly the same values in all frames of reference; recall the remarks at the beginning of this section. We may write Dirac's Eq. (5.53) in Feynman's slash notation as

$$\left[-\mathrm{i}\partial\!\!\!/ + \frac{m_e c}{\hbar}\right]\Psi = 0 \tag{5.67}$$

The basic algebraic structure of the Dirac matrices given in Eq. (5.52) reads

$$\gamma^\mu\gamma^\nu + \gamma^\nu\gamma^\mu = 2g^{\mu\nu}\,\mathbf{1}_4 \tag{5.68}$$

which defines a Clifford algebra. Other representations of the Dirac equation — like the Weyl representation in Eq. (5.38) — can be obtained through a similarity transformation,

$$\tilde{\gamma}^\mu = M\gamma^\mu M^{-1} \tag{5.69}$$

with M being an arbitrary non-singular matrix.

5.3 Solution of the Free-Electron Dirac Equation

5.3.1 Particle at Rest

After having derived a truly relativistic quantum mechanical equation for a freely moving electron (i.e., in the absence of external electromagnetic fields), we now derive its solutions. It is noteworthy from a conceptual point of view that the solution of the field-free Dirac equation can in principle be pursued in two ways: (i) one could directly obtain the solution from the (full) Dirac equation (5.23) for the electron moving with constant velocity $\boldsymbol{v}$ or (ii) one could aim for the solution for an electron at rest — which is particularly easy to obtain — and then Lorentz transform the solution according to Eq. (5.55) to an inertial frame of reference which moves with constant velocity $(-\boldsymbol{v})$ with respect to the frame of reference that observes the electron at rest.

Since the existence of these options is indeed conceptually important, we shall introduce the solutions for a particle at rest first, in order to understand how the Dirac equation simplifies in such a frame of reference. In a frame of reference in which the electron is observed at rest, no kinetic energy operators contribute to the energy expectation value and the spatial derivatives (corresponding to the linear 3-momentum operator $\boldsymbol{p}$) can thus be neglected. The only term which survives in the Hamiltonian H^D of Eq. (5.23) is the rest energy term proportional to $m_e c^2$. The Dirac equation (5.23) then reads

$$\mathrm{i}\hbar\frac{\partial}{\partial t}\Psi = \left[\beta m_e c^2\right]\Psi \tag{5.70}$$

Its four solutions are

$$\Psi_1^{(+)} = \exp\left(-\mathrm{i}\frac{m_e c^2}{\hbar}t\right)\begin{pmatrix}1\\0\\0\\0\end{pmatrix}, \quad \Psi_2^{(+)} = \exp\left(-\mathrm{i}\frac{m_e c^2}{\hbar}t\right)\begin{pmatrix}0\\1\\0\\0\end{pmatrix} \tag{5.71}$$

$$\Psi_1^{(-)} = \exp\left(\mathrm{i}\frac{m_e c^2}{\hbar}t\right)\begin{pmatrix}0\\0\\1\\0\end{pmatrix}, \quad \Psi_2^{(-)} = \exp\left(\mathrm{i}\frac{m_e c^2}{\hbar}t\right)\begin{pmatrix}0\\0\\0\\1\end{pmatrix} \tag{5.72}$$

as can easily be verified by insertion into Eq. (5.70). Note that the β matrix entries control the sign on the right hand side of Eq. (5.70) in the case of a negative coefficient in the exponential. The solutions in Eqs. (5.71) and (5.72) are normalized to one.

If we consider the eigenvalue equation of the Dirac Hamiltonian, we may write the energy expectation value for a particle at rest evaluated with the eigensolutions from Eq. (5.71),

$$\begin{aligned} E^{(+)} &= \langle \Psi_{1,2}^{(+)} | H^D | \Psi_{1,2}^{(+)} \rangle \\ &= \exp\left(\mathrm{i}\frac{m_e c^2}{\hbar}t\right)\left[m_e c^2\right]\exp\left(-\mathrm{i}\frac{m_e c^2}{\hbar}t\right) = m_e c^2 \end{aligned} \tag{5.73}$$

and evaluated with the eigensolutions in Eq. (5.72),

$$\begin{aligned} E^{(-)} &= \langle \Psi_{1,2}^{(-)} | H^D | \Psi_{1,2}^{(-)} \rangle \\ &= \exp\left(-\mathrm{i}\frac{m_e c^2}{\hbar}t\right)\left[-m_e c^2\right]\exp\left(\mathrm{i}\frac{m_e c^2}{\hbar}t\right) = -m_e c^2 \end{aligned} \tag{5.74}$$

Note that the spatial integration has been omitted here since it reduces to unity by suitable normalization of the plane wave spinors (cf. box normalization in nonrelativistic quantum mechanics).

We understand that Dirac's 4×4 β matrix divides the four solutions of Eqs. (5.71) and (5.72) into two classes. One class contains two eigenvectors which describe a spin-1/2-particle with a positive rest energy of $m_e c^2$ as one would have expected based on the derivation of the Dirac equation presented here. The second set of solutions in Eq. (5.72) describes a spin-1/2-particle with negative energy $-m_e c^2$. While the positive energy of the upper two solutions are physically meaningful, the negative energy of the solutions in Eq. (5.72) requires an explanation, which we will consider in section 5.5 after having derived the solutions for freely *moving* particles. Recall that we have encountered negative energies already in the context of the Klein–Gordon equation and that the Dirac equation was constructed so as to reproduce the results of the Klein–Gordon equation for freely moving particles and hence also for particles at rest. The fact that the β matrix distinguishes between positive and negative energy solutions in the case of block-diagonal Hamiltonians, i.e. those without contributions on the off-(block)diagonal places, will be met again in chapter 12 when we study such operators in the presence of external electrostatic potentials.

5.3.2 Freely Moving Particle

We now proceed to derive the Dirac states for a freely moving electron of mass m_e. Note that the charge of the fermion does not enter the Dirac equation for this fermion being at rest or moving freely with constant velocity $\boldsymbol{v}$. The solutions in Eqs. (5.71) and (5.72) may now be subjected to a general Lorentz boost as given by Eq. (3.81) into an inertial frame of reference moving relatively to the previous one, in which the fermion is at rest, with velocity $(-\boldsymbol{v})$. This option, namely that the solutions for a complicated kinematic problem can be obtained from those of a simple kinematic problem in a suitably chosen frame of reference by a Lorentz transformation, cannot be overemphasized from a conceptual point of view. However, instead of this Lorentz transformation a direct solution of the Dirac Eq. (5.23) is easier. For this purpose we choose an ansatz of plane waves,

$$\Psi(x) = u(\boldsymbol{p}) \exp\left[-\mathrm{i}\,\frac{p \cdot x}{\hbar}\right] \tag{5.75}$$

with the scalar product for 4-vectors, p and x, given by

$$p \cdot x = Et - \boldsymbol{p} \cdot \boldsymbol{r} \tag{5.76}$$

as defined in Eq. (3.115). Note that the $\boldsymbol{p}$ in this case denotes the momentum eigenvalue and not the momentum operator. In section 5.2.2 the standard representation of the 4×4 Dirac matrices $\boldsymbol{\alpha} = (\alpha_1, \alpha_2, \alpha_3) = (\alpha_x, \alpha_y, \alpha_z)$ was derived, which allows us to write Dirac's equation Eq. (5.23) for a freely moving fermion in compact notation as

$$[c\,\boldsymbol{\alpha} \cdot \hat{\boldsymbol{p}} + \beta m_e c^2]\Psi \;=\; \mathrm{i}\hbar\frac{\partial}{\partial t}\Psi \tag{5.77}$$

The eigenstate Ψ necessarily contains four components and is called a 4-spinor or simply a spinor. Due to the block structure of the matrices α^i in Eqs. (5.37), the spinor is often split into an upper and a lower 2-spinor, Ψ^L and Ψ^S,

$$\Psi \;=\; \begin{pmatrix} \Psi_1 \\ \Psi_2 \\ \Psi_3 \\ \Psi_4 \end{pmatrix} \;=\; \begin{pmatrix} \Psi^L \\ \Psi^S \end{pmatrix} \tag{5.78}$$

(where L denotes "large" and S "small", respectively, for reasons explained at a later stage). Then, the Dirac equation may be cast in split notation,

$$c\,\boldsymbol{\sigma}\cdot\hat{\boldsymbol{p}}\,\Psi^S + m_e c^2 \Psi^L = \mathrm{i}\hbar\frac{\partial}{\partial t}\Psi^L \tag{5.79}$$

$$c\,\boldsymbol{\sigma}\cdot\hat{\boldsymbol{p}}\,\Psi^L - m_e c^2 \Psi^S = \mathrm{i}\hbar\frac{\partial}{\partial t}\Psi^S \tag{5.80}$$

For the sake of completeness, we shall also present the Dirac equation explicitly in the form of four coupled first-order partial differential equations,

$$c\hat{p}_z\Psi_3 + c(\hat{p}_x - \mathrm{i}\,\hat{p}_y)\Psi_4 + m_e c^2\Psi_1 = \mathrm{i}\hbar\frac{\partial}{\partial t}\Psi_1 \tag{5.81}$$

$$c(\hat{p}_x + \mathrm{i}\,\hat{p}_y)\Psi_3 - c\hat{p}_z\Psi_4 + m_e c^2\Psi_2 = \mathrm{i}\hbar\frac{\partial}{\partial t}\Psi_2 \tag{5.82}$$

$$c\hat{p}_z\Psi_1 + c(\hat{p}_x - \mathrm{i}\,\hat{p}_y)\Psi_2 - m_e c^2\Psi_3 = \mathrm{i}\hbar\frac{\partial}{\partial t}\Psi_3 \tag{5.83}$$

$$c(\hat{p}_x + \mathrm{i}\,\hat{p}_y)\Psi_1 - c\hat{p}_z\Psi_2 - m_e c^2\Psi_4 = \mathrm{i}\hbar\frac{\partial}{\partial t}\Psi_4 \tag{5.84}$$

For the solution of these equations, we now determine $u(\boldsymbol{p})$ in Eq. (5.75). By comparison with Eqs. (5.71) and (5.72) we note that $u(\boldsymbol{p})$ is a 4-vector and, in particular, $u(\boldsymbol{p} = \boldsymbol{0})$ is identical to either of the four 4-vectors in those equations. In the general case of nonvanishing momentum we obtain the solution for $u(\boldsymbol{p})$ by insertion of the ansatz of Eq. (5.75) into Eq. (5.77),

$$[c\,\boldsymbol{\alpha}\cdot\hat{\boldsymbol{p}} + \beta m_e c^2]u(\boldsymbol{p})\exp\left[-\mathrm{i}\frac{p\cdot x}{\hbar}\right] = \mathrm{i}\hbar\frac{\partial}{\partial t}u(\boldsymbol{p})\exp\left[-\mathrm{i}\frac{p\cdot x}{\hbar}\right] \tag{5.85}$$

After multiplication of Eq. (5.85) by $\exp\left[+\mathrm{i}\,(p\cdot x)/\hbar\right]$ from the left we obtain

$$\begin{aligned} c\,\exp\left[+\mathrm{i}\frac{p\cdot x}{\hbar}\right]\boldsymbol{\alpha}\cdot\hat{\boldsymbol{p}}\exp\left[-\mathrm{i}\frac{p\cdot x}{\hbar}\right]u(\boldsymbol{p}) + \beta m_e c^2 u(\boldsymbol{p}) \\ = \mathrm{i}\hbar\exp\left[+\mathrm{i}\frac{p\cdot x}{\hbar}\right]\frac{\partial}{\partial t}u(\boldsymbol{p})\exp\left[-\mathrm{i}\frac{p\cdot x}{\hbar}\right] \end{aligned} \tag{5.86}$$

Recalling Eq. (5.76) allows us to write

$$c\,\exp\left[+\mathrm{i}\frac{p\cdot x}{\hbar}\right]\boldsymbol{\alpha}\cdot\hat{\boldsymbol{p}}\,\exp\left[-\mathrm{i}\frac{p\cdot x}{\hbar}\right] = c\,\boldsymbol{\alpha}\cdot\boldsymbol{p} \tag{5.87}$$

since we have for any direction i

$$\hat{p}_i\,\exp\left[-\mathrm{i}\frac{p\cdot x}{\hbar}\right] = \mathrm{i}\hbar\frac{\partial}{\partial x_i}\exp\left[-\mathrm{i}\frac{p\cdot x}{\hbar}\right] = p_i\exp\left[-\mathrm{i}\frac{p\cdot x}{\hbar}\right] \tag{5.88}$$

For the right hand side of Eq. (5.86) we have

$$\mathrm{i}\hbar\,\exp\left[+\mathrm{i}\frac{p\cdot x}{\hbar}\right]\frac{\partial}{\partial t}\exp\left[-\mathrm{i}\frac{p\cdot x}{\hbar}\right] = \mathrm{i}\hbar\left[-\frac{\mathrm{i}}{\hbar}E\right] = E \tag{5.89}$$

so that we finally obtain

$$\left[c\,\boldsymbol{\alpha}\cdot\boldsymbol{p}+\beta m_e c^2\right]u(\boldsymbol{p}) \;=\; Eu(\boldsymbol{p}) \tag{5.90}$$

where we have utilized the fact that the partial derivative of $u(\boldsymbol{p})$ with respect to spatial coordinates is zero, $\hat{\boldsymbol{p}}u(\boldsymbol{p}) = -i\hbar\nabla u(\boldsymbol{p}) = 0$. It must be emphasized that $\boldsymbol{p}$ in Eq. (5.90) denotes the eigenvalue of the momentum 3-vector of Eq. (5.77) — in order to highlight this difference, the momentum operator in Eqs. (5.77) and (5.87) is explicitly marked by a hat.

We can write Eq. (5.90) in split notation as

$$\begin{pmatrix} m_e c^2 - E & c\,\boldsymbol{\sigma}\cdot\boldsymbol{p} \\ c\,\boldsymbol{\sigma}\cdot\boldsymbol{p} & -m_e c^2 - E \end{pmatrix}\begin{pmatrix} u^L(\boldsymbol{p}) \\ u^S(\boldsymbol{p}) \end{pmatrix} = \begin{pmatrix} 0 \\ 0 \end{pmatrix} \tag{5.91}$$

which yields relations between upper and lower components,

$$u^L = \left[\frac{c\,\boldsymbol{\sigma}\cdot\boldsymbol{p}}{E-m_e c^2}\right]u^S \tag{5.92}$$

$$u^S = \left[\frac{c\,\boldsymbol{\sigma}\cdot\boldsymbol{p}}{E+m_e c^2}\right]u^L \tag{5.93}$$

If we insert the upper Eq. (5.92) into the lower Eq. (5.93),

$$u^S = \left[\frac{c\,\boldsymbol{\sigma}\cdot\boldsymbol{p}}{E+m_e c^2}\right]\left[\frac{c\,\boldsymbol{\sigma}\cdot\boldsymbol{p}}{E-m_e c^2}\right]u^S \tag{5.94}$$

we need to postulate that

$$\left[\frac{c\,\boldsymbol{\sigma}\cdot\boldsymbol{p}}{E+m_e c^2}\right]\left[\frac{c\,\boldsymbol{\sigma}\cdot\boldsymbol{p}}{E-m_e c^2}\right] = 1 \tag{5.95}$$

in order to guarantee consistency. The latter equation, however, allows us to determine the energy E to be,

$$E = \pm\sqrt{c^2\boldsymbol{p}^2 + m_e^2 c^4} \tag{5.96}$$

which is also consistent with what has already been said regarding the free-particle Klein–Gordon equation. We may now choose the ansätze

$$u_1^L = \begin{pmatrix} 1 \\ 0 \end{pmatrix} \quad \text{and} \quad u_2^L = \begin{pmatrix} 0 \\ 1 \end{pmatrix} \tag{5.97}$$

and obtain then u^S from Eq. (5.93), where the scalar product of the Pauli matrices and the vector of momentum eigenvalues is

$$\boldsymbol{\sigma}\cdot\boldsymbol{p} = \begin{pmatrix} p_z & p_x - ip_y \\ p_x + ip_y & -p_z \end{pmatrix} \tag{5.98}$$

so that we may write the resulting 4-spinors $u(\boldsymbol{p})$ as

$$u_1^{(+)}(\boldsymbol{p}) = \mathcal{N} \begin{pmatrix} 1 \\ 0 \\ \dfrac{cp_z}{E+m_ec^2} \\ \dfrac{c(p_x+\mathrm{i}p_y)}{E+m_ec^2} \end{pmatrix}, \quad u_2^{(+)}(\boldsymbol{p}) = \mathcal{N} \begin{pmatrix} 0 \\ 1 \\ \dfrac{c(p_x-\mathrm{i}p_y)}{E+m_ec^2} \\ \dfrac{-cp_z}{E+m_ec^2} \end{pmatrix} \tag{5.99}$$

If, however, we choose u^S to be

$$u_1^S = \begin{pmatrix} 1 \\ 0 \end{pmatrix} \quad \text{and} \quad u_2^S = \begin{pmatrix} 0 \\ 1 \end{pmatrix} \tag{5.100}$$

and then obtain u^L from Eq. (5.92), so that we may write the resulting 4-spinors $u(\boldsymbol{p})$ as

$$u_1^{(-)}(\boldsymbol{p}) = \mathcal{N} \begin{pmatrix} \dfrac{cp_z}{E-m_ec^2} \\ \dfrac{c(p_x+\mathrm{i}p_y)}{E-m_ec^2} \\ 1 \\ 0 \end{pmatrix}, \quad u_2^{(-)}(\boldsymbol{p}) = \mathcal{N} \begin{pmatrix} \dfrac{c(p_x-\mathrm{i}p_y)}{E-m_ec^2} \\ \dfrac{-cp_z}{E-m_ec^2} \\ 0 \\ 1 \end{pmatrix} \tag{5.101}$$

corresponding to the negative energy eigenvalue $E = -|E|$ of Eq. (5.96). The normalization constant $\mathcal{N}$ can be chosen such that

$$u_i^{(\pm),\dagger}(\boldsymbol{p}) \cdot u_j^{(\pm)}(\boldsymbol{p}) = \delta_{ij} \tag{5.102}$$

$$u_i^{(\pm),\dagger}(\boldsymbol{p}) \cdot u_j^{(\mp)}(\boldsymbol{p}) = 0 \tag{5.103}$$

$\forall i, j \in \{1, 2\}$. For i=j, Eq. (5.102) reads for any of the free-particle spinors given above,

$$u_i^{(\pm),\dagger} \cdot u_i^{(\pm)} = \mathcal{N}^2 \left[1 + \frac{c^2\boldsymbol{p}^2}{(E \pm m_ec^2)^2} \right] \stackrel{!}{=} 1 \tag{5.104}$$

which yields for the squared normalization constant,

$$\begin{aligned} \mathcal{N}^2 &= \frac{1}{1+\dfrac{c^2\boldsymbol{p}^2}{(E\pm m_ec^2)^2}} = \frac{(E\pm m_ec^2)^2}{(E\pm m_ec^2)^2 + c^2\boldsymbol{p}^2} \\ &\stackrel{(5.96)}{=} \frac{(E\pm m_ec^2)^2}{(E\pm m_ec^2)^2 + E^2 - m_e^2c^4} = \frac{(E\pm m_ec^2)^2}{(E\pm m_ec^2)^2 + (E+m_ec^2)(E-m_ec^2)} \\ &= \frac{(E\pm m_ec^2)}{(E\pm m_ec^2)+(E\mp m_ec^2)} = \frac{E\pm m_ec^2}{2E} \end{aligned} \tag{5.105}$$

where we note that both the counter and the denominator of the fraction either carry a global (+) or a global (−) sign for the $u_i^{(+)}$ and $u_i^{(-)}$ solutions, respectively, so that the normalization constant is the same for all four spinors $u_i^{(\pm)}(\boldsymbol{p})$,

$$\mathcal{N} = \sqrt{\frac{|E| + m_e c^2}{2|E|}} \tag{5.106}$$

The four solutions belong to two different kinds of eigenvalues. While $u_1^{(+)}$ and $u_2^{(+)}$ times the exponential yield positive eigenvalues, the solutions $u_1^{(-)}$ and $u_2^{(-)}$ times the exponential possess negative energy eigenvalues. One may verify by direct substitution that these four spinors satisfy the free-particle Dirac equation and that they are orthogonal to one another.

Since the positive- and negative-energy solutions for all possible $\boldsymbol{p}$ form a complete orthonormal basis set, we may write the most general free-particle wave function as a superposition of the basis spinors,

$$\begin{aligned} \Psi(x) = \Psi(\boldsymbol{r}, t) &= \mathcal{N} \left(\sum_{\boldsymbol{p}} \sum_{\nu=1,2} c_{\boldsymbol{p},\nu}^{(+)} u_\nu^{(+)}(\boldsymbol{p}) \exp\left[\mathrm{i} \frac{\boldsymbol{p} \cdot \boldsymbol{r} - |E| t}{\hbar} \right] \right. \\ &\left. - \sum_{\boldsymbol{p}} \sum_{\nu=1,2} c_{\boldsymbol{p},\nu}^{(-)} u_\nu^{(-)}(\boldsymbol{p}) \exp\left[\mathrm{i} \frac{\boldsymbol{p} \cdot \boldsymbol{r} + |E| t}{\hbar} \right] \right) \end{aligned} \tag{5.107}$$

with the expansion coefficients $c_{\boldsymbol{p},\nu}^{(\pm)}$ for this wave packet to be determined from a Fourier expansion of Ψ at $t = 0$.

5.3.3 The Dirac Velocity Operator

For later convenience, it is instructive to investigate the relativistic definition of the velocity operator in j direction as defined by the Heisenberg picture in Eq. (4.44),

$$\dot{x}^k = \frac{\mathrm{i}}{\hbar} \left[H^D, x^k \right] \tag{5.108}$$

For our freely moving Dirac electron, the commutator reads,

$$\left[H^D, x^k \right] = \left[c\boldsymbol{\alpha} \cdot \boldsymbol{p}, x^k \right] \stackrel{(4.56)}{=} \left[c\alpha^j p_j, x^k \right] = c\alpha^j \underbrace{\left[p_j, x^k \right]}_{-\mathrm{i}\hbar \delta^k{}_j} = -\mathrm{i}\hbar c \alpha^k \tag{5.109}$$

which then yields

$$\dot{x}^k = c\alpha^k \tag{5.110}$$

(compare the definition of the nonrelativistic velocity operator in section 4.3.7). The k-th component of $\boldsymbol{\alpha}$ scaled by the speed of light c yields the velocity operator $\dot{x}^k$ in k-th direction. Since the direction k has been picked arbitrarily, the vector $\boldsymbol{\alpha}$ of Dirac matrices can be identified with the velocity operator $\dot{\boldsymbol{r}}$ if it is measured in units of c. We may anticipate here that this result also holds in the case of a system of many electrons discussed later in chapter 8 because all additional operators that are introduced into the many-electron Hamiltonian commute with the position operator x^k of a given electron except for those containing its conjugate momentum component p_k, which is for the electron under consideration the same kinetic energy term that survives in Eq. (5.109). In section 5.4 we will meet a different approach to 'deduce' the very same velocity operator on the basis of the correspondence principle.

The fact that we have just identified the Dirac matrices $\boldsymbol{\alpha}$ as being proportional to the velocity operator leads, however, to a couple of paradoxes [71]:
a) Since the eigenvalues of α^k are either $+1$ or -1, the velocity can take only values of $\pm c$ although the electron is not massless.
b) Furthermore, the measurement of the velocity in two different directions k and l is incompatible since the Dirac matrices for these directions do not commute, $[\alpha^k, \alpha^l] \neq 0$, although the momenta p_k and p_l do commute and can hence be measured with arbitrary precision simultaneously.
c) The velocity operator can be chosen not to be a constant of the motion although we are studying a freely moving particle.
d) Since the Dirac Hamiltonian H^D and any component α^k of $\boldsymbol{\alpha}$ do not commute, the energy eigenfunctions are not simultaneous eigenfunctions of the velocity operator. The plane-wave solutions of the freely moving particle become only eigenfunctions of α^k in the case of massless particles which can be easily verified.
e) If we compute the expectation value of the velocity operator in k-direction,

$$\langle \alpha^k \rangle = \int d^3r \Psi^\dagger(\boldsymbol{r},t) \alpha^k \Psi(\boldsymbol{r},t) \tag{5.111}$$

we obtain with the most general free-particle solution for a wave packet given in Eq. (5.107) of the last section,

$$\begin{aligned} \langle \alpha^k \rangle \;=\; & \sum_{\boldsymbol{p}} \sum_{\nu=1,2} \frac{c p_k}{|E|} \left(|c^{(+)}_{\boldsymbol{p},\nu}|^2 - |c^{(-)}_{\boldsymbol{p},\nu}|^2 \right) - \sum_{\boldsymbol{p}} \sum_{\nu,\mu=1,2} \frac{m_e c^2}{|E|} \\ & \times \left\{ c^{(-),\star}_{\boldsymbol{p},\mu} c^{(+)}_{\boldsymbol{p},\nu} u^{(-),\dagger}_{\mu}(\boldsymbol{p})\, \alpha^k u^{(+)}_{\nu}(\boldsymbol{p}) \exp\left[-2\mathrm{i}\frac{Et}{\hbar}\right] \right. \\ & \left. - c^{(+),\star}_{\boldsymbol{p},\nu} c^{(-)}_{\boldsymbol{p},\mu} u^{(+),\dagger}_{\nu}(\boldsymbol{p})\, \alpha^k u^{(-)}_{\mu}(\boldsymbol{p}) \exp\left[2\mathrm{i}\frac{Et}{\hbar}\right] \right\} \end{aligned} \tag{5.112}$$

We see that the first two (time-independent) terms represent the group velocity of the wave packet constructed by the positive- and by the negative-energy solutions, respectively. The third and fourth terms, however, are time-dependent and represent rapid oscillations. A similar observation can be made for the expectation value of the position operator x^k. The oscillatory behavior is called *Zitterbewegung* and is solely due to the interference of positive- and negative-energy solutions as is evident from the last two terms of the right hand side of Eq. (5.112). Hence, the Zitterbewegung vanishes if the wave packet has contributions either solely from the positive-energy or solely from the negative-energy solutions.

5.4
Dirac Electron in External Electromagnetic Potentials

The covariant form of the Dirac equation of a freely moving particle, Eq. (5.53), allows us to incorporate arbitrary external electromagnetic fields. These fields can then be used to describe the interaction of electrons with light. But it must be noted that the treatment of light is then purely classical. A fully quantized description of light and matter on an equal footing is introduced only in quantum electrodynamics and discussed in chapter 7.

The only guiding principle for the derivation of the field-dependent Dirac equation for our single electron interacting with external electromagnetic fields is that the resulting Dirac equation is still Lorentz covariant. But this is easily achieved if the Dirac equation in covariant form is used as a starting point. It is this form which allows us to introduce additional fields and fix possible phases solely on the basis of the symmetry principle, namely that the form of the equation has to be conserved under Lorentz transformations. The standard way which preserves Lorentz covariance is the minimal coupling procedure, which we have already encountered in the classical context in section 3.4.3.2. In the following we shall introduce the minimal coupling first and then consider the question whether this approach is unique.

For the Dirac equation with external electromagnetic fields in covariant form it is appropriate to define suitable 4-quantities. In this way, time and spatial coordinates are re-unified for the relativistic space-time framework. The components of the (linear) 4-momentum read

$$p^0 = p_0 = i\hbar\frac{\partial}{\partial(ct)} \quad \text{and} \quad p^i = -p_i = i\hbar\frac{\partial}{\partial x_i} \quad \text{with } i \in \{1,2,3\} \tag{5.113}$$

and those of the 4-potential are given by $A^\mu = (\phi, \boldsymbol{A})$. Note that the vector potential $\boldsymbol{A} = (A^1, A^2, A^3)$ contains the contravariant components of the 4-potential. According to Eq. (5.53) minimal coupling then requires the following substitution

$$i\hbar\partial_\mu \longrightarrow i\hbar\partial_\mu - \frac{q_e}{c}A_\mu \tag{5.114}$$

which may also be written in time-like and space-like components as

$$i\hbar\frac{\partial}{\partial t} \longrightarrow i\hbar\frac{\partial}{\partial t} - q_e\phi \tag{5.115}$$

$$i\hbar\frac{\partial}{\partial x^i} \longrightarrow i\hbar\frac{\partial}{\partial x^i} - \frac{q_e}{c}A_i \quad \text{with } i \in \{1,2,3\} \tag{5.116}$$

Now, we have a unified substitution pattern at hand, which also comprises the time-like coordinates. Substitution of Eq. (5.114) in the field-free Dirac equation as written in Eq. (5.53) yields the covariant form of the Dirac equation with external electromagnetic fields,

$$\left[-\gamma^{\mu}\left(i\hbar\partial_{\mu} - \frac{q_e}{c}A_{\mu}\right) + m_e c\right]\Psi = 0 \tag{5.117}$$

This equation can be transformed to a more familiar form by separating spatial and time coordinates

$$\gamma^0\left(i\hbar\partial_0 - \frac{q_e}{c}A_0\right)\Psi = \left[-\gamma^i\left(i\hbar\partial_i - \frac{q_e}{c}A_i\right) + m_e c\right]\Psi \tag{5.118}$$

Multiplying this result by γ^0 given by Eq. (5.50) yields

$$i\hbar\frac{\partial}{\partial t}\Psi = \left[c\,\boldsymbol{\alpha}\cdot\left(\boldsymbol{p} - \frac{q_e}{c}\boldsymbol{A}\right) + \beta m_e c^2 + q_e\phi\right]\Psi \tag{5.119}$$

where we emphasize again that $\boldsymbol{A} = (A^1, A^2, A^3)$ contains the contravariant components of the vector potential $A^i = -A_i$ so that the sign in front of the $q_e/c\boldsymbol{A}$ term had to be changed.

We thus understand that the structure of the product $\gamma^0\gamma^i = \alpha^i$ determines where the components of the vector potential A^i enter the Dirac Hamiltonian. This, of course, depends on the representation of the Dirac matrices γ^{μ} chosen in section 5.2.2. However, the *scalar* potential $A_0 = \phi$ always enters the Dirac Hamiltonian on the diagonal, because of $\gamma^0\gamma^0 = \mathbf{1}_4$ by virtue of Eq. (5.51) in any representation. Apart from Lorentz covariance being obviously preserved, invariance under gauge transformations,

$$\Psi(x) \longrightarrow \exp\left[-i\frac{q_e}{\hbar c}\chi(x)\right]\Psi(x) \tag{5.120}$$

$$A_{\mu}(x) \longrightarrow A_{\mu}(x) + \partial_{\mu}\chi(x) \tag{5.121}$$

is also preserved by minimal coupling [with $x = (ct, \boldsymbol{r})$].

5.4.1 Kinematic Momentum

The connection of the canonical momentum operator $\boldsymbol{p}$ with the effect of external vector potentials $\boldsymbol{A}$ on a moving electron with charge $q_e = -e$ is often simply written as

$$\boldsymbol{p} \longrightarrow \boldsymbol{p} - \frac{q_e}{c}\boldsymbol{A} = \boldsymbol{p} + \frac{e}{c}\boldsymbol{A} \equiv \boldsymbol{\pi} \tag{5.122}$$

where we introduced the mechanical-momentum operator $\boldsymbol{\pi}$ (also called kinematic or linear momentum in section 2.4.6). Then, the familiar shorthand notation for the Dirac electron in external electromagnetic fields gives

$$i\hbar\frac{\partial}{\partial t}\Psi = \left[c\,\boldsymbol{\alpha}\cdot\boldsymbol{\pi} + \beta m_e c^2 - e\phi\right]\Psi \tag{5.123}$$

It is important to note that the kinematic momentum components do not commute,

$$[\pi_i, \pi_j] = \frac{q}{c}i\hbar\sum_{k=1}^{3}\varepsilon_{ijk}B_k \tag{5.124}$$

in analogy to the classical Poisson bracket given in Eq. (2.162). Only in the case of a vanishing vector potential are kinematic and canonical momenta identical; compare Eq. (4.56).

5.4.2
Electromagnetic Interaction Energy Operator

Finally, we may note that the Dirac Eq. (5.119) can be re-arranged to yield an expression for the interaction energy operator $\hat{V}$,

$$i\hbar\frac{\partial}{\partial t}\Psi = \left[c\,\boldsymbol{\alpha}\cdot\boldsymbol{p} + \beta m_e c^2 + \underbrace{q_e\phi - q_e\,\boldsymbol{\alpha}\cdot\boldsymbol{A}}_{\equiv\hat{V}}\right]\Psi \tag{5.125}$$

by comparison with the classical expression for the interaction energy for a particle moving in an electromagnetic field from Eq. (2.157),

$$V = q\phi - \frac{q}{c}\dot{\boldsymbol{r}}\cdot\boldsymbol{A} \tag{5.126}$$

we see that they are equivalent in the sense of the correspondence principle if we have the following substitution for the classical velocity,

$$\dot{\boldsymbol{r}} \stackrel{\text{c.p.}}{\longrightarrow} c\boldsymbol{\alpha} \tag{5.127}$$

This is exactly the expression for the Dirac velocity operator which we derived in section 5.3.3 from the Heisenberg equation of motion. In section 8.1, this result for the electromagnetic interaction energy will become important for the interpretation of the relativistic electron–electron interaction.

5.4.3
Nonrelativistic Limit and Pauli Equation

In this section, we derive the two-component Pauli equation from the Dirac equation in external electromagnetic fields. It is also desirable to recover the

Schrödinger equation in order to see the connection between relativistic theory and nonrelativistic quantum mechanics. For this purpose, we rewrite Eq. (5.123) in split notation

$$i\hbar\frac{\partial}{\partial t}\begin{pmatrix}\Psi^L\\ \Psi^S\end{pmatrix} = c\begin{pmatrix}(\sigma\cdot\pi)\Psi^S\\ (\sigma\cdot\pi)\Psi^L\end{pmatrix} + m_e c^2\begin{pmatrix}\Psi^L\\ -\Psi^S\end{pmatrix} + V\begin{pmatrix}\Psi^L\\ \Psi^S\end{pmatrix} \tag{5.128}$$

Since the lowest possible nonrelativistic energy of a free particle in Schrödinger quantum mechanics is zero instead of $+m_e c^2$ as in Dirac's theory, we need to shift the origin of the energy scale by $-m_e c^2$. Formally this corresponds to replacing V by $V - m_e c^2$,

$$i\hbar\frac{\partial}{\partial t}\begin{pmatrix}\Psi^L\\ \Psi^S\end{pmatrix} = c\begin{pmatrix}(\sigma\cdot\pi)\Psi^S\\ (\sigma\cdot\pi)\Psi^L\end{pmatrix} - 2m_e c^2\begin{pmatrix}0\\ \Psi^S\end{pmatrix} + V\begin{pmatrix}\Psi^L\\ \Psi^S\end{pmatrix} \tag{5.129}$$

We now focus on the small component and write the lower part of Eq. (5.129) as

$$\left(i\hbar\frac{\partial}{\partial t} + 2m_e c^2 - V\right)\Psi^S = c(\sigma\cdot\pi)\Psi^L \tag{5.130}$$

For nonrelativistic energies, the energy $E \rightarrow i\hbar\partial/\partial t$ and the potential V are small compared to the rest energy $m_e c^2$ so that we may approximate Ψ^S by

$$\Psi^S \approx \frac{\sigma\cdot\pi}{2m_e c}\Psi^L \tag{5.131}$$

Eq. (5.131) will become important in chapter 10 as the so-called *kinetic-balance* condition. It shows that the lower component of the spinor Ψ^S is by a factor of $1/c$ smaller than Ψ^L (for small linear momenta), which is the reason why Ψ^L is also called the *large component* and Ψ^S the *small component*. In the limit $c \rightarrow \infty$, the small component vanishes, $\Psi^S(\boldsymbol{r},t) = 0$.

Inserting the kinetic-balance condition Eq. (5.131) into the remaining upper component of Eq. (5.129),

$$i\hbar\frac{\partial}{\partial t}\Psi^L = \frac{(\sigma\cdot\pi)(\sigma\cdot\pi)}{2m_e}\Psi^L + V\Psi^L \tag{5.132}$$

and utilizing Dirac's relation derived in appendix D.2,

$$(\sigma\cdot\pi)(\sigma\cdot\pi) = \pi^2 - \frac{q_e\hbar}{c}\sigma\cdot\boldsymbol{B} \tag{5.133}$$

finally yields

$$\frac{1}{2m_e}\left(\boldsymbol{p} - \frac{q_e}{c}\boldsymbol{A}\right)^2\Psi^L - \frac{q_e\hbar}{2m_e c}(\sigma\cdot\boldsymbol{B})\Psi^L + V\Psi^L = i\hbar\frac{\partial}{\partial t}\Psi^L \tag{5.134}$$

which is the Pauli equation known from nonrelativistic quantum mechanics. If we rewrite

$$\begin{aligned}\left(p - \frac{q_e}{c}A\right)^2 &= p^2 - \frac{q_e}{c}(p \cdot A + A \cdot p) + \frac{q_e^2}{c^2}A^2 \\ &= p^2 - 2\frac{q_e}{c}p \cdot A + \frac{q_e^2}{c^2}A^2 \end{aligned} \tag{5.135}$$

where we utilized $A \cdot p = p \cdot A - (p \cdot A)$ and the fact that the divergence of the vector potential vanishes in Coulomb gauge $(p \cdot A) = -i\hbar(\nabla \cdot A) = -i\hbar(\text{div}A) = 0$, see Eq. (2.131). Note that the term linear in the momentum operator, $(2qe_e/c)p \cdot A$, is *imaginary*. The mixed term can now be expressed in terms of the angular momentum coupled to a constant and homogeneous magnetic field B=curl$A$,

$$2p \cdot A \overset{(2.156)}{=} p \cdot (B \times r) = (r \times p) \cdot B \overset{(4.87)}{=} l \cdot B \tag{5.136}$$

and we can rewrite the Pauli equation with the definition of the spin operator, Eq. (4.139), as

$$\left[\frac{p^2}{2m_e} + \frac{q_e^2 A^2}{2m_e c^2} \underbrace{- \frac{q_e}{2m_e c}(l + 2s)}_{\equiv \mu} \cdot B + V\right]\Psi^L = i\hbar\frac{\partial}{\partial t}\Psi^L \tag{5.137}$$

where we understand that the magnetic moment μ of the Pauli electron, which interacts with the external magnetic field B, is generated by the angular and spin momenta,

$$\mu = \mu_{\text{orbit}} + \mu_{\text{spin}} = -\frac{\mu_B}{\hbar}l - 2\frac{\mu_B}{\hbar}s = -\mu_B l - g\mu_B s \tag{5.138}$$

Here, we introduced two important quantities: (i) the *Bohr magneton* $\mu_B = e\hbar/2m_e c$ and the *gyromagnetic ratio* of the electron, $g = 2$, which is also known as the *Landé factor* and which is already very well reproduced by the Dirac and Pauli theories.

The operator quadratic in the vector potential, $q_e^2 A^2/2m_e c^2$, is often called the *diamagnetic term*, while the remaining magnetic-potential operators linear in B (or A, respectively) are called *paramagnetic terms*.

The nonrelativistic limit for infinite speed of light, $c \to \infty$, immediately recovers the Schrödinger equation. Obviously, in the nonrelativistic limit all magnetic interactions are eliminated. If magnetic phenomena are to be studied in a nonrelativistic theory, the Pauli equation has to be employed. To answer the question in which cases the Schrödinger equation might be useful, we recall the argument used earlier, namely that V is small compared to $m_e c^2$. If V is huge, the Schrödinger equation will no longer be a valid approximation. We shall see in section 6.9 for which nuclear charges this will be the case.

It is important to note that everything that has been said in this section transfers directly to the many-electron Dirac equation in chapter 9 and all subsequent chapters.

5.5
Interpretation of Negative-Energy States: Dirac's Hole Theory

Dirac considered his equation a success not only because of its Lorentz covariance fulfilled even in the presence of external electromagnetic fields, but also because it predicted the spin properties of the electron (for instance, it explained the fine-structure splitting in the spectroscopy of hydrogen atoms; see next chapter). The negative-energy states, however, bothered him [72], and he did not simply regard them as a mathematical artefact to be ignored in a physical context. The latter would have been rather unsatisfactory considering the fact that *all* energy eigenvalues are potentially measurable according to the principles of quantum theory. In the context of molecular science, we will nevertheless omit the negative-energy states — though this is done in a very sophisticated way (see chapter 11). However, we shall discuss the physical significance of the negative-energy states in this section now.

Part (a) on the left hand side of Figure 5.1 depicts the situation which emerged after solution of the Dirac equation for a freely moving fermion in section 5.3. Since, in principle, the negative-energy states are not bounded from below and hence states of infinite negative energy are accessible by a fermion, this situation obviously causes problems for energy conservation and the stability of the world as we observe it (even if we neglect interactions through potential energy terms for the moment). We may anticipate here that the influence of external potential as created by the attractive electrostatic interaction of electrons with atomic nuclei does not change this situation. Potentials, which are attractive for electrons, induce electronic bound states close to m_ec^2 as we shall derive in chapter 6 (this situation is anticipated in sketch (b) in Figure 5.1).

In 1930, Dirac presented the first partially satisfactory solution for the dilemma caused by the free-fermion states of negative energy [72]. He writes: "The most stable states for an electron (i.e., the states of lowest energy) are those with negative energy and very high velocity. All the electrons in the world will tend to fall into these states with emission of radiation. The Pauli exclusion principle, however, will come into play and prevent more than one electron going into any one state. Let us assume there are so many electrons in the world that all the most stable states are occupied, or, more accurately, that *all the states of negative energy are occupied except perhaps a few of small velocity*. Any electrons with positive energy will now have very little chance of jumping into negative-energy states and will therefore behave as electrons that are observed to behave in the laboratory. We shall have an infinite number of electrons in negative-energy states, and indeed an infinite number per unit volume all over the world, but if their distribution is exactly uniform we should expect them to be completely unobservable. *Only the small departures from exact uniformity, brought about by some of the negative-energy states being unoccupied, can we hope to observe.*" [italics as in the original paper]. Thus, he assumed that all these states have to be regarded as occupied for the vacuum ground state, establishing the Dirac sea of negative continuum states. Because of the Pauli exclusion principle these states are then no longer available for positive-energy electrons, which yields an explanation for the observed stability of the hydrogen atom. Moreover, Dirac's interpretation involves the existence of a new type of elementary particle, namely the charge-conjugated fermion of the same mass as the electron, also called the anti-particle of the electron or anti-electron or *positron*. This is predicted by a hole theory, which assumes that excitation of an negative-energy electron is possible as illustrated in sketch (c) on the right hand side of Figure 5.1. Of course, the amount of energy required is of the order of $2m_ec^2$, which is larger than 37500 hartree in Hartree atomic units, i.e., of the order of 100 million kJ mol^{-1}. The excited electron leaves a hole in the Dirac sea, which for the sake of charge conservation is positively charged and can be interpreted as a new particle, the positron. Once provided, the energy can be liberated by recombination of electron and positron under emission of electromagnetic radiation.

Thus, electron–positron pair creation is predicted and interpreted as the excitation of a negative-energy sea electron into the positive continuum, which requires at least an energy of $2m_ec^2$. This energy is beyond the energy range which is of importance in the molecular sciences. When Dirac proposed his hole theory in 1930, the positron had not yet been discovered experimentally. Therefore, he was led to identify the positively charged particle, which arises by excitation of a sea electron, as the only known positively charged particle at that time, namely the proton — despite its larger mass [72]. This connection between electron and proton was suspected by several researchers like, for instance, Hermann Weyl, but it was Dirac who considered this option with great caution [72]. The most remarkable success of Dirac's theory of the electron is that this charge-conjugated particle was a couple of years later observed as the positron by the American physicist Carl David Anderson.

Of course, Dirac was well aware of the many difficulties arising from the electron–positron interpretation — from the mass-dissymmetry if the negative-energy electron was a proton to the fact that he has actually created a many-particle theory which, as a true many-particle theory, would have been difficult to study. We shall have a closer look into these difficulties below.

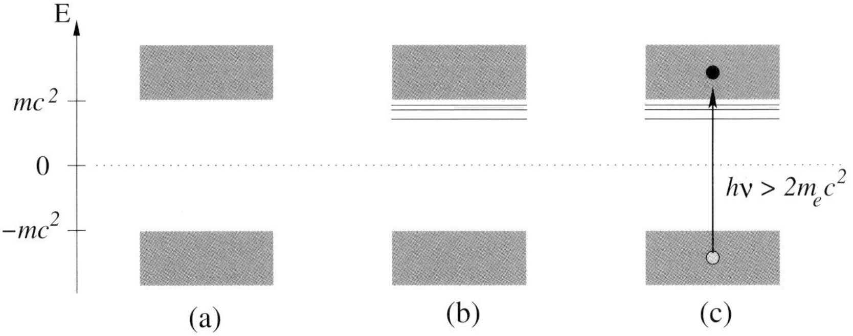

Figure 5.1 Sketch of the spectrum of the Dirac Hamiltonian for a free electron (a) and a bound electron in a Coulombic potential attractive for electrons [see chapter 6] (b). In (c) creation of an electron–positron pair is illustrated according to Dirac's hole theory, where all negative-energy states are assumed to be occupied for the vacuum state.

Though Dirac's hole theory could account for an acceptable interpretation of the negative continuum states for the very first time, it is still plagued by a series of inconsistencies incompatible with a fundamental physical theory from today's point of view. First of all, it crucially demands the fermionic nature of the electrons since it heavily relies on the Pauli principle in order to guarantee that the negative-energy states are *in*accessible. Hole theory could thus not cope with negative-energy states for bosonic fields, which do, however, occur in *any* relativistic theory, e.g., in the framework of the Klein–Gordon equation describing spinless mesons. In addition, Dirac's postulate of an occupied sea of negative-energy states is by no means a necessary consequence of the Dirac equation. Even worse, it is not even formally related to this mathematical equation at all, but rather an additional, metaphysical *ad hoc* assumption necessary for the restoration of a meaningful physical picture. By contrast, a fundamental theory of electrons and positrons should comprise all ingredients for a consistent physical interpretation in its fundamental mathematical framework. In addition, hole theory cannot cope with situations like radioactive β^+-decay either, where only a positron but no electron is created.

Furthermore, starting from the Dirac equation, the assumptions of hole theory lead automatically to a formulation for an infinite number of particles occupying the Dirac sea, and the actual number of electrons and positrons cannot be deduced from hole theory. This is a most disturbing aspect of Dirac's hole theory, namely that it is actually a *many*-particle theory. So far we have considered only a single fermion in the universe and set up an equation to describe its motion. Now, we face a conceptual generalization, which in turn requires a description of the motion of infinitely many fermions. This leads to the inclusion of additional interaction operators in the Dirac equation, and it is this ineraction of electrons which makes life difficult for molecular scientists, as we shall discuss in parts III and IV of this book.

Obviously, the situation with Dirac's theory of the electron requires revision and extension, which allows us to keep the virtues of Dirac's equation and to find a formulation which is more consistent with respect to the conceptual difficulties just mentioned. This will lead to the formulation of quantum electrodynamics, as we shall see in chapter 7. We then understand that the original Dirac equation must not be considered as a quantum mechanical wave equation, but rather as the equation of motion for a classical fermionic field, which might subsequently be quantized by the methods of second quantization (meaning quantization of fields). Then, in quantum field theory no negative-energy states and related pathologies show up due to the beneficial protocol of normal ordering. It is interesting to note that even today, when hole theory and quantum field theory are compared, situations have been found where they do not only differ from a conceptual point of view, but lead to different results in the presence of external perturbations [73,74]. However, before we delve deeper into these issues, we first consider the Dirac electron in an external field as an essential prerequisite of all later chapters.

Further Reading

F. Schwabl, [46]. *Advanced Quantum Mechanics.*

In his very useful book on advanced topics in quantum mechanics Schwabl presents a very extensive discussion of relativistic quantum mechanics as needed for the molecular sciences (rather than with a strong focus on particle physics). It can be recommended to readers who are starting to learn about relativistic quantum mechanics as well as to those who want to quickly look up certain topics.

P. Strange, [75]. *Relativistic Quantum Mechanics — with Applications in Condensed Matter and Atomic Physics.*

This book is an excellent introduction and an advanced text at the same time. It will not fail the beginner and provides a lot of additional material necessary to better understand relativistic quantum mechanics applied to spherically symmetric systems, i.e., to atoms, but also to solid state physics.

B. Thaller, [76]. *The Dirac Equation.*

Thaller's book is a comprehensive discussion of the Dirac equation from a rather mathematical point of view. Ten years after publication of this work, Thaller contributed a review-like article [77] to Schwerdtfeger's collection of review articles on relativistic quantum mechanics [5].

J. J. Sakurai, [71]. *Advanced Quantum Mechanics.*

Like Sakurai's general introduction to quantum theory [16], this advanced volume also offers plenty of insight condensed into a single volume. It can be warmly recommended.

M. E. Rose, [78]. *Relativistic Electron Theory.*

Rose's book is a classic text, which is still an excellent reference for many fundamental issues regarding the Dirac equation.

6
The Dirac Hydrogen Atom

The simplest system of relevance for molecular sciences is the hydrogen atom. The quantum mechanical equation of motion for an electron in the attractive (external) potential of a positively charged atomic nucleus can be solved analytically. From these analytical results many important consequences follow, which make the description of many-electron atoms and molecules feasible (like the choice of proper basis functions for the expansion of molecular spinors).

6.1
Separation of Electronic Motion in a Nuclear Central Field

In the previous chapter we derived the Dirac equation for an electron moving in an arbitrary electromagnetic potential. We next proceed step-wise toward a many-electron theory which incorporates any kind of electromagnetic interactions. The most simple example is the hydrogen atom consisting of two interacting particles: the electron and the proton.

The quantum mechanical description of these two fermions is straightforward in a formal sense considering the one-particle Dirac Hamiltonians for electron e and proton p plus the interaction operator V_{ep},

$$H = h_e^D \otimes \mathbf{1} + \mathbf{1} \otimes h_p^D + V_{ep} \tag{6.1}$$

Note that the derivation of the Dirac equation in chapter 5 holds for any freely moving spin-1/2 fermion and hence also for the proton; we come back to such general two-particle Hamiltonians at the beginning of chapter 8. In the field-free Dirac Hamiltonian only the rest mass determines which fermion is considered. Accordingly, the total wave function of the hydrogen atom reads

$$\Psi = \psi_e \otimes \psi_p \tag{6.2}$$

It will turn out in the following treatment that these equations need not be made more explicit since various approximations (clamped nucleus, approximation of V_{ep} by the simple electrostatic Coulomb interaction) lead to a simplified expression for Hamiltonian, energy eigenvalue and wave function, which

Relativistic Quantum Chemistry. Markus Reiher and Alexander Wolf

ISBN: 978-3-527-31292-4

are still in remarkable agreement with experiment and adopt the form of the Schrödinger solutions for the hydrogen atom in the limit of $c \rightarrow \infty$.

In order to achieve this simplification of the relativistic treatment, we consider an isolated system of the electron and the atomic nucleus, which are both treated as point charges without additional properties (i.e., in particular, the possible occurrence of a spin of the nucleus is left aside for the discussion here). Their interaction is to be described by a scalar potential energy operator, which reads

$$V(r) = q_e\phi = \frac{q_e\, q_{nuc}}{|\boldsymbol{r} - \boldsymbol{R}|} = -\frac{Ze^2}{r} \equiv V_{nuc}(r) \tag{6.3}$$

in Gaussian units, where r denotes the distance between the electron's position $\boldsymbol{r}$ and the nucleus coordinate $\boldsymbol{R}$. Z is the nuclear charge number. Since we assume an instantaneous electrostatic interaction only, contributions from a vector potential or retardation effects are neglected (compare sections 3.5 and 8.1 for a discussion on how these restrictions can be overcome for two interacting charges). Hence, for this hydrogen atom 'model' Dirac's equation is still the relativistic exact one. But a 'real' hydrogen atom is far more complex. However, the fact that we consider a point-like nucleus allows us to treat all hydrogen-like atoms on the same footing, i.e., those atoms which feature any nuclear charge, $q_{nuc}=(+Ze)$, but only a single electron. Restricting the discussion to this special model of hydrogen-like atoms is justified if electrostatic point interactions dominate (and they do as we shall see in chapters 8 and 9).

The fact that we do not treat the proton (or the atomic nucleus) and the electron on the same footing is already evident from the fact that we consider only one coordinate in space-time, namely the coordinate of the electron $x = (ct, \boldsymbol{r})$. The eigen time of the proton does not occur and we will also eliminate the spatial coordinate of the proton, $\boldsymbol{R}$, which occurs in the electron–proton distance $r = |\boldsymbol{r} - \boldsymbol{R}|$ in the following.

To further simplify the problem we will consider the nucleus to be at rest so that the kinetic energy operator for the nucleus can be neglected. Then, the nucleus position may be chosen such that it coincides with the origin of the coordinate system, which yields a purely electronic coordinate $\boldsymbol{r}$ and its distance to the nucleus r. This fixed nucleus approximation will be used extensively in the molecular case in later chapters. It is usually justified by the consideration that already the proton mass is almost 2000 times larger than the electron's rest mass m_e. Hence, the electron should move much faster than the atomic nucleus so that the nucleus can be considered at rest (in a classical picture). Of course, if the atomic nucleus acquires significant kinetic energy (as can be the case for fast moving nuclei or for nuclei in highly excited rotational states of molecules) this approximation may no longer be valid. Then, a separation of electronic and nucleus coordinates will be mandatory. In nonrelativistic theory, this is achieved by a coordinate transformation, where the separation of the center of mass yields a Schrödinger equation for the remaining relative coordinate. In turn, a reduced mass is introduced. In Dirac theory, however, this is not straightforward as the kinetic energy operators for electron and nucleus are not described on the same footing: while the Dirac kinetic energy operator is to be taken for the electron, the appropriate choice for the nucleus, which may be fermionic or bosonic, is not straighforward.

We therefore include the effect of the atomic nucleus by introducing the scalar potential energy operator of Eq. (6.3) into the Dirac equation for external fields in Eq. (5.119) with $\boldsymbol{A} = 0$ to obtain the equation of motion for Dirac hydrogen-like atoms,

$$[c\,\boldsymbol{\alpha}\cdot\boldsymbol{p} + \beta m_e c^2 + V(r)]\Psi(t,\boldsymbol{r}) = \mathrm{i}\hbar\frac{\partial}{\partial t}\Psi(t,\boldsymbol{r}) \tag{6.4}$$

Effects of moving nuclei have been investigated, of course, because they cannot be neglected in high-precision calculations (see, e.g., the work by Yennie *et al.* [79–81]). By construction, we understand that the effect of the proton is 'external' to the electron, which is why the additional scalar potential $V(r)$ is also called the external potential. It can hardly be overemphasized that the effect of the electrostatic charge of the atomic nucleus is considered here to be its only relevant physical property. We thus neglect the fact that it may have a spin which would contribute a magnetic vector potential to the set of potentials the electron interacts with. In chapter 15 we come back to this issue when we consider molecular properties. At this stage it is only important to note that the inclusion of such additional interaction types results in comparatively tiny effects on the energy of the system, which is dominated by the electrostatic forces. For this reason it is usually neglected from the very beginning, although the complete interaction pattern would be much more involved. This is particularly true as a 'fully' relativistic theory of the isolated system 'nucleus plus electron' would require a relativistically consistent description beyond the instantaneous Coulombic interactions. We come back to this issue in the context of the interaction of two electrons in chapter 8, which is the only case for which relativistic corrections to a Coulomb-type interaction of two point charges are usually considered.

For the solution of the Dirac hydrogen atom defined by Eq. (6.4) we first note that the space and time variables of the electron are well separated — as shown in section 4.2 for the general case — so that we may use the ansatz,

$$\Psi(t,\boldsymbol{r}) = \exp(-\mathrm{i}Et/\hbar)\Psi(\boldsymbol{r}) \tag{6.5}$$

Note that this ansatz does not violate the Lorentz covariance of Eq. (6.4) because after having performed a Lorentz transformation to a new set of coordinates $(ct', \boldsymbol{r}')$ one would choose an ansatz as in Eq. (6.5) but for the new coordinates t' and $\boldsymbol{r}'$. Of course, the new time coordinate t' may then be expressed by the *old* time and spatial coordinates t and $\boldsymbol{r}$ as governed by the Lorentz transformation.

The ansatz of Eq. (6.5) now allows us to separate the coordinates in Eq. (6.4),

$$\begin{aligned}[c\,\boldsymbol{\alpha}\cdot\boldsymbol{p}+\beta m_e c^2+V(r)]\exp\left(-\mathrm{i}\frac{Et}{\hbar}\right)\Psi(\boldsymbol{r}) &= \mathrm{i}\hbar\frac{\partial}{\partial t}\exp\left(-\mathrm{i}\frac{Et}{\hbar}\right)\Psi(\boldsymbol{r}) \\ &= E\exp\left(-\mathrm{i}\frac{Et}{\hbar}\right)\Psi(\boldsymbol{r}) \qquad (6.6)\end{aligned}$$

Since the time-dependent and the spatial part of the equation each have to be equal to the separation constant E, we may write Eq. (6.6) as two different equations and integrate out the part of the wave function $\Psi(t,\boldsymbol{r})$ independent of the coordinates in which the operator is expressed,

$$\underbrace{[c\,\boldsymbol{\alpha}\cdot\boldsymbol{p}+\beta m_e c^2+V(r)]}_{\equiv h^D}\Psi(\boldsymbol{r}) = E\Psi(\boldsymbol{r}) \qquad (6.7)$$

$$\mathrm{i}\hbar\frac{\partial}{\partial t}\exp\left(-\mathrm{i}\frac{Et}{\hbar}\right) = E\exp\left(-\mathrm{i}\frac{Et}{\hbar}\right) \qquad (6.8)$$

of which the latter is already solved by the ansatz of Eq. (6.5). Hence, we now have to find $\Psi(\boldsymbol{r})$ by solving Eq. (6.7), for which we introduced a shorthand notation, $h^D\Psi(\boldsymbol{r}) = E\Psi(\boldsymbol{r})$.

When Dirac proposed his equation for the electron [66], he already considered the case of a central scalar field and studied the relation to Schrödinger's wave equation and its extensions including spin by Pauli and Darwin. The solution of the Dirac central-field problem, however, was due to Darwin [70] and Gordon [82] who in turn recovered Sommerfeld's energy expression but, of course, as an energy eigenvalue.

The solution of Eq. (6.7) can be conveniently accomplished in a new set of spatial coordinates. Since the potential energy operator $V(r)$ depends on a radial distance coordinate and hence not on any orientational coordinate, solution of Eq. (6.7) should be attempted in polar coordinates (r,ϑ,φ). The transformation equations from Cartesian coordinates to spherical coordinates and its inverse transformation have already been introduced in section 4.4, Eqs. (4.89)–(4.94).

6.2
Schrödinger Hydrogen Atom

Since one is often more familiar with nonrelativistic Schrödinger quantum mechanics for atoms because of its simpler structure, we recall its basic ingredients, which serve two purposes. For one, the similarities to the Dirac theory of atoms become evident. Second, we will come back to the Schrödinger theory when considering the nonrelativistic limit in section 6.6. In nonrelativistic theory, the Schrödinger equation for the hydrogen atom,

$$\underbrace{\left[-\frac{\hbar^2}{2m_e}\Delta + V(r)\right]}_{=h^{nr}} \Psi^{nr}(\boldsymbol{r}) = E^{nr}\Psi^{nr}(\boldsymbol{r}) \tag{6.9}$$

is to be transformed to spherical coordinates. Hence, one seeks an expression for the Laplacian operator,

$$\Delta = \Delta(x,y,z) = \frac{\partial^2}{\partial x^2} + \frac{\partial^2}{\partial y^2} + \frac{\partial^2}{\partial z^2} \tag{6.10}$$

in polar coordinates $\Delta = \Delta(r,\vartheta,\varphi)$. Although tedious and lengthy, the coordinate transformation for the second derivatives is standard and straighforward [83]. Its important result is that the Laplacian in spherical coordinates can be written such that the square $\boldsymbol{l}^2$ of the (orbital) angular momentum operator $\boldsymbol{l}$ of Eq. (4.87) appears and collects all angular coordinates,

$$\begin{aligned}\Delta(r,\vartheta,\varphi) &= \frac{1}{r^2}\frac{\partial}{\partial r}\left(r^2\frac{\partial}{\partial r}\right) + \frac{1}{r^2}\times \\ &\quad \underbrace{\left[\frac{1}{\sin\vartheta}\frac{\partial}{\partial\vartheta}\left(\sin\vartheta\frac{\partial}{\partial\vartheta}\right) + \frac{1}{\sin^2\vartheta}\frac{\partial^2}{\partial\varphi^2}\right]}_{\equiv -\boldsymbol{l}^2/\hbar^2} = \frac{1}{r}\frac{\partial^2}{\partial r^2}r - \frac{1}{r^2}\frac{\boldsymbol{l}^2}{\hbar^2}\end{aligned} \tag{6.11}$$

Hence, the radial and angular coordinates can be separated,

$$\left[-\frac{\hbar^2}{2m_e}\left(\frac{1}{r}\frac{\partial^2}{\partial r^2}r - \frac{\boldsymbol{l}^2}{\hbar^2 r^2}\right) + V(r)\right]\Psi^{nr}(\boldsymbol{r}) = E^{nr}\Psi^{nr}(\boldsymbol{r}) \tag{6.12}$$

and the eigenfunctions of $\boldsymbol{l}^2$, i.e., the spherical harmonics $Y_{lm}(\vartheta,\varphi)$ of Eq. (4.113), can be used as the angular dependent part in an ansatz for the total stationary state $\Psi^{nr}(\boldsymbol{r})$,

$$\Psi^{nr}(\boldsymbol{r}) \quad\longrightarrow\quad \Psi^{nr}_{nlm}(\boldsymbol{r}) = R^{nr}_{nl}(r)Y_{lm}(\vartheta,\varphi) = \frac{P^{nr}_{nl}(r)}{r}{}_{lm}(\vartheta,\varphi) \tag{6.13}$$

Eq. (6.12) can also be written in terms of a radial momentum operator p_r

$$\left[\frac{p_r^2}{2m_e} + \frac{\boldsymbol{l}^2}{2m_e r^2} + V(r)\right]\Psi^{nr}(\boldsymbol{r}) = E^{nr}\Psi^{nr}(\boldsymbol{r}) \tag{6.14}$$

which we have already derived in chapter 4 when we derived the expression for the kinetic energy operator of a rotating particle in Eq. (4.103). In section

4.4 we showed that a radial momentum operator, Eq. (4.101), can be deduced to become

$$p_r \equiv -i\hbar \frac{1}{r}\frac{\partial}{\partial r} r = -i\hbar \left(\frac{\partial}{\partial r} + \frac{1}{r} \right) \tag{6.15}$$

so that the squared radial momentum operator reads

$$p_r^2 = -\hbar^2 \left[\frac{\partial^2}{\partial r^2} + \frac{2}{r}\frac{\partial}{\partial r} \right] = -\hbar^2 \left[\frac{1}{r}\frac{\partial^2}{\partial r^2} r \right] \tag{6.16}$$

[compare Eq. (4.100)].

It remains to determine the radial function $R_{nl}(r)$. The radial equation to be solved for $R_{nl}^{nr}(r)$ and $P_{nl}^{nr}(r)$, respectively, is then obtained by multiplying Eq. (6.12) by $\langle Y_{lm}(\vartheta, \varphi)| \cdot$ from the left, i.e., by multiplication with the complex conjugate $Y_{lm}^{\star}(\vartheta, \varphi)$ and subsequent integration over all angular degrees of freedom ϑ and φ. The resulting equation,

$$\left[-\frac{\hbar^2}{2m_e} \left(\frac{1}{r}\frac{\partial^2}{\partial r^2} r - \frac{l(l+1)}{r^2} \right) + V(r) \right] R_{nl}^{nr}(r) = E_{nl}^{nr} R_{nl}^{nr}(r) \tag{6.17}$$

can be rewritten after multiplication by r as,

$$\left[-\frac{\hbar^2}{2m_e} \left(\frac{\partial^2}{\partial r^2} - \frac{l(l+1)}{r^2} \right) + V(r) \right] P_{nl}^{nr}(r) = E_{nl}^{nr} P_{nl}^{nr}(r) \tag{6.18}$$

where the new function $P_{nl} = rR_{nl}$ is convenient for the definition of the radial density; see below. The additional (repulsive) potential energy term $+\hbar^2 l(l+1)/(2m_e r^2)$ is called the centrifugal potential because of its close resemblence to the classical expression for the rotational energy, $\boldsymbol{l}^2/2mr^2$. However, it solely arises from the operator for the *kinetic* energy. The nonrelativistic centrifugal potential depends on the angular momentum quantum number l and thus vanishes for totally symmetric s-shells (where $l = 0$). After the solution of Eq. (6.17), the energy of the nonrelativistic quantum state turns out to depend only on the principal quantum number n. The energy takes a particularly simple form in Hartree atomic units, namely

$$E^{nr} = E_{nl}^{nr} = E_n^{nr} = -\frac{Z}{2n^2} \tag{6.19}$$

The most important aspect to note here is that the angular momentum l enters the nonrelativistic Hamiltonian in a way which allows us to classify the nonrelativistic quantum states according to the irreducible representations of the group of rotations, i.e., with respect to the angular momentum quantum number l. To be more precise, we note that $\boldsymbol{l}^2$ and h^{nr} commute,

$$[h^{nr}, \boldsymbol{l}^2] = 0 \tag{6.20}$$

and thus there exists a set of common eigenfunctions.

6.3 Total Angular Momentum

In the next section, we investigate how angular momentum enters the Dirac equation for hydrogen-like atoms. For this we need to consider the *total* angular momentum operator $\boldsymbol{j}$ introduced in section 4.4.4. The four-dimensional structure of the Dirac equation must be taken into account, and we are advised to generalize the vectorial sum to obtain an operator, which is defined on the space of four-component spinors, i.e., which is a 3-vector of 4×4 matrices,

$$\boldsymbol{j}_4 = \boldsymbol{j}\mathbf{1}_2 \overset{(4.145)}{=} (\boldsymbol{l}\mathbf{1}_2 + \boldsymbol{s})\,\mathbf{1}_2 = \boldsymbol{l}\mathbf{1}_4 + \boldsymbol{s}\mathbf{1}_2 \overset{(4.139)}{=} \boldsymbol{l}\mathbf{1}_4 + \frac{\hbar}{2}\boldsymbol{\sigma}\mathbf{1}_2 = \boldsymbol{l}\mathbf{1}_4 + \frac{\hbar}{2}\boldsymbol{\Sigma} \quad (6.21)$$

with $\boldsymbol{l}\mathbf{1}_4 \equiv (l_x\mathbf{1}_4, l_y\mathbf{1}_4, l_z\mathbf{1}_4)$ accordingly. The vector of 4×4 matrices $\boldsymbol{\Sigma}$ is the generalization of $\boldsymbol{\sigma}$ and reads explicitly,

$$\boldsymbol{\Sigma} \equiv \begin{pmatrix} \boldsymbol{\sigma} & 0 \\ 0 & \boldsymbol{\sigma} \end{pmatrix} = \left[\begin{pmatrix} \sigma_x & 0 \\ 0 & \sigma_x \end{pmatrix}, \begin{pmatrix} \sigma_y & \mathbf{0} \\ 0 & \sigma_y \end{pmatrix}, \begin{pmatrix} \sigma_z & 0 \\ 0 & \sigma_z \end{pmatrix} \right] \quad (6.22)$$

The commutation relations of the total angular momentum components, Eq. (4.111), transfer to the four-dimensional case

$$[j_i, j_j] = i\hbar \sum_{k=1}^{3} \epsilon_{ijk}\, j_k \quad \Longrightarrow \quad [j_{4,i}, j_{4,j}] = i\hbar \sum_{k=1}^{3} \epsilon_{ijk}\, j_{4,k} \quad (6.23)$$

The Dirac Hamiltonian for hydrogen-like atoms, h^D, and the square as well as any component of the total angular momentum operator $\boldsymbol{j}_4$ commute,

$$[h^D, \boldsymbol{j}_4^2] = 0, \quad (6.24)$$

$$[h^D, j_{4,i}] = 0, \text{ with } i \in \{x, y, z\} \quad (6.25)$$

which is straighforward to show by inserting explicit expressions for the operators in Eqs. (6.24) and (6.25) and resolving the commutator. Therefore, h^D and $\boldsymbol{j}_4^2$ possess the same eigenstates. But note that h^D and $\boldsymbol{l}^2$ do not commute. Thus, the eigenstates of h^D cannot be classified according to l, but to j, the total angular quantum number. These common eigenfunctions of h^D and $\boldsymbol{j}_4^2$ are then only simultaneous eigenfunctions of *one* additional component of $\boldsymbol{j}_4$ and not of the other two because the components $j_{4,i}$ do not commute.

6.4
Separation of Angular Coordinates in the Dirac Hamiltonian

From the Dirac Hamiltonian for hydrogen-like atoms in split notation,

$$h^D = \begin{pmatrix} m_e c^2 + V(r) & c\boldsymbol{\sigma}\cdot\boldsymbol{p} \\ c\boldsymbol{\sigma}\cdot\boldsymbol{p} & -m_e c^2 + V(r) \end{pmatrix} \tag{6.26}$$

it is clear that we require an expression of $(\boldsymbol{\sigma}\cdot\boldsymbol{p})$ in polar coordinates as the potential energy operator is already represented in these new coordinates and the constant terms on the diagonal are not affected by a coordinate transformation.

6.4.1
Spin–Orbit Coupling

The structure of the Dirac Hamiltonian in Eq. (6.26) demands an analysis of the action of the operator $\boldsymbol{\sigma}\cdot\boldsymbol{p}$ on the two components of the 4-spinor given in Eq. (6.51). We rewrite the product $(\boldsymbol{\sigma}\cdot\boldsymbol{p})$ to generate the orbital angular momentum operator in analogy to the nonrelativistic case above. According to Eq. (4.87) we understand that a way must be found to create a vector product from position and linear momentum, $\boldsymbol{r}\times\boldsymbol{p}$. Recalling Eq. (4.167), we may rewrite the operator product $(\boldsymbol{\sigma}\cdot\boldsymbol{p})$ as

$$\boldsymbol{\sigma}\cdot\boldsymbol{p} = \mathbf{1}_2\cdot(\boldsymbol{\sigma}\cdot\boldsymbol{p}) = \left(\frac{\boldsymbol{\sigma}\cdot\boldsymbol{r}}{r}\right)\left(\frac{\boldsymbol{\sigma}\cdot\boldsymbol{r}}{r}\right)(\boldsymbol{\sigma}\cdot\boldsymbol{p}) = \frac{1}{r^2}(\boldsymbol{\sigma}\cdot\boldsymbol{r})(\boldsymbol{\sigma}\cdot\boldsymbol{r})(\boldsymbol{\sigma}\cdot\boldsymbol{p}) \tag{6.27}$$

which produces after exploitation of Dirac's relation

$$(\boldsymbol{\sigma}\cdot\boldsymbol{r})(\boldsymbol{\sigma}\cdot\boldsymbol{p}) = \boldsymbol{r}\cdot\boldsymbol{p} + \mathrm{i}\boldsymbol{\sigma}\cdot(\boldsymbol{r}\times\boldsymbol{p}) \tag{6.28}$$

the orbital angular momentum operator $\boldsymbol{l}$

$$\boldsymbol{\sigma}\cdot\boldsymbol{p} = \left(\frac{\boldsymbol{\sigma}\cdot\boldsymbol{r}}{r}\right)\frac{1}{r}\left[\boldsymbol{r}\cdot\boldsymbol{p} + \mathrm{i}\boldsymbol{\sigma}\cdot(\boldsymbol{r}\times\boldsymbol{p})\right] = \frac{1}{r^2}(\boldsymbol{\sigma}\cdot\boldsymbol{r})\left[\boldsymbol{r}\cdot\boldsymbol{p} + \mathrm{i}(\boldsymbol{\sigma}\cdot\boldsymbol{l})\right] \tag{6.29}$$

The scalar product $(\boldsymbol{r}\cdot\boldsymbol{p})$ has already been derived in Eq. (4.97) in chapter 4; it reads

$$\boldsymbol{r}\cdot\boldsymbol{p} = -\mathrm{i}\hbar r\frac{\partial}{\partial r} \tag{6.30}$$

From Eq. (4.161) we understand that Eq. (6.29) already features a separation into radial r and angular (ϑ, φ) variables. All angular variables are contained in the operator product $(\boldsymbol{\sigma}\cdot\boldsymbol{l})$, which is essentially the spin–orbit coupling operator known from the Pauli approximation. The remaining operators can all be expressed by the radial variable r alone.

From Eq. (6.29) it is clear that we will require the eigenstates and eigenvalues of the spin–orbit coupling operator $(\boldsymbol{\sigma}\cdot\boldsymbol{l})$. Recalling the definitions of the spin operator $\boldsymbol{s} = \hbar\boldsymbol{\sigma}/2$ and of the total angular momentum $\boldsymbol{j} = \boldsymbol{l}\mathbf{1}_2 + \boldsymbol{s}$ as well as the equality $(\boldsymbol{l} \cdot \boldsymbol{s}) = (\boldsymbol{s} \cdot \boldsymbol{l})$ we can formulate an operator identity

$$\boldsymbol{j}^2 = \boldsymbol{l}^2\mathbf{1}_2 + 2(\boldsymbol{l} \cdot \boldsymbol{s}) + \boldsymbol{s}^2 = \boldsymbol{l}^2\mathbf{1}_2 + \hbar(\boldsymbol{\sigma} \cdot \boldsymbol{l}) + \boldsymbol{s}^2 \tag{6.31}$$

which then allows us to deduce an eigenvalue equation of $(\boldsymbol{\sigma} \cdot \boldsymbol{l})$,

$$\begin{aligned}(\boldsymbol{\sigma} \cdot \boldsymbol{l})\chi_{jm_j}^{(\pm)} &= \frac{1}{\hbar}\left(\boldsymbol{j}^2 - \boldsymbol{l}^2 - \boldsymbol{s}^2\right)\chi_{jm_j}^{(\pm)} \\ &= \hbar\left[j(j+1) - l(l+1) - \frac{3}{4}\right]\chi_{jm_j}^{(\pm)}\end{aligned} \tag{6.32}$$

where the second equality follows from Eqs. (4.156)–(4.158). Taking into account the two cases $j = l + 1/2$ for $\chi_{jm_j}^{(+)}$ and $j = l - 1/2$ for $\chi_{jm_j}^{(-)}$ we then may write

$$(\boldsymbol{\sigma} \cdot \boldsymbol{l})\chi_{jm_j}^{(+)} = \hbar l\,\chi_{jm_j}^{(+)} = \hbar\left[-1 + \left(j + \frac{1}{2}\right)\right]\chi_{jm_j}^{(+)} \tag{6.33}$$

$$(\boldsymbol{\sigma} \cdot \boldsymbol{l})\chi_{jm_j}^{(-)} = -\hbar(l+1)\chi_{jm_j}^{(-)} = \hbar\left[-1 - \left(j + \frac{1}{2}\right)\right]\chi_{jm_j}^{(-)} \tag{6.34}$$

as one may verify by inserting the expressions for the angular momentum coupled Pauli spinors of Eqs. (4.155) and (4.154).

6.4.2 Relativistic Azimuthal Quantum Number Analog

Eqs. (6.33) and (6.34) suggest that adding $\hbar\mathbf{1}_2$ to the operator on the left hand side of the equations simplifies the eigenvalue as $(-\hbar\mathbf{1}_2)$ then drops out. We can therefore define the operator $\boldsymbol{k}$,

$$\boldsymbol{k} \equiv \boldsymbol{\sigma} \cdot \boldsymbol{l} + \hbar\mathbf{1}_2 \tag{6.35}$$

that fulfills the eigenvalue equation

$$\boldsymbol{k}\chi_{jm_j}^{(\pm)} = \pm\hbar\left(j + \frac{1}{2}\right)\chi_{jm_j}^{(\pm)} \equiv \hbar\kappa^{(\pm)}\chi_{jm_j}^{(\pm)} \tag{6.36}$$

in which we introduced a new quantum number κ. From the equation above we conclude that

$$\kappa^{(+)} = j + \frac{1}{2} \tag{6.37}$$

is the eigenvalue to $\chi_{jm_j}^{(+)}$, while

$$\kappa^{(-)} = -\kappa^{(+)} = -j - \frac{1}{2} \tag{6.38}$$

is the eigenvalue to $\chi_{jm_j}^{(-)}$. Hence, we may classify the pair of Pauli spinors for a given set of quantum numbers (j, m_j) equally well by the new quantum number κ,

$$\chi_{\kappa m_j} \equiv \chi_{jm_j}^{(+)}, \tag{6.39}$$

$$\chi_{-\kappa m_j} \equiv \chi_{jm_j}^{(-)} \tag{6.40}$$

The quantum number κ will turn out to play an analogous role in the Dirac theory of atoms that is played by l in the nonrelativistic theory. In the 1930s, Swirles introduced a special notation to assign all different quantum numbers of different quantum mechanical states $\Psi_i(\boldsymbol{r})$ to the familiar spectroscopic notation for the states of hydrogen-like atoms like s ≡sharp, p ≡principal, d ≡diffuse (see Table 6.1 for an overview).

Table 6.1 Quantum numbers and shell labels after Swirles [84]. The last line 'degeneracy' anticipates the number of spinors with the same energy eigenvalue, a fact that is actually already imprinted in the ansatz for the 4-spinor in Eq. (6.51).

Shell	$s_{1/2}$	$p_{1/2}$	$p_{3/2}$	$d_{3/2}$	$d_{5/2}$	$f_{5/2}$	$f_{7/2}$	$g_{7/2}$	$g_{9/2}$
Symbol	s	$\bar{p}$	p	$\bar{d}$	d	$\bar{f}$	f	$\bar{g}$	g
κ	−1	1	−2	2	−3	3	−4	4	−5
l	0	1	1	2	2	3	3	4	4
$j = \|\kappa\| - 1/2$	1/2	1/2	3/2	3/2	5/2	5/2	7/2	7/2	9/2
Parity ($[-1]^l$)	+	−	−	+	+	−	−	+	+
Degeneracy ($2\|\kappa\| = 2j + 1$)	2	2	4	4	6	6	8	8	10

6.4.3 Four-Dimensional Generalization

In analogy to the two-dimensional Eq. (6.27) we can also study the product $(\boldsymbol{\alpha} \cdot \boldsymbol{r})(\boldsymbol{\alpha} \cdot \boldsymbol{p})$,

$$(\boldsymbol{\alpha} \cdot \boldsymbol{r})(\boldsymbol{\alpha} \cdot \boldsymbol{p}) = \mathbf{1}_2 \otimes [\boldsymbol{r} \cdot \boldsymbol{p} + i\boldsymbol{\sigma} \cdot (\boldsymbol{r} \times \boldsymbol{p})] \tag{6.41}$$

instead of $(\boldsymbol{\alpha} \cdot \boldsymbol{p})$ of Eq. (6.7). In close analogy to the nonrelativistic hydrogen atom in Eq. (6.14) we introduce the radial momentum p_r of Eq. (4.101). With Eqs. (4.97) and (4.101) we can write

$$\boldsymbol{r} \cdot \boldsymbol{p} = r p_r + i\hbar \tag{6.42}$$

Eq. (6.41) can then be written as

$$(\boldsymbol{\alpha} \cdot \boldsymbol{r})(\boldsymbol{\alpha} \cdot \boldsymbol{p}) = \mathbf{1}_2 \otimes [r p_r \mathbf{1}_2 + i(\hbar \mathbf{1}_2 + \boldsymbol{\sigma} \cdot \boldsymbol{l})] \tag{6.43}$$

In view of Eq. (6.27) it is convenient to define the scalar product of $\boldsymbol{\alpha}$ with the unit vector $\hat{\boldsymbol{r}}$ in $\boldsymbol{r}$ direction,

$$\alpha_r \equiv \boldsymbol{\alpha} \cdot \hat{\boldsymbol{r}} = \frac{(\boldsymbol{\alpha} \cdot \boldsymbol{r})}{r} \tag{6.44}$$

which is the four-dimensional generalization of $(\boldsymbol{\sigma} \cdot \hat{\boldsymbol{r}})$ from section 4.4.4,

$$\alpha_r = \frac{1}{r} \begin{pmatrix} 0 & \boldsymbol{\sigma} \cdot \boldsymbol{r} \\ \boldsymbol{\sigma} \cdot \boldsymbol{r} & 0 \end{pmatrix} = \begin{pmatrix} 0 & \boldsymbol{\sigma} \cdot \hat{\boldsymbol{r}} \\ \boldsymbol{\sigma} \cdot \hat{\boldsymbol{r}} & 0 \end{pmatrix} \tag{6.45}$$

Its square equals the 4×4 unit matrix $\mathbf{1}_4$,

$$(\alpha_r)^2 = \frac{1}{r^2} \begin{pmatrix} (\boldsymbol{\sigma} \cdot \boldsymbol{r})(\boldsymbol{\sigma} \cdot \boldsymbol{r}) & 0 \\ 0 & (\boldsymbol{\sigma} \cdot \boldsymbol{r})(\boldsymbol{\sigma} \cdot \boldsymbol{r}) \end{pmatrix} \overset{(4.167)}{=} \mathbf{1}_4 \tag{6.46}$$

This can be utilized to simplify Eq. (6.43),

$$\begin{aligned} (\boldsymbol{\alpha} \cdot \boldsymbol{r})(\boldsymbol{\alpha} \cdot \boldsymbol{p}) &= \alpha_r r(\boldsymbol{\alpha} \cdot \boldsymbol{p}) \\ &= \mathbf{1}_2 \otimes (r p_r \mathbf{1}_2 + \mathrm{i}[\hbar \mathbf{1}_2 + \boldsymbol{\sigma} \cdot \boldsymbol{l}]) \qquad \Big| \cdot \frac{\alpha_r}{r} \qquad (6.47) \\ \Rightarrow (\boldsymbol{\alpha} \cdot \boldsymbol{p}) &= \alpha_r \left\{ p_r \mathbf{1}_4 + \frac{\mathrm{i}}{r}[\hbar \mathbf{1}_4 + \mathbf{1}_2 \otimes (\boldsymbol{\sigma} \cdot \boldsymbol{l})] \right\} \qquad (6.48) \end{aligned}$$

We define the 4×4 analog of $\boldsymbol{k}$ in Eq. (6.35) as $\boldsymbol{K}$,

$$\boldsymbol{K} \equiv \beta(\hbar \mathbf{1}_4 + \boldsymbol{\Sigma} \cdot \boldsymbol{l}) = \beta[\mathbf{1}_2 \otimes (\hbar \mathbf{1}_2 + \boldsymbol{\sigma} \cdot \boldsymbol{l})] \overset{(6.35)}{=} \beta(\mathbf{1}_2 \otimes \boldsymbol{k}) = \begin{pmatrix} \boldsymbol{k} & 0 \\ 0 & -\boldsymbol{k} \end{pmatrix} \tag{6.49}$$

One may wonder why it is necessary to introduce the β factor. But the eigenvectors of $\boldsymbol{K}$ are only identical to the angular parts of the spinor with this definition to ensure that the parity is the same of the upper and lower components of the spinor in Eq. (6.51) [see also Eq. (4.164)]. Recalling the eigenvalue equation for $\boldsymbol{k}$ given in Eq. (6.36), the eigenvectors of $\boldsymbol{K}$ are obtained as a combination of a pair of Pauli spinors for given (j, m_j),

$$\boldsymbol{K} \begin{pmatrix} \chi_{\kappa m_j} \\ \chi_{-\kappa m_j} \end{pmatrix} = \begin{pmatrix} \boldsymbol{k} & 0 \\ 0 & -\boldsymbol{k} \end{pmatrix} \begin{pmatrix} \chi_{\kappa m_j} \\ \chi_{-\kappa m_j} \end{pmatrix} = \hbar\kappa \begin{pmatrix} \chi_{\kappa m_j} \\ \chi_{-\kappa m_j} \end{pmatrix} \tag{6.50}$$

6.4.4 Ansatz for the Spinor

Because of the general 2×2 block structure of the Hamiltonian h^D in Eq. (6.26), an appropriate ansatz for the stationary state $\Psi(\boldsymbol{r})$ is

$$\Psi(\boldsymbol{r}) = \begin{pmatrix} F_i(r) \chi_{j m_j}(\vartheta, \varphi) \\ \mathrm{i} G_i(r) \chi'_{j m_j}(\vartheta, \varphi) \end{pmatrix} \tag{6.51}$$

whose radial, $F_i(r)$ and $G_i(r)$, and angular, $\chi_{jm_j}(\vartheta, \varphi)$ and $\chi'_{jm_j}(\vartheta, \varphi)$, components are determined by solving Eq. (6.7). The angular components are the Pauli spinors already encountered in section 4.4.4. With this ansatz the radial functions $F_i(r)$ and $G_i(r)$ turn out to be real once the Dirac equation is solved — as we shall see below — and we anticipated that the transformed Dirac operator for hydrogen-like atoms is indeed separable in r and (ϑ, φ). The eigenstates of total angular momentum introduced in Eqs. (4.147) and (4.159) turned out to become the angular part in the product ansatz for the 4-spinor $\Psi(\boldsymbol{r})$ according to our analysis in the previous section. Since *both* are eigenstates for a given (j, m_j) but of *different* azimuthal quantum number, i.e., $\chi^{(+)}_{jm_j}$ and $\chi^{(-)}_{jm_j}$, we understand that the 4-spinor $\Psi(\boldsymbol{r})$ can then only be an eigenstate of $\boldsymbol{j}_4^2$ and $j_{4,z}$ and not of $\boldsymbol{l}^2\mathbf{1}_4$. The subsequent derivations will show that the ansatz is correct as it allows us to obtain the general solution of the Dirac equation for hydrogen-like atoms.

6.5
Radial Dirac Equation for Hydrogen-Like Atoms

For the solution of the Dirac equation for hydrogen-like atoms, Eq. (6.7), we have succeeded in showing that the angular (and spin) variables can be separated from the radial-dependent part and that the eigenstates of total angular momentum are the required angular-dependent functions in the ansatz for the 4-spinor given in Eq. (6.51). In the last step, we now derive and solve the radial equation, from which we then determine the yet unknown radial functions $F_i(r)$ and $G_i(r)$.

Rewriting Eq. (6.48) yields

$$(\boldsymbol{\alpha} \cdot \boldsymbol{p}) = \boldsymbol{\alpha}_r p_r + \frac{\mathrm{i}}{r} \boldsymbol{\alpha}_r \beta \boldsymbol{K} \tag{6.52}$$

since $\beta^2 = \mathbf{1}_4$ and the Dirac Hamiltonian follows then as

$$h^D = c\boldsymbol{\alpha}_r p_r + \frac{\mathrm{i}c}{r}\boldsymbol{\alpha}_r \beta \boldsymbol{K} + m_e c^2 \beta + V(r)\mathbf{1}_4 \tag{6.53}$$

$$= \begin{pmatrix} m_e c^2 + V(r) & c\left(\frac{\boldsymbol{\sigma} \cdot \boldsymbol{r}}{r}\right)\left[p_r + \frac{\mathrm{i}}{r}\boldsymbol{k}\right] \\ c\left(\frac{\boldsymbol{\sigma} \cdot \boldsymbol{r}}{r}\right)\left[p_r + \frac{\mathrm{i}}{r}\boldsymbol{k}\right] & -m_e c^2 + V(r) \end{pmatrix} \tag{6.54}$$

where we omitted additional $\mathbf{1}_2$ unit matrices to indicate the $2{\times}2$ structure of all terms in the matrix. Recall that the operator $(\boldsymbol{\sigma} \cdot \boldsymbol{r}/r) = (\boldsymbol{\sigma} \cdot \hat{\boldsymbol{r}})$ depends solely on the angular variables φ and ϑ as highlighted by Eq. (4.161).

With the definition of the radial momentum p_r in Eq. (4.101) the Hamiltonian h^D can now be expressed in terms of a radial coordinate r only,

$$h^D = \begin{pmatrix} m_e c^2 + V(r) & -\mathrm{i}c\left(\frac{\boldsymbol{\sigma}\cdot\boldsymbol{r}}{r}\right)\left[\frac{\hbar}{r}\frac{\partial}{\partial r}r - \frac{\boldsymbol{k}}{r}\right] \\ -\mathrm{i}c\left(\frac{\boldsymbol{\sigma}\cdot\boldsymbol{r}}{r}\right)\left[\frac{\hbar}{r}\frac{\partial}{\partial r}r - \frac{\boldsymbol{k}}{r}\right] & -m_e c^2 + V(r) \end{pmatrix} \tag{6.55}$$

while all angular and spin degrees of freedom are collected in the operator $\boldsymbol{k}$. We now exploit the fact that we have identified the eigenfunctions of $\boldsymbol{k}$ in the preceding section and rewrite the ansatz of Eq. (6.51) as

$$\Psi_{n\kappa m_j}(\boldsymbol{r}) = \begin{pmatrix} F_i(r)\chi^{(\pm)}_{jm_j}(\vartheta,\varphi) \\ \mathrm{i}G_i(r)\boldsymbol{\sigma}\cdot\hat{\boldsymbol{r}}\chi^{(\pm)}_{jm_j}(\vartheta,\varphi) \end{pmatrix} \stackrel{(6.62)}{=} \begin{pmatrix} F_{n\kappa}(r)\chi_{\kappa m_j}(\vartheta,\varphi) \\ \mathrm{i}G_{n\kappa}(r)\chi_{-\kappa m_j}(\vartheta,\varphi) \end{pmatrix} \tag{6.56}$$

where we introduced the imaginary unit i to account for the imaginary units on the off-diagonal in Eq. (6.53) so that $F_{n\kappa}(r)$ and $G_{n\kappa}(r)$ are *both* real functions. Without loss of generality we assign a single pair of radial functions, $F_{n\kappa}(r)$ and $G_{n\kappa}(r)$, to a given set of $(2j+1)$-degenerate spherical spinors $\chi^{(\pm)}_{jm_j}(\vartheta,\varphi)$ with given (j, m_j), which is called *equivalence restriction.*

6.5.1 Radial Functions and Orthonormality

In close analogy to the nonrelativistic theory, it is advantageous (as we shall see in the following) to introduce a second set of radial functions, $P_i = P_{n_i\kappa_i}$ and $Q_i = Q_{n_i\kappa_i}$, as

$$\Psi_i(\boldsymbol{r}) \to \Psi_{n_i\kappa_i m_{j(i)}}(\boldsymbol{r}) = \begin{pmatrix} \frac{P_{n_i\kappa_i}(r)}{r}\chi_{\kappa_i m_{j(i)}}(\vartheta,\varphi) \\ \mathrm{i}\frac{Q_{n_i\kappa_i}(r)}{r}\chi_{-\kappa_i m_{j(i)}}(\vartheta,\varphi) \end{pmatrix} \tag{6.57}$$

The subscripts in Eq. (6.57) differentiate between different sets of radial functions belonging to different states $\Psi_i(\boldsymbol{r}) = \Psi_i(r,\vartheta,\varphi)$ with the composite index $i = \{n_i, \kappa_i, m_{j(i)}\}$.

The orthonormality of the spherical harmonics Y_{lm} and of the spin eigenvectors ρ_{sm_s} (with $s = 1/2$ and $m_s = \pm 1/2$) ensures orthonormality of the Pauli spinors,

$$\begin{aligned} \langle\chi_{\kappa_i m_{j(i)}}|\chi_{\kappa_j m_{j(j)}}\rangle &= \int_0^{\pi}\sin\vartheta\,\mathrm{d}\vartheta\int_0^{2\pi}\mathrm{d}\varphi\,\chi^{\dagger}_{\kappa_i m_{j(i)}}(\vartheta,\varphi)\cdot\chi_{\kappa_j m_{j(j)}}(\vartheta,\varphi) \\ &= \delta_{\kappa_i\kappa_j}\delta_{m_{j(i)}m_{j(j)}} = \delta_{ij} \end{aligned} \tag{6.58}$$

Accordingly, the 4-spinors of Eq. (6.57) are orthonormal,

$$\begin{aligned}
\langle \Psi_i | \Psi_j \rangle &= \langle \Psi_{n_i \kappa_i m_{j(i)}} | \Psi_{n_j \kappa_j m_{j(j)}} \rangle \\
&= \langle F_{n_i \kappa_i} \chi_{\kappa_i m_{j(i)}} | F_{n_j \kappa_j} \chi_{\kappa_j m_{j(j)}} \rangle + \langle G_{n_i \kappa_i} \chi_{-\kappa_i m_{j(i)}} | G_{n_j \kappa_j} \chi_{-\kappa_j m_{j(j)}} \rangle \\
&= \langle F_{n_i \kappa_i} | F_{n_j \kappa_j} \rangle \langle \chi_{\kappa_i m_{j(i)}} | \chi_{\kappa_j m_{j(j)}} \rangle + \langle G_{n_i \kappa_i} | G_{n_j \kappa_j} \rangle \langle \chi_{-\kappa_i m_{j(i)}} | \chi_{-\kappa_j m_{j(j)}} \rangle \\
&= \left[\langle F_{n_i \kappa_i} | F_{n_j \kappa_j} \rangle + \langle G_{n_i \kappa_i} | G_{n_j \kappa_j} \rangle \right] \delta_{\kappa_i \kappa_j} \delta_{m_{j(i)} m_{j(j)}} \\
&= \delta_{n_i n_j} \delta_{\kappa_i \kappa_j} \delta_{m_{j(i)} m_{j(j)}} = \delta_{ij} \qquad (6.59)
\end{aligned}$$

if we require the radial functions to be orthonormalized according to

$$\begin{aligned}
\langle F_{n_i \kappa_i} | F_{n_j \kappa_j} \rangle + \langle G_{n_i \kappa_i} | G_{n_j \kappa_j} \rangle &= \int_0^\infty \left[F_i(r) F_j(r) + G_i(r) G_j(r) \right] r^2 \mathrm{d}r \\
&= \int_0^\infty \left[P_i(r) P_j(r) + Q_i(r) Q_j(r) \right] \mathrm{d}r = \delta_{ij} = \delta_{n_i n_j} \delta_{\kappa_i \kappa_j} \qquad (6.60)
\end{aligned}$$

(recall that the volume element $\mathrm{d}^3 r = \mathrm{d}x\mathrm{d}y\mathrm{d}z$ becomes in spherical coordinates $r^2 \mathrm{d}r \sin\vartheta \mathrm{d}\vartheta \mathrm{d}\varphi$).

6.5.2
Radial Eigenvalue Equations

With the above ansatz for the spinor we can rewrite the stationary Dirac equation for hydrogen-like atoms with the Hamiltonian of the form of Eq. (6.55) as,

$$h^D \Psi_i = \begin{pmatrix} (V + m_e c^2) F_i \chi_{\kappa_i m_{j(i)}} - \mathrm{i} c\hbar \left(\frac{\boldsymbol{\sigma} \cdot \boldsymbol{r}}{r} \right) \chi_{-\kappa_i m_{j(i)}} \left[\frac{1}{r} \frac{\mathrm{d}}{\mathrm{d}r} r - \frac{\kappa_i}{r} \right] \mathrm{i} G_i \\ -\mathrm{i} c\hbar \left(\frac{\boldsymbol{\sigma} \cdot \boldsymbol{r}}{r} \right) \chi_{\kappa_i m_{j(i)}} \left[\frac{1}{r} \frac{\mathrm{d}}{\mathrm{d}r} r + \frac{\kappa_i}{r} \right] F_i + (V - m_e c^2) \mathrm{i} G_i \chi_{-\kappa_i m_{j(i)}} \end{pmatrix} \qquad (6.61)$$

where we utilized the fact that the Pauli spinors are eigenfunctions of $\boldsymbol{k}$. Now, when $(\boldsymbol{\sigma} \cdot \hat{\boldsymbol{r}})$ operates on the Pauli spinors $\chi_{\kappa m_j}(\vartheta, \varphi)$ we know from Eq. (4.174) that this results in

$$(\boldsymbol{\sigma} \cdot \hat{\boldsymbol{r}}) \chi_{\kappa m_j}(\vartheta, \varphi) = -\chi_{-\kappa m_j}(\vartheta, \varphi) \qquad (6.62)$$

Replacing κ by $-\kappa$ or, equivalently, multiplying Eq. (6.62) by $(\boldsymbol{\sigma} \cdot \hat{\boldsymbol{r}})$ from the left and recalling Eq. (4.167) yields after multiplication by (-1)

$$-\chi_{\kappa m_j}(\vartheta, \varphi) = (\boldsymbol{\sigma} \cdot \hat{\boldsymbol{r}}) \chi_{-\kappa m_j}(\vartheta, \varphi) \qquad (6.63)$$

With the help of Eqs. (6.62) and (6.63) we then obtain

$$h^D \Psi_i = \begin{pmatrix} (V + m_e c^2) F_i \chi_{\kappa_i m_j(i)} - c\hbar \chi_{\kappa_i m_j(i)} \left[\frac{1}{r} \frac{\mathrm{d}}{\mathrm{d}r} r - \frac{\kappa_i}{r} \right] G_i \\ \mathrm{i} c\hbar \chi_{-\kappa_i m_j(i)} \left[\frac{1}{r} \frac{\mathrm{d}}{\mathrm{d}r} r + \frac{\kappa_i}{r} \right] F_i + (V - m_e c^2) \mathrm{i} G_i \chi_{-\kappa_i m_j} \end{pmatrix} \qquad (6.64)$$

which can be written as

$$h^D \Psi_i = \begin{pmatrix} \left\{ (V + m_e c^2) F_i - c\hbar \left[\frac{1}{r}\frac{\mathrm{d}}{\mathrm{d}r} r - \frac{\kappa_i}{r} \right] G_i \right\} \chi_{\kappa_i m_j(i)} \\ \left\{ c\hbar \left[\frac{1}{r}\frac{\mathrm{d}}{\mathrm{d}r} r + \frac{\kappa_i}{r} \right] F_i + (V - m_e c^2) G_i \right\} \mathrm{i}\chi_{-\kappa_i m_j(i)} \end{pmatrix} \stackrel{!}{=} E_i \Psi_i \quad (6.65)$$

We may now either multiply the last equation by $\langle (\chi_{\kappa_i m_j(i)}, -\mathrm{i}\chi_{-\kappa_i m_j(i)})|$ or simply recognize that

$$(V + m_e c^2) F_i(r) - c\hbar \left[\frac{1}{r}\frac{\mathrm{d}}{\mathrm{d}r} r - \frac{\kappa_i}{r} \right] G_i(r) = E_i F_i(r) \quad (6.66)$$

$$c\hbar \left[\frac{1}{r}\frac{\mathrm{d}}{\mathrm{d}r} r + \frac{\kappa_i}{r} \right] F_i(r) + (V - m_e c^2) G_i(r) = E_i G_i(r) \quad (6.67)$$

to finally arrive at radial equations. This set of equations can be written in the form of a matrix eigenvalue problem, which takes a particularly simple form with the second set of radial functions defined in Eq. (6.57),

$$\begin{pmatrix} V(r) + m_e c^2 & c\hbar \left[-\frac{\mathrm{d}}{\mathrm{d}r} + \frac{\kappa_i}{r} \right] \\ c\hbar \left[\frac{\mathrm{d}}{\mathrm{d}r} + \frac{\kappa_i}{r} \right] & V(r) - m_e c^2 \end{pmatrix} \begin{pmatrix} P_i(r) \\ Q_i(r) \end{pmatrix} = E_i \begin{pmatrix} P_i(r) \\ Q_i(r) \end{pmatrix} \quad (6.68)$$

6.5.3
Solution of the Coupled Dirac Radial Equations

The matrix Eq. (6.68) may also be written explicitly for each composite index $i = \{n_i, \kappa_i, m_{j(i)}\}$ as a set of two coupled ordinary differential equations of first order,

$$[V(r) + m_e c^2] P_i(r) + c\hbar \left[-\frac{\mathrm{d}}{\mathrm{d}r} + \frac{\kappa_i}{r} \right] Q_i(r) = E_i P_i(r) \quad (6.69)$$

$$c\hbar \left[\frac{\mathrm{d}}{\mathrm{d}r} + \frac{\kappa_i}{r} \right] P_i(r) + [V(r) - m_e c^2] Q_i(r) = E_i Q_i(r) \quad (6.70)$$

As ordinary differential equations, they can be solved analytically. We proceed to solve the two coupled radial equations and present the analytic expression for the spinor of the ground state. For this purpose, we introduce the substitutions,

$$a^{\pm} \equiv \frac{m_e c^2 \pm E}{c\hbar} = \frac{m_e c}{\hbar} \pm \frac{E}{c\hbar} \quad (6.71)$$

$$z \equiv \frac{Ze^2}{c\hbar} \quad (6.72)$$

The definition of z depends on the explicit choice of the electron–nucleus attraction potential energy. If it is of the form of point particle interactions as in

Eq. (6.3), $V(r) = -Ze^2/r$, the choice for z is as given above. In general, one may write $z \equiv -v_{-1}$ (compare section 6.9). These substitutions help to rewrite the system of coupled differential equations as,

$$\left[\frac{\mathrm{d}}{\mathrm{d}r} - \frac{\kappa}{r}\right] Q(r) \quad + \quad \left[\frac{z}{r} - a^-\right] P(r) = 0 \tag{6.73}$$

$$\left[\frac{\mathrm{d}}{\mathrm{d}r} + \frac{\kappa}{r}\right] P(r) \quad - \quad \left[\frac{z}{r} + a^+\right] Q(r) = 0 \tag{6.74}$$

in which we dropped the state index i for the different states for the sake of brevity. If the upper equation is now differentiated and substituted in the lower (and vice versa), normalizable solutions for large r are obtained if the two radial functions behave like $\exp(-r)$, so that the following ansatz for the radial functions is appropriate,

$$P(r) \equiv p(r) \exp(-\rho r) \tag{6.75}$$

$$Q(r) \equiv q(r) \exp(-\rho r) \tag{6.76}$$

Next, we study the short-range behavior, $r \to 0$, for which we assume that the radial functions may be expanded in a Taylor series around the origin,

$$p(r) = r^\alpha \sum_{i=0}^{\infty} p_i r^i \tag{6.77}$$

$$\Rightarrow P(r) = p(r)e^{-\rho r} = \left(r^\alpha \sum_{i=0}^{\infty} p_i r^i\right)\left(\sum_{i=0}^{\infty} \frac{\rho^i}{i!} r^i\right) = r^\alpha \sum_{i=0}^{\infty} a_i r^i \tag{6.78}$$

$$q(r) = r^\alpha \sum_{i=0}^{\infty} q_i r^i \tag{6.79}$$

$$\Rightarrow Q(r) = q(r)e^{-\rho r} = \left(r^\alpha \sum_{i=0}^{\infty} q_i r^i\right)\left(\sum_{i=0}^{\infty} \frac{\rho^i}{i!} r^i\right) = r^\alpha \sum_{i=0}^{\infty} b_i r^i \tag{6.80}$$

with α being a yet unknown exponent. Both series expansions feature the same unknown exponent α, i.e., both radial functions start with the same exponent, otherwise the first coefficients a_0 and b_0 would be vanishing, as one may verify by considering this case explicitly along the lines presented in the following (we will come back to this problem later in this chapter when we discuss the effect of finite nuclear charge distributions on the radial functions).

Insertion of the ansätze of Eqs. (6.75) and (6.76) into Eqs. (6.73)–(6.74) and multiplication by $\exp(+\rho r)$ yields

$$\frac{\mathrm{d}}{\mathrm{d}r} q(r) - \rho q(r) - \frac{\kappa}{r} q(r) \quad + \quad \left[\frac{z}{r} - a^-\right] p(r) = 0, \tag{6.81}$$

$$\frac{\mathrm{d}}{\mathrm{d}r} p(r) - \rho p(r) + \frac{\kappa}{r} p(r) \quad - \quad \left[\frac{z}{r} + a^+\right] q(r) = 0 \tag{6.82}$$

and then employing the series expansions for $p(r)$ and $q(r)$ and their derivatives we obtain

$$\begin{aligned}\sum_{i=0}^{\infty}(\alpha+i)q_i r^{\alpha+i-1} &- \rho\sum_{i=0}^{\infty}q_i r^{\alpha+i} - \kappa\sum_{i=0}^{\infty}q_i r^{\alpha+i-1} \\ &+ z\sum_{i=0}^{\infty}p_i r^{\alpha+i-1} - a^-\sum_{i=0}^{\infty}p_i r^{\alpha+i} = 0 \end{aligned} \tag{6.83}$$

$$\begin{aligned}\sum_{i=0}^{\infty}(\alpha+i)p_i r^{\alpha+i-1} &- \rho\sum_{i=0}^{\infty}p_i r^{\alpha+i} + \kappa\sum_{i=0}^{\infty}p_i r^{\alpha+i-1} \\ &- z\sum_{i=0}^{\infty}q_i r^{\alpha+i-1} - a^+\sum_{i=0}^{\infty}q_i r^{\alpha+i} = 0 \end{aligned} \tag{6.84}$$

Multiplication by $r^{1-\alpha}$ finally yields

$$\sum_{i=0}^{\infty}(\alpha+i)q_i r^i - \rho\sum_{i=0}^{\infty}q_i r^{i+1} - \kappa\sum_{i=0}^{\infty}q_i r^i + z\sum_{i=0}^{\infty}p_i r^i - a^-\sum_{i=0}^{\infty}p_i r^{i+1} = 0 \tag{6.85}$$

$$\sum_{i=0}^{\infty}(\alpha+i)p_i r^i - \rho\sum_{i=0}^{\infty}p_i r^{i+1} + \kappa\sum_{i=0}^{\infty}p_i r^i - z\sum_{i=0}^{\infty}q_i r^i - a^+\sum_{i=0}^{\infty}q_i r^{i+1} = 0 \tag{6.86}$$

At the short-range limit, $r \to 0$, only those terms survive which are independent of r,

$$\alpha q_0 - \kappa q_0 + z p_0 = (\alpha-\kappa)q_0 + z p_0 = 0 \quad \Rightarrow \quad q_0 = -\frac{z}{\alpha-\kappa}p_0 \tag{6.87}$$

$$\alpha p_0 + \kappa p_0 - z q_0 = (\alpha+\kappa)p_0 - z q_0 = 0 \quad \Rightarrow \quad q_0 = \frac{\alpha+\kappa}{z}p_0. \tag{6.88}$$

For non-zero a_0 and b_0, these two equations determine the exponent α,

$$-\frac{z}{\alpha-\kappa} = \frac{\alpha+\kappa}{z} \quad \Rightarrow \quad \alpha = \sqrt{\kappa^2 - z^2} = \sqrt{\kappa^2 - Z^2e^4/(\hbar^2c^2)} \tag{6.89}$$

which is a *nonintegral* real number, a fact that will become important in the discussion of the short-range behavior of the radial functions in the later part of this chapter. In principle, also the negative square root would be allowed, but in view of Eqs. (6.77) and (6.79) we select the positive root. The only unknowns left in the latter equations are the coefficients a_i and b_i for which we must find a recursion relation. Eqs. (6.85) and (6.86) are most suitable for this purpose as they can be written as

$$[(\alpha-\kappa)q_0 + zp_0] + \sum_{i=1}^{\infty}\left[(\alpha+i-\kappa)q_i - \rho q_{i-1} + zp_i - a^- p_{i-1}\right] r^i = 0, \tag{6.90}$$

$$[(\alpha+\kappa)p_0 - zq_0] + \sum_{i=1}^{\infty}\left[(\alpha+i+\kappa)p_i - \rho p_{i-1} - zq_i - a^+ q_{i-1}\right] r^i = 0 \tag{6.91}$$

whose solutions are obtained if all coefficients vanish independently,

$$(\alpha + i - \kappa)q_i - \rho q_{i-1} + z p_i - a^- p_{i-1} = 0 \quad (6.92)$$
$$(\alpha + i + \kappa)p_i - \rho p_{i-1} - z q_i - a^+ q_{i-1} = 0 \quad (6.93)$$

[the first terms for r^0 in brackets on the left hand side of Eqs. (6.90) and (6.91) vanish, of course, owing to the choice of α according to Eqs. (6.87) and (6.88)]. From the last set of equations recurrence relations can be derived by rewriting them as

$$\rho(\alpha + i - \kappa)q_i - \rho^2 q_{i-1} + \rho z p_i - \rho a^- p_{i-1} = 0 \quad (6.94)$$
$$a^-(\alpha + i + \kappa)p_i - a^- \rho p_{i-1} - a^- z q_i - (a^- a^+)q_{i-1} = 0 \quad (6.95)$$

and subtracting both

$$\left[\rho(\alpha + i - \kappa) + a^- z\right] q_i - \left[\rho^2 - (a^- a^+)\right] q_{i-1} + \left[\rho z - a^-(\alpha + i + \kappa)\right] p_i = 0 \quad (6.96)$$

eliminates the p_{i-1} term. We could equally well choose to eliminate q_{i-1}, which would provide a second equation,

$$\left[a^+(\alpha + i - \kappa) + \rho z\right] q_i - \left[(a^+ a^-) - \rho^2\right] p_{i-1} + \left[a^+ z - \rho(\alpha + i + \kappa)\right] p_i = 0 \quad (6.97)$$

We can now fix the coefficient ρ in the exponent of the ansatz in Eqs. (6.75)–(6.76),

$$\rho^2 = a^+ a^- \quad (6.98)$$
$$\Rightarrow \rho = \sqrt{a^+ a^-} \quad (6.99)$$
$$= \sqrt{\left(\frac{m_e c^2 + E}{c\hbar}\right)\left(\frac{m_e c^2 - E}{c\hbar}\right)} = \frac{\sqrt{m_e^2 c^4 - E^2}}{c\hbar}$$

such that the p_{i-1} and q_{i-1} terms, respectively, vanish in both equations. Now, Eqs. (6.96) and (6.97) become identical [multiply, for instance Eq. (6.97) by ρ/a^+ and utilize Eq. (6.99) to transform Eq. (6.97) into Eq. (6.96)]. Hence, we have obtained a relation between the coefficients of the series expansions for $p(r)$ and $q(r)$,

$$q_i = \frac{-\left[\rho z - a^-(\alpha + i + \kappa)\right]}{\rho(\alpha + i - \kappa) + a^- z} p_i = \frac{-\left[a^+ z - \rho(\alpha + i + \kappa)\right]}{a^+(\alpha + i - \kappa) + \rho z} p_i \quad (6.100)$$

In order to obtain a square-integrable representation of the total 4-spinor, the series expansions for the radial functions $p(r)$ and $q(r)$ must terminate, and we are thus advised to study the long-range behavior of the relation in Eq. (6.100) (note that we proceed in the same way as one would in order to solve the nonrelativistic radial equation of Schrödinger's H atom).

In the limit $r \to \infty$ the natural number i approaches infinity so that

$$\begin{aligned} q_i &= p_i \lim_{i\to\infty} \frac{-a^+z + \rho(\alpha + i + \kappa)}{a^+(\alpha + i - \kappa) + \rho z} \\ &= p_i \lim_{i\to\infty} \left(\frac{-\frac{a^+z}{i} + \frac{\rho\alpha}{i} + \frac{\rho i}{i} + \frac{\rho\kappa}{i}}{\frac{a^+\alpha}{i} + \frac{a^+ i}{i} - \frac{a^+\kappa}{i} + \frac{\rho z}{i}} \right) \\ &= p_i \left(\frac{\rho}{a^+} \right) \quad \text{for } i \to \infty \end{aligned} \tag{6.101}$$

Accordingly, Eqs. (6.92) and (6.93) would become

$$iq_i - \rho q_{i-1} + zp_i - a^- p_{i-1} = 0 \tag{6.102}$$

$$ip_i - \rho p_{i-1} - zq_i - a^+ q_{i-1} = 0 \tag{6.103}$$

in the limit of large i taken for each term in the equations separately (also note that the terms p_{i-1} and q_{i-1} do not vanish because they would also depend on i as the equations imply). A relation between a_i and q_{i-1} for large i can be obtained for the upper Eq. (6.102) by replacing p_i and p_{i-1} by q_i and q_{i-1} through Eq. (6.101) for both values of i,

$$iq_i - \rho q_{i-1} + z\left(\frac{a^+}{\rho}\right) q_i - a^- \left(\frac{a^+}{\rho}\right) q_{i-1} = 0 \tag{6.104}$$

which can be simplified to

$$\left(i + z\frac{a^+}{\rho}\right) q_i - \left(\rho + a^- \frac{a^+}{\rho}\right) q_{i-1} = 0 \tag{6.105}$$

and then reads

$$q_{i-1} = -\frac{1}{2}\left(-\frac{i}{\rho} - \frac{za^+}{\rho^2}\right) q_i \approx \frac{i}{2\rho} q_i \quad \text{(for large } i\text{)} \tag{6.106}$$

Hence, we have

$$q_i \approx \frac{2\rho}{i} q_{i-1} \quad \text{for large } i \tag{6.107}$$

Analogously we would obtain for the second set of coefficients

$$p_i \approx \frac{2\rho}{i} p_{i-1} \quad \text{for large } i \tag{6.108}$$

The series expansions for $p(r)$ and $q(r)$ in Eqs. (6.77) and (6.79) can therefore be written in the large-i limit as,

$$p(r) \xrightarrow{\text{for large } i} \sum_{j=0}^{\infty} \frac{(2\rho)^j}{j!} r^j = \exp(2\rho r) \tag{6.109}$$

$$q(r) \xrightarrow{\text{for large } i} \sum_{j=0}^{\infty} \frac{(2\rho)^j}{j!} r^j = \exp(2\rho r) \tag{6.110}$$

Obviously, these functions do not vanish at large r as they should in order to yield normalizable spinors. To recover square-integrable spinors and thus square-integrable radial functions, these series expansions *must* truncate. Let us assume that this is the case for some $(i-1) = n_r$. Then, the recursion relations of Eqs. (6.92) and (6.93) read for the vanishing coefficients to follow, $p_{n_r} = q_{n_r} = 0$,

$$-\rho q_{n_r} - a^- p_{n_r} = 0 \quad \Leftrightarrow \quad q_{n_r} = -\frac{a^-}{\rho} p_{n_r} \tag{6.111}$$

$$-\rho p_{n_r} - a^+ q_{n_r} = 0 \quad \Leftrightarrow \quad p_{n_r} = -\frac{a^+}{\rho} q_{n_r} \tag{6.112}$$

Equating Eq. (6.111) and Eq. (6.100) for $i = n_r$ yields

$$-\frac{a^-}{\rho} p_{n_r} = \frac{-a^+ z + \rho(\alpha + n_r + \kappa)}{a^+(\alpha + n_r - \kappa) + \rho z} p_{n_r} \tag{6.113}$$

which is to be fulfilled for any p_{n_r}, and hence

$$\begin{aligned}
-\frac{a^-}{\rho}\left[a^+(\alpha + n_r - \kappa) + \rho z\right] &= -a^+ z + \rho(\alpha + n_r + \kappa) \\
\overset{(6.98)}{\Leftrightarrow} \quad -\rho(\alpha + n_r - \kappa) - a^- z &= -a^+ z + \rho(\alpha + n_r + \kappa) \\
\overset{(6.72)}{\Rightarrow} \quad (a^+ - a^-)\frac{Ze^2}{c\hbar} &= 2\rho(\alpha + n_r)
\end{aligned} \tag{6.114}$$

6.5.4
Energy Eigenvalue, Quantization and the Principal Quantum Number

From Eq. (6.114) we can deduce with the help of Eqs. (6.71) and (6.99),

$$E\frac{Ze^2}{c\hbar} = \sqrt{m_e^2 c^4 - E^2}(\alpha + n_r) \tag{6.115}$$

an expression for the energy eigenvalue,

$$E^2 \frac{Z^2 e^4}{c^2 \hbar^2} = \left[m_e^2 c^4 - E^2\right](\alpha + n_r)^2$$

$$\Rightarrow E^{\pm} = \pm m_e c^2 \left[1 + \frac{Z^2 e^4}{c^2 \hbar^2 (\alpha + n_r)^2}\right]^{-\frac{1}{2}} \tag{6.116}$$

Because of $\alpha = \sqrt{\kappa^2 - Z^2 e^4/(\hbar^2 c^2)}$, Eq. (6.89), we note that those pairs of spinors with the same absolute value $|\kappa|$ lead to the same energy and are thus degenerate. By construction, the integer n_r, which is called the *radial quantum number* can take any values which are allowed for the summation index i in the equations above,

$$n_r = 0, 1, 2, 3, \ldots \quad , \quad n_r \in \mathbf{N}_0 \tag{6.117}$$

The quantum number n_r is related to κ as we shall see in the following. From the truncation condition Eq. (6.111) we deduce for the smallest possible value of $n_r = 0$,

$$\frac{q_0}{p_0} = -\frac{a^-}{\rho} = -\frac{m_e c^2 - E}{c\sqrt{m_e^2 c^4 - E^2}} < 0 \text{ in the case of } E = E^+ \tag{6.118}$$

while we can rewrite the short-range representation of the differential equation in Eq. (6.87),

$$\frac{q_0}{p_0} = -\frac{z}{\alpha - \kappa} = -\frac{Ze^2}{c\left[\sqrt{\kappa^2 - Z^2 e^4/(\hbar^2 c^2)} - \kappa\right]} = \begin{cases} < 0 & \text{if} \quad \kappa < 0 \\ > 0 & \text{if} \quad \kappa > 0 \end{cases} \tag{6.119}$$

Consequently, since both equations, Eq (6.118) and Eq. (6.119), hold at the same time, we must reject all positive κ values for $n_r = 0$. The *principal quantum number* n can be defined as

$$n = n_r + |\kappa| = n_r + j + \frac{1}{2} \tag{6.120}$$

so that the energy eigenvalue of Eq. (6.116) for the electronic bound states in Dirac hydrogen-like atoms reads

$$E^+ = m_e c^2 \left\{1 + \left(\frac{Ze^2/(\hbar c)}{\sqrt{\kappa^2 - Z^2 e^4/(\hbar^2 c^2)} + n - |\kappa|}\right)^2\right\}^{-\frac{1}{2}} \tag{6.121}$$

or in Hartree atomic units

$$E_{n|\kappa|} \equiv E^+ = c^2 \left\{1 + \left(\frac{Z/c}{\sqrt{\kappa^2 - Z^2/c^2} + n - |\kappa|}\right)^2\right\}^{-\frac{1}{2}} \tag{6.122}$$

Hence the energy depends on the pair $(n, |\kappa|) = (n_i, |\kappa_i|)$, while a shell i, for which we use a single set of radial functions $(P_i, Q_i) = (P_{n_i\kappa}, Q_{n_i\kappa})$ owing to

Table 6.2 Explicit list of angular momentum quantum numbers (see also Table 6.1).

κ	± 1	± 2	...	$\pm(n-1)$	$\pm n$
j	$1/2$	$3/2$	...	$n-3/2$	$n-1/2$
$j+1/2$	1	2	...	$n-1$	n
l	0	1	...	$n-2$	$n-1$
	1	2	...	$n-1$	

Table 6.3 Quantum numbers sorted by shells and assigned to shell and term symbols.

n	n_r	κ	$\|\kappa\|$	j	l	Shell	Term nL_j
1	0	-1	1	$1/2$	0	$1s$	$1S_{1/2}$
2	1	-1	1	$1/2$	0	$2s$	$2S_{1/2}$
	1	$+1$	1	$1/2$	1	$2\bar{p}$	$2P_{1/2}$
	0	-2	2	$3/2$	1	$2p$	$2P_{3/2}$
3	2	-1	1	$1/2$	0	$3s$	$3S_{1/2}$
	2	$+1$	1	$1/2$	1	$3\bar{p}$	$3P_{1/2}$
	1	-2	2	$3/2$	1	$3p$	$3P_{3/2}$
	1	$+2$	2	$3/2$	2	$3\bar{d}$	$3D_{3/2}$
	0	-3	3	$5/2$	2	$3d$	$3D_{5/2}$
4	3	-1	1	$1/2$	0	$4s$	$4S_{1/2}$
	3	$+1$	1	$1/2$	1	$4\bar{p}$	$4P_{1/2}$
	2	-2	2	$3/2$	1	$4p$	$4P_{3/2}$
	2	$+2$	2	$3/2$	2	$4\bar{d}$	$4D_{3/2}$
	1	-3	3	$5/2$	2	$4d$	$4D_{5/2}$
	1	$+3$	3	$5/2$	3	$4\bar{f}$	$4F_{5/2}$
	0	-4	4	$7/2$	3	$4f$	$4F_{7/2}$
5	4	-1	1	$1/2$	0	$5s$	$5S_{1/2}$
	...	...	...	...	...	...	...

the equivalence restriction, is defined by (n_i, κ_i). Tables 6.2 and 6.3 list possible combinations of the different quantum numbers.

From the general expression of Eq. (6.122) the ground state energy E_{11} for $(n, \kappa) = (1, -1)$ can be derived (in Hartree atomic units),

$$E_0 \equiv E_{11} = c^2\sqrt{1 - Z^2/c^2} \tag{6.123}$$

6.5.5
The Four-Component Ground State Wave Function

The ground state is doubly degenerate with the two eigen solutions,

$$\Psi_{1(-1)1/2} = \frac{[2m_e(Z\alpha)]^{3/2}}{\sqrt{4\pi}} \sqrt{\frac{1+\alpha_1}{2\Gamma(1+2\alpha_1)}} [2m_e(Z\alpha)r]^{\alpha_1-1} \times e^{-m_e(Z\alpha)r} \begin{pmatrix} 1 \\ 0 \\ \frac{\mathrm{i}(1-\alpha_1)}{Z\alpha}\cos\vartheta \\ \frac{\mathrm{i}(1-\alpha_1)}{Z\alpha}\sin\vartheta e^{\mathrm{i}\varphi} \end{pmatrix} \quad (6.124)$$

$$\Psi_{1(-1)(-1/2)} = \frac{[2m_e(Z\alpha)]^{3/2}}{\sqrt{4\pi}} \sqrt{\frac{1+\alpha_1}{2\Gamma(1+2\alpha_1)}} [2m_e(Z\alpha)r]^{\alpha_1-1} \times e^{-m_e(Z\alpha)r} \begin{pmatrix} 0 \\ 1 \\ \frac{\mathrm{i}(1-\alpha_1)}{Z\alpha}\sin\vartheta e^{\mathrm{i}\varphi} \\ \frac{-\mathrm{i}(1-\alpha_1)}{Z\alpha}\cos\vartheta \end{pmatrix} \quad (6.125)$$

where $\alpha_1 = \sqrt{1 - Z^2\alpha^2}$. Here, α is the fine-structure constant, $\alpha = e^2/(\hbar c)$. $\Gamma(x)$ is the Gamma function resulting from the normalization condition. It is often more convenient to write the above spinors in terms of the Bohr radius, $a_0 = \hbar^2/(m_e e^2)$, instead of the electron rest mass m_e [note, in particular, that $m_e \alpha = \hbar/(ca_0)$].

The *most general* expressions involve confluent hypergeometric functions for the one-electron states [48]. Additional expressions for the analytic solutions of the energetically lowest lying states for the radial functions of hydrogen-like Dirac atoms can be found elsewhere [48, 85, 86].

6.6
The Nonrelativistic Limit

Before we investigate in more detail the eigenpairs of the Dirac equation for hydrogen-like atoms, i.e., the energy eigenvalue and the corresponding set of radial functions, we shall look into the relation of the Dirac and Schrödinger hydrogen atoms. In section 6.3, we have already recalled some elements of the nonrelativistic theory of atoms. Now, we transfer the general discussion of obtaining the nonrelativistic limit, $c \to \infty$, presented in section 5.4.3 to the case of atoms. Since the radial and angular coordinates may be treated separately, we derive the nonrelativistic radial equation, Eq. (6.17), from the

Dirac radial equations, Eqs. (6.69) and (6.70), and study the nonrelativistic limit of the Pauli spinors separately.

For $c \rightarrow \infty$ (this means instant action of forces) the radial Dirac equations take the form of the radial Schrödinger equation for a particle moving in a central field $V(r)$. If we write Eqs. (6.69) and (6.70) — after subtraction of $-m_e c^2$, division of Eq. (6.70) by c and introduction of $E_i' \equiv E_i - m_e c^2$ — as

$$V(r)P_i(r) - \hbar\left[\frac{\mathrm{d}}{\mathrm{d}r} - \frac{\kappa_i}{r}\right][cQ_i(r)] = E_i' P_i(r) \tag{6.126}$$

$$\hbar\left[\frac{\mathrm{d}}{\mathrm{d}r} + \frac{\kappa_i}{r}\right]P_i(r) + \frac{V(r)}{c}Q_i(r) - 2m_e c Q_i(r) = \frac{E_i'}{c}Q_i(r) \tag{6.127}$$

we may rearrange the second equation,

$$cQ_i(r) = \frac{1}{2m_e}\left\{\hbar\left[\frac{\mathrm{d}}{\mathrm{d}r} + \frac{\kappa_i}{r}\right]P_i(r) + \frac{V(r) - E_i'}{c}Q_i(r)\right\} \tag{6.128}$$

insert it into the first,

$$V(r)P_i(r) - \hbar\left[\frac{\mathrm{d}}{\mathrm{d}r} - \frac{\kappa_i}{r}\right]\left(\frac{1}{2m_e}\left\{\hbar\left[\frac{\mathrm{d}}{\mathrm{d}r} + \frac{\kappa_i}{r}\right]P_i(r) + \frac{V(r) - E_i'}{c}Q_i(r)\right\}\right) = E_i' P_i(r) \tag{6.129}$$

and take the limit $c \rightarrow \infty$ so that $P_i(r) \rightarrow P_i^{nr}(r)$

$$V(r)P_i^{nr}(r) - \frac{\hbar^2}{2m_e}\left[\frac{\mathrm{d}}{\mathrm{d}r} - \frac{\kappa_i}{r}\right]\left[\frac{\mathrm{d}}{\mathrm{d}r} + \frac{\kappa_i}{r}\right]P_i^{nr}(r) = E_i' P_i^{nr}(r) \tag{6.130}$$

We can resolve the brackets in the latter equation,

$$\left[\frac{\mathrm{d}}{\mathrm{d}r} - \frac{\kappa_i}{r}\right]\left[\frac{\mathrm{d}}{\mathrm{d}r} + \frac{\kappa_i}{r}\right] = \left[\frac{\mathrm{d}^2}{\mathrm{d}r^2} - \frac{\kappa_i}{r^2} - \frac{\kappa_i^2}{r^2}\right] = \left[\frac{\mathrm{d}^2}{\mathrm{d}r^2} - \frac{\kappa_i + \kappa_i^2}{r^2}\right] \tag{6.131}$$

and then arrive at the Schrödinger equation for hydrogen-like atoms, Eq. (6.9),

$$\left(-\frac{\hbar^2}{2m_e}\frac{\mathrm{d}^2}{\mathrm{d}r^2} + \frac{\kappa_i(\kappa_i + 1)\hbar^2}{2m_e r^2} + V(r)\right)P_i^{nr}(r) = E_i' P_i^{nr}(r) \tag{6.132}$$

from which it is evident how the mass m_e in the Schrödinger kinetic energy arises from the rest mass m_e of the energy-shifted Dirac equation. Eq. (6.132) holds for both choices of the quantum number κ_i,

$$\kappa_i = \begin{cases} l_i \\ -(l_i + 1) \end{cases} \tag{6.133}$$

and, hence, yields in both cases the centrifugal potential $+l_i(l_i+1)\hbar^2/(2m_e r^2)$ for shell i.

In the derivation presented above we subtracted $m_e c^2$ from both radial Dirac equations in order to remove any c-dependence from the upper equation, which then allowed us to easily evaluate the limit $c \to \infty$. A similar effect was also induced in the first discussion on the nonrelativistic limit in section 5.4.3.

The consideration of the nonrelativistic limit of the Dirac energy eigenvalue for the hydrogen-like atom with a Coulombic potential for the electron–nucleus attraction, Eq. (6.3), demonstrates the effect of subtracting the rest energy $m_e c^2$ and leads us to a discussion of the reference energy in the following section 6.7.

The energy eigenvalue in Eq. (6.122) in Hartree atomic units ('$\hbar = e = m_e = 4\pi\epsilon_0 = 1$') may be expanded in a Taylor series,

$$E_i = E_{n_i|\kappa_i|} = c^2 \left\{ 1 - \frac{(Z/c)^2}{2n^2} - \frac{(Z/c)^4}{2n^3} \left[\frac{1}{j+\frac{1}{2}} - \frac{3}{4n} \right] + O\left(\frac{Z^6}{c^6} \right) \right\} \tag{6.134}$$

which allows us to most easily consider the nonrelativistic limit for large c,

$$\begin{aligned} E_i(c \to \infty) &= \lim_{c\to\infty} \left[c^2 - \frac{Z^2}{2n^2} - \frac{Z^2(Z/c)^2}{2n^3} \left[\frac{1}{j+\frac{1}{2}} - \frac{3}{4n} \right] + O\left(\frac{Z^6}{c^4} \right) \right] \\ &= c^2 - \frac{Z^2}{2n^2} \end{aligned} \tag{6.135}$$

From this equation we understand that the nonrelativistic Schrödinger energy eigenvalue for hydrogen-like atoms with point-like Coulomb nucleus, Eq. (6.19), is obtained only *after subtraction of the rest energy*,

$$E_i' = -\frac{Z^2}{2n^2} = E_i(c \to \infty) - m_e c^2 \tag{6.136}$$

where we reintroduced the rest mass m_e for the sake of convenience. The nonrelativistic wave function can then either be obtained by a limiting process $c \to \infty$ applied to the Dirac eigen states or by solving the nonrelativistic radial equation derived above.

6.7
Choice of the Energy Reference and Matching Energy Scales

From section 6.6 we understand that the nonrelativistic energy is not the first term of the series expansion, but the rest energy $m_e c^2$ is. Therefore, the energy eigenvalues will differ by $m_e c^2$. We will later see that this result transfers to the many-electron case. Since a constant W can be added to the time-independent Dirac equation, Eq. (6.7),

$$h^D\Psi(\boldsymbol{r}) + W\Psi(\boldsymbol{r}) = E\Psi(\boldsymbol{r}) + W\Psi(\boldsymbol{r}) \tag{6.137}$$

$$\Rightarrow \left[h^D + W\right]\Psi(\boldsymbol{r}) = E'\Psi(\boldsymbol{r}) \quad \text{with } E' \equiv E + W \tag{6.138}$$

we may choose an energy reference $W \equiv m_e c^2 \mathbf{1}_4$, which matches the non-relativistic limit for $c \to \infty$ exactly. Of course, one may equally well add a constant to the Schrödinger equation in order to match the zero energy level of the Dirac equation. This is, however, not done as the energy takes very large values then — apart from historical reasons and because of the importance of Schrödinger-type quantum mechanics in quantum chemistry. One may also argue that the Schrödinger energy scale is convenient as it yields negative energies for electronic bound states and positive energies for freely moving electrons in continuum states.

The general Dirac equation, Eq. (5.119), for an electron in an external scalar potential, i.e., $\boldsymbol{A} = 0$, reads after shifting the zero energy level

$$\left[c\,\boldsymbol{\alpha}\cdot\boldsymbol{p} + (\beta - \mathbf{1}_4)m_e c^2 + q_e\Phi\right]\Psi = \mathrm{i}\hbar\frac{\partial}{\partial t}\Psi \tag{6.139}$$

The radial Dirac equations, Eqs. (6.69) and (6.70), for hydrogen-like atoms then become,

$$V(r)P_i(r) + c\hbar\left[-\frac{\mathrm{d}}{\mathrm{d}r} + \frac{\kappa_i}{r}\right]Q_i(r) = E_i P_i(r) \tag{6.140}$$

$$c\hbar\left[\frac{\mathrm{d}}{\mathrm{d}r} + \frac{\kappa_i}{r}\right]P_i(r) + [V(r) - 2m_e c^2]Q_i(r) = E_i Q_i(r) \tag{6.141}$$

and, according to Eq. (6.137), we can immediately write the energy eigenvalue for Dirac hydrogen-like atoms in a Coulombic potential energy $V(r) = -Ze^2/r$ as

$$E'_{n|\kappa|} \equiv E_{n|\kappa|} - m_e c^2 \tag{6.142}$$

$$= m_e c^2\left[\left\{1 + \left(\frac{Ze^2/(\hbar c)}{\sqrt{\kappa^2 - Z^2e^4/(\hbar^2c^2)} + n - |\kappa|}\right)^2\right\}^{-\frac{1}{2}} - 1\right] \tag{6.143}$$

instead of going through the derivation again.

In the following chapters, the by $m_e c^2$-shifted energy is usually used. Hence, all bound electronic states, which appeared at positive energies prior to the energy shift, now show up at negative energy. Nonetheless, it is common to still call these solutions *positive-energy solutions*, while the artificial 'positronic' continuum states that originally and still appear at negative energies are called *negative-energy* solutions.

6.8 Wave Functions and Energy Eigenvalues in the Coulomb Potential

In this section, we present graphical representations of the components and discuss some important properties like the number of nodes and the energy eigenvalues.

6.8.1 Features of Dirac Radial Functions

While the mathematical expressions for the radial functions are the more complicated the larger the principal quantum number is, the graphical representation of basic features can easily be sketched as in the nonrelativistic case. By convention, the large components $P_i(r)$ are chosen to start with positive values from the origin of the r coordinate.

As early as 1931, White presented pictures of the states of the Dirac hydrogen atom [87]. All graphical representations of radial functions $P_i(r)$ and $Q_i(r)$ in this chapter have been produced with a computer program for numerical atomic structure calculations [88]. The mathematical and algorithmic approach will be described in detail in chapter 9. Figure 6.1 provides ground state radial functions of some hydrogen-like atoms. The radial functions $P_i(r)$ and $Q_i(r)$ take zero values at the origin because of the r-weighting according to Eq. (6.57). Consequently, the weak singularity of $F_i(r)$ and $G_i(r)$ at the origin is not visible.

Figure 6.1 also shows the increasing size of the small component $Q_{1s}(r)$ with nuclear charge Ze. For superheavy atoms — say, for $Z > 100$ — the small component is no longer small. For $Z = c \approx 137$, $P_{1s}(r)$ and $Q_{1s}(r)$ are of similar magnitude in absolute value but different sign at any distance r.

An important feature of the radial functions of the 4-spinor is the number of (radial) nodes, i.e., the number of positions where these functions are zero. This number is determined by the polynomial prefactor to the exponential function of the radial functions. Hence, the number of nodes solely depends on the quantum numbers that are used to classify the spinor. $P_{n\kappa}(r)$ always has $(n - l - 1)$ radial nodes like its nonrelativistic analog, the Schrödinger radial function $P_{nl}(r)$. However, $Q_{n\kappa}(r)$ has as many nodes as $P_{n\kappa}(r)$ for negative values of κ and one additional node for positive values of κ_i [89, p. 754]. Interestingly enough, all nodes of $P_{n\kappa}(r)$ lie before (from the origin's point of view) the absolute extremum of the function. However, the $Q_{n\kappa}(r)$ function has some nodes behind its absolute extremum.

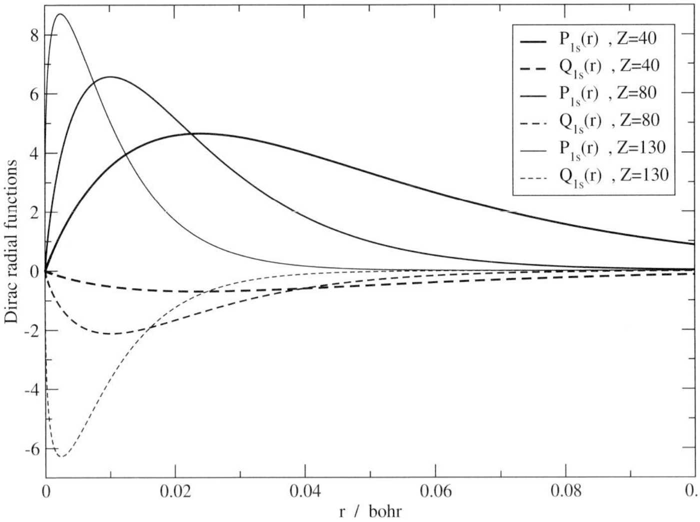

Figure 6.1 Radial functions $P_{1s}(r)$ and $Q_{1s}(r)$, in Hartree atomic units, of the ground state Dirac spinor of the hydrogen-like atoms with $Z = 40, 80, 130$, i.e., for Zr^{39+}, Hg^{79+}, and 130^{129+}. With increasing nuclear charge the radial functions contract. The small component $Q_{1s}(r)$ is small indeed for not too large nuclear charges, but its overall 'size' increases with Z in such a way that $P_{1s}(r)$ and $Q_{1s}(r)$ are of similar 'size' (but different sign) for $Z = 130$, i.e., if $Z \approx c$ in atomic units.

Finally, we should note that spin–orbit coupling is the reason for the splitting of the p shells into $p_{1/2}$ and $p_{3/2}$, the d shells into $d_{3/2}$ and $d_{5/2}$ and so on. The effect on the radial functions is, however, not very pronounced for small and intermediate nuclear charges Ze. We may refer here to Figure 9.3 of a later chapter, where the effect is highlighted for the shells of a Rn cation. Moreover, we will come back to this issue in the context of chemical bonding in chapter 16 (see also the discussion of spin–orbit splitting in atoms in section 16.1). The effect of spin–orbit splitting is reduced for increasing principal quantum number n (compare the energy eigenvalues given in Table 6.4).

6.8.2
Spectrum of Dirac Hydrogen-like Atoms with Coulombic Potential

The energy eigenvalues of the Dirac hydrogen atom were derived in Eq. (6.122). While the corresponding expression for the Schrödinger hydrogen atom takes the simple form $-Z^2/(2n^2)$ in Hartree atomic units, this is no longer the case for the Dirac energy eigenvalues. Especially, the energy eigenvalues are not simply scaled with the squared nuclear charge number Z^2. Hence, division by Z^2 does not yield a constant energy independent of Z in the case of Dirac hydrogen-like atoms (compare also Figure 6.6).

Table 6.4 provides a list of the energy eigenvalues for the hydrogen atom (i.e., $Z = 1$) in Dirac's and Schrödinger's theory. It has already been discussed in section 6.7 that the energy reference must be shifted in order to make the spectra comparable. Hence, the energy E_i' shifted by $m_e c^2$ can be directly compared to the Schrödinger eigenvalue E_i^{nr}. As one would expect, the energy difference between both eigenvalues is rather small for the light hydrogen atom. In general, the difference between E_i' and E_i^{nr} is small for small Z. It is dramatically enlarged with increasing nuclear charge number Z, and we come back to this point in section 6.9 as this effect depends on the form of the electron–nucleus attraction potential (compare again Figure 6.6).

The energy difference is also the smaller the larger n and j are. Thus, the largest difference when compared with the energy eigenvalue of the Schrödinger equation is found for $1s_{1/2}$, the next largest for $2s_{1/2}$ and $2p_{1/2}$, and so on. Also, the two p, the two d and the two f etc. shells behave rather differently. But note that Table 6.4 does not resolve the degenerate states explicitly. For example, the $2S_{1/2}$ and the $2P_{1/2}$ state are represented by the same entry for the pair $(n, j) = (2, 1/2)$.

Table 6.4 Energy eigenvalues for the Dirac and Schrödinger hydrogen atom in Hartree atomic units. The Dirac eigenvalue E_i and the shifted energy E_i' were calculated for $c = 137.037$. Then $m_e c^2$ takes a value of 18779.139369 in Hartree atomic units for this value of the speed of light. The difference between E_i' and E_i^{nr} is rather small and can only be enhanced for larger values of the nuclear charge number Z. The difference is the larger the smaller n and j are.

n	j	E_i	E_i'	E_i^{nr}
1	1/2	18778.639362343	-0.500006656	−0.50000000
2	1/2	18779.014366920	-0.125002080	−0.12500000
	3/2	18779.014368584	-0.125000416	
3	1/2	18779.083812705	-0.055556295	−0.05555555
	3/2	18779.083813198	-0.055555802	
	5/2	18779.083813362	-0.055555638	
4	1/2	18779.108118662	-0.031250338	−0.03125000
	3/2	18779.108118870	-0.031250130	
	5/2	18779.108118939	-0.031250061	
	7/2	18779.108118974	-0.031250026	
5	1/2	18779.119368819	-0.020000181	−0.02000000
	3/2	18779.119368925	-0.020000075	
	5/2	18779.119368961	-0.020000039	
	7/2	18779.119368979	-0.020000021	
	9/2	18779.119368989	-0.020000011	

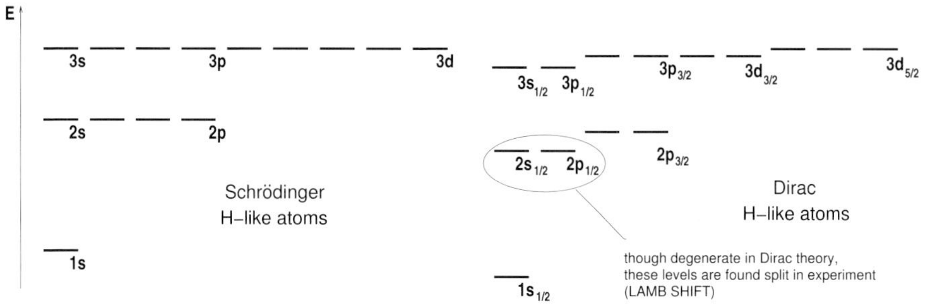

Figure 6.2 Bound electronic ground and excited states of one-electron Schrödinger (left) and Dirac (right) atoms. Note the so-called accidental degeneracy in the Schrödinger case: all energies depend only on the principal quantum number n and are the same for the l quantum number. This degeneracy is lifted in the Dirac case (fine-structure splitting due to coupling of spin and orbital momenta). Note that the positive-energy *continuum* states (and in the case of the Dirac hydrogen-like atoms also the negative-energy continuum states) are not depicted.

Figure 6.2 compares the ground and excited bound-state energies of Schrödinger and Dirac hydrogen-like atoms. Apart from the lowering of the energies due to kinematic relativistic effects, shells of different j quantum number split, which is called *fine-structure splitting*. This effect is therefore due to *spin–orbit coupling* because the different j values in a given (n, l) shell originate from the coupling of spin and angular momentum of that shell. On the other hand, for a given principal quantum number n, the energies are not split. This is called accidental degeneracy and can be understood by group-theoretical means [90,91].

Of course, the complete spectrum also features continuum states at positive energy and, in the Dirac case, infinitely many negative-energy states are obtained as already anticipated in Figure 5.1. *Infinitely many negative-energy continuum states (see section 5.3) always show up whenever the spectrum of a Hamiltonian is investigated that contains the free-particle Dirac Hamiltonian — independently of the particular form of the potential energy operator V.* Experimentally it is known that the $ns_{1/2}$ and the $np_{1/2}$ shells are not degenerate as predicted by Dirac's theory. Instead, these two energy levels split slightly, which is called the *Lamb shift*. This issue can only be resolved in the theory of quantum electrodynamics, a theory introduced in chapter 7.

6.8.3
Radial Density and Expectation Values

Owing to the radial symmetry of atoms, all angular- and spin-dependent terms can be treated analytically so that these degrees of freedom can be in-

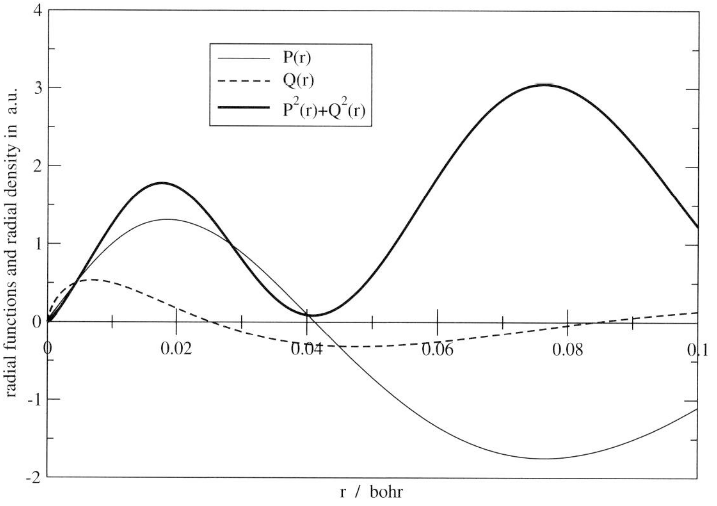

Figure 6.3 Magnification of the excited-state radial density, in Hartree atomic units (a.u.), of a super-heavy hydrogen-like atom. The electron density distribution and both radial functions are shown. Note the *non*-occurrence of nodes in the radial density (the high nuclear charge guarantees that this effect is magnified).

tegrated out. Only the radial-dependent terms remain in expectation values. This still holds in the case of many-electron atoms as we shall see in chapter 9.

If we can transform a one-electron operator $\hat{o}$ to spherical coordinates in such a way that

$$\hat{o}(x,y,z) \quad \longrightarrow \quad \hat{o}(r,\vartheta,\varphi) = \hat{o}(r)\cdot\hat{o}(\vartheta,\varphi) \tag{6.144}$$

we avoid the introduction of new symbols and distinguish the different operators solely by their coordinate dependence.

After integration of the angular- and spin-dependent terms,

$$\begin{aligned}\langle \Psi_{n_i\kappa_i m_{j(i)}}|\hat{o}|\Psi_{n_j\kappa_j m_{j(j)}}\rangle &= \langle F_{n_i\kappa_i}|\hat{o}(r)|F_{n_j\kappa_j}\rangle\langle \chi_{\kappa_i m_{j(i)}}|\hat{o}(\vartheta,\varphi)|\chi_{\kappa_j m_{j(j)}}\rangle \\ &+\langle G_{n_i\kappa_i}|\hat{o}(r)|G_{n_j\kappa_j}\rangle\langle \chi_{-\kappa_i m_{j(i)}}|\hat{o}(\vartheta,\varphi)|\chi_{-\kappa_j m_{j(j)}}\rangle\end{aligned} \tag{6.145}$$

we can reduce the full expectation value to radial integrals,

$$\begin{aligned}\langle \Psi_{n_i\kappa_i m_{j(i)}}|\hat{o}|\Psi_{n_j\kappa_j m_{j(j)}}\rangle &= A_{\kappa_i m_{j(i)},\kappa_j m_{j(j)}}\int_0^\infty r^2\,\mathrm{d}r\,F^\star_{n_i\kappa_i}(r)\hat{o}(r)F_{n_j\kappa_j}(r) \\ &+A_{-\kappa_i m_{j(i)},-\kappa_j m_{j(j)}}\int_0^\infty r^2\,\mathrm{d}r\,G^\star_{n_i\kappa_i}(r)\hat{o}(r)G_{n_j\kappa_j}(r)\end{aligned} \tag{6.146}$$

where the result of the angular integration is represented by the number $A_{\kappa_i m_{j(i)}}$ and $A_{-\kappa_i m_{j(i)}}$. If $\hat{o}(r)$ is a multiplicative operator and both states are equal, i.e., $i = j$, the radial integrals can be simplified to read

$$\int_0^\infty r^2 \mathrm{d}r\, \hat{o}(r) \left[F^2_{n_i\kappa_i}(r) + G^2_{n_i\kappa_i}(r)\right] = \int_0^\infty \mathrm{d}r\, \hat{o}(r) \left[P^2_{n_i\kappa_i}(r) + Q^2_{n_i\kappa_i}(r)\right] = \int_0^\infty \mathrm{d}r\, \hat{o}(r)\, \rho_i(r) \qquad (6.147)$$

where we utilized the fact that the radial functions are real-valued by construction, so that we may simply employ their square. Additionally, we assume that $A_{\kappa_i m_{j(i)}, \kappa_j m_{j(j)}} = A_{-\kappa_i m_{j(i)}, -\kappa_j m_{j(j)}}$. In the last step we introduced the radial density $\rho_i(r)$ of state $\Psi_i(\boldsymbol{r})$,

$$\rho_i(r) \equiv P^2_{n_i\kappa_i}(r) + Q^2_{n_i\kappa_i}(r) \qquad (6.148)$$

Since the radial functions $P_i(r)$ and $Q_i(r)$ do not, in general, feature nodes at the same radial position, the resulting radial density will not show any nodes — apart from the one at the origin $r = 0$. This is in sharp contrast to the nonrelativistic radial density $\rho_i^{nr}(r)$, which is the square of the nonrelativistic radial function $P_i^{nr}(r)$. Hence, $\rho_i^{nr}(r)$ and $P_i^{nr}(r)$ possess nodes at the same values of r. Figure 6.3 displays the radial functions P_i and Q_i as well as the resulting radial density for a hydrogen-like atom with nuclear charge Ze in an interval of the radial coordinate r. The large nuclear charge in this one-electron atom ensures that the Dirac radial density can be distinguished from zero. If the small component is small, which is the case for small nuclear charge numbers Z, the radial density $\rho_i(r)$ approaches values very close to zero if $P_i(r)$ is zero.

It is occasionally argued that the missing nodes in the Dirac radial density explain how 'electrons get across the nodes' [92] as there are no longer any nodes. However, one must keep in mind that this is a somewhat artificial question as time has been eliminated from the stationary Dirac and Schrödinger equations. The question remains why the density dramatically decreases at certain radial distance intervals, at which the large component features radial nodes, whether there is a true node or only a very small density.

Finally, we note that analytic expressions are available for the radial moments $\langle r^k \rangle$ of hydrogen-like Dirac atoms [93].

6.9
Finite Nuclear Size Effects

So far, we have assumed that the interaction of an electron with an atomic nucleus is an instantaneous electrostatic monopole–monopole interaction of the moving electron with the point charge of a nucleus chosen to be at rest, which can be described by a Coulombic potential energy operator introduced in Eq. (6.3). The larger the nuclear charge $+Ze$, however, the larger the size of the nucleus becomes. Hence, heavy atomic nuclei are not well represented by a point charge. Replacing a point-like charge by an extended charge distribution has the important mathematical consequence that the resulting potential is no longer singular at the origin like Ze/r — from which the attractive potential energy operator $-Ze^2/r$ results — but features finite values for $r = 0$. Asymptotically, a finite positive charge distribution yields the same $\propto 1/r$ potential as that yielded by a point charge.

Three questions need to be answered now: (i) What is the effect of this subtle and pronounced short-range difference in the potential energy operator on energies and other observables, (ii) how to choose a function to describe the positive nuclear charge distribution and (iii) does it make a difference whether Dirac or Schrödinger quantum mechanics are considered? Regarding the last question, we may anticipate that the effect of a finite nuclear charge on total energies in the nonrelativistic framework hardly makes any difference, while the effect is pronounced in the Dirac framework (see section 9.8.4).

It must also be emphasized that *electronic* structure theory for molecular science benefits from a description of the external potential generated by atomic nuclei which are as structureless as possible so that different isotopes of an element are not distinguished. Otherwise, electronic structure calculations would have to be carried out for all possible permutations of isotopes in a molecule. However, the effect on the (electronic) wave function and on the total electronic energy would be negligible, which is the reason why we usually reject detailed modeling of different isotopic nuclei in a molecule. For the same reason, we rejected weak interactions brought about by higher electrostatic multipole moments or by nuclear spin in section 5.4. Such weak nuclear electromagnetic and hyperfine effects may be treated via perturbation theory at a later stage (compare chapter 15).

Theoretical nuclear physics does not provide a unique model function for the positive charge distribution derived from quantum chromodynamics. That is why there is a certain degree of arbitrariness in the choice of such functions. In the spherically symmetric case, the radial Poisson equation

$$\frac{1}{r}\frac{\mathrm{d}^2}{\mathrm{d}r^2} r\phi_{nuc}(r) = +4\pi\rho(r) \tag{6.149}$$

relates the nuclear charge distribution to a potential function $\phi_{nuc}(r)$. The Laplacian Δ of the three-dimensional Poisson equation yields a radial derivative and a centrifugal term as in the case of the kinetic energy of the Schrödinger hydrogen atom discussed the beginning of this chapter. Here, however, the centrifugal term vanishes because $l = 0$, as can be understood from a multipole expansion of our nuclear charge density defined such that only the

s-symmetric monopole contribution survives, $\rho_{nuc}(\boldsymbol{r}) = \rho(r)Y_{00}(\vartheta,\varphi)$. The potential function corresponding to a given model nuclear charge density distribution ρ_{nuc} is then obtained after integration of the radial Poisson equation

$$-r\phi_{nuc}(r) = 4\pi \int_0^r \rho_{nuc}(u)\, u^2\, \mathrm{d}u + 4\pi r \int_r^\infty \rho_{nuc}(u)\, u\, \mathrm{d}u \tag{6.150}$$

(in anticipation, observe the similarity of the radial Poisson equation with its solution for $\phi_{nuc}(r)$ and of the general potential functions in many-electron atoms in section 9.3.4; they are equivalent for s-symmetric densities where the centrifugal term from the Laplacian Δ vanishes).

Many model potentials $\phi_{nuc}(r)$ have been used [94] but two have become most important in electronic structure calculations. These are the homogeneous and the Gaussian charge distributions. The homogeneously or uniformly charged sphere is a simple model for the finite size of the nucleus. It is piecewise defined, because the positive charge distribution is confined in a sphere of radius R. The total nuclear charge $+Ze$ is uniformly distributed over the nuclear volume $4\pi R^3/3$,

$$\rho_{hom}(r) = \begin{cases} 3Ze/(4\pi R^3) & ; \quad r \leq R \\ 0 & ; \quad r > R \end{cases} \tag{6.151}$$

The radius R has to be fixed empirically and may be understood as the "size" of the nucleus. This charge density distribution leads through Eq. (6.150) and multiplication by $q_e = -e$ to the homogeneous electron–nucleus potential energy operator

$$V_{hom}(r) = \begin{cases} -\dfrac{Ze^2}{2R}\left[3 - \dfrac{r^2}{R^2}\right] & ; \quad r \leq R \\ -Ze^2/r & ; \quad r > R \end{cases} \tag{6.152}$$

provided that the charge density distribution is normalized to the total charge of the nucleus,

$$4\pi \int_0^\infty \rho_{hom}(u)\, u^2\, \mathrm{d}u = +Ze \tag{6.153}$$

Hence, outside the spherical nucleus, for $r > R$, the ordinary Coulomb attraction governs the electron–nucleus interaction.

The Gaussian and Fermi charge density distributions are continuously defined, meaning that the nuclear charge is not exactly zero even at a large distance from the center of the nucleus. The Gaussian charge density distribution,

$$\rho_{\text{Gauss}}(r) = \rho_{\text{Gauss},0} \exp\left(-\frac{r^2}{R^2}\right) \tag{6.154}$$

with $\rho_{\text{Gauss},0}$ fixed through the normalization condition,

$$\rho_{\text{Gauss},0} = \frac{Ze}{4\pi} \frac{1}{\sqrt{\pi}R^3} \tag{6.155}$$

yields through Eq. (6.150) and multiplication by $q_e = -e$ the following electron–nucleus potential energy

$$V_{\text{Gauss}}(r) = -\frac{Ze^2}{r} \text{erf}\left(\frac{r}{R}\right) \tag{6.156}$$

where erf(x) denotes the error function.

6.9.1
Consequences of the Nuclear Charge Distribution

We have already seen above that the choice of a point-like atomic nucleus limits the Dirac theory to atoms with a nuclear charge number $Z \leq c$, i.e., $Z_{\text{max}} \approx 137$. A nonsingular electron–nucleus potential energy operator allows us to overcome this limit if an atomic nucleus of finite size is used. In relativistic electronic structure calculations on atoms — and thus also for calculations on molecules — it turned out that the effect of different finite-nucleus models on the total energy is comparable but quite different from the energy of a point-like nucleus (compare also section 9.8.4).

We approach the effect of finite nuclear charge models from a formal perspective and introduce a general electron–nucleus potential energy V_{nuc}, which may be expanded in terms of a Taylor series around the origin,

$$V_{nuc} = \sum_{k=-1}^{\infty} v_k r^k \tag{6.157}$$

This Taylor series is required to analyze the radial Dirac equation, Eq. (6.68), close to the origin in order to determine the origin behavior of the solution functions. Obviously, in the case of a point-like nucleus we have $v_{-1} = -Ze^2$ and $v_{k \geq 0} = 0$. For the sake of simplicity, we may assume only radial-symmetric nuclear charge distributions, which is an approximation to atomic nuclei that may be of prolate or oblate shape. In order to study the effect of the chosen nuclear charge distribution on the radial functions of the eigen spinor, we again assume that the radial functions are analytic at the origin as in Eqs. (6.78) and (6.80) and determine the first exponent of this series expansion α by solving the two coupled radial Dirac equations using only the first terms of the series expansion for the electron–nucleus potential in Eq. (6.157),

$$V_{nuc}(r) = v_{-1} r^{-1} + v_0 + O(r) \tag{6.158}$$

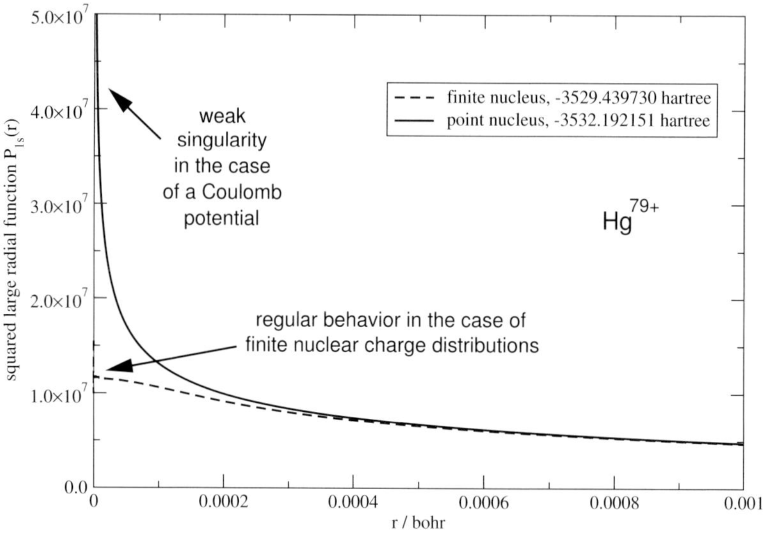

Figure 6.4 Origin behavior of the squared large component radial function $F_{1s}^2(r \to 0) \propto r^{2\alpha} a_0^2 + O(r^{2\alpha+2})$ in Hartree atomic units calculated for a point nucleus with Coulomb potential and for a finite nucleus. Note the weak singularity in the case of the point-like nucleus due to the nonintegral exponent α.

with

$$v_{-1} = \begin{cases} -Ze^2 & \text{for point-like nuclear charges} \\ 0 & \text{for finite-size nuclei} \end{cases} \tag{6.159}$$

We obtain from the Dirac equation for the hydrogen atom for the coefficients of the $r^{\alpha-1}$ term

$$\frac{v_{-1}}{\hbar c} a_0 + (\kappa_i - \alpha) b_0 = 0 \tag{6.160}$$

$$\frac{v_{-1}}{\hbar c} b_0 + (\kappa_i + \alpha) a_0 = 0 \tag{6.161}$$

which yields

$$\alpha = \sqrt{\kappa^2 - v_{-1}^2/(\hbar^2 c^2)} = \begin{cases} \sqrt{\kappa^2 - Z^2 e^4/(\hbar^2 c^2)} & \text{point-like nuclei} \\ |\kappa| & \text{finite-size nuclei} \end{cases} \tag{6.162}$$

[recall that we have derived the expression for point nuclei already in Eq. (6.89)]. We thus understand that it makes an important qualitative difference whether a singular Coulomb potential or a finite attraction potential is employed in the description of a relativistic Dirac atom. While for point-like nuclei the first exponent of the series expansion is *not* integral, it is well-behaved in the case of the potential of a finite nucleus. Interestingly, this qualitative difference of the short-range behavior of the radial functions does not occur in the case of nonrelativistic Schrödinger atoms (because the differential equation is of second order in this case so that the lowest powers in the radial function's series expansion are solely determined by the second derivative — and by the centrifugal potential if $l > 0$ — rather than by the nuclear potential).

The result of the different origin behavior of the radial functions transfers to the many-electron case and creates substantial drawbacks for numerical methods applied to solve the radial equation. Numerical solution methods always require finite higher derivatives which then become singular at the origin. We come back to this point in chapter 9.

6.9.2
Spinors in External Scalar Potentials of Varying Depth

In view of the qualitatively different origin behavior of the radial functions and thus of the spinor, it is instructive to investigate the consequences in greater detail. Figure 6.4 displays the squared large-component radial functions for the highly charged mercury hydrogen-like ion. The ground state energies of this one-electron ion are also included in this figure. As one would expect for a potential which is not singular at the origin as the finite-nucleus potential is, the total energy is not as negative for the finite-nucleus ion than it is in the case of a point-like nucleus. Again, all figures in this section were produced with a numerical atomic structure program [88]. However, an analytic solution for Schrödinger hydrogen-like atoms is available for the homogeneously charged sphere potential.

The radial functions of the ground state are well defined even for $Z > c$ in the case of a finite nucleus because $\alpha_0 = |\kappa_0|$ does not depend on the nuclear charge number Z, while it does in the case of a point nucleus ($\alpha_i = \sqrt{\kappa_i^2 - Z^2/c^2}$ in Hartree atomic units). Consequently, the standard Coulomb model for the electron–nucleus attraction can only be employed for atoms with $Z \leq c$ (see dashed line in Figure 6.5); we may study any atom theoretically if we employ a finite-nucleus model. Figure 6.5 presents the resulting ground state energies.

As the ground state energy monotonically decreases with increasing nuclear charge $+Ze$, this energy will penetrate the Dirac sea of positronic states. From Figure 6.5 we understand that this will be the case at $Z \approx 170$. The highest energy positronic state is denoted by a thick horizontal line. Of course, there are infinitely many positronic states below this line, which are not explicitly displayed in Figure 6.5. Clearly, the exact form and slope of the energy depending on Z depends on the finite-nucleus model potential chosen, but on

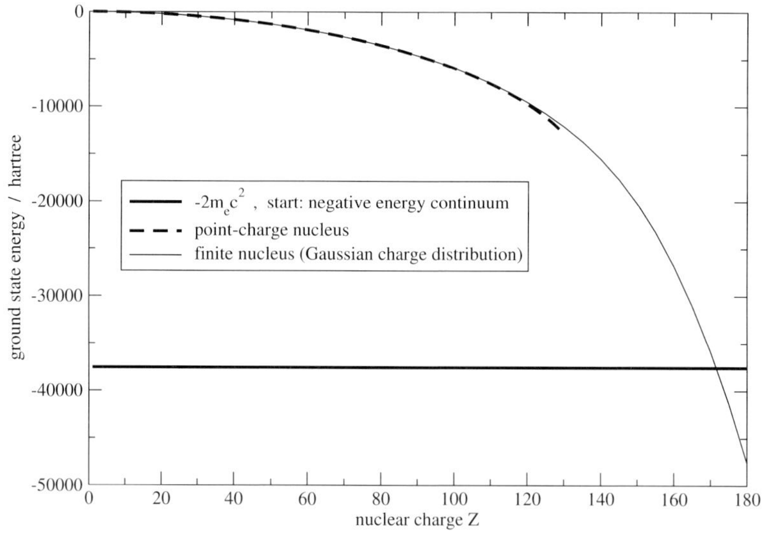

Figure 6.5 Dependence of the total ground state energies of hydrogen-like atoms E'_{1s} shifted by $-m_ec^2$ on the nuclear charge $+Ze$. The onset of the negative-energy state, i.e., of the so-called positronic states at $-2m_ec^2 \approx -37600$ hartree is indicated by a horizontal line. In the case of a Coulombic potential of a point-like nucleus the dashed line stops at $Z = c \approx 137$ as analytical radial functions are defined only up to this point as the exponent in Eq. (6.89) would become imaginary. By contrast, it is possible to proceed to even higher nuclear charge numbers for finite-size nuclei (see text for explanations). Then, we observe that the electronic ground states enter the negative continuum at $-2m_ec^2$ for $Z > 170$.

the scale given in Figure 6.5 all standard model potentials (i.e., all reasonable model potentials) would not deviate from one another.

Since the nonrelativistic energy eigenvalue divided by Z^2 is a constant, it is convenient to plot the ground state total energy scaled by $1/Z^2$. The result is shown in Figure 6.6. Of course, now the constant highest-energy positronic state at $-2m_ec^2$ hartree is no longer a straight line if divided by Z^2. As hydrogen-like atoms with $Z > c$ can be investigated if a finite nucleus is assumed (which is, of course, well justified), we may revisit the situation depicted in Figure 6.1 and also plot the radial functions of the ground state spinor for a one-electron atom with $Z = 170$ (Figure 6.7). We already understood that the so-called 'large' and 'small' component's radial functions for a point-like nucleus with $Z = c$ are identical. Hence, this (historically old) notion might become somewhat disturbing and has therefore been replaced by 'upper' and 'lower' components.

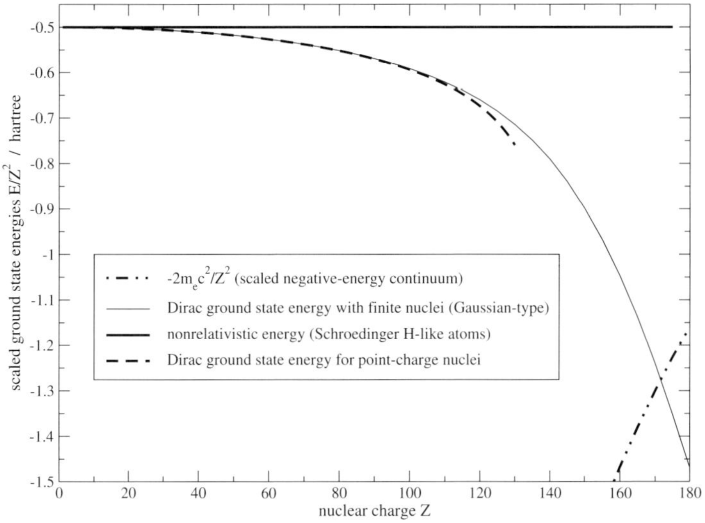

Figure 6.6 Comparison of ground state energies E'_{1s}/Z^2 scaled by $1/Z^2$ obtained for hydrogen-like atoms from Schrödinger quantum mechanics (horizontal line on top at -0.5 hartree), from Dirac theory with a Coulomb potential from a point-like nucleus (dashed line) and from Dirac theory with a finite nuclear charge distribution of Gaussian form (thin black line). The highest energy of the 'positronic' continuum states, $-2m_ec^2$, appears as a thick black line, which is bent because of the $1/Z^2$ scaling.

Finally, it may be anticipated that all that has been said with respect to finite-nucleus effects for hydrogen-like atoms also holds for many-electron atoms to be discussed in chapter 9. This is possible as we shall see because one-electron functions are determined by solution of coupled first-order differential equations of the type of the Dirac equation for a single electron. In the case of spherically symmetric many-electron atoms, these equations reduce to radial equations that are very much akin to those of the Dirac hydrogen atom. The origin behavior of the radial functions as discussed in this section then transfers directly to the many-electron case if the electron–electron interaction does not contribute any additional coefficient to the r^{-1} term.

6.10 Momentum Space Representation

So far only the position-space formulation of the (stationary) Dirac Eq. (6.7) has been discussed, where the momentum operator $\boldsymbol{p}$ acts as a derivative op-

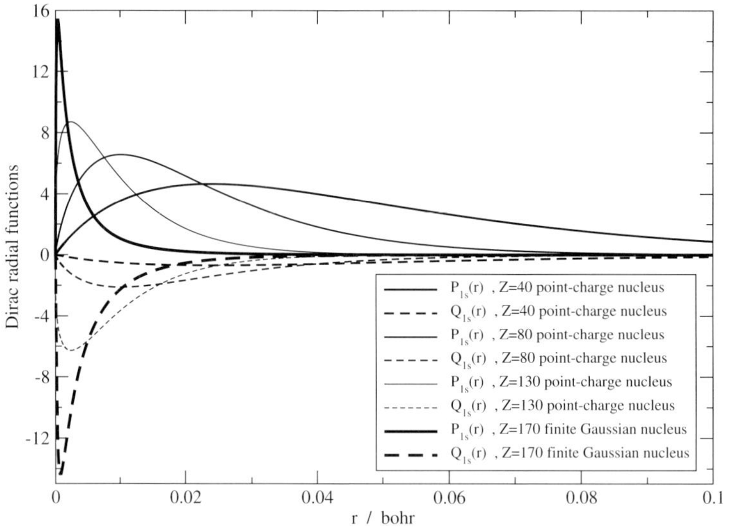

Figure 6.7 This figure comprises the radial functions shown already in Figure 6.1, in Hartree atomic units, plus those obtained for Gaussian nuclear charge distribution for the one-electron atom with $Z = 170$. Only in the case of a finite nucleus, quantum mechanical states of atoms with nuclear charges $Z > c$ are defined ($c \approx 137.037$ in Hartree atomic units).

erator on the 4-spinor Ψ. However, for later convenience in the context of elimination and transformation techniques (chapters 11–12), the Dirac equation is now given in momentum-space representation. Of course, a momentum-space representation is the most suitable choice for the description of extended systems under periodic boundary conditions, but we will later see that it gains importance for unitarily transformed Dirac Hamiltonians in chapters 11 and 12. We have already encountered such a situation, namely when we discussed the square-root energy operator in Eq. (5.4), which cannot be evaluated if $\boldsymbol{p}$ takes the form of a differential operator.

After Fourier transformation (see appendix E), the momentum-space representation of the Dirac equation is given by

$$h^D \Psi(\boldsymbol{p}) = \left[c\,\boldsymbol{\alpha}\cdot\boldsymbol{p} + \beta\, m_e c^2 + V) \right] \Psi(\boldsymbol{p}) = E\,\Psi(\boldsymbol{p}) \tag{6.163}$$

which looks formally very similar to the original Eq. (6.7). However, the external potential is no longer a local (multiplicative) operator, but is now — owing to the Fourier transformation — defined as an integral operator

$$V\,\Psi(\boldsymbol{p}) = \int \frac{d^3p'}{(2\pi\hbar)^3}\, V(\boldsymbol{p}-\boldsymbol{p}')\,\Psi(\boldsymbol{p}') \tag{6.164}$$

on the spinor Ψ. The integral kernel of V is the Fourier transform of $V(\boldsymbol{r})$, i.e.,

$$V(\boldsymbol{p}-\boldsymbol{p}') = \int d^3r \exp\left[-\frac{i}{\hbar}(\boldsymbol{p}-\boldsymbol{p}')\cdot\boldsymbol{r}\right] V(\boldsymbol{r}) = -\frac{4\pi Z e^2 \hbar^2}{|\boldsymbol{p}-\boldsymbol{p}'|^2} \tag{6.165}$$

where the last equality only holds for a Coulomb potential $V(\boldsymbol{r}) = -Ze^2/r$. Thus, the four coupled differential Eqs. (5.81)–(5.84) have been transformed into four coupled integral Eqs. (6.163).

The solution of the Dirac equation does not need to be carried out explicitly as demonstrated for the position-space form in the beginning of this chapter. Instead the spinors are conveniently obtained by a Fourier transformation of the position-space solution functions. The Fourier transform of the spinor reads

$$\Psi(\boldsymbol{p}) = \int d^3r \exp\left[-\frac{i}{\hbar}\boldsymbol{p}\cdot\boldsymbol{r}\right] \Psi(\boldsymbol{r}) \tag{6.166}$$

where the same symbol Ψ has been used throughout to denote both the position- and momentum-space Dirac spinor for the sake of brevity.

Further Reading

F. Schwabl, [46]. *Advanced Quantum Mechanics.*

Schwabl provides many details on how the Dirac hydrogen atom can be solved, though some essential steps are treated very briefly. However, the material in Schwabl's book is still more extensive than in many of the new publications on the subject. Also, the reader will find a more detailed comparison to the Klein–Gordon equation in external electromagnetic fields. But note that Schwabl chooses a different sign convention for the spherical spinors with effect on almost all equations of the Dirac hydrogen atom compared to those derived here.

H. A. Bethe, E. E. Salpeter, [48,95]. *Quantum Mechanics of One- and Two-Electron Atoms.*

This classic text is *the* standard reference for the Dirac theory of hydrogen- and helium-like atoms. Bethe had already published a review article on one- and two-electron atoms in *Handbuch der Physik* in 1933 [39], which is quite readable — especially because it demonstrates the early difficulties with a quantum theory of bound-state electrons and their interactions.

T. P. Das, [85]. *Relativistic Quantum Mechanics of Electrons.*

Das' book contains quite a lot of material on the hydrogen atom; in particular, explicit expressions for the excited state wave functions of the hydrogen atom are tabulated which we recommend for further reference. Also, alternative derivations to those presented in this volume can be found. For instance, the action of $(\boldsymbol{\sigma}\cdot\hat{\boldsymbol{r}})$ on spherical spinors is presented in a different manner than in our chapter 4.

D. Andrae, [94]. Review article: *Finite Nuclear Charge Density Distributions in Electronic Structure Calculations for Atoms and Molecules.*

This review article presents a good overview on all finite-nuclear charge distributions which have been invented to model (mainly radial symmetric) extended nuclear charges. The number of such models is huge and this review provides a very useful classification of the many different empirical approaches.

Part III

Four-Component Many-Electron Theory

7
Quantum Electrodynamics

Quantum electrodynamics is the fundamental physical theory which obeys the principles of special relativity and allows us to describe the mutual interactions of electrons and photons. It is intrinsically a many-particle theory, though much too complicated from a numerical point of view to be the basis for the theoretical framework of the molecular sciences. Nonetheless, it is the basic theory of chemistry and its essential concepts, and ingredients are introduced in this chapter.

7.1
Elementary Quantities and Notation

This chapter is itended to comprise a compact introduction to quantum electrodynamics (QED). The purpose of this presentation is to provide some basic understanding, which then allows us to better justify all approximations made in the following chapters in order to devise a many-electron theory which yields numerically accurate results for physical observables and allows us to perform actual calculations on computer hardware.

7.1.1
Lagrangian for Electromagnetic Interactions

From today's point of view it is fair to say that QED appears to be the final theory for all electromagnetic processes for which other fundamental interactions, i.e., weak, strong, and gravitational forces, may be neglected. This perspective is supported by innumerable low- and high-energy scattering experiments of charged particles and photons and spectral analyses [96]. It can account for nearly all observed phenomena from macroscopic scales down to about 10^{-13} cm, and is based on the classical Lagrangian density

$$\begin{aligned}\mathcal{L}_{\text{QED}} &= \mathcal{L}_{\text{rad}} + \mathcal{L}_{\text{mat}} + \mathcal{L}_{\text{int}} \\ &= -\frac{1}{16\pi}F^{\mu\nu}F_{\mu\nu} + \bar{\psi}\big(\mathrm{i}\hbar c\partial\!\!\!/ - \tilde{m}c^2\big)\psi - \frac{1}{c}j_\mu(A^\mu + A^{\mu,\text{ext}}) \quad (7.1)\end{aligned}$$

Relativistic Quantum Chemistry. Markus Reiher and Alexander Wolf

ISBN: 978-3-527-31292-4

The first term ($\mathcal{L}_{\text{rad}}$) describes the electromagnetic degrees of freedom based on the 4-potential A^μ, the second ($\mathcal{L}_{\text{mat}}$) the Dirac matter field ψ, whose excitations may later, i.e., after quantization, be interpreted as electrons and positrons, and the last term ($\mathcal{L}_{\text{int}}$) accounts for the interaction between the former two. To write such a Lagrangian goes back to the early days of (the old) quantum electrodynamical theory [70].

The gauge field $A^\mu = (\phi, \boldsymbol{A})$ contains the scalar and vector potentials and defines the antisymmetric electromagnetic field tensor $F^{\mu\nu}$ already introduced in section 3.4. A^{ext} comprises, as before, external, i.e., non-dynamical, electromagnetic potentials in addition to the radiation field. In accordance with section 2.4 the dynamical electric and magnetic fields can be obtained as

$$\boldsymbol{E} = -\nabla\phi - \frac{1}{c}\frac{\partial \boldsymbol{A}}{\partial t} \quad \text{and} \quad \boldsymbol{B} = \nabla \times \boldsymbol{A} \tag{7.2}$$

The coupling between the four components of the Dirac spinor field is accomplished by the mathematical structure of a four-dimensional Clifford algebra (see also chapter 5) [97] defined by the anticommutation relation $\{\gamma^\mu, \gamma^\nu\} = 2g^{\mu\nu}$. The coupling between the large and small components of the Dirac spinor ψ in the standard representation of the Clifford algebra is only due to the odd terms containing γ occurring in the Lagrangian, a fact that will be extensively exploited in the following chapters.

In order to cast the expressions in the simplest possible form, Feynman's slash-notation $\not{\partial} = \gamma^\mu \partial_\mu$, which we introduced in section 5.2.4, has been employed in Eq. (7.1). The definition of the adjoint spinor field

$$\bar{\psi} = \psi^\dagger \gamma^0 = \left(\psi_1^*, \psi_2^*, -\psi_3^*, -\psi_4^*\right) \tag{7.3}$$

specifies the Dirac Lagrangian $\mathcal{L}_{\text{mat}}$ completely. The interaction term contains the charge–current density

$$j^\mu = (c\varrho, \boldsymbol{j}) = \tilde{e}c\bar{\psi}\gamma^\mu\psi \tag{7.4}$$

which is given in an explicitly covariant form (recall also section 3.4.1.1). The charge density and current density include the multiplication by charge as compared to Eqs. (5.45) and (5.46), and are hence no longer particle densities (but we refrain from adding a subscript to j^μ in order to indicate this). Both the mass $\tilde{m} > 0$ and the charge $\tilde{e} < 0$ of the electron are not the physically observable quantities but the bare parameters that are still subject to a renormalization procedure according to the rules of quantum field theory.

7.1.2
Lorentz and Gauge Symmetry and Equations of Motion

The Lagrangian $\mathcal{L}_{\text{QED}}$ describes both electromagnetic and fermionic degrees of freedom and the electromagnetic interaction between photons and matter simultaneously, that is, the photon and the Dirac field are treated as dynamical variables. It has all symmetry properties necessary for a fundamental physical theory; it is Lorentz covariant as well as gauge invariant. Lorentz covariance, i.e., invariance in form under arbitrary Lorentz transformations $\Lambda^{\mu}{}_{\nu}$ is manifest by inspection of the Lagrangian given by Eq. (7.1), which only contains Lorentz scalars, i.e., quantities where all relativistic 4-indices are completely contracted and summed over. The upper or lower position of a Lorentz index comprises the contra- or covariant behavior, respectively, of the corresponding quantity under Lorentz transformations. The transformation properties of scalar and vector potentials are extensively discussed in section 3.4.

The second fundamental symmetry property of the Lagrangian is established by local gauge transformations of the form

$$A_{\mu}(x) \longrightarrow A'_{\mu}(x) = A_{\mu}(x) - \partial_{\mu}\chi(x) \tag{7.5}$$

$$\psi(x) \longrightarrow \psi'(x) = \exp\left[-\frac{\mathrm{i}\tilde{e}}{\hbar c}\chi(x)\right]\psi(x) \tag{7.6}$$

with $\chi(x)$ being an arbitrary gauge function, which leaves the action

$$S = \frac{1}{c}\int \mathrm{d}^4x\, \mathcal{L}_{\mathrm{QED}} = \int \mathrm{d}^3r\, \mathrm{d}t\, \mathcal{L}_{\mathrm{QED}} \tag{7.7}$$

i.e., the space–time integral over the Lagrangian, invariant. Note that gauge invariance is even a property of the first two terms of the Lagrangian of Eq. (7.1) itself, whereas the gauge transformation of the interaction term $\mathcal{L}_{\mathrm{int}}$ gives rise to an additional term $\tilde{e}cj_{\mu}\partial^{\mu}\chi$, whose integral over all space–time will vanish after integration by parts because of the continuity equation given below, cf. Eq. (3.187) and the discussion in section 3.4.1. The special form of the gauge transformation of the spinor ψ, which simply describes rotations in the complex plane, is the reason for calling QED an Abelian $U(1)$ gauge theory. The Hamiltonian principle of least action, $\delta S \overset{!}{=} 0$ for arbitrary infinitesimal variations of the dynamical variables A^{μ} and ψ, yields the coupled equations of motion for the electromagnetic and Dirac fields,

$$\partial_{\mu}F^{\mu\nu} = \frac{4\pi}{c}j^{\nu} \tag{7.8}$$

$$\left(\mathrm{i}\hbar c\partial\!\!\!/ - \tilde{m}c^2\right)\psi = \tilde{e}\left(A\!\!\!/ + A\!\!\!/^{\mathrm{ext}}\right)\psi \tag{7.9}$$

Eq. (7.8) is the most general covariant form of the inhomogeneous Maxwell equations, which immediately imply the continuity equation $\partial_{\mu}j^{\mu} = \dot{\varrho} + \mathrm{div}\boldsymbol{j} = 0$ of section 5.2.3, and Eq. (7.9) is the covariant time-dependent Dirac equation in the presence of external electric and magnetic fields. The homogeneous Maxwell equations are automatically satisfied by the sole existence of the 4-potential A^{μ}. In this context it is very important to note that the Dirac equation is interpreted as a classical Euler–Lagrange equation for the spinor field ψ rather than as a quantum-mechanical wave equation.

7.2
Classical Hamiltonian Description

7.2.1
Exact Hamiltonian

The transition to a Hamiltonian formulation of this field theory requires the definition of the conjugate momenta

$$\pi = \frac{\partial \mathcal{L}_{\mathrm{QED}}}{\partial \dot{\psi}} = i\hbar\, \psi^{\dagger} \quad \text{and} \quad \Pi_{\mu} = \frac{\partial \mathcal{L}_{\mathrm{QED}}}{\partial \dot{A}^{\mu}} = \frac{1}{4\pi c} F_{\mu 0} \tag{7.10}$$

and is achieved by a Legendre transformation (cf. section 2.2). After some tedious algebraic manipulations and taking advantage of Eq. (7.2) the final Hamiltonian density is given as

$$\begin{aligned}\mathcal{H}_{\mathrm{QED}} &= \Pi_{\mu} \dot{A}^{\mu} + \pi \dot{\psi} - \mathcal{L}_{\mathrm{QED}} = \frac{1}{8\pi}(\boldsymbol{E}^2 + \boldsymbol{B}^2) + \frac{1}{4\pi}\boldsymbol{E} \cdot \nabla \phi \\ &\quad - \bar{\psi}\left(i\hbar c \gamma^{k} \partial_{k} - \tilde{m} c^2\right)\psi + \tilde{e}\bar{\psi}\gamma^{\mu}\psi(A_{\mu} + A_{\mu}^{\mathrm{ext}})\end{aligned} \tag{7.11}$$

Although this expression is no longer manifestly Lorentz or gauge invariant, all physical observables like energies, field strengths, transition amplitudes, etc. to be deduced from this Hamiltonian are Lorentz and gauge invariant. A subsequent integration over all space yields the classical Hamiltonian

$$\begin{aligned}H_{\mathrm{QED}} &= \int d^3r\, \mathcal{H}_{\mathrm{QED}} = \int d^3r \left\{ \frac{1}{8\pi}(\boldsymbol{E}^2 + \boldsymbol{B}^2) \right. \\ &\quad \left. + \psi^{\dagger}\left[c\,\boldsymbol{\alpha} \cdot \left(- i\hbar\nabla - \frac{\tilde{e}}{c}\boldsymbol{A} - \frac{\tilde{e}}{c}\boldsymbol{A}^{\mathrm{ext}}\right) + \beta \tilde{m} c^2 + \tilde{e}\phi^{\mathrm{ext}}\right]\psi \right\}\end{aligned} \tag{7.12}$$

$$= \int d^3r \left\{ \frac{1}{8\pi}(\boldsymbol{E}^2 + \boldsymbol{B}^2) + \psi^{\dagger} h^{D}\, \psi \right\} \tag{7.13}$$

where integration by parts of the second term of Eq. (7.11) and Gauss' law [$\nu = 0$ in Eq. (7.8)] have been employed; for later convenience we switched to the standard notation of the Dirac matrices as of Eqs. (5.37) and (5.50). The Hamiltonian given by Eq. (7.12) immediately exhibits the familiar form of the interaction of the Dirac field with the transverse vector potential $\boldsymbol{A}$ and external fields and also contains the rest energy and kinetic contributions of the electron. These terms are conveniently summarized by the quantity h^D in Eq. (7.13), which can be interpreted as the familiar one-electron Dirac Hamiltonian.

7.2.2
The Electron–Electron Interaction

Having derived the explicit form of the familiar one-particle Dirac Hamiltonian from the fundamental Lagrangian of QED, one may wonder where the equivalents of the two-electron interactions are hidden. Unfortunately, the derivation of these two-electron terms, being incorporated in the term of the electromagnetic field energy, i.e., the first term of Eq. (7.12), is far more subtle, and is most easily performed in Coulomb gauge. It has to be emphasized that for all developments so far no special gauge has been chosen and the most general framework has been presented. For the rest of this chapter, however, we fix the gauge uniquely by imposing Coulomb's gauge condition, $\text{div}\boldsymbol{A} = 0$, on the electromagnetic fields, which significantly simplifies Gauss' law, i.e., the 0-component of Eq. (7.8) to

$$\triangle\phi = -4\pi\varrho = -4\pi\tilde{e}\psi^{\dagger}\psi \tag{7.14}$$

which is just the familiar Poisson equation of electrostatics. It is explicitly solved by the Coulomb potential

$$\phi(\boldsymbol{r},t) = A^0(\boldsymbol{r},t) = \int d^3r' \frac{\varrho(\boldsymbol{r}',t)}{|\boldsymbol{r}-\boldsymbol{r}'|} = \int d^3r' \frac{\tilde{e}\psi^{\dagger}(\boldsymbol{r}',t)\psi(\boldsymbol{r}',t)}{|\boldsymbol{r}-\boldsymbol{r}'|} \tag{7.15}$$

which, of course, is the origin of the name "Coulomb gauge" (compare also section 2.4 for the choice of gauge). In Coulomb gauge, the scalar potential is given by the *instantaneous* Coulomb potential and hence completely determined by the charge density ρ, cf. section 2.4.4.2, and thus no retardation effects occur from the scalar potential. However, since electromagnetic waves (and thus photons) can only be exchanged with the finite speed of light c, in Coulomb gauge retardation effects will occur for the vector potential $\boldsymbol{A}$, cf. Eq. (2.141). Unfortunately, retardation effects are not easy to handle in quantum theory and are best approached from classical electrodynamics (see section 3.5). We do not delve into this point here (but come back to it later in section 8.1.4), because this aspect would obstruct the presentation of the basics of QED here. In section 7.3, we also discuss implications of this special form of the scalar potential ϕ with respect to the quantization procedure.

After having fixed the gauge, the analysis of the two-electron interaction requires a decomposition of the electric field $\boldsymbol{E}$ into longitudinal (irrotational) and transverse (solenoidal) components,

$$\boldsymbol{E} = \boldsymbol{E}_l + \boldsymbol{E}_t \quad \text{with} \quad \boldsymbol{E}_l = -\nabla\phi \quad \text{and} \quad \boldsymbol{E}_t = -\frac{1}{c}\frac{\partial \boldsymbol{A}}{\partial t} \tag{7.16}$$

Consequently, the complete electromagnetic field energy, which comprises *all* two-electron effects, may be written as

$$\frac{1}{8\pi}\int d^3r\,\left(\boldsymbol{E}^2+\boldsymbol{B}^2\right) = \frac{1}{8\pi}\int d^3r\,\boldsymbol{E}_l^2 + \frac{1}{8\pi}\int d^3r\,\left(\boldsymbol{E}_t^2+\boldsymbol{B}^2\right) \tag{7.17}$$

where the mixed term $\boldsymbol{E}_l \cdot \boldsymbol{E}_t$ vanishes in Coulomb gauge, as is seen after a further integration by parts. This clearly demonstrates that electromagnetic interactions between two charged particles are mediated by the exchange of (virtual) photons, i.e., the quanta of the electromagnetic field.

The first term on the right hand side of Eq. (7.17) represents the instantaneous Coulomb interaction between the charge distributions,

$$\frac{1}{8\pi}\int d^3r\,\boldsymbol{E}_l^2(\boldsymbol{r},t) \stackrel{(7.14)}{=} \frac{1}{2}\int d^3r\,\phi(\boldsymbol{r},t)\varrho(\boldsymbol{r},t)$$

$$\stackrel{(7.15)}{=} \frac{1}{2}\int d^3r d^3r'\,\psi^\dagger(\boldsymbol{r},t)\psi^\dagger(\boldsymbol{r}',t)\frac{\tilde{e}^2}{|\boldsymbol{r}-\boldsymbol{r}'|}\psi(\boldsymbol{r},t)\psi(\boldsymbol{r}',t) \tag{7.18}$$

$$= \frac{1}{2}\int d^3r d^3r'\,\frac{\varrho(\boldsymbol{r},t)\,\varrho(\boldsymbol{r}',t)}{|\boldsymbol{r}-\boldsymbol{r}'|} \tag{7.19}$$

where integration by parts and the Coulomb gauge condition have been employed so that no retardation occurs for the zeroth component of the 4-current [see Eqs. (2.138) and (2.139)]. *It is thus only the longitudinal component of the electric field which mediates the instantaneous Coulomb interaction.* Magnetic interactions, retardation effects due to the finite speed of light, and radiation corrections are contained in the second term on the right hand side of Eq. (7.17). In Coulomb gauge, its expansion in powers of $1/c$ yields the leading terms

$$\frac{1}{8\pi}\int d^3r\,\left(\boldsymbol{E}_t^2+\boldsymbol{B}^2\right) = \frac{1}{2}\int d^3r d^3r'\,\psi^\dagger(\boldsymbol{r})\psi^\dagger(\boldsymbol{r}')\left\{-\frac{\tilde{e}^2}{2|\boldsymbol{r}-\boldsymbol{r}'|}\left(\boldsymbol{\alpha}\cdot\boldsymbol{\alpha}' + \right.\right.$$

$$\left.\left. + \frac{\left(\boldsymbol{\alpha}\cdot(\boldsymbol{r}-\boldsymbol{r}')\right)\left(\boldsymbol{\alpha}'\cdot(\boldsymbol{r}-\boldsymbol{r}')\right)}{|\boldsymbol{r}-\boldsymbol{r}'|^2} + \ldots\right)\right\}\psi(\boldsymbol{r})\psi(\boldsymbol{r}') \tag{7.20}$$

which can be identified as the so-called Breit interaction. Note the resemblance to the classical Darwin interaction derived in section 3.5.5; we will present an explicit derivation and analysis in sections 8.1.4 and 8.1.5.

Higher-order terms occurring in the expansion of Eq. (7.20) are not given here since they do not play any role for quantum chemistry as we shall discuss in the subsequent chapters. The expressions for the two-electron interactions derived here, i.e., Eqs. (7.18) and (7.20) are neither gauge nor Lorentz invariant, since we explicitly broke these symmetries by choosing one special gauge and the truncation of the expansion. For chemical purposes, however, they are an excellent approximation to the exact interaction. It should be emphasized that so far no quantum theory has yet been developed. Eq. (7.12) is just the classical Hamiltonian of the relativistic $U(1)$ gauge field theory of electrodynamics interacting with a dynamical fermionic field. In order to interpret the excitations of these fields as physical particles, i.e., photons, electrons, and positrons, quantization of this theory has to be achieved.

7.3 Second-Quantized Field-Theoretical Formulation

Unfortunately, quantization of this gauge theory is not as straightforward as the development of the theory up to this point. As it is reflected by the mere existence of a gauge symmetry transformation, gauge theories necessarily comprise redundant degrees of freedom, which have to be removed by a suitable gauge fixing procedure before consistent quantization may be achieved. This is most easily seen by application of the *Coulomb, radiation*, or *transverse* gauge, $\mathrm{div}\boldsymbol{A} = 0$, which has been introduced before. Because of Eq. (7.15), the scalar potential ϕ is then no longer an independent dynamical variable, but uniquely determined by the charge distribution ϱ. The remaining spatial components of the vector potential are subject to the Coulomb gauge condition. Hence, only two independent field components survive, reflecting the fact that real photons possess only two *transversal* polarization states.

Taking all this into account, consistent quantization of this gauge field theory may be achieved, e.g., by a constrained canonical procedure [98, 99] or the manifestly covariant Gupta–Bleuler formalism, which employs an indefinite metric at the expense of an easy physical interpretation [100–102]. In the following the basic ideas of the canonical procedure will briefly be presented.

According to the fundamental principles of quantum mechanics, the variables of the theory, i.e., the gauge and Dirac field amplitudes A^μ and ψ_α, respectively, are promoted to operators, which have to satisfy the canonical equal time commutation or anticommutation relations with their conjugate momenta, Π_μ and π_α, respectively. It is this procedure of upgrading classical field amplitudes to quantum mechanical (field) operators which is somewhat misleadingly dubbed "second quantization". One should be aware of the fact that actually a classical field is quantized exactly once by this procedure, rather than an already quantum mechanical object being quantized again. The conjugate momenta are given by Eq. (7.10), and yield the fundamental relations

$$\{\psi_\alpha(\boldsymbol{r},t), \pi_\beta(\boldsymbol{r}',t)\} = i\hbar\,\delta_{\alpha\beta}\,\delta^{(3)}(\boldsymbol{r}-\boldsymbol{r}')\,, \quad \alpha,\beta = 1,\ldots,4 \tag{7.21}$$

$$\left[A^i(\boldsymbol{r},t), \Pi_j(\boldsymbol{r}',t)\right] = i\hbar\,\delta^{\mathrm{tr}}_{ij}(\boldsymbol{r}-\boldsymbol{r}')\,, \quad i,j = 1,2,3 \tag{7.22}$$

where, as before, curly brackets $\{,\}$ denote an anticommutator and normal brackets $[,]$ a commutator. Note that the *transverse* or *divergenceless* δ-

distribution

$$\delta^{\rm tr}_{ij}(\boldsymbol{r}-\boldsymbol{r}') = \left(\delta_{ij} - \frac{\partial_i\partial_j}{\nabla^2}\right)\delta^{(3)}(\boldsymbol{r}-\boldsymbol{r}') \tag{7.23}$$

$$= \int \frac{d^3p}{(2\pi\hbar)^3}\, e^{i\boldsymbol{p}\cdot(\boldsymbol{r}-\boldsymbol{r}')/\hbar}\left(\delta_{ij} - \frac{p_i p_j}{\boldsymbol{p}^2}\right) \tag{7.24}$$

has been employed in Eq. (7.22) in order to satisfy Gauss' law. Eq. (7.22) establishes a non-local commutation relation between the gauge field A^i and its conjugate momentum Π_i, which seem to lead to acausal behavior. However, the commutation relations between all physical observables, e.g., the components of the electric and magnetic fields, are strictly local and hence causal. All commutators or anticommutators, respectively, between other *independent* variables vanish. The scalar potential ϕ, however, is no longer an independent variable since it is completely determined by the Dirac field ψ according to Eq. (7.15). Its commutation behavior is thus completely determined by Eq. (7.21) and given by

$$\left[\phi(\boldsymbol{r},t), \psi_\alpha(\boldsymbol{r}',t)\right] = -\frac{\tilde{e}}{|\boldsymbol{r}-\boldsymbol{r}'|}\,\psi_\alpha(\boldsymbol{r}',t) \tag{7.25}$$

By following the ideas of canonical quantization, both manifest Lorentz covariance and gauge invariance of the theoretical description have been sacrificed. All physical quantities like transition amplitudes and S-matrix elements, however, will be Lorentz covariant and independent of the chosen gauge.

The last necessary ingredient toward a well-defined and consistent quantum field theory of radiation interacting with a fermionic matter field is the introduction of normal-ordered products of field operators. The necessity for this step is easily realized by consideration of the vacuum expectation value of Gauss' law,

$$\langle 0 \,|\, {\rm div}\boldsymbol{E}(\boldsymbol{r},t) \,|\, 0\rangle = 4\pi\tilde{e}\,\langle 0 \,|\, \psi^\dagger(\boldsymbol{r},t)\psi(\boldsymbol{r},t) \,|\, 0\rangle = 4\pi\tilde{e}\sum_n \left|\langle 0 \,|\, \psi^\dagger \,|\, n\rangle\right|^2 \tag{7.26}$$

where the sum runs over a complete, i.e., infinite set of states. Thus, even the vacuum would feature a divergent charge density. This completely unphysical and ridiculous result is sometimes traced back to the infinite negative charge of the sea of negative energy states, which are all supposed to be occupied according to Dirac's hypothesis from 1930 [72] in order to prevent even more unphysical processes like the production of infinite amounts of energy by an electron falling down the negative continuum (compare section 5.5). It is one of the greatest benefits of the formalism of second quantization that it accounts for a consistent and sensible description of physical processes without relying on Dirac's hole theory.

Thus, all operator products containing field operators have to be replaced by normal products, i.e., products where *normal ordering* is guaranteed. This implies that all positive-frequency components occurring in the momentum space expansion of field operators are always on the right hand side of the components with negative frequency. As a consequence, the vacuum expectation value of Gauss' law vanishes as it should. The Hamiltonian is then given as

$$H_{\text{QED}} = \int d^3r : \psi^\dagger(x) h^D(x) \psi(x) : + \frac{1}{8\pi} \int d^3r : \frac{1}{c^2} \dot{\boldsymbol{A}}(x)^2 + \left(\text{curl}\boldsymbol{A}(x)\right)^2 : + \frac{1}{2} \int d^3r d^3r' \frac{:\varrho(\boldsymbol{r},t)\,\varrho(\boldsymbol{r}',t):}{|\boldsymbol{r}-\boldsymbol{r}'|} \tag{7.27}$$

and has the desired property of variational stability, i.e., it is bounded from below by the ground state energy

$$\langle 0 \,|\, H_{\text{QED}} \,|\, 0 \rangle = 0 \tag{7.28}$$

Our discussion of the QED Hamiltonian is deliberately brief (and somewhat imprecise) because it would otherwise require a very extensive presentation of especially the interaction part of matter and photon fields. We come back to this issue in section 8.1.5 before we derive the electromagnetic interaction energies in semi-classical many-electron theory sufficient for molecular science. However, for more details on the QED Hamiltonian we refer to Ref. [32, p. 283 ff.].

Though the charge density $:\varrho(x):$ is an operator not commuting with the Hamiltonian, the total charge

$$Q = \int d^3r :\varrho(\boldsymbol{r},t): = \tilde{e} \int d^3r :\psi^\dagger(\boldsymbol{r},t)\psi(\boldsymbol{r},t): \tag{7.29}$$

is a constant of the motion, $[H_{\text{QED}}, Q] = 0$. Consequently, only those particle creation and annihilation processes are permitted which leave the total charge invariant and give rise only to fluctuations of the charge density. Therefore, only electron–positron pairs may be created out of the vacuum, but never a single charged fermion on its own. This even holds in the case of the β^+-decay mentioned in 5.5, where a proton decays to yield a neutron, a positron and a neutrino. The charge of the decaying proton is conserved in the emitted positron. However, this process is governed by electro-weak forces and, hence, requires the more fundamental theory of electro-weak interactions. Of course, for neutral photons, the quanta of the electromagnetic field, no comparable restriction holds, and they may be created or annihilated in arbitrary number, depending on the details of the electromagnetic interaction. As a consequence of these results, the particle number is no longer a conserved quantity, in sharp contrast to nonrelativistic theories. Finally, the classical Euler–Lagrange equations for the Maxwell and the Dirac fields, i.e., the coupled nonlinear system of equations given by Eqs. (7.8) and (7.9), can be transferred to the quantum world. The only necessary modification, which unfortunately increases the level of complexity significantly, is the replacement of field products by their normal-ordered counterparts,

$$\partial_\mu F^{\mu\nu} = 4\pi\tilde{e} :\bar{\psi}(x)\gamma^\nu \psi(x): \tag{7.30}$$

$$\left(i\hbar c \partial\!\!\!/ - \tilde{m}c^2\right)\psi(x) = \tilde{e} :A\!\!\!/ \psi(x): + \tilde{e} A\!\!\!/^{\text{ext}}\psi(x) \tag{7.31}$$

Similarly to the classical setup presented before, these equations describe the time evolution of the field operators and hence *all* dynamical processes in Fock space, which can be considered a generalization of the Hilbert-space concept for arbitrary particle numbers (see section 8.6.4 for a more precise definition).

7.4
Implications for the Description of Atoms and Molecules

So far the fundamental framework of quantum electrodynamics has been presented. Starting with the Lagrangian of a classical Abelian gauge field theory the corresponding Hamiltonian of the quantized theory has been derived employing the techniques of second quantization. The resulting theory of QED properly describes all interaction processes of electrons and positrons with photons and the external potentials, covering a huge class of phenomena from atomic and molecular bound states or excitation spectra to high-energy scattering results. Developed in final form in the late 1940s [103–115], puzzling problems like the accurate calculation of the electron's gyromagnetic ratio or the Lamb shift, i.e., the splitting of the $2s_{1/2}$ and $2p_{1/2}$ lines of hydrogen due to radiative corrections, self-energy, and vacuum polarization could be accomplished for the first time using renormalization techniques. Even today it is still the numerically most accurate theory ever developed by theoretical physics, and being the first fully relativistic gauge field theory it has served as a prototype for the invention of other fundamental theories for weak, strong, and gravitational interactions.

All interactions relevant for chemistry, i.e., the description of atoms, molecules, and solids are perfectly described by this theory or its generalization to electroweak interactions. However, the solution of its equations of motion (7.30) and (7.31) or even the much simpler calculation of stationary ground states or energy expectation values is far out of reach for all systems under investigation by chemistry. Even today, calculations explicitly taking care of only some of the simplest QED effects are hardly feasible for an atom with more than a couple of electrons [116–118].

However, the energies necessary for pair-creation processes are magnitudes larger than the energy scale relevant to chemistry and the molecular sciences.

Hence, QED is the fundamental theory of chemistry describing perfectly well all phenomena that *do not play any* role for chemistry. It is therefore highly desirable for chemical applications to integrate out irrelevant degrees of freedom, i.e., the possibility of pair creation and the interaction with the radiation field, from the very beginning in order to arrive at a theory for a fixed number of electrons only. As a consequence, this simplification not only makes the theoretical framework fit much better to most chemical questions focused on the electronic structure of molecules, but is also an essential prerequisite for further approximations towards a computationally feasible approach.

Moreover, employing the techniques of second quantization prohibits the direct interpretation of the field operators ψ as usual quantum mechanical wave functions, since superpositions of states with variable numbers of particles are not compatible with the simple probabilistic interpretation of the wave function. In order to restore this feature it has become the standard procedure of quantum chemistry to return to a first quantized formulation based on suitable generalizations of the original Dirac equation Eq. (7.9), which will be the subject of the next chapter.

Further reading

Our introduction to quantum electrodynamics has been kept rather brief. The reason for this decision is the fact that, although it is the fundamental theory for molecular science, a semi-classical version (chapter 8), in which the electromagnetic fields are not considered quantized, has turned out to be a suitable quantum theory for chemistry. Therefore, we simply provide for further study an extensive list of references to excellent books on quantum electrodynamics instead. While the first two books consider the results of the theory from an interpretive point of view, the subsequent volumes present the theory in all technical detail.

P. Teller, [119]. *An Interpretive Introduction to Quantum Field Theory.*

In this wonderful book, Teller works out the basic notions of quantum field theory including second quantization and their meaning. It is a great book to start with as an introduction to quantum field theory and will serve everyone as a good basis for further study. The book is very readable with an emphasis on the concepts of field theory.

R. P. Feynman, [120]. *QED — The strange theory of light and matter.*

This popular science book by one of the inventors of quantum electrodynamics is a marvelous introduction into the concepts of the theory. Although the mathematical apparatus is not easy to grasp (compare also his 'technical' monograph [121]), Feynman managed to present the essential ideas in the clearest possible language.

S. Weinberg, [99]. *The Quantum Theory of Fields. Volume I. Foundations.*

Weinberg succeeded in providing a readable presentation of his subject. He starts from the very beginning and requires surprisingly little previous knowledge of quantum mechanics. His book is an excellent read.

S. S. Schweber, [122]. *An Introduction to Relativistic Quantum Field Theory.*

This classic text features a careful presentation without lacking any details and explains all foundations of QED. It can also be warmly recommended for more advanced topics of QED, like Feynman diagrams, the Furry picture etc.

J. D. Bjorken, S. D. Drell, [123]. *Relativistic Quantum Fields.*

The textbook by Bjorken and Drell is very popular for good reason and can be highly recommended for further reference. The book is accompanied by a more introductory volume on relativistic quantum mechanics by the same authors [65].

A. I. Akhiezer, V. B. Berestetskii, [124]. *Quantum Electrodynamics.*

This book is another classic text. It is well worth reading and contains many details that are surprisingly easy to grasp. For an important example, compare the discussion of the retarded electron–electron interaction starting from the scattering matrix of two charges that finally leads to a part of an integral kernel known as the Breit operator.

L. D. Landau, E. M. Lifshitz, [125]. *Course of Theoretical Physics: 4. Quantum Electrodynamics.*

The fourth volume of Landau and Lifshitz's famous course on Theoretical Physics is devoted to quantum electrodynamics. It also covers the problem of electron–electron scattering and the two-electron bound state problem (Breit equation) from this point of view. Since our presentation of QED was rather brief in this chapter, we recommend the volume from Landau and Lifshitz's course for further reference.

F. Gross, [32]. *Relativistic Quantum Mechanics and Field Theory.*

We include Gross' valuable book in this list as an example of a modern presentation of QED, which by contrast to many such modern accounts does not start from modern gauge theories right away, but which includes a lot of basic background material so that the field theory appears in the broader context in which it belongs. The contents list is well arranged and allows the reader to quickly obtain an overview of the subject.

T. P. Das, [85]. *Relativistic Quantum Mechanics of Electrons.*

In the last part of his classic book, Das also provides some material regarding the quantum electrodynamical treatment of the two-electron problem.

8
First-Quantized Dirac-Based Many-Electron Theory

In the preceding chapter quantum electrodynamics was established as the well-defined and exact theory for matter subject to electromagnetic interactions. However, we noted that the second-quantized formalism with variable particle numbers is not necessarily the most appropriate one for the molecular sciences. Pair-creation processes cannot occur in the energy range relevant to molecular systems and hence have not to be taken into account if we can benefit from their neglect. And in fact, a consistent and numerically reliable many-electron theory can be developed starting from a first-quantized picture of fixed particle numbers and neglecting the quantization of electromagnetic fields. This chapter presents the basic framework to be further detailed in the subsequent chapters.

In the molecular sciences it is most appropriate to adopt a pragmatic attitude toward the Dirac equation in order to set up a theory which closely resembles nonrelativistic many-electron theory. We will see that we can afford quite a number of approximations designed such that the numerical effect on physical observables still resembles that of a truly relativistic many-electron theory. Hence, we proceed from the fundamental physical principles of Einstein's special theory of relativity to approximations of different degree. As a matter of fact this is exactly the program of relativistic quantum chemistry that we shall start to develop in this chapter.

Quantum mechanical calculations in the molecular sciences do not necessarily involve a variation of the number of particles (especially not through pair creation and annihilation processes). This even holds true in the case of particle exchange processes as the reactants involved can be described in a fixed-particle-number framework. For example, a reductant can be treated together with the molecule to be reduced as a whole system such that the number of electrons remains constant during the reduction process. Also, the energy of liberated electrons can be considered zero, and thus such electrons can be neglected from one step to the next in a reaction sequence. This is, for instance, useful for ionization processes, where the released electron is considered to be at rest and features zero energy at infinite distance so that it makes no contribution to the Hamiltonian of the ionized system. There is therefore a need to proceed from QED to a computationally more appropriate albeit less

Relativistic Quantum Chemistry. Markus Reiher and Alexander Wolf

ISBN: 978-3-527-31292-4

rigorous theory. Still, this transition is to be made in a consistent manner so that all essential numerical effects on observables can be captured. In order to obey this boundary condition of accuracy while striving at feasibility we start with the next logical step following chapter 6 which is the one from the one-electron system to the two-electron case in a first-quantized theory.

The Hamiltonian of the two-electron atom already features all pair-interaction operators that are required to describe a system with an arbitrary number of electrons and nuclei. Hence, the step from one to two electrons is much larger than from two to an arbitrary number of electrons. For the latter we are well advised to benefit from the development of nonrelativistic quantum chemistry, where the many-electron Hamiltonian is exactly known, i.e., where it is simply the sum of all kinetic energy operators according to Eq. (4.48) plus all electrostatic Coulombic pair interaction operators.

In a classical picture we may say that because of the nonpolarizability of the elementary *point* particles in a many-electron system, we can employ the classical electrostatic pair interaction energy for point charges in Eq. (2.147) — rather than the more general one in Eq. (2.148) — for the 'deduction' of a suitable electrostatic pair interaction operator guided by the correspondence principle. Unfortunately, the situation is more complicated because of magnetic interactions as is elaborated in the following section.

8.1 Two-Electron Systems and the Breit Equation

In this section, we derive the electromagnetic potential energy operator for two interacting electrons (or for two charges in general). The basic principles for our derivation are the inclusion of non-quantized electromagnetic fields and the correspondence principle. The derivation can, of course, be embedded into the more general picture of quantum electrodynamics. However, it appears that because of the fast development of the modern theory of quantum electrodynamics in the late 1940s, the consistent presentation of the semi-classical theory important for chemistry has not attracted much attention since those times. In the preface to the fourth edition of his textbook from 1957, Dirac comments on the situation as follows [55]: *The main change from the third edition is that the chapter on quantum electrodynamics has been rewritten. [...] In present-day high-energy physics the creation and annihilation of charged particles is a frequent occurrence. A quantum electrodynamics which demands conservation of the number of charged particles is therefore out of touch with physical reality. So I have replaced it by a quantum electrodynamics which includes creation and annihilation of electron–positron pairs. This involves abandoning any close analogy with classical electron theory, but provides a closer description of nature. It seems that the classical concept of an electron is no longer a useful model in physics, except possibly for elementary theories that are restricted to low-energy phenomena.* Exactly these latter low-energy phenomena are the ones that we have to deal with in molecular science.

Presentations of the semi-classical theory are quite rare. The textbook by Mott and Sneddon [126] from 1948 is an example where this has been attempted in some detail, although — after having derived the quantum mechanical energy expression for the retarded electromagnetic interaction of two electrons — the authors come to the conclusion that ... *it will be appreciated that the derivation does not depend on any very consecutive argument* [126, p. 339]. In the following we will see that the situation is actually not that bad.

The first step toward a practical 'relativistic' many-electron theory in the molecular sciences is the investigation of the two-electron problem in an external field which we meet, for instance, in the helium atom. Salpeter and Bethe derived a relativistic equation for the two-electron bound-state problem [95,127–130] rooted in quantum electrodynamics, which features two separate times for the two particles. If we assume, however, that an absolute time is a good approximation, we arrive at an equation first considered by Breit [68, 131, 132]. The Bethe–Salpeter equation as well as the Breit equation hold for a 16-component wave function. From a formal point of view, these 16 components arise when the two four-dimensional one-electron Hilbert spaces are joined by direct multiplication to yield the two-electron Hilbert space.

8.1.1
Dirac Equation Generalized for Two Bound-State Electrons

According to the Dirac equation for an electron in external electromagnetic fields, Eq. (5.119), we should make the following ansatz for the two-electron system including the external potential energy, V_{nuc}, of resting nuclei,

$$\left\{\left[-i\hbar\frac{\partial}{\partial t_1}+c\,\boldsymbol{\alpha}_1\cdot\left(\boldsymbol{p}_1-\frac{q_1}{c}\boldsymbol{A}^{(2)}\right)+\beta_1 m_e c^2+\right.\right.$$
$$\left.q_1\phi^{(2)}+V_{nuc}(r_1)\right]\otimes\mathbf{1}_4+\mathbf{1}_4\otimes\left[-i\hbar\frac{\partial}{\partial t_2}+c\,\boldsymbol{\alpha}_2\cdot\left(\boldsymbol{p}_2-\frac{q_2}{c}\boldsymbol{A}^{(1)}\right)\right.$$
$$\left.\left.+\beta_2 m_e c^2+q_2\phi^{(1)}+V_{nuc}(r_2)\right]\right\}\Psi(ct_1,\boldsymbol{r}_1,ct_2,\boldsymbol{r}_2)=0 \quad (8.1)$$

with the vector and scalar potentials felt by one electron and generated by the other, respectively. Here, two (16×16)-matrices result after evaluation of the tensor products and are then added. This equation was built up from the two separate Dirac equations of the general type in Eq. (5.119),

$$\left[-i\hbar\frac{\partial}{\partial t_1}+c\,\boldsymbol{\alpha}_1\cdot\left(\boldsymbol{p}_1-\frac{q_1}{c}\boldsymbol{A}^{(2)}\right)+\beta_1 m_e c^2+q_1\phi^{(2)}+V_{nuc}(r_1)\right]\psi_1=0 \quad (8.2)$$

$$\left[-i\hbar\frac{\partial}{\partial t_2} + c\,\boldsymbol{\alpha}_2 \cdot \left(\boldsymbol{p}_2 - \frac{q_2}{c}\boldsymbol{A}^{(1)} \right) + \beta_2 m_e c^2 + q_2\phi^{(1)} + V_{nuc}(r_2) \right] \psi_2 = 0 \quad (8.3)$$

valid for each electron. It is important to understand that each of these two equations is exact: these individual Dirac equations describe any of the two electrons in some external electromagnetic field (here, generated by the other electron) which does not, in principle, require any reference to the specific sources. If these equations could be solved exactly, the exact one-electron wave functions would become available. Hence, the *exact* total wave function consists then of '$4 \otimes 4$=16' components arranged in vector form according to

$$\Psi(ct_1, \boldsymbol{r}_1, ct_2, \boldsymbol{r}_2) = \psi_1(ct_1, \boldsymbol{r}_1) \otimes \psi_2(ct_2, \boldsymbol{r}_2) = \begin{pmatrix} \psi_1^{L_1}(ct_1, \boldsymbol{r}_1)\psi_2(ct_2, \boldsymbol{r}_2) \\ \psi_1^{L_2}(ct_1, \boldsymbol{r}_1)\psi_2(ct_2, \boldsymbol{r}_2) \\ \psi_1^{S_1}(ct_1, \boldsymbol{r}_1)\psi_2(ct_2, \boldsymbol{r}_2) \\ \psi_1^{S_2}(ct_1, \boldsymbol{r}_1)\psi_2(ct_2, \boldsymbol{r}_2) \end{pmatrix} \quad (8.4)$$

where we denoted the two upper large components of ψ_1 as L_1 and L_2, while the lower small components carry S_1 and S_2 superscripts, accordingly. Apart from the fact that the Pauli principle might not be fulfilled for this wave function (an issue that is resolved in depth later in this chapter), the ansatz is exact for the one-particle states ψ_1 and ψ_2 of Eqs. (8.2) and (8.3). These solution functions are, however, usually *not* known although it is common in the physics literature to assume that they are known or that efficient approximations are available, which is also no trivial task. We may regard the observation that *approximate* one-electron states dramatically reduce the accuracy of quantum mechanical expectation values calculated from them as *the* contribution of quantum chemistry to many-electron physics. As we shall see later in this chapter (and also in the numerical examples of chapters 9 and 16), a reliable approximation of a total wave function is very involved — even for helium. The accuracy of an energy expectation value must not be judged solely on the basis of the terms included in the Hamiltonian as long as the wave function is not represented accurately enough. This must also be kept in mind when, for instance, Breit's original results on the validity of two-electron operators are examined.

Two problems need to be addressed next. First, what to do with the two different time derivatives and, second, how are the scalar and vector potentials of the electrons to be chosen. With respect to the former issue, we adopt the single absolute time frame, $t_1, t_2 \rightarrow t$ of nonrelativistic theory. This reduces the two time derivatives to one with respect to the new absolute time t,

$$\frac{\partial}{\partial t_1} \otimes \mathbf{1}_4 + \mathbf{1}_4 \otimes \frac{\partial}{\partial t_2} \equiv \frac{\partial}{\partial t_1}\mathbf{1}_4 \otimes \mathbf{1}_4 + \mathbf{1}_4 \otimes \frac{\partial}{\partial t_2}\mathbf{1}_4 \rightarrow \frac{\partial}{\partial t}(\mathbf{1}_4 \otimes \mathbf{1}_4) \quad (8.5)$$

so that we may write

$$\left[\left(-\mathrm{i}\hbar\frac{\partial}{\partial t_1}\right)\otimes \mathbf{1}_4 + \mathbf{1}_4 \otimes \left(-\mathrm{i}\hbar\frac{\partial}{\partial t_2}\right)\right]\psi_1(ct_1,\boldsymbol{r}_1)\otimes\psi_2(ct_2,\boldsymbol{r}_2) \longrightarrow \cdots$$

$$\cdots \longrightarrow -\mathrm{i}\hbar\left[\frac{\partial\psi_1(t,\boldsymbol{r}_1)}{\partial t}\psi_2(t,\boldsymbol{r}_2) + \psi_1(t,\boldsymbol{r}_1)\frac{\partial\psi_2(t,\boldsymbol{r}_2)}{\partial t}\right]$$

$$= -\mathrm{i}\hbar\frac{\partial}{\partial t}\left[\psi_1(t,\boldsymbol{r}_1)\psi_2(t,\boldsymbol{r}_2)\right] \quad (8.6)$$

where we choose to write the sum of Kronecker products as a sum of ordinary products so that our ansatz can finally be cast in the form

$$\left\{\left[c\,\boldsymbol{\alpha}_1\cdot\left(\boldsymbol{p}_1 - \frac{q_1}{c}\boldsymbol{A}^{(2)}\right) + \beta_1 m_e c^2 + q_1\phi^{(2)}\right.\right.$$
$$\left.+V_{nuc}(r_1)\right] + \left[c\,\boldsymbol{\alpha}_2\cdot\left(\boldsymbol{p}_2 - \frac{q_2}{c}\boldsymbol{A}^{(1)}\right) + \beta_2 m_e c^2 + q_2\phi^{(1)}\right.$$
$$\left.\left.+V_{nuc}(r_2)\right]\right\}\Psi(t,\boldsymbol{r}_1,\boldsymbol{r}_2) = \mathrm{i}\hbar\frac{\partial}{\partial t}\Psi(t,\boldsymbol{r}_1,\boldsymbol{r}_2) \quad (8.7)$$

for which the wave function can now be written as being constructed from ordinary one-electron product states

$$\Psi(t,\boldsymbol{r}_1,\boldsymbol{r}_2) = \psi_1(t,\boldsymbol{r}_1)\psi_2(t,\boldsymbol{r}_2) \quad (8.8)$$

This product ansatz is, of course, also not exactly known as we shall discuss later in this chapter. In chemistry, explicit and sufficiently accurate many-electron calculations are the aim. The solution functions ψ_1 and ψ_2 of Eqs. (8.2) and (8.3) are not known because the structure of the potentials is too complicated even in the case of only two electrons — as we shall see — to allow for an analytic solution. We will see in the later sections 8.5 – 8.7 how to deal with this problem by invoking a mean-field approach.

8.1.2
The Gaunt Operator for Unretarded Interactions

The next problem is to choose for one electron the scalar and vector potentials generated by the other electron. We approach this issue step by step and first study the *unretarded* classical scalar potential created by electron 2,

$$\phi_{2,\mathrm{unret.}}(\boldsymbol{r}_1,\boldsymbol{r}_2) = \frac{q_2}{r_{12}} \quad (8.9)$$

This expression is obviously in accord with the standard Coulomb law, but could also be derived from the most general Eq. (3.208) in the limit of infinite speed of light, i.e., after expansion and truncation of the absolute value of the denominator in that equation. With this expression for the scalar potential, we obtain the unretarded expression for the corresponding vector potential of the moving electron 2 from Eq. (3.238),

$$A_{2,\text{unret.}}(r_1, r_2, \dot{r}_2) = \frac{\dot{r}_2}{c}\phi_{2,\text{unret.}} = \frac{\dot{r}_2}{c}\frac{q_2}{r_{12}} \tag{8.10}$$

Corresponding expressions for the fields produced by electron 1 are obtained by exchanging the subscripts 1 and 2. In order to find the quantum mechanical operator expressions, we invoke the only guiding principle that we are familiar with in quantum mechanics, namely the correspondence principle. The velocities $\dot{r}_1$ and $\dot{r}_2$ are then to be substituted by the velocity operators $c\alpha_1$ and $c\alpha_2$ of Dirac's theory derived in section 5.3.3. Note that we switch from an explicit trajectory picture, which has been the basis of all derivations in section 3.5, to the operator formulation in quantum mechanics and solely trust in the correspondence principle that this is the correct way to obtain a trajectory-free quantum picture of the interaction of charged moving particles.

Moreover, one might also think about introducing momentum operators instead of velocity operators. Obviously, that would yield quite different expressions since no α matrices are then introduced at all. However, Eq. (8.7) gives us a hint whether we should use the velocity operator or the momentum operator since it contains via the minimal-coupling expression terms like $(\alpha_i A^{(j)})$, which will only be symmetric in the two coordinates i and j if we introduce the velocity operator and *not* the momentum operator. Recall that we have encountered this line of reasoning already in section 5.4.2.

Consequently, we choose the unretarded electromagnetic potential operators to become

$$\hat{\phi}_{i,\text{unret.}}(r_i, r_j) = \frac{q_i}{r_{ij}} \quad \text{and} \quad \hat{A}_{i,\text{unret.}}(r_1, r_2) = \alpha_i \frac{q_i}{r_{ij}} \tag{8.11}$$

where we used the hat notation only once to indicate the difference from the classical Eqs. (8.9) and (8.10). These (unretarded) electromagnetic field operators can now be employed in Eq. (8.7). However, now it is important to avoid double counting. Since the electromagnetic fields are no longer retarded, their effect is felt instantaneously by both electrons at the *same* time, and we would count the interaction twice in Eq. (8.7) for any instant of time in our absolute frame. Therefore, we are advised to account for this problem and divide the terms that carry the potentials by two to produce

$$\Bigg\{\Bigg[c\,\alpha_1 \cdot \Bigg(p_1 - \frac{q_1}{c}\underbrace{\frac{1}{2}\alpha_2\frac{q_2}{r_{12}}}_{=A^{(2)}}\Bigg) + \beta_1 m_e c^2 + q_1\underbrace{\frac{1}{2}\frac{q_2}{r_{12}}}_{=\phi^{(2)}}$$

$$+V_{nuc}(r_1)\Big] + \Big[c\,\boldsymbol{\alpha}_2\cdot\Big(\boldsymbol{p}_2 - \frac{q_2}{c}\underbrace{\frac{1}{2}\boldsymbol{\alpha}_1\frac{q_1}{r_{12}}}_{=\boldsymbol{A}^{(1)}}\Big) + \beta_2 m_e c^2 + q_2\underbrace{\frac{1}{2}\frac{q_1}{r_{12}}}_{=\phi^{(1)}}$$

$$+V_{nuc}(r_2)\Big]\Big\}\Psi(t,\boldsymbol{r}_1,\boldsymbol{r}_2) = \mathrm{i}\hbar\frac{\partial}{\partial t}\Psi(t,\boldsymbol{r}_1,\boldsymbol{r}_2) \qquad (8.12)$$

This equation can now easily be rearranged into a compact form,

$$\Big[c\,\boldsymbol{\alpha}_1\cdot\boldsymbol{p}_1 + c\,\boldsymbol{\alpha}_2\cdot\boldsymbol{p}_2 + \beta_1 m_e c^2 + \beta_2 m_e c^2 - q_1 q_2\frac{\boldsymbol{\alpha}_1\boldsymbol{\alpha}_2}{r_{12}}$$

$$+\frac{q_1 q_2}{r_{12}} + V_{nuc}(r_1) + V_{nuc}(r_2)\Big]\Psi(t,\boldsymbol{r}_1,\boldsymbol{r}_2) = \mathrm{i}\hbar\frac{\partial}{\partial t}\Psi(t,\boldsymbol{r}_1,\boldsymbol{r}_2) \qquad (8.13)$$

In order to understand the connection between minimal coupling, which we have just exploited, and a quantum mechanical approach that starts from the *field-free* Dirac operators and simply adds the interaction energy as an operator, we start again with the classical expression for the *unretarded* interaction energy of two moving charges. This classical energy for *nonretarded* fields follows directly from Eq. (3.211) where $\gamma(\dot{\boldsymbol{r}}_2)$ approaches 1 as c increases so that we can then employ the *unretarded* scalar potential of Eq. (8.9). Inserting the unretarded scalar potential into Eq. (3.211) for $c \to \infty$ yields the *unretarded* classical interaction energy

$$V_{12} = \frac{q_1 q_2}{r_{12}}\left(1 - \frac{\dot{\boldsymbol{r}}_1\cdot\dot{\boldsymbol{r}}_2}{c^2}\right) \qquad (8.14)$$

where the first term is the electrostatic Coulomb energy, while the second is due to the magnetic interaction of the moving electrons. If we now substitute the velocities by their quantum analogs as before,

$$\hat{V}_{12} = \frac{q_1 q_2}{r_{12}}\left(1 - \boldsymbol{\alpha}_1\boldsymbol{\alpha}_2\right) \qquad (8.15)$$

we easily identify the standard Coulomb operator as the first term and then the so-called *Gaunt operator*,

$$G_0(1,2) = -q_1 q_2\frac{\boldsymbol{\alpha}_1\boldsymbol{\alpha}_2}{r_{12}} \overset{(4.139)}{=} -\frac{q_1 q_2}{r_{12}}\frac{4}{\hbar^2}\begin{pmatrix} 0 & \boldsymbol{s}_1\boldsymbol{s}_2 \\ \boldsymbol{s}_1\boldsymbol{s}_2 & 0 \end{pmatrix} \qquad (8.16)$$

We stress that this interaction enters the quantum mechanical Hamiltonian because of the *unretarded* vector potential. Hence, we understand that the *instantaneous magnetic interaction* of moving electrons shows up in four-component first-quantized Dirac theory as the interaction of the electron's two spin momenta as expressed by the spin operators of the two electons on the right hand side of the equation above. This interaction term was considered extensively by Gaunt for the description of spin–spin interactions in the triplet state of helium [133,134].

It must be noted here that the interaction energy operator of Eq. (8.15) already comprises all magnetic interactions — though in an approximate manner — *of the two electrons, and the vector potentials of the moving electrons do not show up as additive terms to the canonical momentum.* The latter, however, should be obvious from our discussion of the Gaunt interaction, where we started from the potentials introduced through minimal coupling and arrived at the scalar potential energy operator (for unretarded interactions). Additional vector potentials may be added to the momentum only if they are truly external to the electronic system.

Moreover, it is important to understand that the operator pair $(\boldsymbol{\alpha}_1\boldsymbol{\alpha}_2)$ is a scalar product of two vectors

$$\boldsymbol{\alpha}_1\boldsymbol{\alpha}_2 = \sum_{i=1}^{3} \alpha_{1,i}\alpha_{2,i} \tag{8.17}$$

where the (4×4)-matrices $\alpha_{1,i}$ and $\alpha_{2,i}$ must not be multiplied according to the rules of matrix multiplication, which is the reason why the central dot of the scalar product has been omitted. Instead, these operators act on the corresponding one-electron state functions. However, we will only later combine them with the proper one-electron state vectors in matrix elements (see section 9.3.3 for an example, but also section 8.5 later in this chapter).

Since the Gaunt term originally stems from the *unretarded* magnetic interaction in classical electrodynamics, the Gaunt operator is also considered to describe (such) magnetic interactions. However, we already know from the discussion of the Darwin interaction energy in section 3.5 that this will not be a consistent approximation. The interaction energy of two moving charges is, however, affected by the retarded electromagnetic fields as in the classical case derived in Eq. (3.249).

8.1.3
The Breit Operator for Retarded Interactions

The operator for the classical potential energy of the *retarded* interaction of the two electrons can again be obtained from the correspondence principle applied to Darwin's approximation of the classical relativistic interaction energy assigned to two moving charges in section 3.5. Therefore, we do not consider the retarded potentials again but start right away with the Darwin interaction energy, which is correct up to $1/c^2$ according to the derivation in section 3.5. This energy is to be used to find an explicit expression for the general electromagnetic interaction energy operator of Eq. (5.125).

As in the case of the unretarded Gaunt interaction, we choose the velocity operator expression $\dot{\boldsymbol{r}} \rightarrow c\boldsymbol{\alpha}$ (instead of the momentum operator) to obtain consistent results as before and obtain the interaction energy of Eq. (3.249) promoted to the operator calculus of quantum theory,

$$\hat{V}_{12} \approx \frac{q_1 q_2}{r_{12}} \underbrace{- \frac{q_1 q_2}{2} \left[\frac{\boldsymbol{\alpha}_1 \boldsymbol{\alpha}_2}{r_{12}} + \frac{(\boldsymbol{r}_{12} \cdot \boldsymbol{\alpha}_1)(\boldsymbol{r}_{12} \cdot \boldsymbol{\alpha}_2)}{r_{12}^3} \right]}_{\equiv B_0(1,2)} \tag{8.18}$$

The term marked as B_0 is the Breit operator (the zero subscript will become clear at a later stage in this section). The approximation sign in Eq. (8.18) indicates that the operator $B_0(1,2)$ contains only retardation effects to leading order as in the Darwin interaction energy, which was the starting point. However, we refrain from assigning a well-defined order in $1/c$ as provided for the Darwin interaction since the Dirac velocity $c\boldsymbol{\alpha}$ cancels some of these prefactors. To be more precise: by comparison with the classical expression of Darwin we see that the relativistic interaction energy operator, Eq. (8.18), does *not* contain the c^{-2} damping factor. Instead, it seems as if the Breit operator in Eq. (8.18) is of the same order in $1/c$ as the Coulomb operator, i.e., of zeroth order.

This is, however, misleading, because the $\boldsymbol{\alpha}_i$ operators have zero block-diagonal entries and, hence, couple only large and small components of the wave function. Matrix elements of the Breit operator B_0 are therefore damped by c^{-2} because of the kinetic balance relation derived in Eq. (5.131). Note that the two-electron Breit operator couples, of course, large and small components from the spinors of two electrons so that the kinetic balance condition hits twice and produces the square of the inverse speed of light. In other words, the fact that the large component always enters energy expectation values multiplied by a small component results in a c^{-2}-damping compared to the Coulomb integrals that contain integrals where the large components are always multiplied by one another but never by a small component. Thus, matrix elements of the Breit operator that are required to evaluate the energy expectation value would again be damped by $1/c^2$ compared to the Coulomb integrals over the first operator in Eq. (8.18).

With Breit's potential energy operator, the stationary quantum mechanical equation for two electrons in the central field of an atomic nucleus reads

$$\Bigg\{ \underbrace{c(\boldsymbol{\alpha} \cdot \boldsymbol{p})_1 + \beta_1 m_e c^2 + V_{nuc}(r_1)}_{\equiv h_1^D} + \underbrace{c(\boldsymbol{\alpha} \cdot \boldsymbol{p})_2 + \beta_2 m_e c^2 + V_{nuc}(r_2)}_{\equiv h_2^D} + \frac{q_1 q_2}{r_{12}} - \frac{q_1 q_2}{2} \left[\frac{\boldsymbol{\alpha}_1 \cdot \boldsymbol{\alpha}_2}{r_{12}} + \frac{(\boldsymbol{r}_{12} \cdot \boldsymbol{\alpha}_1)(\boldsymbol{r}_{12} \cdot \boldsymbol{\alpha}_2)}{r_{12}^3} \right] \Bigg\} \Psi(\boldsymbol{r}_1, \boldsymbol{r}_2) = E \Psi(\boldsymbol{r}_1, \boldsymbol{r}_2) \tag{8.19}$$

The Breit interaction implements the effect of retardation as its classical Darwin analog was derived for this case. We should emphasize that the specific form of this retardation term contains a certain degree of arbitrariness due to the choice of the gauge function in the derivation of the Darwin interaction, which was solely designed to produce a symmetric form of the interaction energy valid for an absolute time scale. The first term of the Breit operator also contains a pair of $\boldsymbol{\alpha}_i$ operators divided by the electron–electron distance, which is therefore associated with this magnetic interaction although it occurs in the *retarded* (though approximate) form of Darwin and is divided by 2 compared to the Gaunt operator in Eq. (8.13). Often, however, only the second term of order r_{12}^{-3} is misleadingly denoted the retardation operator. In order to extract the contribution solely due to retardation,

$$B_0(1,2) = G_0(1,2) + B_{\text{ret.}}(1,2) \tag{8.20}$$

we may subtract the Gaunt operator and obtain the true *retardation operator*

$$B_{\text{ret.}}(1,2) = \frac{q_1 q_2}{2}\left[\frac{\boldsymbol{\alpha}_1\boldsymbol{\alpha}_2}{r_{12}} - \frac{(\boldsymbol{r}_{12}\cdot\boldsymbol{\alpha}_1)(\boldsymbol{r}_{12}\cdot\boldsymbol{\alpha}_2)}{r_{12}^3}\right] \tag{8.21}$$

Though this retardation operator includes products of the Dirac matrices for the two interacting particles, $\boldsymbol{\alpha}_1\boldsymbol{\alpha}_2$, magnetic interactions mediated via the vector potential seem to be included. However, already the derivation of the classical retarded interaction in section 3.5 clearly shows that all vector potential terms inherently feature a $1/c^2$ prefactor, as in Eq. (3.211), so that the vector potential contribution and, hence, the magnetic contribution of the interaction enters these equations unretarded [135]. This is also clear from the rigorous collection of terms after a Taylor expansion around time t_1 in Eq. (3.230).

Including only the Gaunt operator in a many-particle Hamiltonian is often preferred in calculations because of its simpler structure, which is comparable in complexity to the instantaneous Coulomb interaction. In the present circumstances, however, it may also be preferred as it is much more well defined than the retardation term whose specific form was determined by the choice of gauge in section 3.5. As noted before, including only the Gaunt interaction would imply that retardation effects are completely omitted and magnetic interaction comes instantaneously into effect. Mann and Johnson [136] found that the Gaunt interaction is about one order of magnitude more important than retardation effects on relative energies of atoms. However, it can also be argued that Gaunt and retardation terms are of the same order and, hence, the complete Breit operator should be included rather than the Gaunt operator.

Gaunt and Breit interaction operators represent potential energies due to magnetic interactions, which one would also assume to play a role in the nonrelativistic many-particle theory given by the Pauli equation, Eq. (5.134). One might argue that in the strict nonrelativistic limit, $c \to \infty$, all magnetic interaction terms vanish in Eq. (5.134) because of the $1/c$ pre-factor [137]. However, one would also expect from any nonrelativistic theory not to neglect magnetic interactions. Hence, although their contribution is small because of the $1/c$ prefactor, we cannot omit the wealth of magnetic phenomena from the nonrelativistic theory and therefore ask about how to include Gaunt- and Breit-type operators in a Pauli-like theory. It seems obvious that this results in expressions where the Dirac matrices $\boldsymbol{\alpha}$ are replaced by Pauli spin matrices $\boldsymbol{\sigma}$. In fact, Gaunt discussed in his early papers this two-component variant involving Pauli rather than Dirac matrices. His work was already based on work by Heisenberg and others in the very early days of quantum mechanics who tried to identify a suitable operator for magnetic interactions.

The interesting aspect, however, is the question of how one would have obtained these expressions starting from Schrödinger theory rather than from Dirac theory. The main difference is that the Dirac velocity operator and the momentum operator are not related whereas the Schrödinger velocity operator is nothing else but the momentum operator divided by the mass, $\dot{\boldsymbol{r}}^{(S)} = \boldsymbol{p}/m$ (cf. section 4.3.7). Also, in the classical case, the Darwin interaction energy can be written using momenta $\boldsymbol{p}$ rather than velocities $\dot{\boldsymbol{r}}$; see Eq. (3.250). Hence, the correspondence principle applied to the Darwin interaction in Eq. (3.250) does not work because no spin–spin interaction terms would then be generated in Dirac theory or if Eq. (3.250) were directly applied in Schrödinger's theory. It is thus interesting to note the importance of the Dirac theory and the benefits of the minimal coupling procedure of section 5.4 in order to acknowledge how the classical electromagnetic potentials of the moving charges enter the Hamiltonian. This is another example of how Dirac's theory nicely introduces all effects of electron spin automatically without any further assumption. The correct expressions for the Pauli Hamiltonian then arise naturally by progressing from four-component Dirac to two-component Pauli theory of section 5.4.3. We will see later in section 13.2 how the Breit operator finally produces the so-called *Breit–Pauli Hamiltonian*. There, it will be demonstrated that the Breit–Pauli Hamiltonian contains an explicit spin–spin interaction term which somewhat resembles the Breit expression though the former contains different pre-factors and an additional global $1/r_{12}^2$ term. But most importantly, we encounter a Darwin-like energy operator, Eq. (13.72), that is like the classical expression of Eq. (3.250) which *explicitly* employs the momentum operators. Hence, the quantum analog of Eq. (3.250) can be thought of as being hidden in the full Breit operator, as it can be shown to emerge from it.

Depending on the type of molecular properties, a Breit operator is often not employed in quantum chemistry. This is because of the c^{-2} damping which reduces the effect of the Breit correction compared to the Coulomb interaction and which makes this term even vanish in the limit of instantaneous interactions, $c \to \infty$. Non-negligible effects may be encountered in accurate calculations of excitation energies and, of course, in spin- or magnetic-field-dependent properties.

Finally, we should again note that the minimal coupling procedure of section 5.4, which allows us to introduce external electromagnetic potentials and, in particular, vector potentials $\boldsymbol{A}$, does *not* need to be employed for the vector potentials that arise from the moving electrons. These magnetic interactions are covered by the Darwin and thus by the Breit interaction — though in an approximate manner (compare section 3.5). The electrons' magnetic fields are thus purely internal and minimal coupling is only required for vector potentials that are truly external to a system of electrons.

The effect of the Breit and Gaunt interactions has been investigated in many studies on atomic systems [136, 138–159] and also on molecular systems [160–163]. However, especially in early molecular calculations, only the Gaunt rather than the Breit operator was used, though they were specified as Breit calculations emphasizing the inclusion of magnetic effects but disregarding the distinction between unretarded and retarded interactions. As one would expect from the discussion so far, the Gaunt interaction is already a small correction to the Coulomb interaction and the retardation correction is then even smaller. Although these magnetic interactions are important — and especially non-negligible in high-resolution spectroscopy — they can usually be neglected in chemical reactions of molecules. For instance, the effect of the Gaunt electron–electron interaction on bond length can be negligible as demonstrated in Ref. [160]: CH_4 [0.0 pm]; SiH_4 [0.0 pm]; GeH_4 [0.1 pm, 326 ppm]; SnH_4 [0.1 pm, 514 ppm]; PbH_4 [0.2 pm, 962 ppm]; — the difference in bond lengths compared to a calculation with only the Coulomb interaction energy operators is given in pm in brackets and the percentage of the change compared to the absolute value of the bond length is given in parts per million (ppm); the wave function was approximated by a single Slater determinant described later in this chapter. The effect of the Gaunt interaction on binding energies in CH_4 is about 0.3 kJ mol^{-1} and increases in the case of PbH_4 to about 2.1 kJ mol^{-1} [160].

To conclude, compared to the exact theory of electrons and photons sketched in chapter 7 and in accordance with Eq. (3.249), we may consider approximate pair interaction terms of Coulomb–Breit type for incorporation into the first-quantized many-electron Hamiltonian in the following section 8.2.

8.1.4 Exact Retarded Electromagnetic Interaction Energy

After having considered the *approximate* retarded interaction, i.e., the Breit interaction, from the point of view of classical electrodynamics, we now follow Rosenfeld [164] and derive the *exact* retarded interaction energy of two moving electrons (note that also Nikolsky [165] has derived the Breit equation starting from Dirac's equation in a quite similar manner). The solution of Eqs. (8.2) and (8.3) can, of course, be written in a formally exact manner for any one-electron Dirac Hamiltonian that does *not* explicitly depend on time as

$$\psi(\boldsymbol{r},t) = \sum_n a_n\, \psi_n(\boldsymbol{r})\, e^{(-\mathrm{i}E_n t/\hbar)} \tag{8.22}$$

with basis functions ψ_n; compare Eqs. (4.3) and (4.22). According to Eq. (5.47), we can write for the 4-current j^μ or, more precisely, for the charge-weighted 4-current $j_c^\mu = q_e j^\mu$ of Eq. (7.4) (cf. also section 3.4.1.1),

$$j_c^\mu(\boldsymbol{r},t) = e^{(-\mathrm{i}\omega_{mn}t)} \Big(\underbrace{c\, q_e \sum_{nm} a_n^\star a_m \psi_n^\dagger \psi_m}_{\rho_c = q_e\rho\,;\ Eq.\ (4.11)} \,,\ \underbrace{q_e c \sum_{nm} a_n^\star a_m \psi_n^\dagger \boldsymbol{\alpha} \psi_m}_{\boldsymbol{j}_c = q_e \boldsymbol{j}\,;\ Eq.\ (5.46)} \Big) \tag{8.23}$$

with the energy difference expressed as

$$E_m - E_n = \hbar\omega_{mn} \tag{8.24}$$

so that $j_c^\mu(\boldsymbol{r},t)$ may be interpreted as a sum over transition 4-currents, which describe transitions between states m and n induced by the potentials of the mutual interaction of the electrons. The scalar and vector potentials of one of the moving charges can then also be obtained as a solution of the four-dimensional Poisson equation [recall Eq. (7.8) and also Eq. (3.163)],

$$\Box A^\mu = +\frac{4\pi}{c} j_c^\mu \tag{8.25}$$

which relates the 4-potential with the 'charged' 4-current of this electron so that the 4-potential is obtained as the (generalized) Poisson integral,

$$A^\mu(\boldsymbol{r}) = \frac{1}{c}\int \mathrm{d}^3r' \frac{j^\mu(\boldsymbol{r}',t')}{|\boldsymbol{r}-\boldsymbol{r}'|} \stackrel{(3.214)}{=} \frac{1}{c}\int \mathrm{d}^3r' \frac{j^\mu(\boldsymbol{r}',t-|\boldsymbol{r}-\boldsymbol{r}'|/c)}{|\boldsymbol{r}-\boldsymbol{r}'|} \tag{8.26}$$

where we explicitly denoted the retardation effect on j_c^μ, compare the analogous Eqs. (2.140) and (2.141). That is, $j_c^\mu(\boldsymbol{r}',t')$ denotes the value of j_c^μ at the retarded time $t-|\boldsymbol{r}-\boldsymbol{r}'|/c$ — compare section 3.5 for the same situation in classical electrodynamics — so that we have for the components of the 4-potential

$$\phi(\boldsymbol{r},t) = q_e\, a_n^\star a_m \sum_{nm} \int \mathrm{d}^3r' \psi_n^\dagger(\boldsymbol{r}') \frac{\exp\left(\mathrm{i}\omega_{nm}\left\{t-|\boldsymbol{r}-\boldsymbol{r}'|/c\right\}\right)}{|\boldsymbol{r}-\boldsymbol{r}'|} \psi_m(\boldsymbol{r}') \tag{8.27}$$

$$\boldsymbol{A}(\boldsymbol{r},t) = q_e\, a_n^\star a_m \sum_{nm} \int \mathrm{d}^3r' \psi_n^\dagger(\boldsymbol{r}') \frac{\boldsymbol{\alpha}\exp\left(\mathrm{i}\omega_{nm}\left\{t-|\boldsymbol{r}-\boldsymbol{r}'|/c\right\}\right)}{|\boldsymbol{r}-\boldsymbol{r}'|} \psi_m(\boldsymbol{r}') \tag{8.28}$$

where the time dependence can now be extracted to yield

$$\phi(\boldsymbol{r},t) = q_e\, a_n^\star a_m\, e^{(\mathrm{i}\omega_{nm}t)} \sum_{nm} \int \mathrm{d}^3 r' \psi_n^\dagger(\boldsymbol{r}') \frac{1}{|\boldsymbol{r}-\boldsymbol{r}'|} e^{\left(\mathrm{i}\omega_{nm}|\boldsymbol{r}-\boldsymbol{r}'|/c\right)} \psi_m(\boldsymbol{r}') \quad (8.29)$$

$$\boldsymbol{A}(\boldsymbol{r},t) = q_e\, a_n^\star a_m\, e^{(\mathrm{i}\omega_{nm}t)} \sum_{nm} \int \mathrm{d}^3 r' \psi_n^\dagger(\boldsymbol{r}') \frac{\boldsymbol{\alpha}}{|\boldsymbol{r}-\boldsymbol{r}'|} e^{\left(\mathrm{i}\omega_{nm}|\boldsymbol{r}-\boldsymbol{r}'|/c\right)} \psi_m(\boldsymbol{r}') \quad (8.30)$$

It is clear that all retardation effects are now coded in the exponential function under the integral. The unretarded potentials are obtained once the exponential vanishes, which is the case if ω_{mn} is set equal to zero as we will investigate below. Then, the total interaction energy expectation value can be generalized from Eq. (2.148), i.e., expressed as the 4-current of electron 1 in the electromagnetic 4-potential generated by electron 2,

$$V_{ee} = \frac{1}{2} \int \mathrm{d}^3 r_1\, j_c^\mu(1) A_\mu(1) = \frac{1}{2} \int \mathrm{d}^3 r_1 \int \mathrm{d}^3 r_2 \frac{j_c^\mu(1) j_{c,\mu}(2)}{r_{12}} \quad (8.31)$$

which can be written as

$$V_{ee} = \frac{1}{2} q_e^2\, a_l^\star\, a_n^\star\, a_m\, a_k\; e^{(\mathrm{i}[\omega_{nm}+\omega_{lk}]t)} \sum_{klmn} V_{lnkm} \quad (8.32)$$

with

$$V_{lnkm} = \int \mathrm{d}^3 r_1 \int \mathrm{d}^3 r_2 \psi_l^\dagger(\boldsymbol{r}_1) \psi_n^\dagger(\boldsymbol{r}_2) \frac{1-\boldsymbol{\alpha}_1\boldsymbol{\alpha}_2}{r_{12}} e^{(\mathrm{i}\omega_{nm} r_{12}/c)} \psi_m(\boldsymbol{r}_2) \psi_k(\boldsymbol{r}_1) \quad (8.33)$$

after insertion of Eqs. (8.23), (8.29) and (8.30). This latter matrix element already contains the *complete and unapproximated* retarded electromagnetic interaction.

Equally well, we could have considered electron 2 in the 4-potential created by electron 1, which would have created a similar expression but with ω_{lk}^2. If we assume that electron 2 changes its state from ψ_m to ψ_n upon interaction, we may associate a change in energy by $\hbar\omega_{nm}^{(2)}$ that must be compensated by the other electron because the total energy must remain unchanged [135]. Hence, we choose

$$\omega_{nm}^{(2)} = -\omega_{lk}^{(1)} \quad (8.34)$$

where we have introduced an additional superscript to denote the particle which undergoes the change associated with an energy measured by the $\omega_{ab}^{(i)}$. Eq. (8.34) restores the symmetry with respect to coordinate exchange as we shall see in more detail below. Moreover, it also makes the time-dependent phase factor in Eq. (8.32) vanish, which is quite convenient.

Occasionally Eq. (8.33) is rewritten as follows. If we split the exponential as

$$
\begin{aligned}
e^{(\mathrm{i}\omega_{nm}r_{12}/c)} &= \frac{1}{2}\left[e^{(\mathrm{i}\omega_{nm}r_{12}/c)} + e^{(\mathrm{i}\omega_{nm}r_{12}/c)}\right] \\
&= \frac{1}{2}\left[e^{(\mathrm{i}\omega_{nm}r_{12}/c)} + e^{-(\mathrm{i}\omega_{lk}r_{12}/c)}\right] \\
&\overset{\omega\equiv|\omega_{nm}|}{=} \frac{1}{2}\left[e^{(\mathrm{i}\omega r_{12}/c)} + e^{-(\mathrm{i}\omega r_{12}/c)}\right] = \cos\left(\frac{\omega r_{12}}{c}\right)
\end{aligned} \tag{8.35}
$$

the integral is explicitly symmetrized and reduced to contain a simple cosine.

The electromagnetic interaction energy will play a significant role in current-density functional theory developed in section 8.8. Then, we will also consider (Pauli) exchange effects which so far have been completely neglected — an approximation in accord with the early work by Møller [166, 167]. For $\omega_{nm} \to 0$ and $\omega_{kl} \to 0$ the interaction energy expectation value reduces to

$$
V_{ee} = \frac{q_e^2}{2} a_l^\star a_n^\star a_m a_k \sum_{klmn} \int \mathrm{d}^3 r_1 \int \mathrm{d}^3 r_2 \psi_l^\dagger(\boldsymbol{r}_1)\psi_n^\dagger(\boldsymbol{r}_2)\frac{1-\boldsymbol{\alpha}_1\boldsymbol{\alpha}_2}{r_{12}}\psi_m(\boldsymbol{r}_2)\psi_k(\boldsymbol{r}_1) \tag{8.36}
$$

Again, we see how the Gaunt interaction emerges with the Gaunt operator in the matrix element (compare the papers by Rosenfeld [164] and Møller [167] which also mark the transition to the modern QED treatment of the problem on which we comment again in section 8.1.5 below). Following Bethe and Fermi [135], we may also recover the Breit interaction from the general expression for the interaction energy matrix element of Eq. (8.33) by expansion of the exponential,

$$
\exp\left(\mathrm{i}\omega_{nm}\frac{r_{12}}{c}\right) = 1 + \mathrm{i}\omega_{nm}\frac{r_{12}}{c} - \omega_{nm}^2\frac{r_{12}^2}{2c^2} + O(c^{-3}) \tag{8.37}
$$

to yield

$$
\begin{aligned}
V_{lnkm} = & \int \mathrm{d}^3 r_1 \int \mathrm{d}^3 r_2\, \psi_l^\dagger(\boldsymbol{r}_1)\, \psi_n^\dagger(\boldsymbol{r}_2) \left\{\frac{1}{r_{12}}\left[1 + \mathrm{i}\omega_{nm}\frac{r_{12}}{c} - \omega_{nm}^2\frac{r_{12}^2}{2c^2}\right.\right. \\
& \left.\left. + O(c^{-3})\right] - \frac{\boldsymbol{\alpha}_1\boldsymbol{\alpha}_2}{r_{12}}\right\} \psi_m(\boldsymbol{r}_2)\, \psi_k(\boldsymbol{r}_1)
\end{aligned} \tag{8.38}
$$

where we already took into account that the vector-potential contribution can be considered as unretarded to this order in c according to what has been found in the preceding section. Eq. (8.38) simplifies to [124, p. 518]

$$
\begin{aligned}
V_{lnkm} = & \int \mathrm{d}^3 r_1 \int \mathrm{d}^3 r_2\, \psi_l^\dagger(\boldsymbol{r}_1)\, \psi_n^\dagger(\boldsymbol{r}_2) \left[\frac{1}{r_{12}} + \mathrm{i}\frac{\omega_{nm}}{c} - \omega_{nm}^2\frac{r_{12}}{2c^2}\right. \\
& \left. - \frac{\boldsymbol{\alpha}_1\boldsymbol{\alpha}_2}{r_{12}} + O(c^{-3})\right] \psi_m(\boldsymbol{r}_2)\, \psi_k(\boldsymbol{r}_1)
\end{aligned} \tag{8.39}
$$

Here, we see that the second term in brackets, which is a constant, vanishes because either we have $\omega_{nn} = (E_n - E_n)/\hbar = 0$ for $n = m$ or it is reduced due to the orthogonality of the eigenstates $i\omega_{nm}/c\langle\psi_n|\psi_m\rangle = 0$ for $n \neq m$. Thus, we have

$$
\begin{aligned}
V_{lnkm} &= \int d^3r_1 \int d^3r_2\, \psi_l^\dagger(\boldsymbol{r}_1)\, \psi_n^\dagger(\boldsymbol{r}_2) \left[\frac{1}{r_{12}} - \omega_{nm}^2 \frac{r_{12}}{2c^2}\right. \\
&\qquad \left. - \frac{\boldsymbol{\alpha}_1 \boldsymbol{\alpha}_2}{r_{12}} + O(c^{-3})\right] \psi_m(\boldsymbol{r}_2)\, \psi_k(\boldsymbol{r}_1)
\end{aligned} \tag{8.40}
$$

According to our derivation, the scaled energy difference in the integral stems from electron 2, which we may explicitly write as $\omega_{nm}^{(2)}$, which in what follows would create an asymmetry in the particle labels [166]. Therefore, since the two interacting electrons are indistinguishable we eventually obtain a symmetric expression if we rewrite the square of the scaled energy difference with Eq. (8.34) as

$$
\omega_{nm}^2 = \omega_{nm}^{(2)} \omega_{nm}^{(2)} = -\omega_{nm}^{(2)} \omega_{lk}^{(1)} \tag{8.41}
$$

which produces a symmetric integrand (compare the choice of gauge by Darwin in the case of classical particles discussed at the end of section 3.5, which can be considered as the classical analog of the symmetrization procedure here). If we now replace the energy eigenvalues of the the stationary reference functions ψ_i by the corresponding operators $h^D(1)$ and $h^D(2)$ we may replace in the integrand,

$$
\begin{aligned}
\omega_{nm}^{(2)} \omega_{lk}^{(1)} r_{12} &\overset{(8.24)}{=} \frac{1}{\hbar^2}\left[E_n^{(2)} E_l^{(1)} - E_m^{(2)} E_l^{(1)} - E_n^{(2)} E_k^{(1)} + E_m^{(2)} E_k^{(1)}\right] r_{12} \\
&\longrightarrow \frac{1}{\hbar^2}\left[h^D(2) h^D(1) r_{12} - h^D(1) r_{12} h^D(2)\right. \\
&\qquad \left. - h^D(2) r_{12} h^D(1) + r_{12} h^D(1) h^D(2)\right] \\
&= \frac{1}{\hbar^2}\left[h^D(2), \left[h^D(1), r_{12}\right]\right] = \frac{1}{\hbar^2}\left[h^D(1), \left[h^D(2), r_{12}\right]\right]
\end{aligned} \tag{8.42}
$$

where the non-commutation of the operators (in contrast to the commuting energy eigenvalues) has been explicitly taken care of by rearranging the one-electron Hamiltonians to the right or to the left of the distance operator r_{12} so that they can operate on the bra or on the ket to produce the correct energy eigenvalue. It must be stressed that this is a somewhat crucial point as we did not explicitly specify whether the reference states — also considered as the unperturbed states if the electron–electron interaction is regarded as a perturbation — are the free-particle states [86] or those in some external electromagnetic field [135] like the external potential of an atomic nucleus; see also the remark on the Furry picture at the end of the next section.

Before we study the final interaction energy matrix element, we note that the double commutator can be resolved since

$$\left[h^D(2), r_{12}\right] = \left[c\boldsymbol{\alpha}_2 \cdot \boldsymbol{p}_2, r_{12}\right] \overset{(4.49)}{=} \frac{\hbar c}{\mathrm{i}}\boldsymbol{\alpha}_2 \cdot (\nabla_2 r_{12}) \overset{(A.21)}{=} -\frac{\hbar c}{\mathrm{i}}\boldsymbol{\alpha}_2 \cdot \frac{\boldsymbol{r}_{12}}{r_{12}} \tag{8.43}$$

where we exploited the fact that all potentials and the rest energy commute with the distance so that only the gradient of the momentum operator survives. With this expression for the commutator, the double commutator requires us to evaluate its gradient

$$\left[c\boldsymbol{\alpha}_1 \cdot \boldsymbol{p}_1, \left[h^D(2), r_{12}\right]\right] = \hbar^2 c^2 \left\{ \sum_{i=1}^{3} \alpha_{1,i}\alpha_{2,i} \left[\frac{1}{r_{12}} - \frac{r_{12,i}^2}{r_{12}^3}\right] - \sum_{i,j \neq i}^{3} \alpha_{1,i}\alpha_{2,j} \frac{r_{12,i}r_{12,j}}{r_{12}^3} \right\} \tag{8.44}$$

and then becomes

$$\frac{1}{\hbar^2}\left[h^D(1), \left[h^D(2), r_{12}\right]\right] = +c^2 \left(\frac{\boldsymbol{\alpha}_1\boldsymbol{\alpha}_2}{r_{12}} - \frac{(\boldsymbol{\alpha}_1 \cdot \boldsymbol{r}_{12})(\boldsymbol{\alpha}_2 \cdot \boldsymbol{r}_{12})}{r_{12}^3} \right) \tag{8.45}$$

The lowest-order term of the interaction matrix element finally reads

$$\begin{aligned} V_{lnkm} \approx & \int d^3r_1 \int d^3r_2\, \psi_l^\dagger(\boldsymbol{r}_1)\, \psi_n^\dagger(\boldsymbol{r}_2) \left[\frac{1}{r_{12}} + \frac{1}{2}\left(\frac{\boldsymbol{\alpha}_1\boldsymbol{\alpha}_2}{r_{12}} \right.\right. \\ & \left.\left. - \frac{(\boldsymbol{\alpha}_1 \cdot \boldsymbol{r}_{12})(\boldsymbol{\alpha}_2 \cdot \boldsymbol{r}_{12})}{r_{12}^3} \right) - \frac{\boldsymbol{\alpha}_1\boldsymbol{\alpha}_2}{r_{12}} \right] \psi_m(\boldsymbol{r}_2)\, \psi_k(\boldsymbol{r}_1) \end{aligned} \tag{8.46}$$

where we did not specify the truncation order, which we know only for the truncated expansion of the exponential but not for the complete integrand, and used an approximation sign instead. This lowest-order term is immediately recognized as being identical to the Gaunt plus retardation matrix element that can be reduced to the one for the Breit operator

$$\begin{aligned} V_{lnkm} \approx & \int d^3r_1 \int d^3r_2\, \psi_l^\dagger(\boldsymbol{r}_1)\, \psi_n^\dagger(\boldsymbol{r}_2) \left[\frac{1}{r_{12}} - \frac{1}{2}\left(\frac{\boldsymbol{\alpha}_1\boldsymbol{\alpha}_2}{r_{12}} \right.\right. \\ & \left.\left. + \frac{(\boldsymbol{\alpha}_1 \cdot \boldsymbol{r}_{12})(\boldsymbol{\alpha}_2 \cdot \boldsymbol{r}_{12})}{r_{12}^3} \right) \right] \psi_m(\boldsymbol{r}_2)\, \psi_k(\boldsymbol{r}_1) \end{aligned} \tag{8.47}$$

as discussed in the preceding section. However, we now see how the time-dependent Dirac states can be utilized to derive a potential energy matrix element that employs only charge density terms, i.e., products of Dirac spinors.

The current density only occurs implicitly as we have obtained a new interaction operator, the Coulomb–Breit operator, for the charge density in an essentially *stationary* picture. Hence, we describe the interaction energy by the stationary states only and have all retardation- and current-density-specific terms hidden in the operator. This picture will be generalized for many-electron systems later in section 8.8.

8.1.5 Breit Interaction from Quantum Electrodynamics

After we have presented the semi-classical derivation of the Breit interaction, in which the electrons are described quantum mechanically while their interaction is considered via classical electromagnetic fields, we should discuss how this interaction can eventually be derived from quantum electrodynamics [168]. Consequently, we revisit the basics already introduced in section 7.2.2.

The interaction of two moving fermions can be expressed as a scattering process of these two charges which then exchange so-called virtual photons of specific angular frequency $\omega = |\omega_{nm}|$. The difficulty is, of course, that the two charges need not be at the same place at a given time in order to be scattered. In classical theory, one electron is scattered at the potential generated by the other electron (taking retardation effects properly into account). In quantum electrodynamics, the first electron is scattered by a virtual photon emitted by the second electron at some earlier time. Naturally, the multitude of scattering processes of electrons and photons known to physics and chemistry — like Coulomb, Compton, Rayleigh, or Raman scattering — can eventually all be reduced to basic events like the one just described; at least in principle. In practice one employs more approximate schemes based on the semi-classical theory derived in this book. Of course, the longitudinal and transverse fields of chapter 7 (see also sections 3.3.2.2 and 3.4.2) that lead to Coulomb and Breit interactions in chapter 7 are then interpreted as the exchange of longitudinal and transverse photons, i.e., of photons with a definite polarization property.

Akhiezer and Berestetskii [124, p. 509–519] present a very readable derivation of an explicit expression for the scattering matrix containing the 4-current of the matter field and an expression for the 4-potential constructed from photon properties. The 4-potential can be expanded in plane waves (analogously to Eqs. (2.120) and (2.121) and subsequently quantized yielding an exponential function containing the angular frequency ω of the exchanged photon [124, p. 159–182]. This angular frequency can also be related to the difference in energy of the one-particle states, which is then much like our presentation given in the last section on the basis of the original papers by Rosenfeld, by Møller and by Bethe and Fermi. The scattering matrix element can then be used to define an effective interaction energy [124, p. 517] which is identical to Eq. (8.32).

The four indices in Eq. (8.33) denote the initial and final states of the two electrons before and after the scattering process. The energy difference $\hbar|\omega_{nm}|$ is equal to the energy of the exchanged (virtual) photon $\hbar\omega$. We may emphasize that the sequence of two Dirac $\boldsymbol{\alpha}_i$ matrices arises from the Dirac 4-currents in the original scattering matrix; see Eq. (8.23) and compare also Eqs. (5.47) and (5.46) for the 4-current of a single particle, which carries the Dirac matrices. In the end, this effective interaction energy leads to the Gaunt and Breit interactions after suitable Taylor series expansions of the exponential function in the integral as demonstrated in the last section. In this way, QED explains the emergence of classical electromagnetic potentials from a second-quantized world.

Hence, there exists a well-defined QED basis to the semi-classical theory. The frequency-dependent interaction energy in Eq. (8.32) is the most general form, and consequently a frequency-dependent form of the Breit interaction which cannot be derived within the semi-classical theory, has been in use and may be considered to be somewhat more fundamental. For this reason, we present it here. The general expression in use is based on a formulation by Bethe and Salpeter [78,95],

$$B_{\omega}(1,2) = -\frac{q_e^2}{2\pi^2}\int \frac{\mathrm{d}^3k}{k^2}\, e^{\mathrm{i}\boldsymbol{k}\cdot\boldsymbol{r}_{12}} \left(\boldsymbol{\alpha}_1\boldsymbol{\alpha}_2 - \frac{(\boldsymbol{\alpha}_1\cdot\boldsymbol{k})(\boldsymbol{\alpha}_2\cdot\boldsymbol{k})}{k^2}\right) \tag{8.48}$$

that explicitly involves the wave vector $\boldsymbol{k}$ of the electromagnetic waves (or photons, respectively). exponential. With the integrals,

$$\frac{1}{2\pi^2}\int \frac{\mathrm{d}^3k}{k^2}\, e^{\mathrm{i}\boldsymbol{k}\cdot\boldsymbol{r}} = \frac{1}{r} \tag{8.49}$$

and

$$\begin{aligned}\frac{1}{2\pi^2}\int \frac{\mathrm{d}^3k}{k^2}\, e^{\mathrm{i}\boldsymbol{k}\cdot\boldsymbol{r}} \frac{(\boldsymbol{a}\cdot\boldsymbol{k})(\boldsymbol{b}\cdot\boldsymbol{k})}{k^2} &= \frac{\mathrm{i}}{2}\left(\boldsymbol{a}\cdot\nabla_{\boldsymbol{r}}\right)\int \frac{\mathrm{d}^3k}{k^2}\, e^{\mathrm{i}\boldsymbol{k}\cdot\boldsymbol{r}} \left(\boldsymbol{b}\cdot\nabla_{\boldsymbol{k}}\right)\frac{1}{k^2} \\ &= \frac{1}{2}\left(\boldsymbol{a}\cdot\nabla_{\boldsymbol{r}}\right)\frac{\boldsymbol{b}\cdot\boldsymbol{r}}{r} = \frac{1}{2r}\left(\boldsymbol{a}\cdot\boldsymbol{b} - \frac{(\boldsymbol{a}\cdot\boldsymbol{r})(\boldsymbol{b}\cdot\boldsymbol{r})}{r^2}\right)\end{aligned} \tag{8.50}$$

where $\boldsymbol{a}$ and $\boldsymbol{b}$ are vectors, we again obtain the frequency-independent Breit operator. However, Bethe and Salpeter note that Breit matrix elements derived from QED do not feature the same order in a perturbation theory compared to the original work by Breit [48, p. 173].

The connection to such a momentum-space representation as given above can also be made by starting from the cosine expression of the retarded interaction derived in the last section — through Eq. (8.35) — which finally produces [167]

$$\frac{\cos\left(\frac{\omega r_{12}}{c}\right)}{r_{12}} = \frac{1}{2\pi^2}\int \mathrm{d}^3k \frac{\exp\left(\mathrm{i}\boldsymbol{k}\cdot\boldsymbol{r}_{12}\right)}{k^2 - \omega^2/c^2} \tag{8.51}$$

Depending on gauge choices, Quiney *et al.* write two expressions became quite popular in atomic physics and quantum chemistry [169]: the interaction operator in Lorentz (Feynman) gauge in configuration-space representation as

$$\hat{V}_{12}^{\mathrm{F}} = \frac{1 - \boldsymbol{\alpha}_1 \boldsymbol{\alpha}_2}{r_{12}} e^{\mathrm{i}\boldsymbol{k}\cdot\boldsymbol{r}_{12}} \tag{8.52}$$

while they have for the same quantity in Coulomb gauge

$$\hat{V}_{12}^{\mathrm{C}} = \frac{1 - \boldsymbol{\alpha}_1 \boldsymbol{\alpha}_2}{r_{12}} e^{\mathrm{i}\boldsymbol{k}\cdot\boldsymbol{r}_{12}} + \frac{(\boldsymbol{\alpha}_1 \cdot \nabla_{\boldsymbol{r}_{12}})(\boldsymbol{\alpha}_2 \cdot \nabla_{\boldsymbol{r}_{12}})}{\boldsymbol{k}^2 r_{12}} \left[e^{\mathrm{i}\boldsymbol{k}\cdot\boldsymbol{r}_{12}} - 1 \right] \tag{8.53}$$

Note the imprinted form of the Breit interaction *plus* an independent exponential retardation factor. Grant and co-workers [169] refer in this context to earlier work by Brown [170] who essentially embeds the Rosenfeld arguments into the QED language with the final result being the Coulomb plus Gaunt operators times the retardation factor as derived in the previous section.

Based on the expression in Coulomb gauge, Eq. (8.53), the quantum-electrodynamics-based expression of the *general* frequency-dependent Breit operator B_ω has been given as [136, 138, 171]

$$\begin{aligned} B_\omega(1,2) &= q_e^2 \left\{ -\boldsymbol{\alpha}_1 \boldsymbol{\alpha}_2 \frac{\cos(\omega r_{12}/c)}{r_{12}} + (\boldsymbol{\alpha}_1 \cdot \nabla_1)(\boldsymbol{\alpha}_2 \cdot \nabla_2) \frac{\cos(\omega r_{12}/c) - 1}{\omega^2 r_{12}/c^2} \right\} \\ &= q_e^2 \left\{ -\boldsymbol{\alpha}_1 \boldsymbol{\alpha}_2 \frac{\cos(2\pi\tilde{\nu} r_{12})}{r_{12}} + (\boldsymbol{\alpha}_1 \cdot \nabla_1)(\boldsymbol{\alpha}_2 \cdot \nabla_2) \frac{\cos(2\pi\tilde{\nu} r_{12}) - 1}{4\pi^2 \tilde{\nu}^2 r_{12}} \right\} \end{aligned} \tag{8.54}$$

where the cosine has been derived from the exponential along the lines of Eqs. (8.35)–(8.35) and the scalar $\boldsymbol{k}^2 = k^2$ was substituted by the square of $k = \omega/c = |\omega_{nm}|/c$. Moreover, $\tilde{\nu}$ is the wave number of the exchanged virtual photon (with frequency ν and wave length λ),

$$\tilde{\nu} = \frac{1}{\lambda} = \frac{\nu}{\lambda\nu} = \frac{\nu}{c} = \frac{\omega}{2\pi c} \tag{8.55}$$

We should emphasize that the form of the general Breit interaction above (reproduced after Mann and Johnson [136]) already contains the frequency-*in*dependent Breit interaction imprinted as we shall see in the following. This is in contrast to the expression given above, which is also the starting point in the monograph by Akhiezer and Berestetskii and starts from the Gaunt-like single pair of Dirac $\boldsymbol{\alpha}$ matrices which we know represents instantaneous magnetic interactions.

In a first-quantized theory, the frequency ν of the virtual photon must not occur, and we therefore seek the limit $c \to \infty$. This limit is identical to the one obtained for $\tilde{\nu} \to 0$ or $\omega \to 0$ as is easily understood from Eq. (8.55), though it is often referred to as the long-wavelength limit $\lambda \to \infty$, which is, of course, just another phrase for the same limit. To analyze this limit we expand the cosine,

$$\cos(2\pi\tilde{\nu}r_{12}) = 1 - \frac{1}{2}(2\pi\tilde{\nu}r_{12})^2 + \frac{1}{24}(2\pi\tilde{\nu}r_{12})^4 + O(\tilde{\nu}^6 r_{12}^6) \tag{8.56}$$

or equivalently

$$\cos\left(\frac{\omega r_{12}}{c}\right) = 1 - \frac{1}{2}\left(\frac{\omega r_{12}}{c}\right)^2 + \frac{1}{24}\left(\frac{\omega r_{12}}{c}\right)^4 + O(c^{-6} r_{12}^6) \tag{8.57}$$

so that we can expand the interaction energy

$$\begin{aligned} B_\omega(1,2) &= q_e^2 \Bigg\{ -\boldsymbol{\alpha}_1\boldsymbol{\alpha}_2 \frac{1}{r_{12}}\left[1 + O(\tilde{\nu}^2 r_{12}^2)\right] + (\boldsymbol{\alpha}_1 \cdot \nabla_1)(\boldsymbol{\alpha}_2 \cdot \nabla_2) \\ &\quad \times \frac{\left[1 - \frac{1}{2}(2\pi\tilde{\nu}r_{12})^2 + \frac{1}{24}(2\pi\tilde{\nu}r_{12})^4 + O(\tilde{\nu}^6 r_{12}^6)\right] - 1}{4\pi^2\tilde{\nu}^2 r_{12}} \Bigg\} \\ &= q_e^2 \left\{ -\boldsymbol{\alpha}_1\boldsymbol{\alpha}_2 \frac{1}{r_{12}} - \frac{1}{2}(\boldsymbol{\alpha}_1 \cdot \nabla_1)(\boldsymbol{\alpha}_2 \cdot \nabla_2) r_{12} + O(c^{-2}) \right\} \end{aligned} \tag{8.58}$$

In the long-wavelength limit we thus obtain,

$$B_0(1,2) = -q_e^2 \left\{ \boldsymbol{\alpha}_1\boldsymbol{\alpha}_2 \frac{1}{r_{12}} + \frac{1}{2}(\boldsymbol{\alpha}_1 \cdot \nabla_1)(\boldsymbol{\alpha}_2 \cdot \nabla_2) r_{12} \right\} \tag{8.59}$$

This equation can be rewritten to adopt the more familiar form of the Breit equation. For this purpose, we employ the following auxiliary expressions

$$\begin{aligned} (\boldsymbol{\alpha}_1 \cdot \nabla_1)(\boldsymbol{\alpha}_2 \cdot \nabla_2) r_{12} &= \sum_{i,j=1}^{3} \left(\alpha_{1,i}\alpha_{2,j} \frac{\partial}{\partial x_{1,i}} \frac{\partial}{\partial x_{2,j}} \right) r_{12} \\ &= -\sum_{i=1}^{3} \alpha_{1,i}\alpha_{2,i} \left(\frac{1}{r_{12}} - \frac{r_{12,i}^2}{r_{12}^3} \right) + \sum_{i,j\neq i}^{3} \alpha_{1,i}\alpha_{2,j} \frac{r_{12,i} r_{12,j}}{r_{12}^3} \\ &= -\frac{\boldsymbol{\alpha}_1\boldsymbol{\alpha}_2}{r_{12}} + \frac{(\boldsymbol{\alpha}_1 \cdot \boldsymbol{r}_{12})(\boldsymbol{\alpha}_2 \cdot \boldsymbol{r}_{12})}{r_{12}^3} \end{aligned} \tag{8.60}$$

— a derivation similar to the one in Eq. (8.44) above. Inserted into Eq. (8.59), this then yields the well-known expression of the Breit interaction operator,

$$B_0(1,2) = -\frac{q_e^2}{2} \left\{ \frac{\boldsymbol{\alpha}_1\boldsymbol{\alpha}_2}{r_{12}} + \frac{(\boldsymbol{\alpha}_1 \cdot \boldsymbol{r}_{12})(\boldsymbol{\alpha}_2 \cdot \boldsymbol{r}_{12})}{r_{12}^3} \right\} \tag{8.61}$$

Because of this relation to quantum electrodynamics, the Breit interaction is often thought to be applicable only in perturbation theory. This is, however, not automatically true in a first-quantized semi-classical theory which we develop now.

As a final remark we may comment on the fact that we need to study the two-electron problem in the attractive external potential of an atomic nucleus, hence, as a bound-state problem. It is immediately seen that this affects, for instance, the expansion in terms of zeroth-order state functions in Eq. (8.22), where bound states of the one-electron problem rather than free-particle states become the basis for the construction of the wave function (and wave-function operators in second quantization). The situation is, however, more delicate than one might think and reference is usually made to the discussion of this issue provided by Furry [172] (Furry picture). Of course, it is of fundamental importance to the QED basis of quantum chemistry. However, as a truly second-quantized QED approach, we abandon it in our semi-classical picture and refer to Schweber for more details [122, p. 566]. Instead, we may adopt from this section only the possibility to include either the Gaunt or the Breit operators in a first-quantized many-particle Hamiltonian.

8.2
Quasi-Relativistic Many-Particle Hamiltonians

Our aim is to develop a many-electron theory for the molecular sciences, for which we started from fundamental physical principles. Then, we realized that for this endeavor we have to compromise these first principles of quantum mechanics if we are to arrive at a stage where actual calculations on molecules and molecular aggregates are feasible. Hence, we proceed to elaborate on the two-particle interactions discussed in the preceding section and go on to derive a formalism, which captures the major part of the effect of special relativity on the numerical values of many-electron observables.

8.2.1
Nonrelativistic Hamiltonian for a Molecular System

Historically, the nonrelativistic many-electron theory was developed and computationally tested first. For molecules containing only light atoms, this approach turned out to be remarkably successful in chemistry. Consequently, we are well advised to be inspired by this success if we look for a relativistic theory that is to be computationally as feasible as its nonrelativistic relative.

A feature of the nonrelativistic external-field-free many-particle Hamiltonian is that its principles of construction are quite simple. First of all, kinetic energy operators for each particle in the system, i.e., for each electron and each atomic nucleus, are summed. Since the elementary particles can be well described as point-like particles, it is not necessary to consider electrostatic multipole terms beyond the electrostatic monopole interactions. That is, since *point* particles cannot be polarized, electrostatic many-body effects on the pair interaction do not arise. The total interaction energy operator can then be obtained through the correspondence principle applied to the classical interaction energy of point charges given in Eq. (2.147). In addition, we may add the retardation and magnetic pair terms as discussed in the previous section for the two-electron case. Often, however, these additional terms due to the magnetic interactions of the moving particles are neglected, which is justified by experience with the numerical results obtained in quantum chemical calculations that show that the electrostatic monopole interaction dominates by far all other terms. In principle, the magnetic interaction terms would have to be introduced also in the framework of nonrelativistic quantum chemistry (as in the Pauli approximation employing the Breit–Pauli Hamiltonian) but are usually neglected for the same reason.

The nonrelativistic many-particle Hamiltonian of a collection of electrons and nuclei in the absence of additional external electromagnetic fields can be written (in Gaussian units) as,

$$H = \sum_{I}^{M} \frac{\boldsymbol{p}_I^2}{2m_I} + \sum_{i}^{N} \frac{\boldsymbol{p}_i^2}{2m_e} + \sum_{J<I}^{M} \frac{Z_I e Z_J e}{|\boldsymbol{R}_I - \boldsymbol{R}_J|} + \sum_{j<i}^{N} \frac{e^2}{|\boldsymbol{r}_i - \boldsymbol{r}_j|} - \sum_{i}^{N}\sum_{I}^{M} \frac{Z_I e^2}{|\boldsymbol{r}_i - \boldsymbol{R}_I|} \tag{8.62}$$

where we use small letters for electronic indices and coordinates while capital letters denote those of the nuclei. Note that restricted double sums have been abbreviated according to,

$$\sum_{i=1}^{N} \sum_{j=i+1}^{N} \longrightarrow \sum_{i<j}^{N} \text{ which can equivalently be written as } \frac{1}{2}\sum_{i\neq j}^{N} \tag{8.63}$$

in order to avoid double counting of the interaction terms.

Of course, the interaction energy operators introduced here are obtained through the correspondence principle applied to the classical interaction energy of point charges given in Eq. (2.147). Owing to the opposite sign of the charges of electrons and nuclei, the electron–nucleus attraction term, i.e., the last term on the right hand side of Eq. (8.62) bears an overall negative sign. As before, we abbreviate the inter-particle distances by a single distance coordinate carrying two indices. The distance between electrons i and j thus reads $r_{ij} \equiv |\boldsymbol{r}_i - \boldsymbol{r}_j|$. In Eq. (8.62), N is the total number of electrons and M the total number of atomic nuclei.

It is important to be aware of the fact that we have largely simplified all interactions present within a molecule or a molecular aggregate. For instance, apart from retardation effects on the electron–electron, electron–nucleus, and nucleus–nucleus interactions, the effect of the weak magnetic field associated with a nuclear spin is completely neglected at this stage. This procedure features two advantages. First, we do not need to consider these additional interaction operators and, second, we obtain a generic model energy for a collection of electrons and nuclei, which is to be understood to be only slightly modulated by additional small interactions like those arising in the presence of a nuclear magnetic moment. Consider a molecule consisting of atomic nuclei of a single element which features two different isotopes of which one possesses a spin and the other one does not. Then, we would have to consider all different distributions of these isotopes over the nuclear positions $\boldsymbol{R}_I$ in a Hamiltonian. Hence, it would be necessary to solve the Schrödinger equation for all these permutations instead of only for one as in the case of the Hamiltonian presented in Eq. (8.62). For many different nuclei, this would soon become highly impractical.

We thus exploit the fact that electrostatic (monopole) interactions dominate all other interactions. It also allows us to treat all other interactions as small perturbations, an issue which then leads to the calculation of molecular properties in the framework of response theory in chapter 15.

8.2.2
First-Quantized Relativistic Many-Particle Hamiltonian

According to the discussion in chapter 7, the Hamiltonian of quantum electrodynamics given by the normal-ordered version of Eq. (7.13) contains many degrees of freedom which are irrelevant for almost all situations of chemical interest. In order to establish a theoretical framework tailor-made for chemistry, where only systems with a fixed number of electrons and nuclei are considered, it is therefore necessary to abandon the principles of second quantization and to revert to a first-quantized formulation explicitly labeling each particle.

For a (quasi-)relativistic many-particle theory, we now need to decide which operators in Eq. (8.62) can be kept and which should be replaced by relativistic operators. Because of the mass of atomic nuclei being at least three orders of magnitude larger than the electron's rest mass, the nuclear kinetic energy operator can be kept as in Eq. (8.62), i.e., as the nonrelativistic Schrödinger-type kinetic energy operators of the M nuclei in the molecular system. Actually, one would have to use Dirac- or Klein–Gordon-type operators instead (in a rigorous tensor product formulation; see also sections 6.1 and 8.4) depending on the spin of the atomic nuclei. This would yield rather complicated expressions for molecular Hamiltonians (and hence for the wave functions) depending on the spin and mass of the particular isotopes involved. Since we employ nonrelativistic kinetic energy operators for the nuclei, we only have to deal with the problem that explicit values for the masses of the isotopes in a molecule need to be specified. From experimental chemistry we know that molecules differing only by their isotope composition show essentially the same behavior in chemical reactions (if we leave kinetic isotope effects aside for the moment). Therefore, we proceed one step further and separate the quantum mechanical description of the electrons completely from the masses of the isotopes in the molecule in section 8.3.

Knowing from the study of the Dirac hydrogen atom in chapter 6 that the Dirac Hamiltonian changes into the Schrödinger Hamiltonian for large values of the speed of light c, we may replace the Schrödinger kinetic energy operators for the electrons in Eq. (8.62) with the one-electron Dirac Hamiltonian of a freely moving particle. We now combine all terms in Eq. (8.62) that depend on the single coordinate of electron i into a single one-electron Hamiltonian $h^D(i)$,

$$h^D(i) = c\,\boldsymbol{\alpha}_i \cdot \left(\boldsymbol{p}_i - \frac{q_e}{c}\boldsymbol{A}_{\text{ext}}\right) + (\beta_i - 1)m_e c^2 + V_{nuc}(\boldsymbol{r}_i) \tag{8.64}$$

of section 5.4 with $\boldsymbol{A}_{\text{ext}} = 0$ for the time being; nonvanishing vector potentials are considered in chapter 15. Thus, $h^D(i)$ comprises the Dirac Hamiltonian for a freely moving electron plus the external electrostatic scalar potential energy for the interaction of electron i with all atomic nuclei

$$V_{nuc}(\boldsymbol{r}_i) = -\sum_I^M \frac{Z_I e^2}{|\boldsymbol{r}_i - \boldsymbol{R}_I|} \tag{8.65}$$

so that the quasi-relativistic many-particle Hamiltonian reads

$$H = \sum_I^M \frac{\boldsymbol{p}_I^2}{2m_I} + \sum_i^N h^D(i) + \sum_{i<j}^N g(i,j) + V_0 \tag{8.66}$$

In Eq. (8.66) the nucleus–nucleus repulsion operators are incorporated in the last term on the right hand side of that equation abbreviated as

$$V_0 \equiv \sum_{J<I}^M \frac{Z_I Z_J e^2}{|\boldsymbol{R}_I - \boldsymbol{R}_J|} \tag{8.67}$$

and the electron–electron repulsion operators have been abbreviated as $g(i,j)$. In the case of the Dirac–Coulomb many-electron Hamiltonian, $g(i,j)$ is to be replaced by

$$g(i,j) \to V_C(i,j) = \frac{e^2}{|\boldsymbol{r}_i - \boldsymbol{r}_j|} \tag{8.68}$$

and in the case of the Dirac–Coulomb–Breit many-electron Hamiltonian

$$g(i,j) = V_C(i,j) + B_\omega(i,j) = \frac{e^2}{|\boldsymbol{r}_i - \boldsymbol{r}_j|} + B_\omega(i,j) \tag{8.69}$$

where also the frequency-independent Breit pair interaction operator $B_0(i,j)$ or the unretarded Gaunt interaction $G_0(i,j)$ may be employed. Although the Breit interaction operator contains the lowest-order retardation and magnetic contributions of *any* two interacting charges, we apply it only for the interactions of two electrons. It is neither considered as an add-on to the instantaneous Coulomb interaction of an electron with the atomic nuclei, V_{nuc}, nor for the mutual instantaneous interactions of the nuclei in V_0 for reasons already mentioned above.

In order to finally calculate binding energies of electrons comparable to the nonrelativistic theory — as outlined in section 6.7 — the origin of the energy scale has been shifted by the rest energy $m_e c^2$ of each single electron in every one-electron operator $h^D(i)$. Thus, the total energy is shifted by $-Nm_e c^2$.

Having left the framework of field theory outlined in chapter 7 and thus having skipped any need for subsequent renormalization procedures, the mass and charge of the electron are now the physically observable quantities, and thus do not bear a tilde on top. In contrast to quantum electrodynamics, the radiation field A^μ is no longer a dynamical degree of freedom in a many-electron theory which closely follows nonrelativistic quantum mechanics. Vector potentials may only be incorporated as external perturbations into the many-electron Hamiltonian of Eq. (8.62). From the QED Eqs. (7.13), (7.19), and (7.20), the Hamiltonian of a system of N electrons and M nuclei is thus described by the many-particle Hamiltonian of Eq. (8.66). In addition, we refer to a common absolute time frame although this will not matter in the following as we consider only the stationary case.

8.2.3
Pathologies of the First-Quantized Formulation

The many-electron Dirac–Coulomb or Dirac–Coulomb–Breit Hamiltonian is neither gauge nor Lorentz invariant as these symmetries have explicitly been broken in section 8.1. Moreover, these relativistic many-particle Hamiltonians feature serious conceptual problems [173], which are solely related to the structure of the one-electron terms, i.e., the Dirac Hamiltonian h^D. Even famous subtleties connected to two-electron processes, e.g., the Brown-Ravenhall disease [174] (see below), also known as continuum dissolution, would not be possible if the Dirac Hamiltonian were bounded from below.

8.2.3.1 Boundedness and Variational Collapse

The unboundedness of the one-electron Dirac Hamiltonian, i.e., the fact that states with infinitely large negative energy are possible, prohibits the use of the operator in any variational approach which aims at making energy expectation values stationary. As we shall see in the following section, additional inconvenient features arise if more than one electron is considered.

In principle, we are only interested in the upper part of the spectrum of the Dirac Hamiltonian, which describes electronic bound states as well as positive-energy continuum states. Hamiltonians which focus on the upper part of the Hamiltonians only so that pair-creation processes can no longer occur are called *no-pair Hamiltonians*. In order to produce such a Hamiltonian, projection operators, $\Lambda_i^{(+)}$, which may be expressed in terms of the sought-for electronic states $\psi_i(\boldsymbol{r})$ as

$$\Lambda_i^{(+)} = |\psi_i(\boldsymbol{r})\rangle\langle\psi_i(\boldsymbol{r})| \tag{8.70}$$

to yield vanishing contributions whenever they hit a 'positronic' state, are introduced [175–178] for the one- and two-electron terms

$$\sum_i^N h^D(i) \quad \longrightarrow \quad \sum_i^N \Lambda_i^{(+)} h^D(i) \Lambda_i^{(+)} \tag{8.71}$$

$$\sum_{i<j}^N g(i,j) \quad \longrightarrow \quad \sum_{i<j}^N \Lambda_i^{(+)} \Lambda_j^{(+)} g(i,j) \Lambda_j^{(+)} \Lambda_i^{(+)} \tag{8.72}$$

In the following, however, we refrain from writing these projectors explicitly, for the sake of brevity but also because they are not explicitly implemented in quantum chemical computer programs. We shall see later how the projection onto 'electronic' states is accomplished. However, Indelicato has demonstrated that a neglect of projection techniques may lead to convergence problems in actual calculations [179, 180].

8.2.3.2 **Continuum Dissolution**

Because of the negative-energy continuum in the spectrum of the one-electron Dirac Hamiltonian $h^D(i)$, a two-electron equation constructed from these building blocks becomes meaningless as, in particular, Brown and Ravenhall noted in their paper [174]. Interestingly, because of their introduction starting with the equation of Breit, Eq. (8.19), this so-called *Brown–Ravenhall disease* — also known as the *continuum dissolution* effect — is rationalized by Brown and Ravenhall for the two-electron equation considering the instantaneous electrostatic Coulomb interaction *only*. Their introductory rationale goes like this. Imagine that the total wave function of Eq. (8.19) would be obtained for *all two-electron potential energy terms neglected*. Then, upon slowly turning on the Coulombic interaction, e^2/r_{12}, the two-electron system can make transitions to states where one electron has any large negative energy, while the other electron is found in the positive-energy continuum such that the total energy of the system remains unchanged. The deeper analysis that follows this argument in Ref. [174] comprises, of course, the complete Breit interaction.

The Brown–Ravenhall disease is the reason why it is often believed that one must not use the Breit operator in first-quantized *variational* calculations; only perturbation-theory calculations should be reasonable. Here, it is important to note that the unperturbed system is the nonrelativistic one, i.e., a two-electron equation with no Dirac one-electron Hamiltonians at all. Of course, according to their rationale, this would already hold true for the pure Dirac–Coulomb Hamiltonian. However, continuum dissolution has never been observed in actual calculations. Of course, one might argue that it has not been particularly intensely searched for because the many-electron wave function is constructed such that negative-energy states are always omitted in actual calculations on molecular systems (as we shall see at the end of this chapter).

For the time being, all quasi-relativistic Hamiltonians in the following may be thought of as being surrounded by proper projection operators onto electronic bound or positive-energy continuum states. Hence, *all* subsequent Hamiltonians are to be understood as no-pair Hamiltonians.

8.2.4
Local Model Potentials for One-Particle QED Corrections

We left the framework of quantum electrodynamics when moving forward from chapter 7 to chapter 8 and have thus willingly given up the possibility of quantitative comparisons with high-resolution spectroscopic measurements in favor of a feasible way to handle the electromagnetic interactions through classical fields. Nevertheless, attempts have been made to re-introduce the numerical effect of quantum electrodynamics on the energy of a quantum mechanical state. This may be achieved by adding terms to each of the one-electron potential energy operators of the Hamiltonian in Eq. (8.66). Such terms may then be carried through all derivations given in this and the following chapters as simple add ons to the one-electron operators of Eq. (8.64). Since QED is able to reproduce the experimentally observed Lamb shift, i.e., the lifting of the degeneracy of $ns_{1/2}$ and $np_{1/2}$ shells, one would require that this qualitative effect is also reproduced by QED model potentials.

Typical QED effects are *self energy* and *vacuum polarization*, which are able to account for the Lamb shift [181]. They arise in a perturbation expansion of the interaction energy, which is often pictorially represented by Feynman diagrams (see the monographs on QED recommended at the end of chapter 7 for details), and thus even higher-order self-energy and vacuum-polarization terms are known. The dominant contribution [182] to the vacuum polarization, which stems from the coupling of the electron–positron field to the Coulomb field of the (point-like) atomic nucleus, arises from the Uehling potential [183], as elaborated by Wichmann and Kroll [184], who provided further corrections. Asymptotically, the Uehling potential diminishes exponentially, while at short distances it behaves like $\ln(r)/r$ with r being the distance from the point charge. Fullerton and Rinker [185] gave expressions (also for finite nuclear charges) that may be efficiently used as additive supplements to the nuclear potential.

In countless studies, such terms have been investigated in detail for atoms (see, for examples, Refs. [186–198, 198–200] and for overviews see Refs. [116, 201]). However, attempts have been made to include them also in molecular calculations in order to assess their magnitude compared to Breit interaction effects [202]. They turn out to be very small, even when compared to the Breit corrections, and therefore provide numerical evidence for the validity of the semi-classical theory of non-quantized electromagnetic fields developed in this book.

8.3
Separation of Nuclear and Electronic Degrees of Freedom: The Born–Oppenheimer Approximation

At this stage, we need to introduce the separation of nuclear and electronic degrees of freedom in close analogy to nonrelativistic quantum mechanics in order to arrive step by step at solvable differential equations. We will then understand how the potential energy surface emerges which allows us to understand chemical processes and which acts as a potential for nuclear motion, i.e., for vibrations. Vibrations may then be described quantum mechanically or even classically. All of these approaches to nuclear motion, which involve quantum dynamical approaches [203,204], approximation methods like the harmonic approximation for vibrational analysis [205–207], or Born–Oppenheimer and Car–Parrinello molecular dynamics based on classical Newtonian dynamics [208, 209], do not have a particular relativistic side in molecular science — apart from the fact that a potential energy surface may require a relativistic theory for its calculation. Of course, having said this, one must keep in mind that these statements refer to numerical experience gained over the years. This does not exclude the necessity for an extensive relativistic treatment which may even include the atomic nuclei in certain very special cases.

In order to simplify the Hamiltonian of Eq. (8.66), we freeze the nuclear motion and thus can neglect the kinetic energy operators for the nuclei. In this 'clamped-nuclei' approximation the remaining *electronic* Hamiltonian reads,

$$H'_{el} = \sum_i^N h^D(i) + \sum_{i<j}^N g(i,j) + V_0 \tag{8.73}$$

where the repulsive energy operator of the clamped nuclei, V_0, is simply added to the electronic Hamiltonian. The eigenfunctions of the many-electron Hamiltonian H'_{el},

$$H'_{el}\Psi_{el,A} = E'_{el,A}\Psi_{el,A} \tag{8.74}$$

are the electronic wave functions $\Psi_{el,A} = \Psi_{el,A}(\{\boldsymbol{r}_i\})$ for the A-th electronic state. These wave functions are defined for fixed nuclear positions, which we denote by a tilde on top of the coordinates, i.e., as $\Psi_{el,A} = \Psi_{el,A}(\{\boldsymbol{r}_i\}, \{\tilde{\boldsymbol{R}}_I\})$, and hence change with changing molecular structure. However, we could generate the full set of electronic wave functions for all positions $\Psi_{el,A}(\{\boldsymbol{r}_i\}, \{\boldsymbol{R}_I\})$ from these point-wise defined functions.

Since the nucleus–nucleus interaction energy operator V_0 is a multiplicative constant with respect to integration over electronic coordinates, it can thus simply be subtracted from the Hamiltonian,

$$H_{el} = H'_{el} - V_0 \tag{8.75}$$

and hence from the electronic energy eigenvalue $E'_{el,A}$,

$$H_{el}\Psi_{el,A} = (E'_{el,A} - V_0)\Psi_{el,A} \equiv E_{el,A}\Psi_{el,A} \tag{8.76}$$

Therefore, we do not refer to this constant shift of the electronic energy in the following chapters and consider the electronic energy shifted by this term, for which we must, however, keep in mind that it depends on the nuclear coordinates and, hence, on the molecular structure. Note also that the eigenfunctions remain unchanged once this shift has been applied. The eigenvalue $E_{el,A}$ is called the electronic energy. It is known only pointwise for fixed nuclear structures $E_{el,A} = E_{el,A}(\{\tilde{\boldsymbol{R}}_I\})$, but since we may solve Eq. (8.76) for any arrangement of the nuclei we may assume that the electronic energy is an analytic function of the nuclear coordinates $E_{el,A} = E_{el,A}(\{\boldsymbol{R}_I\})$. Any local minimum of this function is then interpreted as a potentially stable chemical structure of a molecule or molecular assembly that may be protected against decomposition by surrounding barriers (first-order saddle points) of various heights.

Having introduced the electronic wave function we may expand the total state in a product basis as

$$\Psi(\{\boldsymbol{r}_i\}, \{\boldsymbol{R}_I\}) = \sum_A \chi_A(\{\boldsymbol{R}_I\})\Psi_{el,A}(\{\boldsymbol{r}_i\}, \{\boldsymbol{R}_I\}) \tag{8.77}$$

where it is convenient to introduce the electronic wave functions $\Psi_{el,A}$ to explicitly depend also on the nuclear coordinates (although we know that this dependence is discrete rather than continuous). The electronic wave functions represent the set of basis functions, and the nuclear wave functions χ_A appear as coordinate-dependent expansion coefficients $\chi_A = \chi_A(\{\boldsymbol{R}_I\})$. This expansion makes sense if we assume that the many-electron basis functions $\Psi_{el,A}$ are known, i.e., if we assume that Eq. (8.76) can be solved, which is to be demonstrated later on in this chapter. Eq. (8.77) is exact as long as the many-electron basis is complete. The fact that the nuclear coordinates are kept frozen is no restriction on the completeness condition. Of course, one may also imagine a complete many-electron basis set with the nuclear coordinates being still dynamical variables (see the geminal basis functions below). Such an expansion would still be exact but it would feature completely different expansion parameters χ'_A compared to the coefficients χ_A in Eq. (8.77). To keep the nuclear coordinates frozen in the basis functions $\Psi_{el,A}$ is, however, the key to their determination and thus to the routine solution of Eq. (8.76) for molecules consisting of an arbitrary number of nuclei. One may anticipate that a routine solution is not possible for the complete set of electronic wave functions but only for a limited number that is thought to be physically relevant. If electronic excitation can be neglected, the electronic ground state, $\Psi_{el,A} = \Psi_{el,0}$ may be sufficient. If, however, one restricts the summation in Eq. (8.77) to only a single electronic state A, an approximation will be introduced since the completeness requirement on the basis functions is no longer fulfilled. The nuclear motion is then confined to a single potential energy surface $E_{el,A} = E_{el,A}(\{\boldsymbol{R}_I\})$ and vibronic coupling cannot be accounted for [210]. If furthermore the kinetic energy operator of the nuclei acting on the electronic wave function is considered to be a small and thus negligible perturbation, we arrive at the so-called *crude adiabatic approximation* [see Eq. (8.84) below].

The expansion in Eq. (8.77) allows us to solve the eigenvalue equation for the full Hamiltonian H,

$$H\Psi = E\Psi \quad \overset{(8.66),(8.73),(8.77)}{\Longrightarrow} \quad \left[\sum_I^M \frac{\boldsymbol{p}_I^2}{2m_I} + H'_{el}\right] \sum_A \chi_A \Psi_{el,A} = E \sum_A \chi_A \Psi_{el,A} \tag{8.78}$$

by multiplication from the left with $\langle \Psi_{el,B}|$ which yields

$$\left\langle \Psi_{el,B} \left| \sum_I^M \frac{\boldsymbol{p}_I^2}{2m_I} \right| \sum_A \Psi_{el,A} \chi_A \right\rangle + \left\langle \Psi_{el,B} \left| H'_{el} \right| \sum_A \Psi_{el,A} \chi_A \right\rangle$$
$$= E \left\langle \Psi_{el,B} \middle| \sum_A \Psi_{el,A} \chi_A \right\rangle \tag{8.79}$$

It must be emphasized that the integration in Eq. (8.79) is over all *electronic* coordinates; the nuclear coordinates remain untouched. The eigenfunctions of Eq. (8.76) are orthogonal and may be chosen to be normalized,

$$\langle \Psi_{el,B} | \Psi_{el,A} \rangle = \delta_{BA} \tag{8.80}$$

as we are free to multiply the electronic Schrödinger eigenvalue equation by an arbitrary factor. This allows us to simplify the Schrödinger equation for the total system to

$$\sum_A \left\langle \Psi_{el,B} \middle| -\sum_I^M \frac{\hbar^2 \Delta_I}{2m_I} \middle| \Psi_{el,A} \chi_A \right\rangle + E'_{el,B} \chi_B = E \chi_B \tag{8.81}$$

For the derivation of this equation we also utilized the electronic Schrödinger equation, Eq. (8.76), for the arbitrarily chosen state B in its integral form $E'_{el,B} = \langle \Psi_{el,B} | H'_{el} | \Psi_{el,B} \rangle$, which, after integration over all electronic coordinates, still explicitly depends on all nuclear coordinates. Moreover, we introduced the explicit expression $-\hbar^2 \Delta_I$ for the square of the momentum operator $\boldsymbol{p}_I^2$ of nucleus I. For the remaining bracket on the left hand side of Eq. (8.81) the product rule needs to be applied twice,

$$\begin{aligned} \Delta_I \Psi_{el,A} \chi_A &= \nabla_I \left[\chi_A \nabla_I \Psi_{el,A} + \Psi_{el,A} \nabla_I \chi_A \right] \\ &= \chi_A \Delta_I \Psi_{el,A} + 2(\nabla_I \chi_A)(\nabla_I \Psi_{el,A}) + \Psi_{el,A} \Delta_I \chi_A \end{aligned} \tag{8.82}$$

in order to arrive at

$$\begin{aligned} -\sum_I^M \frac{\hbar^2 \Delta_I}{2m_I} \chi_B + (E'_{el,B} - E)\chi_B = \sum_A \Bigg[& \chi_A \left\langle \Psi_{el,B} \middle| \sum_I^M \frac{\hbar^2 \Delta_I}{2m_I} \middle| \Psi_{el,A} \right\rangle \\ & + \sum_I^M (\nabla_I \chi_A) \cdot \left\langle \Psi_{el,B} \middle| \frac{\hbar^2 \nabla_I}{m_I} \middle| \Psi_{el,A} \right\rangle \Bigg] \end{aligned} \tag{8.83}$$

If we neglect the so-called non-adiabatic terms on the right hand side, we arrive at an eigenvalue equation,

$$\left[-\sum_I^M \frac{\hbar^2 \Delta_I}{2m_I} + E'_{el,B}(\{\boldsymbol{R}_I\}) \right] \chi_B(\{\boldsymbol{R}_I\}) = E \chi_B(\{\boldsymbol{R}_I\}) \tag{8.84}$$

which is the nuclear Schrödinger equation in Born–Oppenheimer approximation. We note that Eq. (8.84) is solved by many eigenpairs $(\chi_{B,K}, E_K)$ for a given potential energy surface $E'_{el,B}$.

In order to emphasize that Eq. (8.84) no longer depends on the coordinates of the electrons, which have been integrated out, we have included the dependence on the nuclear coordinates explicitly. In this way, the electronic energy enters the nuclear Schrödinger equation as a potential energy called the potential energy surface. Since we can solve the electronic Schrödinger equation, Eq. (8.76), only for a pre-assumed, fixed molecular structure, $E_{el,B}(\{\boldsymbol{R}_I\})$ is known only point-wise. Solving the nuclear Schrödinger equation is well described in monographs on nonrelativistic quantum mechanics and we may refer, for example, to Refs. [45,204,205]. Starting from Eq. (8.84), which describes all nuclear motions of the molecular system under consideration, the translational motion of the center of mass can be separated and internal coordinates can be introduced [211].

In the following, we will now be concerned with solving for the *electronic* Schrödinger equation, whose solution we assumed to be possible in the preceding paragraphs, since it is this equation for which a (pseudo-)relativistic approach is important when its numerical effect on the electronic energy hypersurface cannot be neglected. Since our first-quantized many-electron Hamiltonian comprises Dirac-like relativistic one-electron (and if necessary two-electron) operators, we consider the calculation of *relativistic* potential energy surfaces to be then employed in studies of molecular structure and nuclear motion.

8.4 Tensor Structure of the Many-Electron Hamiltonian and Wave Function

The solution of the electronic Schrödinger equation, Eq. (8.76), requires the integration of a complicated partial differential equation depending on $3N$ electronic coordinates as variables. The complexity of this problem increases with the number of electrons in the molecule. In order to establish a general solution strategy it is mandatory to first study the underlying formal framework of many-particle quantum mechanics, namely that of a linear vector space — the Hilbert space (cf. section 4.1). The N-electron Hilbert space, which 'hosts' the total quantum mechanical state vector, is then constructed by direct multiplication of the one-electron Hilbert spaces.

In particular, it is important to realize that an N-electron state is not obtained by simple multiplication of one-particle states but through a direct product (i.e., tensor or Konecker product). Standard quantum chemical approaches, however, do not require this tensor product formulation and hence the formal structure is usually abandoned (compare section 8.5). An exception is the emerging density matrix renormalization group (DMRG) algorithm [212–216], which explicitly constructs the N-electron Hamiltonian from direct products of terms operating on one-electron states of a one-particle Hilbert space. We will also encounter unitary transformation techniques for the Dirac–Coulomb Hamiltonian in chapter 12, for which it is mandatory to be aware of this underlying structure in order to break down a unitary transformation for a total state into one- and two-electron terms.

According to the standard representation of the Dirac theory, the one-electron operators considered in chapters 5 and 6 are (4×4)-matrices acting on the one-particle Hilbert space

$$^1\mathcal{H} = \{\phi(\boldsymbol{r}) \in \mathbb{C}^4 : \phi \text{ is square integrable}\} \tag{8.85}$$

For many-electron Hamiltonians defined as in Eq. (8.73) the situation is more involved. The Hilbert space of an N-electron system is the tensor product (or direct product) of the N one-particle Hilbert spaces $^1\mathcal{H}$, i.e.,

$$^N\mathcal{H} = {}^1\mathcal{H}(1) \otimes {}^1\mathcal{H}(2) \otimes \cdots \otimes {}^1\mathcal{H}(N) \tag{8.86}$$

where the number in parentheses labels the corresponding electron. Since $^N\mathcal{H}$ comprises normalizable spinors with 4^N components, also the operators acting on $^N\mathcal{H}$ no longer feature a simple (4×4) structure, but are $(4^N \times 4^N)$-dimensional quantities. Consequently, the one-particle Dirac operator for the i-th electron, $h^D(i)$, is actually given by

$$h^D(i) \equiv \mathbf{1}(1) \otimes \mathbf{1}(2) \otimes \cdots \otimes \mathbf{1}(i-1) \otimes h^D(i) \otimes \mathbf{1}(i+1) \otimes \cdots \otimes \mathbf{1}(N) \tag{8.87}$$

while the two-electron term, $g(i,j)$, reads

$$V_C(i,j) = \frac{1}{r_{ij}} [\mathbf{1}(1) \otimes \cdots \otimes \mathbf{1}(i) \otimes \cdots \otimes \mathbf{1}(j) \otimes \cdots \otimes \mathbf{1}(N)] \tag{8.88}$$

where each factor on the right hand side of Eqs. (8.87) and (8.88) is still a (4×4)-operator. Its action on all but the i-th electron is described by the identity operator, i.e., it has no action on these electrons, and only electron number i is affected in a non-trivial fashion by this operator. The Breit operator is represented similarly. For the sake of brevity, we only write the Gaunt operator,

$$G_0(i,j) = -\frac{1}{r_{ij}} [\mathbf{1}(1) \otimes \cdots \otimes \boldsymbol{\alpha}(i) \otimes \cdots \otimes \boldsymbol{\alpha}(j) \otimes \cdots \otimes \mathbf{1}(N)] \tag{8.89}$$

Any two-electron entries for not necessarily neighboring i and j thus enter the proper places in the 2^N super-matrix Hamiltonian. Because of Eq. (8.86) a total quantum mechanical state Ψ possesses the structure

$$\Psi = \mathcal{A}\left[\psi_1(1) \otimes \psi_2(2) \otimes \cdots \otimes \psi_N(N)\right] \tag{8.90}$$

where the ψ_i denote one-particle states and $\mathcal{A}$ is the antisymmetrizer

$$\mathcal{A} = \frac{1}{\sqrt{N!}} \sum_{p=1}^{N!} (-1)^p \mathcal{P}_p \tag{8.91}$$

that takes care of the Pauli principle by explicit implementation of the sign changes upon exchange of any pair of electronic coordinates. Even numbers p are assigned to those permutation operators $\mathcal{P}_p$ that are generated by an even number of pair permutations and odd p's denote those generated by an odd number of pair permutations. The antisymmetrizer also carries the normalization constant for the resulting linear combination of direct products of one-particle states.

Usually, the exact one-particle states are *not* known as they would be the eigenstates defined by the most general single-electron Dirac equation, Eq. (5.119). This equation can, however, not be solved analytically because even in the absence of additional fields that are *external* to the electronic system the remaining scalar potential energy operators due to the other electrons in the system — even when the vector potentials due to electrons' magnetic interactions (Gaunt and Breit) are neglected — represent one-particle equations much too complicated to be solved for the one-electron states ψ_i (recall section 8.1.1). However, we may circumvent this problem and expand each of the unknown one-particle states in a complete set $\{\phi_k\}$ of one-particle basis functions so that the total state reads

$$\Psi = \mathcal{A}\left(\sum_{k_1} c_{k_1}\phi_{k_1}(1) \otimes \sum_{k_2} c_{k_2}\phi_{k_2}(2) \otimes \cdots \otimes \sum_{k_N} c_{k_N}\phi_{k_N}(N)\right) \tag{8.92}$$

Since summation and direct multiplication commute, we may write the last equation as

$$\Psi = \sum_{k_1,k_2,\ldots,k_N} \underbrace{c_{k_1}c_{k_2}\cdots c_{k_N}}_{\equiv C_{I(k_1,k_2,\ldots,k_N)}} \mathcal{A}\left(\phi_{k_1}(1) \otimes \phi_{k_2}(2) \otimes \cdots \otimes \phi_{k_N}(N)\right) \tag{8.93}$$

which can be abbreviated as

$$\Psi = \sum_I C_I \Theta_I \tag{8.94}$$

if we introduce the composite index $I{=}I(k_1,k_2,\ldots,k_N)$ and the antisymmetrized direct product Θ_I as

$$\Theta_I = \Theta_{k_1,k_2,\ldots,k_N}(\boldsymbol{r}_1,\boldsymbol{r}_2,\ldots,\boldsymbol{r}_N) \equiv \mathcal{A}\left(\psi_{k_1}(\boldsymbol{r}_1) \otimes \psi_{k_2}(\boldsymbol{r}_2) \otimes \cdots \otimes \psi_{k_N}(\boldsymbol{r}_N)\right) \tag{8.95}$$

Eq. (8.94) represents an exact expression for the quantum mechanical state of a many-electron system. Note that the expansion coefficients C_I of the N-particle basis states are directly related to the expansion coefficients of the one-particle states c_{k_i}. It is sufficient to know either of them (and this fact is related to the observation made below that a full configuration interaction wave function does not require the optimization of orbitals; see the next section).

8.5
Approximations to the Many-Electron Wave Function

Having identified a suitable Hamiltonian for a practical many-electron theory, namely the Dirac–Coulomb Hamiltonian (or the somewhat more sophisticated Dirac–Coulomb–Breit Hamiltonian), we now need to solve the corresponding eigenvalue Eq. (8.76). For this purpose we again follow the prescriptions given by the nonrelativistic theory for many-electron systems. This section is devoted to a brief sketch of the standard and well-established approach to deriving an approximation to the many-electron wave function Ψ_{el} by means of a product of functions depending only on the coordinates of a single electron (independent particle model). Of course, this ansatz for the many-electron state does not take into account that the motion of the electrons is actually correlated. In quantum chemistry, there are other, more specialized approaches around, which we skip here for the sake of brevity. However, *relativistic* quantum chemistry is first of all about the proper choice of operators, while the wave function is approximated by standard techniques that have almost all been described on the same footing in the very sophisticated monograph by Helgaker, Jørgensen and Olsen [217]. We refer the reader to this monograph for many details on various electronic structure methods and to textbooks on quantum chemistry in general.

8.5.1
The Independent-Particle Model

Solving the electronic Schrödinger equation means that we need to determine a very complicated function, $\Psi_{el} = \Psi_{el}(\{\boldsymbol{r}_i\})$, whose variables are all electronic coordinates. This implies that the solution will be the more complicated the more electrons are involved in our first-quantized description. In order to simplify the problem and to find a suitable starting point, we introduce a product ansatz,

$$\theta_{k_1,k_2,\ldots,k_N}(\boldsymbol{r}_1,\boldsymbol{r}_2,\ldots,\boldsymbol{r}_N) = \psi_{k_1}(\boldsymbol{r}_1)\psi_{k_2}(\boldsymbol{r}_2)\ldots\psi_{k_N}(\boldsymbol{r}_N) \tag{8.96}$$

where each of the N electrons is associated with a function ψ_i depending on the electron's coordinate $\boldsymbol{r}_i$. To be more precise, the choice for the function must be a 4-spinor in order to match the 4×4 structure of the one-electron Dirac Hamiltonians h^D, which serve as one-electron operators in the electronic Hamiltonian H_{el} through Eq. (8.66). Also, Eq. (8.96) is the pragmatic analog of Eq. (8.95) (except for the antisymmetrization) where all direct products have been written as ordinary multiplications for the sake of convenience, which is justified if the many-electron Hamiltonian is written as in Eq. (8.73).

Note that it is clear from the beginning that such a product ansatz of one-electron spinors, which is usually called the Hartree product, is not a proper choice considering the electron–electron interaction terms $g(i,j)$, which couple the electronic coordinates in the electronic Hamiltonian H_{el}. Apart from this, the Hartree product of spinors does not fulfill the Pauli principle. To satisfy the Pauli principle, we implement it into the ansatz for the wave function using the antisymmetrizer of Eq. (8.91). This is convenient because we do not want to give up the simple structure of the Hartree product $\theta_{k_1,k_2,\ldots,k_N}$. After antisymmetrization, a linear combination of Hartree products with permutated coordinates results. In total, $N!$ permutations are possible of which all that can be constructed from an even number of pairwise exchanged coordinates carry a positive sign while those set up from an odd number of pair permutations carry a negative sign. The resulting linear combination of Hartree products changes sign whenever any two coordinates are exchanged, as this would introduce an additional pair permutation per Hartree product. To systematically construct all $N!$ Hartree products one simply uses the antisymmetrizer $\mathcal{A}$ introduced in Eq. (8.91). The antisymmetrizer also guarantees the normalization of the resulting linear combination of Hartree product. The *antisymmetrized* Hartree product Θ_K then reads

$$\Theta_K = \mathcal{A}\theta_{k_1,k_2,\ldots,k_N}(\boldsymbol{r}_1, \boldsymbol{r}_2, \ldots, \boldsymbol{r}_N) \qquad (8.97)$$

where K denotes the specific set $\{k_1, k_2, \ldots, k_N\}$ of one-electron spinors in the Hartree product $\theta_K \equiv \theta_{k_1,k_2,\ldots,k_N}$. Θ_K can always be written as a determinant — the so-called Slater determinant — for a given set of spinors

$$\Theta_K(1,\ldots,N) = \frac{1}{\sqrt{N!}} \begin{vmatrix} \psi_{k_1}(\boldsymbol{r}_1) & \psi_{k_1}(\boldsymbol{r}_2) & \cdots & \psi_{k_1}(\boldsymbol{r}_N) \\ \psi_{k_2}(\boldsymbol{r}_1) & \psi_{k_2}(\boldsymbol{r}_2) & \cdots & \psi_{k_2}(\boldsymbol{r}_N) \\ \vdots & \vdots & & \vdots \\ \psi_{k_N}(\boldsymbol{r}_1) & \psi_{k_N}(\boldsymbol{r}_2) & \cdots & \psi_{k_N}(\boldsymbol{r}_N) \end{vmatrix} \qquad (8.98)$$

The Slater determinant represents the simplest approximation to an electronic ground state, and its orbitals may be optimized to approximate the ground state as closely as possible within the independent particle model. We shall see later in this chapter how the optimization of the orbitals can be done within (Dirac–)Hartree–Fock theory by relying on the variational principle.

8.5.2
Configuration Interaction

Given a set of one-electron functions ψ_{k_i} (orbitals) which contains more functions than there are electrons, more than one determinant can be constructed. If the orbitals are physically meaningful, e.g., if they correspond to occupied hydrogen-like orbitals that can be called the electronic configuration of the system under study, we use the notion *electronic configuration* synonymously with *Slater determinant*. All possible Slater determinants form an N-particle basis set, into which the quantum mechanical state Ψ_A can be expanded. This expansion is called *configuration interaction*.

Not all of these determinants are eigenfunctions of the total squared spin operator S^2, but one may construct linear combinations of determinants that are [218, 219]. Then, the complexity of the problem can be reduced by definition of these so-called configuration state functions (CSF),

$$\Phi_I = \sum_K \Theta_K B_{KI} \tag{8.99}$$

which are superpositions of Slater determinants with (analytically) known coefficients B_{KI}. The exact total electronic wave function is obtained by expansion into the complete (infinite) set of CSFs constructed from a complete set of one-electron spinors,

$$\Psi_A = \sum_{I=0}^{\infty} \Phi_I C_{IA} \tag{8.100}$$

which is the so-called full configuration interaction (FCI) wave function and resembles the formally exact representation of the state in Eq. (8.94), of course. Note that the subscript *el* has been skipped in this and all following equations referring to the total electronic state for the sake of brevity. For the FCI expansion in Eq. (8.100) the set of orthonormal spinors may be chosen arbitrarily. Any complete set of one-electron functions can be used to calculate the exact many-electron wave function in an FCI approach.

The expansion coeeficients C_{IA} for state A are called CI coefficients. Hence, electronic ground and excited states A can be described on equal footing within a CI-type approach by different sets of expansion parameters $\boldsymbol{C}_A = \{C_{IA}\}$. Later in this chapter, we will see that these coefficients can be determined by diagonalization of the proper Hamiltonian matrix, which in turn yields the coefficient vectors $\boldsymbol{C}_A$ for as many electronic states as there are basis functions Φ_I.

In approximate CI methods, the set of many-particle basis functions Φ_I is restricted and not infinitely large, i.e., it is usually not complete. Then, the many-particle basis is usually constructed systematically from a given reference basis function (like the Slater determinant, which is constructed to approximate the ground state of a many-electron system in (Dirac–)Hartree–Fock theory).

In this systematic construction process, spinors which enter the reference Slater determinant (these are the so-called occupied spinors) are substituted by new orthogonal spinors (which are, for instance, the virtual spinors that are obtained in a Dirac–Fock–Roothaan calculation; see chapter 10). One then usually distinguishes sets of singly substituted many-electron functions $\Phi_{o_i}^{v_i}$, where an occupied spinor o_i has been substituted by a virtual spinor v_i, from doubly substituted $\Phi_{o_i o_j}^{v_i v_j}$ and so on. The $\Phi_{o_i}^{v_i}$ are also called single-excited determinants, the $\Phi_{o_i o_j}^{v_i v_j}$ then double-excited determinants etc. It is important to understand that these excitations are not to be confused with the excited states of quantum mechanics; they simply denote a substitution pattern. Eq. (8.100) can be rewritten in terms of the excitation hierarchy as

$$\begin{aligned}\Psi_A &= \Phi_0 C_{0A} + \sum_{(o_i v_i)}^{N} \Phi_{o_i}^{v_i} C_{(o_i v_i)A} + \sum_{(o_i v_i),(o_j v_j)}^{N} \Phi_{o_i o_j}^{v_i v_j} C_{(o_i v_i, o_j v_j)A} + \\ &\quad \cdots + \sum_{(o_i v_i),\ldots,(o_N v_N)}^{N} \Phi_{o_i \ldots o_N}^{v_i \ldots v_N} C_{(o_i v_i, \ldots, o_N v_N)A} \end{aligned} \tag{8.101}$$

Note that this presentation of the approximation of the relativistic many-electron wave function closely follows the systematic construction of electronic wave functions in nonrelativistic quantum chemistry. Consequently, a truncated CI expansion that takes only all single substitutions into account is called CI-Singles (in short CIS), one that also includes double substitutions is called CI-Singles-Doubles (in short CISD), and so forth. A configuration interaction wave function restricted to single and double substitutions is a computationally feasible though necessarily approximate scheme. The ansatz for the CISD wave function simply reads

$$\Psi_A^{\text{CISD}} = \Phi_0 C_{0A} + \sum_{(o_i v_i)}^{N} \Phi_{o_i}^{v_i} C_{(o_i v_i)A} + \sum_{(o_i v_i),(o_j v_j)}^{N} \Phi_{o_i o_j}^{v_i v_j} C_{(o_i v_i, o_j v_j)A} \tag{8.102}$$

A truncated CI expansion, however, suffers from a lack of size consistency. This means that the electronic energy of two identical molecules separated at infinite distance is not equal to twice the energy of the monomer if all energies are obtained from a truncated-CI calculation. To correct for the size consistency error extrapolation schemes that estimate the contribution of discarded excitations to the electronic energy have been developed [220].

To improve on the selection of important configurations, a reference space consisting of Slater determinants (or symmetry-adapted configuration state functions, respectively) with large CI coefficients can be defined from which single and double excitations are then generated. This defines the so-called multi-reference configuration-interaction (MRCI) approach [221,222]. In analogy to Eq. (8.102), the MRCI wave function can be written as

$$\Psi_A^{\text{MRCI}} = \sum_{\nu} \Phi_{\nu}^{\text{ref}} C_{\nu A} + \sum_{\nu,(o_i v_i)}^{N} \Phi_{\nu, o_i}^{\text{ref}, v_i} C_{(o_i v_i)A} + \sum_{\nu,(o_i v_i),(o_j v_j)}^{N} \Phi_{\nu, o_i o_j}^{\text{ref}, v_i v_j} C_{(o_i v_i, o_j v_j)A} \tag{8.103}$$

where the single Hartree–Fock reference determinant Φ_0 has been replaced by a linear combination of selected N-electron basis functions $\{\Phi_\nu^{\text{ref}}\}$.

In those cases where different configurations are close in energy a single electronic configuration like the Hartree–Fock determinant is not a good starting point for the CI expansion because of the biased set of orbitals optimized for only one electronic configuration. The multi-configurational self-consistent field (MCSCF) approach avoids this drawback. The MCSCF wave function replaces the single Slater determinant of Hartree–Fock theory by a linear combination of determinants. A subsequent optimization procedure is then required for the expansion coefficients $\boldsymbol{C} = \{C_\nu\}$ as well as for the orbitals ψ_i from which the Slater determinants (or CSFs) are constructed. This procedure ensures that an optimal set of orbitals is obtained for a truncated CI expansion. The MCSCF approximation to the exact electronic wave function then reads

$$\Psi_A^{\text{MCSCF}} = \sum_{\nu \in S} \Phi_\nu^{\text{opt}} C_{\nu A} \tag{8.104}$$

where S represents a selected subspace and $\Phi_\nu^{\text{opt}} = \Phi_\nu^{\text{opt}}(\{\psi_i^{\text{opt}}\})$ indicates that the molecular orbitals for the pre-selected set of Slater determinats are optimized for the MCSCF expansion. By constrast, the single reference CI methods solely employ Hartree–Fock-type orbitals. The first generation of MCSCF methods optimized CI coefficients and orbitals alternately until convergence was achieved [223], which can often cause convergence problems. The second generation of MCSCF methods [224,225] employs second derivatives for the *simultaneous* optimization of CI coefficients and orbitals (usually expressed as state-transfer and orbital rotation parameters [217]). Although this is much more computer time demanding, it finally pays off because of fast convergence. The most successful variant among the MCSCF techniques is the complete-active-space self-consistent-field (CASSCF) method [226,227], which constructs the reference space of determinants as an FCI in a *restricted* orbital space, the so-called active space. However, the size of the active space suffers from the same restrictions as the one in an FCI calculation. Moreover, the contribution of the neglected virtual orbitals to the correlation energy is often not negligible. In order to account for this contribution, a perturbation theory to second order (CASPT2) [228,229] is needed if chemical accuracy of, say, 4 kJ mol^{-1} on relative energies is envisaged.

In general, MCSCF approaches are a suitable means to take into account the so-called static electron correlation resulting from nearly degenerate orbitals. However, they suffer from a lack of dynamic electron correlation, which is determined by those configurations in the FCI expansion that have smaller CI coefficients and which are omitted in the MCSCF expansion by construction. It is important to understand that the terms *static* and *dynamic* are widely used but to a certain degree arbitrarily defined as they only refer to the relative ratio of CI coefficients in the FCI expansion. Many-body perturbation theory — formulated as Møller–Plesset [230,231] or CASPT2 theory — offers an efficient way to take into account dynamic correlation and to improve the results from Hartree–Fock and MCSCF calculations, respectively. However, two major drawbacks of perturbation theory must not be forgotten: (i) The energy corrections may eventually become very large and an increased absolute value of the energy does not imply that the energy has been improved as this can only be guaranteed in variational approaches. (ii) A convergence order by order is also not guaranteed [232–234].

While we will discuss details of the MCSCF approach in combination with the simple Hartree–Fock model in the next chapter for atoms and in chapter 10 for molecules, we will not give many details on the closely related CI approach and refer the reader to the literature for more extensive presentations of the formalism and for technical aspects. A formalism for relativistic four-component CI calculations was devised and applied to Hg and Pb atoms in the early 1980s [235,236]. Later a complete open shell CI (COSCI) and a relativistic variant of the restricted active space CI method were implemented [237–239] in the pioneering four-component electronic structure program for molecules, MOLFDIR [240–242]. Later, a direct CI module for the treatment of large determinant expansions [243,244] was included in the now well progressing DIRAC code [245–247].

8.5.3
Detour: Explicitly Correlated Wave Functions

The two-electron interaction operators $g(i,j)$ in the many-electron Hamiltonian are the reason why a product ansatz for the electronic wave function $\Psi_{el}(\{\boldsymbol{r}_i\})$ that separates the coordinates of the electrons is not the proper choice if the exact function is to be obtained in a single product of one-electron functions. Instead, the electronic coordinates are coupled and the motion of electrons is correlated. It is therefore natural to assume that a suitable ansatz for the many-electron wave function requires functions that depend on two coordinates. Work along these lines has got a long history [248–254]. The problem, however, is then what functional form to choose for these functions. First attempts in molecular quantum mechanics used simply terms that are linear in the interelectronic distance $r_{ij} = |\mathbf{r}_i - \mathbf{r}_j|$ [255–259], while it could be shown that an exponential ansatz is more efficient [260–263].

It turns out that the functional form of a Gaussian distribution is very suitable when it comes to the evaluation of many-electron integrals (compare chapter 10). Therefore, a Gaussian geminal,

$$\phi_{ij} = \phi_{ij}(\alpha_{ij}, \mathbf{r}_i, \mathbf{r}_j) = \exp\left(-\alpha_{ij} r_{ij}^2\right) \tag{8.105}$$

represents an appropriate form for practical applications. Such a two-electron basis function then enters an antisymmetrized product ansatz for the N-electron basis function,

$$\Phi_I = \mathcal{A}\left[\prod_{j>i}^{N} \phi_{ij}(\alpha_{ij}, \mathbf{r}_i, \mathbf{r}_j)\right] = \mathcal{A}\left[\phi_{11}\phi_{12}\phi_{13}\cdots\phi_{NN}\right], \tag{8.106}$$

composed from $N(N-1)/2$ different Gaussian geminals. This N-electron function can be written in a very compact form,

$$\Phi_I = \mathcal{A}\left[\prod_{j>i}^{N} \exp\left(-\alpha_{ij}^{(I)} r_{ij}^2\right)\right] = \mathcal{A}\exp\left(-\sum_{j>i}^{N} \alpha_{ij}^{(I)} r_{ij}^2\right) \tag{8.107}$$

which is a single Gaussian function with a quite complex exponent. However, the exponent is actually much simpler to handle than one might think. General expressions for matrix elements are available [264, 265] that require only a little generalization of what will be presented for one-electron Gaussians in chapter 10.

Despite its explicit dependence on the interelectronic coordinates r_{ij} this N-electron function is not an exact approximation to the N-electron wave function, as the Gaussian form has been arbitrarily chosen. Therefore, a CI-type expansion cannot be avoided. However, it can be expected that this expansion will converge much faster than the one discussed above. Still, the yet unknown set of parameters $\{\alpha_{ij}^{(I)}\}$, which parametrizes an N-electron basis function of the type in Eq. (8.107), needs to be determined. If these are determined based on a variational principle, i.e., the parameters are determined such that the energy is stationary when they undergo an infinitesimally small variation (see below), it is guaranteed that any geminal-type test function yields an energy above the true energy. In other words, the energy expectation value which is lowest is also the best approximation to the exact total electronic energy.

The geminal ansatz still requires more effort than the standard one-electron approach of the independent particle model. It is therefore usually restricted to small molecules for feasibility reasons. As an example how the nonlinear optimization problem can be handled we refer to the stochastic variational approach [266]. However, the geminal ansatz as presented above has the useful feature that all elementary particles can be treated on the same footing. This means that we can actually use such an ansatz for total wave functions without employing the Born–Oppenheimer approximation, which utilizes the fact that nuclei are much heavier than electrons. Hence, electrons and nuclei can be treated on the same footing [266–268] and even mixed approaches are possible, where protons and electrons are treated in the external field of heavier nuclei [269–272]. The integrals required for the matrix elements are hardly more complicated than those over one-electron Gaussians [264, 265, 273].

8.5.4
Orthonormality Constraints and Total Energy Expressions

In order to calculate expectation values for a wave function of the structure given in Eq. (8.100), it is convenient to introduce orthonormality constraints on the one-electron spinors

$$\langle \psi_i | \psi_j \rangle = \delta_{ij} \tag{8.108}$$

Then, we can benefit largely from factors that become zero upon resolution of an expectation value over two Slater determinants because of the orthogonality of two different spinors. Also, the Slater determinants turn out to be orthonormal then,

$$\langle \Theta_K | \Theta_L \rangle = \delta_{KL} \tag{8.109}$$

Additionally, orthonormal Slater determinants result in orthonormal CSFs,

$$\langle \Phi_I | \Phi_J \rangle = \delta_{IJ} \tag{8.110}$$

The orthonormality restriction for the spinors, Eq. (8.108), may be split into two parts in the case of atoms. The product ansatz for the spinor automatically yields orthonormal angular parts (coupled spherical harmonics; cf. chapter 9). But these do not contain information about the principal quantum numbers in the composite indices i and j. For this reason, the restriction to orthonormal spinors results in the orthonormality restriction for radial functions

$$\langle P_i | P_j \rangle + \langle Q_i | Q_j \rangle = \delta_{ij} \tag{8.111}$$

This holds true only if $\kappa_i = \kappa_j$ since the spin and angular parts guarantee orthogonality in all other cases. If one wants to calculate excited electronic states using the same set of radial functions optimized for both the ground state and the excited state within the same symmetry, the orthonormality requirement for these states leads to orthogonal vectors of CI coefficients,

$$\langle \Psi_{el,A} | \Psi_{el,B} \rangle = \sum_I C_{IA}^\star C_{IB} = \delta_{AB} \tag{8.112}$$

The electronic energy $E_{el,A}$ can then be written as,

$$\begin{aligned} E_{el,A} &= \langle \Psi_{el,A} \mid H_{el} \mid \Psi_{el,A} \rangle \\ &= \sum_{IJ} C_{IA}^\star C_{JA} \sum_{KL} B_{KI}^\star B_{LJ} \, \langle \Theta_K \mid H_{el} \mid \Theta_L \rangle \\ &= \sum_{IJ} C_{IA}^\star C_{JA} \sum_{KL} B_{KI}^\star B_{LJ} \, \langle \mathcal{A}\theta_K \mid H_{el} \mid \mathcal{A}\theta_L \rangle \end{aligned} \tag{8.113}$$

Utilizing the self-adjointness of the antisymmetrizer in step (A), the commutation relation $[\mathcal{A}^\dagger, H_{el}] = 0$ in (B), and its idempotency in step (C) (compare Ref. [274]), we obtain

$$\begin{aligned}
E_{el,A} &= \sum_{IJ} C^\star_{IA} C_{JA} \sum_{KL} B^\star_{KI} B_{LJ} \, \langle \theta_K \mid \mathcal{A}^\dagger H_{el} \mathcal{A} \mid \theta_L \rangle \\
&\overset{(A)}{=} \sum_{IJ} C^\star_{IA} C_{JA} \sum_{KL} B^\star_{KI} B_{LJ} \, \langle \theta_K \mid \mathcal{A} H_{el} \mathcal{A} \mid \theta_L \rangle \\
&\overset{(B)}{=} \sum_{IJ} C^\star_{IA} C_{JA} \sum_{KL} B^\star_{KI} B_{LJ} \, \langle \theta_K \mid H_{el} \mathcal{A}^2 \mid \theta_L \rangle \\
&\overset{(C)}{=} \sum_{IJ} C^\star_{IA} C_{JA} \sum_{KL} B^\star_{KI} B_{LJ} \, \langle \theta_K \mid H_{el} \sqrt{N!} \mathcal{A} \mid \theta_L \rangle
\end{aligned} \tag{8.114}$$

If we may now introduce the explicit expression of Eq. (8.91) for the antisymmetrizer,

$$\begin{aligned}
E_{el,A} &= \sum_{IJ} C^\star_{IA} C_{JA} \sum_{KL} B^\star_{KI} B_{LJ} \sum_p (-1)^p \, \langle \theta_K \mid H_{el} \mid \mathcal{P}_p \theta_L \rangle \\
&= \sum_{IJ} C^\star_{IA} C_{JA} \sum_{KL} B^\star_{KI} B_{LJ} \sum_p (-1)^p \times \\
&\quad \left[\left\langle \theta_K \middle| \sum_{i=1}^{N} h(i) \middle| \mathcal{P}_p \theta_L \right\rangle + \left\langle \theta_K \middle| \frac{1}{2} \sum_{j \neq i}^{N} g(i,j) \middle| \mathcal{P}_p \theta_L \right\rangle \right]
\end{aligned} \tag{8.115}$$

we note that only a single pair permutation $\mathcal{P}_{ij}$ of coordinates yields a nonvanishing contribution in the two-electron part of the expectation value,

$$\begin{aligned}
E_{el,A} &= \sum_{IJ} C^\star_{IA} C_{JA} \sum_{KL} B^\star_{KI} B_{LJ} \left[\left\langle \theta_K \middle| \sum_{i=1}^{N} h(i) \middle| \theta_L \right\rangle \right. \\
&\quad \left. + \left\langle \theta_K \middle| \frac{1}{2} \sum_{j \neq i}^{N} g(i,j) \left\{ 1 - \mathcal{P}_{ij} \right\} \middle| \theta_L \right\rangle \right]
\end{aligned} \tag{8.116}$$

The energy expression derived so far cannot be cast in a more elegant form at this stage since we need to be able to resolve the matrix elements over Hartree products K and L. Fortunately, these are easy to evaluate because of the fact that the total integral separates into products of integrals over individual electronic coordinates,

$$\begin{aligned}
\left\langle \theta_K \middle| \sum_{i_1 i_2 \ldots i_n}^{N} o(i_1, i_2, \ldots, i_n) \middle| \theta_L \right\rangle &= N(N-1) \cdots (N-n+1) \\
&\times \langle \psi_{k_1}(1) \psi_{k_2}(2) \ldots \psi_{k_n}(n) \mid o(1,2,\ldots,n) \mid \psi_{l_1}(1) \psi_{l_2}(2) \ldots \psi_{l_n}(n) \rangle \\
&\times \langle \psi_{k_{n+1}} | \psi_{l_{n+1}} \rangle \times \cdots \times \langle \psi_{k_N} | \psi_{l_N} \rangle
\end{aligned} \tag{8.117}$$

where the formal equivalence of integrals over different electronic coordinates has been extensively exploited so that we may select the first n coordinates for which we evaluate the matrix element over the n-electron operator $o(1,2,\ldots,n)$.

The matrix element on the left hand side of Eq. (8.117) is only different from zero if $k_a = l_a$ for $a \neq i,j,\ldots,n$ while k_i can be different from l_i, k_j can be different from $l_j,\ldots$ and k_n can be different from l_n. Hence, all matrix elements over Hartree products vanish as soon as the two Hartree products K and L differ by more than n orbitals since they cannot be included in the integral over the n-electron operator $o(1,2,\ldots,n)$ and thus produce vanishing overlap integrals.

Fortunately, the many-electron Hamiltonian contains only one- and two-electron operators, and thus only seven different cases need to be distinguished. For one-electron operators we obtain

$$\left\langle \theta_K \middle| \sum_i^N o(i) \middle| \theta_L \right\rangle = \begin{cases} \sum_i^N \langle \psi_{k_i} | o | \psi_{k_i} \rangle & \text{if } K = L \\ \langle \psi_{k_j} | o | \psi_{l_j} \rangle & \forall\, k_i = l_i \text{ but one } j, k_j \neq l_j \\ 0 & \text{else} \end{cases} \tag{8.118}$$

and for two-electron operators

$$\left\langle \theta_K \middle| \sum_{ij}^N o(i,j) \middle| \theta_L \right\rangle = \begin{cases} \frac{1}{2} \sum_{ij} (k_i k_j || k_i k_j) & \text{if } K = L \\ \sum_j (k_i k_j || l_i k_j) & \forall\, k_a = l_a \text{ but one } i, k_i \neq l_i \\ (k_i k_j || l_k l_l) & \forall\, k_a = l_a \text{ but two } i,j, k_i \neq l_i \\ & \text{and} k_j \neq l_j \\ 0 & \text{else} \end{cases} \tag{8.119}$$

where we introduced the shorthand notation

$$\begin{aligned} (k_i k_j || l_k l_l) &= \langle \psi_{k_i}(1) \psi_{k_j}(2) \mid o(1,2) \mid \psi_{l_k}(1) \psi_{l_l}(2) \rangle \\ &\quad - \langle \psi_{k_i}(1) \psi_{k_j}(2) \mid o(1,2) \mid \psi_{l_l}(1) \psi_{l_k}(2) \rangle \end{aligned} \tag{8.120}$$

The equations above are known as the Slater–Condon rules for the evaluation of many-electron matrix elements. Note that the summation signs on the right hand side run over orbital indices instead of electronic coordinates since the indistinguishability of the electrons has been exploited.

Only in the simplest model case when the CI expansion comprises only a single determinant, which is known as the Hartree–Fock approximation, can the energy expression be given in a compact form at this stage,

$$E_{el,0}^{\text{DHF}} = \sum_{i}^{N} h_i + \frac{1}{2}\sum_{ij}^{N}(J_{ij} - K_{ij}) \tag{8.121}$$

after having evaluated all matrix elements for $I = J$ and $K = L$ according to the Slater–Condon rules. The superscript DHF stands for Dirac–Hartree–Fock which denotes the fact that the Hartree–Fock expression utilizes the Dirac one-electron Hamiltonian and we introduced the one-electron integrals h_i,

$$h_i \equiv \langle \psi_i(1) \mid h^D(1) \mid \psi_i(1)\rangle \tag{8.122}$$

and the two-electron Coulomb integrals,

$$J_{ij} \equiv \langle \psi_i(1)\psi_j(2) \mid g(1,2) \mid \psi_i(1)\psi_j(2)\rangle \tag{8.123}$$

and exchange integrals,

$$K_{ij} \equiv \langle \psi_i(1)\psi_j(2) \mid g(1,2) \mid \psi_j(1)\psi_i(2)\rangle \tag{8.124}$$

8.6 Second Quantization for the Many-Electron Hamiltonian

The general energy expression derived so far takes a rather clumsy form and requires the explicit reference to the Slater–Condon rules for the evaluation of integrals. In this section we utilize *second quantization* [275], which is an integral tool in any quantum field theory (see chapter 7) and thus covered in textbooks on QED as a new language in quantum chemistry. Second quantization will allow us to write compact and elegant expressions for expectation values that automatically incorporate the Slater–Condon rules through the invocation of matrix elements over new operators, called creation and annihilation operators. The introduction of these operators can be done in close connection to the independent particle model of Hartree–Fock theory, i.e., as acting on Slater determinants. As a consequence, integrals over one-electron functions show up as parameters in front of annihilation and creation operators, but they are only well defined once the first-quantized many-electron Hamiltonian has been chosen. In the context of electronic structure theory, second quantization mainly enters as a bookkeeping scheme rather than in its original field quantization sense. Nonetheless, we will be able to describe different numbers of electrons on the same footing in Fock space.

8.6.1
Creation and Annihilation Operators

The essential new operators in second quantization are the creation operator (the *creator*) [46,223,276]

$$a_q^\dagger \Theta_K = \begin{cases} 0 & \text{if } q \in \{p_1, ..., p_N\} \\ (-1)^{N-k} \Theta_{K'}(p_1, ..., p_{k-1}, q, p_k, ..., p_N) & \text{if } p_{k-1} < q < p_k \end{cases} \tag{8.125}$$

and the annihilation operator (the *annihilator*)

$$a_q \Theta_K = \begin{cases} 0 & \text{if } q \notin \{p_1, ..., p_N\} \\ (-1)^{N-k} \Theta_{K''}(p_1, ..., p_{k-1}, p_{k+1}, ..., p_N) & \text{if } q = p_k \end{cases} \tag{8.126}$$

Creation and annihilation operators create and annihilate spinors, respectively, in normalized Slater determinants $\Theta_K(p_1, ..., p_N)$. The resulting Slater determinants, $\Theta_{K'}(p_1, ..., p_{k-1}, q, p_k, ..., p_N)$ and $\Theta_{K''}(p_1, ..., p_{k-1}, p_{k+1}, ..., p_N)$, are $(N+1)$-electron and $(N-1)$-electron basis functions, respectively. The creators and annihilators operate on columns of the determinants while maintaining the correct order of the spinors with respect to their orbital indices. Accordingly, their definition accounts for the sign changes in the determinant. In the case of the annihilator, the N-th row and the k-th column are annihilated resulting in a prefactor of $(-1)^{N-k}$. In order to retain the normalization for $\Theta_{K'}(p_1, ..., p_{k-1}, q, p_k, ..., p_N)$ and $\Theta_{K''}(p_1, ..., p_{k-1}, p_{k+1}, ..., p_N)$ multiplication by $\sqrt{1/(N+1)}$ and by $\sqrt{N}$, respectively, is implicitly required. The creator–annihilator formalism allows us to conveniently expand the N-electron Slater determinant with respect to its N-th row [223],

$$\Theta_K = \frac{1}{\sqrt{N}} \sum_p^{n'} \psi_p(N) \left[a_p \Theta_K\right] = \frac{1}{\sqrt{N}} (-1)^{N-k} \sum_p^{n'} \psi_p(N)\, \Theta_{K''} \tag{8.127}$$

and the resulting determinants with respect to the $(N-1)$-th row,

$$\Theta_K = \frac{1}{\sqrt{N(N-1)}} \sum_{p,q}^{n'} \psi_q(N-1)\, \psi_p(N) \left[a_q a_p \Theta_K\right] \tag{8.128}$$

and so forth. The label n' denotes the maximum number of one-particle functions, i.e., spinors (or spin orbitals in nonrelativistic theory). Hence, these equations do not hold for a set of functions which has been reduced in number for symmetry reasons (like the equivalence restriction of chapter 6).

Both types of operators do not commute but fulfill the following anticommutation rules

$$\{a_p^\dagger, a_q\} = a_p^\dagger a_q + a_q a_p^\dagger = \delta_{pq} \tag{8.129}$$

$$\left\{a_p^\dagger, a_q^\dagger\right\} = a_p^\dagger a_q^\dagger + a_q^\dagger a_p^\dagger = 0 \tag{8.130}$$

$$\{a_p, a_q\} = a_p a_q + a_q a_p = 0 \tag{8.131}$$

and are mutually adjoint. This allows us to work with annihilation operators in matrix elements (cf. Eqs. (8.135) and (8.139)) only.

8.6.2
Reduction of Determinantal Matrix Elements to Matrix Elements Over Spinors

To calculate expectation values, we have to get rid of the determinantal wave function in matrix elements over a many-electron operator $\hat{O}$ like the many-electron Hamiltonian and reduce them to computable one-electron and two-electron matrix elements. This can be done easily using creators and annihilators since we can expand a Slater determinant according to Eq. (8.127) and obtain in cases of one-electron operators

$$\langle \Psi_A | \hat{O}(1) | \Psi_B \rangle = \sum_{IJ} C^{\star}_{IB} \langle \Phi_I | \hat{O}(1) | \Phi_J \rangle C_{JA} \tag{8.132}$$

$$\langle \Phi_I | \hat{O}(1) | \Phi_J \rangle = \sum_{KL} B^{\star}_{KI} \langle \Theta_K | \hat{O}(1) | \Theta_L \rangle B_{LJ} \tag{8.133}$$

$$\langle \Theta_K | \hat{O}(1) | \Theta_L \rangle = \frac{1}{N} \sum_{pq}^{n'} \langle \psi_p | \hat{O}(1) | \psi_q \rangle_1 t^{KL}_{pq} \tag{8.134}$$

Additionally, we have introduced

$$t^{KL}_{pq} = \langle \Theta_K | a^{\dagger}_p a_q | \Theta_L \rangle = \langle a_p \Theta_K | a_q \Theta_L \rangle = \begin{cases} \pm 1 & \text{if } a_p \Theta_K = \pm a_q \Theta_L \\ 0 & \text{else} \end{cases} \tag{8.135}$$

Remember and that the factor $1/N$ results from the reduction of the original N-electron Slater determinant to a Slater determinant with $(N-1)$ electrons for which we must reduce the normalization constant from $1/\sqrt{N!}$ to $1/\sqrt{(N-1)!}$. Analogously, the two-electron operator is resolved according to

$$\langle \Psi_A | \hat{O}(1,2) | \Psi_B \rangle = \sum_{IJ} C^{\star}_{IB} \langle \Phi_I | \hat{O}(1,2) | \Phi_J \rangle C_{JA} \tag{8.136}$$

$$\langle \Phi_I | \hat{O}(1,2) | \Phi_J \rangle = \sum_{KL} B^{\star}_{KI} \langle \Theta_K | \hat{O}(1,2) | \Theta_L \rangle B_{LJ} \tag{8.137}$$

$$\langle \Theta_K | \hat{O}(1,2) | \Theta_L \rangle = \frac{1}{N(N-1)} \sum_{pqrs}^{n'} \langle \psi_p \langle \psi_r | \hat{O}(1,2) | \psi_s \rangle_2 \psi_q \rangle_1 T^{KL}_{pqrs} \tag{8.138}$$

where we grouped the coordinates of electrons 1 and 2 in a double-bracket notation and defined

$$\begin{aligned} T^{KL}_{pqrs} &= \langle \Theta_K | a^{\dagger}_p a^{\dagger}_r a_s a_q | \Theta_L \rangle \\ &= \langle a_r a_p \Theta_K | a_s a_q \Theta_L \rangle = \begin{cases} \pm 1 & \text{if } a_r a_p \Theta_K = \pm a_s a_q \Theta_L \\ 0 & \text{else} \end{cases} \end{aligned} \tag{8.139}$$

At this stage, it is important to understand that the matrix elements containing products of creation and annihilation operators can be easily evaluated owing to the turnover rule as employed in Eqs. (8.135) and (8.139) because they are nothing else but overlap integrals of many-electron basis functions. Thus, their absolute value is either one or zero depending on whether or not the resulting $(N-1)$-electron and $(N-2)$-electron basis functions, $a_p\Theta_K$, $a_q\Theta_L$ and $a_r a_p\Theta_K$, $a_s a_q\Theta_L$ are identical (apart from the sign) or not.

8.6.3 Many-Electron Hamiltonian and Energy

In second quantization we may now rewrite the many-electron Hamiltonian as

$$H_{el}^{s.q.} = \sum_{ij} h_{ij}\, a_i^\dagger a_j + \frac{1}{2}\sum_{ijkl} g_{ijkl}\, a_i^\dagger a_j^\dagger a_l a_k \tag{8.140}$$

with the (generalized) one-electron integrals,

$$h_{ij} \equiv \langle \psi_i(1) \mid h^D(1) \mid \psi_j(1)\rangle \tag{8.141}$$

and the two-electron integrals,

$$g_{ijkl} \equiv \langle \psi_i(1)\psi_j(2) \mid g(1,2) \mid \psi_k(1)\psi_l(2)\rangle \tag{8.142}$$

The resulting total electronic energy expression,

$$E_{el,A} = \langle \Psi_{el,A} \mid H_{el}^{s.q.} \mid \Psi_{el,A}\rangle \tag{8.143}$$

may now be conveniently simplified by defining

$$\gamma_{ij}^A = \sum_{IJ} C_{IA}^\star C_{JA} \sum_{KL} B_{KI}^\star B_{LJ} t_{ij}^{KL} \tag{8.144}$$

and

$$\Gamma_{ijkl}^A = \sum_{IJ} C_{IA}^\star C_{JA} \sum_{KL} B_{KI}^\star B_{LJ} T_{ijkl}^{KL} \tag{8.145}$$

so that we may write

$$E_{el,A} = \sum_{ij}^{n'} \gamma_{ij}^A h_{ij} + \frac{1}{2}\sum_{ijkl}^{n'} \Gamma_{ijkl}^A g_{ijkl} \tag{8.146}$$

where n' denotes the total number of one-electron states taken into account. This last equation represents the most general expression for the electronic energy of a many-electron system with a given total number of electrons. The

γ_{ij}^A and Γ_{ijkl}^A are called *structure factors* [277] or *coupling coefficients* [224]. They are the one- and two-electron density matrices.

In the case of the simplest approximation of the many-electron wave function, which is a single Slater determinant that manifests the basis of the Hartree–Fock model, the electronic energy simplifies to

$$E_{el,0}^{\mathrm{DHF}} = \sum_i^N h_{ii} + \frac{1}{2} \sum_{ij}^N (J_{ijij} - K_{ijji}) \tag{8.147}$$

with $n' = N$. This expression has already been given in Eq. (8.121) but we are now using generalized two-electron Coulomb integrals,

$$J_{ij} \rightarrow J_{ijij} \quad \equiv \quad g_{ijij} \tag{8.148}$$

and generalized exchange integrals,

$$K_{ij} \rightarrow K_{ijji} \quad \equiv \quad g_{ijji} \tag{8.149}$$

8.6.4 Fock Space and Occupation Number Vectors

Up to this point we have tailored the second-quantization formalism in close connection to the independent-particle picture introduced before. However, the formalism can be generalized in an even more abstract fashion. For this we introduce so-called *occupation number vectors*, which are state vectors in Fock space. Fock space is a mathematical concept that allows us to treat variable particle numbers (although this is hardly exploited in quantum chemistry; see for an exception the Fock-space coupled-cluster approach mentioned in section 8.9). Accordingly, it represents loosely speaking all Hilbert spaces for different but fixed particle numbers and can therefore be formally written as a *direct sum* of N-electron Hilbert spaces,

$$\mathcal{F} = \bigoplus_{N=0}^{\infty} {}^N\mathcal{H} \tag{8.150}$$

each of them being constructed from a tensor product of one-electron Hilbert spaces according to Eq. (8.86).

The occupation number vector can be written as a sequence of 0's and 1's according to the occupation of a particular one-electron state

$$\Psi \equiv |i_1, i_2, i_3, \ldots, i_\infty\rangle \quad \forall i_j \in \{0, 1\} \tag{8.151}$$

In particular, we may generate any reference state from the vacuum state,

$$|vac\rangle = |0, 0, 0, \ldots, 0\rangle \tag{8.152}$$

by application of a sequence of creation operators $a_{i_j}^\dagger$ that create the desired electronic configuration.

What has been said so far assumed exactly known one-particle states. These are, however, not known in quantum chemistry: the orbitals are only approximations and, hence, a CI-like expansion of the state results (compare section 8.5). However, the formalism can still be elegantly employed in chemistry [217, 278] as a bookkeeping scheme that also takes care of the Slater–Condon rules as demonstrated above. Consequently, the total state is then to be expressed in a CI-like manner as a superposition of many different occupation number vectors. As the one-electron states are now basis functions rather than one-electron spinors being a solution of Eq. (5.119), the creators and annihilators operate on these basis states, i.e., include them or remove them from a basis state in occupation-number representation. Nonetheless, it is said that an electron is 'created' or 'annihilated'.

Finally, in expectation values sequences of annihilation and creation operators stemming from the second-quantized Hamiltonian and from the states in bra and ket of the full bra-ket must be evaluated for which rules like Wick's theorem, which simply implements the anticommutation relations of operator pairs to obtain a relation to normal ordered operator products, can be beneficial [32, 279].

8.6.5
Fermions and Bosons

Finally, we should note that all that has been said so far is valid for fermionic annihilation and creation operators only. In the case of bosons these operators need to fulfill commutation relations instead of the anticommutation relations. The fulfillment of anticommutation and commutation relations corresponds to Fermi–Dirac and Bose–Einstein statistics, respectively, valid for the corresponding particles. There thus exists a well-established connection between statistics and spin properties of particles. It can be shown [32], for instance, that Dirac spinor fields fulfill anticommutation relations after having been quantized (actually, this result is the basis for the antisymmetrization simply postulated in section 8.5). Hence, in occupation number representation each state can only be occupied by one fermion because attempting to create a second fermion in state i, which has already been occupied, gives zero if anticommutation symmetry holds,

$$\left\{a_i^\dagger, a_i^\dagger\right\} = 2a_i^\dagger a_i^\dagger = 0 \tag{8.153}$$

This fact is thus the basis of Pauli's principle, which is therefore also termed *Pauli's exclusion principle*. In addition, we have the Fermi–Dirac statistics for other states (in occupation number representation)

$$|1_{i_1}, 1_{i_2}\rangle = a^\dagger_{i_1} a^\dagger_{i_2} |00\rangle = -a^\dagger_{i_2} a^\dagger_{i_1} |00\rangle = -|1_{i_2}, 1_{i_1}\rangle \tag{8.154}$$

8.7
Derivation of Effective One-Particle Equations

Although we have already derived the most general expression for the electronic energy of a many-electron system in Eq. (8.146) we are not yet in a position to explicitly evaluate this expression as we know neither the spinors nor the CI coefficients. In this section we derive equations to determine both. For this purpose we consider the energy expression of Eq. (8.146) as an optimization problem. Hence, we rely on the variational principle to determine the minimum of the total electronic energy with respect to the two sets of parameters, i.e., the spinors and the CI coefficients.

8.7.1
The Minimax Principle

In order to determine those parameters that parametrize our trial wave function (i.e., the one-electron spinors and the CI coefficients), we would like to employ the variational principle. The unboundedness of the one-electron Dirac operator, however, prohibits its use and we are in desperate need of a solution to this problem. In the 1980s, many authors discussed the issue of how a basis set expansion of the one-electron spinors affects the variational stability (we come back to this particular issue in chapter 10) and how this is related to the need to choose projection operators as discussed in section 8.2.3.1.

A formal solution is the so-called *minimax principle* [280], which states that the problem of variational collapse is avoided by determining the minimum of the electronic energy with respect to the large component of the spinor, while guaranteeing a maximum of the energy with respect to the small component. How such a saddle point may look has been depicted already by Schwarz and Wechsel-Trakowski [173] (see also Ref. [281]). For such basis-set expansion techniques it turned out to be decisive to fulfill the kinetic-balance condition also for the basis functions (see again chapter 10 for further reference). Interestingly, fully numerical four-component calculations have already been carried out around 1970 without encountering variational collapse. Loosely speaking, in numerical approaches it is possible to search for optimized spinors in the vicinity of the nonrelativistic solution (controllable by the spinor energy) as we shall see in chapter 9.

In view of what has been said in section 8.2.3.1, one may wonder how the projection operators are applied in variational procedures to ensure variational stability. Surprisingly, there is no *explicit* representation of the projection operators. They may be thought of as being hidden in the algorithmic structure of the actual variational procedure employed. In the fully numerical approach (chapter 9) it is the search for the electronic bound-state solutions; in the basis-set approach (chapter 10) the projectors of Eq. (8.70) may be thought of as being applied when the Fock equation is diagonalized to produce electronic bound and continuum states that are *orthogonal* to the 'positronic' states. Moreover, in the latter approach the 'positronic' states are then also not considered for the construction of the density matrix (cf. section 10.4.2 later). Although in this latter approach the 'positronic' negative-energy states are omitted from the density matrix, they nevertheless are optimized (the bound-state solutions are obtained as being orthogonal to the positronic states). We will see later in this section that the solution for a spinor requires an iterative scheme, for which one may think of the implicitly applied projection operators as being also optimized in the iterations.

Nonetheless, one may wonder what happens if the positronic states are included in the wave function, i.e., if they become occupied in a basis-set representation of the one-electron spinors. As we shall see in chapter 10, the molecular spinors are expanded for practical reasons in a set of Gaussian-type functions regardless of whether they correspond to positive or negative energies. These Gaussians are atom-centered basis functions, which would artificially lead to a strong localization of electrons in negative-energy states close to the atomic nuclei (and not everywhere in space as Dirac's hole theory would demand). The electron–electron Coulombic repulsion operators would then finally produce a highly artificial energy spectrum of one-electron states. Apart from this reason why one should *not* occupy the negative-energy states in quantum chemical calculations on molecules, it is also not desired to do so because the methods devised in the following chapters of this book are tailored to describe positive-energy states only.

After the wealth of numerical studies in the 1980s, rigorous results on the minimax principle were aimed at in the late 1990s and beginning of the new millennium [282–290]. Other studies aim at rewriting the Dirac operator such that the variational principle can be applied without precautions [291–294]. In this book, however, we will describe in chapters 9 and 10 the numerically stable approaches that have been well established in atomic and molecular physics, and then turn to elimination procedures for the small components which will naturally produce variationally stable Hamiltonians.

8.7.2 Variation of the Energy Expression

The projectors onto positive energy states are usually not explicitly implemented into a computer program or into the equations that it is to solve. Therefore, we ignore these projectors and perform the variation irrespective of the fact that variational collapse may occur. This is solely justified by the result, for which we then *a posteriori* determine how the variational collapse is avoided — along with what has just been anticipated in the preceding section.

8.7.2.1 Variational Conditions

Starting from the total electronic energy

$$E_{el,A}[\Psi_A] = E_{el,A}[\{C_{IA}\}, \{\psi_i\}] = \langle \Psi_A \mid H_{el} \mid \Psi_A \rangle \tag{8.155}$$

we determine the parameters $\{C_{IA}\}$ and $\{\psi_i\}$ of the wave function from the variational condition,

$$\delta E[\Psi_A] = \langle \delta\Psi_A \mid H_{el} \mid \Psi_A \rangle + \langle \Psi_A \mid H_{el} \mid \delta\Psi_A \rangle \stackrel{!}{=} 0 \tag{8.156}$$

The variational condition must be fulfilled by each parameter individually,

$$\frac{\partial E_{el,A}[\{C_{KA}\}, \{\psi_k\}]}{\partial C_{IA}} \stackrel{!}{=} 0 \tag{8.157}$$

$$\frac{\delta E_{el,A}[\{C_{KA}\}, \{\psi_k\}]}{\delta \psi_i} \stackrel{!}{=} 0 \tag{8.158}$$

where we have now denoted the sets of CI coefficients and orbitals as $\{C_{KA}\}$ and $\{\psi_k\}$, respectively, in order to facilitate the differentiation with respect to C_{IA} and ψ_i. However, if we want to apply the general energy expression of Eq. (8.146) we may not vary this expression freely as it was derived under the assumption that the spinors are orthonormal. The boundary condition of orthonormality imposed on the set of spinors $\{\psi_i\}$ leads to orthonormal configuration state functions Φ_I and thus to a normalized state Ψ_A (see section 8.5.4).

8.7.2.2 The CI Eigenvalue Problem

The CI eigenvalue equation, from which the CI coefficients (and the total electronic energies of the ground and excited states) are calculated, can also be derived from the general expectation value, which includes the normalization condition,

$$E_{el,A} = \frac{\langle \Psi_A \mid H_{el} \mid \Psi_A \rangle}{\langle \Psi_A \mid \Psi_A \rangle} = \frac{\sum_{KL} C^{\star}_{KA} C_{LA} \langle \Phi_K \mid H_{el} \mid \Phi_L \rangle}{\sum_{KL} C^{\star}_{KA} C_{LA} \langle \Phi_K \mid \Phi_L \rangle} \tag{8.159}$$

For the denominator of Eq. (8.159) we obtain,

$$\sum_{KL} C_{KA}^{\star} C_{LA} \langle \Phi_K \mid \Phi_L \rangle = \sum_K C_{KA}^2 \tag{8.160}$$

considering that the CSFs are orthonormal, Eq. (8.110), for orthonormal orbitals. We may now have for the stationarity condition,

$$\frac{\partial E_{el,A}[\{C_{KA}\}]}{\partial C_{IA}} = \frac{\partial}{\partial C_{IA}} \frac{\sum_{KL} C_{KA}^{\star} C_{LA} \langle \Phi_K \mid H_{el} \mid \Phi_L \rangle}{\sum_K C_{KA}^2} \stackrel{!}{=} 0 \tag{8.161}$$

which after application of the quotient rule

$$\left(\frac{u}{v}\right)' = u'\frac{1}{v} + u\left(\frac{1}{v}\right)' = \frac{u'v - v'u}{v^2} = \left(\frac{u'}{v}\right) - \left(\frac{v'}{v}\right)\frac{u}{v} \tag{8.162}$$

yields

$$\left(\frac{\sum_K C_{KA}^{\star} \langle \Phi_K \mid H_{el} \mid \Phi_I \rangle}{\sum_K C_{KA}^2}\right) - \left(\frac{C_{IA}^{\star}}{\sum_K C_{KA}^2}\right) E_{el,A} + c.c. \ = 0 \tag{8.163}$$

where $c.c.$ denotes the complex conjugate expression,

$$c.c. \equiv \left(\frac{\sum_L C_{LA} \langle \Phi_I \mid H_{el} \mid \Phi_L \rangle}{\sum_K C_{KA}^2}\right) - \left(\frac{C_{IA}}{\sum_K C_{KA}^2}\right) E_{el,A}. \tag{8.164}$$

Since both expressions are connected by complex conjugation, Eq. (8.163) is fulfilled if $c.c. = 0$, which we can write after multiplication by the denominator as

$$\sum_L \langle \Phi_I \mid H_{el} \mid \Phi_L \rangle C_{LA} = E_{el,A} C_{IA} \tag{8.165}$$

Such an equation holds for any CI coefficient. Considering the equations for *all* CI coefficients at once, we may write them in matrix form as

$$\boldsymbol{H}\boldsymbol{C}_A = E_{el,A}\boldsymbol{C}_A \tag{8.166}$$

This equation represents the CI eigenvalue equation for electronic state A. We can construct as many orhtogonal CI vectors $\boldsymbol{C}_I$ for different electronic states I as there are CSF basis functions. All of these equations can be combined to a single CI eigenvalue equation valid for all these electronic states,

$$\boldsymbol{H}\boldsymbol{C} = \boldsymbol{E}_{el}\boldsymbol{C} \tag{8.167}$$

where $\boldsymbol{E}_{el}$ is a diagonal matrix that contains the total electronic energies on the diagonal. Instead of differentiating the explicitly normalized CI energy eigenvalue, the CI matrix eigenvalue equation may also be derived by employing a

Lagrange multiplier approach, which then guarantees the orthonormality of the CSFs via explicit constraints added to the energy functional after multiplication with Lagrange multipliers.

For the solution of the CI matrix eigenvalue equations a diagonalization procedure is needed. Only a triangular part of the symmetric Hamiltonian matrix $\boldsymbol{H}$ needs to be set up. The matrix eigenvalue equation can then be solved for the CI coefficients (their eigenvectors) by a Householder transformation and a subsequent QL diagonalization of the resulting tridiagonal matrix. This method works well for $\boldsymbol{H}$ matrices with a dimension that is not too large (say, at most 10^4) and it yields all eigenvectors. For larger matrices, however, the Davidson subspace iteration techniques must be employed since $\boldsymbol{H}$ could not even be kept in the memory of a computer. Davidson's algorithm is designed for the calculation of a few eigenvectors of a large CI matrix by an iterative procedure. Handling these large eigenvalue problems only became possible with the invention of subspace iteration techniques such as the Lanczos scheme [295]. Davidson improved on these developments by introducing a preconditioner for diagonal dominant CI matrices [296–302]. Currently, it is possible to handle $\boldsymbol{H}$ matrices up to the dimension 10^{10}.

8.7.3
Self-Consistent Field Equations

To determine a set of spinors ψ_i for a given expression of the total electronic energy, we again apply the variational principle (see chapter 4). Thus, the energy as a functional of the spinors $E_{el,a}[\{\psi_i\}]$ is minimized. This variation must be carried out under the constraint that the orbitals remain orthonormal. Therefore we define a Lagrange functional L as

$$L[\{\psi_i\},\{\epsilon_{ij}\}] = E_{el,A}[\{\psi_i\}] - \sum_{ij}^{N} \epsilon_{ij}(\langle\psi_i \mid \psi_j\rangle - \delta_{ij}) \tag{8.168}$$

with ϵ_{ij} being the Lagrangian multipliers. For stationarity, the variation δL with respect to *all* its parameters must equal zero. However, we assume the CI coefficients to be fixed, which is no severe restriction as we can optimize orbitals and CI coefficients alternately (for a simultaneous optimization approach see the MCSCF procedure discussed in chapter 10). Thus, we shall determine

$$\delta_{\psi_i} L[\{\psi_i\},\{\epsilon_{ij}\}] = \delta_{\psi_i} E_{el,A} - \sum_{j}^{N} \epsilon_{ij}(\langle\delta\psi_i \mid \psi_j\rangle + \langle\psi_j \mid \delta\psi_i\rangle) \stackrel{!}{=} 0 \tag{8.169}$$

and

$$\frac{\partial L[\{\psi_i\},\{\epsilon_{ij}\}]}{\partial \epsilon_{ij}} = \langle\psi_i \mid \psi_j\rangle - \delta_{ij} \stackrel{!}{=} 0 \tag{8.170}$$

It is easily seen that the partial differentiation of $L[\{\psi_i\},\{\epsilon_{ij}\}]$ with respect to ϵ_{ij} as expressed in Eq. (8.170) simply leads to the orbital-orthonormality condition that needs to be simultaneously fulfilled with Eq. (8.169).

If we were to assume a basis set expansion for the spinor of the type of a linear combination of atomic orbitals (LCAO) we could differentiate the Lagrangian functional directly and would obtain equations in matrix form (compare the Dirac–Hartree–Roothaan equations in chapter 10). Here, we proceed in a more general way and proceed with the general method of variations. The variation of any of the matrix elements over an operator o containing ψ_i in $L[\{\psi_i\},\{\epsilon_{ij}\}]$ may be written as the limit for infinitely small variations of a given orbital ψ_i as

$$\begin{aligned} & \delta\left[\langle\psi_i|o|\psi_j\rangle + \langle\psi_j|o|\psi_i\rangle + \langle\psi_i|o|\psi_i\rangle\right] \\ = & \lim_{\delta\psi_i\to 0}\left[\frac{\langle\psi_i+\delta\psi_i|o|\psi_j\rangle + \langle\psi_j|o|\psi_i+\delta\psi_i\rangle + \langle\psi_i+\delta\psi_i|o|\psi_i+\delta\psi_i\rangle}{(\psi_i+\delta\psi_i)-\psi_i}\right. \\ & \left. - \frac{\langle\psi_i|o|\psi_j\rangle + \langle\psi_j|o|\psi_i\rangle + \langle\psi_i|o|\psi_i\rangle}{(\psi_i+\delta\psi_i)-\psi_i}\right] \end{aligned} \tag{8.171}$$

Resolving the term in brackets allows us to write this limit as

$$\begin{aligned} & \lim_{\delta\psi_i\to 0}\frac{\langle\delta\psi_i|o|\psi_j\rangle + \langle\psi_j|o|\delta\psi_i\rangle + \langle\delta\psi_i|o|\psi_i\rangle + \langle\psi_i|o|\delta\psi_i\rangle + \langle\delta\psi_i|o|\delta\psi_i\rangle}{\delta\psi_i} \\ = & \lim_{\delta\psi_i\to 0}\frac{\langle\delta\psi_i|o|\psi_j\rangle + \langle\psi_j|o|\delta\psi_i\rangle + \langle\delta\psi_i|o|\psi_i\rangle + \langle\psi_i|o|\delta\psi_i\rangle}{\delta\psi_i} \end{aligned} \tag{8.172}$$

where only the linear variations survive the limiting process. Now, the complex conjugate terms are again abbreviated as "*c.c.*" and we finally obtain for the variation of a matrix element of a general operator o that contains ψ_i

$$\begin{aligned} \delta\left[\langle\psi_i|o|\psi_j\rangle + \langle\psi_j|o|\psi_i\rangle + \langle\psi_i|o|\psi_i\rangle\right] = & \lim_{\delta\psi_i\to 0}\frac{\langle\delta\psi_i|o|\psi_j\rangle + \langle\delta\psi_i|o|\psi_i\rangle}{\delta\psi_i} \\ & + c.c. = 0 \end{aligned} \tag{8.173}$$

The variation of the Lagrangian functional with respect to the spinors can now be written as

$$\begin{aligned} \delta L[\{\psi_i\}] = & \sum_j^N \gamma_{ij}\langle\delta\psi_i \mid h^D \mid \psi_j\rangle + \frac{1}{2}\sum_{jkl}^N \Gamma_{ijkl} \times \\ & \left[\langle\delta\psi_i(1)\psi_j(2)|g(1,2)(1-P_{12})|\psi_k(1)\psi_l(2)\rangle\right. \end{aligned}$$

$$+\langle\psi_j(1)\delta\psi_i(2)|g(1,2)(1-P_{12})|\psi_k(1)\psi_l(2)\rangle\Big]$$

$$-\sum_j^N \epsilon_{ij}\langle\delta\psi_i|\psi_j\rangle + c.c. = 0 \tag{8.174}$$

After rearranging and adjusting the summation indices we obtain

$$\begin{aligned}\delta L[\{\psi_i\}] &= \sum_j^N \Big[\gamma_{ij}\langle\delta\psi_i \mid h^D \mid \psi_j\rangle \\ &\quad + \sum_{kl}^N \Gamma_{ikjl}\langle\delta\psi_i(1)\psi_k(2)|g(1,2)(1-P_{12})|\psi_j(1)\psi_l(2)\rangle \\ &\quad -\epsilon_{ij}\langle\delta\psi_i|\psi_j\rangle\Big] + c.c. = 0\end{aligned} \tag{8.175}$$

This equation holds for any variation of $\delta\psi_i$ so that we may require the remaining integrand to be zero,

$$\sum_j^N \left[\gamma_{ij}h^D\psi_j + \sum_{kl}^N \Gamma_{ikjl}\langle\psi_k(2)|g(1,2)(1-P_{12})|\psi_j(1)\psi_l(2)\rangle_2 - \epsilon_{ij}\psi_j\right] = 0 \tag{8.176}$$

at any point in space. $\langle\ldots\rangle_2$ denotes integration over the dynamical variables of electron 2. It is convenient to introduce the Fock operator f_{ij},

$$f_{ij}(\boldsymbol{r}) = \gamma_{ij}h^D(\boldsymbol{r}) + \sum_{kl}^N \Gamma_{ikjl}\left[J_{kl}(\boldsymbol{r}) - K_{kl}(\boldsymbol{r})\right] \tag{8.177}$$

with the Coulomb operator J_{kl}

$$J_{kl}\psi_j \equiv \langle\psi_k(2)|g(1,2)|\psi_l(2)\rangle_2\psi_j(1) \tag{8.178}$$

and the exchange operator K_{kl}

$$K_{kl}\psi_j \equiv \langle\psi_k(2)|g(1,2)|\psi_j(2)\rangle_2\psi_l(1) \tag{8.179}$$

so that we can write the stationarity condition as

$$\sum_j^N f_{ij}\psi_j = \sum_j^N \epsilon_{ij}\psi_j \tag{8.180}$$

This equation is the so-called self-consistent field (SCF) equation which we may rearrange for spinor ψ_i to become

$$\left[f_{ii} - \epsilon_{ii}\right]\psi_i = \sum_{j,j\neq i}^N \left[\epsilon_{ij} - f_{ij}\right]\psi_j \tag{8.181}$$

The solution functions, ψ_i, of all N SCF equations enter the definition of the Fock operator through the Coulomb and exchange integrals. Therefore, the SCF equations cannot be solved simply and an iterative procedure must be devised in order to converge a guess of functions until self-consistency is reached, hence the name (see section 8.7.5 below for further details). Note that in order to fulfill Eq. (8.170) we need to orthonormalize the solution functions (in every iteration step).

8.7.4
Dirac–Hartree–Fock Equations

The derivation has been quite general so far, i.e., we have derived the most general form of the self-consistent field equations that one may use to optimize a set of one-electron spinors for truncated CI expansions in the MCSCF approach.

In the case of a single Slater determinant employed for the approximation of the electronic state Ψ_A, all f_{ij} for $i \neq j$ vanish and the self-consistent field equations simplify to

$$f_{ii}\psi_i = \sum_j^N \epsilon_{ij}\psi_j \tag{8.182}$$

which is also called the (Dirac–)Hartree–Fock equation for spinor ψ_i. The Fock operator then reads

$$f_{ii} = h^D + \sum_k^N [J_{kk} - K_{kk}] \tag{8.183}$$

where for the one-electron density matrix elements $\gamma_{ij} = \delta_{ij}$ holds — i.e., only the diagonal entries survive and become occupation numbers in the Hartree–Fock case. Similarly, the general two-electron structure factor Γ_{ikjl} simplifies and can be omitted here.

It can be (quite easily) proved that the Fock operator f_{ii} is invariant under unitary transformations [274, p. 274–275]. Hence, we may diagonalize the matrix of Lagrangian multipliers $\boldsymbol{\epsilon} = \{\epsilon_{ij}\}$,

$$U^\dagger \boldsymbol{\epsilon} U = \tilde{\boldsymbol{\epsilon}}_{diag} \tag{8.184}$$

and obtain the *canonical* (Dirac–)Hartree–Fock equations,

$$f_{ii}(\boldsymbol{r})\tilde{\psi}_i(\boldsymbol{r}) = \tilde{\epsilon}_{ii}\tilde{\psi}_i(\boldsymbol{r}) \tag{8.185}$$

which are written as pseudo-eigenvalue equations. Note, however, that the exchange operator is a nonlocal operator which produces an 'inhomogeneous' contribution after application to $\tilde{\psi}_i(\boldsymbol{r})$.

The canonical spinors $\tilde{\psi}_i$,

$$\tilde{\psi}_i = \sum_j^N U_{ij} \psi_j \tag{8.186}$$

are linear combinations of the old spinors chosen in such a way that the matrix of Lagrangian multipliers becomes diagonal. In the following we will drop the tilde sign as it is usually clear from the context what type of orbital is being considered.

As a consequence of these considerations we note that the spinors are not unique because two sets of spinors that are related by a unitary transformation yield the same total electronic energy. Since these sets feature spinors of different spatial form, unitary transformations which effect a spatial localization of spinors are applied in quantum chemistry to provide one-electron functions that resemble the qualitative picture of orbitals widely invoked in chemistry [303].

Solving the canonical Hartree–Fock equations for a particular molecule directly yields the canonical spinors of that molecule. How the equations can be solved is discussed from a technical point of view in the next section, while their explicit solution procedures are presented in chapters 9 and 10. The terms "Dirac–Hartree–Fock", "Dirac–Fock", and "four-component Hartree–Fock" are used synonymously in research papers.

The Coulomb and exchange operators depend on the occupied spinors only. The electron–electron interaction is then represented by these two operators in a mean-field manner, which is the reason why the SCF equations are also called mean-field equations. In the case of an atom, only these two operators make the difference to the exactly solvable Dirac hydrogen atom. Since $\gamma_{ij} = \delta_{ij}$ in the Hartree–Fock case, the kinetic energy operator, $c\boldsymbol{\alpha} \cdot \boldsymbol{p}$ and the electron–nucleus potential energy operator V_{nuc} are exactly the same. In case of a molecule, the one-electron part h^D in f_{ii} differs from the Dirac Hamiltonian for the freely moving electron or from the Dirac hydrogen atom only in the explicit form of the external potential. In the general case of a molecule, the external potential contains all Coulombic interaction operators of the electron with the nuclei of the molecule as given by Eq. (8.65). It is important to note that the (geometric) structure of a molecule enters the self-consistent field equations directly through the electron–nucleus interaction (and somewhat indirectly through the spatial form of the spinors in the two-electron terms).

The relativistic four-component Hartree–Fock approach is handled analogously to the well-established nonrelativistic Hartree–Fock theory [274, 279]. Their most significant difference is the fact that the kinetic energy operators are of different form in both theories. However, in the nonrelativistic limit $c \rightarrow \infty$, the one-electron part h^D of the four-component Fock operator becomes identical to the one-component one-electron operator h^S of the (nonrelativistic) Schrödinger-type Fock operator,

$$h^S = -\hbar^2 \frac{\Delta}{2m_e} + V_{ext} = -\hbar^2 \frac{\Delta}{2m_e} + V_{nuc} \tag{8.187}$$

as can always be seen numerically in four-component calculations for unphysically large values of the speed of light. The reasons have already been discussed in section 5.4.3, where the small components of the 4-spinor ψ_i vanish for infinite speed of light c. In addition, the remaining two large components become the well-known uncoupled α- and β-spin orbitals of nonrelativistic many-electron theory.

In general, a single Slater determinant will not be a good approximation to the total electronic wave function, although the Hartree–Fock energy can approach $> 95\%$ of the exact electronic energy. Unfortunately, it turns out that the missing $< 5\%$ are decisive to relative energies in molecular science. It is therefore common to define the Hartree–Fock error as the difference between the exact and the Dirac–Hartree–Fock energy $E_{el}^{\rm DHF}$,

$$\Delta E_{corr} = E_{el,0} - E_{el}^{\rm DHF} \tag{8.188}$$

This energy difference ΔE_{corr} is called the *correlation energy*. Since all energies on the right hand side depend on the nuclear coordinates, which have been kept frozen in the electronic Schrödinger equation, the correlation energy also depends on the molecular structure, $\Delta E_{corr} = \Delta E_{corr}(\{\boldsymbol{R}_I\})$.

Because of the neglected electron correlation, four-component Dirac–Hartree–Fock calculations are as inaccurate as nonrelativistic Hartree–Fock calculations (with respect to the quality of the approximation to the electronic wave function). Attempts to reduce the correlation energy, i.e., the Hartree–Fock error, are often denoted by the phrase to 'include electron correlation'. Note that the Dirac–Hartree–Fock approach yields an approximation to the electronic ground state. Excited states cannot be treated — only in crude models. The reason for this is that a CI-like method allows one to obtain the excited states orthogonal to the ground state in a variational procedure. Technically speaking, the CI vector of an excited state is orthonormalized to the CI vector of the ground state after solution of the CI matrix eigenvalue problem.

8.7.5
The Relativistic Self-Consistent Field

The SCF equations are operator equations in which the operator depends on the solution functions. We have already anticipated that they therefore must be solved iteratively. However, the situation is not as straightforward as it might appear. The reason for this is that the simple idea — (1) to guess some spinors, (2) to calculate the one-electron integrals and with the guess-spinors the two-electron integrals, (3) and then to construct the Fock operator and solve the SCF equation for this approximate Fock operator to obtain a new set of spinors, which can then be employed again in step (2) until convergence is reached — does not yield convergence except for the simplest few-electron cases.

A simple (though not very efficient) solution is to use simple convergence control schemes which only admix the new solutions to the starting orbital set via a flexible parameter. Much more efficient schemes have been developed like the direct inversion of the iterative subspace (DIIS) by Pulay *et al.* [304–306], in which the new guess spinors for the next iteration step are expanded into a couple of previous solutions. This expansion can even very conveniently be done with the (approximate) Fock operators.

As discussed above, depending on the representation of the spinors, negative-energy states are ignored (if obtained), and hence projection operators are implicitly taken into account. Also, the spinor guess is no obstacle to these calculations. Many different types are available and quite efficient. To name only a few, (i) one may use the so-called core-Hamiltonian guess, which is obtained by solving the SCF equation neglecting the two-electron operators in the first step, (ii) or apply hydrogen-like functions with effective nuclear charge numbers in the exponent, (iii) or solve the one-electron equation with a model potential for the electron–electron interaction that does not depend on the spinors, or (iv) solve a secular equation with (parametrized) integrals as in semi-empirical or (extended-)Hückel methods [274,279].

The solution of the SCF equations involves quite a number of technical tricks decisive for actual calculations. In the end, they represent an optimization problem that can be tackled in many different ways. An interesting aspect to mention, though, is that usually only the first derivative (the linear variation) is considered. The set of spinors obtained produces a stationary energy but it is not guaranteed that this is a true minimum rather than a saddle point. Nevertheless, this possibility is usually not tested unless if a special type of MCSCF calculations is carried out which we introduce in section 10.6.

To conclude, we now have derived the equations to determine the one-electron spinors. The next two chapters will focus on how they can be solved. The explicit solution depends on the system under consideration. For atoms, numerical techniques were derived quite early on, and we will demonstrate how these methods work in principle in chapter 9. Then, chapter 10 considers the case of general molecules. But before we proceed to these chapters, we need to close the present chapter by introducing two important approaches to incorporating electron-correlation effects in quantum chemical calculations: density functional and coupled-cluster theory.

8.8 Relativistic Density Functional Theory

An interesting approach to the quantum mechanical description of many-electron systems like atoms, molecules and solids is based on the idea that it should be possible to find a quantum theory that solely refers to observable quantities. Instead relying on a wave function, such a theory should be based on the electron density. In this section, we introduce the basic concepts of this density functional theory (DFT) from fundamental relativistic principles. The equations that need to be solved within DFT are similar in structure to the SCF one-electron equations. For this reason, the focus here is on selected conceptual issues of relativistic DFT. From a practical and algorithmic point of view, most contemporary DFT variants can be considered as an improved model compared to the Hartree–Fock method, which is the reason why this section is very brief. Also, nonrelativistic DFT already embraces a large number of conjectures and proofs and it is not yet a completed theory. On the contrary, nonrelativistic DFT is still under significant development. This holds especially true for the relation to four-component DFT. The scope of this book is far too limited to elaborate on all the delicate issues and relations between nonrelativistic DFT, spin-DFT and current-density functional theory. For elaborate accounts that also address the many formal difficulties arising in the context of DFT we therefore refer the reader to excellent monographs devoted to the subject [307–309].

From what has been said already with respect to the variational collapse and the minimax principle, it is clear from the beginning that the standard derivation of the Hohenberg–Kohn theorems [310], which are the fundamental theorems of DFT based on the variational principle, must be modified compared to nonrelativistic theory [307–309]. Also, we already know that the electron density is only the zeroth component of the 4-current, and we anticipate that the relativistic, i.e., the fundamental version of DFT must rest on the 4-current and that various variants may be derived afterwards. The main issue of nonrelativistic DFT for practical applications is the choice of the exchange–correlation functional [311], an issue of equal importance in relativistic DFT [312,313] but beyond the scope of this volume.

While DFT calculations are currently of too low accuracy to be of value in atomic structure theory, their simple Hartree–Fock-like algorithmic structure makes them an ideal workhorse for molecular calculations which do not require ultimate accuracy (see section 10.4.3 for a list of references to the original research literature).

It should be mentioned that DFT is in particularly widespread use in solid-state physics because electron–correlation effects are efficiently included via additional potentials in the SCF equations (see Ref. [314] for a formulation of Dirac–Hartree–Fock equations for solids). Such potentials can be approximated quite accurately because already the free-electron gas is often a good approximation to describe the distribution of electrons in a solid. By contrast, it is more involved to consider electronic correlations in solids via the excitation hierarchies of *ab initio* methods because of the extended structure of a solid and the huge number of one-particle states resulting from this fact.

Because of the translational symmetry of a solid, plane waves as introduced in section 5.3.2 are the optimum basis functions for the representation of periodic one-particle states and electronic densities. A basis expansion can benefit from plane waves as well as from functions that are capable of reproducing the nodal structure of one-electron functions in atomic cores. For instance, the modified augmented plane wave (MAPW) method [315] combines four-component plane waves with four-component spinors for bound states, from which, finally, an expansion into spherical waves is subtracted.

8.8.1 Electronic Charge and Current Densities for Many Electrons

Within Dirac theory for a *single* electron, we may write the probability 4-current j^μ of section 5.2.3 as

$$j^\mu = c\,(\gamma^0\Psi^\dagger)\gamma^\mu\Psi \tag{8.189}$$

with the γ^μ matrices as defined in section 5.2.4. The charge-current density $j_{c,\mu}$ of an electron follows after multiplication by the negative elementary charge $q_e = -e$,

$$j_c^\mu = q_e c\,(\gamma^0\Psi^\dagger)\gamma^\mu\Psi \tag{8.190}$$

as already discussed in section 5.2.3 for the field-free case. The 4-current contains the density and the current density. We have already seen in chapter 4 how both can be deduced from the continuity equation — also for a many-electron system.

The expression for the 4-current holds also in the presence of external magnetic fields. In order to introduce this dependence explicitly, we first write the covariant form of the field-dependent Dirac equation in Eq. (5.117) as

$$\Psi = \frac{1}{m_e c}\gamma^\nu\left(i\hbar\,\partial_\nu - \frac{q_e}{c}A_\nu\right)\Psi \tag{8.191}$$

and, accordingly, its complex conjugate as

$$\Psi^\dagger = \frac{1}{m_e c}\gamma^\nu\left(-i\hbar\,\partial_\nu - \frac{q_e}{c}A_\nu\right)\Psi^\dagger \tag{8.192}$$

If we now write Eq. (8.190) somewhat artificially in a split form

$$j_c^\mu = \frac{q_e c}{2}\left[(\gamma^0\Psi^\dagger)\gamma^\mu\Psi + (\gamma^0\Psi^\dagger)\gamma^\mu\Psi\right] \tag{8.193}$$

we may replace half of the spinors in this expression by using Eqs. (8.191) and (8.192) and arrive at

$$j_c^\mu = \frac{q_e}{2m_e}\left[(\gamma^0\Psi^\dagger)\gamma^\mu\gamma^\nu\left(i\hbar\,\partial_\nu - \frac{q_e}{c}A_\nu\right)\Psi + \left(-i\hbar\,\partial_\nu - \frac{q_e}{c}A_\nu\right)(\gamma^0\Psi^\dagger)\gamma^\nu\gamma^\mu\Psi\right] \tag{8.194}$$

We may now split the 4-current into two parts,

$$j^\mu = j^{(1),\mu} + j^{(2),\mu} \tag{8.195}$$

which shall sort the terms where the summation index ν coincides or does not coincide with μ, respectively. Then, we can exploit the properties of the γ^μ matrices given in Eq. (5.51), and the two parts of the charge-current read

$$j_c^{(1),\mu} = -\frac{q_e}{2m_e}i\hbar\left([\partial_\nu(\gamma^0\Psi^\dagger)]\Psi - (\gamma^0\Psi^\dagger)\partial_\nu\Psi\right) - \frac{q_e^2}{m_e c}A_\nu(\gamma^0\Psi^\dagger)\Psi \tag{8.196}$$

and

$$j_c^{(2),\mu} = \frac{q_e}{2m_e}i\hbar\left((\gamma^0\Psi^\dagger)\gamma^\mu\gamma^\nu\partial_\nu\Psi - [\partial_\nu(\gamma^0\Psi^\dagger)]\gamma^\nu\gamma^\mu\Psi\right) - \frac{q_e^2}{c}A_\nu(\gamma^0\Psi^\dagger)\underbrace{[\gamma^\mu\gamma^\nu + \gamma^\nu\gamma^\mu]}_{=0;\ \text{Eq. (5.52)}}\Psi \tag{8.197}$$

which is the so-called *Gordon decomposition* of the 4-current.

Important for molecular science is now how the *many-electron* density and current density can be defined. Following the lines of reasoning introduced in chapter 4, we derive the density and current-density expressions for the many-electron system according to Ehrenfest's theorem employing the many-electron Dirac–Coulomb–(Breit) Hamiltonian

$$\frac{\partial}{\partial t}\langle\hat{\rho}_{\boldsymbol{r}}\rangle = \frac{i}{\hbar}\left\langle\left[\sum_{i=1}^{N}(c\boldsymbol{\alpha}\cdot\boldsymbol{p}_i), \hat{\rho}_{\boldsymbol{r}}\right]\right\rangle \tag{8.198}$$

Note that all multiplicative two-electron interaction operators cancel in the commutator so that the definition of density and current density does not depend on the choice for the approximate electron–electron interaction. The right hand side of the above equation can be simplified to become

$$\left\langle\left[\sum_{i=1}^{N}(c\boldsymbol{\alpha}\cdot\boldsymbol{p}_i), \hat{\rho}_{\boldsymbol{r}}\right]\right\rangle \overset{(4.74)}{=} \left\langle\left[\sum_{i=1}^{N}(c\boldsymbol{\alpha}\cdot\boldsymbol{p}_i), \sum_{j=1}^{N}\delta(\boldsymbol{r}_j - \boldsymbol{r})\right]\right\rangle = \left\langle\sum_{i=1}^{N}\left[c\boldsymbol{\alpha}\cdot\boldsymbol{p}_i, \delta(\boldsymbol{r}_i - \boldsymbol{r})\right]\right\rangle \tag{8.199}$$

so that Eq. (8.198) reads

$$\frac{\partial}{\partial t}\langle \hat{\rho}_{\boldsymbol{r}}\rangle = \frac{\mathrm{i}}{\hbar}\left\langle \sum_{i=1}^{N} \left[c\boldsymbol{\alpha}\cdot\boldsymbol{p}_i, \delta(\boldsymbol{r}_i - \boldsymbol{r})\right] \right\rangle \tag{8.200}$$

By noting that the electrons are physically indistinguishable, we may write

$$\frac{\partial}{\partial t}\langle \hat{\rho}_{\boldsymbol{r}}\rangle = \frac{\mathrm{i}}{\hbar}N\left\langle \left[c\boldsymbol{\alpha}\cdot\boldsymbol{p}_1, \delta(\boldsymbol{r}_1 - \boldsymbol{r})\right] \right\rangle \tag{8.201}$$

This equation can be rewritten to finally yield the current density of an N-electron system if we substitute $\boldsymbol{p}_i = -\mathrm{i}\hbar\nabla_i$,

$$\begin{aligned}\frac{\partial}{\partial t}\langle \hat{\rho}_{\boldsymbol{r}}\rangle &= cN\left\langle \Psi \middle| \boldsymbol{\alpha}\cdot\nabla_1\delta(\boldsymbol{r}_1-\boldsymbol{r}) - \delta(\boldsymbol{r}_1-\boldsymbol{r})\boldsymbol{\alpha}\cdot\nabla_1 \middle| \Psi\right\rangle \\ &= -cN\Big(\left\langle \nabla_1\Psi \middle| \boldsymbol{\alpha}\delta(\boldsymbol{r}_1-\boldsymbol{r}) \middle| \Psi\right\rangle + \left\langle \Psi \middle| \delta(\boldsymbol{r}_1-\boldsymbol{r})\boldsymbol{\alpha} \middle| \nabla_1\Psi\right\rangle\Big)\end{aligned} \tag{8.202}$$

where we exploited the antihermiticity of the ∇-operator or, in other words, an integration by parts utilizing that $\Psi^\dagger\Psi$ vanishes at the boundaries at infinity ($\pm\infty$). We may write the Ehrenfest equation now by recalling the sum rule as

$$\frac{\partial}{\partial t}\langle \hat{\rho}_{\boldsymbol{r}}\rangle = -\nabla\cdot cN\left\langle \Psi \left| \boldsymbol{\alpha}\delta(\boldsymbol{r}_1-\boldsymbol{r}) \right| \Psi\right\rangle \tag{8.203}$$

Note that we have exchanged the differentiation with respect to $\boldsymbol{r}_1$ in the integral by a differentiation of the integral with respect to $\boldsymbol{r}$ and that the time in Eq. (8.204) is the (absolute) time for the change of the total electron density (individual time-like coordinates for individual electrons are not considered). From the last equation we obtain the most general form of a continuity equation for an N-electron system,

$$\frac{\partial}{\partial t}\langle \hat{\rho}_{\boldsymbol{r}}\rangle = -\nabla\cdot c\left\langle \Psi \left| \boldsymbol{\alpha}\hat{\rho}_{\boldsymbol{r}} \right| \Psi\right\rangle \tag{8.204}$$

We may define the N-electron current-density as

$$\boldsymbol{j} \equiv c\left\langle \Psi \left| \boldsymbol{\alpha}\hat{\rho}_{\boldsymbol{r}} \right| \Psi\right\rangle \tag{8.205}$$

Density and current density are now defined for arbitrary wave functions Ψ or proper approximations to it. In the case of a single Slater determinant (SD) containing a set of N orbitals $\{\psi_i\}$ both read

$$\rho^{\mathrm{SD}}(\boldsymbol{r}) = \sum_i^N \psi_i^\dagger(\boldsymbol{r})\cdot\psi_i(\boldsymbol{r}) \tag{8.206}$$

$$\boldsymbol{j}^{\mathrm{SD}}(\boldsymbol{r}) = c\sum_i^N \psi_i^\dagger(\boldsymbol{r})\cdot\boldsymbol{\alpha}\cdot\psi_i(\boldsymbol{r}) \tag{8.207}$$

As one would demand, in the simple case of a single electron, where $\Psi \rightarrow \psi(\boldsymbol{r})$, Eq. (8.204) reduces to the one-particle continuity equation,

$$\frac{\partial}{\partial t}\left(\psi^{\dagger}\cdot\psi\right) = -\nabla\cdot c\left(\psi^{\dagger}\cdot\boldsymbol{\alpha}\cdot\psi\right) \tag{8.208}$$

To summarize, the electron density and the current density can be expressed in terms of one-electron spinors, and consequently we may benefit a lot from all that has already been elaborated in section 8.1 about the interaction of two electrons.

8.8.2
Current-Density Functional Theory

The basic idea of DFT is to deduce a theory that provides quantum mechanical observables solely on the basis of the electron density or, more precisely, on the basis of the 4-current rather than on the relativistic wave function. Naturally, one would like to base this deduction on QED [313]. However, in the spirit of this book, we would be satisfied by a 'first-quantized' semi-classical theory.

Hohenberg and Kohn [310] derived two theorems in the realm of nonrelativistic quantum mechanics, which state that the energy can be solely calculated from the electron density and that any trial density produces an energy, which is an upper bound to the desired exact energy. The proofs are quite straightforward and not involved. However, they also provoked scepticism, which inspired rigorous work on the mathematical foundations of DFT. We can, however, not delve deeper into these issues here.

Rajagopal and Callaway [316] gave a relativistic generalization of the Hohenberg–Kohn theorem, in which the 4-current is treated like the electron density in the original nonrelativistic formulation of Kohn and Hohenberg. The ground state electronic energy is then written as a functional of the 4-current,

$$E_{el}[j^{\mu}] = T[j^{\mu}] + V_{nuc}[j^{\mu}] + J[j^{\mu}] + E_{XC}[j^{\mu}] \tag{8.209}$$

where $T[j^{\mu}]$ is the kinetic energy functional, $V_{nuc}[j^{\mu}]$ is the external potential energy of the interaction of the electronic 4-current j^{μ} with all nuclei, and $J[j^{\mu}]$ is the classical Coulomb repulsion energy of the electrons as already deduced for two electrons in Eq. (8.31), while $E_{XC}[j^{\mu}]$ denotes an exchange–correlation current-density functional, which contains all quantum, i.e., exchange and correlation effects. For the sake of simplicity, we assume that no external electromagnetic potentials are present other than the monopole potential of the atomic nuclei. Unfortunately, $E_{XC}[j^{\mu}]$ exists but is not known in analytical form, which is the reason why present-day DFT is hampered by more or less accurate approximations to $E_{XC}[j^{\mu}]$.

8.8.3
The Four-Component Kohn–Sham Model

We have already seen in section 8.1 that (i) a Dirac electron with electromagnetic potentials created by all other electrons [see Eqs. (5.119) and (8.2)] cannot be solved analytically, which is the reason, why the total wave function as given in Eq. (8.4) cannot be calculated, and also that (ii) the electromagnetic interactions may be conveniently expressed through the 4-currents of the electrons as given in Eq. (8.31) for the two-electron case. Now, we seek a one-electron Dirac equation, which can be solved exactly so that a Hartree-type product becomes the *exact* wave function of this system. Such a separation requires in order to be exact (after what has been said in section 8.5), a Hamiltonian, which is a sum of strictly local operators. The local interaction terms may be extracted from a 4-current based interaction energy such as the one in Eq. (8.31). Of course, we need to take into account Pauli exchange effects that have been omitted in section 8.1.4, and we also need to take care of electron correlation effects. This leads us to the Kohn–Sham (KS) model of DFT.

Usually, the Kohn–Sham model is introduced differently, as outlined below, based on physical ideas rather than on rigorous mathematical foundations, which have been given though we cannot reproduce them because of the limited scope of this book. One may think that both electron density and current density depending only on a single set of coordinates might provide simpler means to calculate physical observables than the employment of complicated electronic wave functions, which after all even depend on N sets of coordinates. The situation is, however, not that simple because one needs to construct such densities. The basic idea of Kohn and Sham [317] is to introduce a surrogate fermionic system, which is *noninteracting*. Hence, it can be represented by a Hamiltonian with at most *local* potential energy operators. Then, the wave function belonging to this 'artificial' fermionic system can be exactly chosen as an antisymmetrized Hartree product. Thus, we define a Kohn–Sham Slater determinant as the exact wave function of the noninteracting fermionic system, from which the electron and current densities are calculated through Eqs. (8.206) and (8.207).

The question now is how the spinors in the Kohn–Sham Slater determinant are obtained in order to compute the densities and how the densities then enter a functional — a function of a function — that allows us to obtain the electronic energy. Under all precautions discussed *in extenso* for the energy expectation value calculated from unbounded operators, a functional derivative may be calculated from the energy functional of Eq. (8.209) similarly to the derivation given in section 8.7. In 1973, Rajagopal and Callaway [316] derived in this way the most general relativistic KS-DFT equations,

$$\left[c\boldsymbol{\alpha}\cdot\left(\boldsymbol{p}-\frac{q_e}{c}\boldsymbol{A}_{\text{eff}}(\boldsymbol{r})\right)+\beta m_e c^2+q_e\phi_{\text{eff}}(\boldsymbol{r})\right]\psi_i(\boldsymbol{r})=\epsilon_i\psi_i(\boldsymbol{r}) \tag{8.210}$$

based on the variational principle. Like the SCF equations of section 8.7, these equations can be used for the determination of the four-component KS orbitals ψ_i. Note that the rest energy $\beta m_e c^2$ has not been shifted by $-m_e c^2$ in this case. The effective potential terms depend on the electronic 4-current j^μ

$$\phi_{\text{eff}}(\boldsymbol{r}) = \phi_{\text{ext}}(\boldsymbol{r}) + \frac{q_e}{c}\int d^3r' \frac{j^0(\boldsymbol{r}')}{|\boldsymbol{r}-\boldsymbol{r}'|} + c\frac{\delta E_{XC}[j^\mu]}{\delta j^0(\boldsymbol{r})} \tag{8.211}$$

$$\boldsymbol{A}_{\text{eff}}(\boldsymbol{r}) - \boldsymbol{A}_{\text{ext}}(\boldsymbol{r}) + \frac{q_e}{c}\int d^3r' \frac{\boldsymbol{j}(\boldsymbol{r}')}{|\boldsymbol{r}-\boldsymbol{r}'|} + c\frac{\delta E_{XC}[j^\mu]}{\delta \boldsymbol{j}(\boldsymbol{r})} \tag{8.212}$$

Besides the exchange–correlation energy functional E_{XC}, we recognize the familiar interaction integrals of Eq. (8.31). The external potentials contain the electron–nucleus interaction as usual. Note especially that the KS equations also resemble the exact Eqs. (8.2) and (8.3), but now for a system of noninteracting fermions!

A Gordon decomposition of the spatial current density provides orbital and spin parts such that the neglect of the former and the introduction of the magnetization $\boldsymbol{m}$,

$$\boldsymbol{m}(\boldsymbol{r}) = \sum_i \psi_i^\dagger(\boldsymbol{r}) \cdot \boldsymbol{\sigma} \cdot \psi_i(\boldsymbol{r}) \tag{8.213}$$

for the latter yields four-component KS-DFT equations [318] which are analogous to their counterparts in nonrelativistic spin-KS-DFT [307]. The magnetization emerges naturally in the framework of relativistic DFT, and it is important to note that the magnetization couples to the vector potential terms.

In the case of the absence of external magnetic fields, Saue and Helgaker [319] emphasized that for four-component quantum chemical calculations the current density would only occur if the Gaunt term of the Breit interaction was considered — which is clear from what has just been said and especially from the discussion in section 8.1.4. Only the current-density would couple to the vector potential, which we assumed to be absent. Accordingly, the description is restricted to the Dirac–Coulomb Hamiltonian, where projection onto the positive-energy states to produce a suitable no-pair operator is to be introduced at a convenient stage, then it will be sufficient to calculate the KS spinors and total energies from the electronic density *alone*, in accord with section 8.1.4. Saue and Helgaker argue [319] that it may be advantageous in view of the approximate nature of present-day functionals to introduce spin-dependent functionals as proposed by MacDonald and Vosko [318] and as applied for a long time in nonrelativistic spin-DFT [307].

8.8.4
Noncollinear Approaches and Collinear Approximations

Also in four-component DFT, spin is no longer a good quantum number because of spin–orbit coupling. It became common to express this fact in the language of solid state physics: *collinear* denotes the existence of an external quantization axis for the spin and can only be justified if spin is a good quantum number. If we restrict a relativistic spin-DFT framework to the z-component of the magnetization, this will mean employing a fixed external quantization axis for the spin, and is then a collinear approach. The z-component of the magnetization is then essentially the same as the *spin density*, which is nothing else but the sum of squared orbitals of α-spin minus the sum of squared β-orbitals in the collinear approach.

However, a spin-operator-dependent Hamiltonian like any of the above mentioned four-component KS Hamiltonians or like the Dirac–Coulomb Hamiltonian in general does not commute with spin operators, so that the collinear approach, which requires spin to be a good quantum number, cannot be applied. In the *noncollinear* approach one may define a spin density which is equal to the length of the magnetization vector and thus always positive. Also, this spin density is invariant under rotations in spin space as the length of the magnetization vector would be invariant. This is a convenient feature regarding the invariance properties of the exchange–correlation contribution to the total DFT energy, as discussed in Ref. [320].

8.9
Completion: The Coupled-Cluster Expansion

So far, we have considered variational wave functions. An important approach to approximate N-electron wave functions is the coupled-cluster (CC) method [321], which is introduced only briefly for reasons of space: for detailed presentations we refer the reader to Refs. [217,322–324]. In practice, the CC wave function is determined by projection rather than variationally. In order to understand this, we first reconsider the CI ansatz and reformulate the FCI wavefunction of Eq. (8.101) in terms of the excitation and de-excitation operators introduced in section 8.6. Using the language of second quantization and assuming the so-called intermediate normalization, $C_0 = 1$, we obtain

$$\Psi^{\mathrm{FCI}} = \Phi_0 + \sum_{ia} C_i^a a_a^\dagger a_i \Phi_0 + \sum_{ijab} C_{ij}^{ab} a_b^\dagger a_j a_a^\dagger a_i \Phi_0 + \sum_{ijkabc} C_{ijk}^{abc} a_c^\dagger a_k a_b^\dagger a_j a_a^\dagger a_i \Phi_0 \cdots \tag{8.214}$$

where we have omitted the state index A. The operator a_i annihilates an electron in the spinor ψ_i, which is occupied in the reference determinant Φ_0. Accordingly, the operator $a_a^\dagger$ creates an electron in the virtual, i.e., in the reference determinant Φ_0 unoccupied spinor ψ_a. The expansion of Eq. (8.214) can be rewritten convieniently as

$$\Psi^{\rm FCI} = (1 + A_1 + A_2 + A_3 + \cdots)\,\Phi_0 = \left(\sum_{I=0}^{N} A_I\right)\Phi_0 \tag{8.215}$$

The A_I are operators that contain the CI coefficients and generate all excitations, i.e., singles,

$$A_1 = \sum_{ia} C_i^a a_a^\dagger a_i \tag{8.216}$$

doubles,

$$A_2 = \sum_{ijab} C_{ij}^{ab} a_a^\dagger a_i a_b^\dagger a_j \tag{8.217}$$

and so on. For the CC formalism we introduce the excitation operator T, which is, in analogy to Eq. (8.215), given by

$$T = T_1 + T_2 + T_3 + \cdots + T_N = \sum_I^N T_I \tag{8.218}$$

Here, T_1 generates all single excitations, T_2 all double excitations and so forth. They are explicitly defined as

$$T_1 = \sum_{ia} t_i^a a_a^\dagger a_i \tag{8.219}$$

$$T_2 = \sum_{ijab} t_{ij}^{ab} a_b^\dagger a_j a_a^\dagger a_i \quad \text{etc.} \tag{8.220}$$

where the so-called cluster amplitudes t_i^a, t_{ij}^{ab}, ... have been introduced in order to distinguish them formally from the CI coefficients of Eq. (8.214). The coupled-cluster approach is in contrast to the linear expansion of a CI ansatz based on an exponential ansatz

$$\Psi^{\rm CC} = \exp(T)\Phi_0 \tag{8.221}$$

The exponential of the operator T is defined by its Taylor series

$$\begin{aligned} \exp(T)\Phi_0 = &\Big(1 + T_1 + T_2 + T_3 + \cdots + \frac{1}{2!}T_1^2 + T_1T_2 + \frac{1}{3!}T_1^3 + \frac{1}{4!}T_1^4 \\ &+ \frac{1}{2!}T_1^2T_2 + T_1T_3 + \frac{1}{2!}T_2^2 + \cdots\Big)\Phi_0 \end{aligned} \tag{8.222}$$

where products of excitation operators appear in contrast with the linear CI ansatz. Examination of the structure of T_1^2,

$$T_1^2 = \sum_{ijab} t_i^a t_j^b a_b^\dagger a_j a_a^\dagger a_i \tag{8.223}$$

which is the simplest representative containing operator products, shows that T_1^2 creates double excitations with cluster amplitudes given by $t_i^a t_j^b$, i.e., a product of cluster amplitudes originally belonging to single excitations. Double excitations are therefore generated not only by T_2, but also by T_1^2. T_2 in Eq. (8.219) features cluster amplitudes of the kind t_{ij}^{ab}, which are denoted *connected* cluster amplitudes, whereas $t_i^a t_j^b$ are called *disconnected* cluster amplitudes. As can be seen on the basis of Eq. (8.214) the linear CI expansion does not provide a counterpart to the disconnected cluster amplitudes. The relation between the CI coefficients C_i^a and C_{ij}^{ab} of Eq. (8.214) and the cluster amplitudes generating single and double excitations can be derived by comparing Eqs. (8.215) and (8.222),

$$C_i^a = t_i^a \tag{8.224}$$

$$C_{ij}^{ab} = t_{ij}^{ab} + t_i^a t_j^b - t_i^b t_j^a \tag{8.225}$$

In principle, the coupled-cluster ansatz for the wave function is exact if the excitation operator in Eq. (8.218) is not truncated. But this defines an FCI approach, which is unfeasible in actual calculations on general many-electron systems. A truncation of the CC expansion at a predefined order in the excitation operator T is necessary from the point of view of computational practice. Truncation after the single and double excitations, for instance, defines the CCSD scheme. However, in contrast with the linear CI ansatz, a truncated CC wave function is still size consistent, because all disconnected cluster amplitudes which can be constructed from a truncated set of connected ones are kept [324]. The maximum excitation in T determines the maximum connected amplitude, and the lower-order excitations in T generate the disconnected cluster amplitudes. However, the nonlinearity of the coupled-cluster ansatz as indicated by Eq. (8.222) prevents the determination of the amplitudes by a variational scheme. For this reason, they are solved by projection (see below).

For actual applications of the cluster ansatz even in its truncated form it is necessary that no infinite sums as in Eq. (8.222) occur. Fortunately, we can benefit from the properties of the creation and annihilation operators. In order to understand how, we write the electronic Schrödinger equation for the coupled-cluster wave function,

$$H_{el} \exp(T)\Phi_0 = E \exp(T)\Phi_0 \tag{8.226}$$

which can be conveniently re-written after multiplication by $\exp(-T)$ from the left,

$$\exp(-T)H_{el}\exp(T)\Phi_0 = E\Phi_0 \tag{8.227}$$

Now, we observe that the transformed Hamiltonian contains only a finite number of terms,

$$\begin{aligned}\exp(-T)H_{el}\exp(T) &= H_{el} + [H_{el},T] + \frac{1}{2!}[[H_{el},T],T] + \frac{1}{3!}[[[H_{el},T],T],T] \\ &\quad + \frac{1}{4!}[[[[H_{el},T],T],T],T]\end{aligned} \tag{8.228}$$

once we have employed the definition, i.e., the expansion of the exponential twice. The latter expansion is known as the Baker–Campbell–Hausdorff formular or, in short, Hausdorff formula. The fact that the Baker–Campbell–Hausdorff formula truncates can be understood as follows. A general creation–annihilation operator pair $a_p^\dagger a_q$ as occurring in the second-quantized electronic Hamiltonian of Eq. (8.140) does not commute with such a pair from the excitation operator in Eq. (8.218),

$$\begin{aligned}a_p^\dagger a_q a_a^\dagger a_i &= \delta_{aq}a_p^\dagger a_i - a_p^\dagger a_a^\dagger a_q a_i \\ &= \delta_{aq}a_p^\dagger a_i - a_a^\dagger a_p^\dagger a_i a_q \\ &= \delta_{aq}a_p^\dagger a_i - \delta_{pi}a_a^\dagger a_q + a_a^\dagger a_i a_p^\dagger a_q\end{aligned} \tag{8.229}$$

so that we obtain for the commutator

$$[a_p^\dagger a_q, a_a^\dagger a_i] = \delta_{aq}a_p^\dagger a_i - \delta_{pi}a_a^\dagger a_q \tag{8.230}$$

Hence, only pairs of creation and annihilation operators survive. Moreover, when the full second-quantized Hamiltonian of Eq. (8.140) enters Eq. (8.228) four such elimination steps leave only the excitation/de-excitation operators of the cluster ansatz for the wave function, like $a_a^\dagger a_i$ which commute with T.

The coupled-cluster equations are then obtained by projection. While the cluster amplitudes are obtained from equations like

$$\begin{aligned}\Big\langle \Phi_{ijk\ldots}^{abc\ldots} \Big| H_{el} + [H_{el},T] + \frac{1}{2!}[[H_{el},T],T] + \frac{1}{3!}[[[H_{el},T],T],T] \\ + \frac{1}{4!}[[[[H_{el},T],T],T],T] \Big| \Phi_0 \Big\rangle = 0\end{aligned} \tag{8.231}$$

the energy can then afterwards be evaluated from the ordinary expectation value

$$E = \langle \exp(T)\Phi_0 | H_{el} | \exp(T)\Phi_0 \rangle$$

$$= \left\langle \Phi_0 \middle| H_{el} + [H_{el}, T] + \frac{1}{2!}[[H_{el}, T], T] + \frac{1}{3!}[[[H_{el}, T], T], T] \right.$$
$$\left. + \frac{1}{4!}[[[[H_{el}, T], T], T], T] \middle| \Phi_0 \right\rangle \qquad (8.232)$$

After having presented some very basic concepts of the coupled-cluster ansatz, we should emphasize that coupled-cluster theory is a broad and highly sophisticated approach. It can, of course, be extended to the calculation of excited states and molecular properties, but for these more detailed treatments the reader is advised to consult the books referred to at the beginning of this section in Refs. [217, 322–324]. The coupled-cluster approach is currently one of the most accurate electronic structure methods for ground-state, close to the minimum structure, single-configuration molecules. However, perturbatively treated triple excitations in the CCSD(T) variant is essential for sufficient accuracy. Single-reference coupled-cluster methods have been implemented not only for the four-component many-electron Hamiltonian introduced in this chapter [325, 326], but are also available for many of the more approximate Hamiltonians to be introduced in chapters 11–14. Since especially the latter requires a mere interfacing of a nonrelativistic coupled-cluster scheme for the wave function with some choice of an approximate relativistic one-electron operator in the many-electron Hamiltonian, we refrain from giving any details here. Instead, we should mention a special coupled-cluster technique, which has proved to be very valuable in four-component calculations.

If the electronic structure of a molecule is not even qualitatively well described by a single electronic configuration, a multi-reference ansatz is also required for the coupled-cluster approach. One option is the so-called Fock-space coupled-cluster (FSCC) approach [327–329]. Details of the method are described (for atoms) in Refs. [154, 155, 330–333]. For instance, the hydrides SnH_4 [334] and CdH [335] have been studied with this approach. The latest development is the so-called Intermediate Hamiltonian Fock-space coupled-cluster method [336, 337], which yields highly accurate results due to an increased size of the model space. Its most convenient feature, however, is its improved convergence.

Further Reading

Many excellent introductory books are available on quantum chemistry and its different approaches to approximating the true wave function of a many-electron system. We do not intend to give a full list of these monographs and textbooks as there are so many of them. Instead, we refer to some prominent examples.

H. A. Bethe, E. E. Salpeter, [48,95]. *Quantum Mechanics of One- and Two-Electron Atoms.*

This classic monograph by Bethe and Salpeter should also be recommended as further reading. Since it considers two-electron, i.e., helium-like atoms, it contains a discussion of the Breit equation. However, this discussion is mostly from the point of view of quantum electrodynamics. It does not rest on Darwin's classical potential energy expression subjected to the correspondence principle. While the former is the more fundamental point of view, in quantum chemistry the latter is sufficient as we neglect or model many physical effects of little importance in chemistry (like radiative corrections) anyway. However, this book also contains much material on how radiative corrections can be considered, if this is desired.

R. McWeeny, [279]. *Methods of Molecular Quantum Mechanics.*

This book represents an excellent account of the basics of nonrelativistic quantum chemistry. All essential concepts of many-electron theory are introduced. It is extremely useful, as relativistic, first-quantized quantum chemistry heavily exploits the historically older nonrelativistic quantum chemistry.

F. Pilar, [274] *Elementary Quantum Chemistry.*

This is a standard quantum chemistry textbook for beginners. It can be recommended to anyone who is just about to start familiarizing himself or herself with the concepts and ideas of quantum chemical many-electron theory.

H. Primas, U. Müller-Herold, [303]. *Elementare Quantenchemie.*

Primas and Müller-Herold have written one of the most concise and precise introductions to nonrelativistic quantum chemistry. Unfortunately, this inspiring text, which reaches beyond the standard technique-dominated presentations, is only available in German.

A. L. Fetter, J. D. Walecka, [338]. *Quantum Theory of Many-Particle Systems.*

The book by Fetter and Walecka is a classic physics text on the quantum mechanical treatment of many-particle systems, which makes extensive use of the language of second quantization.

9
Many-Electron Atoms

The spherical symmetry of the attractive nuclear potential of atoms in the absence of additional fields allows one to formulate the many-electron problem in the radial coordinate only. It is instructive to study the difference between these many-electron equations and the single-electron Dirac hydrogen atom from chapter 6. In addition, analytical results can be obtained for the unkown atomic spinors in the short- and long-range limit that also serve well for spherical averages in the case of molecules. Last but not least, atomic spinors are also a good starting point for establishing approximations to molecular spinors discussed in the following chapter.

In the preceding chapter we introduced the basic framework of an essentially first-quantized relativistic theory for many-electron systems like atoms, molecules or molecular aggregates. Chapter 8 ended with general one-electron mean-field equations to be solved for the one-electron spinors from which the approximation to the total electronic wave function is constructed. The explicit solution of such mean-field equations depends on the system to be investigated. In the case of isolated atoms the radial symmetry of the external potential can be exploited to a large extent as we shall see in this chapter. As a consequence, the mean-field partial differential equations become sets of coupled ordinary differential equations of first order, which can be conveniently solved numerically on a mesh of grid points as demonstrated in the following sections. The next chapter is then devoted to the case of general molecules.

At first sight, the simplest many-electron systems are many-electron atoms. However, the spherical symmetry of atoms implies that all Dirac one-particle states can be classified according to the quantum numbers of total angular momentum because the Hamiltonian and the square of the total angular momentum operator commute. Consequently, symmetry can be extensively exploited in the set-up of the many-electron configuration state functions. Moreover, this leads to a fully analytical treatment of all parts of the formalism which depend on angular and spin degrees of freedom, so that highly accurate numerical methods can be employed for the solution of the one-particle self-consistent field equations. In fully numerical schemes, all functions are represented on a grid of points distributed in space. We will see in the follow-

Relativistic Quantum Chemistry. Markus Reiher and Alexander Wolf

ISBN: 978-3-527-31292-4

ing that many results obtained for hydrogen-like atoms in chapter 6 transfer to the case of many-electron atoms. Of course, significant extensions are necessary, and the most important result will be that the radial equations cannot be solved analytically for the many-electron case.

Important landmarks in the history of relativistic atomic structure can be found in the literature [188, 339, 340], of which we mention only a few explicitly. Relativistic theory of atoms with more than one electron started as early as in 1935 when Bertha Swirles [84] applied the Hartree–Fock formalism to the Dirac equation. Because of the lack of computers in those days, only a few calculations could be carried out. The situation changed in the 1960s when Grant used Racah's tensor algebra [341–344] for the analytic integration over all angular terms and derived a general expression for the total electronic energy of a closed-shell atom in the central-field approximation. From this expression he deduced self-consistent field equations for the determination of spinors [89, 345, 346]. In the late 1960s pioneering Dirac–Hartree–Fock and four-component DFT calculations were performed on heavy and super-heavy atoms [347–350]. In 1973, Desclaux [351] calculated accurate spinor energies, total energies, and other expectation values fully numerically for nearly all neutral atoms of the periodic table in the Dirac–Hartree–Fock model for closed shells and configuration averages. This study was later repeated and extended by Visscher and Dyall in the mid 1990s [352].

Our capabilities to investigate many-electron systems depend on the availability of computer programs for the solution of SCF-type equations, which are complicated coupled differential equations. Therefore, we should recall some of the cornerstones of these technical developments. Desclaux published a program for multi-configuration SCF calculations in 1975 [353] that was continuously extended [179, 354, 355]. Then, in 1980 Grant and collaborators published their multi-configuration SCF code [356, 357] which was later reorganized to form the GRASP package [358]. The numerical solution methods they used are comparable to those applied in Desclaux's code [359]. Parpia *et al.* extended the GRASP program to facilitate large-scale CI calculations [360]). Additionally, work has been done on relativistic basis set calculations for atoms within the past decades (see, for examples, Refs. [142, 145, 148, 152–155, 330–333, 361–365]). Basis set expansions of the four-component atomic spinors are more difficult to control with respect to accuracy when compared with numerical grid-based methods, but they provide virtual atomic spinors in the diagonalization of the SCF equations, which can immediately be employed for CI-type wave functions in electron correlation methods. Moreover, such expansion techniques are an appropriate starting point for the molecular theory and are therefore discussed in chapter 10.

In the context of numerical solution methods, we should also mention the special case of *linear* molecules for which fully numerical methods are feasible and efficient. Because of their high point group symmetry, $D_{\infty h}$ and $C_{\infty v}$, the total electronic wave function is still an eigenfunction of one component of the total angular momentum operator. Exploiting the symmetry of this special class of molecules allows us to efficiently use purely numerical methods as well. Fully numerical grid-based approaches have been tested for Dirac–Fock calculations on diatomics [366–369]. The finite-element method has also been tested for Dirac–Fock and four-component Kohn–Sham calculations for few-electron diatomics [370–374]. These numerical techniques are able to provide accurate Dirac–Hartree–Fock results and thus yield reference data for comparisons with more approximate basis-set approaches (to be discussed in the next chapter). The limits of such numerical techniques for molecules, however, are the facts that (i) four-component numerical schemes for molecules with more than two atoms have not yet been established (such schemes have mainly been developed for nonrelativistic calculations [375–379]), (ii) molecular structure optimizations and general gradient techniques will be clumsy because gradients and approximates to Hessian matrices must be calculated numerically, (iii) virtual orbitals, which are needed for correlated methods, are not obtained automatically by contrast to basis-set approaches, and, finally, (iv) convergence of iterative solution methods is usually a troublesome issue in fully numerical methods.

After this short (and unavoidably selective) overview and introduction we shall now work out the theory of relativistic atomic structure calculations.

9.1 Transformation of the Many-Electron Hamiltonian to Polar Coordinates

We have seen in chapter 8 that the many-electron Dirac–Coulomb(–Breit) Hamiltonian can be written as

$$\hat{H}_{el} = \sum_{i}^{N} h^D(i) + \sum_{j<i}^{N} g(i,j) \tag{9.1}$$

consisting of 4×4 one- and two-particle operators, $h^D(i)$ and $g(i,j)$, respectively (the nucleus–nucleus repulsion term V_0 of Eq. (8.66) vanishes in the case of a single atom, of course). The nucleus is considered to be 'clamped' and its coordinate is chosen to coincide with the origin of the coordinate system. The operators $h^D(i)$ are of the form of the Dirac one-electron Hamiltonian of the hydrogen atom. They contain operators for the kinetic energy of each electron and its attraction by the atomic nucleus. Each $h^D(i)$ depends on one set of electron coordinates i only. In the case of atoms, the nucleus–electron attraction operator is a single term and $h^D(i)$ is identical in form to the Dirac Hamiltonian of hydrogen-like atoms and ions. Its explicit form in polar coordinates has already been derived for hydrogen-like atoms in Eq. (6.55) of chapter 6. The 4×4 matrices $g(i,j)$ account for the electron–electron interaction.

In order to exploit the spherical symmetry of an atom, the Hamiltonian needs to be transformed to polar coordinates first. This transformation was given for $h^D(i)$ Eq. (6.65). But the electron–electron interaction operators $g(i,j)$ also need to be transformed to polar coordinates in order to allow us to separate the radial from the angular coordinates in the ansatz for the one-electron functions like the one in Eq. (6.51).

9.1.1
Comment on Units

A comment on units is appropriate. In Hartree atomic units, all equations adopt a rather short form when all constants that take a value of one are omitted ('$\hbar$=e=m_e=$4\pi\epsilon_0$=1'). In this book we employ Gaussian units, so that only $4\pi\epsilon_0$=1 holds. The interaction energy operators in atoms contribute many terms, and to multiply all of them by the product of the interacting charges would make the equations quite clumsy. In the rest of this chapter we therefore leave out the product of the charges $q_e^2 = e^2$, which would be equal to one in Hartree atomic units anway. In any unit system used in this book, the speed of light c adopts a value determined by experiment. In atomic units this is approximately 137.04 a.u. Of course, any quantum mechanical result will then always depend on the value chosen for the speed of light and for actual calculations this should be given with sufficiently many significant figures after the decimal point (see section 9.8).

9.1.2
Coulomb Interaction in Polar Coordinates

The simplest form of the two-electron interaction operator, $g(i,j) = 1/r_{ij}$, will be considered first. It is sufficient to pick out any two electronic coordinates 1 and 2 as in $1/r_{12}$ because of the indistinguishability of the electrons (cf. section 8.5). The $1/r_{12}$ Coulomb operator,

$$\frac{1}{r_{12}} = \frac{1}{|\boldsymbol{r}_1 - \boldsymbol{r}_2|} = \frac{1}{\sqrt{(\boldsymbol{r}_1 - \boldsymbol{r}_2)^2}} \tag{9.2}$$

can be rewritten as

$$\frac{1}{r_{12}} = \frac{1}{\sqrt{\boldsymbol{r}_1^2 + \boldsymbol{r}_2^2 - 2\boldsymbol{r}_1\boldsymbol{r}_2}} = \frac{1}{\sqrt{r_1^2 + r_2^2 - 2r_1 r_2 \cos\vartheta_{12}}} \tag{9.3}$$

which can then be expressed as

$$\frac{1}{r_{12}} = \frac{1}{r_>}\sqrt{1 + \left(\frac{r_<}{r_>}\right)^2 - 2\left(\frac{r_<}{r_>}\right)\cos\vartheta_{12}} \tag{9.4}$$

where $r_< = \min\{r_1, r_2\}$ and $r_> = \max\{r_1, r_2\}$. We will later see in section 9.3.4 how 'the larger r', $r_>$, and 'the smaller r', $r_<$, are resolved in calculations. Taylor series expansion [380]

$$\frac{1}{\sqrt{1+x^2-2x\cos\vartheta}} = 1+(\cos\vartheta)x+\frac{1}{2}(3\cos\vartheta^2-1)x^2+\cdots \tag{9.5}$$

$$= \sum_{\nu=0}^{\infty} x^{\nu} P_{\nu}(\cos\vartheta) \tag{9.6}$$

results in a sum of Legendre polynomials $P_\nu(\cos\vartheta)$ and can be utilized to obtain

$$\frac{1}{r_{12}} = \sum_{\nu=0}^{\infty} \frac{r_<^{\nu}}{r_>^{\nu+1}} P_\nu(\cos\vartheta_{12}) \tag{9.7}$$

From the addition theorem in Eq. (4.125),

$$P_\nu(\cos\vartheta_{12}) = \frac{4\pi}{2\nu+1} \sum_{m=-\nu}^{\nu} Y^{\star}_{\nu m}(\vartheta_1,\varphi_1) Y_{\nu m}(\vartheta_2,\varphi_2) \tag{9.8}$$

it follows that

$$\frac{1}{r_{12}} = \sum_{\nu=0}^{\infty} \frac{r_<^{\nu}}{r_>^{\nu+1}} \frac{4\pi}{2\nu+1} \sum_{m=-\nu}^{\nu} Y^{\star}_{\nu m}(\vartheta_1,\varphi_1) Y_{\nu m}(\vartheta_2,\varphi_2) \tag{9.9}$$

Hence, the inverse of the interelectronic distance can be conveniently separated in polar coordinates. This represents the multipole expansion of the two-electron interaction. It may already be anticipated that it will be important to be able to solve integrals containing the angular two-electron operator $Y^{\star}_{\nu m}(\vartheta_1,\varphi_1) Y_{\nu m}(\vartheta_2,\varphi_2)$ analytically. We shall come back to this issue in section 9.3.2.

9.1.3
Breit Interaction in Polar Coordinates

By contrast to the transformation of standard Coulomb interaction operators to polar coordinates, this task is much more involved in the case of the Breit interaction discussed in section 8.1. The Breit operator is symmetric with respect to an exchange of the full electron coordinate sets of electrons 1 and 2. This symmetry is partially lost if the radial terms are considered separately, i.e., when only two variables, r_1 and r_2, remain. Grant and Pyper [138] derived a decomposition of the frequency-dependent Breit operator in spherical coordinates,

$$B_\omega(1,2) = B_\omega^{(1)}(1,2) + B_\omega^{(2)}(1,2) \tag{9.10}$$

with

$$B_\omega^{(1)}(1,2) = \sum_L \frac{4\pi}{2L+1} \sum_{\nu=L-1}^{L+1} v_{\nu L} V_\nu(1,2) \boldsymbol{\alpha}_1 Y_{L\nu}(\vartheta_1,\varphi_1) \boldsymbol{\alpha}_2 Y_{L\nu}(\vartheta_2,\varphi_2) \quad (9.11)$$

$$\begin{aligned} B_\omega^{(2)}(1,2) &= \sum_L w_L \frac{4\pi}{2L+1} \\ &\times \{ W^\omega_{L-1,L+1,L}(1,2) \boldsymbol{\alpha}_1 Y_{L(L-1)}(\vartheta_1,\varphi_1) \boldsymbol{\alpha}_2 Y_{L(L+1)}(\vartheta_2,\varphi_2) \\ &+ W^\omega_{L+1,L-1,L}(1,2) \boldsymbol{\alpha}_1 Y_{L(L+1)}(\vartheta_1,\varphi_1) \boldsymbol{\alpha}_2 Y_{L(L-1)}(\vartheta_2,\varphi_2) \} \end{aligned} \quad (9.12)$$

Recall the frequency-dependent Breit operator of section 8.1.5 where ω is the angular frequency of the exchanged (virtual) photon,

$$\omega = \frac{E_\omega}{\hbar} = 2\pi\nu = \frac{2\pi c}{\lambda} \quad (9.13)$$

where we explicitly included the reduced Planck constant $\hbar$, which takes the value of one in Hartree atomic units. Following Grant and Pyper, we have

$$w_L = -\frac{\sqrt{L(L+1)(2L-1)(2L+3)}}{(2L+1)^2} \quad (9.14)$$

as well as

$$v_{L-1,L} = -\frac{L+1}{2L+1} \quad \text{and} \quad v_{LL} = 1 \quad \text{and} \quad v_{L+1,L} = -\frac{L}{2L+1} \quad (9.15)$$

The radial dependent operators $V_\nu(1,2)$ and $W_{\nu,\nu',L}(1,2)$ can be expressed in terms of spherical Bessel functions, $j_\nu(x)$ and $n_\nu(x)$ [381]

$$V_\nu(1,2) = -[\nu]\omega j_\nu(\omega r_<) n_\nu(\omega r_>) \quad (9.16)$$

$$W^\omega_{\nu-1,\nu+1,\nu}(1,2) = \begin{cases} [\nu]\omega j_{\nu-1}(\omega r_1) n_{\nu+1}(\omega r_2) + \dfrac{[\nu]^2}{\omega^2} \dfrac{r_1^{\nu-1}}{r_2^{\nu+2}} \text{ if } r_1 \le r_2 \\ [\nu]\omega n_{\nu-1}(\omega r_1) j_{\nu+1}(\omega r_2) \text{ if } r_1 > r_2 \end{cases} \quad (9.17)$$

with the abbreviations

$$[i,j,...] = (2i+1)(2j+1)\cdots \;\Rightarrow\; [\nu] = (2\nu+1) \quad (9.18)$$

As before in section 9.1.2, $r_< = \min\{r_1, r_2\}$ and accordingly we have $r_> = \max\{r_1, r_2\}$.

Because of the importance of Darwin's expression for the classical electromagnetic interaction of two moving charges (section 3.5), we are particularly interested in the frequency-independent radial form of the Breit operator. This represents the consistent interaction term to approximately include the retarded electromagnetic interaction of the electrons in our semi-classic formalism that describes only the elementary particles (electrons) quantum mechanically. In this long-wavelength limit, $\omega \to 0$, the radial operator $V_\nu(1,2)$ in Eq. (9.16) becomes $U_\nu(1,2)$ — already known from the Coulomb case in Eq. (9.9) — and can be conveniently split as

$$U_\nu(1,2) = \frac{r_<^\nu}{r_>^{\nu+1}} = \bar{U}_\nu(1,2) + \bar{U}_\nu(2,1) \tag{9.19}$$

with

$$\bar{U}_\nu(1,2) = \begin{cases} \dfrac{r_1^\nu}{r_2^{\nu+1}} & \text{if } r_1 < r_2 \\ 0 & \text{if } r_1 > r_2 \end{cases} \tag{9.20}$$

The radial operator in Eq. (9.17) reads in the long-wavelength limit, $\omega \to 0$,

$$W^\omega_{\nu-1,\nu+1,\nu}(1,2) \;\to\; W^0_{\nu-1,\nu+1,\nu}(1,2) = \frac{[\nu]}{2}\left\{\bar{U}_{\nu+1}(1,2) - \bar{U}_{\nu-1}(1,2)\right\} \tag{9.21}$$

$$W^\omega_{\nu+1,\nu-1,\nu}(1,2) \;\to\; W^0_{\nu+1,\nu-1,\nu}(1,2) = \frac{[\nu]}{2}\left\{\bar{U}_{\nu+1}(2,1) - \bar{U}_{\nu-1}(2,1)\right\} \tag{9.22}$$

with the $\bar{U}_\nu(1,2)$ operators as defined above.

It turned out [158] that a special symmetrized form of the Breit operator leads to more simplified matrix elements contributing to the electronic energy expression. The total Breit operator in its ω-dependent form as well as in the long-wavelength limit is symmetric with respect to an interchange of the electron coordinates 1 and 2. For $B_0^{(1)}(1,2)$ this symmetry holds even for the radial and angular coordinates independently. However, this is not the case for the $B_0^{(2)}(1,2)$ term, where we have $B_0^{(2)}(1,2) = B_0^{(2)}(2,1)$ but not independently for the radial part and the angular and spin part.

By introducing the tensor operator

$$X_{L\nu}(i) \equiv \sqrt{\frac{4\pi}{2L+1}}\,\boldsymbol{\alpha}_i Y_{L\nu}(\vartheta_i,\varphi_i) \tag{9.23}$$

as an abbreviation and merging the double sum of Eq. (9.11) into a single one, we can write for the first term of the Breit operator

$$\begin{aligned} B_0^{(1)}(1,2) = \sum_\nu U_\nu(1,2)\Big[&-\frac{\nu+2}{2\nu+3}X_{(\nu+1)\nu}(1)X_{(\nu+1)\nu}(2) \\ &+X_{\nu\nu}(1)X_{\nu\nu}(2) - \frac{\nu-1}{2\nu-1}X_{(\nu-1)\nu}(1)X_{(\nu-1)\nu}(2)\Big] \end{aligned} \tag{9.24}$$

For a symmetric expression of $B_0^{(2)}(1,2)$ with the terms in the sum again ordered according to the index ν of $U_\nu(1,2)$, we investigate the relation

$$U_\nu(1,2)T_\nu(1,2) = \bar{U}_\nu(1,2)\bar{T}_\nu(1,2) + \bar{U}_\nu(2,1)\bar{T}_\nu(2,1) \tag{9.25}$$

where the operator $\bar{T}_\nu(1,2)$,

$$\begin{aligned}\bar{T}_\nu(1,2) \quad = \quad & \{w_{\nu-1}[\nu-1]X_{(\nu-1)(\nu-2)}(1)X_{(\nu-1)\nu}(2) \\ & - w_{\nu+1}[\nu+1]X_{(\nu+1)\nu}(1)X_{(\nu+1)(\nu+2)}(2)\}\end{aligned} \tag{9.26}$$

contains all angular dependent terms of $B_0^{(2)}(1,2)$. The identity

$$U_\nu(1,2)T_\nu(1,2) = \frac{U_{\nu,+}(1,2)T_{\nu,+}(1,2) + U_{\nu,-}(1,2)T_{\nu,-}(1,2)}{2} \tag{9.27}$$

with

$$U_{\nu,\pm}(1,2) \quad = \quad \bar{U}_\nu(1,2) \pm \bar{U}_\nu(2,1) \tag{9.28}$$

$$T_{\nu,\pm}(1,2) \quad = \quad \bar{T}_\nu(1,2) \pm \bar{T}_\nu(2,1) \tag{9.29}$$

splits the operator product $U_\nu(1,2)T_\nu(1,2)$ into a symmetric and an antisymmetric part. The frequency-independent operator $B_0^{(2)}(1,2)$ can now be written explicitly as

$$\begin{aligned} B_0^{(2)}(1,2) = -\frac{1}{4}\sum_\nu \Big[& U_{\nu,+}(1,2)\,\{w_{\nu+1}[\nu+1]\,\big[X_{(\nu+1)\nu}(1)X_{(\nu+1)(\nu+2)}(2) \\ & +X_{\nu(\nu+2)}(1)X_{(\nu+1)\nu}(2)\big] - w_{\nu-1}[\nu-1]\,\big[X_{(\nu-1)(\nu-2)}(1)X_{(\nu-1)\nu}(2) \\ & \qquad +X_{(\nu-1)\nu}(1)X_{(\nu-1)(\nu-2)}(2)\ \big]\} \\ & +U_{\nu,-}(1,2)\,\{w_{\nu+1}[\nu+1]\,\big[X_{(\nu+1)\nu}(1)X_{(\nu+1)(\nu+2)}(2) \\ & -X_{(\nu+1)(\nu+2)}(1)X_{(\nu+1)\nu}(2)\big] - w_{\nu-1}[\nu-1]\,\big[X_{(\nu-1)(\nu-2)}(1)X_{(\nu-1)\nu}(2) \\ & \qquad - X_{(\nu-1)\nu}(1)X_{(\nu-1)(\nu-2)}(2)\big]\}\Big] \end{aligned} \tag{9.30}$$

where $[\nu+1] = 2\nu+3$ and $[\nu-1] = 2\nu-1$. The factor 1/4 in this Eq. (9.30) originates from the original expression for the operator $B_0^{(2)}(1,2)$, Eqs. (9.21) and (9.22), and from the identity in Eq. (9.27).

9.1.4
Atomic Many-Electron Hamiltonian

After having transformed all operators of the many-electron atomic Hamiltonian in Eq. (9.1) to polar coordinates, we may write the Dirac–Coulomb

Hamiltonian for an atom explicitly as

$$\hat{H}_{el} = \sum_{i}^{N} \begin{pmatrix} m_e c^2 + V_{nuc}(r_i) & -ic\left(\frac{\boldsymbol{\sigma}\cdot\boldsymbol{r}_i}{r_i}\right)\left[\frac{\hbar}{r_i}\frac{\partial}{\partial r_i}r_i - \frac{k_i}{r_i}\right] \\ -ic\left(\frac{\boldsymbol{\sigma}\cdot\boldsymbol{r}_i}{r_i}\right)\left[\frac{\hbar}{r_i}\frac{\partial}{\partial r_i}r_i - \frac{k_i}{r_i}\right] & -m_e c^2 + V_{nuc}(r_i) \end{pmatrix}$$
$$+\sum_{j<i}^{N}\sum_{\nu=0}^{\infty}\frac{r_<^{\nu}}{r_>^{\nu+1}}\frac{4\pi}{2\nu+1}\sum_{m=-\nu}^{\nu}Y_{\nu m}^{\star}(\vartheta_i,\varphi_i)Y_{\nu m}(\vartheta_j,\varphi_j) \tag{9.31}$$

where the external potential $V_{nuc}(r_i)$ is either a simple Coulombic potential energy operator or represents a finite-nucleus model as introduced in section 6.9. In the former case it simply reads $-Ze^2/r_i$ with Z being the nuclear charge number of the atom under consideration and r_i being the distance of electron i to the nucleus. In accord with the Born–Oppenheimer approximation the nucleus is assumed to be infinitely heavy and its coordinate is chosen to be the origin of the coordinate system. If the energy-level shift of $-m_e c^2$ per electron as discussed in section 6.7 is to be considered the entries on the diagonal of the one-electron operators are shifted by $-m_e c^2$ each so that the rest energy no longer appears in their upper left block. For the sake of brevity we skip the addition of the Breit interaction terms in polar coordinates and thus refrain from presenting the explicit form of the Dirac–Coulomb–Breit Hamiltonian for atoms.

9.2 Atomic Many-Electron Wave Function and jj-Coupling

Since the structure of the *many-electron* Dirac Hamiltonian operator for atoms and thus the one-electron Fock operator of self-consistent field equations still allows us to separate radial and angular degrees of freedom (as in the case of the one-electron atoms solved analytically in chapter 6), the coordinates in the 4-spinor's components may be separated into radial and spherical spinor parts,

$$\psi_i \longrightarrow \psi_{n_i\kappa_i m_{j(i)}}(\boldsymbol{r}) = \frac{1}{r}\begin{pmatrix} P_{n_i\kappa_i}(r)\chi_{\kappa_i m_{j(i)}}(\vartheta,\varphi) \\ \mathrm{i}Q_{n_i\kappa_i}(r)\chi_{-\kappa_i m_{j(i)}}(\vartheta,\varphi) \end{pmatrix} \tag{9.32}$$

Because the one-electron operators are identical in form to the one-electron operator in hydrogen-like systems, we use for the independent particle model of Eq. (8.96) for the basis of the many-electron wave function a product consisting of N such hydrogen-like spinors. This ansatz allows us to treat the nonradial part analytically. The radial functions remain unknown. In principle, they may be expanded into a set of known basis functions, but we focus in this chapter on numerical methods, which can be conveniently employed for the one-dimensional radial problem that arises after integration of all angular and spin degrees of freedom.

In addition, we adopt the equivalence restriction which limits the number of radial functions. We choose only one radial function for all possible $m_{j(i)}$ values corresponding to one κ_i (compare Table 6.1) as the one-electron Fock operator will depend only on κ_i — since the radial Dirac Hamiltonian for hydrogen-like atoms depends only on κ_i — and is thus the same for the $2|\kappa_i|=2j+1$ degenerate states. In the following, we will use the indices p, q, r, s to denote the individual atomic spinors that enter the Hartree product, while the indices i, j, k, l label the nondegenerate sets of spinors and are thus shell indices. Hence, we again exploit the degeneracy of atomic spinors through the equivalence restriction. The orthonormality requirement of Eq. (8.108) translates onto the radial functions and simply reads

$$\langle P_{n_i\kappa_i}(r)|P_{n_j\kappa_j}(r)\rangle + \langle Q_{n_i\kappa_i}(r)|Q_{n_j\kappa_j}(r)\rangle = \delta_{ij} \tag{9.33}$$

for shells i and j (the spherical 2-spinors $\chi_{\kappa_i m_{j(i)}}(\vartheta, \varphi)$ and $\chi_{-\kappa_i m_{j(i)}}(\vartheta, \varphi)$ are already orthogonal; compare also chapter 6). In the last equation, the integration is only over the radial coordinate r with volume element $r^2\,dr$.

Spinors defined according to Eq. (9.32) finally enter the Slater determinants of Eq. (8.98). In general, for a given set of shell functions and occupation numbers, more than one Slater determinant must be constructed for the superposition to yield a CSF which is an eigenfunction of the scalar angular momentum operators $\hat{\boldsymbol{J}}^2$ and $\hat{J}_z$. These superpositions can be built up analytically because of the analytic treatment of the angular and spin parts of the spinors. The set of spinors in that Slater determinant or CSF which contributes most to the CI expansion of the total electronic *ground* state Ψ_A is called the *electronic configuration* of an atom.

In spherical symmetry the conservation of the total electronic angular momentum is maintained and the corresponding squared operator $\hat{\boldsymbol{J}}^2$ must commute with the electronic Dirac Hamiltonian. Therefore, the eigenfunctions of the Dirac Hamiltonian can be classified by the angular momentum quantum numbers. Further, a construction scheme for the generation of many-electron total angular momentum eigenfunctions out of one-electron eigenfunctions, the *jj*-coupling scheme, can be used to generate the analytic form of the total wave function for an electronic state A.

The simplest way to construct a total electronic wave function for given J (and M_J) is to couple two angular momenta at a time. Hence, once the first two one-electron total angular momenta are coupled, the resulting one is coupled with the next one-electron total angular momentum and so forth until N coupled one-electron functions set up a CSF of well-defined J and M_J. However, more elegant ways to construct total angular momentum eigenfunctions exist. In the 1940s Racah introduced a tensor algebra for this purpose [341–344]. In this Racah algebra the angular momentum coupling to final J and M_J is achieved in terms of parental states, coefficients of fractional parentage, seniority numbers and coefficients of fractional grand parentage [41, 382–385]. Here, we choose the pedestrian way of successive coupling of angular momenta step by step — although this is quite clumsy in practice if a large set of CSF basis functions is to be constructed. We can, however, benefit from the fact that most shells within a given CSF are in fact closed shells. Usually, only a few shells are open, mostly those in the valence region. Therefore, we can largely exploit the fact that all electrons in closed shells couple to a $^1S_0$ state.

Obviously, to generate eigenfunctions of $\hat{J}_z = \hat{j}_z(1) + \hat{j}_z(2)$ one can multiply the two angular momentum eigenfunctions of the uncoupled representation to get the eigenvalue $M_J = m_{j(1)} + m_{j(2)}$. Since for two momenta $i = 1$ and $i = 2$ holds $-j_i \leq m_{j(i)} \leq j_i$, we can write $M_J = -J, -J+1, ..., J$ where $J = j_1 + j_2, j_1 + j_2 - 1, ..., |j_1 - j_2|$. For the generation of orthonormal eigenfunctions of $\hat{\boldsymbol{J}}^2$ we employ the Clebsch–Gordan sum introduced in section 4.4

$$|JM_J\rangle = \sum_{m_j} |j_1 m_j\rangle \, |j_2(M_J - m_j)\rangle \, \langle j_1 j_2 m_j (M_J - m_j)|j_1 j_2 J M_J\rangle \tag{9.34}$$

Recall section 4.4.2 and Eq. (4.136) for the definition of the Clebsch–Gordan coefficients (vector coupling coefficients) and Wigner-3j-symbols. Of course, the Clebsch–Gordan coefficients finally define the B_{KI} expansion coefficients of the pre-contracted CSF in Eq. (8.99).

For shell occupations $1, 2j$ (one hole) and $2j+1$ (closed-shell) only **one** Slater determinant represents the total symmetry (j is the total angular momentum for an electron in a particular shell: $j = l \pm s$). Furthermore, for shell occupations 2 and $2j - 1$ (two holes) Slater determinants result which are independent of the coupling scheme (for $|\kappa| > 1$). Two such Slater determinants, $\Phi_1(n)$ and $\Phi_2(m)$, with n and m electrons can be coupled

$$\begin{aligned}\Phi(n+m; JM_J) &= \sum_{M_J'} \Phi_1(n; J_1(M_J - M_J')) \, \Phi_2(m; J_2 M_J') \\ &\quad \times \langle J_1 J_2 (M_J - M_J') M_J | J M_J\rangle\end{aligned} \tag{9.35}$$

The Slater determinants are multiplied as follows

$$||\psi_{i_1} \cdots \psi_{i_n}|| \times ||\psi_{j_1} \cdots \psi_{j_m}|| \Rightarrow ||\psi_{i_1} \cdots \psi_{i_n} \psi_{i_{(n+1)}} \cdots \psi_{i_{(n+m)}}|| \tag{9.36}$$

to give an $(n+m)$-dimensional antisymmetrized and normalized Slater determinant $(i_1 < i_2 < \cdots < i_{(n+m)})$.

Compared to nonrelativistic CSF expansions classified according to their LS symmetry, the situation becomes more complicated for relativistic expansions, since usually many more relativistic configuration state functions (CSFs) have to be used to represent the a given nonrelativistic CSF analog. This has been discussed for LS- and jj-coupled atomic CSFs [386,387].

9.3 One- and Two-Electron Integrals in Spherical Symmetry

From the discussion in section 8.5.4 it is clear that we need to be able to calculate three- and six-dimensional integrals, the so-called one- and two-electron integrals in order to determine expectation values for a many-electron atom. According to the ansatz for the atomic spinors in Eq. (9.32) we proceed to analytically separate the angular-dependent parts leaving one- and two-dimensional radial integrals to be solved. It must be stressed that the analytic result of the angular integration depends crucially on the choice of phase of the Pauli spinors in Eqs. (4.154) and (4.155) as already emphasized in section 4.4.4 so that a given analytic expression might require adjustment accordingly.

9.3.1 One-Electron Integrals

The integrals over the coordinates of a single electron

$$h^D_{pq} = \langle \psi_p | h^D | \psi_q \rangle \tag{9.37}$$

can be evaluated considering the one-electron operator of Eq. (6.53). The integration over the spin and angular degrees of freedom results in the integrated scalar product of the 2-component spherical spinors, i.e., $\langle \chi_{\kappa_p m_{j(p)}} | \chi_{\kappa_q m_{j(q)}} \rangle = \delta_{pq}$. We then obtain for the original one-electron Dirac Hamiltonian

$$\begin{aligned} h^D_{pq} = \delta_{\kappa_p \kappa_q} \delta_{m_{j(p)} m_{j(q)}} \Bigg[& m_e c^2 \langle P_{n_p\kappa_p} | P_{n_q\kappa_q} \rangle - m_e c^2 \langle Q_{n_p\kappa_p} | Q_{n_q\kappa_q} \rangle \\ & + \langle P_{n_p\kappa_p} | V_{\text{nuc}} | P_{n_q\kappa_q} \rangle + \langle Q_{n_p\kappa_p} | V_{\text{nuc}} | Q_{n_q\kappa_q} \rangle \\ -c\hbar \left\langle P_{n_p\kappa_p} \middle| \frac{\mathrm{d}}{\mathrm{d}r} - \frac{\kappa_q}{r} \middle| Q_{n_q\kappa_q} \right\rangle & + c\hbar \left\langle Q_{n_p\kappa_p} \middle| \frac{\mathrm{d}}{\mathrm{d}r} + \frac{\kappa_q}{r} \middle| P_{n_q\kappa_q} \right\rangle \Bigg] \end{aligned} \tag{9.38}$$

or after the shift in energy by the rest energy $m_e c^2$ for one electron,

$$\begin{aligned} h^D_{pq} - m_e c^2 \delta_{pq} = \delta_{\kappa_p \kappa_q} \delta_{m_{j(p)} m_{j(q)}} \Bigg[& - 2 m_e c^2 \langle Q_{n_p\kappa_p} | Q_{n_q\kappa_q} \rangle \\ & + \langle P_{n_p\kappa_p} | V_{\text{nuc}} | P_{n_q\kappa_q} \rangle + \langle Q_{n_p\kappa_p} | V_{\text{nuc}} | Q_{n_q\kappa_q} \rangle \\ -c\hbar \left\langle P_{n_p\kappa_p} \middle| \frac{\mathrm{d}}{\mathrm{d}r} - \frac{\kappa_q}{r} \middle| Q_{n_q\kappa_q} \right\rangle & + c\hbar \left\langle Q_{n_p\kappa_p} \middle| \frac{\mathrm{d}}{\mathrm{d}r} + \frac{\kappa_q}{r} \middle| P_{n_q\kappa_q} \right\rangle \Bigg] \end{aligned} \tag{9.39}$$

which is evident if we start directly from the shifted one-electron Hamiltonian as in Eq. (6.139). The energy shift thus only affects the diagonal matrix elements as can also be derived by starting from Eq. (9.38) and noting that

$$\begin{aligned} m_e c^2 \langle P_p | P_q \rangle - m_e c^2 \langle Q_p | Q_q \rangle &= m_e c^2 \langle P_p | P_q \rangle - m_e c^2 \langle Q_p | Q_q \rangle \\ &\quad + \underbrace{m_e c^2 \langle Q_p | Q_q \rangle - m_e c^2 \langle Q_p | Q_q \rangle}_{0} \\ &= m_e c^2 \underbrace{\langle P_p | P_q \rangle + \langle Q_p | Q_q \rangle}_{\delta_{pq} \text{ after Eq. (9.33)}} - 2 m_e c^2 \langle Q_p | Q_q \rangle \end{aligned} \tag{9.40}$$

with the shorthand notation $P_{n_p \kappa_p} = P_p$ and so forth. The sum of all radial integrals,

$$\sum_{pq}^{M'} h^D_{pq} = \sum_{ij}^{M} h^D_{ij} \delta_{\kappa_i \kappa_j} \sum_{m_{j(p)}=-j_i}^{j_i} \sum_{m_{j(q)}=-j_j}^{j_j} \delta_{m_{j(p)} m_{j(q)}} \tag{9.41}$$

is now reduced where we used the composite indices i and j for all equal $\{n_p \kappa_p\}$ and $\{n_q \kappa_q\}$ pairs of shells. The radial integral for the two shells reads (with shifted energy reference)

$$\begin{aligned} h^D_{ij} &= -2 m_e c^2 \langle Q_{n_i \kappa_i} | Q_{n_j \kappa_j} \rangle + \langle P_{n_i \kappa_i} | V_{\text{nuc}} | P_{n_j \kappa_j} \rangle + \langle Q_{n_i \kappa_i} | V_{\text{nuc}} | Q_{n_j \kappa_j} \rangle \\ &\quad - c\hbar \left\langle P_{n_i \kappa_i} \left| \frac{\mathrm{d}}{\mathrm{d}r} - \frac{\kappa_j}{r} \right| Q_{n_j \kappa_j} \right\rangle + c\hbar \left\langle Q_{n_i \kappa_i} \left| \frac{\mathrm{d}}{\mathrm{d}r} + \frac{\kappa_j}{r} \right| P_{n_j \kappa_j} \right\rangle \end{aligned} \tag{9.42}$$

Note that the differential volume element for the integrations in terms of the radial functions $P_{n\kappa}(r)$ and $Q_{n\kappa}(r)$ is dr and not r^2dr (see chapter 6). Moreover, there will be M'^2 (pq)-pairs if M' is the total number of spinors while there are only M^2 (ij)-pairs where M is the total number of pairs of radial functions (i.e. the number of shells).

9.3.2
Electron–Electron Coulomb Interaction

In the following, the sets of spatial coordinates appearing in the two-electron integrals will be distinguished by 1 and 2. To evaluate Coulomb- and exchange-integrals over the coordinates 1 and 2 of any two electrons

$$g_{pqrs} = \left\langle \psi_p(1) \left\langle \psi_r(2) \left| 1/r_{12} \right| \psi_s(2) \right\rangle_2 \psi_q(1) \right\rangle_1 \tag{9.43}$$

we replace the spinors by their product ansatz and the operator by Eq. (9.9) and get

$$g_{pqrs} = \sum_{\nu=0}^{\infty} \frac{4\pi}{2\nu+1} \sum_{m=-\nu}^{\nu} \int_0^{\infty} \mathrm{d}r_1 \int_0^{\infty} \mathrm{d}r_2 \frac{r_<^{\nu}}{r_>^{\nu+1}}$$

$$\times \left\{ P_{n_p\kappa_p}(r_1)P_{n_q\kappa_q}(r_1) \int_{\Omega_1} \chi^{\dagger}_{\kappa_p m_{j(p)}}(\Omega_1)\chi_{\kappa_q m_{j(q)}}(\Omega_1)Y^{\star}_{\nu m}(\Omega_1)\mathrm{d}\Omega_1 \right.$$
$$\left. + Q_{n_p\kappa_p}(r_1)Q_{n_q\kappa_q}(r_1) \int_{\Omega_1} \chi^{\dagger}_{-\kappa_p m_{j(p)}}(\Omega_1)\chi_{-\kappa_q m_{j(q)}}(\Omega_1)Y^{\star}_{\nu m}(\Omega_1)\mathrm{d}\Omega_1 \right\}$$
$$\times \left\{ P_{n_r\kappa_r}(r_2)P_{n_s\kappa_s}(r_2) \int_{\Omega_2} \chi^{\dagger}_{\kappa_r m_{j(r)}}(\Omega_2)\chi_{\kappa_s m_{j(s)}}(\Omega_2)Y_{\nu m}(\Omega_2)\mathrm{d}\Omega_2 \right.$$
$$\left. + Q_{n_r\kappa_r}(r_2)Q_{n_s\kappa_s}(r_2) \int_{\Omega_2} \chi^{\dagger}_{-\kappa_r m_{j(r)}}(\Omega_2)\chi_{-\kappa_s m_{j(s)}}(\Omega_2)Y_{\nu m}(\Omega_2)\mathrm{d}\Omega_2 \right\} \quad (9.44)$$

The integration symbol $\int_{\Omega}\mathrm{d}\Omega = \int_0^{\pi}\sin\vartheta\mathrm{d}\vartheta\int_0^{2\pi}\mathrm{d}\varphi$ denotes the integration over all angular coordinates collected in Ω. Each integral over these angular degrees of freedom Ω can be written explicitly as

$$\int_{\Omega} \chi^{\dagger}_{\kappa_p m_{j(p)}}(\Omega)\chi_{\kappa_q m_{j(q)}}(\Omega)Y^{(\star)}_{\nu m}(\Omega)d\Omega$$
$$= \int_{\Omega}\left[Y^{\star}_{l_p(m_{j(p)}-\frac{1}{2})}(\Omega)\langle l_p\,\tfrac{1}{2}\,(m_{j(l)}-\tfrac{1}{2})\,\tfrac{1}{2}|j_l\,m_{j(l)}\rangle \right.$$
$$Y_{l_q(m_{j(q)}-\frac{1}{2})}(\Omega)\langle l_q\,\tfrac{1}{2}\,(m_{j(q)}-\tfrac{1}{2})\,\tfrac{1}{2}|j_q\,m_{j(q)}\rangle$$
$$+ Y^{\star}_{l_p(m_{j(p)}+\frac{1}{2})}(\Omega)\langle l_p\,\tfrac{1}{2}\,(m_{j(p)}+\tfrac{1}{2})\,\tfrac{1}{2}|j_p\,m_{j(p)}\rangle$$
$$\left. Y_{l_q(m_{j(q)}+\frac{1}{2})}(\Omega)\langle l_q\,\tfrac{1}{2}\,(m_{j(q)}+\tfrac{1}{2})\,\tfrac{1}{2}|j_q\,m_{j(q)}\rangle \right] Y^{(\star)}_{\nu m}(\Omega)d\Omega \quad (9.45)$$

which can be written as two integrals of the type [58, p. 11]

$$\int Y^{\star}_{l'm'}(\Omega)Y_{LM}(\Omega)Y_{lm}(\Omega)d\Omega = (-1)^{m'}\sqrt{\frac{(2l'+1)(2L+1)(2l+1)}{4\pi}}$$
$$\times \begin{pmatrix} l' & L & l \\ -m' & M & m \end{pmatrix}\begin{pmatrix} l' & L & l \\ 0 & 0 & 0 \end{pmatrix} \quad (9.46)$$

(see, for instance, Ref. [16, p. 216] for a derivation). In 1961, Grant [345] introduced the following notation for these integrals,

$$\int_{\Omega} \chi^{\dagger}_{\kappa_p m_{j(p)}}(\Omega)\;\chi_{\kappa_q m_{j(q)}}(\Omega)\quad Y^{\star}_{\nu m}(\Omega)d\Omega$$
$$= \sqrt{\frac{2\nu+1}{4\pi}}\quad d^{\nu}_{j_p m_{j(p)} j_q m_{j(q)}}\,\delta_{m(m_{j(q)}-m_{j(p)})} \quad (9.47)$$
$$\int_{\Omega} \chi^{\dagger}_{\kappa_r m_{j(r)}}(\Omega)\;\chi_{\kappa_s m_{j(s)}}(\Omega)\quad Y_{\nu m}(\Omega)d\Omega$$
$$= \sqrt{\frac{2\nu+1}{4\pi}}\quad d^{\nu}_{j_s m_{j(s)} j_r m_{j(r)}}\,\delta_{m(m_{j(r)}-m_{j(s)})} \quad (9.48)$$

Note that the complex conjugation of $Y_{\nu m}(\Omega)$ results in a different order of indices in the coefficients $d^{\nu}_{jm_j j' m_{j'}}$. Grant's coefficients are given by [345]

$$\begin{aligned} d^{\nu}_{jm_j j' m_{j'}} &= \sqrt{(2j+1)(2j'+1)}\;(-1)^{m_{j'}+1/2} \\ &\quad \times \begin{pmatrix} j' & \nu & j \\ -m_{j'} & m & m_j \end{pmatrix} \begin{pmatrix} j' & \nu & j \\ 1/2 & 0 & -1/2 \end{pmatrix} \end{aligned} \tag{9.49}$$

and are thus numbers that can be readily calculated from Wigner $3j$-symbols. These coefficients are the relativistic analogs of the Gaunt coefficients in nonrelativistic atomic structure theory. They are different from zero, i.e. $d^{\nu}_{jm_j j' m_{j'}} \neq 0$, only if

$$m = m_{j'} - m_j \quad ; \quad |j - j'| \leq \nu \leq j + j' \tag{9.50}$$

and

$$j + j' + \nu = \begin{cases} \text{even} & \text{if } \operatorname{sgn}(\kappa) \neq \operatorname{sgn}(\kappa') \\ \text{odd} & \text{if } \operatorname{sgn}(\kappa) = \operatorname{sgn}(\kappa') \end{cases} \tag{9.51}$$

due to the so-called triangle condition for the Wigner $3j$-symbols [58]. Integration over all angular coordinates, Ω_1 and Ω_2, simplifies Eq. (9.44) to yield

$$\begin{aligned} g_{pqrs} &= \sum_{\nu=0}^{\infty} \int_0^{\infty} dr_1 \int_0^{\infty} dr_2 \frac{r_<^{\nu}}{r_>^{\nu+1}} \\ &\quad \times \{P_{n_p\kappa_p}(r_1)P_{n_q\kappa_q}(r_1) + Q_{n_p\kappa_p}(r_1)Q_{n_q\kappa_q}(r_1)\} \\ &\quad \times \{P_{n_r\kappa_r}(r_2)P_{n_s\kappa_s}(r_2) + Q_{n_r\kappa_r}(r_2)Q_{n_s\kappa_s}(r_2)\} \\ &\quad \times \delta_{(m_{j(q)}-m_{j(p)})(m_{j(r)}-m_{j(s)})} d^{\nu}_{j_p m_{j(p)} j_q m_{j(q)}} d^{\nu}_{j_s m_{j(s)} j_r m_{j(r)}} \end{aligned} \tag{9.52}$$

The total number of original integrals is thereby reduced after carrying out the integrations over the angular parts,

$$\begin{aligned} \sum_{pqrs}^{M'} g_{pqrs} &= \sum_{ijkl}^{M} \sum_{\nu=\nu_{min}}^{\nu_{max}} g_{ijkl\nu} \sum_{m_{j(p)}=-j_i}^{j_i} \sum_{m_{j(q)}=-j_j}^{j_j} \sum_{m_{j(r)}=-j_k}^{j_k} \sum_{m_{j(s)}=-j_l}^{j_l} \\ &\quad \times d^{\nu}_{j_i m_{j(p)} j_j m_{j(q)}} d^{\nu}_{j_l m_{j(s)} j_k m_{j(r)}} \delta_{(m_{j(q)}-m_{j(p)})(m_{j(r)}-m_{j(s)})} \end{aligned} \tag{9.53}$$

with the abbreviation

$$g_{ijkl\nu} = \langle P_i | U_{kl\nu}(r_1) | P_j \rangle_{r_1} + \langle Q_i | U_{kl\nu}(r_1) | Q_j \rangle_{r_1} \tag{9.54}$$

Every $g_{ijkl\nu}$ consists of four integrals. They add up to four times as many integrals as in nonrelativistic atomic structure theory [388]. Furthermore, the sum over m was skipped since there exists only one possible combination, $m_{j(q)} - m_{j(p)} = m_{j(r)} - m_{j(s)}$, with nonvanishing $3j$-symbols. The $U_{kl\nu}(r)$ potential functions are defined as

$$U_{kl\nu}(r_1) = \left\langle P_k(r_2) \left| \frac{r_<^\nu}{r_>^{\nu+1}} \right| P_l(r_2) \right\rangle_{r_2} + \left\langle Q_k(r_2) \left| \frac{r_<^\nu}{r_>^{\nu+1}} \right| Q_l(r_2) \right\rangle_{r_2} \tag{9.55}$$

which is to be understood as

$$\begin{aligned} U_{kl\nu}(r_1) &= \frac{1}{r_1^{\nu+1}} \int_0^{r_1} \mathrm{d}r_2\, r_2^\nu \left[P_k(r_2)P_l(r_2) + Q_k(r_2)Q_l(r_2)\right] \\ &\quad + r_1^\nu \int_{r_1}^{\infty} \mathrm{d}r_2 \frac{1}{r_2^{\nu+1}} \left[P_k(r_2)P_l(r_2) + Q_k(r_2)Q_l(r_2)\right] \end{aligned} \tag{9.56}$$

For Eq. (9.52) and Eq. (9.53), respectively, the following four selection rules apply (one is the parity selection rule and the others are those derived from Eq. (9.47), Eq. (9.48), and Eq. (9.51)):

1. $m_{j(p)} + m_{j(r)} = m_{j(q)} + m_{j(s)}$
2. $|j - j'| \le \nu \le j + j'$ where $\begin{cases} |j - j'| = \max\{|j_p - j_q|, |j_r - j_s|\} \\ j + j' = \min\{j_p + j_q, j_r + j_s\} \end{cases}$
3. $j_p + j_q + \nu$ $\begin{cases} \text{even} & \text{if } \operatorname{sgn}(\kappa_p) \neq \operatorname{sgn}(\kappa_q) \\ \text{odd} & \text{if } \operatorname{sgn}(\kappa_p) = \operatorname{sgn}(\kappa_q) \end{cases}$
4. $j_r + j_s + \nu$ $\begin{cases} \text{even} & \text{if } \operatorname{sgn}(\kappa_r) \neq \operatorname{sgn}(\kappa_s) \\ \text{odd} & \text{if } \operatorname{sgn}(\kappa_r) = \operatorname{sgn}(\kappa_s) \end{cases}$

These rules truncate the infinite sum over ν.

9.3.3
Electron–Electron Frequency-Independent Breit Interaction

For the calculation of the general frequency-independent Breit integral,

$$b_{0,pqrs} = \langle \psi_p(1) \, \langle \psi_r(2) \, |B_0(1,2)| \, \psi_s(2) \rangle_2 \, \psi_q(1) \rangle_1 \tag{9.57}$$

we require an analytic expression for the integration over spin and angular parts of one electronic coordinate in close analogy to the preceding Coulomb case. For the sake of brevity we fall back to the analytic results for the integration over a single $X_{ba}(1)$ operator of Eq. (9.23)

$$\begin{aligned} \langle \psi_p(1) \, |X_{ba}(1)| \, \psi_q(1) \rangle &= (-1)^{2j_p - m_p + 1/2} \sqrt{(2j_p+1)(2j_q+1)} \\ \times \Pi(l_p + l_q + a) &\begin{pmatrix} j_p & b & j_q \\ -m_p & (m_p - m_q) & m_q \end{pmatrix} \begin{pmatrix} j_p & b & j_q \\ 1/2 & 0 & -1/2 \end{pmatrix} \\ \times \, \mathrm{i} &\left[E_1^a(\kappa_p, \kappa_q, b) P_p(r_1) Q_q(r_1) - E_{-1}^a(\kappa_p, \kappa_q, b) Q_p(r_1) P_q(r_1) \right] \end{aligned} \tag{9.58}$$

which has been derived by Grant and Pyper [138]. In order to keep the above equation as compact as possible, we introduce

$$E^a_\beta(\kappa_p,\kappa_q,b) = \begin{cases} \dfrac{b+\beta(\kappa_q-\kappa_p)}{\sqrt{b(2b-1)}} & ;\ a=b-1 \\ \dfrac{\beta(\kappa_p+\kappa_q)}{\sqrt{b(b+1)}} & ;\ a=b \\ \dfrac{-b-1+\beta(\kappa_q-\kappa_p)}{\sqrt{(b+1)(2b+3)}} & ;\ a=b+1 \end{cases} \tag{9.59}$$

and

$$\Pi(n) = \begin{cases} 0 & \text{if } n \text{ even} \\ 1 & \text{if } n \text{ odd} \end{cases} \tag{9.60}$$

The integral for the second coordinate set is of the same form. Finally, we obtain for the general Breit matrix element over one-electron spinors [158]

$$\begin{aligned} b_{0,pqrs} = & \sqrt{[j_p,j_q,j_r,j_s]}\delta_{(m_p-m_q)(m_s-m_r)}(-1)^{m_p+m_s} \\ & \times \sum_\nu \Pi(l_p+l_q+\nu)\Pi(l_r+l_s+\nu) \\ & \times\{ A_{1+}(p,q,r,s;\nu)\langle P_p(r)|W_{rs\nu}(r)|Q_q(r)\rangle \\ & + A_{1-}(p,q,r,s;\nu)\langle Q_p(r)|W_{sr\nu}(r)|P_q(r)\rangle \\ & + A_{2-}(p,q,r,s;\nu)\langle P_p(r)|W_{sr\nu}(r)|Q_q(r)\rangle \\ & + A_{2+}(p,q,r,s;\nu)\langle Q_p(r)|W_{rs\nu}(r)|P_q(r)\rangle \\ & + B_{1+}(p,q,r,s;\nu)\langle P_p(r)|V_{rs\nu}(r)|Q_q(r)\rangle \\ & + B_{1-}(p,q,r,s;\nu)\langle Q_p(r)|V_{sr\nu}(r)|P_q(r)\rangle \\ & + B_{2-}(p,q,r,s;\nu)\langle P_p(r)|V_{sr\nu}(r)|Q_q(r)\rangle \\ & + B_{2+}(p,q,r,s;\nu)\langle Q_p(r)|V_{rs\nu}(r)|P_q(r)\rangle\} \end{aligned} \tag{9.61}$$

with the newly defined Breit potential functions

$$W_{rs\nu}(r_1) = \langle P_r(r_2)|U_{\nu,+}(1,2)|Q_s(r_2)\rangle_{r_2} \tag{9.62}$$

$$V_{rs\nu}(r_1) = \langle P_r(r_2)|U_{\nu,-}(1,2)|Q_s(r_2)\rangle_{r_2} \tag{9.63}$$

(the operator definitions are as in section 9.1.3) and the prefactors given by

$$A_{ab}(p,q,r,s;\nu) = {}^W F^{\nu+1}_{ab} + {}^W G^{\nu+1}_{ab} + {}^W F^{\nu}_{1} + {}^W F^{\nu-1}_{ab} + {}^W G^{\nu-1}_{ab} \tag{9.64}$$

$$B_{ab}(p,q,r,s;\nu) = {}^V G^{\nu+1}_{ab} + {}^V G^{\nu-1}_{ab} \tag{9.65}$$

Table 9.1 Coefficients ${}^{W}F^{c}_{ab}$. The index a can take the values 1 and 2, b stands for $+$ or $-$ to replace the $\pm$ sign in the table entries while c is $\nu+1$, ν, or $\nu-1$. The resulting five possible coefficients are given in the last two lines. Each of these values has to be multiplied by the quotient [prefactor] in the first line containing H^{ν}_{pqrs} symbols defined in Eq. (9.66).

a	c: $\nu+1\,;\ \nu\geq 0$	$\nu\,;\ \nu\geq 0$	$\nu-1\,;\ \nu\geq 1$
[prefactor]	$\frac{-(\nu+2)\,H^{\nu+1}_{pqrs}}{(\nu+1)(2\nu+1)(2\nu+3)}\times$	$\frac{KK'\,H^{\nu}_{pqrs}}{\nu(\nu+1)}\times$	$\frac{(1-\nu)\,H^{\nu-1}_{pqrs}}{(2\nu-1)\nu(2\nu+1)}\times$
1	$(\nu+1\pm K)(\nu+1\pm K')$	1	$(-\nu\pm K)(-\nu\pm K')$
2	$(-\nu-1\pm K)(\nu+1\pm K')$		$(\nu\pm K)(-\nu\pm K')$

Table 9.2 Coefficients ${}^{W}G^{c}_{ab}$ ($\circ\ \rightarrow\ +$) and ${}^{V}G^{c}_{ab}$ ($\circ\ \rightarrow\ -$). The general symbol $\circ$ has to be replaced by $+$ for the ${}^{W}G^{c}_{ab}$ and by $-$ for the ${}^{V}G^{c}_{ab}$. The coefficients are to be read as in Table 9.1 (including the indices a, b, and c).

a	c: $\nu+1\,;\ \nu\geq 0$	$\nu-1\,;\ \nu\geq 1$
[prefactor]	$\frac{H^{\nu+1}_{pqrs}}{4(2\nu+3)}\times$	$\frac{-H^{\nu-1}_{pqrs}}{4(2\nu-1)}\times$
1	$\big[(\nu+1\pm K)(-\nu-2\pm K')$ $\circ(-\nu-2\pm K)(\nu+1\pm K')\big]$	$\big[(\nu-1\pm K)(-\nu\pm K')$ $\circ(-\nu\pm K)(\nu-1\pm K')\big]$
2	$\big[(-\nu-1\pm K)(-\nu-2\pm K')$ $\circ(\nu+2\pm K)(\nu+1\pm K')\big]$	$\big[(-\nu+1\pm K)(-\nu\pm K')$ $\circ(\nu\pm K)(\nu-1\pm K')\big]$

The coefficients ${}^{W}F^{c}_{ab}$, ${}^{W}G^{c}_{ab}$, and ${}^{V}G^{c}_{ab}$ (Tables 9.1–9.2), whose spinor dependence has been skipped for brevity, contain products of Wigner $3j$-symbols

$$
\begin{aligned}
H^{\nu}_{pqrs} &= \begin{pmatrix} j_p & \nu & j_q \\ -m_p & (m_p-m_q) & m_q \end{pmatrix}\begin{pmatrix} j_p & \nu & j_q \\ 1/2 & 0 & -1/2 \end{pmatrix} \\
&\quad\times\begin{pmatrix} j_r & \nu & j_s \\ -m_r & (m_r-m_s) & m_s \end{pmatrix}\begin{pmatrix} j_r & \nu & j_s \\ 1/2 & 0 & -1/2 \end{pmatrix}
\end{aligned} \tag{9.66}
$$

and the abbreviations $K=\kappa_q-\kappa_p$, $K'=\kappa_s-\kappa_r$. For later convenience we introduce the following abbreviations

$$
\begin{aligned}
G_{n\pm pqrs\nu} &= \sqrt{[j_p,j_q,j_r,j_s]}\,\delta_{(m_p-m_q)(m_s-m_r)}(-1)^{m_p+m_s} \\
&\quad\times\Pi(l_p+l_q+\nu)\Pi(l_r+l_s+\nu)\, g_{n\pm pqrs\nu}
\end{aligned} \tag{9.67}
$$

Here, $G_{n\pm pqrs\nu}$ is an abbreviation for either $C_{n\pm pqrs\nu}$ (then, $g_{n\pm pqrs\nu}$ becomes $c_{n\pm pqrs\nu}$) or $D_{n\pm pqrs\nu}$ (where $g_{n\pm pqrs\nu}$ must be replaced by $d_{n\pm pqrs\nu}$).

$$c_{1+pqrs\nu} = \langle P_p(r)|W_{rs\nu}(r)|Q_q(r)\rangle, \quad c_{1-pqrs\nu} = \langle Q_p(r)|W_{sr\nu}(r)|P_q(r)\rangle \quad (9.68)$$
$$c_{2-pqrs\nu} = \langle P_p(r)|W_{sr\nu}(r)|Q_q(r)\rangle, \quad c_{2+pqrs\nu} = \langle Q_p(r)|W_{rs\nu}(r)|P_q(r)\rangle \quad (9.69)$$
$$d_{1+pqrs\nu} = \langle P_p(r)|V_{rs\nu}(r)|Q_q(r)\rangle, \quad d_{1-pqrs\nu} = \langle Q_p(r)|V_{sr\nu}(r)|P_q(r)\rangle \quad (9.70)$$
$$d_{2-pqrs\nu} = \langle P_p(r)|V_{sr\nu}(r)|Q_q(r)\rangle, \quad d_{2+pqrs\nu} = \langle Q_p(r)|V_{rs\nu}(r)|P_q(r)\rangle \quad (9.71)$$

With these coefficients we obtain the simplified expression for the Breit matrix element

$$b_{0,pqrs} = \sum_{\nu=\nu_{min}}^{\nu_{max}} \sum_{n\pm} \left[C_{npqrs\nu} c_{n\pm pqrs\nu} + D_{npqrs\nu} d_{n\pm pqrs\nu} \right] \quad (9.72)$$

The integrals in Eqs. (9.61) and (9.72) can be written in an even more compact form [158].

9.3.4
Calculation of Potential Functions

The two-electron terms, Coulomb as well as Breit integrals, both require an inner integration over the second electronic coordinate which is in our presentation hidden in the potential functions. To define these potential functions is useful as they allow one to utilize a differential equation [158] for their calculation rather than explicit, straightforward integration. The latter is to be avoided as an integration over, say, one-thousand grid points in the second radial coordinate r_2 then needs to be carried out for each of the one-thousand grid points in the first radial coordinate r_1. With the potential functions, however, equations analogous to the Poisson equation from classical electrostatics are recovered. Instead, one can calculate the potential functions according to Hartree's method from two coupled differential equations of first order [389].

The potential functions $U_{kl\nu}(r)$ of Eq. (9.55) are required for the electron–electron Coulomb interaction. We recall that they may be written as

$$U_{kl\nu}(r_1) = \frac{1}{r_1^{\nu+1}} \int_0^{r_1} dr_2 \rho_{kl}(r_2) r_2^\nu + r_1^\nu \int_{r_1}^\infty dr_2 \rho_{kl}(r_2) \frac{1}{r_2^{\nu+1}} \quad (9.73)$$

where we used the generalized density-like function $\rho_{kl}(r)$

$$\rho_{kl}(r) = P_k(r) P_l(r) + Q_k(r) Q_l(r) \quad (9.74)$$

as an abbreviation for the product of radial functions. These integrals have to be evaluated for every value of r_1 on a numerical grid, which would soon become computationally impractical as noted above. Fortunately, there are more efficient ways for computing these integrals which can be easily derived by virtue of the following definitions

$$Y_{kl\nu}(r) = rU_{kl\nu}(r) \tag{9.75}$$

$$Z_{kl\nu}(r) = \frac{1}{r^{\nu+1}}\int_0^r d\tau\rho_{kl}(\tau)\tau^\nu \tag{9.76}$$

9.3.4.1 **First-Order Differential Equations**

The differentiation of $Y_{kl\nu}(r)$ in Eq. (9.75)

$$\frac{\mathrm{d}}{\mathrm{d}r}Y_{kl\nu}(r) = U_{kl\nu}(r) + r\frac{\mathrm{d}}{\mathrm{d}r}U_{kl\nu}(r) \tag{9.77}$$

yields

$$\begin{aligned}\frac{\mathrm{d}}{\mathrm{d}r}Y_{kl\nu}(r) &= U_{kl\nu}(r) + r\left\{-\frac{\nu+1}{r^{\nu+2}}\int_0^r \mathrm{d}\tau\rho_{kl}(\tau)\tau^\nu + \frac{\rho_{kl}(r)}{r}\right.\\ &\quad \left. +\nu r^{\nu-1}\int_r^\infty \mathrm{d}\tau\frac{\rho_{kl}(\tau)}{\tau^{\nu+1}} - \frac{\rho_{kl}(r)}{r}\right\} \qquad (9.78)\\ &= -\frac{\nu}{r^{\nu+1}}\int_0^r \mathrm{d}\tau\rho_{kl}(\tau)\tau^\nu + (\nu+1)r^\nu\int_r^\infty \mathrm{d}\tau\frac{\rho_{kl}(\tau)}{\tau^{\nu+1}} \qquad (9.79)\end{aligned}$$

because of $\frac{\mathrm{d}}{\mathrm{d}r}\int_a^r f(s)\mathrm{d}s = \frac{\mathrm{d}}{\mathrm{d}r}[F(r) - F(a)] = f(r)$. With Eq. (9.76), the equation above can be easily reduced to

$$\frac{\mathrm{d}}{\mathrm{d}r}Y_{kl\nu}(r) = -(2\nu+1)Z_{kl\nu}(r) + (\nu+1)U_{kl\nu}(r) \tag{9.80}$$

The same can be done for the $Z_{kl\nu}(r)$ functions

$$\frac{\mathrm{d}}{\mathrm{d}r}Z_{kl\nu}(r) = -\frac{\nu+1}{r}Z_{kl\nu}(r) + \frac{\rho_{kl}(r)}{r} \tag{9.81}$$

Eqs. (9.80) and (9.81) are formally identical to the ones obtained for the nonrelativistic case which are known to be an improper choice for the determination of $U_{kl\nu}(r)$ [388, p. 233].

9.3.4.2 **Derivation of the Radial Poisson Equation**

Instead, we investigate the second derivative of $Y_{kl\nu}(r)$,

$$\begin{aligned}\frac{\mathrm{d}^2}{\mathrm{d}r^2}Y_{kl\nu}(r) &= -\frac{\nu+1}{r^2}Y_{kl\nu}(r) + \frac{\nu+1}{r}\frac{\mathrm{d}}{\mathrm{d}r}Y_{kl\nu}(r)\\ &\quad -(2\nu+1)\left\{-\frac{\nu+1}{r}Z_{kl\nu}(r) + \frac{\rho_{kl}(r)}{r}\right\} \qquad (9.82)\end{aligned}$$

which can be simplified and reordered to become

$$\left(\frac{\mathrm{d}^2}{\mathrm{d}r^2} - \frac{\nu(\nu+1)}{r^2}\right)Y_{kl\nu}(r) = -\frac{2\nu+1}{r}\rho_{kl}(r) \tag{9.83}$$

This second-order differential equation is known as the radial Poisson equation. The left hand side of Eq. (9.88) is formally identical to the one used in the nonrelativistic case.

The boundary conditions are $Y_{kl\nu}(0) = 0$ and $Y_{kl\nu}(\infty) = 0$ (except in the case where $\nu = 0$ and $k = l$ for the Coulomb interaction; see below) and ν taking values from $|j_k - j_l|$ to $j_k + j_l$ with the restriction that $\nu + j_k + j_l$ is even if $\text{sgn}(\kappa_k) \neq \text{sgn}(\kappa_l)$ and that it is odd if $\text{sgn}(\kappa_k) = \text{sgn}(\kappa_l)$. The differential equation is to be solved as a two-point boundary problem.

9.3.4.3 **Breit Potential Functions**

To evaluate the two-electron integrals within the CI and SCF equations the functions $W_{kl\nu}(r)$ and $V_{kl\nu}(r)$ can be calculated in close analogy to the Coulomb case [158]. With the definition of a general 'density' distribution

$$\rho_{kl}(r) = P_k(r)Q_l(r) \tag{9.84}$$

we have for the potential functions

$$W_{kl\nu}(r) = \frac{1}{r_1^{\nu+1}} \int_0^{r_1} d\,r_2 \rho_{kl}(r_2) r_2^\nu + r_1^\nu \int_{r_1}^\infty d\,r_2 \frac{\rho_{kl}(r_2)}{r_2^{\nu+1}} \tag{9.85}$$

$$V_{kl\nu}(r) = -\frac{1}{r_1^{\nu+1}} \int_0^{r_1} d\,r_2 \rho_{kl}(r_2) r_2^\nu + r_1^\nu \int_{r_1}^\infty d\,r_2 \frac{\rho_{kl}(r_2)}{r_2^{\nu+1}} \tag{9.86}$$

Again, there is an efficient way for computing these integrals [158] which can be easily shown using the additional definition

$$Y^S_{kl\nu}(r) = r S_{kl\nu}(r)\,, \text{with } S = W \text{ or } V \tag{9.87}$$

The second derivative of $Y^S_{kl\nu}(r)$ may be rewritten as a Poisson equation for the determination of $Y^U_{kl\nu}(r) = rW_{kl\nu}(r)$,

$$\left(\frac{d^2}{dr^2} - \frac{\nu(\nu+1)}{r^2}\right) Y^U_{kl\nu}(r) = -\frac{2\nu+1}{r}\rho_{kl}(r) \tag{9.88}$$

and as a Poisson-type equation for the determination of $Y^V_{kl\nu}(r) = rV_{kl\nu}(r)$,

$$\left(\frac{d^2}{dr^2} - \frac{\nu(\nu+1)}{r^2}\right) Y^V_{kl\nu}(r) = -\frac{\rho_{kl}(r)}{r} - 2\rho'_{kl}(r) \tag{9.89}$$

with the derivative with respect to r defined as

$$\rho'_{kl}(r) = P'_k(r)Q_l(r) + P_k(r)Q'_l(r) \tag{9.90}$$

9.4 Total Expectation Values

The expansion of the wave function in a basis of Slater determinants allows us to reduce matrix elements of any kind to one- and two-electron integrals as demonstrated for the energy expectation value for a general many-electron system in section 8.5.4. Following the discussion of the one- and two-electron matrix elements in the preceding section we can now write the total electronic energy for any atomic configuration interaction wave function in terms of radial integrals only. Expectation values for other atomic property operators can be derived in the very same way.

We should emphasize that the shell indices in the following expression strictly refer to occupied bound-state spinors. Neither positive-energy nor negative-energy spinors are taken into account and the finally derived SCF equations are solved for these states, i.e., the solution algorithms are designed such that the relativistic solutions are searched for in the vicinity of the nonrelativistic solutions so that it is guaranteed that no variational collapse occurs. This is an implicit protocol to implement the explicit projection operators discussed in section 8.7.1.

9.4.1 General Expression for the Electronic Energy

From the atomic many-electron Hamiltonian we calculate the total electronic energy of an electronic state A according to Eq. (8.113). Then along the lines of what has been presented in chapter 8 we get for the expectation value of an atomic Dirac–Coulomb Hamiltonian [277]

$$
\begin{aligned}
E_{el,A} &= \sum_{IJ} C_{IA} C_{JA} \left[\sum_{ij}^{n} h_{ij} \delta_{\kappa_i \kappa_j} \sum_{m_{j(p)}=-j_i}^{j_i} \sum_{m_{j(q)}=-j_j}^{j_j} \delta_{m_{j(p)} m_{j(q)}} \sum_{KL} B_{KI} B_{LJ} t_{pq}^{KL} \right. \\
&\quad + \frac{1}{2} \sum_{ijkl}^{n} \sum_{\nu} \left(\langle P_i(r) | U_{kl\nu}(r) | P_j(r) \rangle + \langle Q_i(r) | U_{kl\nu}(r) | Q_j(r) \rangle \right) \\
&\quad \times \sum_{m_{j(p)}=-j_i}^{j_i} \sum_{m_{j(q)}=-j_j}^{j_j} \sum_{m_{j(r)}=-j_k}^{j_k} \sum_{m_{j(s)}=-j_l}^{j_l} \left\{ \delta_{(m_{j(q)}+m_{j(s)})(m_{j(p)}+m_{j(r)})} \right. \\
&\quad \left. \left. \times d^{\nu}_{j_i m_{j(p)} j_j m_{j(q)}} d^{\nu}_{j_l m_{j(s)} j_k m_{j(r)}} \sum_{KL} B_{KI} B_{LJ} T^{KL}_{pqrs} \right\} \right]
\end{aligned}
\tag{9.91}
$$

where the integration over all angular degrees of freedom has already been carried out. Additionally, we have substituted the integrals over spinors by integrals over radial functions for which we must adjust the summation boundary n' of section 8.6.3 to n ($n \leq n'$ but usually $n \ll n'$), where n is now the total number of shells. The CI coefficients C_{IA} as well as the CSF expansion coefficients B_{KI} have been chosen to be real without loss of generality. The energy expression above can then be simplified as in section 8.6.3 by introducing the structure factors,

$$\gamma_{ij}^{IJ} = \delta_{\kappa_i\kappa_j} \sum_{m_{j(p)}=-j_i}^{j_i} \sum_{m_{j(q)}=-j_j}^{j_j} \delta_{m_{j(p)}m_{j(q)}} \sum_{KL} B_{KI}B_{LJ}t_{pq}^{KL} \tag{9.92}$$

and

$$\Gamma_{ijkl\nu}^{IJ} = \sum_{m_{j(p)}=-j_i}^{j_i} \sum_{m_{j(q)}=-j_j}^{j_j} \sum_{m_{j(r)}=-j_k}^{j_k} \sum_{m_{j(s)}=-j_l}^{j_l} \left\{ \delta_{(m_{j(q)}+m_{j(s)})(m_{j(p)}+m_{j(r)})} \right. \\ \left. \times d^{\nu}_{j_i m_{j(p)} j_j m_{j(q)}} d^{\nu}_{j_l m_{j(s)} j_k m_{j(r)}} \sum_{KL} B_{KI}B_{LJ}T_{pqrs}^{KL} \right\} \tag{9.93}$$

to yield [277]

$$E_{el,A} = \sum_{IJ} C_{IA}C_{JA} \left[\sum_{ij}^{n} h_{ij}\gamma_{ij}^{IJ} + \frac{1}{2}\sum_{ijkl}^{n}\sum_{\nu} g_{ijkl\nu}\Gamma_{ijkl\nu}^{IJ} \right] \tag{9.94}$$

Of course, the CI coefficients can be absorbed in the structure factors

$$\gamma_{ij} = \sum_{IJ} C_{IA}C_{JA}\gamma_{ij}^{IJ} \tag{9.95}$$

$$\Gamma_{ijkl\nu} = \sum_{IJ} C_{IA}C_{JA}\Gamma_{ijkl\nu}^{IJ} \tag{9.96}$$

so that the electronic energy reads

$$E_{el,A} = \sum_{ij}^{n} h_{ij}\gamma_{ij} + \frac{1}{2}\sum_{ijkl}^{n}\sum_{\nu} g_{ijkl\nu}\Gamma_{ijkl\nu} \tag{9.97}$$

In this way, the CI coefficients finally enter the SCF equations, which is necessary to couple spinor and CI coefficient optimizations in MCSCF calculations. For the sake of brevity we have skipped the state index A for the structure factors $\gamma_{ij}^{(A)} \rightarrow \gamma_{ij}$ and $\Gamma_{ijkl\nu}^{(A)} \rightarrow \Gamma_{ijkl\nu}$. Note also that the two-electron structure factors $\Gamma_{ijkl\nu}^{IJ}$ and $\Gamma_{ijkl\nu}$, respectively, depend on the additional subscript ν introduced via the expansion of the $1/r_{12}$ operator in Eq. (9.9).

9.4.2
Breit Contribution to the Total Energy

The Breit interaction changes the total energy,

$$E_{el,A}^{CB} = E_{el,A} + E_{el,A}^{B} \tag{9.98}$$

There are two possibilities to calculate $E^{CB}_{el,A}$, namely as $\langle E_{el} + E^{B}_{el,A} \rangle$ where the Breit contribution is added to the total energy with radial functions determined self-consistently using only the Coulomb interaction, i.e. added as a first-order perturbation, or where both Coulomb and Breit operators are used for the self-consistent calculation of the radial functions. However, in both cases the analytical form of the matrix element is the same so that it is not necessary to distinguish between these cases here.

$E^{B}_{el,A}$ is the contribution resulting from the frequency-independent Breit interaction

$$E^{B}_{el,A} = \left\langle \Psi_{el,A} \left| \frac{1}{2} \sum_{i \neq j} B_0(i,j) \right| \Psi_{el,A} \right\rangle = \sum_{IJ} C^{\star}_{I} C_J \left\langle \Phi_I \left| \frac{1}{2} \sum_{i \neq j} B_0(i,j) \right| \Phi_J \right\rangle \quad (9.99)$$

expanded as before in terms of configuration state functions Φ_I which can be written as minimum linear combinations of Slater determinants Θ_R with analytically known expansion coefficients B_{RI} (see section 9.2). To reduce this matrix element to radial integrals we proceed as in the case of the Coulomb electron–electron interaction. A matrix element over CSFs reads

$$\begin{aligned} \left\langle \Phi_I \left| \frac{1}{2} \sum_{i \neq j} B_0(i,j) \right| \Phi_J \right\rangle &= \frac{1}{2} \sum_{pqrs} b_{0,pqrs} \sum_{RS} B_{RI} T^{RS}_{pqrs} B_{SJ} \\ &= \frac{1}{2} \sum_{ijklv} \sum_{n\pm} \left[C^{IJ}_{n\pm ijklv} c^{n}_{ijklv} + D^{IJ}_{n\pm ijklv} d^{n}_{ijklv} \right] \end{aligned} \quad (9.100)$$

with the CSF-index-dependent coefficients

$$G^{IJ}_{n\pm ijklv} = \sum_{m_p=-j_i}^{j_i} \sum_{m_q=-j_j}^{j_j} \sum_{m_r=-j_k}^{j_k} \sum_{m_s=-j_l}^{j_l} G_{n\pm pqrsv} \sum_{RS} B_{RI} B_{SJ} T^{RS}_{pqrs} \quad (9.101)$$

where G^{IJ}_{nijklv} is a substitute for either A^{IJ}_{nijklv} or B^{IJ}_{nijklv}, respectively. The coefficients from Eqs. (9.101) contain all analytically known coefficients in the expression for the frequency-independent Breit contribution. Finally, these coefficients enter the final Breit structure factors

$$\Gamma_{n\pm ijklv} = \sum_{IJ} C_{IA} C_{JA} \; C^{IJ}_{n\pm ijklv} \text{ and } \Delta_{n\pm ijklv} = \sum_{IJ} C_{IA} C_{JA} \; D^{IJ}_{n\pm ijklv} \quad (9.102)$$

so that $E^{B}_{el,A}$ can now be written as depending on shell indices

$$E^{B}_{el,A} = \frac{1}{2} \sum_{ijklvn\pm} \Gamma_{n\pm ijklv} c_{n\pm ijklv} + \frac{1}{2} \sum_{ijklvn\pm} \Delta_{n\pm ijklv} \, d_{n\pm ijklv} \quad (9.103)$$

where the radial integrals to be determined are given in Eqs. (9.68)–(9.71). The sums in Eq. (9.103) can be restricted because of symmetry relations [158].

9.4.3
Dirac–Hartree–Fock Total Energy of Closed-Shell Atoms

The energy expectation values studied so far were all written for the most general ansatz for the total state, which is a multi-configuration wave function that also includes the full configuration interaction case. The independent-particle model, which is the basis of Hartree–Fock theory, leads to a single Slater determinant to approximate the quantum mechanical many-electron state. The energy expression for this most simple model wave function may be derived either along the lines of the derivation of the multi-configuration energy expression or by simplification of the final ground-state energy expression noting that all C_{IA} and B_{RI} coefficients vanish except for $C_{10} = B_{11} = 1$ and that most matrix elements over creation and annihilation operators vanish.

In accordance with the equivalence restriction a shell is classified by one pair of quantum numbers n and κ. A shell comprises $2|\kappa|$ atomic spinors and is thus $2|\kappa|$-fold degenerate. A closed shell is characterized by the fact that all of these degenerate spinors enter the single Slater determinant in Dirac–Hartree–Fock theory.

In closed-shell atoms, the m_j dependence of the structure factors vanishes so that much simpler equations are obtained. The expectation value of the total ground state energy can be written in cases of atoms with closed shells for the electronic state A as (cf. Refs. [89] and [24, p. 152]),

$$
\begin{aligned}
E_{el,0}^{\mathrm{DHF}} = & \sum_i d_i \left[c\hbar \left\langle P_i \left| -\frac{\mathrm{d}}{\mathrm{d}r} + \frac{\kappa_i}{r} \right| Q_i \right\rangle + \left\langle P_i \middle| V_{nuc} \middle| P_i \right\rangle \right. \\
& \left. + c\hbar \left\langle Q_i \left| \frac{\mathrm{d}}{\mathrm{d}r} + \frac{\kappa_i}{r} \right| P_i \right\rangle + \left\langle Q_i \left| V_{nuc} - 2m_e c^2 \right| Q_i \right\rangle \right] \\
& + \frac{1}{2} \sum_i d_i(d_i - 1) \left(\langle P_i | U_{ii0} | P_i \rangle + \langle Q_i | U_{ii0} | Q_i \rangle \right) \\
& + \sum_{ij, i>j} d_i d_j \left(\langle P_i | U_{jj0} | P_i \rangle + \langle Q_i | U_{jj0} | Q_i \rangle \right) \\
& - \frac{1}{4} \sum_i \sum_{\nu>0} d_i^2 A_{ii\nu} \left(\langle P_i | U_{ii\nu} | P_i \rangle + \langle Q_i | U_{ii\nu} | Q_i \rangle \right) \\
& - \frac{1}{2} \sum_{ij, i>j} \sum_{\nu} d_i d_j A_{ij\nu} \left(\langle P_i | U_{ij\nu} | P_j \rangle + \langle Q_i | U_{ij\nu} | Q_j \rangle \right)
\end{aligned}
\tag{9.104}
$$

with the potential energy functions $U_{ij\nu}$ defined as before, and the coefficient

$$
A_{ij\nu} = 2 \begin{pmatrix} j_i & \nu & j_j \\ 1/2 & 0 & -1/2 \end{pmatrix}^2
$$

resulting from the integration over angular coordinates. $d_i = 2|\kappa_i|$ denotes the occupation number of the i-th shell. The sums over ν run from $\nu = |j_i - j_j|$ to $\nu = j_i + j_j$ (constraints: $j_i + j_j + \nu$ has to be even if sgn(κ_i)$\neq$sgn(κ_j) and $j_i + j_j + \nu$ has to be odd if sgn(κ_i)=sgn(κ_j)).

A simplified closed-shell expression for the Breit interaction can be derived as well [158]. For open-shell atoms, an average of configurations yields a similar total energy expression except for the intra-shell exchange energy [24, p. 164].

9.5 General Self-Consistent-Field Equations and Atomic Spinors

In section 8.7 it was discussed how the one-electron spinors can be determined from the variational principle yielding self-consistent field equations. In the atomic case, these equations reduce to coupled ordinary differential equations because we have integrated out all angular degrees of freedom already at the stage of the total energy expectation value. The SCF equations can be derived either from the self-consistent field Eqs. (8.181) by carrying out the transformation to polar coordinates and integration over all angular variables afterwards or by explicit variation of the total energy expressions derived in the preceding section along the lines of the general formulation in chapter 8.

What remains is the calculation of the two radial functions $P_{n_i\kappa_i}(r)$ and $Q_{n_i\kappa_i}(r)$. For this task the coupled first-order differential SCF equations have to be solved. The explicit form of these equations depends on the many-electron Hamiltonian used to calculate the total electronic energy. For the Dirac–Coulomb Hamiltonian in combination with the general multi-configuration wave function these equations turn out to read [277]

$$\begin{pmatrix} V_i^P(r) - \epsilon_{ii} & A_i^\dagger(r) \\ A_i(r) & V_i^Q(r) - \epsilon_{ii} \end{pmatrix} \begin{pmatrix} P_i(r) \\ Q_i(r) \end{pmatrix} = \begin{pmatrix} X_i^P(r) \\ X_i^Q(r) \end{pmatrix} \tag{9.105}$$

with the kinetic and potential terms

$$A_i(r) = \gamma_{ii} c\hbar \left(\frac{\mathrm{d}}{\mathrm{d}r} + \frac{\kappa_i}{r} \right) \implies A_i^\dagger(r) = \gamma_{ii} c\hbar \left(-\frac{\mathrm{d}}{\mathrm{d}r} + \frac{\kappa_i}{r} \right) \tag{9.106}$$

$$V_i^P(r) = \gamma_{ii} V_{\mathrm{nuc}}(r) + \sum_{kl\nu} \Gamma_{iikl\nu} U_{kl\nu}(r) \tag{9.107}$$

$$V_i^Q(r) = V_i^P(r) - \gamma_{ii} 2 m_e c^2 \tag{9.108}$$

and the inhomogeneities

$$
\begin{aligned}
X_i^P(r) &= -\sum_{j,j\neq i}\left[\left(\sum_{klv}\Gamma_{ijklv}U_{klv}(r)+\gamma_{ij}V_{\text{nuc}}(r)-\epsilon_{ij}\right)P_j(r)\right.\\
&\quad\left.+\gamma_{ij}c\hbar\left(-\frac{\mathrm{d}}{\mathrm{d}r}+\frac{\kappa_j}{r}\right)Q_j(r)\right] \qquad (9.109)\\
X_i^Q(r) &= -\sum_{j,j\neq i}\left[\left(\sum_{klv}\Gamma_{ijklv}U_{klv}(r)+\gamma_{ij}(V_{\text{nuc}}(r)-2m_ec^2)-\epsilon_{ij}\right)Q_j(r)\right.\\
&\quad\left.+\gamma_{ij}c\hbar\left(\frac{\mathrm{d}}{\mathrm{d}r}+\frac{\kappa_j}{r}\right)P_j(r)\right] \qquad (9.110)
\end{aligned}
$$

These SCF equations contain the r variable only and are thus one-dimensional, which makes them particularly accessible by numerical solution methods. The corresponding Dirac–Coulomb–*Breit* SCF equations [158] can be obtained in an analogous way from the energy expression that includes the Breit term. It is, however, interesting to note that some of the Breit potential energy terms enter the *off-diagonal* and are thus added to $A_i^{(\dagger)}(r)$. This is a consequence of the off-diagonal, so-called 'odd' nature of the Breit interaction that couples large and small components (compare also chapter 12).

The inhomogeneous SCF Eqs. (9.105) for *all* shells may be cast into the form of a homogeneous multiparameter super-equation. Its solution, however, is not as straightforward as the sequential solution of the individual inhomogeneous equations [390, 391].

The CI expansion coefficients, which enter the structure factors γ_{ij} and Γ_{ijkl}, are obtained as eigenvectors of the CI eigenvalue problem introduced in section 8.7 for a CI-type wave function.

The non-diagonal Lagrangian multipliers ϵ_{ij} of Eq. (9.105) can be calculated by multiplication with $(P_j(r), Q_j(r))$ and integration to get

$$
\begin{aligned}
\epsilon_{ij} &= \sum_n\Big[\langle P_j(r)|A_n^\dagger(r)|Q_n(r)\rangle+\langle Q_j(r)|A_n(r)|P_n(r)\rangle\\
&\quad+\langle P_j(r)|-\gamma_{ii}-V_{\text{nuc}}+\sum_{klv}\Gamma_{inklv}U_{klv}|P_n(r)\rangle\\
&\quad\langle Q_j(r)|-\gamma_{ii}(-V_{\text{nuc}}+2m_ec^2)+\sum_{klv}\Gamma_{inklv}U_{klv}|Q_n(r)\rangle\Big] \qquad (9.111)
\end{aligned}
$$

This equation does not have to be evaluated for all possible ϵ_{ij}. The condition $\epsilon_{ij} = \epsilon_{ji}$ proved by Hinze [223] allows one to control the numerical solution of the SCF equations and is actually an identical solution condition to obtain the MCSCF spinors. It can be fulfilled (in terms of machine precision and numerical accuracy) only if the SCF iteration is converged. On the other hand, a large discrepancy $\epsilon_{ij} \neq \epsilon_{ji}$ could result in convergence problems. Of course, Eq. (9.111) has to be adjusted if the Breit interaction enters the two-electron interaction terms and hence modifies the potential functions. These changes, however, are straightforward [158].

9.5.1
Dirac–Hartree–Fock Equations

Because of the reduced number of terms in the Dirac–Hartree–Fock energy, the self-consistent-field equations for the single-determinant approximation of Hartree–Fock theory contain comparatively few potential and inhomogeneity functions, although they adopt the same form as the most general SCF equations discussed so far. The variation of the Lagrangian functional for the closed-shell case yields the Dirac–Hartree–Fock equations, which are equal to Eqs. (9.105) if (i) the one-electron structure factor γ_{pp} calculated or a given atomic spinor p (not for the corresponding shell i) is set equal to one in Eqs. (9.106) and (9.108),

$$\gamma_{pp} \quad \rightarrow \quad 1 \tag{9.112}$$

and if (ii) the following modifications are taken into account

$$\begin{aligned}
V_i^P(r) \quad &\rightarrow \quad V_{\mathrm{nuc}}(r) + \sum_j d_j U_{jj0}(r) - \sum_{\nu=0} \frac{d_i}{2} A_{ii\nu} U_{ii\nu}(r) \\
X_i^P(r) \quad &\rightarrow \quad \sum_{j,j\neq i} \left[\sum_\nu \frac{d_j}{2} A_{ij\nu} U_{ij\nu}(r) P_j(r) + \epsilon_{ij} \delta_{\kappa_i,\kappa_j} P_j(r) \right] \\
X_i^Q(r) \quad &\rightarrow \quad \sum_{j,j\neq i} \left[\sum_\nu \frac{d_j}{2} A_{ij\nu} U_{ij\nu}(r) Q_j(r) + \epsilon_{ij} \delta_{\kappa_i,\kappa_j} Q_j(r) \right]
\end{aligned}$$

with the summation indices j running over all shells *occupied* in the Hartree–Fock–Slater determinant. The occupation numbers are given by $d_i = 4l_i + 2$ as introduced in section 9.4.3. The one-electron structure factors γ_{ii} for the shells (not for the spinors p of that shell) then play the role of generalized occupation numbers. The closed-shell equations with the frequency-independent Breit interaction contain, like the more general Breit MCSCF equations, additive terms on the off-diagonal (as well as to the inhomogeneity) [158].

9.5.2
Comparison of Atomic Hartree–Fock and Dirac–Hartree–Fock Theories

The nonrelativistic Hartree–Fock theory (abbreviated HF in the equations to follow) formally resembles Dirac–Hartree–Fock (DHF) theory for the Dirac–Coulomb Hamiltonian. Of course, for large c of, say, 10^4 to 10^5 a.u. they even yield the same results. For this reason we shall make an explicit comparison of both in this section. The total energy for closed-shell atoms after integration over all angular and spin coordinates is in both cases given by

$$E_{el,0} = \sum_i d_i h_{ii} + \frac{1}{2} \sum_{ij} d_i d_j \left[\int_0^\infty dr\, \rho_{ii} U_{jj0} - \frac{1}{2} \sum_\nu A_{ij\nu} \int_0^\infty dr\, \rho_{ij} U_{ji\nu} \right] \quad (9.113)$$

if we introduce the general one-electron integrals h_{ii},

$$h_{ii} = \begin{cases} h_{ii}^S = \langle P_i(r) | h^S(r) | P_i(r) \rangle \\ h_{ii}^D = \langle (P_i(r), Q_i(r)) | h^D(r) | (P_i(r), Q_i(r)) \rangle \end{cases} \quad (9.114)$$

over the Schrödinger and Dirac one-electron radial Hamiltonians, $h^S(r)$ and $h^D(r)$, respectively. These one-electron Hamiltonians are explicitly given by

$$h^S(r) = -\frac{\hbar^2}{2m_e} \frac{d^2}{dr^2} + \frac{l_i(l_i+1)\hbar^2}{2m_e r^2} + V_{nuc}(r) \quad (9.115)$$

$$h^D(r) = \begin{pmatrix} V_{nuc}(r) & A_i^\dagger(r) \\ A_i(r) & V_{nuc}(r) - 2m_e c^2 \end{pmatrix} \quad (9.116)$$

The relativistic variant of the energy expression of Eq. (9.113) was introduced in section 9.4.3, while the nonrelativistic energy is not derived in this book, and we may refer the reader to the presentation by others [24,388]. Of course, its derivation follows the very same principles presented in this chapter for the Dirac-based many-electron Hamiltonian, and all expressions of Hartree–Fock theory would also be obtained upon taking the nonrelativistic limit, $c \to \infty$ (compare also section 6.6).

For the sake of convenience, we defined a generalized radial density $\rho_{ij}(r)$,

$$\rho_{ij}(r) = \begin{cases} P_i(r)P_j(r) & \text{for Hartree–Fock} \\ P_i(r)P_j(r) + Q_i(r)Q_j(r) & \text{for Dirac–Hartree–Fock} \end{cases} \quad (9.117)$$

which allows us to write the two-electron terms for both theories in a formally identical way.

In the case of Hartree–Fock theory, the summation over ν in Eq. (9.113) runs from $|l_i - l_j|$ to $l_i + l_j$, with $\nu + l_i + l_j$ even. In the relativistic Dirac–Hartree–Fock framework, the summation is for all ν from $|j_i - j_j|$ to $j_i + j_j$, with $\nu + j_i + j_j$ odd if $\text{sgn}(\kappa_i) = \text{sgn}(\kappa_j)$, and $\nu + j_i + j_j$ even if $\text{sgn}(\kappa_i) \neq \text{sgn}(\kappa_j)$. The symmetry coefficients are given as squared Wigner 3j-symbols

$$A_{ij\nu}^{\text{HF}} = \begin{pmatrix} l_i & \nu & l_j \\ 0 & 0 & 0 \end{pmatrix}^2 \quad \text{and} \quad A_{ij\nu}^{\text{DHF}} = 2 \begin{pmatrix} j_i & \nu & j_j \\ \frac{1}{2} & 0 & -\frac{1}{2} \end{pmatrix}^2 \quad (9.118)$$

For closed-shell atoms the occupation numbers d_i are given by

$$d_i = \begin{cases} 4l_i + 2 & \text{for Hartree–Fock} \\ 2|\kappa_i| = 2j_i + 1 & \text{for Dirac–Hartree–Fock} \end{cases} \tag{9.119}$$

(compare the degeneracy of shells in Table 6.1).

Variation of the Hartree–Fock total electronic energy expectation value with respect to the orbitals yields the radial Hartree–Fock equations [392, 393] that can be written as homogeneous differential equations

$$\left(-\frac{\hbar^2}{2m_e} \frac{\mathrm{d}^2}{\mathrm{d}r^2} + \frac{l_i(l_i+1)\hbar^2}{2m_e r^2} + V_{nuc} + W_i^S \right) P_i(r) = \epsilon_i P_i(r) \tag{9.120}$$

where we introduced the pseudo-local Hartree–Fock electron–electron interaction potential

$$\begin{aligned} W_i^S(r) &\equiv \sum_j d_j \left[U_{jj0}(r) - \frac{1}{2} \sum_\nu A_{ij\nu}^{\mathrm{HF}} U_{ji\nu}(r) \frac{P_j(r)}{P_i(r)} \right] \\ &= V_i(r) + \frac{X_i^S(r)}{P_i(r)} \end{aligned} \tag{9.121}$$

The superscript S indicates the nonrelativistic Hartree–Fock approach based on the Schrödinger one-electron operator. The Hartree–Fock equations are written in a somewhat artificially homogeneous fashion because of the devision of $X_i^S(r)$ by $P_i(r)$, and inhomogeneities arise from the j-shell contributions through $P_j(r)$ (and therefore the Hartree–Fock interaction potentials are pseudo-local). Potential singularities due to the nodes of the $P_i(r)$ functions are ignored.

The pseudo-local presentation of the self-consistent-field Hartree–Fock equation nicely shows how closely these one-particle equations formulated for a many-electron atom resemble the Schrödinger equation for one-electron atoms. Hence, the SCF equation for a given orbital of given angular momentum l is formally like the Schrödinger equation for the electron in the hydrogen atom in the same angular momentum state l. The main difference is the occurrence of the mean-field potential $W_i^S(r)$, which comprises the interaction with all other electrons through the radial functions which enter $W_i^S(r)$. It is also this pseudo-local mean-field potential of the electron–electron interaction which prevents the analytical solution of the Hartree–Fock equation considering simply that $W_i^S(r)$ depends on all solution functions.

In close analogy to the nonrelativistic Hartree–Fock approach, we rewrite the relativistic Dirac–Hartree–Fock equations

$$\begin{pmatrix} V_{nuc} + W_i^P & A_i^\dagger \\ A_i & V_{nuc} + W_i^Q - 2m_e c^2 \end{pmatrix} \begin{pmatrix} P_i(r) \\ Q_i(r) \end{pmatrix} = \epsilon_i \begin{pmatrix} P_i(r) \\ Q_i(r) \end{pmatrix} \tag{9.122}$$

with pseudo-local Dirac–Hartree–Fock interaction potentials

$$\begin{aligned} W_i^P(r) &\equiv \sum_j d_j \left[U_{jj0}(r) - \frac{1}{2} \sum_\nu A_{ij\nu}^{\mathrm{DHF}} U_{ji\nu}(r) \frac{P_j(r)}{P_i(r)} \right] \\ &= V_i(r) + \frac{X_i^P(r)}{P_i(r)} \end{aligned} \tag{9.123}$$

$$\begin{aligned} W_i^Q(r) &\equiv \sum_j d_j \left[U_{jj0}(r) - \frac{1}{2} \sum_\nu A_{ij\nu}^{\mathrm{DHF}} U_{ji\nu}(r) \frac{Q_j(r)}{Q_i(r)} \right] \\ &= V_i(r) + \frac{X_i^Q(r)}{Q_i(r)} \end{aligned} \tag{9.124}$$

whereby the functions $V_i(r)$ and $X_i^{(P,Q)}(r)$ and their nonrelativistic analogs will be analyzed in the following. These quasi-local self-consistent-field potentials $W_i^{(S,P,Q)}(r)$ contain two different parts, $V_i(r)$ and $X_i^{(S,P,Q)}(r)$. For the homogeneous part, $V_i(r)$, which represents the Coulomb interaction and the correction for the self-interaction, we get

$$V_i(r) = \sum_j d_j U_{jj0}(r) - \frac{1}{2} d_i \sum_\nu A_{ii\nu} U_{ii\nu}(r) \tag{9.125}$$

for both Hartree–Fock and Dirac–Hartree–Fock potentials. The inhomogeneous part, which originates from the exchange interaction only, is in the nonrelativistic Hartree–Fock framework

$$X_i^S(r) = -\frac{1}{2} \sum_{j \neq i} d_j \sum_\nu A_{ij\nu}^{\mathrm{HF}} U_{ji\nu}(r) P_j(r) \tag{9.126}$$

while we obtain in the relativistic case

$$X_i^P(r) = -\frac{1}{2} \sum_{j, j \neq i} d_j \sum_\nu A_{ij\nu}^{\mathrm{DHF}} U_{ji\nu}(r) P_j(r) \tag{9.127}$$

$$X_i^Q(r) = -\frac{1}{2} \sum_{j, j \neq i} d_j \sum_\nu A_{ij\nu}^{\mathrm{DHF}} U_{ji\nu}(r) Q_j(r) \tag{9.128}$$

As in the case of nonrelativistic Hartree–Fock theory, the Dirac–Hartree–Fock equations (and, of course, also any other form of self-consistent-field equations like the MCSCF equations) in pseudo-local form resemble the quantum mechanical equation of the Dirac hydrogen atom. Again, it is the mean-field potential functions $W_i^P(r)$ and $W_i^Q(r)$ which prohibit an analytical solution as they depend on all solution functions. Still, we are able to use the same symmetry labels for the spinors of the many-electron atom as we used for the Dirac hydrogen atom. The κ- or j-symmetry of an atomic spinor of a given many-electron atom or ion thus corresponds to the ground or excited state of the Dirac hydrogen-like atom with same nuclear charge number Z (or, more generally speaking, with the same electron–nucleus potential V_{nuc}). Of course, all this is due to the spherical symmetry.

9.5.3
Relativistic and Nonrelativistic Electron Densities

The spherically averaged electron density is an important quantity [394] as it can be defined even for an atom in a molecule. It follows from the general definition of the electron ground-state density $\rho(\boldsymbol{r})$, which is to be calculated from the total electronic ground state wave function Ψ_0, Eq. (4.9). For a representation in polar coordinates $\boldsymbol{r} = (r, \varphi, \vartheta)$ we define a *radial* electron density $D(r)$ such that

$$\int_0^\infty \mathrm{d}r D(r) \equiv N \tag{9.129}$$

Comparing this definition with the integration of $\rho(\boldsymbol{r})$ over all space, Eq. (4.10)

$$\int_{-\infty}^{\infty} \mathrm{d}^3 r\, \rho(\boldsymbol{r}) = \int_0^\infty r^2 \mathrm{d}r \int_0^{2\pi} \int_0^{\pi} \sin\vartheta\, \mathrm{d}\vartheta\, \mathrm{d}\varphi\, \rho[\boldsymbol{r}(\boldsymbol{r})] = N \tag{9.130}$$

we find

$$D(r) = r^2 \int_0^{2\pi} \int_0^{\pi} \sin\vartheta\, \mathrm{d}\vartheta\, \mathrm{d}\varphi\, \rho(\boldsymbol{r}) \tag{9.131}$$

In order to define a density which addresses an atom in a general molecule as well as a spherically symmetric atom, we need to define a spherically averaged density $\overline{\rho}(r)$ that yields the total (electronic) charge in a spherical shell $\overline{\rho}_C$ with inner radius r and thickness $\mathrm{d}r$, i.e., $D(r)\mathrm{d}r$, divided by the volume of this shell ($4\pi r^2 \mathrm{d}r$),

$$\overline{\rho}(r) = \frac{D(r)\mathrm{d}r}{4\pi r^2 \mathrm{d}r} = \frac{D(r)}{4\pi r^2} \quad \rightarrow \quad \overline{\rho}_C(r) = q_e\, \overline{\rho}(r) = \tag{9.132}$$

so that integration over all shells, $\int \overline{\rho}(r)\, 4\pi r^2 \mathrm{d}r = N$, still yields the total number of electrons. This spherically averaged electron density $\overline{\rho}(r)$ is the quantity one should consider for the investigation of atomic properties when discussing the electron density along a radial ray starting from the nucleus of an isolated atom or of an atom in a molecule. In general, one may write the electron density in the language of second quantization as

$$\rho(\boldsymbol{r}) = \sum_{pq}^{k} \gamma_{pq} \psi_p^\dagger(\boldsymbol{r}) \psi_q(\boldsymbol{r}) \tag{9.133}$$

for a relativistic spinor as well as for a nonrelativistic orbital. Here, k is the total number of such one-electron states and $\{\gamma_{pq}\}$ is the first-order density matrix (structure factor for spinors) as before. The density matrix elements may be written for both cases as

$$\gamma_{pq} = \left\langle \Psi_0 \left| \hat{E}_{pq} \right| \Psi_0 \right\rangle = \left\langle a_p \Psi_0 | a_q \Psi_0 \right\rangle \tag{9.134}$$

with the excitation operator $\hat{E}_{pq}$ of second quantization

$$\hat{E_{pq}} = a_p^\dagger a_q \tag{9.135}$$

and operates on orbitals, spin orbitals or spinors, respectively. Expressing Ψ_0 in Eq. (9.134) by the linear combination of configuration state functions Φ_{IA} from Eq. (8.100) then yields for the ground state $A = 0$

$$\gamma_{pq} = \sum_{IJ} C_{I0} C_{J0} \left\langle \Phi_I \left| \hat{E}_{pq} \right| \Phi_J \right\rangle = \sum_{IJ} C_{I0} C_{J0} \left\langle a_p \Phi_I | a_q \Phi_J \right\rangle \tag{9.136}$$

where we assumed the CI expansion coefficients C_{I0} to be real. In Hartree–Fock theory with its single basis function Φ_{10} this simplifies to a single summation with $C_{I0} = C_{J0} = 1$ and the diagonal density matrix elements become the occupation numbers of spinor states

$$\gamma_{pq} \to \gamma_{pp} = \left\langle \Phi_1 \left| \hat{E}_{pp} \right| \Phi_1 \right\rangle = \begin{cases} 0 \text{ or } 2 & \text{nonrelativistic} \\ 0 \text{ or } 1 & \text{four-component} \end{cases} \tag{9.137}$$

due to the orthogonality of the one-electron functions ψ_i from which Φ_{10} is constructed. This illustrates explicitly that the integral described by the brackets in Eq. (9.136) using the annihilation operators to annihilate one orbital from Φ_{10} is equivalent to the integration over all but one electronic coordinate in Eq. (4.9). In addition, we should emphasize that in the nonrelativistic theory the last equation refers to the special case of a closed-shell electronic structure such that the excitation operator can be written for two spin states,

$$E_{pq} = a_{p\alpha}^\dagger a_{q\alpha} + a_{p\beta}^\dagger a_{q\beta}. \tag{9.138}$$

If, in the case of atoms, the angular degrees of freedom are integrated out analytically, we switch from spinor indices p, q to shell indices i, j and arrive at an explicit expression for $D(r)$,

$$D(r) = \sum_{ij} \gamma_{ij} D_{ij}(r) \times \left\{ \begin{array}{c} \left\langle Y_{l_i m_i} | Y_{l_j m_j} \right\rangle \\ \left\langle \chi_{\kappa_i m_{j,i}} | \chi_{\kappa_j m_{j,j}} \right\rangle \end{array} \right\} = \sum_i \gamma_{ii} D_{ii}(r) \tag{9.139}$$

where now the shell structure factors γ_{ij} need to be employed, and then the generalized radial shell density $D_{ij}(r)$ reads

$$D_{ij}(r) = \begin{cases} P_i(r) P_j(r) & \text{nonrelativistic} \\ P_i(r) P_j(r) + Q_i(r) Q_j(r) & \text{four-component} \end{cases} \tag{9.140}$$

The spherically averaged density then reads

$$\bar{\rho}(r) = (4\pi r^2)^{-1} D(r) = \begin{cases} \sum_{ij} \dfrac{\gamma_{ij}}{4\pi r^2} P_i(r)P_j(r) & \text{nonrelativistic} \\ \sum_{ij} \dfrac{\gamma_{ij}}{4\pi r^2} \left[P_i(r)P_j(r) + Q_i(r)Q_j(r)\right] & \text{four-component} \end{cases} \tag{9.141}$$

and is given in particles per bohr^3. In the special case of closed-shell molecules with N electrons represented by a single Slater determinant the general expression for $\rho(\boldsymbol{r})$ simplifies in the nonrelativistic theory to

$$\rho(\boldsymbol{r}) = \sum_{i=1}^{N/2} 2\psi_i^\star(\boldsymbol{r})\psi_i(\boldsymbol{r}) \tag{9.142}$$

and in case of the four-component relativistic theory to

$$\rho(\boldsymbol{r}) = \sum_{i=1}^{N} \psi_i^\dagger(\boldsymbol{r})\psi_i(\boldsymbol{r}) \tag{9.143}$$

if no additional symmetries (like spherical or point group symmetry) are exploited. In the case of atoms with nsh subshells, the equivalence restriction of one and the same radial function per subshell (nl) or $(n\kappa)$, respectively, allows us to simplify this equation even further, and we obtain for the radial and the spherically averaged densities

$$\bar{\rho}(r) = \sum_{i}^{nsh} d_i \frac{D_{ii}(r)}{4\pi r^2} \tag{9.144}$$

with the subshell occupation numbers as in Eq. (9.119).

9.6
Analysis of Radial Functions and Potentials at Short and Long Distances

Analytic knowledge on otherwise not exactly treatable many-electron equations is only available at the boundaries of the coordinate system, namely at short distances in close proximity to the nucleus and asymtoptically, i.e., at very large distances. Spinors turn out to be exponentially decreasing asymptotically (compare also chapter 6), which is connected to the normalization condition. Numerical solution methods always imply that accuracy must not be compromised at the boundaries, and special attention needs to be paid to this issue. Therefore, it is important to gain any analytic knowledge available on the otherwise numerically determined parameters, the radial functions, of a calculation.

The analytic results can be generalized to the molecular case if spherical averaging is applied to the electronic wave function of the molecule. In 1957, Kato derived analytic properties for the exact many-electron wave function with a focus on their behavior at an atomic nucleus with Coulomb singularity [395]. Some of these results were later considered from the point of view of a relativistic first-quantized many-electron theory [396].

9.6.1 Short-Range Behavior of Atomic Spinors

The radial functions are the only unknown functions in any ansatz for the atomic total wave function. Their calculation through the solution of self-consistent field equations just introduced is necessarily approximate — either through basis set expansion techniques or through representation on a mesh of grid points. Therefore, it is worthwhile to extract from the self-consistent-field equations any analytical knowledge possible on the atomic spinors. The resulting analytical knowledge is also of significance for any approximate scheme for nonspherical systems (like molecules) which may even lack radial symmetry that is then reintroduced through spherical averaging (as in the case of atomic centers in molecules). Usually, such information is obtained in the limiting cases for small and large radial distances.

We assume that the radial functions are analytic at the origin,

$$P_i(r) = r^{\alpha_i} \sum_{k=0}^{\infty} a_{k,i} r^k \tag{9.145}$$

$$Q_i(r) = r^{\alpha_i} \sum_{k=0}^{\infty} b_{k,i} r^k \tag{9.146}$$

and determine the first exponent of this series expansion α_i by solving the SCF equation for the ith shell using series expansions for the coefficient functions

$$V_i^P(r) = v_{-1,i} r^{-1} + v_{0,i} + O(r^k)\,,\ (k \in \mathbf{R}) \tag{9.147}$$

$$X_i^{P,Q}(r) = x_{-1,i} r^{-1} + x_{0,i} + O(r^k)\,,\ (k \in \mathbf{R}) \tag{9.148}$$

with

$$v_{-1,i} = v_{-1} = \begin{cases} -Z & \text{for point-like nuclei} \\ 0 & \text{for finite-sized nuclei} \end{cases} \tag{9.149}$$

We have already seen that the potential functions $U_{klv}(r)$ can contribute only to the r^j terms (with $j \geq 0$) in section 9.6.1.2. Here we assume that the inhomogeneity vanishes at the origin, i.e. $x_{-1} = x_0 = 0$. We obtain from the SCF equations for the coefficients of the $r^{\alpha_i - 1}$ term

$$(v_{-1}/c)\, a_0 + (\kappa_i - \alpha_i) b_0 = 0 \tag{9.150}$$

$$(v_{-1}/c)\, b_0 + (\kappa_i + \alpha_i) a_0 = 0 \tag{9.151}$$

which yields (in Hartree atomic units)

$$\alpha_i = \sqrt{\kappa_i^2 - (v_{-1})^2/c^2} = \begin{cases} \sqrt{\kappa_i^2 - Z^2/c^2} & \text{point-like nuclei} \\ |\kappa_i| & \text{finite-sized nuclei} \end{cases} \tag{9.152}$$

For point-like nuclei the first exponent of the series expansion is *not* integral. This creates substantial drawbacks for the numerical methods, which always require finite higher derivatives. The result, Eq. (9.152), has one important consequence for finite-sized nuclei, seen immediately when writing down Eqs. (9.150) and (9.151) explicitly for this case,

$$(\kappa_i - \alpha_i) b_{0,i} = 0 \tag{9.153}$$

$$(\kappa_i + \alpha_i) a_{0,i} = 0 \tag{9.154}$$

which yield

$$b_{0,i} = 0 \quad \text{if } \kappa_i > 0 \tag{9.155}$$

$$a_{0,i} = 0 \quad \text{if } \kappa_i < 0 \tag{9.156}$$

This result has consequences for the origin corrections in the discretized SCF equations of numerical solution methods (see below).

9.6.1.1 Cusp-Analogous Condition at the Nucleus

In the Dirac many-electron case, Eqs. (9.150) and (9.151) allow us to determine the ratio of the first coefficients in the series expansion of $P_i(r)$ and $Q_i(r)$ in the point-like nucleus case:

$$\frac{b_{0,i}^{(r)}}{a_{0,i}^{(r)}} = \frac{(\alpha_i + \kappa_i)c}{Z} \tag{9.157}$$

This ratio is comparable to Kato's nonrelativistic cusp condition [395], which is the ratio of the derivative of the nonrelativistic radial function to itself (note that the $Q_i(r)$ function can be regarded as the derivative of the $P_i(r)$ function), which can be used to check the accuracy of the computed radial functions.

9.6.1.2 Coulomb Potential Functions

For later convenience we study the short-range series expansion

$$Y_{kl\nu}(r) = r^{\nu+1}\langle r^{-(\nu+1)}\rangle + r^{\alpha_k+\alpha_l} \sum_{i=1}^{\infty} d_i^{(r)} r^i \tag{9.158}$$

and the asymptotic long-range series expansion

$$Y_{kl\nu}(r) \sim r^{-\nu} \sum_{i=0}^{\infty} d_i^{(r)} r^{-i} \tag{9.159}$$

which are known to be valid for the potential functions $Y_{kl\nu}(r)$. Note that Eq. (9.158) is fulfilled for all corresponding densities $\rho_{kl}(r)$ in Eq. (9.74)! The integral representation

$$Y_{kl\nu}(r) = \frac{1}{r^\nu}\int_0^r d\tau\,\tau^\nu \rho_{kl}(\tau) + r^{\nu+1}\int_r^\infty d\tau\,\tau^{-(\nu+1)}\rho_{kl}(\tau) \quad (9.160)$$

related to Eq. (9.88) is useful for the analysis, though not a very good choice for the actual determination of $Y_{kl\nu}(r)$, as dicussed in section 9.3.4. The integral representation can be used to analyze the limit for $r \to \infty$

$$\lim_{r\to\infty} Y_{kl\nu}(r) = \left\{\int_0^\infty d\tau\ \tau^\nu \rho_{kl}(\tau)\right\} \lim_{r\to\infty}\frac{1}{r^\nu} \quad (9.161)$$

which reduces to

$$\lim_{r\to\infty} Y_{kl\nu}(r) = \langle \rho_{kl}(r)\rangle \ ,\ \nu = 0 \quad (9.162)$$

$$\left[\Rightarrow \lim_{r\to\infty} Y_{kl\nu}(r) = \langle r^0\rangle_{kl} \ ,\ \nu = 0 \text{ in the Coulomb case}\right] \quad (9.163)$$

$$\lim_{r\to\infty} Y_{kl\nu}(r) = 0 \ ,\ \nu \geq 1 \quad (9.164)$$

(and is immediately fulfilled for closed numerical grids (cf. section 9.7.2)).

In practice, Eq. (9.88) may be treated only over a finite range of the radial variable $r \in [0, r_{max}]$ ($r_{max} \leq \infty$) provided that r_{max} is a sufficiently large number that causes $P_k(r_{max}) \approx 0$ and $Q_k(r_{max}) \approx 0$. With these constraints a tolerable error will be introduced due to the exponentially decaying behavior of the two radial functions. The potential functions $Y_{kl\nu}(r)$ require more attention, since a finite r_{max} leads to the modified upper boundary condition

$$Y_{kl\nu}(r_{max}) \approx \frac{1}{r_{max}^\nu}\int_0^{r_{max}} d\tau\,\tau^\nu\rho_{kl}(\tau) \approx \frac{1}{r_{max}^\nu}\int_0^\infty d\tau\,\tau^\nu\rho_{kl}(\tau) \quad (9.165)$$

as derived from Eq. (9.159). The correct limiting values result when r_{max} approaches infinity. Eq. (9.165) indicates that the $Y_{kl\nu}(r)$ functions do not vanish at the boundary for a finite r_{max}.

9.6.2 Origin Behavior of Interaction Potentials

To analyze the short-range behavior of the potentials for closed-shell atoms [397], we make use of the series expansion of the radial functions around the origin of the coordinate system. The origin behavior of the Hartree–Fock radial functions can be shown to be regular at the origin,

$$P_i^{\rm HF}(r) = r^{l_i+1}\sum_{k=0}^{\infty} a_{k,i}^{\rm HF} r^k \quad (9.166)$$

by considering the short-range limit of the nonrelativistic radial differential equation, Eq. (9.120), analogously to what has been done for the relativistic case above. We adopt the short-range behavior of the pair of Dirac–Hartree–Fock radial functions according to Eqs. (9.145) and (9.146) but include the superscript "DHF" in order to make the functions and expansion parameters distinguishable from the nonrelativistic case above,

$$P_i^{\text{DHF}}(r) = r^{\alpha_i} \sum_{k=0}^{\infty} a_{k,i}^{\text{DHF}} r^k \quad \text{and} \quad Q_i(r) = r^{\alpha_i} \sum_{k=0}^{\infty} b_{k,i} r^k \tag{9.167}$$

(see the derivation in section 9.6.1 above).

According to Eqs. (9.125)–(9.128), we need to evaluate the behavior of the potential functions $U_{ij\nu}(r)$,

$$\begin{aligned} \lim_{r_1 \to 0} U_{ij\nu}^{\text{HF}}(r_1) &= \lim_{r_1 \to 0} \Bigg[r_1^{k-1} \left(\frac{a_{0,i}^{\text{HF}} a_{0,j}^{\text{HF}}}{\nu + k} - \frac{a_{0,i}^{\text{HF}} a_{0,j}^{\text{HF}}}{k - \nu - 1} \right) \\ &\quad + r_1^{\nu} \int_0^{\infty} dr_2 \frac{\rho_{ij}^{\text{HF}}(r_2)}{r_2^{\nu+1}} + O(r_1^k) \Bigg] \end{aligned} \tag{9.168}$$

with $k = 3 + l_i + l_j \geq 3$. The second integral in Eq. (9.160) was evaluated according to

$$\lim_{\epsilon \to 0} \int_{\epsilon}^{\infty} dr f(r) = \lim_{\epsilon \to 0} \left[\int_0^{\infty} dr f(r) - \int_0^{\epsilon} dr f(r) \right]$$

with $f(r)$ being a radial-coordinate dependent function so that the ϵ-independent expectation value $\langle f(r) \rangle$ enters the expression. The boundary ϵ is chosen such that a Taylor series expansion for the integrand $f(r)$ converges. This series expansion is to be constructed from the series expansions of the radial functions.

In an analogous way, we obtain

$$\begin{aligned} \lim_{r_1 \to 0} U_{ij\nu}^{\text{DHF}}(r_1) &= \lim_{r_1 \to 0} \Bigg[r_1^{m-1} \left(a_{0,i}^{\text{DHF}} a_{0,j}^{\text{DHF}} + b_{0,i} b_{0,j} \right) \left(\frac{1}{\nu + m} - \frac{1}{m - \nu - 1} \right) \\ &\quad + r_1^{\nu} \int_0^{\infty} dr_2 \frac{\rho_{ij}^{\text{DHF}}(r_2)}{r_2^{\nu+1}} + O(r_1^m) \Bigg] \end{aligned} \tag{9.169}$$

with $m = 1 + \alpha_i + \alpha_j \geq 1$. Here we have used the power series expansions Eqs. (9.166) and (9.167), respectively. In both equations the first term vanishes at the origin, and from the second term we get contributions only if $\nu = 0$. Therefore we obtain the result

$$\lim_{r_1 \to 0} U_{ij\nu}(r_1) = \begin{cases} 0 & ; \quad \nu \neq 0 \\ \displaystyle\int_0^{\infty} dr_2 \frac{\rho_{ij}(r_2)}{r_2} & ; \quad \nu = 0 \end{cases} \tag{9.170}$$

valid for Hartree–Fock and Dirac–Hartree–Fock potential functions.

9.6.3
Short-Range Electron–Electron Coulomb Interaction

Since the constraint $\nu = 0$ in Eq. (9.170) can only be fulfilled if $l_i = l_j$ or $j_i = j_j$, respectively, and the symmetry coefficients are then given as (cf. e.g. [58])

$$A_{ii0} = \frac{2}{d_i} \tag{9.171}$$

the value of the homogeneous part of Hartree–Fock and Dirac–Hartree–Fock mean-field potentials at the origin is

$$V_i(0) = \lim_{r\to 0} V_i(r) = \sum_j d_j \int_0^\infty \mathrm{d}s\, \frac{\rho_{jj}(s)}{s} - \int_0^\infty \mathrm{d}s\, \frac{\rho_{ii}(s)}{s} \tag{9.172}$$

However, the short-range behavior of the inhomogeneous part $X_i^R(r)/R_i(r)$ (with R being $\{S,P,Q\}$ and R_i the corresponding radial function) of the electron–electron interaction potentials is different in nonrelativistic and relativistic theory because of the exponents in the series expansions of the radial functions, Eqs. (9.166) and (9.167). This difference has its origin in the structure of the differential equations Eq. (9.120), which are of second order, and Eq. (9.122), which are coupled first-order differential equations. The way in which the structure of the differential equations determines the exponents in the radial function's series expansions has been shown above. In the case of Dirac–Hartree–Fock theory, these exponents additionally depend on the type of model used for the nucleus, i.e., point nucleus or finite nucleus. In contrast to Dirac–Hartree–Fock theory, the exponents are independent of the electron–nucleus interaction model in the Hartree–Fock theory.

9.6.4
Exchange Interaction at the Origin

The origin behavior of $X_i^R(r)/R_i(r)$ is derived in the following for the nonrelativistic and the relativistic Hartree–Fock theory. For the determination of $X_i^S(0)/P_i(0)$, we evaluate products of potential functions $U_{ij\nu}(r)$ and ratios of radial functions $P_j(0)/P_i(0)$,

$$\begin{aligned}\lim_{r_1\to 0}\left[U_{ij\nu}(r_1)\frac{P_j(r_1)}{P_i(r_1)}\right] &= \lim_{r_1\to 0}\left[r_1^{m-1}\left(\frac{a_{0,j}^{\mathrm{HF}2}}{\nu+l_i+l_j+3} - \frac{a_{0,j}^{\mathrm{HF}2}}{l_i+l_j-\nu+2}\right)\right.\\ &\quad \left. +\; r_1^{\nu+l_j-l_i}\int_0^\infty \frac{\rho_{ij}^{\mathrm{HF}}(r_2)}{r_2^{\nu+1}}\mathrm{d}r_2 \frac{a_{0,j}^{\mathrm{HF}}}{a_{0,i}^{\mathrm{HF}}} + O(r_1^m)\right]\end{aligned} \tag{9.173}$$

with $m = 3 + 2l_j$ where we used the series expansion Eq. (9.166) and the definition of the potential functions $U_{kl\nu}$. Only the last term yields nonvanishing contributions for short distances r, provided $\nu = l_i - l_j$ and $l_i \geq l_j$ (otherwise this term always vanishes, since negative values for ν are not allowed by the selection rules). The former constraint can easily be understood, since for $\nu > l_i - l_j$ we have $\lim_{r_1 \to 0} (r_1^{\nu+l_j-l_i}) = 0$, and $\nu < l_i - l_j$ is not allowed because of the selection rules. Finally we get

$$\lim_{r \to 0} \left[U_{ij\nu}(r) \frac{P_j(r)}{P_i(r)} \right] = \begin{cases} 0 & ; \quad \nu \neq l_i - l_j \\ \left[\int_0^\infty \frac{\rho_{ij}^{\text{HF}}(s)}{s^{\nu+1}} \text{d}s \right] \frac{a_{0,j}^{\text{HF}}}{a_{0,i}^{\text{HF}}} & ; \quad \nu = l_i - l_j \end{cases} \tag{9.174}$$

and we are now able to evaluate $X_i^S(r)/P_i(r)$ in Eq. (9.197) for $r = 0$,

$$\lim_{r \to 0} \left[\frac{X_i^S(r)}{P_i(r)} \right] = -\frac{1}{2} \sum_{j;\{j \neq i, l_j \leq l_i\}} d_j A_{ij(l_i - l_j)} \left[\int_0^\infty \frac{\rho_{ij}^{\text{HF}}(s)}{s^{l_i - l_j + 1}} \text{d}s \right] \frac{a_{0,j}^{\text{HF}}}{a_{0,i}^{\text{HF}}} \tag{9.175}$$

to arrive at Eq. (9.197). The exchange interaction contributions to the mean-field potential in Dirac–Hartree–Fock theory at the origin require us to consider different cases because of different first exponents in the series expansions for the radial functions resulting from different nucleus models as derived in Eq. (9.152). In the case of finite nuclei, either a_0^{DHF} or b_0 in Eq. (9.167) may be zero, dependent on $\text{sgn}(\kappa_i)$. Thus, one must distinguish between the following two cases:

1. $\kappa > 0, \alpha = \kappa$; this yields

$$a_0^{\text{DHF}} = 0 \quad \Rightarrow \quad k_{min} = 1 \quad \text{and} \quad b_0 \neq 0 \quad \Rightarrow \quad m_{min} = 0 \tag{9.176}$$

2. $\kappa < 0, \alpha = -\kappa$; this yields

$$a_0^{\text{DHF}} \neq 0 \quad \Rightarrow \quad k_{min} = 0 \quad \text{and} \quad b_0 = 0 \quad \Rightarrow \quad m_{min} = 1 \tag{9.177}$$

Here we introduced k_{min} and m_{min}, which denote the lowest indices k and m in Eq. (9.167) with non-zero values for the coefficients a_k^{DHF} and b_m. Both k_{min} and m_{min} are zero in case of a point nucleus. In analogy to the Hartree–Fock case, we therefore obtain

$$\lim_{r_1 \to 0} \left[U_{ij\nu}(r_1) \frac{P_j(r_1)}{P_i(r_1)} \right] = \lim_{r_1 \to 0} \left[r_1^{\beta_{ij}^P} \left(\int_0^\infty \text{d}r_2 \frac{\rho_{ij}^{\text{DHF}}(r_2)}{r_2^{\nu+1}} \right) \frac{a_{k_{j,min},j}^{\text{DHF}}}{a_{k_{i,min},i}^{\text{DHF}}} \right] \tag{9.178}$$

$$\lim_{r_1 \to 0} \left[U_{ij\nu}(r_1) \frac{Q_j(r_1)}{Q_i(r_1)} \right] = \lim_{r_1 \to 0} \left[r_1^{\beta_{ij}^Q} \left(\int_0^\infty \text{d}r_2 \frac{\rho_{ij}^{\text{DHF}}(r_2)}{r_2^{\nu+1}} \right) \frac{b_{m_{j,min},j}}{b_{m_{i,min},i}} \right] \tag{9.179}$$

where we already skipped the terms vanishing in the short-range limit. The exponents β are defined by the following expressions,

$$\beta_{ij}^{P} = \nu + (\alpha_j + k_{j,min}) - (\alpha_i + k_{i,min}) \tag{9.180}$$

$$\beta_{ij}^{Q} = \nu + (\alpha_j + m_{j,min}) - (\alpha_i + m_{i,min}) \tag{9.181}$$

Obviously, in the case of an external potential of a point nucleus, both expressions are equal, $\beta_{ij}^{P} = \beta_{ij}^{Q} = \beta_{ij}$. The next task is the determination of the lowest possible values $\beta_{ij,min}$, since only $\beta_{ij,min} = 0$ results in regular, nonvanishing contributions at the origin.

In the presence of the external potential of a point nucleus, the exponents α_i are nonintegral, while the allowed values of ν are always integers. Therefore, additional cases must be considered in order to determine those which lead to nonvanishing contributions at the origin in Eqs. (9.178) and (9.179), i.e., which yield exponents $\beta_{ij} \not> 0$.

a) $|\kappa_i| = |\kappa_j|$: The lowest possible value for ν is $\nu_{min} = 0$, so that

$$\beta_{ij,min} = 0 \tag{9.182}$$

b) $|\kappa_i| < |\kappa_j|$: Here we have $\nu_{min} = j_j - j_i = |\kappa_j| - |\kappa_i|$, and therefore

$$\beta_{ij,min} = |\kappa_j| - |\kappa_i| - \left(|\kappa_i| \sqrt{1 - \frac{Z^2}{c^2\kappa_i^2}} - |\kappa_j| \sqrt{1 - \frac{Z^2}{c^2\kappa_j^2}} \right) \tag{9.183}$$

$$\approx 2|\kappa_j| - 2|\kappa_i| + \frac{Z^2}{2c^2} \left(\frac{1}{|\kappa_i|} - \frac{1}{|\kappa_j|} \right) > 0 \tag{9.184}$$

We may recall from chapter 6 that the quantum numbers j_i and κ_i are related by $j_i = |\kappa_i| - 1/2$ and $\nu_{min} = |j_i - j_j| = ||\kappa_i| - |\kappa_j||$. In the last transformation, we utilized the Taylor series expansion $\sqrt{1+x} = 1 + x/2 + O(x^2)$, which converges for $|x| < 1$.

c) $|\kappa_i| > |\kappa_j|$: This leads to electron–electron interaction potential singularities, since $\nu_{min} = |\kappa_i| - |\kappa_j|$, from which we obtain

$$\beta_{ij,min} = |\kappa_i| - |\kappa_j| - \left(|\kappa_i| \sqrt{1 - \frac{Z^2}{c^2\kappa_i^2}} - |\kappa_j| \sqrt{1 - \frac{Z^2}{c^2\kappa_j^2}} \right) \tag{9.185}$$

$$\approx \frac{Z^2}{2c^2} \left(\frac{1}{|\kappa_i|} - \frac{1}{|\kappa_j|} \right) < 0 \tag{9.186}$$

This shows that all electron–electron interaction potentials $W_i^P(r), W_i^Q(r)$ (see Eqs. (9.123) and (9.124)), in which contributions with $|\kappa_i| > |\kappa_j|$ occur, i.e., all electron–electron interaction potentials except those with the minimal $|\kappa_i| = 1$, behave nonregularly at the origin.

Only for $|\kappa_i| = 1$ do we find a regular short-range behavior of the electron–electron interaction potentials for the case of a point nucleus. In this case, we find nonvanishing contributions only for $|\kappa_j| = |\kappa_i| = 1$, where $\nu_{min} + j_i + j_j = 2j_i$ is odd. Thus, these contributions occur only, if $\text{sgn}(\kappa_i) = \text{sgn}(\kappa_j)$ and therefore $\kappa_i = \kappa_j = \pm 1$.

In the presence of an external potential of a finite nucleus, the values $\beta_{ij,min}$ are different for large and small components in accordance with Eqs. (9.180) and (9.181). They are dependent on $\text{sgn}(\kappa)$, which can be seen by means of Eqs. (9.176) and (9.177). Writing Eqs. (9.180) and (9.181) explicitly for all possible combinations $\{\text{sgn}(\kappa_i), \text{sgn}(\kappa_j)\}$, we obtain

$$\beta_{ij}^P = \begin{cases} \nu + |\kappa_j| - |\kappa_i| & ; \quad \text{sgn}(\kappa_i) = \text{sgn}(\kappa_j) \\ \nu + |\kappa_j| - |\kappa_i| - 1 & ; \quad \kappa_i > 0 \wedge \kappa_j < 0 \\ \nu + |\kappa_j| - |\kappa_i| + 1 & ; \quad \kappa_i < 0 \wedge \kappa_j > 0 \end{cases} \tag{9.187}$$

$$\beta_{ij}^Q = \begin{cases} \nu + |\kappa_j| - |\kappa_i| & ; \quad \text{sgn}(\kappa_i) = \text{sgn}(\kappa_j) \\ \nu + |\kappa_j| - |\kappa_i| - 1 & ; \quad \kappa_i < 0 \wedge \kappa_j > 0 \\ \nu + |\kappa_j| - |\kappa_i| + 1 & ; \quad \kappa_i > 0 \wedge \kappa_j < 0 \end{cases} \tag{9.188}$$

It is useful to regard the lowest possible values ν_{min}, since $\nu \geq |j_i - j_j|$ has to be fulfilled as well as the constraint concerning the sum $\nu + j_i + j_j$, so that

$$\nu_{min} = \begin{cases} ||\kappa_i| - |\kappa_j|| & ; \quad \text{sgn}\,(\kappa_i) = \text{sgn}\,(\kappa_j) \\ ||\kappa_i| - |\kappa_j|| + 1 & ; \quad \text{sgn}\,(\kappa_i) \neq \text{sgn}\,(\kappa_j) \end{cases} \tag{9.189}$$

since $|j_i - j_j| + j_i + j_j = ||\kappa_i| - |\kappa_j|| + |\kappa_i| + |\kappa_j| - 1$ is always odd. Combining Eqs. (9.187), (9.188), and (9.189) yields

$$\beta_{ij,min}^P = \begin{cases} ||\kappa_i| - |\kappa_j|| + |\kappa_j| - |\kappa_i| & ; \quad \text{sgn}(\kappa_i) = \text{sgn}(\kappa_j) \\ ||\kappa_i| - |\kappa_j|| + |\kappa_j| - |\kappa_i| & ; \quad \kappa_i > 0 \wedge \kappa_j < 0 \\ ||\kappa_i| - |\kappa_j|| + |\kappa_j| - |\kappa_i| + 2 & ; \quad \kappa_i < 0 \wedge \kappa_j > 0 \end{cases} \tag{9.190}$$

$$\beta_{ij,min}^Q = \begin{cases} ||\kappa_i| - |\kappa_j|| + |\kappa_j| - |\kappa_i| & ; \quad \text{sgn}(\kappa_i) = \text{sgn}(\kappa_j) \\ ||\kappa_i| - |\kappa_j|| + |\kappa_j| - |\kappa_i| & ; \quad \kappa_i < 0 \wedge \kappa_j > 0 \\ ||\kappa_i| - |\kappa_j|| + |\kappa_j| - |\kappa_i| + 2 & ; \quad \kappa_i > 0 \wedge \kappa_j < 0 \end{cases} \tag{9.191}$$

These expressions may be investigated subject to the absolute values of the κ quantum numbers:

$$||\kappa_i| - |\kappa_j|| + |\kappa_j| - |\kappa_i| = \begin{cases} 0 & ; \quad |\kappa_i| \geq |\kappa_j| \\ 2|\kappa_j| - 2|\kappa_i| > 0 & ; \quad |\kappa_i| < |\kappa_j| \end{cases} \tag{9.192}$$

Thus, the lowest exponents $\beta_{ij,min}$ are for the case of a finite nucleus

$$\beta_{ij,min}^P = \begin{cases} 0 & ; \quad \text{sgn}(\kappa_i) = \text{sgn}(\kappa_j) \wedge |\kappa_i| \geq |\kappa_j| \\ 0 & ; \quad \kappa_i > 0 \wedge \kappa_j < 0 \wedge |\kappa_i| \geq |\kappa_j| \\ \beta_{ij,min}^P > 0 & ; \quad \text{otherwise} \end{cases} \tag{9.193}$$

$$\beta^{Q}_{ij,min} = \begin{cases} 0 & ; \quad \mathrm{sgn}(\kappa_i) = \mathrm{sgn}(\kappa_j) \wedge |\kappa_i| \geq |\kappa_j| \\ 0 & ; \quad \kappa_i < 0 \wedge \kappa_j > 0 \wedge |\kappa_i| \geq |\kappa_j| \\ \beta^{Q}_{ij,min} > 0 & ; \quad \text{otherwise} \end{cases} \tag{9.194}$$

It can easily be seen that in contrast to the point-nucleus case, there exist no singularities in the electron–electron interaction potentials in the case of a finite nucleus. Hence, an analytical expression for the electron–electron interaction potentials at the origin can be determined for a finite nucleus, and for a point nucleus as far as shells with $|\kappa_i| = 1$ are concerned.
To evaluate the inhomogeneous terms in Eqs. (9.199) and (9.201) we have to carry out the summation in Eqs. (9.127) and (9.128), and obtain in the short-range limit

$$\lim_{r\to 0}\left[\frac{X_i^P(r)}{P_i(r)}\right] = \begin{cases} -\dfrac{1}{2}\sum\limits_{j,j\neq i} \delta_{(\beta^P_{ij,min},0)} d_j A^{\mathrm{DHF}}_{ij\nu_{min}} \left(\displaystyle\int_0^\infty \mathrm{d}s\, \frac{\rho^{\mathrm{DHF}}_{ij}(s)}{s^{\nu_{min}+1}}\right) \dfrac{a^{\mathrm{DHF}}_{k_{j,min},j}}{a^{\mathrm{DHF}}_{k_{i,min},i}} & \text{finite} \\ -\dfrac{1}{2}\sum\limits_{j,j\neq i} \delta_{(\kappa_i,\kappa_j)} d_j A^{\mathrm{DHF}}_{ij0} \left(\displaystyle\int_0^\infty \mathrm{d}s\, \frac{\rho^{\mathrm{DHF}}_{ij}(s)}{s}\right) \dfrac{a^{\mathrm{DHF}}_{0,j}}{a^{\mathrm{DHF}}_{0,i}} & \text{point, } |\kappa_i| = 1 \\ -\infty \quad \text{point, } |\kappa_i| \neq 1 \end{cases} \tag{9.195}$$

$$\lim_{r\to 0}\left[\frac{X_i^Q(r)}{Q_i(r)}\right] = \begin{cases} -\dfrac{1}{2}\sum\limits_{j,j\neq i} \delta_{(\beta^Q_{ij,min},0)} d_j A^{\mathrm{DHF}}_{ij\nu_{min}} \left(\displaystyle\int_0^\infty \mathrm{d}s\, \frac{\rho^{\mathrm{DHF}}_{ij}(s)}{s^{\nu_{min}+1}}\right) \dfrac{b_{m_{j,min},j}}{b_{m_{i,min},i}} & \text{finite} \\ -\dfrac{1}{2}\sum\limits_{j,j\neq i} \delta_{(\kappa_i,\kappa_j)} d_j A^{\mathrm{DHF}}_{ij0} \left(\displaystyle\int_0^\infty \mathrm{d}s\, \frac{\rho^{\mathrm{DHF}}_{ij}(s)}{s}\right) \dfrac{b_{0,j}}{b_{0,i}} & \text{point, } |\kappa_i| = 1 \\ -\infty \quad \text{point, } |\kappa_i| \neq 1 \end{cases} \tag{9.196}$$

where 'point' denotes a point-like atomic nucleus whereas 'finite' denotes a nucleus of finite size with corresponding nonsingular electron–nucleus potential.

9.6.5 Total Electron–Electron Interaction at the Nucleus

In the analysis that leads to Eq. (9.152) it is assumed that the electron–electron interaction potential functions do not behave like $1/r$ or are even more singular at the origin. Since in the relativistic case the point-like nucleus can only be applied for $Z \leq c$, the prefactor in Eq. (9.186) is always ≤ 1. The term in parentheses in Eq. (9.186) is always larger than -1, such that $\beta > -1$. Typical values for the Zn atom are $\beta \approx -0.02$. Therefore, in case of singular behavior of the Dirac–Hartree–Fock electron–electron interaction potentials, the series expansions of the radial functions remain unaffected by the weak singularity.

Once the homogeneous part $V_i(r)$ of the total Hartree–Fock interaction potential has been added to the inhomogeneous term $X_i^S(r)/P_i(r)$, whose short-range behavior is given by Eq. (9.175), we get

$$\begin{aligned} W_i^S(0) &= \lim_{r\to 0}\left[V_i(r) + \frac{X_i^S(r)}{P_i(r)}\right] = \left\langle R^{-1}\right\rangle - \int_0^\infty \mathrm{d}s\, \frac{\rho_{ii}^{\mathrm{HF}}(s)}{s} \\ &\quad -\frac{1}{2}\sum_{j;\{j\neq i, l_j\leq l_i\}} d_j A_{ij(l_i-l_j)}^{\mathrm{HF}} \left(\int_0^\infty \mathrm{d}s\, \frac{\rho_{ij}^{\mathrm{HF}}(s)}{s^{l_i-l_j+1}}\right) \frac{a_{0,j}^{\mathrm{HF}}}{a_{0,i}^{\mathrm{HF}}} \end{aligned} \qquad (9.197)$$

where we introduced the expectation value for the many-electron radial operator $R^{-1} = \sum_i^N r_i^{-1}$,

$$\left\langle R^{-1}\right\rangle = \sum_i d_i \int_0^\infty \mathrm{d}r\, \frac{\rho_{ii}(r)}{r} \qquad (9.198)$$

which is the shell-independent part of $W_i^S(0)$ originating solely from the Coulomb interaction. Because of the series expansions of the radial functions we get different expressions for the radial density $\rho_{ij}^{\mathrm{DHF}}(r)$ depending on the nuclear model used. This leads to an origin behavior of the exchange contributions to the electron–electron interaction potentials, for which different cases have to be considered. The result, Eq. (9.197), can also be obtained for open-shell atoms [397].

Analogously to the derivation of Eq. (9.197), we have for the Dirac–Hartree–Fock potentials by combining the inhomogeneities of Eqs. (9.195)–(9.196) with the homogeneous part

$$W_i^P(0) = \lim_{r\to 0}\left[V_i^P(r) + \frac{X_i^P(r)}{P_i(r)}\right] \qquad (9.199)$$

$$\begin{aligned} &= \langle R^{-1}\rangle - \int_0^\infty \mathrm{d}s\, \frac{\rho_{ii}^{\mathrm{DHF}}(s)}{s} \\ &\quad - \begin{cases} \dfrac{1}{2}\sum_{j,j\neq i} \delta_{(\beta_{ij,min}^P,0)} d_j A_{ij(l_i-l_j)}^{\mathrm{DHF}} \left(\int_0^\infty \mathrm{d}s\, \dfrac{\rho_{ij}^{\mathrm{DHF}}(s)}{s^{\nu_{min}+1}}\right) \dfrac{a_{k_{j,min},j}^{\mathrm{DHF}}}{a_{k_{i,min},i}^{\mathrm{DHF}}} & \text{finite} \\ \dfrac{1}{2}\sum_{j,j\neq i} \delta_{(\kappa_i,\kappa_j)} d_j A_{ij0}^{\mathrm{DHF}} \left(\int_0^\infty \mathrm{d}s\, \dfrac{\rho_{ij}^{\mathrm{DHF}}(s)}{s}\right) \dfrac{a_{0,j}^{\mathrm{DHF}}}{a_{0,i}^{\mathrm{DHF}}} & \text{point, } |\kappa_i| = 1 \\ \infty \quad \text{point, } |\kappa_i| \neq 1 \end{cases} \end{aligned}$$

(9.200)

$$W_i^Q(0) = \lim_{r\to 0}\left[V_i^P(r) + \frac{X_i^Q(r)}{Q_i(r)}\right] \tag{9.201}$$

$$= \langle R^{-1}\rangle - \int_0^\infty ds\,\frac{\rho_{ii}^{\mathrm{DHF}}(s)}{s}$$
$$- \begin{cases} \frac{1}{2}\sum_{j,j\neq i}\delta_{(\beta^Q_{ij,min},0)}d_jA^{\mathrm{DHF}}_{ij(l_i-l_j)}\left(\int_0^\infty ds\,\frac{\rho_{ij}^{\mathrm{DHF}}(s)}{s^{\nu_{min}+1}}\right)\frac{b_{m_{j,min},j}}{b_{m_{i,min},i}} & \text{finite}\\ \frac{1}{2}\sum_{j,j\neq i}\delta_{(\kappa_i,\kappa_j)}d_jA^{\mathrm{DHF}}_{ij0}\left(\int_0^\infty ds\,\frac{\rho_{ij}^{\mathrm{DHF}}(s)}{s}\right)\frac{b_{0,j}}{b_{0,i}} & \text{point, } |\kappa_i| = 1\\ \infty \quad \text{point, } |\kappa_i|\neq 1 \end{cases} \tag{9.202}$$

The third index of the symmetry coefficients results from the possible values of j_i, j_j, and κ_i, κ_j, respectively, under the restrictions included in the Kronecker delta. It can be seen from these equations that the main contribution to the electron–electron interaction potentials at the origin is given by the $\langle R^{-1}\rangle$ expectation value, which by far dominates the above expressions. Additional contributions can be written in terms of $\langle r^{-n}\rangle$ matrix elements. These small corrections, which cause the shell dependence of the electron–electron interaction potentials, result from the local self-interaction term and the non-local exchange interaction only, while the part that concerns the Coulomb interaction (self-interaction included) is equal for all shells.

Furthermore, it is interesting to analyze for the Dirac–Hartree–Fock interaction potentials the connection between $W_i^P(r)$ and $W_i^Q(r)$. In general, these two components are equal in their long-range behavior (see section 9.6.6 for details), but they may be different in the short-range limit. Nevertheless, these electron–electron interaction potentials are identical at the origin in some special cases for the point nucleus; considering the cusp-analogous condition of Eq. (9.157) for the case of a point nucleus already discussed in section 9.6.1.1,

$$\frac{b_{0,i}}{a_{0,i}^{\mathrm{DHF}}} = \frac{(\alpha_i + \kappa_i)c}{Z} \tag{9.203}$$

it follows that

$$\frac{a_{0,i}^{\mathrm{DHF}}}{a_{0,j}^{\mathrm{DHF}}} = \frac{b_{0,i}}{b_{0,j}} \tag{9.204}$$

if $\kappa_i = \kappa_j$. Therefore, we obtain from Eqs. (9.200) and (9.202)

$$W_i^P(0) = W_i^Q(0) \tag{9.205}$$

provided $|\kappa_i| = 1$. This demostrates that $W_i^P(r)$ and $W_i^Q(r)$ are equal at the origin for $s_{1/2}$ and $p_{1/2}$ shells in a point-like nuclear potential. For $|\kappa_i| > 1$,

$W_i^P(r)$ and $W_i^Q(r)$ are singular at the origin. Note that Eq. (9.203) does not hold for the case of a finite nucleus in general. Graphical representations of the interaction potentials can be found in Ref. [397], which is the reference on which this section is based.

9.6.6 Asymptotic Behavior of the Interaction Potentials

The long-range behavior of the electron–electron interaction potentials has been the subject of many investigations [398–401] and has already been discussed with classical argument by Condon and Shortley [38, p. 159/160]. For large values of r, the electron–electron interaction potentials are dominated by the $1/r$-decay of the potential functions $U_{ij\nu}$ with $\nu = 0$ since

$$\lim_{r\to\infty} U_{ij\nu}(r) = \langle \rho_{ij}(r) r^{\nu} \rangle \lim_{r\to\infty} r^{-\nu-1} \tag{9.206}$$

Thus, the homogeneous contributions to the electron–electron interaction potentials in the long-range limit may be written as

$$\lim_{r\to\infty} V_i(r) = \frac{1}{r}\sum_j d_j \langle \rho_{jj} \rangle - \frac{d_i}{2} A_{ii0} \langle \rho_{ii} \rangle = \frac{1}{r}\left[\sum_j d_j - 1\right] \tag{9.207}$$

which holds for both the Hartree–Fock and the Dirac–Hartree–Fock potentials. The $-1/r$ contribution in Eq. (9.207) results from the self-interaction term of the exchange integrals because the potential V_i contains *all* homogeneous contributions. In density functional theory, the Coulomb part of the potential contains a self-interaction term, which must be compensated by the potential from the exchange density functional. Therefore, the asymptotic $-1/r$ behavior of this exchange potential can be exploited as a boundary condition for approximate exchange density functionals [402, 403].

The inhomogeneous part is determined by

$$\lim_{r\to\infty} \frac{X_i^S(r)}{P_i(r)} = -\frac{1}{2}\sum_{j,j\neq i} d_j A_{ij\nu_{min}} \langle \rho_{ij} \rangle \lim_{r\to\infty} \left\{ \frac{P_j(r)}{r^{1+\nu_{min}} P_i(r)} \right\} = 0 \tag{9.208}$$

within Hartree–Fock theory and by

$$\lim_{r\to\infty} \frac{X_i^P(r)}{P_i(r)} = \frac{1}{2}\sum_{j,j\neq i} d_j A^{\text{DHF}}_{ij\nu_{min}} \langle \rho_{ij} \rangle \lim_{r\to\infty} \left\{ \frac{P_j(r)}{r^{1+\nu_{min}} P_i(r)} \right\} = 0 \tag{9.209}$$

$$\lim_{r\to\infty} \frac{X_i^Q(r)}{Q_i(r)} = \frac{1}{2}\sum_{j,j\neq i} d_j A^{\text{DHF}}_{ij\nu_{min}} \langle \rho_{ij} \rangle \lim_{r\to\infty} \left\{ \frac{Q_j(r)}{r^{1+\nu_{min}} Q_i(r)} \right\} = 0 \tag{9.210}$$

within Dirac–Hartree–Fock theory. The vanishing of contributions from the inhomogeneities $X_i^{(S,P,Q)}(r)$ comes from the use of orthonormal radial functions yielding $\langle \rho_{ij} \rangle = 0$. Thus, there are only contributions from the homogeneous part, which is equal in both cases, and we finally obtain

$$\lim_{r \to \infty} W_i^{(S,P,Q)}(r) = \frac{\sum_j d_j - 1}{r} = \frac{N-1}{r} \tag{9.211}$$

As can be seen from the last section, this expression holds for closed- and open-shell systems. If we add the electron–nucleus potential, this is just the potential of the ion, which the electron leaves behind, i.e., $-(Z - N + 1)/r$.

9.7 Numerical Discretization and Solution Techniques

In this chapter we have seen how all angular dependencies can be integrated out analytically because of the spherical symmetry of the central field potential of an atom. We are now left with the task to determine the yet unknown radial functions, for which we derived the self-consistent field equations based upon the variational minimax principle. We now address the numerical solution of these equations. The mean-field potential in the set of coupled SCF equations is the reason why we cannot solve them analytically. Hence, the radial functions need to be approximated in some way in order to make these equations solvable within the particular approximation chosen. While basis set expansion techniques are one option and have been employed [142, 145, 148, 152–155, 330–333, 361–365] we consider this approach only in the next chapter and focus instead on the very accurate fully numerical techniques here.

In order to benefit from this accuracy and from controllable truncation errors, finite difference methods for the discretization of all radial functions on an equidistant mesh of grid points with step size h are advantageous. The radial functions and all components of the self-consistent-field equations are then known only at these grid points, and the differential equations are transformed into difference equations. However, solving the resulting difference equations on an equidistant grid in the variable r is not efficient. For small distances many grid points are needed for a satisfactory description of all changes of the radial functions (i.e., for their extrema and nodes) whereas fewer are needed for large distances where these functions decay exponentially. A high density of points near the origin would thus lead to an equally high density in the outer regions. To avoid this one introduces a transformation to a new radial variable, which we may call s.

Before we continue, we note that all equations in this section on numerical methods will be given in Hartree atomic units with $\hbar = e = m_e = 4\pi\epsilon_0 = 1$, which we adopt for the sake of brevity. The advantage is that numerically calculated energy expectation values are consistently obtained in the energy unit of this a.u. system, i.e. in hartrees, and can be easily transformed to, for instance, kJ mol^{-1} by the conversion factor of 2625.5 kJ mol^{-1} hartree^{-1}. In Hartree atomic units, only the speed of light remains as a fundamental constant in the equations whose value depends on the currently accepted value. This is an important difference from nonrelativistic atomic structure theory, where all equations can be solved without using any data from experiment. Accordingly, accurate numerical results obtained by solving nonrelativistic atomic SCF equations stand once and for all, while analogous relativistic equations are affected by the value chosen for the speed of light. However, the situation is not that bad since this dependence is not very pronounced and since the speed of light is already quite accurately known. For this reason, Visscher and Dyall [352] argued that all calculation may choose a fixed value of c irrespective of tiny future corrections. Then, calculated results are perfectly comparable which is of paramount importance to computational quantum chemistry. The accepted value for the speed of light is $c = 137.0359895$ a.u.

9.7.1
Variable Transformations

A suitable variable transformation to a new variable $s(r)$ should fold many grid points to small r values. Obviously, such a transformation function should be of the form shown in Figure 9.1. All equations involved in the algorithms must then be transformed to the new variable $s = s(r)$ where the equidistant grid with step size h is generated. Equidistant grid points in this new variable s correspond to points in the old variable r which lie closer together in the short-range than in the asymptotic region to account for the structure of the coefficient functions in Eqs. (9.105) and (9.88). This leads to important criteria that need to be fulfilled by a suitable variable transformation [404, 405] expressed as a monotonous and bijective function $s(r)$. The function $s = s(r)$ must be differentiable twice,

$$\frac{ds}{dr} = w^2 \quad \text{and} \quad \frac{d^2s}{dr^2} = \frac{d}{dr}w^2 \tag{9.212}$$

where we have introduced an abbreviation for the first derivative. Additionally, we assume the existence of the second derivative $w'' = d^2w/ds^2$. In order to use the analysis in the r-dependent description and to take advantage of the theory of ordinary linear differential equations to the largest possible extent we will focus on transformation functions where origins match, $s(r{=}0){=}0$. This can always be achieved. A further requirement for simplicity without loss of generality is that all grid points in the new variable s will be in the interval $[0, 1]$. In the following sections we will discuss the general form of these transformation functions (including two typical choices) followed by the application of the transformation to the differential equations and integrals encountered.

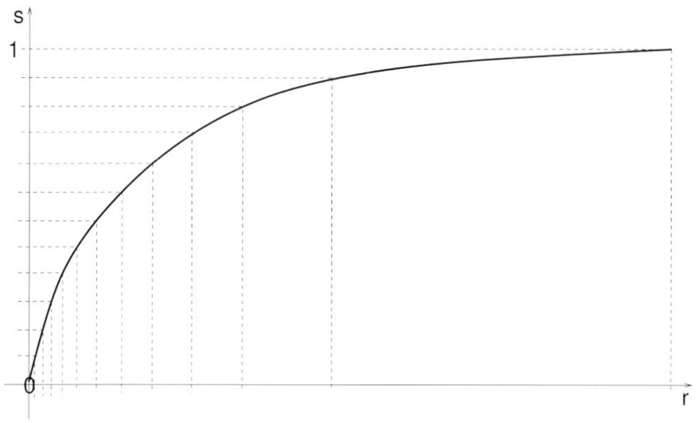

Figure 9.1 Illustration of how to generate an equidistant grid in a new variable s, which corresponds to a widening of the grid the larger the original variable r gets.

Andrae *et al.* [404, 405] have shown that a suitable variable transformation that meets the above criteria may be chosen arbitrarily out of general classes of transformation functions. An appropriate general form of the variable transformation and its inverse can be given by

$$s(r) = \frac{1}{T}\left[t(r) - t_0\right], \qquad r(s) = \bar{t}(Ts + t_0) \tag{9.213}$$

with $T = t(r_{max}) - t_0$ and $t_0 = t(r = 0)$. $\bar{t}$ denotes the inverse function of t. This general formulation permits, for instance, the combination of two functions to generate a very useful grid for electron-scattering calculations. Here one would like to use a grid point distribution similar to those used in atomic structure calculations and, additionally, a different grid at larger distances to be able to describe the wave function of the moving electron. Another application of this useful formulation is the combination of two different grids to ensure that enough grid points are inside the nucleus if a finite-nucleus model is chosen.

9.7.2
Explicit Transformation Functions

One can think of an infinite number of transformation functions which could be used in atomic structure calculations, but two of them have been used widely in atomic structure calculations. These are the rational and the logarithmic grid:

9.7.2.1 **The Logarithmic Grid**

This grid,

$$s(r) = \frac{1}{T}\left[\ln(r+b) - \ln b\right] \;\Rightarrow\; r(s) = b\exp(Ts) - b \tag{9.214}$$

is an open grid which means that it must be normalized using

$$T = \ln(r_{max} + b) - \ln b \tag{9.215}$$

to get $s(r_{max}) = 1$. A similar grid is frequently used in atomic structure calculations [388, 389, 406] (without substracting $\ln b$). The choice of a maximum radius r_{max} means enclosure of the atom in a sphere. The logarithmic transformation yields the following two scaling functions:

$$w^2 = \frac{ds}{dr} = \frac{1}{T[r+b]} = \frac{1}{Tb[\exp(Ts)]} \tag{9.216}$$

$$w'' = \frac{d^2 w}{ds^2} = T^2 \frac{w}{4} \tag{9.217}$$

9.7.2.2 **The Rational Grid**

The rational grid,

$$s(r) = \frac{r}{r+b} \;\Rightarrow\; r(s) = \frac{bs}{1-s} \tag{9.218}$$

is a closed grid which does not need a specification of a maximum radial distance. It has the useful property that the second derivative of the square root of the w^2 function vanishes, i.e. $w'' = 0$. The w^2 scaling function itself is given as

$$w^2 = \frac{ds}{dr} = \frac{(1-s)^2}{b} \tag{9.219}$$

9.7.3
Transformed Equations

Having introduced the variable transformation of arbitrary form, we may now briefly consider the form of the Poisson and self-consistent-field equations in the new variable s. Regarding the CI eigenvalue problem of Eq. (8.167), which can be solved by diagonalization techniques discussed in section 8.7.2, the CSF matrix elements must be transformed to the new variable s. Of course, the calculation of any expectation values requires the same procedure. Such matrix elements over many-particle basis functions and operators are finally broken down to integrals over spinors (e.g., through the application of Slater–Condon rules), and thus we need to consider the transformation of radial integrals in the original r-variable to the new variable s,

$$\begin{aligned}
&\int_0^\infty dr \ \left[P_i(r)\hat{o}(r)P_j(r) + Q_i(r)\hat{o}(r)Q_j(r)\right] \\
&= \int_0^1 w^{-2} ds \left\{P_i[s(r)]\hat{o}[s(r)]P_j[s(r)] + Q_i[s(r)]\hat{o}[s(r)]Q_j[s(r)]\right\} \quad (9.220)
\end{aligned}$$

where $\hat{o}[r(s)]$ is a multiplicative one-electron operator, e.g. $c\kappa_j/r$ or $V_{nuc}(r)$. In the case of differential operators as occurring in $A_i(r)$ or $A_i^\dagger(r)$ we have

$$\begin{aligned}
&\int_0^\infty dr \ \left[Q_i(r)\frac{dP_j(r)}{dr} - P_i(r)\frac{dQ_j(r)}{dr}\right] \\
&= \int_0^1 ds \left\{Q_i[s(r)]\frac{dP_j[s(r)]}{ds} - P_i[s(r)]\frac{dQ_j[s(r)]}{ds}\right\} \quad (9.221)
\end{aligned}$$

The w^{-2} term vanishes here because of the transformed first derivative $d/dr = w^2\, d/ds$. The upper boundary in the original r-variable was set to infinity but may equally well be chosen as some r_{max} provided this is large enough to ensure a negligibly small truncated integral element. The integrals can now be solved numerically with a standard method like the trapezoidal rule [407,408] or the Simpson formula (cf. appendix F).

9.7.3.1 **SCF Equations**

We proceed on in order to formulate the differential Eqs. (9.105) in the variable s. We substitute $d/dr = w^2(d/ds)$ and get the transformed equations in normal form

$$\begin{aligned}
&\left(\frac{d}{ds} + \frac{\kappa_i}{w^2 r(s)}\right) P_i[s(r)] \\
&\qquad + \frac{1}{cw^2}\left[V_i^P[s(r)] \quad -2c^2 - \epsilon_i\right] Q_i[r(s)] \quad = \quad \frac{1}{cw^2} X_i^Q[s(r)] \quad (9.222) \\
&\left(\frac{d}{ds} - \frac{\kappa_i}{w^2 r(s)}\right) Q_i[s(r)] \\
&\qquad - \frac{1}{cw^2}\left[V_i^P[s(r)] \qquad - \epsilon_i\right] P_i[s(r)] \quad = \quad -\frac{1}{cw^2} X_i^P[s(r)] \quad (9.223)
\end{aligned}$$

The boundary conditions that must be met by the transformed components are that both components must vanish at $s(r=0) = 0$ and $s(r_{max}) = 1$ (with $r_{max} \le \infty$).

9.7.3.2 **Regular Solution Functions for Point-Nucleus Case**

As was pointed out in section 9.6 the Eqs. (9.105) with a singular electron-nucleus potential of Coulomb type yield nonanalytic solution functions. To obtain regular functions for the point-like nuclei the operator identity

$$\frac{d}{dr} = r^{-\gamma_i} \frac{d}{dr} r^{\gamma_i} - \frac{\gamma_i}{r} \tag{9.224}$$

with

$$\gamma_i = |\kappa_i| - \alpha_i = |\kappa_i| - \sqrt{\kappa_i^2 - (Z/c)^2} \tag{9.225}$$

suggested by Biegler-König can be used. The power γ_i depends on the quantum number κ_i and is chosen such that it cancels out the real power α_i in the short-range series expansion and replaces it by the integral number $|\kappa_i|$. A generalized distance variable x is chosen in Eq. (9.224) to indicate that the identity may be introduced in the old variable r or in the new variable s. As an illustration we give an explicit formulation of the transformed SCF equations

$$\begin{aligned}\left(\frac{d}{ds} + \frac{\kappa_i - \gamma_i}{w^2 r(s)}\right) & [r(s)]^{\gamma_i} P_i[r(s)] \\ & + \frac{1}{cw^2}\left[V_i[r(s)] - 2c^2 - \epsilon_i\right] \quad [r(s)]^{\gamma_i} Q_i[r(s)] = \frac{[r(s)]^{\gamma_i}}{cw^2} X_i^Q[r(s)]\end{aligned} \tag{9.226}$$

$$\begin{aligned}\left(\frac{d}{ds} - \frac{\kappa_i + \gamma_i}{w^2 r(s)}\right) & [r(s)]^{\gamma_i} Q_i[r(s)] \\ & - \frac{1}{cw^2}\left[V_i[r(s)] - \epsilon_i\right] \quad [r(s)]^{\gamma_i} P_i[r(s)] = -\frac{[r(s)]^{\gamma_i}}{cw^2} X_i^P[r(s)]\end{aligned} \tag{9.227}$$

where we introduced the operator identity in the original variable r and wrote $r(s)$ in order to denote that a grid point will be determined in s so that the corresponding value for r is then fixed. For $\gamma_i = 0$ these equations pass into the original Eqs. (9.222) and (9.223) for nonsingular potentials, i.e., in finite nucleus cases, which yield solution functions that are automatically analytic at the origin. The equations above are of the following general form in the new variable s,

$$\frac{dy_U}{ds} + F_{UU}(s)y_U(s) + F_{UL}(s)y_L(s) = G(s) \tag{9.228}$$

$$-\frac{dy_L}{ds} + F_{LU}(s)y_U(s) + F_{LL}(s)y_L(s) = H(s) \tag{9.229}$$

whose discretization on the equidistant mesh of grid points have to be considered.

9.7.3.3 **Poisson Equations**

To derive the transformed differential equation for the two-electron potential functions $Y_{klv}([r(s)])$ from Eq. (9.88) we take advantage of the operator identity

$$\frac{d^2}{dr^2} = w^2 \frac{d}{ds}\left(w^2 \frac{d}{ds}\right) = w^2 \left[\frac{d^2}{ds^2} w - \left(\frac{d^2 w}{ds^2}\right)\right] \tag{9.230}$$

to obtain

$$\left\{ \frac{d^2}{ds^2} - \frac{1}{w} \left(\frac{d^2 w}{ds^2} \right) - \frac{\nu(\nu+1)}{w^4[r(s)]^2} \right\} wY_{kl\nu}[r(s)] = -\frac{2\nu+1}{w^3 r(s)} \rho_{kl}[r(s)] \quad (9.231)$$

The left-hand side of this equation is, of course, identical in form to the non-relativistic case. At the lower boundary, the transformed potential functions $(wY_{kl\nu})[r(s)]$ must vanish at $s(r=0)=0$ while the upper boundary value is given with Eq. (9.161) by

$$(wY_{kl\nu})[s(r_{max})] = (wY_{kl\nu})[1] \approx \frac{w(r_{max})}{r^{\nu}_{max}} \int_0^1 ds \rho_{kl}[r(s)] \quad (9.232)$$

If r_{max} approaches infinity the upper boundary value will be zero provided that the weight function w goes to zero as well (otherwise the case $k=l$ and $\nu=0$ must be treated separately). Of course, analogous formulae hold if the Breit interaction is included.

9.7.4
Numerical Solution of Matrix Equations

We are now prepared for the difficult task of solving the transformed SCF equations. This involves the process of expressing the equations in such a way that it can be dealt with on a computer. In addition, we already know that the equations cannot be solved in one shot because the operators depend on the solution and a self-consistent-field approach is envisaged, for which tailored convergence accelerators are needed.

Finite-difference methods operating on a grid consisting of equidistant points $(\{x_i\}, x_i = ih + x_0)$ are known to be one of the most accurate techniques available [409]. Additionally, on an equidistant grid all discretized operators appear in a simple form. The uniform step size h allows us to use the Richardson extrapolation method [407,410] for the control of the numerical truncation error. Many methods are available for the discretization of differential equations on equidistant grids and for the integration (quadrature) of functions needed for the calculation of expectation values.

Simple discretization schemes use derivatives of Lagrangian interpolation polynomials that approximate the function known only at the grid points $\{x_i\}$. These schemes consist of tabulated numbers multiplied by the function's values at m contiguous grid points and are referred to as "m-point-formulae" by Bickley [411] (cf. Ref. [381, p.914]). For an acceptable truncation error $O(h^t)$, $t=4$ or higher, m is larger than t which leads to an extended amount of computation.

We will use the standard Numerov scheme for second-order differential equations without a first derivative [407, 412] and an analogous scheme for first-order differential equations (both are derived in appendix F). It is possible to derive rather complicated discretization formulae using only three contiguous points for second-order differential equations that contain a first derivative [413] (be aware of some misprints in this reference). These techniques have the advantage that only three contiguous grid points are needed for a truncation error of order h^4 (i.e., $m = 3$, $t = 4$) but we have to pay the price of a nondiagonal coefficient function containing the potentials, which is formally equivalent to a second derivative of the potential.

After choosing an appropriate variable transformation function $s(r)$ as well as the number of inner grid points n the differential equations have to be discretized at the chosen points. Because of the normalization of the range of s to the unit interval, these grid points s_k and the corresponding step size h are given by the general formula

$$s_k = s_0 + kh = kh \; , \quad k = 0, 1, ..., n, n+1 \; , \quad h = \frac{s(r_{max}) - s_0}{n+1} = \frac{1}{n+1} \tag{9.233}$$

A total of $n + 2$ grid points results (n inner points plus two boundary points $s(0) = s_0 = 0$ and $s(r_{max}) = s_{n+1} = 1$). However, the numerical work can be restricted almost completely to the n inner grid points. The discretization of s allows us to calculate the corresponding grid points in r and the discrete values of the weight function w — both depending, of course, on the chosen explicit variable transformation. All additional data like coefficient functions of the differential equations can then be calculated in a general way (using the vectors s, r and w). Additional grid-specific information may be required depending on the finite difference method to be used.

In the following sections, implications resulting from the general treatment of the variable transformations are formulated for a couple of finite difference methods of an acceptable numerical truncation error of order h^4. The accuracy of the data produced by a proper computer implementation may be checked and improved by the Richardson extrapolation.

The previous sections have shown that we are concerned with linear ordinary first- and second-order differential equations which have to be solved numerically. Their discretized forms — the finite-difference equations — can be written as matrix equations which can be solved using linear algebra techniques. For convenience, we write matrices (in addition to vectors) in this section in bold face and break with our convention employed in the preceding chapters. All matrix equations (except those of eigenvalue type) can be solved via LU decomposition with pivoting followed by a backsubstitution step as described in Ref. [408]. In the case of a vanishing determinant of the U matrix, an additional iteration step improves the accuracy of the method. The LU decomposition of the matrix operator on the right hand side into a *Lower* and an *U*pper triangular matrix is possible,

$$\boldsymbol{Ay} = \boldsymbol{g} \;\Rightarrow\; \boldsymbol{LUy} = \boldsymbol{g} \tag{9.234}$$

even for almost singular matrices: zero elements are replaced by very small numbers. Since equations for triangular matrices,

$$\boldsymbol{LUy} = \boldsymbol{g} \;\Rightarrow\; \boldsymbol{Lz} \equiv \boldsymbol{g} \tag{9.235}$$

can be solved easily for $\boldsymbol{z}$ by simply starting with the *first* row, we may then use backsubstitution starting in the *last* row,

$$\boldsymbol{Uy} = \boldsymbol{z} \tag{9.236}$$

to obtain the desired solution $\boldsymbol{y}$. The obtained solution can then be improved via inverse iteration (cf. Refs. [414, p. 46] and [388, p. 268-270]).

9.7.5 Discretization and Solution of the SCF equations

The coupled first-order differential Eqs. (9.228) and (9.229) may be written in discretized form [405]

$$\left.\frac{\mathrm{d}y_U}{\mathrm{d}s}\right|_{s=s_k} + F_{UU}(s_k)y_U(s_k) + F_{UL}(s_k)y_L(s_k) \;=\; G(s_k) \tag{9.237}$$

$$-\left.\frac{\mathrm{d}y_L}{\mathrm{d}s}\right|_{s=s_k} + F_{LU}(s_k)y_U(s_k) + F_{LL}(s_k)y_L(s_k) \;=\; H(s_k) \tag{9.238}$$

These equations are fulfilled at every grid point s_k and may be abbreviated as

$$y'_{U,k} + F_{UU,k}\, y_{U,k} + F_{UL,k}\, y_{L,k} \;=\; G_k \tag{9.239}$$

$$-y'_{L,k} + F_{LU,k}\, y_{U,k} + F_{LL,k}\, y_{L,k} \;=\; H_k \tag{9.240}$$

The application of a finite-difference method transforms the first derivatives $y'_{(U,L),k}$ at a grid point s_k into a linear combination of function values $y_{(U,L),m}$ at contiguous points s_m around s_k. This yields two linear difference equations for every inner grid point s_k. The resulting set of linear equations can be combined into a $2n \times 2n$ matrix equation

$$\boldsymbol{Ay} = \boldsymbol{g} \tag{9.241}$$

which is in general inhomogeneous, i.e. $\boldsymbol{g} \neq \boldsymbol{0}$. The elements of the vector $\boldsymbol{y} = (y_1, ..., y_l, ..., y_{(2n)})^{\top}$ are the values of the two functions y_U and y_L still to be determined. The indices U and L are omitted in the vector in order to indicate that the values of the functions may be sorted in such a way that the matrix $\boldsymbol{A}$

becomes as simple as possible. A single first-order differential equation leads to a matrix A which will be tri-diagonal due to the discretization of the first derivative. The elements of the band matrix A are determined by the values $F_{UU,k}$, $F_{UL,k}$, $F_{LU,k}$, and $F_{LL,k}$ of the corresponding coefficient functions and the finite difference approximation of the first derivative (see below for explicit formulae). The elements of the inhomogeneity vector $\boldsymbol{g} = (g_1, ..., g_l, ..., g_{(2n)})^\top$ depend on the values G_k and H_k of the inhomogeneity functions $G(s)$ and $H(s)$. These elements have to be sorted in the same way as it was done for the vector $\boldsymbol{y}$, to account for the resulting row interchanges in the matrix A. Since this analysis deals with eigenvalue equations, the matrix A can be split

$$A = T + V + \epsilon_i S \tag{9.242}$$

related to the kinetic energy $\langle T \rangle$, to the potential energy $\langle V \rangle$, and a metric S. It follows from Eqs. (9.222) and (9.223) that the elements of the metric consist of values of the function $1/(cw^2)$ at successive grid points.

Transformed solution functions $y = y_{L,U} = r^{\gamma_i} z_{L,U}$ occur for point-like nuclei in electronic structure calculations of atoms, but it is not necessary to introduce these functions explicitly. In general, the solution functions may be written as $y = vz$. Eq. (9.241) is then converted to

$$\boldsymbol{A}\boldsymbol{y} = \boldsymbol{A}\boldsymbol{V}\boldsymbol{z} = \widetilde{\boldsymbol{A}}\boldsymbol{z} = \boldsymbol{g} \tag{9.243}$$

by introducing the diagonal weight matrix V whose diagonal consists of the values of the function $v = r^{\gamma_i}$ at all inner grid points. The vector z represents the set of function values of $P_i[r(s)]$ and $Q_i[r(s)]$ in their s-dependent form.

For the discretization of the first-order SCF equations a scheme can be developed [405] that yields the required order for the truncated series expansion terms similar to the well-known standard Numerov method for second-order differential equations [392,412]. In this scheme the first derivative at the point s_k is approximated by a difference formula which requires values at three contiguous points. This formula may be derived analogously to the deduction of the Numerov formula (see appendix F) and is comparable to the Simpson integration formula. As for the Numerov method, we have to pay a price for this low-point approximation to the first derivative. The procedure affects all remaining terms in the differential equation (namely the potential energy and inhomogeneity) because it also requires their first derivative, which was again approximated through three successive points. Finally we obtain a tri-diagonal band matrix representation for each differential equation of the coupled system Eqs. (9.226) and (9.227).

For *one* of the coupled pair of self-consistent-field equations we may write the matrix A and the inhomogeneity vector g with the following definition of matrix and vector elements

$$d_k = \frac{4h}{3} F_k, \quad e_k^{\pm} = \pm 1 + \frac{h}{3} F_k, \quad g_k = \frac{h}{3}(G_{k-1} + 4G_k + G_{k+1}) \tag{9.244}$$

as

$$
A = \begin{pmatrix}
d_1^* & e_2^{+*} & f & & & & & & \\
e_1^- & d_2 & e_3^+ & & & & & & \\
 & e_2^- & d_3 & e_4^+ & & & & & \\
 & & \ddots & \ddots & \ddots & & & & \\
 & & & e_{k-1}^- & d_k & e_{k+1}^+ & & & \\
 & & & & \ddots & \ddots & \ddots & & \\
 & & & & & e_{n-3}^- & d_{n-2} & e_{n-1}^+ & \\
 & & & & & & e_{n-2}^- & d_{n-1} & e_n^+ \\
 & & & & & & & e_{n-1}^- & d_n
\end{pmatrix}, \quad g = \begin{pmatrix} g_1^* \\ g_2 \\ g_3 \\ \vdots \\ g_k \\ \vdots \\ g_{n-2} \\ g_{n-1} \\ g_n^* \end{pmatrix} \tag{9.245}
$$

Those elements marked by an asterisk require additional corrections to conserve the numerical accuracy. In the first line a nonvanishing contribution from the origin has been incorporated into the elements of A by setting $d_1^* = d_1 + \Delta d$ and $e_2^{+*} = e_2^+ + \Delta e$ if $|\kappa_i| = 1$ (for both equations of the coupled system). To obtain an origin correction of order h^4 it may be necessary to introduce a third contribution f in Eq. (9.245). There are three ways to avoid this additional contribution. Simple matrix manipulations concerning the first two lines can be used to cancel f but this will change all elements of A and g in the first row. The second possibility may be introduced earlier if one requires an origin correction at only two points with order h^4. The disadvantage of both of these procedures is that the eigenvalue ϵ_i is needed so that the partition in Eq. (9.242) is no longer possible. The third way is to subtract fy_3 from the first element of the inhomogeneity g_1^*. However, this may cause problems in iterative solution methods if the inhomogeneity vector g is zero, i.e. in homogeneous equations. The corrections are determined from the requirement that the linear combination $\Delta dy_1 + \Delta ey_2 + fy_3$ shall be equal to the contribution of the origin within the desired error $O(h^4)$ or better.

No contribution from the origin occurs if $G_0 = 0$ or $H_0 = 0$ in the first element g_1^* of the inhomogeneity vector. It is assumed that the inhomogeneities vanish at the upper boundary, $G_{n+1} = 0$ and $H_{n+1} = 0$, which leads to a modified last element g_n^*. In addition, a nonvanishing contribution $e_{n+1}^+ y_{n+1}$ from the last grid point may also be included in g_n^* but this does not occur in the case under study here.

Note that we get two equations of the same type as Eq. (9.245) for our coupled system of differential equations, which can be combined into a single hepta-diagonal matrix equation that further allows us to determine the two radial functions simultaneously! To derive this hepta-diagonal system we apply Eq. (F.25) (see appendix F) to the transformed coupled Eqs. (9.226) and (9.227) and arrange all $2 \times n$ discretized equations into a matrix equation. The first two rows of this matrix equation must be corrected for origin contributions in cases where $|\kappa| = 1$.

Because of the implicit use of the first derivative of the coefficient functions, the Simpson-type discretization method may be problematic if piecewise-defined potential terms are present (for some finite-nucleus models like the homogeneous charge distribution of Eq. (6.151)). For such special cases, it would be more advantageous to use a scheme which employs a diagonal matrix representation of the effective potential (cf. Ref. [405] for details).

The matrix equation obtained needs to be subjected to a sophisticated solution algorithm for the radial functions as well as for the Lagrange multipliers to be implemented in a computer program.

9.7.6
Discretization and Solution of the Poisson Equations

Before we discuss in detail the numerical discretization scheme used for the Poisson equation, which by the way is very similar to the discretization of the radial Schrödinger equation given in Eq. (9.120) [404], we sum up some of its general features. We consider the general form of the radial Poisson equation, Eq. (9.231), at one of the grid points s_k,

$$\left.\frac{d^2 y}{ds^2}\right|_{s=s_k} + F(s_k)\, y(s_k) = G(s_k) \tag{9.246}$$

which may be abbreviated as

$$y_k'' + F_k\, y_k = G_k \tag{9.247}$$

A numerical treatment requires an approximation for the second derivative. Application of a finite difference method transforms the second derivative y_k'' at a grid point s_k into a linear combination of function values y_m at contiguous grid points s_m around the point s_k. The differential equation at the grid point s_k thus becomes a linear difference equation in the unknown function values y_m at successive grid points. The resulting set of linear equations can be combined into an $(n \times n)$ matrix equation, which is inhomogeneous in this case,

$$\boldsymbol{A}\boldsymbol{y} = \boldsymbol{g} \tag{9.248}$$

The elements of the vector $\boldsymbol{y} = (y_1, \ldots, y_k, \ldots, y_n)^\top$ are the function values to be determined. The elements of the band matrix $\boldsymbol{A}$ are determined by the values F_k of the coefficient function $F(s)$ and by the finite difference approximation to the second derivative. The elements of the inhomogeneity vector $\boldsymbol{g} = (g_1, \ldots, g_k, \ldots, g_n)^\top$ depend analogously on the values G_k of the inhomogeneity function $G(s)$. After the setup of the required matrices and of the inhomogeneity vector, the resulting matrix equations can be solved by LU decomposition and backsubstitution.

The discretization scheme, which leads to an error $O(h^4)$ for second-order differential equations (without first derivative) with the lowest number of points in the difference equation, is the method frequently attributed to Numerov [407, 412]. It can be efficiently employed for the transformed Poisson Eq. (9.231). In this approach, the second derivative at grid point s_k is approximated by the second central finite difference at this point, corrected to order h^4, and requires values at three contiguous points (see appendix F for details). Finally, we obtain tri-diagonal band matrix representations for both the second derivative and the coefficient function of the differential equation. The resulting matrix $\boldsymbol{A}$ and the inhomogeneity vector $\boldsymbol{g}$ are then

$$\boldsymbol{A} = \begin{pmatrix} d_1^* & e_2^* & & & & & & \\ e_1 & d_2 & e_3 & & & & & \\ & e_2 & d_3 & e_4 & & & & \\ & & \ddots & \ddots & \ddots & & & \\ & & & e_{k-1} & d_k & e_{k+1} & & \\ & & & & \ddots & \ddots & \ddots & \\ & & & & e_{n-3} & d_{n-2} & e_{n-1} & \\ & & & & & e_{n-2} & d_{n-1} & e_n \\ & & & & & & e_{n-1} & d_n \end{pmatrix}, \quad \boldsymbol{g} = \begin{pmatrix} g_1^* \\ g_2 \\ g_3 \\ \vdots \\ g_k \\ \vdots \\ g_{n-2} \\ g_{n-1} \\ g_n^* \end{pmatrix} \tag{9.249}$$

with the matrix and vector elements given by

$$d_k = 10e_k - 12, \quad e_k = 1 + \frac{h^2}{12}F_k, \quad g_k = \frac{h^2}{12}(G_{k-1} + 10G_k + G_{k+1}) \tag{9.250}$$

except for those with an asterisk. In the first line, a nonvanishing contribution from the origin, the point $s_0 = 0$, resulting from the term $e_0 y_0 = (h^2/12)F_0 y_0$, must be incorporated into the remaining two entries by setting $d_1^* = d_1 + \Delta d$ and $e_2^* = e_2 + \Delta e$. The corrections Δd and Δe are determined from the requirement that the linear combination $\Delta d\, y_1 + \Delta e\, y_2$ shall be equal to the contribution from the origin within the desired error of order h^4 or better. In the first element g_1^* of the inhomogeneity vector, no contribution from the origin occurs in most cases where $G_0 = 0$. The inhomogeneity was assumed to have vanished at the upper boundary, $G_{n+1} = 0$, which leads to a modified last element g_n^*. Additionally, a nonvanishing contribution $e_{n+1} y_{n+1}$ from the last grid point may sometimes be included in g_n^*. Such a contribution from the last grid point may occur only in the case of the radial Poisson equations, because of the long-range behavior of the two-electron potential functions. This contribution $e_{n+1}y_{n+1}$ must be calculated separately for each variable transformation used. A general expression for this term, as is available for the origin contribution e_0y_0 (see below), has not yet been derived. This nonvanishing contribution from the upper boundary is the reason why some of the 'closed' variable transformations must not be used over the full range $r \in [0, \infty]$ within the Numerov scheme, since the contribution $e_{n+1}y_{n+1}$ becomes infinite.

In some cases, where $\alpha_i + \alpha_j - 1 < 0$ so that we get $G_0 \to \infty$, it is not possible to correct the Numerov scheme for the Poisson equation [404, 405]. In these cases one can use a five-point central difference formula [411] (also with truncation error of order h^4) which leaves the coefficient function on the diagonal. However, this five-point formula leads to difficulties at the boundaries, where values at grid points s_{-1} (at the lower boundary) and s_{n+2} (at the upper boundary) would be needed, which lie outside the range of definition.

The discretized Poisson equations can easily be solved on a computer using the established techniques of LU decomposition and subsequent backsubstitution as described in Ref. [408]. The left-hand side of the Poisson equations only needs to be LU decomposed once, while the right-hand side changes in every SCF cycle. Therefore, the backsubstitution needs to be done in every SCF cycle (as well as the additional iteration step — if necessary — for accuracy improvement).

9.7.7
Extrapolation Techniques and Other Technical Issues

Finite difference methods allow the use of techniques which extrapolate to step size $h \to 0$ (i. e., the exact solution) and control the numerical truncation error [407, 408]. This can be done for every numerically calculated quantity F if we assume an analytic behavior of F,

$$F(h) = F(0) + Ah^t + O(h^{t+1}) \tag{9.251}$$

where t is the order of the truncation error connected to the chosen numerical method. If F is known for three different step sizes, the quantities $F(0)$ (the extrapolated result for step size zero), A, and t can be calculated. Since all numerical methods are usually employed with an order $t \geq 4$, this can be used to check this 'theoretical order'. Note that the extrapolated value $F(0)$ will be correct to order $t + 1$ and we gain only one or two figures in accuracy compared to the result calculated with the largest number of grid points.

Multigrid methods with control of the numerical truncation error are very useful for the solution of matrix equations since they start with a small number of grid points, use extrapolation techniques similar to Richardson's and reduce the step size until the result is accurate enough. The method of Bulirsch and Stoer (cf. [408, p. 718–725] and [415, p. 288–324]), for instance, consists essentially of three ideas. The calculated values for a given step size h are analytic functions of h — which is, of course, fulfilled here. The analytic expression can always be approximated as a rational function, i.e., a quotient of two polynomials in h, which is the basis for the rational extrapolation [408, p. 104-107]. The applied numerical method must be of even order in all higher corrections in the truncation term in order to gain two orders at a time [408, p. 717].

The SCF equations are now solved for a given number of points. Then, mid-points are added to this starting grid and the equations are solved again. This process is repeated until the rational extrapolation leads to a sufficiently small degree of error. A 3-point formula [381, p. 914] without origin correction for the first derivative, which is only of second order in h, can be applied for the discretization. This is possible since the numerical error is controlled by rational extrapolation. Its advantage is that the discretization matrix in the equation to be solved in the SCF procedure is only penta-diagonal and its elements are easily computed. Additionally, the Bulirsch–Stoer method could also be used in cases where one would like to use a diagonal representation for the coefficient functions of the differential equations for reasons of numerical stability.

Since the SCF equations can only be solved iteratively, a good starting approximation is needed to set up the potentials $U_{kl\nu}(r)$ in the Fock operator. The quality of this approximation determines the convergence behavior of the SCF cycles. There are numerous possibilities to generate an initial set of radial functions of which some will be mentioned below.

Of course, one may always use hydrogen-like spinors with a screened effective nuclear charge number in the exponential to generate starting functions for the first Fock operator guess. More elaborate approaches, however, employ models for the electron–electron interaction to replace all spinor-dependent $U_{kl\nu}(r)$ potential functions in the SCF equations. They may be used, then, to generate the first approximation for the radial functions. Systematic electron–electron model potentials have been devised based on the work by Green and collaborators (see Ref. [416] and references therein). With these potentials *homogeneous* Dirac–Hartree–Fock equations can be devised which can be discretized with the methods just discussed for the SCF equations. Then, LU decomposition of the matrix and inverse iteration solve the discretized equations for some estimate of the spinor energies.

Another model potential can be obtained utilizing the statistical theory for the electron distribution in an atom due to the studies done by Thomas and Fermi [211, p. 145-156]. The Thomas–Fermi potential can be seen as the simplest potential possible within the framework of the density functional theory of section 8.8. This close connection to DFT shows that exchange and correlation functionals can easily be introduced into the program code despite numerical problems that could arise because of the complicated structure of exchange–correlation functionals.

Numerical discretization methods pose an interesting consequence for fully numerical Dirac–Hartree–Fock calculations. These grid-based methods are designed to directly calculate only those radial functions on a given set of mesh points that occupy the Slater determinant. It is, however, not possible to directly obtain any 'excess' radial functions that are needed to generate new CSFs as excitations from the Dirac–Hartree–Fock Slater determinant. Hence, one cannot directly start to improve the Dirac–Hartree–Fock results by methods which capture electron correlation effects based on excitations that start from a single Slater determinant as reference function. This is very different from basis-set expansion techniques to be discussed for molecules in the next chapter. The introduction of a one-particle basis set provides so-called virtual spinors automatically in a Dirac–Hartree–Fock–Roothaan calculation, which are not produced by the direct and fully numerical grid-based approaches.

In short CI expansions, one may set up explicitly those CSFs which allow us to assign a correlating radial function to a given radial function of the Dirac–Hartree–Fock Slater determinant. This correlating function has got some well-defined properties; for instance, the virtual radial function and the one to be correlated should 'live' in the same spatial region. However, this can create additional technical difficulties for the guess of those radial functions which have been introduced to account for the correlation of a particular shell in the CI expansions with more than one CSF. The above-mentioned model potentials are not the best choice, and additional adjustments to them are necessary so that it can be certain that the shells to be correlated live in the same radial space.

9.8
Results for Total Energies and Radial Functions

In this last section, we shall present some examples of results of numerical atomic structure calculations to demonstrate properties of radial functions and the magnitude of specific effects like the choice of the finite nucleus model or the inclusion of the Breit operator. It should be emphasized that the reliability of numerical calculations is solely governed by the affordable length of the CI expansion of the many-electron wave function since the numerical solution techniques allow us to determine spinors with almost arbitrary accuracy. Expansions with many tens of thousands of CSFs can be routinely handled (with the basis-set techniques of chapter 10, expansions of billions of CSFs are feasible via subspace iteration techniques [296]).

As an illustrative case study, we present results for the fine-structure splitting induced by spin–orbit coupling in magnesium-like ions obtained from a very small CI expansion. The fine-structure splitting $^3P_0 -^3 P_1$ of the magnesium-like ions Sc^{9+} and Se^{22+} is calculated with the Dirac–Coulomb Hamiltonian employing the numerical methods described in this chapter [88] (c=137.0359895, logarithmic grid with 400 grid points, grid parameter $b = 7.5 \times 10^{-5}$ and a maximum distance $r_{max} = 14$ bohr). The orbitals are optimized in an MCSCF-OL (optimal-level) scheme, in which the orbitals have been optimized for each state separately. These data are compared to an earlier study by Das and Grant [141].

Table 9.3 CI coefficients for important configurations that contribute to the $^3P_0$ and $^3P_1$ states of the magnesium-like ions Sc^{9+} and Se^{22+}.

		$^3P_0$			$^3P_1$			
ion	Z	CSF	Ref. [88]	Ref. [141]	CSF	LS analog	Ref. [88]	Ref. [141]
Sc^{9+}	21	3s3p	0.99787	0.998	3s3p	$^3P_1$	0.83528	0.835
		3p3d	0.06518	0.065	3s3p	$^1P_1$	−0.54586	−0.546
					3p3d	$^3D_1$	0.03060	0.030
					3p3d	$^3P_1$	0.04942	0.049
					3p3d	$^1P_1$	−0.03110	−0.031
Se^{22+}	34	3s3p	0.99871	0.999	3s3p	$^3P_1$	0.90874	0.909
		3p3d	0.05069	0.050	3s3p	$^1P_1$	−0.41343	−0.414
					3p3d	$^3D_1$	0.03654	0.036
					3p3d	$^3P_1$	0.04300	0.043
					3p3d	$^1P_1$	−0.00924	−0.009

Table 9.4 Fine-structure splitting $\Delta E =^3 P_0 -^3 P_1$ of the magnesium-like ions Sc^{9+} and Se^{22+} in cm^{-1}. In addition, we have included results from an elaborate many-body perturbation theory (MBPT) study (including the Breit interaction) by Johnson and co-workers [417]. The experimental data have been taken from Refs. [418] and [419]. (See Refs. [419] and [417] for further references to the literature on Mg-like ions.)

Z	MCSCF-OL [88]	MCSCF-EAL [141]	MBPT [417]	Exp.
21	2100	1963	1971	2006 [418]
34	19030	18598	18375	18384 [419]

While the CI coefficients given in Table 9.3 are equal, the results for the fine-structure splitting (Table 9.4) differ since the total electronic energies in Ref. [141] have been corrected for the Breit interaction within perturbation theory. Moreover, an MCSCF-EAL (extended average level) scheme was employed for the orbital optimization by Das and Grant (cf. Ref. [420]). The Breit contribution changes the numerical value by $\approx 2-5\%$. The direct comparison with the experimental data demonstrates the accuracy that can be expected from such a short CI expansion.

9.8.1 Electronic Configurations and the Aufbau Principle

After having developed the theory for atomic structure calculations we are now in a position to discuss which shells are occupied in the minimum-energy (Hartree–Fock) Slater determinant or, more generally, in those Slater determinants which carry a significant weight in the CI expansion of the total wave function. In qualitative discussions, these electronic configurations play an important role. However, it must be understood that the unambiguous assignment of electronic configurations to quantum mechanical states requires well-separated orbital energies leading to a dominant configuration in the CI expansion.

It is clear, that the concept of electron configurations as a synonym for a total state is on shaky ground if the orbital energies are close. This occurs in particular if excited states are studied where no unique electron configuration can be identified in the CI expansion because many carry similar weight. It also occurs for heavy atoms because of their large number of electrons requiring orbitals in the CSFs with similar orbital energies. In quantum chemistry this situation has been called *static electron correlation*, and in chemistry it leads to rather meaningless discussions on which orbital is to be considered occupied for a qualitative picture.

Hence, close-lying orbital energies lead to CI expansions without any prominent CI coefficient that would allow us to assign the electron configuration, i.e., the set of orbitals which occupy the Slater determinant associated with that CI coefficient. Accordingly, the famous Aufbau principle, which assigns an electronic configuration *without* carrying out any calculation, is only valid for well-separated orbital energies. Since the orbital energies get closer and closer for large quantum numbers as depicted in Figure 9.2, the Aufbau principle works reliably only for comparatively light atoms and for closed shells.

9.8.2 Radial Functions

In the case of hydrogen-like atoms we have already noted that the number of nodes of the radial functions depends on the quantum numbers (see section 6.8.1). $P_{n_i\kappa_i}(r)$ always has $(n_i - l_i - 1)$ radial nodes (as many as its nonrelativistic limit, the radial function $P_{n_il_i}(r)$, has). $Q_{n_i\kappa_i}(r)$ has as many nodes as $P_{n_i\kappa_i}(r)$ for negative values of κ_i and one additional node for positive values

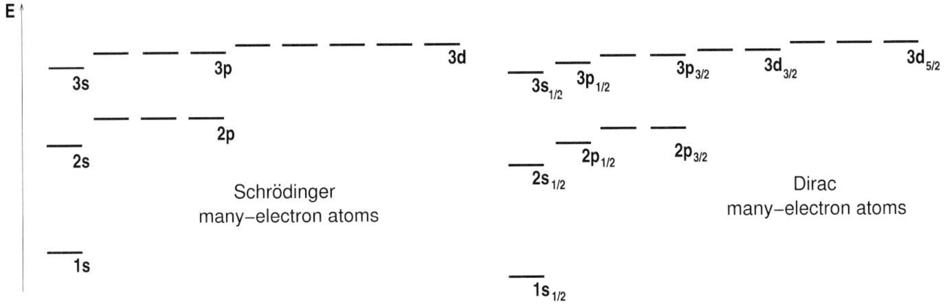

Figure 9.2 Sketch of orbital energies of many-electron Schrödinger (left) and Dirac (right) atoms. Note the lifting of the accidental degeneracy in the Schrödinger case (see Figure 6.2): all energies are now also different for different l quantum number because of the electron–electron interaction terms in the SCF equations. Observe also the general energy lowering in the Dirac case and the non-degeneracy of all shells that are distinguished by n and κ.

of κ_i [89]. Interestingly enough, all nodes of $P_{n_i\kappa_i}(r)$ lie before (from the origin's point of view) the absolute extremum of the function. Moreover, radial functions of MCSCF wave functions may have more (the so-called 'spurious') nodes [388]. These nodes can be found only behind the absolute extremum, although it spurious nodes were also reported near the origin [388].

Figure 9.3 presents graphical representations of the two sets of radial functions, $P_{n_i\kappa_i}(r)$ and $Q_{n_i\kappa_i}(r)$, obtained for a single Slater determinant approximation to the electronic wave function of the krypton atom and the krypton-like radon ion. The sets of radial functions for both atomic systems are very similar, of course. However, the positively charged radon ion naturally features contracted radial functions, which already dropped to values close to zero at radial distances less than 1 bohr. Due to the normalization condition of Eq. (9.33) imposed on the radial functions, the functional values of the radial functions in the highly charged cation of Rn are increased compared to those of neutral Kr. Furthermore, Figure 9.3 clearly shows the pairs of spin–orbit-split radial functions; compare especially the $3p_{1/2}$ and the $3p_{3/2}$ as well as the $4p_{1/2}$ and the $4p_{3/2}$ shell functions. The difference within each pair of radial functions is more pronounced for the charged cation with larger nuclear charge Z compared to the lighter Kr atom. From the figure it is also clear that the small-component radial functions $Q_{n_i\kappa_i}(r)$ are more compact and short-ranged than the large-component functions. This, however, depends on the nuclear charge number Z as can be seen for the heavy cation where the radial extension of the small-component radial functions is more extended compared to the set of large-component radial functions.

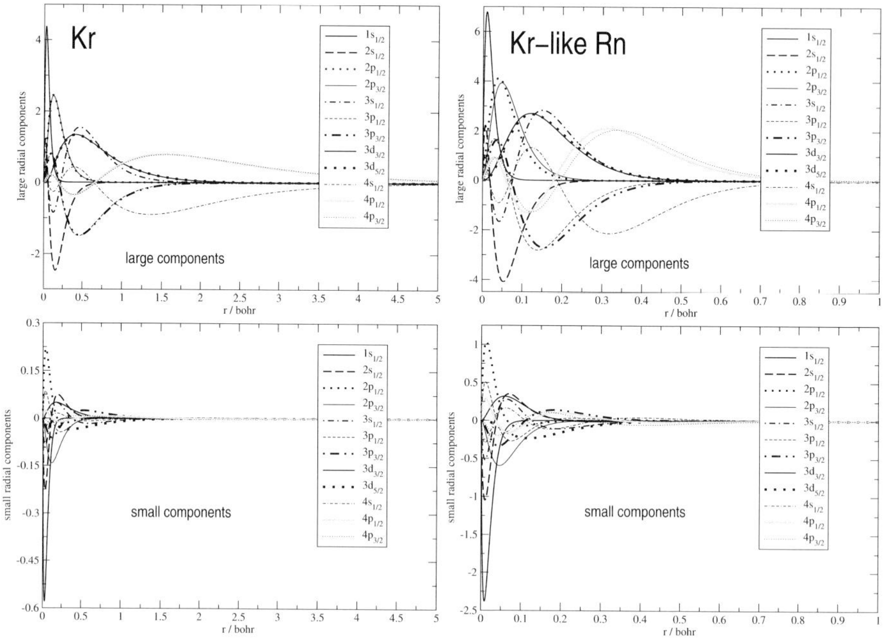

Figure 9.3 Dirac–Fock radial functions of the shells occupied in the Hartree–Fock ground state determinant for the neutral rare gas Kr as well as for the positively charged Kr-like Rn. The large components $P_i(r)$ of these shells are depicted in the upper graphs, while the graphs at the bottom show the small components $Q_i(r)$. Note the number of nodes in these functions and that the small components $Q_i(r)$ of a shell i are short-ranged. Moreover, the spin–orbit-coupling induced splitting of the radial shells is visible for the p-shell of Kr-like Rn (top right graph), but not for Kr. The radial functions have been obtained with a numerical atomic structure program [88] that implements all methodology developed in this chapter.

9.8.3
Effect of the Breit Interaction on Energies and Spinors

The Breit interaction becomes important for the calculation of fine structure splittings and for highly charged ions. An example has already been given in this section for magnesium-like ions. For further examples we may refer to the literature (see also the references given in section 8.1).

The effect of the Breit interaction on the wave function can be conveniently studied by comparing radial moments $\langle r \rangle$ of shells calculated with Dirac–Coulomb and Dirac–Coulomb–Breit Hamiltonians. The effect is not very large as can be seen for Li- and Be-like ions in Figure 9.4. From the plot we note that the effect of the Breit interaction on the radial functions is small, but increases linearly with the nuclear charge number Z. Moreover, the four different models to describe the positive nuclear charge distribution (point-like, exponential, Gaussian shaped and Fermi) can hardly be distinguished.

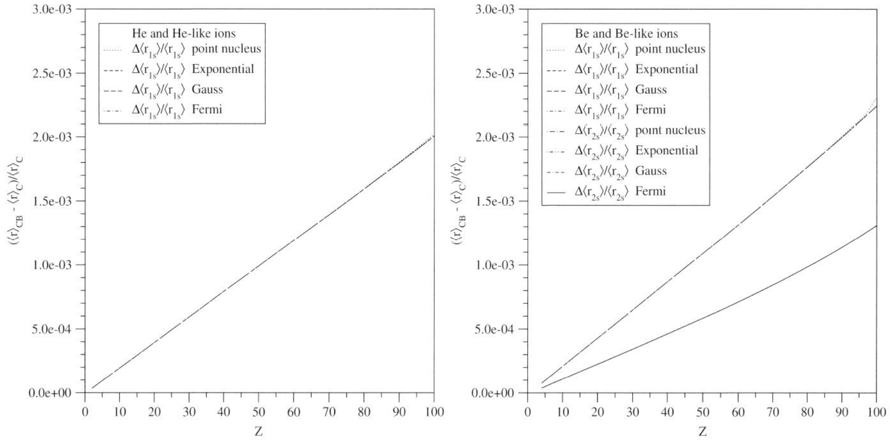

Figure 9.4 Effect of the frequency-independent Breit interaction on $1s_{1/2}$ and $2s_{1/2}$ radial functions for He- and Be-like cations as measured by the first radial momentum $\langle r \rangle = \int_0^\infty r^2 \mathrm{d}r \, r[P_i^2(r) + Q_i^2(r)]$. For the plot, the difference between radial functions obtained with the Dirac–Coulomb and the Dirac–Coulomb–Breit Hamiltonian has been divided by the radial moment obtained with the Dirac–Coulomb Hamiltonian in order to extract the relative effect of the shell functions that strongly contract with increasing nuclear charge $+Ze$. The data have been taken from Ref. [159] and have been corrected for the kink which showed up for the $2s_{1/2}$ shell of point-nucleus Be ions for large nuclear charges, as that turned out to be a numerical artifact. The radial functions were determined in variational closed-shell Dirac–Fock calculations.

9.8.4
Effect of the Nuclear Charge Distribution on Total Energies

In section 6.9 we already introduced finite-size models of the atomic nucleus and analyzed their effect on the eigenstates of the Dirac hydrogen atom. This analysis has been extended in the previous sections to the many-electron case. It turned out that neither the electron–electron interaction potential functions nor the inhomogeneities affect the short-range behavior of the shell functions already obtained for the one-electron case. Table 9.5 now provides the total electronic energies calculated for the hydrogen atom and some neutral many-electron atoms obtained for different nuclear potentials provided by Visscher and Dyall [352], who also provided a list of recommended finite-nucleus model parameters recommended for use in calculations in order to make computed results comparable.

As can be seen from the table, the effect on the total energy is notable when switching from the singular potential of a point nucleus to a finite-nucleus potential. The finite-nucleus potentials, however, not differ very much. Most important is the effect on relative energies. Here we may note that the effect is observable but small for moderate nuclear charges; it becomes larger for superheavy atoms (see Ref. [421] for a comparative study).

Table 9.5 Total ground state energies E'_0 of the hydrogen atom and some closed-(sub)shell atoms in Hartree atomic units reproduced after Visscher and Dyall [352].

	configuration	$E'_{0,\text{point-like}}$	$E'_{0,hom}$	$E'_{0,\text{Fermi}}$	$E'_{0,\text{Gauss}}$
H	$1s^1$	−0.5000066566	−0.5000066561	−0.5000066556	−0.5000066561
He	$1s^2$	−2.8618133422	−2.8618133228	−2.8618133156	−2.8618133228
Be	$[\text{He}]2s^2$	−14.575892267	−14.575891698	−14.575891699	−14.575891698
Ne	$[\text{He}]2s^2 2p^6$	−128.69196952	−128.69193054	−128.69193055	−128.69193055
Mg	$[\text{Ne}]3s^2$	−199.93515947	−199.93506693	−199.93506695	−199.93506696
Ar	$[\text{Ne}]3s^2 3p^6$	−528.68445072	−528.68376234	−528.68376260	−528.68376282
Ca	$[\text{Ar}]4s^2$	−679.71124970	−679.71016054	−679.71016104	−679.71016145
Zn	$[\text{Ar}]4s^2 3d^{10}$	−1794.6217017	−1794.6129678	−1794.6129740	−1794.6129834
Kr	$[\text{Zn}]4p^6$	−2788.8848401	−2788.8605625	−2788.8605828	−2788.8606235
Sr	$[\text{Kr}]5s^2$	−3178.1124394	−3178.0798789	−3178.0799081	−3178.0799692
Cd	$[\text{Kr}]5s^2 4d^{10}$	−5593.4455479	−5593.3182982	−5593.3184461	−5593.3188373
Xe	$[\text{Cd}]5p^6$	−7447.1627620	−7446.8940386	−7446.8943892	−7446.8954397
Ba	$[\text{Xe}]6s^2$	−8135.9845174	−8135.6431155	−8135.6435767	−8135.6450112
Yb	$[\text{Xe}]6s^2 4f^{14}$	−14069.298848	−14067.664150	−14067.666905	−14067.677260
Hg	$[\text{Xe}]6s^2 4f^{14} 5d^{10}$	−19653.650425	−19648.849250	−19648.858221	−19648.896156
Rn	$[\text{Hg}]6p^6$	−23611.192831	−23602.005520	−23602.023307	−23602.104246

Further Reading

Excellent monographs on atomic structure theory are available. The material presented in the preceding section is, however, self-contained and by and large complementary to what is presented in the following list of monographs.

I. P. Grant, [47]. *Relativistic Quantum Theory of Atoms and Molecules — Theory and Computation.*

In a recent voluminous monograph, Grant presented relativistic atomic structure theory from the perspective of all developments by himself and his collaboratos in the second half of the 20th century. The total electronic wave function is elegantly constructed in terms of Racah algebra instead of the pedestrian way chosen for the sake of simplicity here. Also, the numerical solution methods for the SCF equations are quite different from the matrix approach presented here.

L. Szasz, [24]. *The Electronic Structure of Atoms.*

Szasz has provided a good presentation of nonrelativistic as well as relativistic atomic structure theory (pretty much on the same level of sophistication as Grant's classic review [89] on relativistic atomic structure).

C. Froese Fischer, T. Brage, P. Jönsson, [389]. *Computational Atomic Structure — An MCHF Approach.*

This monograph presents atomic structure theory from the nonrelativistic perspective with an emphasis on calculations. Relativistic effects are considered by quasi-relativistic Hamiltonians, and the Dirac many-electron case is only addressed in the appendix. However, the book provides a good presentation of the general philosophy and strategy in numerical atomic structure theory. Prior to this monograph, Froese Fischer published a now classic book on numerical nonrelativistic Hartree–Fock theory in the 1970s [388]. Another classic text on this subject was delivered by Hartree in the 1950s [406].

W. R. Johnson, [42]. *Atomic Structure Theory.*

The book by Johnson covers nonrelativistic and relativistic atomic structure theory, providing many details on angular momentum coupling as well as a bridge to numerical solution methods.

R. D. Cowan, [382]. *The Theory of Atomic Structure and Spectra.*

For those interested in learning more about atomic spectroscopy, the book by Cowan is a sophisticated presentation and can be recommended. During the long history of this topics, of course, many monographs have been published (see Refs. [385, 422] for further examples).

10
General Molecules and Molecular Aggregates

After having derived a relativistic four-component Hamiltonian operator in chapter 8 — often called a 'fully relativistic' Hamiltonian although fundamental requirements of special relativity have been given up (no absolute time and Lorentz covariance) — and having employed it for spherically symmetric atoms, it is now appropriate to proceed to the case of general molecules. Here, we need to introduce suitable approximations for the one-electron functions of molecular systems. Results and inventions of the historically earlier developed nonrelativistic theory are transferred to the relativistic regime. However, a couple of subtleties like variational collapse, kinetic balance and the role of the negative-energy states need to be taken care of.

The general SCF equations derived in chapter 8 are valid for any many-electron system. The differentiation between an atom and a molecule is done in this formalism through the explicit choice of the external potential energy V_{nuc} which enters the one-electron operators. It is the spatial symmetry of V_{nuc} that determines according to which symmetry the quantum mechanical states and hence the one-electron functions can be classified. Then, the symmetry determines the way of how to solve the SCF equations. That is, in the case of an atom the spherically symmetric V_{nuc} allowed us to treat all angular degrees of freedom analytically. In the case of molecules, the point group symmetry of V_{nuc} will be important instead of the continuous group of three-dimensional rotations in the case of atoms, which allowed us to treat all angular variables analytically.

In this chapter the case of *general* molecules — meaning molecules of arbitrary structure and thus arbitrary external nuclear potential — is considered. We will understand how the numerical solution of mean-field equations for many-electron atoms helps us to solve the molecular problem. The key element is the introduction of analytically known one-particle basis functions rather than an elaborated numerical treatment on a three-dimensional spatial grid, which is possible but not desirable (a fact that will become most evident in section 10.5).

Technically, the four-component quantum chemical methods for molecules will turn out to be very much like those known from nonrelativistic quantum chemistry. Historically, the latter based on Schrödinger's one-electron Hamil-

Relativistic Quantum Chemistry. Markus Reiher and Alexander Wolf

ISBN: 978-3-527-31292-4

tonian were developed first. Four-component molecular quantum chemistry adapted all techniques developed for one-component nonrelativistic theory. Therefore, the focus in this chapter will be on the principle ideas of molecular quantum chemistry and on the peculiarities of the relativistic formulation.

Although Kim [423] had already applied the basis-set expansion technique of Roothaan [424] and Hall [425] to the Dirac one-electron equation in 1967, technical problems of variational stability had to be overcome before the first four-component Dirac–Hartree–Fock calculations could be carried out in the early 1980s [426–428]. Especially, the kinetic-balance problem was in the center of these technical issues [429]. All four-component molecular electronic structure computer programs work like their nonrelativistic relatives because of the formal similarity of the theories, where one-electron Schrödinger operators are replaced by four-component Dirac operators, enforcing a four-component spinor basis. Obviously, the spin symmetry must be treated in a different way. It is replaced by time-reversal symmetry being the basis of Kramers' theorem, as we shall see in this chapter. In addition, point group symmetry is replaced by the theory of double groups (see also section 10.4.4 in this chapter), since spatial and spin coordinates cannot be treated separately.

Today, several computer programs for *ab initio* four-component molecular electronic structure calculations have been established for routine calculations and are still under constant development. Although the theory of one-electron basis-set expansions introduced into the equations of chapter 8 still needs to be developed in this chapter, we may anticipate some developments on molecular relativistic calculations from a historical perspective. The pioneering Dutch program MOLFDIR developed in the mid 1980s was the first general program for routine four-component molecular calculations of high accuracy. On the basis of the Dirac–Hartree–Fock kernel of the package [240–242], correlation methods were introduced. A complete open-shell CI and a relativistic variant of the restricted active space CI method were implemented [237–239]. As a size-consistent correlation method, a coupled-cluster module was developed [325, 326]. In turn, this coupled-cluster module also provides total energies from Møller–Plesset perturbation theory to second order, i.e., from MP2 calculations. Cartesian Gaussian basis functions, that are discussed in the next section in Eqs. (10.14) and (10.15), have been chosen for the representation of the molecular spinors. The magnetic part of the frequency-independent Breit interaction could be evaluated either self-consistently or as a first-order perturbation, and the program even supported all double groups which can be constructed from the point group O_h and its subgroups.

The development of MOLFDIR came to an end in 2001 and the developers of this program joined forces with a new Scandinavian program, DIRAC, that emerged in the mid 1990s [430]. DIRAC contains an elegant implementation of Dirac–Hartree–Fock theory as a direct SCF method [245] in terms of quaternion algebra [246,247]. For the treatment of electron correlation, MP2 has been implemented in a direct Kramers-restricted RMP2 version [431]. Additionally, the formalism of a relativistic second-generation MCSCF procedure has also been developed within this framework [244, 432]. This MCSCF module accounts for near degeneracies which occur more frequently than in nonrelativistic theory because of the spin-orbit splitting of shells. The MCSCF program has been supplemented by a direct CI module for the treatment of large determinant expansions [243].

Besides these widely applied computer programs, new developments took place and deserve special attention. The UTCHEM program emerged in the new millennium as a competitive and efficient tool which provides many different wave-function methods [433–439]. Another program is BERTHA, that can perform Dirac–Hartree–Fock and RMP2 calculations [163, 440, 441]. The two-electron integrals are evaluated using the well-established McMurchie–Davidson algorithm. Spherical Gaussians — as introduced in Eqs. (10.8) and (10.9) below — are used as basis functions. The evaluation of matrix elements for electromagnetic properties within the quantum electrodynamical framework has been described [442–446].

We should mention that relativistic electronic structure calculations on solid-state systems have been carried out as well. However, our focus here will not be on the special case of translational symmetry as we rather concentrate on molecules and molecular aggregates. The general principles may, of course, be transferred to the special case of crystals, and we may refer the reader for further details to Refs. [314, 315, 447–454], where four-component and also approximate relativistic Hamiltonians are considered.

10.1 Basis Set Expansion of Molecular Spinors

In order to describe the one-electron spinors of molecules that enter the Slater determinants to approximate the total electronic wave function it is natural to be inspired by the fact that molecules are composed of atoms. It should be noted that this at first sight obvious fact is not obvious at all as we describe only the ingredients, i.e., the elementary particles (electrons and atomic nuclei) and not individual atoms of a molecule. To define what an atom in a molecule shall be is not a trivial issue in view of the continuous total electron distribution $\rho(\boldsymbol{r})$ of a molecule. Nonetheless, we may utilize what we have just learned about many-electron atoms and go back to the old idea of a linear combination of atomic orbitals (LCAO).

For the relativistic description of molecules this means that each molecular spinor $\psi_i(\boldsymbol{r})$ entering a Slater determinant is to be expanded in a set of four-component atomic spinors $\psi_k^{(at)}(\boldsymbol{r})$,

$$\psi_i(\boldsymbol{r}) = \sum_k^{m'} d_{ik} \psi_k^{(at)}(\boldsymbol{r}) \tag{10.1}$$

Each atomic spinor $\psi_k^{(at)}(\boldsymbol{r}) = \psi_k^{(at)}(\boldsymbol{r}, \boldsymbol{R}_A)$ has its center at the position of the nucleus $\boldsymbol{R}_A$ of some atom A. In a first step, we include only those atomic spinors $\psi_k^{(at)}(\boldsymbol{r})$ which would be considered in an atomic Dirac–Hartree–Fock calculation on every atom of the molecule. Of course, if a given atom occurs more than once in the molecule, a set of atomic spinors of this atom is to be placed at every position where a nucleus of this type of atom occurs in the molecule. The number of basis spinors m' is then smallest for such a minimal basis set. In this case, it can be calculated as the number of shells s per atom times the degeneracy d of these shells times the number of atoms M in the molecule, $m' = s(A) \times d(s) \times M$.

In practice, the number of basis spinors m' increases even further for a minimal basis set because we have no analytic expression for the atomic spinors of a many-electron atom available. Hence, the atomic spinors themselves need to be expanded in terms of known basis functions $\phi_\mu^{(a)}(\boldsymbol{r}, \boldsymbol{R}_A)$.

$$\psi_k^{(at)}(\boldsymbol{r}, \boldsymbol{R}_A) = \sum_\mu^{m'_k} \begin{pmatrix} b_{k\mu}^{(1)} \phi_\mu^{(1)}(\boldsymbol{r}, \boldsymbol{R}_A) \\ b_{k\mu}^{(2)} \phi_\mu^{(2)}(\boldsymbol{r}, \boldsymbol{R}_A) \\ b_{k\mu}^{(3)} \phi_\mu^{(3)}(\boldsymbol{r}, \boldsymbol{R}_A) \\ b_{k\mu}^{(4)} \phi_\mu^{(4)}(\boldsymbol{r}, \boldsymbol{R}_A) \end{pmatrix} \tag{10.2}$$

where we now considered individual expansion parameters for the different components of the atomic spinor. Without loss of generality, we assume the same number m'_k of expansion coefficients for all components (if m'_k is larger than necessary for some component, some $b_{k\mu}^{(a)}$ will simply be zero). In addition, we distinguish functions $\phi_\mu^{(a)}$ for different components by the superscript '(a)' (though one may simply choose one basis set for all components). The expansion coefficients $b_{k\mu}^{(a)}$ can be determined either by a fit to numerically computed atomic spinors or directly by the same basis-set methods developed here for molecules (cf. section 10.4 below). The final basis set expansion for a molecular spinor can then be written as,

$$\psi_i(\boldsymbol{r}) = \sum_k^{m'} \sum_\mu^{m'_k} d_{ik} \begin{pmatrix} b_{k\mu}^{(1)} \phi_\mu^{(1)}(\boldsymbol{r}, \boldsymbol{R}_A) \\ b_{k\mu}^{(2)} \phi_\mu^{(2)}(\boldsymbol{r}, \boldsymbol{R}_A) \\ b_{k\mu}^{(3)} \phi_\mu^{(3)}(\boldsymbol{r}, \boldsymbol{R}_A) \\ b_{k\mu}^{(4)} \phi_\mu^{(4)}(\boldsymbol{r}, \boldsymbol{R}_A) \end{pmatrix} \equiv \sum_\mu^{m} \begin{pmatrix} c_{i\mu}^{(1)} \phi_\mu^{(1)}(\boldsymbol{r}, \boldsymbol{R}_A) \\ c_{i\mu}^{(2)} \phi_\mu^{(2)}(\boldsymbol{r}, \boldsymbol{R}_A) \\ c_{i\mu}^{(3)} \phi_\mu^{(3)}(\boldsymbol{r}, \boldsymbol{R}_A) \\ c_{i\mu}^{(4)} \phi_\mu^{(4)}(\boldsymbol{r}, \boldsymbol{R}_A) \end{pmatrix} \tag{10.3}$$

Accordingly, the total number of basis functions is $4 \times m$ with $m = m' \times m'_k$. A useful feature of the linear combination of atom-centered atomic basis functions is the fact that the basis functions can be classified according to their l or κ symmetry, respectively (see below).

In principle, the number of basis spinors m needs to be infinitely large for a complete basis and hence for an exact representation of a molecular spinor in this basis. For practical reasons, however, it must be as small as possible in order to keep the computational effort as low as possible. The basis set size should thus be small but still allow for sufficiently accurate calculations. In order to achieve this, we need to exploit the physics of the problem to the largest extent; a procedure in which the LCAO idea is the first step. Hence, we emphasize that the expansion of Eq. (10.2) provides an optimum description of an atomic spinor. Keeping in mind the idea of a minimal basis set constructed of atomic spinors, we may well freeze the coefficients $b^{(a)}_{k\mu}$ and thus reduce the number m back to the smaller number m'. Then, only the m' coefficients d_{ik} are to be determined rather than the $4m$ $c_{i\mu}$ coefficients. To use fixed $b^{(a)}_{k\mu}$ in a molecular calculation is known as using a 'contracted' basis set. Various variants of such contractions are known but we shall not delve deeper into such purely technical issues. Basis functions that have not been contracted are called primitives or primitive basis functions.

A minimal basis set is certainly not suitable for a description of atoms in molecules. Chemically speaking, molecules are characterized by interacting atoms which polarize each other through additional external fields that distort the spherical symmetry of an atom. If the spherical symmetry of an atom is distorted by an external field then the one-electron functions require a partial wave expansion introducing higher-angular-momentum spherical harmonics and thus higher-angular-momentum basis functions in order to finally account for the polarization of the electron density. These higher-angular-momentum basis functions are called polarization functions, accordingly.

Needless to say, contraction coefficients that have been optimized for a given relativistic, pseudo-relativistic or nonrelativistic scheme are usually not transferable, especially not in highly accurate calculations. Hence, if this is not taken into account, significant accuracy reductions may be observed. Also, the nonrelativistic limit may no longer be well defined in a contracted basis set if one expects to obtain the nonrelativistic result by simply employing a value for the speed of light which is, say, five orders of magnitude larger than usual. The relativistic and nonrelativistic electronic energies are no longer of the same quality but are affected by finite-basis errors of different magnitude.

10.1.1 Kinetic Balance

From Eqs. (5.130) and (5.131) in chapter 5 we know that the kinetic balance condition relates the small and large 2-spinors. In the stationary case, in which the time-dependence of the Dirac equation drops out and a separation constant ϵ comes in according to $(i\hbar)\,\partial/\partial t \rightarrow \epsilon$, we may write Eq. (5.130) as

$$2m_e c^2 \psi^S = c(\boldsymbol{\sigma} \cdot \boldsymbol{\pi})\psi^L + V\psi^S - \epsilon\psi^S \tag{10.4}$$

so that the small component can be expressed as a function of the large component,

$$\psi^S = \frac{c}{2m_e c^2 - V + \epsilon}(\boldsymbol{\sigma} \cdot \boldsymbol{\pi})\psi^L \approx \frac{\boldsymbol{\sigma} \cdot \boldsymbol{\pi}}{2m_e c}\psi^L \tag{10.5}$$

where we assumed $2m_e c^2 \gg -V + \epsilon$ in the last step. Because of the formal similarity of the time-independent Dirac equation and the one-electron SCF Eqs. (8.185), we may instantaneously write the kinetic balance condition for the molecular spinors ψ_i,

$$\psi_i^S = -\frac{c}{2m_e c^2 - W + \epsilon_i}(\boldsymbol{\sigma} \cdot \boldsymbol{\pi})\psi_i^L \approx \frac{\boldsymbol{\sigma} \cdot \boldsymbol{\pi}}{2m_e c}\psi_i^L \tag{10.6}$$

where we now assumed that $2m_e c^2 \gg -W + \epsilon_i$ in the last step. where ϵ_i is the orbital energy and W is the total interaction potential of the Fock operator, i.e., the potential containing the electron–nucleus and electron–electron mean-field interaction. Note that in the case of a vanishing external vector potential $\boldsymbol{\pi}$ simply reduces to the ordinary canonical momentum operator $\boldsymbol{p}$. This boundary condition must, of course, be fulfilled by any basis set representation of the molecular spinors as well.

In 1967, Kim [423] formulated relativistic SCF equations by expanding the unknown spinors into a set of known basis functions. However, he did not take special precautions so that the relation between the upper and lower two components is also satisfied by the basis functions. This was taken into account later by Lee and McLean in 1982 [428]. In the case of finite basis sets, which are used for the representation of the one-electron spinors, the basis sets for the small component must be restricted such as to maintain 'kinetic balance' [429]. A hierarchy of approximations is used in practice for this purpose [455]. A natural choice is to apply 'atomic balance'.

In the numerical solution of the SCF orbital equations kinetic balance restrictions are not required, as this condition will be satisfied exactly. However, in the numerical solution of MCSCF equations for purely correlating orbitals difficulties may arise if the 'orbital energy' ϵ_i becomes too negative [180, 194, 386, 456]. Here it is suggested to use projection operators to eliminate the functions which correspond to the negative continuum.

10.1.2
Special Choices of Basis Functions

The choice and generation of basis sets has been addressed by many authors [147,149,440,457–466]. While we consider here only the basic principles of basis-set construction, we should note that this is a delicate issue as it determines the accuracy of a calculation. Therefore, we refer the reader to the references just given and to the review in Ref. [467]. In Ref. [462] it is stressed that the selection of the number of basis functions used for the representation of a shell $\{n_i\kappa_i\}$ should not be made on the grounds of the nonrelativistic shell classification $\{n_il_i\}$ but on the natural basis of j quantum numbers resulting in basis sets of similar size for, e.g., $s_{1/2}$ and $p_{1/2}$ shells, while the $p_{3/2}$ basis may be chosen to be smaller. As a consequence, if, for instance, $p_{1/2}$ and $p_{3/2}$ shells are treated on the $\{n_il_i\}$ footing, the number of contracted basis functions may be doubled (at least in principle). The ansatz which has been used most frequently for the representation of molecular one-electron spinors is a basis expansion into Gauss-type spinors.

The expansion of four-component one-electron functions into a set of global basis functions can be done in several ways independently of the particular choice of the type of the basis functions. For instance, four independent expansions may be used for the four components as sketched above. However, one might also relate the expansion coefficients of the four components to each other. In contrast to these expansions, the molecular spinors can also be expressed in terms of 2-spinor expansions. This latter ansatz reflects the structure of the one-electron Fock operator and is therefore very efficient (for a detailed discussion compare Ref. [468]) The expansion in terms of 2-spinor basis functions reads

$$\psi_i(\boldsymbol{r}) = \begin{pmatrix} \sum_\mu c_{i\mu}^{(L)} \phi_\mu^L(\boldsymbol{r}_A) \\ \sum_\mu c_{i\mu}^{(S)} \phi_\mu^S(\boldsymbol{r}_A) \end{pmatrix} \tag{10.7}$$

with the spherical two-component basis functions

$$\phi_\mu^L(\boldsymbol{r}_A) = \frac{P_\mu(r_A)}{r_A}\chi_{\kappa_\mu m_\mu}(\vartheta_A, \varphi_A) \tag{10.8}$$

$$\phi_\mu^S(\boldsymbol{r}_A) = \mathrm{i}\frac{Q_\mu(r_A)}{r_A}\chi_{-\kappa_\mu m_\mu}(\vartheta_A, \varphi_A) \tag{10.9}$$

resembling the atomic spinors of chapter 9. The subscript A denotes that the electronic coordinate is to be taken relative to the atomic nucleus A at which it is centered, $\boldsymbol{r}_A = \boldsymbol{r} - \boldsymbol{R}_A$. Again, the two-component spherical spinors $\chi_{\kappa_\mu m_\mu}(\vartheta_A, \varphi_A)$ are those already introduced in chapters 4 and 6. Since we exploit the (2×2) super-structure of the Fock operator, we reduce the number of coefficients to be determined from four to two per basis spinor. The radial parts may also be chosen analogously to those of hydrogen-like atoms and then read

$$P_\mu^{\rm STO}(r_A) = N_\mu^L \, r_A^{l_\mu+1} \exp\left(-\zeta_\mu r_A\right) \tag{10.10}$$

$$Q_\mu^{\rm STO}(r_A) = N_\mu^S \, \left[(\kappa_\mu + l_\mu + 1) - 2\zeta_\mu r_A\right] r_A^{l_\mu} \exp\left(-\zeta_\mu r_A\right) \tag{10.11}$$

Hence, the factors ζ_μ in the exponents are the only adjustable parameters of these basis functions (note that the ζ_μ are usually called the 'exponents' of the basis functions). A basis set is thus characterized solely by the set of 'exponents' $\{\zeta_\mu\}$ (and by the set of contraction coefficients $b_{k\mu}$; see above). Since these exponents enter the basis functions in a nonlinear fashion, they are usually fixed prior to a calculation so that only linear optimization techniques need to be employed for the determination of the $c_{i\mu}^{(L)}$ and $c_{i\mu}^{(S)}$ coefficients.

We will later in this chapter understand that the use of ordinary exponential functions (so-called Slater-type functions) does not lead to solvable SCF equations because of the two-electron integrals that cannot be calculated analytically. It turned out that this problem can be avoided if Gauss-type radial basis functions of the form

$$P_\mu^{\rm GTO}(r_A) = N_\mu^L \, r_A^{l_\mu+1} \exp\left(-\zeta_\mu r_A^2\right) \tag{10.12}$$

$$Q_\mu^{\rm GTO}(r_A) = N_\mu^S \, \left[(\kappa_\mu + l_\mu + 1) - \zeta_\mu r_A\right] r_A^{l_\mu} \exp\left(-\zeta_\mu r_A^2\right) \tag{10.13}$$

are used. The small component's radial function has been fixed according to the kinetic balance condition in both cases.

Because of the different short- and long-range behavior of Gaussians compared to Slater-type functions, more Gaussian functions are needed to represent an atomic orbital properly. While Slater functions, i.e., ordinary exponential functions, have a nonvanishing slope at the origin, Gaussians feature a zero slope at the origin and asymptotically they decay faster than Slater functions. In order to mimic the Slater-type behavior of the atomic spinors of a many-electron atom more Gaussians are required and the number of basis functions per shell m_k', and thus the total number m, increases.

As an alternative to the spherical Gaussian basis sets introduced so far, Cartesian Gaussian functions,

$$\phi_\mu^L(\boldsymbol{r}_A) = N_\mu^L \, x^{\alpha_\mu} y^{\beta_\mu} z^{\gamma_\mu} \exp\left(-\zeta_\mu^L r_A^2\right) \tag{10.14}$$

$$\phi_\mu^S(\boldsymbol{r}_A) = N_\mu^S \, x^{\alpha_\mu} y^{\beta_\mu} z^{\gamma_\mu} \exp\left(-\zeta_\mu^S r_A^2\right) \tag{10.15}$$

may be used. The sum of the exponents α_μ, β_μ, and γ_μ is related to the angular quantum number. The approximate kinetic balance condition is to be fulfilled here as well, which means that first derivatives of Cartesian primitive Gaussians,

$$\frac{\partial}{\partial x} x^{\alpha_\mu} \exp\left(-\zeta_\mu r_A^2\right) = \left[\alpha_\mu x^{\alpha_\mu - 1} - 2\zeta_\mu x^{\alpha_\mu+1}\right] \exp\left(-\zeta_\mu r_A^2\right) \tag{10.16}$$

have to be evaluated to associate the exponents ζ_μ^S for the small component with those of the large component ζ_μ^L. In the case of contractions of Gaussian basis functions the kinetic balance condition becomes even less rigorous. For these cases an atomic balance procedure has been developed [469,470].

Because of the derivative operator $\boldsymbol{p} = -i\hbar\nabla$ every large component Gaussian basis spinor ϕ_i^L gives rise to two small-component basis functions with the same exponent. That is, starting from a p_x-type Gaussian basis function for the large component, for example, the kinetic balance requirement produces an s-type and a d-type basis function for the small component. Therefore, the small-component basis has to comprise at least all spinors originating by differentiation of the large-component basis. As a consequence, the small-component basis will contain functions with higher angular momentum than the large-component basis in order to represent both upper and lower parts of the Dirac Hamiltonian with equal quality for electronic solutions. The small-component basis can thus become twice as large as the large-component basis, as sketched in Figure 10.1, resulting in a significantly higher computational cost for a brute force SCF procedure. While, for instance, the MOLFDIR program generates in this way a huge basis set for the small component (though this component is not that important in terms of its contribution to expectation values), a modern program like UTCHEM circumvents this problem. If we rewrite the SCF equations, Eqs. (8.185), for a given molecular spinor ψ_i in split notation as

$$\begin{pmatrix} W & c\,\boldsymbol{\sigma}\cdot\boldsymbol{p} \\ c\,\boldsymbol{\sigma}\cdot\boldsymbol{p} & W - 2m_ec^2 \end{pmatrix} \begin{pmatrix} \psi_i^L \\ \psi_i^S \end{pmatrix} = \epsilon_i \begin{pmatrix} \psi_i^L \\ \psi_i^S \end{pmatrix} \tag{10.17}$$

where W again collects V_{nuc} and the electron–electron mean-field potential energy terms (note that these equations already incorporate the energy level shift by $-m_ec^2$), we may insert the approximate kinetic balance condition and obtain

$$\begin{pmatrix} W & c\,\boldsymbol{\sigma}\cdot\boldsymbol{p} \\ c\,\boldsymbol{\sigma}\cdot\boldsymbol{p} & W - 2m_ec^2 \end{pmatrix} \begin{pmatrix} \psi_i^L \\ \dfrac{\boldsymbol{\sigma}\cdot\boldsymbol{p}}{2m_ec}\psi_i^L \end{pmatrix} = \epsilon_i \begin{pmatrix} \psi_i^L \\ \dfrac{\boldsymbol{\sigma}\cdot\boldsymbol{p}}{2m_ec}\psi_i^L \end{pmatrix} \tag{10.18}$$

Hence, we absorb the $(\boldsymbol{\sigma}\cdot\boldsymbol{p})$-dependent prefactor into the operator

$$\begin{pmatrix} W & c\,\boldsymbol{\sigma}\cdot\boldsymbol{p}\dfrac{\boldsymbol{\sigma}\cdot\boldsymbol{p}}{2m_ec} \\ c\,\boldsymbol{\sigma}\cdot\boldsymbol{p} & (W - 2m_ec^2)\dfrac{\boldsymbol{\sigma}\cdot\boldsymbol{p}}{2m_ec} \end{pmatrix} \begin{pmatrix} \psi_i^L \\ \psi_i^L \end{pmatrix} = \epsilon_i \begin{pmatrix} \psi_i^L \\ \dfrac{\boldsymbol{\sigma}\cdot\boldsymbol{p}}{2m_ec}\psi_i^L \end{pmatrix} \tag{10.19}$$

and use the same basis sets for large and small components. Note that the remaining prefactor on the right-hand side would then later be absorbed in a modified overlap matrix (compare section 10.4).

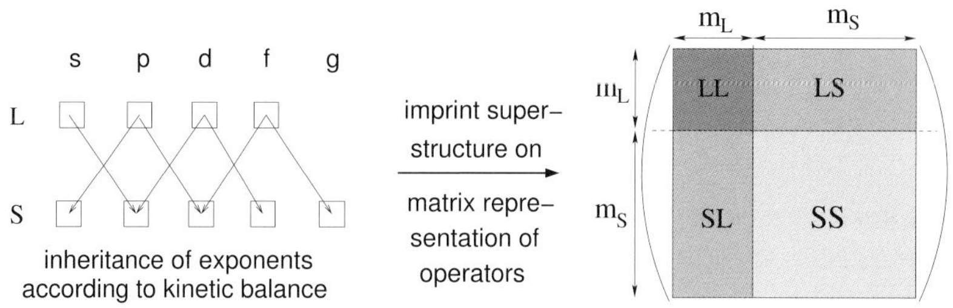

Figure 10.1 Generation of small component's exponents from large component's exponents through kinetic balance explained in Eq. (10.16) (left). L and S denote the large- and small-component basis functions and s, p, d, f, g represent their angular momentum symmetry. The resulting super-block structure of the matrix representation of one-electron operators is depicted on the right hand side. The size of the upper-left LL-block corresponds to the nonrelativistic situation. The lighter the color of a block the more demanding is its evaluation in actual calculations.

The procedure not to apply the $(\boldsymbol{\sigma} \cdot \boldsymbol{p})$-operator onto the basis functions but to consider it explicitly and rewrite the one-electron equation is a major trick which has been dicsussed by many authors and which is intimately connected with the modified Dirac equation and the so-called infinite-order quasi-relativistic methods discussed in some detail in section 14.1.

Apart from the expansion into Gauss-type functions, the use of Slater-type functions has been discussed [471], although the analytic evaluation of integrals becomes as hopeless as in the nonrelativistic theory. Therefore, these Slater-type functions are only a good choice for atoms, linear molecules, or four-component density functional calculations, where integrals over the total electron density are evaluated numerically.

10.2 Dirac–Hartree–Fock Electronic Energy in Basis Set Representation

In the preceding chapter on isolated, spherically symmetric atoms, we chose to derive CI and SCF equations from the most general expression of the total electronic energy in which the total state is expanded in a CSF many-particle basis set. Here, we could proceed in the same way but start instead with the simple Dirac–Hartree–Fock theory for the sake of convenience. Its extension we discuss later in this chapter.

The starting point for our discussion is the Dirac–Hartree–Fock energy expression in Eq. (8.121) which we may write as

$$E_{el,0}^{\mathrm{DHF}} = \sum_{i}^{N} \langle \psi_i | h^D | \psi_i \rangle + \frac{1}{2} \sum_{ij}^{N}$$

$$\times \Big[\langle \psi_i(1)\psi_j(2) \mid g(1,2) \mid \psi_i(1)\psi_j(2) \rangle - \langle \psi_i(1)\psi_j(2) \mid g(1,2) \mid \psi_j(1)\psi_i(2) \rangle \Big] \tag{10.20}$$

This expression is formally the same for its nonrelativistic one-component Hartree–Fock relative,

$$E_{el,0}^{\text{HF}} = \sum_i^N \langle \psi_i | h^S | \psi_i \rangle + \frac{1}{2} \sum_{ij}^N \times \Big[\langle \psi_i(1)\psi_j(2) \mid g(1,2) \mid \psi_i(1)\psi_j(2) \rangle - \langle \psi_i(1)\psi_j(2) \mid g(1,2) \mid \psi_j(1)\psi_i(2) \rangle \Big] \tag{10.21}$$

where h^S denotes the Schrödinger one-electron operator containing the kinetic energy operator for a single electron, $\boldsymbol{p}^2/(2m_e)$ and the one-electron potential V_{nuc}. The N (nonrelativistic) molecular (spin) orbitals ψ_i are constructed as a product of a one-component spatial orbital and one of the two spin functions, α or β, respectively. The one-electron spin functions play a rather formal role in nonrelativistic quantum chemistry and are simply integrated out to yield the restricted Hartree–Fock energy for closed-shell electronic structures,

$$E_{el,0}^{\text{HF}} = 2 \sum_i^{N/2} \langle \psi_i | h^S | \psi_i \rangle + \frac{1}{2} \sum_{ij}^{N/2} \times \Big[2\langle \psi_i(1)\psi_j(2) \mid g(1,2) \mid \psi_i(1)\psi_j(2) \rangle - \langle \psi_i(1)\psi_j(2) \mid g(1,2) \mid \psi_j(1)\psi_i(2) \rangle \Big] \tag{10.22}$$

which finally reduces the number of orbitals to be calculated by one half because each spatial orbital ψ_i is associated once with an α-spin function and once with a β-spin function in closed-shell cases (for convenience, we use the same symbols for spin and spatial orbitals as the upper summation limit indicates the kind of orbital to be employed). If an ordinary one-component basis set is now introduced for the spatial orbitals $\psi_i(\boldsymbol{r})$,

$$\psi_i(\boldsymbol{r}) = \sum_\mu^m c_{i\mu} \phi_\mu(\boldsymbol{r}) \tag{10.23}$$

the energy expression reads

$$E_{el,0}^{\text{HF}} = 2 \sum_i^{N/2} \sum_{\mu\nu}^m c_{i\mu}^\star c_{i\nu} \langle \phi_\mu | h^S | \phi_\nu \rangle + \frac{1}{2} \sum_{ij}^{N/2} \sum_{\mu\nu\lambda\kappa}^m$$

$$\times\left[2c_{i\mu}^{\star}c_{i\lambda}c_{j\nu}^{\star}c_{j\kappa}\langle\phi_\mu(1)\phi_\nu(2)\mid g(1,2)\mid\phi_\lambda(1)\phi_\kappa(2)\rangle\right.$$
$$\left.-c_{i\mu}^{\star}c_{j\lambda}c_{j\nu}^{\star}c_{i\kappa}\langle\phi_\mu(1)\phi_\nu(2)\mid g(1,2)\mid\phi_\lambda(1)\phi_\kappa(2)\rangle\right] \tag{10.24}$$

which can be written in compact form as

$$E_{el,0}^{\mathrm{HF}} = 2\sum_{\mu\nu}^{m} D_{\mu\nu}^{nr}\langle\phi_\mu|h^S|\phi_\nu\rangle + \frac{1}{2}\sum_{\mu\nu\lambda\kappa}^{m}\left\{\left[2D_{\mu\lambda}^{nr}D_{\nu\kappa}^{nr} - D_{\mu\kappa}^{nr}D_{\nu\lambda}^{nr}\right]\right.$$
$$\left.\times\langle\phi_\mu(1)\phi_\nu(2)\mid g(1,2)\mid\phi_\lambda(1)\phi_\kappa(2)\rangle\right\} \tag{10.25}$$

if the (nonrelativistic) closed-shell density matrix $\boldsymbol{D} = \{\boldsymbol{D}^{nr}\}$ for (one-component) molecular orbitals is defined as,

$$D_{\mu\nu}^{nr} = \sum_{i=1}^{N} c_{i\mu}^{\star}c_{i\nu} = 2\sum_{i=1}^{N/2} c_{i\mu}^{\star}c_{i\nu} \tag{10.26}$$

The one-electron integrals over the predefined basis functions need to be calculated and can be arranged in the form of an m-dimensional matrix $\boldsymbol{h} = \{h_{\mu\nu}\}$ with its elements defined by

$$h_{\mu\nu} = \langle\phi_\mu|h^S|\phi_\nu\rangle \tag{10.27}$$

Observe that the first sum can be written in terms of the newly defined matrices as

$$\sum_{\mu\nu}^{m} D_{\mu\nu}^{nr}\langle\phi_\mu|h^S|\phi_\nu\rangle = \sum_{\mu}^{m}\sum_{\nu}^{m} D_{\mu\nu}^{nr}h_{\mu\nu}^{S} = \sum_{\mu}^{m}\sum_{\nu}^{m} D_{\mu\nu}^{nr}h_{\nu\mu}^{S} = Tr(\boldsymbol{D}\cdot\boldsymbol{h}) \tag{10.28}$$

Introducing, however, a basis set expansion as of Eqs. (10.2) and (10.3) in the *relativistic* case for the molecular spinors ψ_i, this leads to a quite complicated expression for the 'simple' Dirac–Hartree–Fock energy of Eq. (10.20), namely to

$$E_{el,0}^{\mathrm{DHF}} = \sum_{i}^{N}\sum_{\mu\nu}^{m}\left[c_{i\mu}^{(1)\star}c_{i\nu}^{(1)}\langle\phi_\mu^{(1)}|V_{nuc}|\phi_\nu^{(1)}\rangle + c_{i\mu}^{(2)\star}c_{i\nu}^{(2)}\langle\phi_\mu^{(2)}|V_{nuc}|\phi_\nu^{(2)}\rangle\right.$$
$$+c_{i\mu}^{(3)\star}c_{i\nu}^{(3)}\langle\phi_\mu^{(3)}|V_{nuc}-2m_ec^2|\phi_\nu^{(3)}\rangle + c_{i\mu}^{(4)\star}c_{i\nu}^{(4)}\langle\phi_\mu^{(4)}|V_{nuc}-2m_ec^2|\phi_\nu^{(4)}\rangle$$
$$+c_{i\mu}^{(1)\star}c_{i\nu}^{(3)}\langle\phi_\mu^{(1)}|cp_z|\phi_\nu^{(3)}\rangle + c_{i\mu}^{(1)\star}c_{i\nu}^{(4)}\langle\phi_\mu^{(1)}|c(p_x-ip_y)|\phi_\nu^{(4)}\rangle$$
$$+c_{i\mu}^{(2)\star}c_{i\nu}^{(3)}\langle\phi_\mu^{(2)}|c(p_x+ip_y)|\phi_\nu^{(3)}\rangle + c_{i\mu}^{(2)\star}c_{i\nu}^{(4)}\langle\phi_\mu^{(2)}|cp_z|\phi_\nu^{(4)}\rangle$$
$$+c_{i\mu}^{(3)\star}c_{i\nu}^{(1)}\langle\phi_\mu^{(3)}|cp_z|\phi_\nu^{(1)}\rangle + c_{i\mu}^{(3)\star}c_{i\nu}^{(2)}\langle\phi_\mu^{(3)}|c(p_x-ip_y)|\phi_\nu^{(2)}\rangle$$

$$
+c_{i\mu}^{(4)\star} c_{i\nu}^{(1} \langle \phi_{\mu}^{(4)} | c(p_x + \mathrm{i}p_y) | \phi_{\nu}^{(1)} \rangle + c_{i\mu}^{(4)\star} c_{i\nu}^{(2)} \langle \phi_{\mu}^{(4)} | cp_z | \phi_{\nu}^{(2)} \rangle \Big]
$$

$$
+\frac{1}{2} \sum_{ij}^{N} \sum_{\mu\nu\lambda\kappa}^{m} \sum_{ab}^{4} \Bigg\{ \left[c_{i\mu}^{(a)\star} c_{i\lambda}^{(a)} c_{j\nu}^{(b)\star} c_{j\kappa}^{(b)} - c_{i\mu}^{(a)\star} c_{j\lambda}^{(a)} c_{j\nu}^{(b)\star} c_{i\kappa}^{(b)} \right]
$$

$$
\times \left\langle \phi_{\mu}^{(a)}(1) \phi_{\nu}^{(b)}(2) \left| \frac{1}{r_{12}} \right| \phi_{\lambda}^{(a)}(1) \phi_{\kappa}^{(b)}(2) \right\rangle \Bigg\} \tag{10.29}
$$

because of the four-component nature of the spinors and the four-dimensional one- and two-electron operators. Especially, the four-dimensional one-electron operator,

$$
h^D = \begin{pmatrix} V_{nuc} & 0 & cp_z & c(p_x - \mathrm{i}p_y) \\ 0 & V_{nuc} & c(p_x + \mathrm{i}p_y) & cp_z \\ cp_z & c(p_x - \mathrm{i}p_y) & V_{nuc} - 2m_e c^2 & 0 \\ c(p_x + \mathrm{i}p_y) & cp_z & 0 & V_{nuc} - 2m_e c^2 \end{pmatrix} \tag{10.30}
$$

needs to be resolved into its components as in Eqs. (5.81)–(5.84). The pair exchange that differentiates Coulomb and exchange integrals is transferred onto the molecular spinor coefficients in Eq. (10.29) to be multiplied with the corresponding two-electron integral over four basis functions each. This type of basis set expansion thus leads to 16 different pairs of molecular spinor coefficient products to be multiplied with two-electron integrals per Coulomb and per exchange integral. If we adopt the more convenient bispinor basis set expansion of Eq. (10.7), only four of these 16 products of pairs survive,

$$
\begin{aligned}
E_{el,0}^{\mathrm{DHF}} = & \sum_{i}^{N} \sum_{\mu\nu}^{m} \Big[c_{i\mu}^{(L)\star} c_{i\nu}^{(L)} \langle \phi_{\mu}^{L} | V_{nuc} | \phi_{\nu}^{L} \rangle + c_{i\mu}^{(S)\star} c_{i\nu}^{(S)} \langle \phi_{\mu}^{S} | V_{nuc} - 2m_e c^2 | \phi_{\nu}^{S} \rangle \\
& + c_{i\mu}^{(L)\star} c_{i\nu}^{(S)} \langle \phi_{\mu}^{L} | c\boldsymbol{\sigma} \cdot \boldsymbol{p} | \phi_{\nu}^{S} \rangle + c_{i\mu}^{(S)\star} c_{i\nu}^{(L)} \langle \phi_{\mu}^{S} | c\boldsymbol{\sigma} \cdot \boldsymbol{p} | \phi_{\nu}^{L} \rangle \Big] \\
& + \frac{1}{2} \sum_{ij}^{N} \sum_{\mu\nu\lambda\kappa}^{m} \Bigg[c_{i\mu}^{(L)\star} c_{i\lambda}^{(L)} c_{j\nu}^{(L)\star} c_{j\kappa}^{(L)} \left\langle \phi_{\mu}^{L}(1) \phi_{\nu}^{L}(2) \left| \frac{1}{r_{12}} \right| \phi_{\lambda}^{L}(1) \phi_{\kappa}^{L}(2) \right\rangle \\
& + c_{i\mu}^{(L)\star} c_{i\lambda}^{(L)} c_{j\nu}^{(S)\star} c_{j\kappa}^{(S)} \left\langle \phi_{\mu}^{L}(1) \phi_{\nu}^{S}(2) \left| \frac{1}{r_{12}} \right| \phi_{\lambda}^{L}(1) \phi_{\kappa}^{S}(2) \right\rangle \\
& + c_{i\mu}^{(S)\star} c_{i\lambda}^{(S)} c_{j\nu}^{(L)\star} c_{j\kappa}^{(L)} \left\langle \phi_{\mu}^{S}(1) \phi_{\nu}^{L}(2) \left| \frac{1}{r_{12}} \right| \phi_{\lambda}^{S}(1) \phi_{\kappa}^{L}(2) \right\rangle \\
& + c_{i\mu}^{(S)\star} c_{i\lambda}^{(S)} c_{j\nu}^{(S)\star} c_{j\kappa}^{(S)} \left\langle \phi_{\mu}^{S}(1) \phi_{\nu}^{S}(2) \left| \frac{1}{r_{12}} \right| \phi_{\lambda}^{S}(1) \phi_{\kappa}^{S}(2) \right\rangle \\
& - c_{i\mu}^{(L)\star} c_{j\lambda}^{(L)} c_{j\nu}^{(L)\star} c_{i\kappa}^{(L)} \left\langle \phi_{\mu}^{L}(1) \phi_{\nu}^{L}(2) \left| \frac{1}{r_{12}} \right| \phi_{\lambda}^{L}(1) \phi_{\kappa}^{L}(2) \right\rangle \\
& - c_{i\mu}^{(L)\star} c_{j\lambda}^{(L)} c_{j\nu}^{(S)\star} c_{i\kappa}^{(S)} \left\langle \phi_{\mu}^{L}(1) \phi_{\nu}^{S}(2) \left| \frac{1}{r_{12}} \right| \phi_{\lambda}^{L}(1) \phi_{\kappa}^{S}(2) \right\rangle
\end{aligned}
$$

$$- c_{i\mu}^{(S)\star} c_{j\lambda}^{(S)} c_{j\nu}^{(L)\star} c_{i\kappa}^{(L)} \left\langle \phi_\mu^S(1)\phi_\nu^L(2) \left| \frac{1}{r_{12}} \right| \phi_\lambda^S(1)\phi_\kappa^L(2) \right\rangle$$
$$- c_{j\mu}^{(S)\star} c_{j\lambda}^{(S)} c_{j\nu}^{(S)\star} c_{i\kappa}^{(S)} \left\langle \phi_\mu^S(1)\phi_\nu^S(2) \left| \frac{1}{r_{12}} \right| \phi_\lambda^S(1)\phi_\kappa^S(2) \right\rangle \Bigg] \quad (10.31)$$

and also the much simpler (2×2) super-structure of the one-electron operator h^D

$$h^D = \begin{pmatrix} V_{nuc} & c\boldsymbol{\sigma}\cdot\boldsymbol{p} \\ c\boldsymbol{\sigma}\cdot\boldsymbol{p} & V_{nuc} - 2m_e c^2 \end{pmatrix} \quad (10.32)$$

can now be exploited. At this stage, we should finally comment on the reason why we have introduced the Coulomb interaction in place of the more general two-electron operator $g(1,2)$. The frequency-independent Breit interaction — and already the simpler spin–spin Gaunt interaction — contain Dirac $\boldsymbol{\alpha}$ matrices, which couple large and small components, for each of the two electronic coordinates. The structure of matrix elements for such off-diagonal, so-called odd operators $\mathcal{O}$,

$$\mathcal{O}(i) = \begin{pmatrix} \mathbf{0} & \mathcal{O}^{LS}(i) \\ \mathcal{O}^{SL}(i) & \mathbf{0} \end{pmatrix} \quad (10.33)$$

is therefore completely different from that of the diagonal Coulomb operator, which is of the form $\langle \phi_\mu^{(L)}(1)\phi_\nu^{(S)}(2)|r_{12}^{-1}|\phi_\lambda^{(L)}(1)\phi_\kappa^{(S)}(2)\rangle$ rather than $\langle \phi_\mu^{(L)}(1)\phi_\nu^{(S)}(2)|\mathcal{O}^{(LS)}(1)\cdot\mathcal{O}^{(SL)}(2)|\phi_\lambda^{(S)}(1)\phi_\kappa^{(L)}(2)\rangle$. If we now define the elements of the m-dimensional relativistic density matrix $\boldsymbol{D} = \{D_{\mu\nu}\}$ with an (LL), (LS), (SL) and (SS) super-structure,

$$D_{\mu\nu}^{(XY)} = \sum_i^N c_{i\mu}^{(X)\star} c_{i\nu}^{(Y)}, \quad \text{and} \quad X,Y \in \{L,S\} \quad (10.34)$$

we can write the Dirac–Hartree–Fock energy in a more compact way as

$$\begin{aligned} E_{el,0}^{\text{DHF}} = & \sum_{\mu\nu}^m \Big[D_{\mu\nu}^{(LL)} \langle \phi_\mu^L | V_{nuc} | \phi_\nu^L \rangle + D_{\mu\nu}^{(SS)} \langle \phi_\mu^S | V_{nuc} - 2m_e c^2 | \phi_\nu^S \rangle \\ & \quad + D_{\mu\nu}^{(LS)} \langle \phi_\mu^L | c\boldsymbol{\sigma}\cdot\boldsymbol{p} | \phi_\nu^S \rangle + D_{\mu\nu}^{(SL)} \langle \phi_\mu^S | c\boldsymbol{\sigma}\cdot\boldsymbol{p} | \phi_\nu^L \rangle \Big] \\ & + \frac{1}{2} \sum_{\mu\nu\lambda\kappa}^m \Big\{ \Big[D_{\mu\lambda}^{(LL)} D_{\nu\kappa}^{(LL)} - D_{\mu\kappa}^{(LL)} D_{\nu\lambda}^{(LL)} \Big] \\ & \quad \times \left\langle \phi_\mu^L(1)\phi_\nu^L(2) \left| \frac{1}{r_{12}} \right| \phi_\lambda^L(1)\phi_\kappa^L(2) \right\rangle \\ & + \Big[D_{\mu\lambda}^{(LL)} D_{\nu\kappa}^{(SS)} - D_{\mu\kappa}^{(LS)} D_{\nu\lambda}^{(SL)} \Big] \\ & \quad \times \left\langle \phi_\mu^L(1)\phi_\nu^S(2) \left| \frac{1}{r_{12}} \right| \phi_\lambda^L(1)\phi_\kappa^S(2) \right\rangle \end{aligned}$$

$$+ \left[D_{\mu\lambda}^{(SS)} D_{\nu\kappa}^{(LL)} - D_{\mu\kappa}^{(SL)} D_{\nu\lambda}^{(LS)}\right]$$
$$\times \left\langle \phi_\mu^S(1)\phi_\nu^L(2) \left| \frac{1}{r_{12}} \right| \phi_\lambda^S(1)\phi_\kappa^L(2) \right\rangle$$

$$+ \left[D_{\mu\lambda}^{(SS)} D_{\nu\kappa}^{(SS)} - D_{\mu\kappa}^{(SS)} D_{\nu\lambda}^{(SS)}\right]$$
$$\times \left\langle \phi_\mu^S(1)\phi_\nu^S(2) \left| \frac{1}{r_{12}} \right| \phi_\lambda^S(1)\phi_\kappa^S(2) \right\rangle \Bigg\} \tag{10.35}$$

The one-electron integrals over the pre-defined basis functions need to be calculated and can be arranged in form of an $(m_L + m_S)$-dimensional matrix $\boldsymbol{h}^D = \{h_{\mu\nu}^D\}$ and its elements still feature the (2×2) super-structure,

$$h_{\mu\nu}^D = \begin{pmatrix} h_{\mu\nu}^{D(LL)} & h_{\mu\nu}^{D(LS)} \\ h_{\mu\nu}^{D(SL)} & h_{\mu\nu}^{D(SS)} \end{pmatrix} = \begin{pmatrix} V_{\mu\nu}^{(LL)} & c\,[\boldsymbol{\sigma}\cdot\boldsymbol{p}]_{\mu\nu}^{(LS)} \\ c\,[\boldsymbol{\sigma}\cdot\boldsymbol{p}]_{\mu\nu}^{(SL)} & V_{\mu\nu}^{(SS)} - 2m_e c^2 S_{\mu\nu}^{(SS)} \end{pmatrix} \tag{10.36}$$

The matrix elements are naturally defined as

$$V_{\mu\nu}^{(LL)} = \langle \phi_\mu^L \mid V_{nuc} \mid \phi_\nu^L \rangle \tag{10.37}$$
$$[\boldsymbol{\sigma}\cdot\boldsymbol{p}]_{\mu\nu}^{(SL)} = \langle \phi_\mu^L \mid \boldsymbol{\sigma}\cdot\boldsymbol{p} \mid \phi_\nu^S \rangle \tag{10.38}$$
$$S_{\mu\nu}^{(SS)} = \langle \phi_\mu^S \mid \phi_\nu^S \rangle \tag{10.39}$$

and so forth. Regarding the two-electron terms we may define in close analogy Coulomb and exchange matrices with entries

$$J_{\mu\lambda}^{(XX)} = \sum_{\nu\kappa} \left[D_{\nu\kappa}^{(LL)} \left\langle \phi_\mu^X(1)\phi_\nu^L(2) \left| \frac{1}{r_{12}} \right| \phi_\lambda^X(1)\phi_\kappa^L(2) \right\rangle + D_{\nu\kappa}^{(SS)} \left\langle \phi_\mu^X(1)\phi_\nu^S(2) \left| \frac{1}{r_{12}} \right| \phi_\lambda^X(1)\phi_\kappa^S(2) \right\rangle \right] ;\ X \in \{L, S\} \tag{10.40}$$

$$J_{\mu\lambda}^{(LS)} = J_{\mu\lambda}^{(SL)} = 0 \tag{10.41}$$

$$K_{\mu\kappa}^{(LL)} = \sum_{\nu\lambda} D_{\nu\lambda}^{(LL)} \left\langle \phi_\mu^L(1)\phi_\nu^L(2) \left| \frac{1}{r_{12}} \right| \phi_\lambda^L(1)\phi_\kappa^L(2) \right\rangle \tag{10.42}$$

$$K_{\mu\kappa}^{(SL)} = \sum_{\nu\lambda} D_{\nu\lambda}^{(SL)} \left\langle \phi_\mu^L(1)\phi_\nu^S(2) \left| \frac{1}{r_{12}} \right| \phi_\lambda^L(1)\phi_\kappa^S(2) \right\rangle \tag{10.43}$$

$$K_{\mu\kappa}^{(LS)} = \sum_{\nu\lambda} D_{\nu\lambda}^{(LS)} \left\langle \phi_\mu^S(1)\phi_\nu^L(2) \left| \frac{1}{r_{12}} \right| \phi_\lambda^S(1)\phi_\kappa^L(2) \right\rangle \tag{10.44}$$

$$K_{\mu\kappa}^{(SS)} = \sum_{\nu\lambda} D_{\nu\lambda}^{(SS)} \left\langle \phi_\mu^S(1)\phi_\nu^S(2) \left| \frac{1}{r_{12}} \right| \phi_\lambda^S(1)\phi_\kappa^S(2) \right\rangle \tag{10.45}$$

Hence, the Coulomb matrix $\boldsymbol{J}$ is block-diagonal, while the exchange matrix $\boldsymbol{K}$ is not,

$$\boldsymbol{J} = \begin{pmatrix} \boldsymbol{J}^{(LL)} & \boldsymbol{J}^{(LS)} \\ \boldsymbol{J}^{(SL)} & \boldsymbol{J}^{(SS)} \end{pmatrix} = \begin{pmatrix} \boldsymbol{J}^{(LL)} & \boldsymbol{0} \\ \boldsymbol{0} & \boldsymbol{J}^{(SS)} \end{pmatrix}, \quad \boldsymbol{K} = \begin{pmatrix} \boldsymbol{K}^{(LL)} & \boldsymbol{K}^{(LS)} \\ \boldsymbol{K}^{(SL)} & \boldsymbol{K}^{(SS)} \end{pmatrix} \tag{10.46}$$

For section 10.4 it is convenient to write the energy as

$$E_{el,0}^{\mathrm{DHF}}[\{\boldsymbol{c}_{i\mu}\}] = \sum_{i=1}^{n} \left(\boldsymbol{c}_i^{(L)}, \boldsymbol{c}_i^{(S)}\right) \cdot \begin{pmatrix} \boldsymbol{E}^{(LL)} & \boldsymbol{E}^{(LS)} \\ \boldsymbol{E}^{(SL)} & \boldsymbol{E}^{(SS)} \end{pmatrix} \cdot \begin{pmatrix} \boldsymbol{c}_i^{(L)} \\ \boldsymbol{c}_i^{(S)} \end{pmatrix} \tag{10.47}$$

with

$$\boldsymbol{E}^{(LL)} = \boldsymbol{V}^{(LL)} + \boldsymbol{J}^{(LL)} - \boldsymbol{K}^{(LL)} \tag{10.48}$$

$$\boldsymbol{E}^{(LS)} = c\,[\boldsymbol{\sigma} \cdot \boldsymbol{p}]^{(LS)} - \boldsymbol{K}^{(LS)} \tag{10.49}$$

$$\boldsymbol{E}^{(SL)} = c\,[\boldsymbol{\sigma} \cdot \boldsymbol{p}]^{(SL)} - \boldsymbol{K}^{(SL)} \tag{10.50}$$

$$\boldsymbol{E}^{(SS)} = \boldsymbol{V}^{(SS)} - 2m_e c^2 \boldsymbol{S}^{(SS)} + \boldsymbol{J}^{(SS)} - \boldsymbol{K}^{(SS)} \tag{10.51}$$

10.3
Molecular One- and Two-Electron Integrals

The reason for the introduction of Gaussian-type basis functions is the fact that two-electron integrals can be solved analytically. All relativistic one- and two-electron integrals can be evaluated using standard techniques that have been developed in nonrelativistic quantum chemistry. For a detailed discussion we may therefore refer to the book by Helgaker, Jørgensen and Olsen [217] and include here only a few general comments.

The number of well-established integral evaluation schemes is not that large. Two examples are the McMurchie–Davidson [472] and the Obara–Saika schemes [473, 474]. All of them have been reviewed in an excellent overview by Lindh [190] to which we also refer regarding the issue on how to efficiently evaluate two-electron integrals over Gaussian basis functions. Lindh emphasizes that since all these approaches aim at the solution of the same types of integrals they are quite similar, and hence the efficient integral calculation depends strongly on the smart implementation of the individual schemes rather than on the choice of a particular scheme. Though some integral programs have been established in the past, such programs are still under constant development, also for four-component molecular calculations as documented in Ref. [475].

10.4 Dirac–Hartree–Fock–Roothaan Matrix Equations

With the basis set expansion of Eq. (10.7) representing a molecular spinor ψ_i, the molecular spinor coefficients $c_{i\mu}$ are the remaining unknowns. The molecular spinor now becomes identical to the set of unknown spinor coefficients. In order to determine these unknowns the variational minimax principle of chapter 8 is invoked. For this procedure, we may again start from the energy expression of section 10.2 and differentiate it or directly insert the basis set expansion of Eq. (10.3) into the SCF Eqs. (8.185). These options are depicted in Figure 10.2. The resulting Dirac–Hartree–Fock equations in basis set representation are called Dirac–Hartree–Fock–Roothaan equations according to the work by Roothaan [424] and Hall [425] on the nonrelativistic analog.

10.4.1 Two Possible Routes for the Derivation

After introducing a set of basis functions, we can rewrite the matrix elements occurring in the total energy as matrix products introduced in section 10.2. For the differentiation of the total electronic energy with respect to the molecular spinor coefficients we write the Lagrangian functional as

$$L\left[\left\{\psi_i^L,\psi_i^S\right\}\right] = E\left[\left\{\psi_i^L,\psi_i^S\right\}\right] - \sum_{ij}^{N}\epsilon_{ij}\left[\langle\psi_i^L|\psi_j^L\rangle + \langle\psi_i^S|\psi_j^S\rangle - \delta_{ij}\right] \qquad (10.52)$$

and then express it in terms of the molecular (bi)spinor coefficients,

$$\begin{aligned} L\left[\left\{c_{i\mu}^{(L)},c_{i\mu}^{(S)}\right\}\right] &= E\left[\left\{c_{i\mu}^{(L)},c_{i\mu}^{(S)}\right\}\right] - \sum_{ij}^{N}\epsilon_{ij}\times \\ &\left[\sum_{\mu\nu}\left(c_{i\mu}^{(L)}c_{j\nu}^{(L)}\langle\phi_i^L|\phi_j^L\rangle + c_{i\mu}^{(S)}c_{j\nu}^{(S)}\langle\phi_i^S|\phi_j^S\rangle\right) - \delta_{ij}\right] \end{aligned} \qquad (10.53)$$

Next, we need to set the first derivative of L with respect to the molecular spinor coefficients equal to zero,

$$\frac{\partial L[\{c_{i\mu}^X\}]}{\partial c_{i\mu}^{(X)}} \stackrel{!}{=} 0 \quad \text{with } X \in \{L,S\} \qquad (10.54)$$

following the derivation in section 8.7 Finally, we obtain for the special case of a basis set expansion,

$$\begin{pmatrix} f^{(LL)} & f^{(LS)} \\ f^{(SL)} & f^{(SS)} \end{pmatrix}\begin{pmatrix} c_i^{(L)} \\ c_i^{(S)} \end{pmatrix} = \epsilon_i \begin{pmatrix} S^{(LL)} & 0 \\ 0 & S^{(SS)} \end{pmatrix}\begin{pmatrix} c_i^{(L)} \\ c_i^{(S)} \end{pmatrix} \qquad (10.55)$$

$$L[\{\psi_i\}] = E_{el,A}[\{\psi_i\}] - \sum_{ij}^{N} \epsilon_{ij}(\langle \psi_i \mid \psi_j \rangle - \delta_{ij}) \quad \text{Eq. (8.168)}$$

$$\xrightarrow[\text{Eq. (10.3)}]{\psi_i = \sum_{\mu}^{m}\{c_{i\mu}^{(a)}\phi_{\mu}^{(a)}\}}$$

$$L[\{c_{i\mu}\}] = E_{el,A}[\{c_{i\mu}\}] - \sum_{ij}^{N} \epsilon_{ij} \times \left(\sum_{\mu\nu}\sum_{a} c_{i\mu}^{(a)} c_{j\nu}^{(a)} \langle \phi_{\mu}^{(a)} \mid \phi_{\nu}^{(a)} \rangle - \delta_{ij}\right) \quad \text{Eq. (10.53)}$$

$$\downarrow \; \delta L[\{\psi_i\}] \overset{!}{=} 0 \quad \text{Eq. (8.169)} \qquad\qquad \downarrow \; \frac{\partial L[\{\psi_i\}]}{\partial c_{i\mu}^{(a)}} \overset{!}{=} 0 \quad \text{Eq. (10.54)}$$

$$\hat{f}_{ii}\psi_i = \epsilon_i \psi_i \quad \text{Eq. (8.185)} \quad \xrightarrow[\text{Eq. (10.3)}]{\psi_i = \sum_{\mu}^{m}\{c_{i\mu}^{(a)}\phi_{\mu}^{(a)}\}} \quad \boldsymbol{fc} = \boldsymbol{Sc\epsilon} \quad \text{Eq. (10.55)}$$

Figure 10.2 Two routes for the derivation of Dirac–Hartree–Fock equations in basis set representation in the lower right corner: the Roothaan equations (recall the caveats of sections 8.2.3 and 8.7.1 required for the application of the variational principle).

These are the relativistic Roothaan equations [423] where

$$\boldsymbol{f}^{(LL)} = \boldsymbol{V}^{(LL)} + \boldsymbol{J}^{(LL)} - \boldsymbol{K}^{(LL)} \tag{10.56}$$

$$\boldsymbol{f}^{(LS)} = c\,[\boldsymbol{\sigma}\cdot\boldsymbol{p}]^{(LS)} - \boldsymbol{K}^{(LS)} \tag{10.57}$$

$$\boldsymbol{f}^{(SL)} = c\,[\boldsymbol{\sigma}\cdot\boldsymbol{p}]^{(SL)} - \boldsymbol{K}^{(SL)} \tag{10.58}$$

$$\boldsymbol{f}^{(SS)} = \boldsymbol{V}^{(SS)} - 2m_e c^2 \boldsymbol{S}^{(SS)} + \boldsymbol{J}^{(SS)} - \boldsymbol{K}^{(SS)} \tag{10.59}$$

This equation may be written in the well-known compact form as a generalized matrix eigenvalue problem,

$$\boldsymbol{fc} = \boldsymbol{Sc\epsilon} \tag{10.60}$$

for all vectors of molecular spinor coefficients collected in the matrix $\boldsymbol{c}$. The orbital energies are then the entries of the diagonal matrix $\boldsymbol{\epsilon} = \{\epsilon_{ii}\} \equiv \{\epsilon_i\}$.

It is important to note that the introduction of a finite basis set produced a Fock matrix where the position dependence has been integrated out — in contrast to the atomic case of chapter 9, where all Coulomb and exchange operators feature a clear position dependence.

10.4.2
Treatment of Negative-Energy States

Similar to nonrelativistic Hartree–Fock theory, the Dirac–Roothaan Eqs. (10.60) are solved iteratively until self-consistency is reached. However, because of the properties of the one-electron Dirac Hamiltonian entering the Fock operator, molecular spinors representing unphysical negative-energy states (recall section 5.5) show up in this procedure. As many of these negative-continuum states are obtained as basis functions for the small component have been employed. The occurrence of these states within four-component SCF theory is an inevitable consequence of the mathematical structure of the Dirac Hamiltonian h^D. We emphasize that it is *not* related to the specific choice of the two-electron interaction potential. As a consequence, the molecular spinor coefficient matrix $\boldsymbol{c}$ comprises m_S states belonging to the negative continuum with spinor energies below $-2m_ec^2$ and m_L positive-energy states above $-m_ec^2$. After a rearrangement of these one-electron states in ascending spinor energy ϵ_i, the coefficient matrix is then given by

$$\boldsymbol{c} = \left(\begin{array}{ccccccccc} c_{11}^{(S)} & \cdots & c_{m_S1}^{(S)} & c_{(m_S+1)1}^{(S)} & \cdots & c_{(m_S+N)1}^{(S)} & \cdots & c_{m1}^{(S)} \\ \vdots & & \vdots & \vdots & & \vdots & & \vdots \\ c_{1m}^{(L)} & \cdots & c_{(m_S)m}^{(L)} & c_{(m_S+1)m}^{(L)} & \cdots & c_{(m_S+N)m}^{(L)} & \cdots & c_{mm}^{(L)} \end{array} \right) \tag{10.61}$$
$$\qquad\qquad \underbrace{\hphantom{c_{11}^{(S)} \cdots c_{m_S1}^{(S)}}}_{m_S} \quad \underbrace{\hphantom{c_{(m_S+1)1}^{(S)} \cdots c_{(m_S+N)1}^{(S)}}}_{N} \quad \underbrace{\hphantom{\cdots c_{m1}^{(S)}}}_{m_L - N}$$

with $m = m_L + m_S$. Accordingly, each column of this matrix contains the $(m_L + m_S)$ expansion coefficients of one molecular spinor ψ_i. The key feature of any four-component SCF protocol is the deletion of the first m_S negative-energy states. They are thus simply disregarded for the construction of the density matrices. As a consequence, only the lowest N electronic bound states are treated as occupied and incorporated into the density matrices,

$$\begin{aligned} D_{\mu\nu}^{(LL)} &= \sum_{i=m_S+1}^{m_S+N} c_{i\mu}^{(L)} c_{i\nu}^{(L)} \quad , \quad D_{\mu\nu}^{(LS)} = \sum_{i=m_S+1}^{m_S+N} c_{i\mu}^{(L)} c_{i\nu}^{(S)} \\ D_{\mu\nu}^{(SL)} &= \sum_{i=m_S+1}^{m_S+N} c_{i\mu}^{(S)} c_{i\nu}^{(L)} \quad , \quad D_{\mu\nu}^{(SS)} = \sum_{i=m_S+1}^{m_S+N} c_{i\mu}^{(S)} c_{i\nu}^{(S)} \end{aligned} \tag{10.62}$$

from which the two-electron mean-field potential of the Fock operator is calculated then. As before, the superscripts (L) and (S) denote the classification of all molecular spinor coefficients according to their membership to either the large or the small component basis set (where we assume that all molecular spinor coefficients are real, which is no restriction of generality for general molecules). This procedure has proved to yield accurate basis-set results for atomic and molecular systems.

10.4.3
Four-Component DFT

A four-component DFT implementation possesses a similar algorithmic structure to that of a Dirac–Hartree–Fock computer program. Accordingly, it was natural to combine pure density functionals with Hartree–Fock-type exchange integrals in Kohn–Sham calculations in order to overcome deficiencies in contemporary approximate exchange–correlation functionals [476,477]. The gradient of these pure exchange–correlation functionals enter the Kohn–Sham Fock operator solely on the block diagonal. Several DFT implementations have been developed and made available for molecular electronic structure calculations [319,434,444–446,478–481].

10.4.4
Symmetry

Considerations of symmetry are beneficial for various reasons. If a suitable group is assigned to a molecular system, all spinors and wave functions can be classified according to its irreducible representations [482–484] and all operators (like, for example, the Fock operator) adopt a block-diagonal form in a symmetry-adapted basis. The latter is, of course, advantageous as the dimension is reduced so that matrix operations — in particular, diagonalization and inversion operations which scale like m^3 if m is the dimension of the matrix — can be carried out more efficiently. Moreover, group-theoretical arguments allow one to determine the symmetry of relevant quantities, which in turn can be helpful in determining vanishing expectation values, as an integral is only different from zero if the integrand transforms like the totally symmetric irreducible representation of the appropriate group (a fact extensively used for the derivation of selection rules for spectroscopic transitions on the basis of Fermi's golden rule; but this fact is also of technical importance).

We have already addressed symmetry issues in preceding chapters. One- and many-electron atoms feature atomic spinors and total states that transform according to the irreducible representation of total angular momentum. These symmetries according to the group of rotations and the corresponding term symbol notation were introduced at the end of chapter 4. There, we already anticipated that symmetries associated with angular momentum can only be exploited for linear molecules. For general molecules, point group symmetry can be exploited. While, in nonrelativistic theories, spin symmetry can always be treated separately and decoupled from point group symmetries, this is no longer the case in relativistic approaches when spin is no longer a good quantum number. Nevertheless we can exploit symmetries as follows.

10.4.5
Kramers' Time Reversal Symmetry

Since spin symmetry is broken through spin–orbit coupling in four-component electronic structure theory, we must seek for a symmetry to replace it. In the case of spherically symmetric atoms this was the *jj*-coupling. Now we also lack rotational symmetry and are in need of a more general concept. This has been discovered by Kramers [485, 486], and is the concept of *time reversal* [52]. In the absence of an external magnetic field, the state of one fermion is doubly degenerate. The Pair of such states is called a *Kramers Pair*. The Kramers time-reversal operator K relates the two states, ψ_i and $\bar{\psi}_i$,

$$K\psi_i = \bar{\psi}_i \tag{10.63}$$

and is defined as [487]

$$K = -\mathrm{i}\begin{pmatrix} \sigma_y & 0 \\ 0 & \sigma_y \end{pmatrix} K_0 \tag{10.64}$$

where K_0 simply performs the complex conjugation as demanded by the equations above. It may be easily shown that this operator possesses the following properties,

$$K^2\psi_i = -\psi_i \quad \text{and} \quad K(a\psi_i) = a^\star\psi_i \tag{10.65}$$

Hence, the relativistic analog of the spin-restriction in nonrelativistic closed-shell Hartree–Fock theory is Kramers-restricted Dirac–Hartree–Fock theory. We should emphasize that our derivation of the Roothaan equation above is the 'pedestrian' way chosen in order to produce this matrix-SCF equation step by step. The most sophisticated formulations are the Kramers-restricted quaternion Dirac–Hartree–Fock implementations [246, 247]. A basis of Kramers pairs, i.e., one adapted to time-reversal symmetry, transforms into another basis under quaternionic unitary transformation [488]. This can be exploited for the optimization of Dirac–Hartree–Fock spinors, but also for MCSCF spinors. In a Kramers one-electron basis, an operator $\boldsymbol{O}$ invariant under time reversal possesses a specific block structure,

$$\boldsymbol{O} = \begin{pmatrix} \boldsymbol{A} & \boldsymbol{B} \\ -\boldsymbol{B}^\star & \boldsymbol{A}^\star \end{pmatrix} \tag{10.66}$$

in which the submatrix $\boldsymbol{A}$ is hermitean, while $\boldsymbol{B}$ is antisymmetric.

10.4.6 Double Groups

Spin and spatial coordinates do not factorize in Dirac-based theories (in contrast to nonrelativistic quantum mechanics). Hence, ordinary point group symmetries are not a proper approach to understanding half-spin systems. Bethe [489] solved this problem by distinguishing between the null rotation and the rotation by 2π in spinor physics, the concepts, however, were introduced much earlier to mathematics (see the book by Altmann [490] for a historical perspective and a detailed account on double groups).

Only for half-spin systems does the double group representation become important, and the identity operation corresponds to a rotation by 4π. Systems with integer spin may be treated according to ordinary point-group symmetry. The symmetry operations that are the elements of a double group are operations in space like rotations, reflections and inversion. Time reversal does not belong to them, though there is a connection to the irreducible representations of the double group [486].

Double group symmetry is employed to the utmost extent in four-component computer programs for molecular calculations in order to increase the efficiency of these codes [487, 491, 492]. We do not have the space to discuss these issues further, but refer to presentations of the relevant character tables as in Ref. [493].

10.5 Analytic Gradients

It is important to note that analytic energy derivatives are the key to molecular structure optimizations and also to the calculation of molecular properties. While the latter issue will be discussed in chapter 15, the former is considered in this section. The major advantage of basis-set-expansion methods is that any kind of derivative can be taken analytically. In fully numerical methods — like those discussed in the preceding chapter — any kind of gradient would have to be evaluated numerically by separate single-point calculations for given distorted values of the variable under consideration.

In chapter 8 we separated electronic from nuclear coordinates. As a consequence, the electronic wave function and hence the total electronic energy are calculated only for a given set of nuclear coordinates $\{\boldsymbol{R}_A\}$ and thus for a predefined molecular structure. From a chemical perspective, mostly those structures are of importance which correspond to minima on the potential energy surface $E_{el,A}(\{\boldsymbol{R}_A\})$. In order to find these minima, first-order derivatives of the total electronic energy of a given state with respect to all nuclear coordinates, the so-called *geometric gradients*, are required in order to define a path of descending energy. Of course, the electronic energy may be expanded in a Taylor series around some reference point, and the gradient is the first correction term to proceed to any other point nearby, higher (geometric) derivatives being useful as well; however, for computational purposes at most approximate second derivatives are also computed in actual calculations.

In the following, we consider only first derivatives of the electronic energy and start with a rather general expression which we differentiate with respect to a general parameter λ that may then be identified with some nuclear coordinate $R_{A,\alpha}$ with $\alpha \in \{x, y, z\}$. In principle, these considerations may be generalized for any other total energy expression like the one of Hartree–Fock or Dirac–Hartree–Fock theory or for any other relativistic many-electron Hamiltonians discussed so far or to be discussed in later chapters. The reader is referred to the comprehensive book by Yamaguchi *et al.* [494] for a detailed presentation of nonrelativistic gradient theory for CI-type wave functions.

The Dirac–Hartree–Fock energy depends on the molecular spinor coefficients $\{c_{i\mu}\}$ and on a given parameter λ that may be, for instance, a nuclear coordinate $R_{A,\alpha}$. A first derivative of this energy may be formulated in the most general form by using the chain rule,

$$\frac{dE_{el,0}^{\text{DHF}}}{d\lambda} = \frac{\partial E_{el,0}^{\text{DHF}}}{\partial\lambda} + \sum_{i\mu} \frac{\partial E_{el,0}^{\text{DHF}}}{\partial c_{i\mu}} \frac{\partial c_{i\mu}}{\partial\lambda} \tag{10.67}$$

An optimized Dirac–Hartree–Fock wave function yields by construction a vanishing partial derivative [compare Eq. (10.54)]

$$\frac{\partial E_{el,0}^{\text{DHF}}}{\partial c_{\mu a}} = 0 \tag{10.68}$$

which implies that the total and partial derivatives with respect to λ coincide,

$$\frac{dE_{el,0}^{\text{DHF}}}{d\lambda} = \frac{\partial E_{el,0}^{\text{DHF}}}{\partial\lambda} \tag{10.69}$$

In the following the focus will be solely on the nuclear-coordinate dependence of the electronic energy, and we choose $\lambda = R_{A,\alpha}$ so that the last equation reads

$$\frac{dE_{el,0}^{\text{DHF}}}{dR_{A,\alpha}} = \frac{\partial E_{el,0}^{\text{DHF}}}{\partial R_{A,\alpha}} \tag{10.70}$$

For the sake of simplicity, we do not distinguish between large and small components in the following expressions in order to demonstrate the principles in a simplest way with less crowded equations (then, the derivation is basically identical to the derivation of the nonrelativistic gradient of Hartree–Fock theory). Since we have two different sets of molecular spinor coefficients connected to large- and small-component basis functions, the differentiation of the Dirac–Hartree–Fock energy is more tedious than the differentiation of the nonrelativistic Hartree–Fock energy. Therefore, we proceed with the derivation of the Hartree–Fock energy for the sake of clarity in order to highlight the principles. The Hartree–Fock energy can be written as

$$E_{el,0}^{\rm HF}(R_{A,\alpha}) = V_0(R_{A,\alpha}) + \sum_{\mu\nu} D_{\mu\nu}(R_{A,\alpha}) h_{\mu\nu}(R_{A,\alpha})$$
$$+\frac{1}{2}\sum_{\mu\nu\kappa\sigma} D_{\mu\nu}(R_{A,\alpha}) D_{\kappa\sigma}(R_{A,\alpha}) (\mu(R_{A,\alpha})\nu(R_{A,\alpha})||\kappa(R_{A,\alpha})\sigma(R_{A,\alpha})) \qquad (10.71)$$

where we employ the atom-centered basis set, which thus also depends on the nuclear positions and hence on the variable $R_{A,\alpha}$. Differentiating the energy expression with respect to λ leads to

$$\begin{aligned} \frac{\partial E_{el,0}^{\rm HF}}{\partial R_{A,\alpha}} &= \sum_{\mu\nu} D_{\mu\nu} \frac{\partial h_{\mu\nu}}{\partial R_{A,\alpha}} + \frac{1}{2} \sum_{\mu\nu\lambda\sigma} D_{\mu\nu} D_{\lambda\sigma} \frac{\partial (\mu\nu||\lambda\sigma)}{\partial R_{A,\alpha}} + \frac{\partial V_0}{\partial R_{A,\alpha}} \\ &+ \sum_{\mu\nu} \frac{\partial D_{\mu\nu}}{\partial R_{A,\alpha}} h_{\mu\nu} + \sum_{\mu\nu\lambda\sigma} \frac{\partial D_{\mu\nu}}{\partial R_{A,\alpha}} D_{\lambda\sigma} (\mu\nu||\lambda\sigma) \end{aligned} \qquad (10.72)$$

The factor $1/2$ in front of the last term disappears because of the product rule and the fact that the pairs of summation indices cover all density matrix elements. The product rule then yields for the derivative of the density matrix

$$\frac{\partial D_{\mu\nu}}{\partial R_{A,\alpha}} = \sum_i^N \left[\frac{\partial c_{i\mu}}{\partial R_{A,\alpha}} c_{i\nu} + c_{i\mu} \frac{\partial c_{i\nu}}{\partial R_{A,\alpha}} \right] = 2 \sum_i^N \frac{\partial c_{i\mu}}{\partial R_{A,\alpha}} c_{i\nu} \qquad (10.73)$$

Now, the last two terms of Eq. (10.72) can be expressed in terms of the molecular spinor coefficients,

$$\begin{aligned} & \sum_{\mu\nu} \frac{\partial D_{\mu\nu}}{\partial R_{A,\alpha}} h_{\mu\nu} + \sum_{\mu\nu\kappa\sigma} \frac{\partial D_{\mu\nu}}{\partial R_{A,\alpha}} D_{\kappa\sigma}(\mu\nu||\kappa\sigma) \\ &= 2 \sum_{\mu\nu} \sum_i^N \frac{\partial c_{i\mu}}{\partial R_{A,\alpha}} h_{\mu\nu} c_{i\nu} + 2 \sum_{\mu\nu\kappa\sigma} \sum_i^N \frac{\partial c_{i\mu}}{\partial R_{A,\alpha}} D_{\kappa\sigma}(\mu\nu||\kappa\sigma) c_{i\nu} \\ &= 2 \sum_{\mu\nu} \sum_i^N \frac{\partial c_{i\mu}}{\partial R_{A,\alpha}} \underbrace{\left[h_{\mu\nu} + \sum_{\kappa\sigma} D_{\kappa\sigma}(\mu\nu||\kappa\sigma) \right]}_{\text{this is an element of the Fock matrix } f_{\mu\nu}} c_{i\nu} \\ &= 2 \sum_{\mu\nu} \sum_i^N \frac{\partial c_{i\mu}}{\partial R_{A,\alpha}} f_{\mu\nu} c_{i\nu} \overset{(10.55)}{=} 2 \sum_i^N \epsilon_i \sum_{\mu\nu} \frac{\partial c_{i\mu}}{\partial R_{A,\alpha}} S_{\mu\nu} c_{i\nu} \end{aligned} \qquad (10.74)$$

Recalling the orthonormality condition of the molecular spinors,

$$\langle \psi_i | \psi_j \rangle = \sum_{\mu\nu} c_{i\mu} S_{\mu\nu} c_{j\nu} = \delta_{ij} \qquad (10.75)$$

and differentiating this expression with respect to $R_{A,\alpha}$ through application of the product rule yields

$$2\sum_{\mu\nu}\frac{\partial c_{i\mu}}{\partial R_{A,\alpha}}S_{\mu\nu}c_{i\nu} = -\sum_{\mu\nu}c_{i\mu}c_{i\nu}\frac{\partial S_{\mu\nu}}{\partial R_{A,\alpha}} \tag{10.76}$$

Eq. (10.76) allows us to substitute the differentiation of molecular spinor coefficients (which we do not want to carry out explicitly as we do not have in general an analytic expression $c_{i\mu} = c_{i\mu}(\lambda)$ available) by a differentiation of overlap integrals of basis functions. As discussed in section 10.3, those derivative integrals can be calculated by standard methods.

Replacing the last two terms of Eq. (10.72) by the last term of Eq. (10.74) and considering Eq. (10.76) finally leads to the following expression for the energy derivative

$$\begin{aligned}\frac{\partial E^{\text{HF}}_{el,0}}{\partial R_{A,\alpha}} &= \sum_{\mu\nu}D_{\mu\nu}\frac{\partial h_{\mu\nu}}{\partial R_{A,\alpha}} + \frac{1}{2}\sum_{\mu\nu\kappa\sigma}D_{\mu\nu}D_{\kappa\sigma}\frac{\partial(\mu\nu||\kappa\sigma)}{\partial R_{A,\alpha}} + \frac{\partial V_0}{\partial R_{A,\alpha}} \\ &- \sum_{\mu\nu}\sum_{i}^{N}\epsilon_i c_{i\mu}c_{i\nu}\frac{\partial S_{\mu\nu}}{\partial R_{A,\alpha}}\end{aligned} \tag{10.77}$$

Because of Eq. (10.77), the gradient expression contains only three types of derivative integrals: differentiated one-electron integrals, differentiated two-electron integrals, and differentiated overlap integrals. It is thus not necessary to know how the variational parameters $\{c_{i\mu}\}$ change with the nuclear coordinates. This simplifies actual calculations significantly. Finally, we should stress again that all integral derivatives can be calculated with the same elegant techniques discussed in section 10.3.

10.6
Post-Hartree–Fock Methods

So far, we have only discussed the four-component basis-set approach in connection with the simplest *ab initio* wave-function model, namely for a single Slater determinant provided by Dirac–Hartree–Fock theory. We know, however, from chapter 8 how to improve on this model and shall now discuss some papers with a specific focus on correlated four-component basis-set methods.

An efficient approach to improve on the Hartree–Fock Slater determinant is to employ Møller–Plesset perturbation theory, which works satisfactorily well for all molecules in which the Dirac–Hartree–Fock model provids a good approximation (i.e., in typical closed-shell single-determinantal cases). The four-component Møller–Plesset perturbation theory has been implemented by various groups [431, 492, 495]. A major bottleneck for these calculations is the fact that the molecular spinor optimization in the SCF procedure is carried out in the atomic-orbital basis set, while the perturbation expressions are given in terms of molecular spinors. Hence, all two-electron integrals required for the second-order Møller–Plesset energy expression must be calculated from the integrals over atomic-orbital basis functions like

$$\langle \psi_i(1)\psi_j(2) \mid g(1,2) \mid \psi_i(1)\psi_j(2) \rangle = \sum_{\mu\lambda\nu\kappa} c^{\star}_{i\mu} c_{i\lambda} c^{\star}_{j\nu} c_{j\kappa} \times \langle \phi_\mu(1)\phi_\nu(2) \mid g(1,2) \mid \phi_\lambda(1)\phi_\kappa(2) \rangle \tag{10.78}$$

This transition to the molecular spinor basis is called four-index transformation for obvious reasons and has been discussed for the four-component case by Esser *et al.* [496] (see also Ref. [437]).

More appropriate than perturbation approaches for improving on the energy are variational approaches (under the specific caveats discussed in chapter 8 with respect to the negative-energy states), because the total electronic energies obtained are much better controlled, an essential property since the exact reference is not known for any interesting many-electron molecule. In particular, we shall address the second-generation MCSCF methods mentioned in chapter 8. For reference to molecular CI and CC theory, please consult sections 8.5.2 and 8.9, respectively.

By contrast to the numerical MCSCF method discussed in the last chapter, the basis-set approach has the convenient advantage that the virtual orbitals come for free by solution of the Roothaan equation. While the fully numerical approaches of chapter 8 do not produce virtual orbitals, as the SCF equations are solved directly for occupied orbitals only and smart bypasses must be devised, this problem does not show up in basis-set approaches. Out of the m basis functions, only N with $N \ll m$ are occupied, while the diagonalization of the matrix Fock operator produces a full set of m orthogonal molecular spinor vectors that can be efficiently employed in the excitation process of any CI-like method.

Moreover, the second-generation MCSCF parametrizes the wave function in a way that enables the *simultaneous* optimization of spinors and CI coefficients, in this context then called orbital or spinor rotation parameters and state transfer parameters, respectively. Then, a Newton–Raphson optimization method is employed which also requires the second derivatives of the MCSCF electronic energy with respect to the molecular spinor coefficients (more precisely, to the orbital rotation parameters) and to the CI coefficients. As we have seen, in Hartree–Fock theory the second derivatives are usually not calculated to confirm that a solution of the SCF procedure has indeed reached a minimum with respect to the large component and not a saddle point. Now, these general MCSCF methods could, in principle, provide such information, though it is often not needed in practice.

The parametrization of the normalized four-component MCSCF state function can be written as [244,432,497]

$$\Psi_A = e^{-\boldsymbol{\kappa}} \sum_I \Phi_I C_{IA} \tag{10.79}$$

where $\exp(-\boldsymbol{\kappa})$ is a unitary matrix, since $\boldsymbol{\kappa}$ is antihermitean, $\boldsymbol{\kappa}^\dagger = -\boldsymbol{\kappa}$. That is,

$$U = e^{-\boldsymbol{\kappa}} \quad \Rightarrow \quad U^\dagger = \left(e^{-\boldsymbol{\kappa}}\right)^\dagger = e^{\boldsymbol{\kappa}} \quad \text{so that} \quad UU^\dagger = \mathbf{1} \tag{10.80}$$

The operator U carries the so-called *orbital rotation* parameters κ_{pq}, which are the means to change, i.e., to optimize the MCSCF orbitals. For the sake of simplicity, we deviate from the definition of the CI wave function used in the research papers quoted above and stick to our equivalent definition of Eq. (8.100). With this definition of the total electronic state, we may write the optimization problem in the form of a Taylor series with the first derivatives (i.e., the gradient)

$$E_I^{(1)} \equiv \frac{\partial E_{el}^{\text{MCSCF}}}{\partial C_I^\star}, \quad E_{pq}^{(1)} \equiv \frac{\partial E_{el}^{\text{MCSCF}}}{\partial \kappa_{pq}^\star} \tag{10.81}$$

and the second derivatives (i.e., the Hessian)

$$E_{IJ}^{(2)} \equiv \frac{\partial E_{el}^{\text{MCSCF}}}{\partial C_I^\star \partial C_J^\star}, \quad E_{I,pq}^{(2)} \equiv \frac{\partial E_{el}^{\text{MCSCF}}}{\partial C_I^\star \partial \kappa_{pq}^\star}, \quad E_{pq,rs}^{(2)} \equiv \frac{\partial E_{el}^{\text{MCSCF}}}{\partial \kappa_{pq}^\star \partial \kappa_{rs}^\star} \tag{10.82}$$

to be taken at some reference set of parameters, which then directly leads to a Newton–Raphson optimization procedure with the Newton step,

$$\begin{pmatrix} C_A \\ \boldsymbol{\kappa} \end{pmatrix} = -\begin{pmatrix} \left\{E_{IJ}^{(2)}\right\} & \left\{E_{I,pq}^{(2)}\right\} \\ \left\{E_{pq,I}^{(2)}\right\} & \left\{E_{pq,rs}^{(2)}\right\} \end{pmatrix}^{-1} \cdot \begin{pmatrix} \left\{E_I^{(1)}\right\} \\ \left\{E_{pq}^{(1)}\right\} \end{pmatrix} \tag{10.83}$$

The optimization with respect to the spinors can be accomplished by obeying the minimax principle, and positronic energy states are allowed to relax in this correlation method (like in Dirac–Hartree–Fock–Roothaan calculations) so that again projection operators which keep the electronic states orthogonal to the positronic states are implicitly taken into account — both concepts have already been discussed in chapter 8.

Although the calculation of the Hessian is quite time consuming, the effort is quickly compensated by the excellent convergence properties of the Newton–Raphson approach [224–226]. This optimization technique solved the convergence problems of first-order MCSCF methods, which optimized orbitals and CI coefficients in an alternating manner (recall chapter 9). Even perturbative improvements of the four-component CASSCF wave function are feasible and have been implemented and investigated [439].

Further Reading

A. Szabo, N. S. Ostlund, [498]. *Modern Quantum Chemistry — Introduction to Advanced Electronic Structure Theory.*

The book by Szabo and Ostlund rests on the one-electron basis-set expansion of the orbitals but does not utilize the abstract language developed by Helgaker, Jørgensen and Olsen in their book mentioned below. It can be recommended as a first step to readers who want to learn more about basis-set methods in nonrelativistic quantum chemistry.

T. Helgaker, P. Jørgensen, J. Olsen, [217]. *Molecular Electronic-Structure Theory.*

This book is a highly sophisticated and advanced monograph on electronic structure theory based on basis set expansions. Throughout, the second-quantized formulation has been adopted, and no other monograph discusses advanced wavefunction-based quantum chemical methods from elaborate second-generation MCSCF to coupled cluster like this ultimate reference. Of course, Hartree–Fock and Møller–Plesset theory as well as all technical aspects (choice of one-particle basis sets, integrals) are also treated in extenso. Considering the fact that relativistic quantum chemical calculations on molecules basically require only to change the one-electron part of the Fock operator if the Dirac–Coulomb many-electron Hamiltonian is chosen, all details that had to be left aside in this chapter — as our focus was solely on the peculiarities of the relativistic formulation — can be obtained from the book by Helgaker, Jørgensen and Olsen.

B. Roos and co-authors, [499, 500]. *Lecture Notes in Quantum Chemistry — European Summer School in Quantum Chemistry.*

A very instructive collection of introductory lectures to many different topics in quantum chemistry (including all those mentioned in this chapter) has been provided by the Lund group. A first collection of these lecture notes has been published by Springer-Verlag [499, 500], but the latest version of these continuously developed notes can only be obtained from the theoretical chemistry group at Lund University (http://www.teokem.lu.se).

D. B. Cook, [501]. *Handbook of Computational Quantum Chemistry.*

There is a strong technical aspect to molecular quantum chemistry because of the complexity of SCF and CI equations and of energy gradient expressions. Cook's book is a good reference to many technical issues to be tackled when the equations are to be implemented in computer programs.

P. von Ragué Schleyer (Editor in Chief), [502]. *Encyclopedia of Computational Chemistry.*

The Encyclopedia is a very valuable reference regarding molecular quantum chemistry. It contains a collection of excellent review articles on almost any topic.

Part IV

TWO-COMPONENT HAMILTONIANS

11
Decoupling the Negative-Energy States

In the preceding chapters we set out from fundamental physical theory to arrive at a suitable theory for calculations on atoms and molecules, which still features four-component one-particle states. However, we noted that not only for small nuclear charges the contribution of the lower components of these spinors are small indeed. Hence, attempts were made to find Hamiltonians which do not require lower components in the corresponding one-particle functions and which thus are more convenient from a conceptual and — if possible — from a computational point of view. The principal options for such an elimination of small components are now introduced.

11.1
Relation of Large and Small Components in One-Electron Equations

Though accurate, first-quantized four-component methods are both computationally demanding and plagued by interpretive problems due to the negative-energy states. From a conceptual point of view it is desirable to decouple the upper and lower components of the Dirac Hamiltonian and to obtain a two-component description for electrons only.

For this purpose, we focus on one-electron operators only, because even the first-quantized many-electron theory reduces to Dirac-like one-electron equations, i.e., to the self-consistent field equations. The Dirac Hamiltonian is then substituted by the four-dimensional Fock operator

$$f = h^D + V \tag{11.1}$$

with h^D being the *field-free* Dirac Hamiltonian derived in chapter 5, and all potential energy terms are collected in $V = V_{nuc} + V_{ee}$, where V_{nuc} contains the point-like or finite-nucleus attraction of an electron with all nuclei in the system and V_{ee} comprises all electron–electron interaction terms whose explicit form depends on the choice of the approximation to the many-electron wave function (compare chapter 8). While h^D also included V_{nuc} in chapter 6, it is now taken as two separate operators (i.e., h^D is the field-free Dirac Hamiltonian) for later convenience.

Relativistic Quantum Chemistry. Markus Reiher and Alexander Wolf

ISBN: 978-3-527-31292-4

11.1.1 Restriction on the Potential Energy Operator

We must emphasize that the derivations in this chapter assume that V is diagonal. Moreover, the upper left and lower right block are identical $V^{LL} = V^{SS}$ and simply abbreviated as V in obviously two-dimensional equations. These restrictions are fulfilled for the Dirac operator of hydrogen-like atoms and, hence, for all one-electron Dirac-like operators where additional potentials can be regarded as a perturbation. In fact, in molecular calculations this is an efficient assumption and is, for instance, made in the one-electron Douglas–Kroll–Hess method discussed later in section 12.5. While the Kohn–Sham Coulomb and exchange–correlation potentials are also diagonal and therefore fulfill the restrictions, this does not hold for SCF-Roothaan-type equations, which feature off-diagonal contributions from the exchange integrals (and, consequently, it does also not hold for *hybrid* density functionals, which contain admixtures of Hartree–Fock exchange). One may, however, include the *block-diagonal* part of the Coulomb and exchange potentials in V. It is obvious that the Breit interaction, which features off-diagonal contributions due to the $\boldsymbol{\alpha}_i$ operators, can also not be handled by the approach sketched in the following.

Nevertheless, we adopt the restriction of a block-diagonal V for the sake of clarity. The more general case considering also off-diagonal potential contributions is then discussed in chapter 15. We now study decoupling of large and small components for one-electron equations only, which can then be generalized to the many-electron case.

11.1.2 The X-Operator Formalism

A key element for the reduction to two-component form is the analysis of the relationship between the large and small components of exact eigenfunctions of the Dirac equation, which we have already encountered in section 5.4.3. This relationship emerges because of the (2×2)-superstructure of the Dirac Hamiltonian, see, e.g., Eq. (5.129), which turned out to be conserved upon derivation of the one-electron Fock-type equations as presented in chapter 8. Hence, because of the (2×2)-superstructure of Fock-type one-electron operators, we may assume that a general relation,

$$\psi^S = X\psi^L \tag{11.2}$$

between the upper and lower components of a 4-spinor in that Fock-type equation holds. Here, X is an undetermined (2×2)-matrix operator, whose properties are the subject of investigation in this section. Of course, from what has been said in the previous chapters we recognize a relation to the kinetic balance condition. But now the focus is on the exact relation between large and small components under explicit consideration of all potential energy operators in the corresponding one-particle equation

$$f\psi = \left[h^D + V\right]\psi = \epsilon\psi \tag{11.3}$$

A state or orbital index has been dropped for the sake of clarity. With $V = 0$ this equation reduces to the free-particle Dirac equation of chapter 5. For $V = -Z/r$ we have the equation for Dirac hydrogen-like atoms for point-like nuclei (see chapter 6), while $V = -Ze^2/r + V_{ee}$ holds for the many-electron atoms of chapter 9 and $V = -\sum_A Z_A e^2/r_A + V_{ee}$ for molecules as in chapter 10, whereby V_{ee} may be a substitute for the mean-field Hartree–Fock potential, a potential that results from a more complex approximation of the wave function or may stem from an energy density functional in the Kohn–Sham equations of section 8.8.

If ψ is an solution of Eq. (11.3) — and V does not contribute off-diagonal terms that would have to be added to $\boldsymbol{\sigma}\cdot\boldsymbol{p}$ — an expression for X can easily be given in closed form

$$X = X(\epsilon) = \left(\epsilon - V + 2m_e c^2\right)^{-1} c\,\boldsymbol{\sigma}\cdot\boldsymbol{p} \tag{11.4}$$

[compare also Eqs. (5.130) and (10.4)]. From this equation we understand that the X-operator depends on the energy eigenvalue ϵ. Thus, each solution of the one-particle equation possesses its own energy-dependent X-operator. Even the countable set of all bound states of the one-particle operator features infinitely many different X-operators. This energy-dependent definition of the X-operator does not serve our purpose to decouple large and small components because the energy eigenvalues need to be determined prior to the construction of X, and hence we have to deal with the negative-energy states again. Nevertheless, the energy-dependent X-operator is a starting point for the so-called *elimination techniques*, in which Eq. (11.2) is applied to substitute the small component in the upper equation of the coupled set in Eq. (11.3). Then, ψ^S is implicitly included in an equation for the large component only. These techniques are considered in greater detail in chapter 13. Here, however, we seek to find an energy-independent X-operator valid for the whole spectrum of the one-particle Hamiltonian.

For an energy-independent X-operator it is necessary to employ the Dirac equation in a different form. Multiplication of the upper of the two Dirac equations in split notation, Eq. (5.79), by X from the left produces a right hand side that reads in stationary form $XE\psi^L$. From Eq. (11.2) we understand that this is identical to $XE\psi^L = E\psi^S$, and hence the two left hand sides of the split Dirac equation become equal,

$$XV\psi^L + Xc\,\boldsymbol{\sigma}\cdot\boldsymbol{p}X\psi^L = c\,\boldsymbol{\sigma}\cdot\boldsymbol{p}\,\psi^L + VX\psi^L - 2m_e c^2 X\psi^L \tag{11.5}$$

X must satisfy this equation for all possible choices of the large components ψ^L, and is thus determined by the nonlinear operator identity [503]

$$X = \frac{1}{2m_ec^2}\left\{c\,\boldsymbol{\sigma}\cdot\boldsymbol{p} - [X,V] - Xc\,\boldsymbol{\sigma}\cdot\boldsymbol{p}X\right\} \tag{11.6}$$

which directly follows from Eq. (11.5).

The solution of this equation is, however, as complex as the solution of the Dirac equation itself, and thus only for a restricted class of potentials, excluding the Coulomb potential, are closed-form solutions for X known [504, 505]. For potentials that are relevant in molecular science it can be solved numerically after a basis set representation for the large component has been introduced (see also chapters 13 and 14).

Eq. (11.6) is a quadratic equation in X and therefore possesses two independent solutions X_+ and X_-, corresponding to positive-energy $(E>-m_ec^2)$ and negative-energy $(E<-m_ec^2)$ spinors, respectively. They originate solely due to the fact that the Dirac one-electron Hamiltonian in Eq. (11.1) describes both positive-energy and negative-energy states on an equal footing (the shift of the energy scale by $-m_ec^2$ does not affect this fact). Since in the case of very strong potentials even electronic solutions may have energies below $-2m_ec^2$ within the negative-energy continuum (see section 6.9), it was suggested to refer to all solutions corresponding to X_+ more rigorously as *class-I* solutions by Heully *et al.* [503]. Accordingly, all solutions connected to X_- are called *class-II* solutions.

Various approximate two-component theories satisfy the kinetic balance relation only to a certain degree. They establish only variationally stable but not variational approaches [455], i.e., the energy expectation value is then bounded from below by a different bound which may be above or below the true ground state energy E_0. The simplest approximation to exact kinetic balance may be obtained in the nonrelativistic limit of Eqs. (11.4) or (11.6),

$$X_+ \xrightarrow{c\to\infty} \frac{\boldsymbol{\sigma}\cdot\boldsymbol{p}}{2m_ec} \quad \text{and} \quad X_- \xrightarrow{c\to\infty} -2m_ec\,(\boldsymbol{\sigma}\cdot\boldsymbol{p})^{-1} \tag{11.7}$$

which is immediately recognized as the familiar (approximate) kinetic balance relation of Eq. (10.6) employed in the four-component basis set approaches discussed in section 10.1. The introduction of the operator X leads to a modified normalization description for the Dirac spinor ψ,

$$\langle\psi|\psi\rangle = \langle\psi^L|\psi^L\rangle + \langle\psi^S|\psi^S\rangle = \langle\psi^L|1 + X^\dagger X|\psi^L\rangle \stackrel{!}{=} 1 \tag{11.8}$$

which can now be expressed in terms of the large component only.

We conclude this section by mentioning that the relationship between the large and small components presented above could equally well be formulated in terms of an operator Y defined by

$$\psi^L = Y\psi^S \tag{11.9}$$

The X- and Y-operators are, of course, closely related to one another, and apart from the trivial relation $X_\pm = Y_\pm^{-1}$ they always satisfy the condition $X_\pm = -Y_\mp^\dagger$. A comprehensive discussion of this connection can be found in Ref. [503]. In molecular science, we are, however, solely interested in class-I (electronic) solutions and thus introduce the definitions $X \equiv X_+$ and $Y \equiv Y_- = -X^\dagger$ in order to keep the notation simple.

11.1.3 Free-Particle Solutions

In order to demonstrate the ambiguity of the X-operator, explicit expressions for both X_+ and X_- are given for the case of a free particle, defined by $V = 0$,

$$X_\pm^{V=0} = \left(\epsilon_\pm + m_e c^2\right)^{-1} c\, \boldsymbol{\sigma}\cdot\boldsymbol{p} \tag{11.10}$$

where

$$\epsilon_\pm = \epsilon_\pm(\boldsymbol{p}) \equiv \pm E_p, \qquad E_p = \sqrt{\boldsymbol{p}^2 c^2 + m_e^2 c^4} > 0 \tag{11.11}$$

is the familiar square root operator of Eq. (5.4) reflecting the relativistic energy–momentum relation. Eq. (11.10) establishes an energy-independent connection between the upper and lower components of free Dirac spinors, since the quantity $\epsilon_\pm$ is to be interpreted as an operator rather than an energy eigenvalue. Also, $X_+^{V=0}$ defines the so-called free-particle Foldy–Wouthuysen transformation to be discussed in detail in section 11.3.

The square-root operator is difficult to evaluate in position space because of the square root to be taken of a differential operator that would represent $\boldsymbol{p}$. We have already discussed this issue in the context of the Klein–Gordon equation in section 5.1.1. Hence, the action of the X-operator is most conveniently studied in momentum space, where the inverse operator may be applied in closed form without expanding the square root.

The four normalized free-particle Dirac eigenspinors of section 5.3 with shifted eigenvalues $\epsilon_\pm - m_e c^2$ may now be written as

$$\psi_{\pm,s}(\boldsymbol{p},t) = \underbrace{\sqrt{\frac{\epsilon_\pm + m_e c^2}{2\epsilon_\pm}} \begin{pmatrix} \rho_s \\ X_\pm^{V=0}\rho_s \end{pmatrix}}_{u_{\pm,s}(\boldsymbol{p})} \exp\left[\frac{\mathrm{i}}{\hbar}\left(\boldsymbol{p}\cdot\boldsymbol{r} - (\epsilon_\pm - m_e c^2)t\right)\right] \tag{11.12}$$

with

$$\rho_1 = \begin{pmatrix} 1 \\ 0 \end{pmatrix} \quad \text{and} \quad \rho_2 = \begin{pmatrix} 0 \\ 1 \end{pmatrix} \tag{11.13}$$

for $s = 1, 2$. These 4-spinors are normalized to unity as usual in quantum chemistry [compare Eq. (8.108)] instead of to $E_p/m_e c^2$, as is often found in the physics literature. After some elementary algebraic manipulations the standard form of the spinors $u_\pm$ of Eqs. (5.99) and (5.101)

$$u_{+,s} = \begin{pmatrix} \sqrt{\dfrac{E_p + m_e c^2}{2E_p}}\, \rho_s \\ \\ \dfrac{c\,\boldsymbol{\sigma}\cdot\boldsymbol{p}}{\sqrt{2E_p(E_p + m_e c^2)}}\, \rho_s \end{pmatrix}, \quad u_{-,s} = \begin{pmatrix} \dfrac{-c\,\boldsymbol{\sigma}\cdot\boldsymbol{p}}{\sqrt{2E_p(E_p + m_e c^2)}}\, \rho_s \\ \\ \sqrt{\dfrac{E_p + m_e c^2}{2E_p}}\, \rho_s \end{pmatrix} \tag{11.14}$$

is recovered if the normalization constant is explicitly taken into account. The above solutions are occasionally found with inverted signs because the eigenvalue equation may be multiplied by -1.

11.2
Closed-Form Unitary Transformation of the Dirac Hamiltonian

Another possibility for the reduction of the 4-spinor to two-component Pauli form is to decouple the Dirac equation by a unitary transformation U to block-diagonal form,

$$f_{bd} = U f U^\dagger = \begin{pmatrix} f_+ & 0 \\ 0 & f_- \end{pmatrix} \tag{11.15}$$

with $UU^\dagger = 1$. Similarly to the Hamiltonian, the spinor is also reduced by this unitary transformation and has only one nonvanishing 2-spinor component,

$$\tilde{\psi} = U\psi = \begin{pmatrix} \tilde{\psi}^L \\ \tilde{\psi}^S \end{pmatrix} \tag{11.16}$$

with $\tilde{\psi}^S = \mathbf{0}$ for class-I solutions and $\tilde{\psi}^L = \mathbf{0}$ for class-II solutions. The unitary transformation guarantees that the two 'effective' operators f_+ and f_- exactly reproduce the entire energy spectrum of the Dirac operator h^D, but without any coupling due to odd terms of the Hamiltonian. Only the spinors are changed — but the unitary transformation also preserves their lengths, i.e., norm. Hence, all electronic eigenvalues and eigenstates of the Dirac Hamiltonian may be obtained variationally by restriction to the simplified eigenvalue problem for f_+, which is no longer plagued by an infinite spectrum of negative-energy eigenvalues that could hamper variational stability.

Of course, what has just been stated for the one-electron Dirac Hamiltonian is also valid for the general one-electron operator in Eq. (11.1). However, the coupling of upper and lower components of the spinor is solely brought about by the off-diagonal $c\boldsymbol{\sigma} \cdot \boldsymbol{p}$ operators of the free-particle Dirac one-electron Hamiltonian and kinetic energy operator, respectively. We shall later see that the occurrence of any sort of potential V will pose some difficulties when it comes to the determination of an explicit form of the unitary transformation U. A universal solution to this problem will be provided in chapter 12 in form of Douglas–Kroll–Hess theory.

Before deriving the explicit form of the matrix U in terms of the operator X it should be mentioned that the spectrum of the Dirac operator h^D is invariant under arbitrary similarity transformations, i.e., non-singular (invertible) transformations U, whether they are unitary or not. But only *unitary* transformations conserve the normalization of the Dirac spinor and leave scalar products and matrix elements invariant. Therefore, a restriction to unitary transformations is inevitable if one is interested in the wave function or in quantities derived from it. Furthermore, the problem of actually carrying out the transformation experiences a great technical simplification by the choice of a unitary transformation, since the inverse transformation U^{-1} would in general hardly be obtained if U were not unitary. It should be recalled that the eigenstates of the transformed one-electron operator are different (in form and thus in spatial shape) for different unitary transformations, while the spectra of the transformed operators are all identical and the norm of the eigenstates is also preserved.

The most general form of the unitary transformation of the operator f is determined by its one-electron fragment h^D, which imprints a block-diagonal super-structure on the unitary matrix U employed,

$$U = \begin{pmatrix} U^{LL} & U^{LS} \\ U^{SL} & U^{SS} \end{pmatrix} \tag{11.17}$$

The requirement $\tilde{\psi}^S = 0$ for positive-energy (class I) solutions leads to the condition

$$[U^{LS} + U^{SS}X]\,\psi^L = 0 \tag{11.18}$$

which has to be satisfied for all 2-spinors ψ^L under consideration. This can only be achieved by the operator identity

$$U^{SL} = -U^{SS}X \tag{11.19}$$

Similarly, the requirement $\tilde{\psi}^L = 0$ for negative-energy (class II) solutions yields together with $X^\dagger = -Y$ the relation

$$U^{LS} = U^{LL}X^\dagger \tag{11.20}$$

Finally, the form of U_{11} and U_{22} is determined by the unitarity condition $UU^\dagger = 1$, which is only satisfied if

$$U^{LL} = e^{i\xi}\left(1 + X^\dagger X\right)^{-1/2} \quad \text{with} \quad \xi \in [0, 2\pi[\tag{11.21}$$

and

$$U^{SS} = e^{i\varphi}\left(1 + XX^\dagger\right)^{-1/2} \quad \text{with} \quad \varphi \in [0, 2\pi[\tag{11.22}$$

where ξ and φ are arbitrary phase factors [506]. Since a global phase is obviously insignificant for the construction of a unitary matrix, we may set $\xi = 0$ without loss of generality and restrict our considerations to the relative phase between U_{11} and U_{22}.

These most general results allow us to rewrite the general form of the unitary matrix U in Eq. (11.17) more explicitly as,

$$U = U(X) = \begin{pmatrix} \left(1 + X^\dagger X\right)^{-1/2} & \left(1 + X^\dagger X\right)^{-1/2} X^\dagger \\ -e^{i\varphi}\left(1 + XX^\dagger\right)^{-1/2} X & e^{i\varphi}\left(1 + XX^\dagger\right)^{-1/2} \end{pmatrix} \tag{11.23}$$

In the literature, two choices for the relative phase φ mainly occur. In their original work [503], where the form of $U = U(X)$ was derived for the first time, Heully and co-workers chose $\varphi = \pi$, which results in an hermitean form of U, while many subsequent publications, e.g., Refs. [455, 507, 508], preferred to choose $\varphi = 0$. In any case, neither the operators nor their spectra are affected by the choice of φ.

With this explicit form of the unitary matrix U, we can calculate the block-diagonal Hamiltonian f_{bd} given by Eq. (11.15). Its components are

$$f_+ = \frac{1}{\sqrt{1 + X^\dagger X}} \left\{ V + c\,\boldsymbol{\sigma}\cdot\boldsymbol{p} X + X^\dagger c\,\boldsymbol{\sigma}\cdot\boldsymbol{p} + X^\dagger (V - 2mc^2) X \right\} \frac{1}{\sqrt{1 + X^\dagger X}} \tag{11.24}$$

and

$$f_- = \frac{1}{\sqrt{1 + XX^\dagger}} \left\{ V - 2mc^2 - c\,\boldsymbol{\sigma}\cdot\boldsymbol{p} X^\dagger - X^\dagger c\,\boldsymbol{\sigma}\cdot\boldsymbol{p} + XVX^\dagger \right\} \frac{1}{\sqrt{1 + XX^\dagger}} \tag{11.25}$$

where the off-diagonal elements of f_{bd} vanish.

If the energy-dependent expression in Eq. (11.4) for the operator X was inserted into Eq. (11.24), the one-electron operator f_+ would be a very complicated function of its own eigenvalues, and the eigenvalue equation

$$f_+(\epsilon)\, \tilde{\psi}^L = \epsilon\, \tilde{\psi}^L \tag{11.26}$$

would have only normalizable solutions if ϵ were a true electronic, i.e., positive-energy eigenvalue of this operator. We have then gained nothing in the case of an energy-dependent X-operator which produces no eigenvalue equation. Only the energy-independent X-operator yields the electronic part of the spectrum in terms of an eigenvalue equation without any reference to the negative-energy states.

Unfortunately, this X-operator and hence the unitary transformation U are not available in closed form, except for a few very special cases, e.g., the free particle or an electron being exposed to a homogeneous magnetic field [505]. It is, however, possible to solve equations of the type shown in Eq. (11.6) purely numerically in a given one-particle basis (see section 11.6 below).

In addition, approximate decoupling schemes can be envisaged in order to arrive at the block-diagonal Hamiltonian of Eq. (11.15), which will be particularly valuable in cases with complicated expressions for the potential V. This can either be achieved by a systematic analytic decomposition of the transformation U into a sequence of unitary transformations, each of which is expanded in an *a priori* carefully chosen parameter. These issues will be addressed in detail in chapter 12, and we shall now stick to the long-known unitary transformation scheme for free particles, namely the free-particle Foldy–Wouthuysen transformation.

11.3 The Free-Particle Foldy–Wouthuysen Transformation

The historically first attempt to achieve the block-diagonalization of the Dirac Hamiltonian h^D is due to Foldy and Wouthuysen and dates back to 1950 [509]. They derived the very important closed-form expressions for both the unitary transformation and the decoupled Hamiltonian for the case of a free particle without invoking something like the X-operator. Because of the discussion in the previous two sections, we can directly write down the final result since the free-particle X-operator of Eq. (11.10) and hence $U_{V=0} \equiv U_0$ are known. With the arbitrary phase of Eq. (11.23) being fixed to zero it is given by

$$U_0 = U(X_+^{V=0}) = A_p \begin{pmatrix} 1 & \boldsymbol{\sigma}\cdot\boldsymbol{P}_p \\ -\boldsymbol{\sigma}\cdot\boldsymbol{P}_p & 1 \end{pmatrix} \tag{11.27}$$

with

$$A_p \equiv \sqrt{\frac{E_p + m_e c^2}{2E_p}}, \quad \boldsymbol{P}_p = \frac{c\,\boldsymbol{p}}{E_p + m_e c^2}, \quad R_p \equiv \boldsymbol{\alpha}\cdot\boldsymbol{P}_p = \frac{c\,\boldsymbol{\alpha}\cdot\boldsymbol{p}}{E_p + m_e c^2} \tag{11.28}$$

where we have already defined a momentum-dependent scalar, R_p, for later convenience. The A_p factors must, of course, not be confused with a component of the vector potential (also not in chapter 12).

In the literature, many forms of the unitary matrix in Eq. (11.27) can be found and are not easy to be recognized as the same — often also because the notation is varied and authors chose not to stick to historically older conventions. To demonstrate how these different forms can be interconverted we may consider some of them explicitly in the following. An often employed version of U_0, identical to the one above, is the exponential form,

$$U_0 = \exp\left(\beta\frac{\boldsymbol{\alpha}\cdot\boldsymbol{p}}{p}\omega(p)\right) \quad \Rightarrow \quad U_0^\dagger = \exp\left(-\beta\frac{\boldsymbol{\alpha}\cdot\boldsymbol{p}}{p}\omega(p)\right) \tag{11.29}$$

with a yet unknown angle $\omega(p)$ for which we anticipated a momentum dependence. Note the imaginary unit in the momentum operator that allows us to directly write the adjoint. After Taylor expansion of the exponential and recalling that $(\boldsymbol{\alpha}\cdot\boldsymbol{p})^2 = \boldsymbol{p}^2$ and $[\beta(\boldsymbol{\alpha}\cdot\boldsymbol{p})]^2 = -\boldsymbol{p}^2$, we obtain an Euler-like formula

$$\exp\left(\pm\beta\frac{\boldsymbol{\alpha}\cdot\boldsymbol{p}}{p}\omega(p)\right) = \cos\left[\omega(p)\right] \pm \frac{\boldsymbol{\alpha}\cdot\boldsymbol{p}}{p}\sin\left[\omega(p)\right] \tag{11.30}$$

in which the exponential is now substituted by sine and cosine functions. The transformed free-particle Dirac Hamiltonian then reads

$$\begin{aligned} h_1 = U_0 h^D U_0^\dagger \stackrel{(11.30)}{=} & \; c\boldsymbol{\alpha}\cdot\boldsymbol{p}\left(\cos\left[2\omega(p)\right] - \frac{m_e c^2}{cp}\sin\left[2\omega(p)\right]\right) \\ & + \beta m_e c^2\left(\cos\left[2\omega(p)\right] + \frac{cp}{m_e c^2}\sin\left[2\omega(p)\right]\right) \end{aligned} \tag{11.31}$$

The angle follows from the condition that the $(\boldsymbol{\alpha}\cdot\boldsymbol{p})$-term on the right hand side, which is the only odd term, shall vanish. This yields

$$\tan\left[(2\omega(p)\right] = \frac{cp}{m_e c} = \frac{cp}{m_e c^2} \tag{11.32}$$

which in turn allows us to obtain explicit expressions for the sine and cosine functions,

$$\begin{aligned} \sin\left[2\omega(p)\right] &= \frac{\tan\left[2\omega(p)\right]}{\sqrt{1+\tan^2\left[2\omega(p)\right]}} = \frac{cp}{\sqrt{m_e^2 c^4 + c^2\boldsymbol{p}^2}} \\ \cos\left[2\omega(p)\right] &= \frac{m_e c^2}{\sqrt{m_e^2 c^4 + c^2\boldsymbol{p}^2}} \end{aligned} \tag{11.33}$$

In order to write the exponential form of Eq. (11.29) explicitly, we write the angle according to Eq. (11.32) as

$$\omega(p) = \frac{1}{2}\arctan\left(\frac{p}{m_e c}\right) \tag{11.34}$$

and, hence, the first given expression for U_0 turns out to be equivalent to the familiar exponential form

$$U_0 = \exp\left(\beta \frac{\boldsymbol{\alpha}\cdot\boldsymbol{p}}{2p} \arctan \frac{p}{m_e c}\right) \overset{(11.30)}{\underset{(11.33)}{=}} A_p \left(\mathbf{1}_4 + \beta R_p \right) \tag{11.35}$$

which was used by Foldy and Wouthuysen in their original work. Its application to the free-particle Dirac Hamiltonian $\hat{h}^D$ with $V = 0$ yields the desired block-diagonal form. However, the free-particle Foldy–Wouthuysen transformation has also proven to be extremely useful in the presence of a scalar potential. As a matter of fact, it is the mandatory starting point for *all* decoupling schemes employing unitary transformations to the Dirac Hamiltonian h^D [510], which may contain *any* potential V, as will be shown in chapter 12, where it is rederived applying the most general framework for the parametrization of U_0.

Since the free-particle Foldy–Wouthuysen transformation can still be performed in closed form even in the presence of a scalar potential V of any form,

$$f_1 = U_0 \left(h^D + V\right) U_0^\dagger = \mathcal{E}_0 + \mathcal{E}_1 + \mathcal{O}_1 \tag{11.36}$$

where the subscripts at each term on the right hand side of this equation denote the order in the scalar potential of the corresponding term. They are given by

$$\mathcal{E}_0 = \beta E_p - m_e c^2 \tag{11.37}$$

$$\mathcal{E}_1 = A_p V A_p + A_p R_p V R_p A_p \tag{11.38}$$

$$\mathcal{O}_1 = \beta A_p \left[R_p , V \right] A_p \tag{11.39}$$

The notation introduced here — which is also used extensively in chapter 12 — reflects the block-diagonal super-structure of the resulting operators. $\mathcal{E}$ denotes a block-diagonal operator with LL and SS blocks, which is called *even*, while $\mathcal{O}$ denotes an operator with entries on the two off-diagonal LS and SL blocks, which is called *odd*. For example, the scalar potential V in the untransformed one-electron operator is an even operator like the rest energy term in h^D. On the other hand, the $c\boldsymbol{\alpha}\cdot\boldsymbol{\sigma}$ term of h^D is an odd operator. Now it is important to understand that the free-particle transformation according to Foldy and Wouthuysen produces a closed-form transformed operator which still contains an odd contribution $\mathcal{O}_1$, while the original untransformed Hamiltonian $(h^D + V)$ features an odd term of zeroth order in the scalar potential that is to be denoted $\mathcal{O}_0^{(0)}$ (the superscript indicates that this is an odd term of the original, untransformed operator). Hence, the free-particle transformation cannot decouple potential-affected one-electron operators, which poses a problem for molecular quantum mechanics to be addressed in chapter 12. For the sake of completeness, we should add that the untransformed operator would read in the new notation

$$f = h^D + V = c\boldsymbol{\alpha} \cdot \boldsymbol{p} + m_e c^2 \beta + V \equiv \mathcal{O}_0^{(0)} + \mathcal{E}_0^{(0)} + \mathcal{E}_1^{(0)} \tag{11.40}$$

This new notation is quite interesting as it emphasizes that the explicit form of the individual operators is of little importance. Moreover, we will understand that higher-order terms can be constructed in a *recursive* manner so that only the explicit expressions in Eqs. (11.37)–(11.39) need to be known (see also chapter 12).

It is crucial to note that all operators containing any power of the momentum operator, e.g. A_p or R_p, do not commute with the scalar potential, since V operates as an integral operator in momentum space according to Eq. (6.164). Alternatively, one may understand that $V(\boldsymbol{r})$ does not commute with $\boldsymbol{p}$ in position space because $\boldsymbol{r}$ does not commute with $\boldsymbol{p}$.

A more explicit formulation of the free-particle Foldy–Wouthuysen transformed Hamiltonian in the presence of scalar potentials, which also highlights the preserved super-structure of this operator, reads

$$f_1 = \begin{pmatrix} f_1^{LL} & f_1^{LS} \\ f_1^{SL} & f_1^{SS} \end{pmatrix} \tag{11.41}$$

with the individual blocks given by

$$f_1^{LL} = E_p - m_e c^2 + A_p\, V A_p + A_p\, \boldsymbol{\sigma}\cdot\boldsymbol{P}_p V \boldsymbol{\sigma}\cdot\boldsymbol{P}_p\, A_p \tag{11.42}$$

$$f_1^{LS} = A_p \left[\boldsymbol{\sigma}\cdot\boldsymbol{P}_p, V\right] A_p = -f_1^{SL} \tag{11.43}$$

$$f_1^{SS} = -E_p - m_e c^2 + A_p\, V A_p + A_p\, \boldsymbol{\sigma}\cdot\boldsymbol{P}_p V \boldsymbol{\sigma}\cdot\boldsymbol{P}_p A_p \tag{11.44}$$

It must be emphasized that these expressions are still exact when compared to the original untransformed operator. The transformed operator f_1 would be completely decoupled if $V = 0$, i.e., if the particle were moving freely, and hence only the kinetic energy operator (apart from the rest energy term) remains. Thus, the even terms of the free-particle Foldy–Wouthuysen transformation already account for all so-called 'kinematic' relativistic effects.

The influence of the small component is to some extent shifted into the Hamiltonian f_1 in such a way that the magnitude of the transformed small component is significantly decreased. This is most easily seen by a restriction to the leading order in $1/c$. Then the relation between the large and small component of the untransformed Dirac spinor is given by

$$\psi^S \sim \frac{\boldsymbol{\sigma}\cdot\boldsymbol{p}}{2m_e c}\,\psi^L \sim \frac{1}{c}\,\psi^L \tag{11.45}$$

After the free-particle Foldy–Wouthuysen transformation the magnitude of the small component

$$\psi_1^S = -\frac{f_1^{SL}}{f_1^{SS} - \epsilon}\psi_1^L \tag{11.46}$$

as compared to the large component has decreased in leading order in $1/c$,

$$\psi_1^S \sim \frac{-p^2\,\boldsymbol{\sigma}\cdot\boldsymbol{p}}{16m^3c^3}\,\psi_1^L \sim \frac{1}{c^3}\,\psi_1^L \tag{11.47}$$

as the Taylor series expansion reveals, and the importance of the lower component has been diminished in the transformed spinor. This development is accompanied by the decrease of the leading order in $1/c$ of the off-diagonal blocks of f_1. The free-particle Foldy–Wouthuysen transformation reduces the leading order of these blocks from c to $1/c$.

11.4 General Parametrization of Unitary Transformations

We have already seen that the free-particle Foldy–Wouthuysen transformation can be expressed in a couple of ways which seem very different at first sight. The only boundary for the explicit choice of a unitary transformation is that the off-diagonal blocks and thus all odd operators vanish. If many different choices are possible — and we will see in the following that this is actually the case — the question arises how are they related and what this implies for the resulting Hamiltonians.

In general, a unitary transformation U can be parametrized by an odd and antihermitean operator W. The antihermiticity of this operator W,

$$W^\dagger = W^{\star,T} = -W \tag{11.48}$$

allows one to easily write the inverse unitary transformation $U^\dagger$ and the property of being 'odd' ensures that we can clearly identify any operator sequence as being in total odd or even later on. Since this is the only mandatory property of W, the antihermitean operator may be chosen freely to serve a given prupose like eliminating the odd contributions in the Dirac Hamiltonian. A well-known example in quantum chemistry is the exponential parametrization,

$$U = \exp(W) \tag{11.49}$$

for which the inverse can be given directly as,

$$U^\dagger = \exp(-W) \tag{11.50}$$

so that

$$U \cdot U^{\dagger} = \exp(W) \cdot \exp(-W) = 1 \tag{11.51}$$

It is now interesting to note that a parametrized unitary transformation — with W being the parameter — can also be expanded into a Taylor series if convergence can be guaranteed. In this sense we can cover all different choices of parametrized unitary transformations in one scheme [511]. The expansion coefficients then characterize all different parametrizations and guarantee that unitarity is fulfilled. This yields infinitely many unitary parametrizations, which are all equivalent with respect to their application in decoupling procedures for the Dirac Hamiltonian. Even if a composite unitary transformation constructed as a sequence of unitary transformations is applied, the product of all resulting Taylor series expansions is just a new Taylor series expansion valid for the overall unitary transformation. Of course, this can easily be seen if one uses the special exponential choice just given,

$$U = U_1 \cdot U_0 = \exp(W_1) \cdot \exp(W_0) = \exp(W_1 + W_0) \equiv \exp(W) \tag{11.52}$$

but it also holds for any sequence where any component is expanded into a series.

11.4.1
Closed-Form Parametrizations

The historically first attempt to parametrize U dates back to 1950 and is due to Foldy and Wouthuysen [509]. It employs an exponential ansatz $U = \exp(W)$. Douglas and Kroll advocated the use of the so-called *square-root* parametrization,

$$U = \sqrt{1 + W^2} + W \tag{11.53}$$

But one may think of additional closed-form parametrizations for U, namely the Cayley-type form [512],

$$U = \frac{2 + W}{2 - W} \tag{11.54}$$

and the McWeeny parametrization [513],

$$U = \frac{1 + W}{\sqrt{1 - W^2}} \tag{11.55}$$

Of course, infinitely many other unitary transformations may be constructed, like

$$U = \frac{a + W}{a - W} \quad \forall a \in \mathbb{R} \tag{11.56}$$

At this stage it is, however, not clear which of these parametrizations should be preferred over the others for application in decoupling procedures and whether they all yield identical block-diagonal Hamiltonians. Furthermore, the four possibilities given above to parametrize unitary transformations differ obviously in their radius of convergence R_c, which is equal to unity for the square root and the McWeeny form, whereas $R_c = 2$ for the Cayley parametrization, and $R_c = \infty$ for the exponential form.

Within any decoupling scheme there are only a few restrictions on the choice of the transformations U. Firstly, they have to be unitary and analytic (holomorphic) functions on a suitable domain of the one-electron Hilbert space $\mathcal{H}$, since any parametrization has necessarily to be expanded in a Taylor series around $W = 0$ for the sake of comparability but also for later application in nested decoupling procedures (see chapter 12). Secondly, they have to permit a decomposition of f_{bd} in even terms of well-defined order in a given expansion parameter of the Hamiltonian (like $1/c$ or V). It is thus possible to parametrize U without loss of generality by a power-series ansatz in terms of an antihermitean operator W, where unitarity of the resulting power series is the only constraint. In the next section this most general parametrization of U is discussed.

11.4.2
Exactly Unitary Series Expansions

The most general ansatz to construct a unitary transformation $U = f(W)$ as an analytic function of an antihermitean operator W is a power series expansion [511],

$$U = a_0 \mathbf{1} + a_1 W + a_2 W^2 + a_3 W^3 + \cdots = a_0 \mathbf{1} + \sum_{k=1}^{\infty} a_k W^k \tag{11.57}$$

which is assumed to converge within a certain domain. Without loss of generality the restriction is imposed that the coefficients a_k are real-valued. To write the power series expansion of the hermitean conjugate transformation $U^\dagger$, is straightforward

$$U^\dagger = a_0 \mathbf{1} - a_1 W + a_2 W^2 - a_3 W^3 + \cdots = a_0 \mathbf{1} + \sum_{k=1}^{\infty} (-1)^k a_k W^k \tag{11.58}$$

after exploiting the antihermiticity $\left(W^\dagger = -W\right)$. In order to guarantee that unitarity is fulfilled, the coefficients a_k have to satisfy a set of constraints, which are determined by

$$\begin{aligned} U\,U^\dagger &= a_0^2 \mathbf{1} + \left(2a_0 a_2 - a_1^2\right) W^2 + \left(2a_0 a_4 + a_2^2 - 2a_1 a_3\right) W^4 \\ &\quad + \left(2a_0 a_6 + 2a_2 a_4 - 2a_1 a_5 - a_3^2\right) W^6 \end{aligned}$$

$$+\left(2a_0a_8 + 2a_2a_6 + a_4^2 - 2a_1a_7 - 2a_3a_5\right)W^8$$
$$+\left(2a_0a_{10} + 2a_2a_8 + 2a_4a_6 - 2a_1a_9 - 2a_3a_7 - a_5^2\right)W^{10}$$
$$+\mathcal{O}\left(W^{12}\right) \stackrel{!}{=} \mathbf{1} \tag{11.59}$$

An important observation is that odd powers of W do not occur in this expression because of the antihermiticity of W. With the requirement that different powers of W be linearly independent, the *unitarity conditions* for the coefficients are obtained. Their explicit form for the first few coefficients reads:

$$a_0 = \pm 1 \tag{11.60}$$
$$a_2 = \tfrac{1}{2}\, a_0\, a_1^2 \tag{11.61}$$
$$a_4 = a_0\left(a_1a_3 - \tfrac{1}{8}a_1^4\right) \tag{11.62}$$
$$a_6 = a_0\left(a_1a_5 + \tfrac{1}{2}a_3^2 - \tfrac{1}{2}a_1^3a_3 + \tfrac{1}{16}a_1^6\right) \tag{11.63}$$
$$a_8 = a_0\left(a_1a_7 + a_3a_5 + \tfrac{3}{8}a_1^5a_3 - \tfrac{3}{4}a_1^2a_3^2 - \tfrac{1}{2}a_1^3a_5 - \tfrac{5}{128}a_1^8\right) \tag{11.64}$$
$$a_{10} = a_0\left(a_1a_9 + a_3a_7 + \tfrac{1}{2}a_5^2 - \tfrac{1}{2}a_1^3a_7 - \tfrac{3}{2}a_1^2a_3a_5 - \tfrac{5}{16}a_1^7a_3 + \tfrac{15}{16}a_1^4a_3^2 + \tfrac{3}{8}a_1^5a_5 - \tfrac{1}{2}a_1a_3^3 + \tfrac{7}{256}a_1^{10}\right) \tag{11.65}$$

The first coefficient a_0 is fixed apart from a global minus sign and can thus always be chosen as $a_0 = 1$. All constraints imposed on lower coefficients a_i, $(i = 0, 2, \ldots, 2k)$ have already been applied to express the condition for the next even coefficient a_{2k+2} in Eqs. (11.60)–(11.65). For example, the dependence of coefficient a_{10} on a_8 was instantaneously resolved by application of the expression for a_8. Therefore all odd coefficients can be chosen *arbitrarily*, and all even coefficients are functions of the lower odd ones, i.e.,

$$a_{2k} = f(a_0, a_1, a_3, a_5, \ldots, a_{2k-1}), \quad \forall k \in \mathbb{N} \tag{11.66}$$

By using the general power series expansion for U all the infinitely many parametrizations of a unitary transformation are treated on an equal footing. However, the question about the equivalence of these parametrizations for application in decoupling Dirac-like one-electron operators needs to be studied. It is furthermore not clear *a priori* whether the antihermitean matrix W can always be chosen in the appropriate way; the mandatory properties of W, i.e., its 'oddness', antihermiticity and behavior as a certain power in the chosen expansion parameter, have to be checked for every single transformation U applied to the untransformed or any pre-transformed Hamiltonian. Since the even expansion coefficients follow from the odd coefficients, the radius of convergence R_c of the power series depends strongly on the choice of the odd coefficients.

As long as exact unitary transformations U, i.e., infinite power series with coefficients a_k satisfying the unitarity conditions given above, are applied to transform the original one-electron operator, the energy eigenvalues of the transformed operator will be exactly the same. Therefore, the eigenvalues $\epsilon = \langle f_{bd} \rangle$ of the completely decoupled operator will certainly not depend on the choice of the odd coefficients a_{2k+1}. All infinitely many different unitary parametrizations derived above are completely equivalent in this sense.

11.4.3
Approximate Unitary and Truncated Optimum Transformations

In actual applications of decoupling transformations the power series expansions of the matrices U_i may always be truncated after a finite number of terms to produce approximately block-diagonal, but efficient, one-electron operators. The eigenvalues will both slightly differ from the exact ones and may depend on the coefficients a_k of Eq. (11.58). It is thus important for approximately unitary transformations to fix the odd coefficients in the best possible way, which would be to minimize the deviation of the eigenvalues of the transformed Hamiltonian (obtained with a truncated expansion for U) from the eigenvalues of the corresponding nth order approximation to the block-diagonal Hamiltonian f_{bd} given by Eq. (11.15). We will see in chapter 12 that it is the fixed finite order of the desired Hamiltonian which allows for a truncation of the parametrizations of the unitary matrices *without* introducing additional errors apart from the truncation of an expansion of the block-diagonal Hamiltonian. All errors due to the truncation procedure will affect only the corrections to the Hamiltonian which are of at least $(n+1)$-th order in the expansion parameter.

Since unitarity is the key property of the transformations it is clear that an optimum truncated transformation is the better suited the better this condition is fulfilled, i.e., the smaller the operator norm $|UU^\dagger - \mathbf{1}|$ is. In the following, this principle will be exploited to determine the optimum parametrization of truncated unitary matrices. If we consider a *truncated* transformation U of the form

$$U \;=\; a_0 \mathbf{1} + a_1 W + a_2 W^2 + O(W^3) \tag{11.67}$$

the minimization of the deviation of $UU^\dagger$ from unity,

$$UU^\dagger - \mathbf{1} = \frac{1}{4} a_1^4 W^4 + O(W^6) \tag{11.68}$$

yields $a_1 = 0$. Eq. (11.68) is to be constructed from the truncated expansion of Eq. (11.67), for which we seek to determine optimum expansion coefficients a_1 and a_2. Therefore, since W^3 and higher powers of W do not occur in the truncated unitary transformation U, they do not contribute to Eq. (11.68).

Because of the unitarity conditions all other coefficients would automatically vanish as well, and U would be the identity transformation. The coefficient a_1 has thus necessarily to be chosen to be different from zero, and since a_1 defines only a simple scaling of W, we may choose $a_1 = 1$.

If the expansion for U is driven to higher orders, the unitarity condition reads

$$UU^\dagger - \mathbf{1} = \left(-a_3^2 + a_1^3 a_3 - \tfrac{1}{8} a_1^6 \right) W^6 + \tfrac{1}{2^6} a_1^2 \left(8a_3 - a_1^3 \right)^2 W^8 \tag{11.69}$$

This expression will in general be minimal if the first term in parentheses vanishes. As a quadratic expression for a_3, it has two solutions. In order to achieve the smallest deviation of U from unitarity possible, we select the smaller of these two solutions, which reads

$$a_3 = \frac{2-\sqrt{2}}{4} a_1^3 \approx 0.14645\, a_1^3 \tag{11.70}$$

With this optimum choice of a_3 we can write Eq. (11.69) as

$$UU^\dagger - \mathbf{1} = \frac{1}{2^6} \left(2\sqrt{2} - 3 \right)^2 a_1^8 W^8 \approx 4.6 \cdot 10^{-4} a_1^8 W^8 \tag{11.71}$$

whose leading order has now been reduced to W^8. Since all other possible unitary parametrizations feature a leading order of W^6, i.e., they will in general lead to a larger deviation of U from unitarity, the transformation with this best choice of the coefficients will be denoted as the *optimum* unitary transformation U^{opt}.

Having previously fixed a_3 according to Eq. (11.70), the transformation U closest to unitarity including the next higher terms of the series expansion of U requires

$$a_5 = \frac{24 - 17\sqrt{2}}{2^6} a_1^5 \approx -6.5 \cdot 10^{-4} a_1^5 \tag{11.72}$$

in order to guarantee unitarity up to terms of leading order W^{10},

$$UU^\dagger - \mathbf{1} = \frac{3}{2^{11}} \left(181 - 2^7\sqrt{2} \right) a_1^{10} W^{10} + \frac{1}{2^{12}} \left(24 - 17\sqrt{2} \right)^2 a_1^{12} W^{12} \tag{11.73}$$

The same ideas may now be repeatedly applied to derive any higher-order terms of the optimum parametrization U^{opt}. This procedure will fix the higher-order odd coefficients a_{2k+1} uniquely. Higher expansion coefficients are given in Table 11.1 and compared with the coefficients of the aforementioned closed-form unitary transformations. Both the exponential and the optimal unitary parametrizations U^{opt} rapidly converge. In general, if the

Table 11.1 Coefficients a_k of the power series expansion of the unitary transformation U for five different parametrizations [506, 511]. The first two coefficients have been fixed to be $a_0 = a_1 = 1$. Note that all coefficients are only given with an accuracy of three digits after the decimal point. R_c is the radius of convergence of the expansions.

	$U=\sqrt{1+W^2}+W$	$U=\exp(W)$	$U^{\rm opt}$
a_2	5.000E−1	5.000E−1	5.000E−1
a_3	0	1.667E−1	1.464E−1
a_4	−1.250E−1	4.167E−2	2.145E−2
a_5	0	8.334E−3	−6.505E−4
a_6	6.250E−2	1.389E−3	−6.505E−4
a_7	0	1.984E−4	4.006E−5
a_8	−3.906E−2	2.480E−5	4.006E−5
a_9	0	2.756E−6	−3.102E−6
a_{10}	2.734E−2	2.756E−7	−3.102E−6
R_c	1	∞	

power series expansion of U is truncated after the term of order W^k, application of the optimum parametrization $U^{\rm opt}$ guarantees that the leading term of $UU^\dagger - \mathbf{1}$ is of order W^{k+4} instead of order W^{k+2} as for all other unitary parametrizations.

For truncated expansions of U the optimal parametrization behaves significantly better than all other choices for the coefficients a_k, as is clearly demonstrated by the deviations of U from unitarity presented in Table 11.2. Truncation of any power series applied in decoupling transformations necessarily requires that the operator norm of W is smaller than unity to guarantee convergence, since otherwise the higher-order terms in W would dominate the expansion. This implies that an even better performance of the optimum unitary parametrization $U^{\rm opt}$ as compared to all other choices for the coefficients is gained than the data in Table 11.2 suggest.

Table 11.2 Lowest-order terms of $UU^\dagger - \mathbf{1}$, for given truncation of U after $\mathcal{O}(W^k)$ for five different types of parametrization [506, 511]. The first two coefficients have been set to unity, $a_0 = a_1 = 1$. All values are rounded and given with an accuracy of only one figure after the decimal point.

	$U=\sqrt{1+W^2}+W$	$U=\exp(W)$	$U^{\rm opt}$
4	$-1.3\text{E}{-1}\,W^6$	$1.4\text{E}{-2}\,W^6$	$4.6\text{E}{-4}\,W^8$
6	$7.8\text{E}{-2}\,W^8$	$3.5\text{E}{-4}\,W^8$	$-2.8\text{E}{-5}\,W^{10}$
8	$-5.5\text{E}{-2}\,W^{10}$	$5.0\text{E}{-6}\,W^{10}$	$2.2\text{E}{-6}\,W^{12}$
10	$4.1\text{E}{-2}\,W^{12}$	$4.6\text{E}{-8}\,W^{12}$	$-1.9\text{E}{-7}\,W^{14}$

11.5
Foldy–Wouthuysen Expansion in Powers of $1/c$

Every transformation method aiming at decoupling of the Dirac Hamiltonian (now written with the energy level shift of $-m_e c^2$ from section 6.7),

$$h^D + V = (\beta - 1)\, m_e c^2 + V + c\,\boldsymbol{\alpha}\cdot\boldsymbol{p} \tag{11.74}$$

$$= \mathcal{E}_0^{(0)} + \mathcal{E}_1^{(0)} + \mathcal{O}_0^{(0)} \tag{11.75}$$

$$= \mathcal{E}_{[-2]}^{(0)} + \mathcal{E}_{[0]}^{(0)} + \mathcal{O}_{[-1]}^{(0)} \tag{11.76}$$

has necessarily to eliminate the odd term $\mathcal{O}_{[-1]}^{(0)} = \mathcal{O}_0^{(0)}$ in the first step. As before, superscripts in parentheses have been chosen to indicate that these terms belong to the initial, untransformed operator written as $h^D + V$ in accordance with Eq. (11.1). In addition, the equation above introduces a new notation for a new order parameter. While the subscripts so far — and also in Eq. (11.75) — indicated the order in the scalar potential V, i.e., a symbolic $\mathcal{E}_i$ or $\mathcal{O}_i$ operator collects all terms that contain V^i, we indicate terms that collect all terms with a given power in $1/c$ by this power written in square brackets as a subscript, as in Eq. (11.76). The two notations are required for the general analysis to be presented in this section and refer to the only two order parameters available, namely V and $1/c$, as we shall see.

Historically, $1/c$ expansions, which allow one to approach the well-known nonrelativisitic limit via $c \to \infty$, were the only ones seriously considered for decades. Therefore, we strictly refer to $1/c$ expansions as Foldy–Wouthuysen expansions. This is in accordance with the original intention of Foldy and Wouthuysen in Ref. [509]. Expansions in terms of the scalar potential V are called Douglas–Kroll–Hess (DKH) expansions.

11.5.1
The Lowest-Order Foldy–Wouthuysen Transformation

In this section we consider the first step of the Foldy–Wouthuysen transformation, which eliminates the first odd operator of order c^{-1}. If we employ the idea presented in Eq. (11.57) for this purpose, we may use the antihermitian operator W to remove the odd term $\mathcal{O}_{[-1]}^{(0)}$. Since this odd term $c\,\boldsymbol{\alpha}\cdot\boldsymbol{p}$ does not depend on the scalar potential, it has to be removed by a unitary transformation which is independent of V in accordance with the closed-form free-particle Foldy–Wouthuysen transformations introduced in section 11.3. The most general ansatz for the initial free-particle transformation U_0 then reads

$$U_{V=0} = U_0 = U_0(W_{[0]}) = a_{0,0}\mathbf{1} + \sum_{k=1}^{\infty} a_{0,k} W_{[0]}^k \tag{11.77}$$

where $W_{[0]}$ is an antihermitean operator which is exactly first order in $1/c$ but independent of V because U_0 is independent of V. If the operator $W_{[0]}$ were of zeroth or even lower order in $1/c$ it would introduce odd terms of lower than minus-first order in the transformed Hamiltonian due to the presence of the term $\mathcal{E}_{[-2]}$, and if it were of second or even higher order in $1/c$, it could not account for eliminating $\mathcal{O}^{(0)}_{[-1]}$. It should be emphasized already at this stage that the subscript in brackets attached to the operator $W_{[k]}$ indicates that the corresponding term is of $(2k+1)$-th order in $1/c$, in contrast to all other even or odd operators where it directly labels the order in $1/c$. Hence, the $W_{[0]}$ is of first order in $1/c$, $W^2_{[0]}$ of second order in $1/c$ and so forth.

The expansion coefficients $a_{0,k}$ have to satisfy the unitarity conditions (11.60)–(11.66). Transformation of the one-electron operator f now yields

$$
\begin{aligned}
f_1 &= U_0 \left(h^D + V\right) U_0^\dagger \\
&= \mathcal{E}_{[-2]} + \mathcal{E}_{[0]} + \mathcal{E}_{[2]} + \underbrace{\mathcal{O}^{(0)}_{[-1]} + a_{0,0}a_{0,1}\left[W_{[0]}, \mathcal{E}_{[-2]}\right]}_{\text{odd}} \\
&\quad + \sum_{k=2}^{\infty} \mathcal{E}^{(1)}_{[2k]} + \sum_{k=1}^{\infty} \mathcal{O}^{(1)}_{[2k-1]}
\end{aligned}
\tag{11.78}
$$

The condition that one has to impose on $W_{[0]}$ in order to eliminate the lowest-order odd term occurring in f_1 then reads [510]

$$
\mathcal{O}^{(1)}_{[-1]} \equiv \mathcal{O}^{(0)}_{[-1]} + a_{0,0}a_{0,1}\left[W_{[0]}, \mathcal{E}_{[-2]}\right] \stackrel{!}{=} 0 \tag{11.79}
$$

Due to the very simple structure of $\mathcal{E}_{[-2]}$, which is only a constant and independent of both $\boldsymbol{p}$ and V, inversion of the commutator is straightforward and yields the odd and antihermitean operator [510]

$$
W_{[0]} = \frac{a_{0,0}}{a_{0,1}}\, \beta\, \frac{c\,\boldsymbol{\alpha}\cdot\boldsymbol{p}}{2m_e c^2} = \frac{a_{0,0}}{a_{0,1}}\, \beta\, \frac{\boldsymbol{\alpha}\cdot\boldsymbol{p}}{2m_e c} \tag{11.80}
$$

which is first order in $1/c$ indeed and turns out to be odd. Because of the odd super-structure of Dirac's $\boldsymbol{\alpha}$ matrices it is clear that $W_{[0]}$ is an *odd* operator.

Eq. (11.77) represents infinitely many different unitary parametrizations U_0 which all decouple the one-electron Hamiltonian f up to zeroth order in $1/c$. The resulting Hamiltonians f_1, which all differ by the specific choice of expansion coefficients $a_{0,k}$, consist of infinitely many terms which can all be assigned a definite order in both $1/c$ and V.

For example, the odd operator $\mathcal{O}^{(1)}_{[1]}$, which was obtained after the free-particle transformation, is first order '[1]' in $1/c$ as well as first order '1' in V,

$$
\mathcal{O}^{(1)}_{[1]} = \left[\beta\, \frac{\boldsymbol{\alpha}\cdot\boldsymbol{p}}{2m_e c}, V\right] \tag{11.81}
$$

Since U_0 is independent of the potential V, all the various Hamiltonians f_1 contain only terms of at most first order in V, albeit arbitrarily high orders in $1/c$. The leading terms of f_1 given by Eq. (11.78) are independent of the chosen parametrization of U_0 — meaning independent of the expansion coefficients $a_{0,k}$ which exactly define any specific choice of unitary transformation. Their explicit form is given by

$$\mathcal{E}_{[-2]} = (\beta - 1)m_e c^2 \tag{11.82}$$

$$\mathcal{E}_{[0]} = V + \beta \frac{p^2}{2m_e} \tag{11.83}$$

$$\mathcal{E}_{[2]} = -\beta \frac{p^4}{8m_e^3c^2} \frac{[\alpha \cdot p[\alpha \cdot p, V]]}{8m_e^2c^2} \tag{11.84}$$

One clearly recovers the rest energy $\mathcal{E}_{[-2]}$ and the nonrelativistic contributions $\mathcal{E}_{[0]}$ as the leading terms of the Foldy–Wouthuysen series given above.

It might be tempting to consider all these infinitely many Hamiltonians f_1 as being equivalent, since they all seem to possess the same spectrum due to the unitary structure of the transformation U_0. However, this understanding would be erroneous since the situation is more subtle. After having chosen a specific set of expansion coefficients $a_{0,k}$ in accordance with the unitarity conditions mentioned above, one has necessarily to check the convergence of both the power series ansatz for U_0 *and* the resulting series expansion of the Hamiltonian f_1. Especially since $W_{[0]}$ is linear in the momentum operator $\boldsymbol{p}$ it is by no means clear whether these series expansions are valid for large momenta, i.e., whether they are related to the original one-electron Hamiltonian at all.

There is, however, one very special parametrization for the transformation U_0 that avoids the expansion of the one-electron operator in any way and is thus free from convergence issues. This particular parametrization yields operators that can be converted into the closed-form free-particle Foldy–Wouthuysen expression defined by Eq. (11.27) or (11.35), which produces the closed-form Hamiltonian f_1 derived in section 11.3. Its expansion coefficients, satisfying all unitarity conditions, are given by

$$a_{0,0} = 1, \quad a_{0,1} = 1, \quad a_{0,2} = \frac{1}{2}, \quad a_{0,3} = \frac{3}{2}, \quad a_{0,4} = \frac{11}{8}$$
$$a_{0,5} = \frac{31}{8}, \quad a_{0,6} = \frac{69}{16}, \quad a_{0,7} = \frac{187}{16}, \quad a_{0,8} = \frac{1843}{128}, \quad a_{0,9} = \frac{4859}{128} \quad \cdots \tag{11.85}$$

The different expressions for U_0 given by Eqs. (11.27) and (11.35) represent well-defined quantities for all momenta $p = |\boldsymbol{p}| \in \mathbb{R}_0^+$, which are free of any singularities. Accordingly, also the free-particle Foldy–Wouthuysen Hamiltonian f_1 obtained by this specific transformation,

$$f_1 = U_0 h^D U_0^\dagger = \mathcal{E}_0 + \mathcal{E}_1 + \mathcal{O}_1^{(1)} \tag{11.86}$$

contains only well-defined expressions valid for all values of the momentum p. *Only* this specific closed-form choice for U_0 facilitates the opportunity of a closed-form evaluation of the Hamiltonian f_1 and abandons the necessity of *any* expansion in $1/c$. However, the legitimacy of this expansion and hence the issue of higher-order Foldy–Wouthuysen transformations will be analyzed in the next section in greater detail. By inspection of the terms occurring in f_1 given by Eq. (11.36) or (11.86), however, there is no doubt about the well-defined behavior of the free-particle Foldy–Wouthuysen transformation U_0 even in the presence of a scalar potential.

In Eq. (11.86), the superscript in parentheses at the odd term has been introduced in order to establish a systematic notation for the derivation of higher-order terms in the next chapter. $\mathcal{O}_1^{(1)}$ is, of course, still given by Eq. (11.39).

As a consequence of this choice for U_0, one arrives at a compact expression for f_1, where each constituent term can only be classified according to its order in V and no longer according to its order in $1/c$. Each term occurring in Eq. (11.36) or (11.86) contains arbitrarily high orders of $1/c$, as is easily seen from a Taylor series expansion in this parameter. If this compact result for f_1 is to be utilized for further decoupling transformations, all terms have necessarily to be classified according to their order in V, which leads to the Douglas–Kroll–Hess protocol to be discussed in the next chapter.

Finally, we should stress that all expressions discussed so far are of $2{\times}2$ super-structure and hence belong to the world of four-component one-electron schemes. In order to obtain the desired two-component operators, which yield or approximate f_+ of Eq. (11.15), we have to neglect all LS, SL and SS blocks in the operators derived so far. Note that *all* unitary matrices U and thus also all antihermitean parameters W feature a $2{\times}2$ super-structure and belong to the four-component regime. Especially, this cannot be avoided as the W operators turned out to be odd and thus couple large and small components of the eigenstates of the one-electron operator.

It has also to be emphasized that the derivation of the initial step of any transformation scheme presented here has employed the most general ansatz possible. Although the final result is the familiar free-particle Foldy–Wouthuysen transformation that has been in use for decades, we stress that our analysis yields this result as a *necessary consequence* obtained from the most general viewpoint [510].

11.5.2
Second-Order Foldy–Wouthuysen Operator: Pauli Hamiltonian

For a spherically-symmetric Coulomb potential $V(r) = -Ze^2/r$, the second-order even term in the Foldy–Wouthuysen expansion, $\mathcal{E}_{[2]}$ in Eq. (11.84), can be cast into the much more familiar form to yield the (relativistic corrections of the) Pauli Hamiltonian,

$$\mathcal{E}_{[2]} = -\beta \frac{\boldsymbol{p}^4}{8m_e^3c^2} + \frac{\hbar^2}{8m_e^2c^2}(\Delta V) + \frac{\hbar}{4m^2c^2}\boldsymbol{\Sigma}\cdot\left[(\nabla V)\times\boldsymbol{p}\right] \quad (11.87)$$

$$= -\beta \frac{\boldsymbol{p}^4}{8m_e^3c^2} - \frac{e\hbar^2}{8m_e^2c^2}\,\mathrm{div}\boldsymbol{E} + \frac{\hbar}{4m_e^2c^2}\frac{1}{r}\frac{\partial V}{\partial r}\boldsymbol{\Sigma}\cdot\boldsymbol{l} \quad (11.88)$$

which are from left to right the mass–velocity, the Darwin, and the spin–orbit coupling term. Note the close analogy to the classical energy expression derived already in Eq. (3.121) of chapter 3. Of course, the nonrelativistic kinetic energy operator as well as the external potential V are not given here because they enter the Pauli Hamiltonian through $\mathcal{E}_{[0]}$ of Eq. (11.83). The angular momentum operator is defined as before in chapters 4 and 6, $\boldsymbol{l} = \boldsymbol{r}\times\boldsymbol{p}$. According to Eq. (2.125), the electric field strength is given by

$$\boldsymbol{E} = -\nabla\left(\frac{V}{q_e}\right) = \frac{\nabla V}{e} \quad \longrightarrow \quad \boldsymbol{E}_{nuc} \stackrel{(A.23)}{=} \frac{Ze\boldsymbol{r}}{r^3} \quad (11.89)$$

which reduces to $\boldsymbol{E}_{nuc}$ in case of a single point-like nucleus. Dirac's even $\boldsymbol{\Sigma}$-matrices are the four-component generalization of Pauli's spin matrices already introduced in Eq. (6.22). We refer to section 13.1 for a more detailed derivation and manipulation of the terms in the Pauli operator.

These familiar approximate relativistic one-electron Hamiltonians, whose expectation values often served to define measures for relativistic effects, are thus only an approximation to the Dirac Hamiltonian of hydrogen-like atoms discussed in chapter 6. Nevertheless, they are also employed in the context of many-electron systems where the one-electron part in f of Eq. (11.1), i.e., $h^D + V_{nuc}$, is approximated solely by the 2×2-analogous expressions for $\mathcal{E}_{[0]} + \mathcal{E}_{[2]}$,

$$(\mathcal{E}_{[0]} + \mathcal{E}_{[2]})_+ = V + \frac{\boldsymbol{p}^2}{2m_e} - \frac{\boldsymbol{p}^4}{8m_e^3c^2} - \frac{e\hbar^2}{8m_e^2c^2}\mathrm{div}\boldsymbol{E} + \frac{\hbar}{4m_e^2c^2}\frac{1}{r}\frac{\partial V}{\partial r}\boldsymbol{\sigma}\cdot\boldsymbol{l} \quad (11.90)$$

which represent a first approximation to f_+ of Eq. (11.15). However, the Pauli Hamiltonian cannot be used in variational calculations as becomes clear in the next section, which discusses the major flaw of any Foldy–Wouthuysen expansion (truncated or not).

11.5.3
Higher-Order Foldy–Wouthuysen Transformations and Their Pathologies

Up to this point we have just been able to decouple a potential-free Dirac Hamiltonian h^D exactly. In the presence of a scalar potential we are able to transform an odd term $\mathcal{O}_0^{(0)}$ into another one $\mathcal{O}_1^{(1)}$.

In this section we now consider the option of getting rid of $\mathcal{O}_1^{(1)}$ or, at least, diminishing it systematically until it vanishes or until its contribution to the tranformed one-electron Hamiltonian can safely be neglected. Of course, we must not forget what has just been stated regarding the convergence of series expansions that may arise. Actually, the $1/c$ expansion of the one-electron Hamiltonian, which we specifically call a Foldy–Wouthuysen expansion, causes problems. Especially the very roots of these problems are often mistakenly identified and are erroneously attributed to the singular behavior of the Coulomb potential near the origin, whereas it is the illegal expansion of the square root operator E_p occurring in the free-particle term $\mathcal{E}_0$ of Eq. (11.37) which is responsible for its insufficiencies as we shall now see.

After having carried out an initial free-particle Foldy–Wouthuysen transformation U_0 one could try to establish a *sequence* of further unitary transformations $U_i = U_i(W_{[i]})$, $(i = 1, 2, 3, \ldots)$ written as

$$U = \cdots U_4 U_3 U_2 U_1 U_0 = \prod_{i=0}^{\infty} U_i \tag{11.91}$$

which would eliminate the respective lowest-order odd term in $1/c$ step by step by suitable choices of the actual expansion parameter $W_{[i]}$. The resulting expression for the expanded one-electron operator can then be formally written as

$$f_{bd} = \sum_{k=-1}^{\infty} \mathcal{E}_{[2k]} = \sum_{k=-1}^{\infty} \begin{pmatrix} \mathcal{E}_{[2k]+} & 0 \\ 0 & \mathcal{E}_{[2k]-} \end{pmatrix} \tag{11.92}$$

where each term features a well-defined order in $1/c$, indicated by the subscript in brackets. As well as the free-particle Foldy–Wouthuysen transformation, this higher-order decoupling scheme was also introduced by Foldy and Wouthuysen in 1950 [509] and has been presented in many textbooks dealing with relativistic quantum mechanics ever since [53,65,95,102]. However, only the lowest-order corrections up to $\mathcal{O}(c^{-2})$ are discussed in these presentations without any warning not to proceed to higher orders in $1/c$. The lowest-order term $\mathcal{E}_0$ is, of course, zeroth-order in V and minus-second-order in $1/c$ due to the rest mass term $(\beta - 1)m_e c^2$, as can be verified by a formal Taylor series expansion of Eq. (11.37) in $1/c$.

It must be stressed again — following the original paper by Foldy and Wouthuysen — that the phrase *Foldy–Wouthuysen transformation* is strictly reserved for denoting a $1/c$ expansion of f_{bd} rather than *any* arbitrary decoupling transformation of the Dirac Hamiltonian as it is sometimes applied in the literature.

According to Eq. (11.92), the decoupled Hamiltonian f_{bd} within the Foldy–Wouthuysen framework is formally given as a series of even terms of well-defined order in $1/c$. In most presentations of the Foldy–Wouthuysen transformation the exponential function parametrization $U_{[i]} = \exp(W_{[i]})$ is applied for each transformation step. However, in the light of the discussion in section 11.4 the specific choice of this parametrization does not matter at all, since one necessarily has to expand U_i into a power series in order to evaluate the Hamiltonian. Consequently, in order to guarantee a most general analysis, the most general parametrization for the Foldy–Wouthuysen transformation should be employed [510]. Thus, U_i is parametrized as a power series expansion in an odd and antihermitean operator $W_{[i]}$, which is of $(2i+1)$-th order in $1/c$, (cf. section 11.4). After n transformation steps, the intermediate, partially transformed Hamiltonian f_n has the following structure,

$$\begin{aligned} f_n &= U_{n-1} f_{n-1} U_{n-1}^\dagger \\ &= \sum_{k=-1}^{2n-1} \mathcal{E}_{[2k]} + \sum_{k=n}^{2n-1} \mathcal{O}_{[2k-1]}^{(n)} + \sum_{k=2n}^{\infty} \left(\mathcal{E}_{[2k]}^{(n)} + \mathcal{O}_{[2k-1]}^{(n)} \right) \end{aligned} \tag{11.93}$$

The next transformation U_n is determined by the odd and antihermitean operator $W_{[n]}$, which has to be chosen as

$$W_{[n]} = \frac{a_{n,0}}{a_{n,1}} \beta \frac{\mathcal{O}_{[2n-1]}^{(n)}}{2m_e c^2} \tag{11.94}$$

in order to eliminate the odd term $\mathcal{O}_{[2n-1]}^{(n+1)} = \mathcal{O}_{[2n-1]}^{(n)} + a_{n,0} a_{n,1} \left[W_{[n]}, \mathcal{E}_{[-2]} \right]$ of f_{n+1}.

Though this Foldy–Wouthuysen procedure may formally be repeated until exact decoupling seems to be achieved, the resulting even terms are highly singular and ill-defined, and are not related to the original Dirac Hamiltonian at all, except for the leading nonrelativistic term $\mathcal{E}_{[0]}$ and to some extent the first relativistic correction $\mathcal{E}_{[2]}$, which could be evaluated perturbatively. The reason for this failure of the higher-order Foldy–Wouthuysen transformation is that it necessarily relies on an illegal $1/c$ expansion of *all* terms occurring in the free-particle Foldy–Wouthuysen Hamiltonian f_1 defined by Eq. (11.36) in order to classify each term. It should be noted that a strict classification of all terms of the expanded block-diagonal Hamiltonian in powers of $1/c$ necessarily requires an expansion of the square-root in E_p. It does not make much sense not to expand E_p but to keep it intact, if all other terms are classified by powers of $1/c$, as the latter terms are then incomplete with respect to any chosen power of $1/c$.

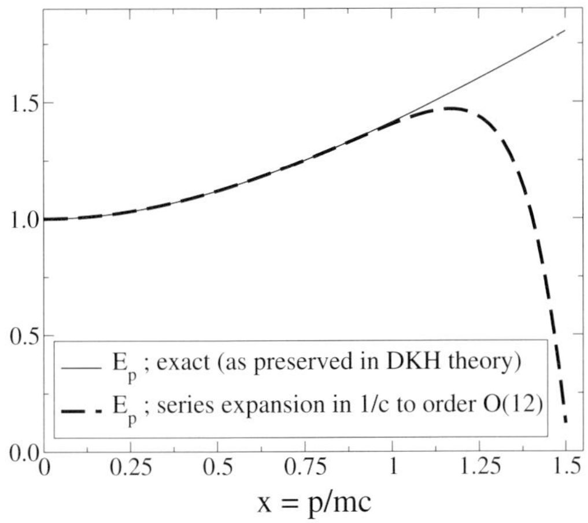

Figure 11.1 Comparison of the exact expression for E_p with the series expansion around the nonrelativistic limit $x \equiv p/m_e c = 0$ up to 12th order $\mathcal{O}(x^{12})$. The series expansions only represent the exact expressions for $x < 1$. Beyond the radius of convergence of the square root of E_p in Eq. (11.11) at $x = 1$ the series expansions and hence the in terms of $1/c$ expanded Hamiltonians are completely ill-defined [510].

In addition, such a power series expansion is, however, only permitted for analytic, i.e., holomorphic functions and must never be extended beyond a singular point. Since the square root occurring in the relativistic energy–momentum relation E_p of Eq. (11.11) possesses branching points at $x \equiv p/m_e c = \pm\text{i}$, *any* series expansion of E_p around the static nonrelativistic limit $x = 0$ is only related to the exact expression for E_p for non-ultrarelativistic values of the momentum, i.e., $|x| < 1$. This is most easily seen by rewriting E_p as

$$E_p = m_e c^2 \sqrt{1 + x^2} \equiv m_e c^2 \exp\left[\frac{1}{2}\ln(1 + x^2)\right] \tag{11.95}$$

which clearly demonstrates that the essential singularities of the logarithm at $x = \pm\text{i}$ confine the domain of convergence of any power series expansion of E_p to $|x| < 1$. For larger momenta, which inevitably arise in the vicinity of the nucleus, the series expansion up to any arbitrary or even infinite order in $p/m_e c$ does not represent the original function E_p (see Figure 11.1). For these momenta, this series does not even converge at all. The singular behavior of the series expansions of E_p, and hence of A_p and $\mathcal{R}_p$, becomes thus the worse the more terms of the expansions are taken into account. If the expansions had not been truncated, they would not even converge at all!

The intrinsic failure of the Foldy–Wouthuysen protocol is thus doubtlessly related to the illegal $1/c$ expansion of the kinetic term E_p, which does not bear any reference to the external potential V. However, in the literature the ill-defined behavior of the Foldy–Wouthuysen transformation has sometimes erroneously been attributed to the singular behavior of the Coulomb potential near the nucleus, and even the existence of the correct nonrelativistic limit of the Foldy–Wouthuysen Hamiltonian is sometimes the subject of dispute. Because of Eqs. (11.82) and (11.83) and the analysis given above, the nonrelativistic limit $c \to \infty$, i.e., $x \to 0$ is obviously well defined, and for positive-energy solutions given by the Schrödinger Hamiltonian $f_{\rm nr} = \boldsymbol{p}^2/2m_e + V$.

11.6
The Infinite-Order Two-Component One-Step Protocol

Alternatively to the non-convergent $1/c$ decoupling attempt, one might wish to refrain from any expansion of the one-electron operator and thus not apply a sequence of unitary transformations. It is possible to envisage a purely numerical procedure in order to derive only a matrix representation of the final block-diagonal Hamiltonian in a given one-particle basis set. Of course, we then have no analytical form of the resulting Hamiltonian available, but this is no obstacle for numerical calculations in molecular science. Such an approach was suggested by Barysz and Sadlej [514–519] and was called the infinite-order two-component (IOTC) method. It explicitly employs the X-operator formalism presented in the preceding sections. The approach is also based on ideas originally presented by Barysz, Sadlej, and Snijders [520] and has been mostly applied to atomic systems [517–519].

In this numerical approach, the free-particle Foldy–Wouthuysen transformation U_0 is a mandatory initial transformation. Then, the sequence of subsequent unitary transformations U_i ($i \geq 1$) of Eq. (12.1) applied to the free-particle Foldy–Wouthuysen Hamiltonian f_1 is united to only one residual transformation step U_1,

$$f_{bd} = U_1 f_1 U_1^\dagger = \begin{pmatrix} f_+ & 0 \\ 0 & f_- \end{pmatrix} \tag{11.96}$$

which is formally parametrized by the operator R, which constitutes the exact relationship between the upper and lower components of the free-particle Foldy–Wouthuysen transformed spinor denoted as ψ_1 for electronic (class-I) solutions [503],

$$\psi_1^S = R\psi_1^L \tag{11.97}$$

The new symbol R has been used for this operator in order to distinguish it from the previously introduced X-operator relating the small and large components of the original, i.e., untransformed Dirac spinor ψ via Eq. (11.2). The matrix U_1 may then be expressed as

$$U_1 = U_1(R) = \begin{pmatrix} \left(1+R^\dagger R\right)^{-1/2} & \left(1+R^\dagger R\right)^{-1/2} R^\dagger \\ -\left(1+RR^\dagger\right)^{-1/2} R & \left(1+RR^\dagger\right)^{-1/2} \end{pmatrix} \tag{11.98}$$

where the arbitrary phase of Eq. (11.23) has been set to zero. Similarly to Eq. (11.6), the requirement of vanishing off-diagonal blocks of f_{bd} leads to a condition imposed on the operator R,

$$R = \left[f_1^{SS}\right]^{-1} \left\{ -f_1^{LS} + R\, f_1^{LL} + R\, f_1^{SL} R \right\} \tag{11.99}$$

After insertion of the explicit expressions for the components of the free-particle Foldy–Wouthuysen Hamiltonian f_1 given by Eqs. (11.36)–(11.39), this equation reads

$$\begin{aligned} E_p R + R E_p &= A_p \left[\sigma\cdot\mathcal{P}_p\, , V\right] A_p + \left[A_p\, V A_p\, , R\right] \\ &\quad + \left[A_p\, \sigma\cdot\mathcal{P}_p\, V \sigma\cdot\mathcal{P}_p\, A_p\, , R\right] + R A_p \left[\sigma\cdot\mathcal{P}_p\, , V\right] A_p R \end{aligned} \tag{11.100}$$

This last equation is the basis for an iterative numerical scheme for a chosen basis set expansion of the two-component spinor with resulting matrix representation of all operators [516]. For this purpose the equation has first to be multiplied by the operator $P_p^{-1} \sigma\cdot\mathcal{P}_p$ from the left in order to reduce it to a computationally feasible form, where

$$P_p = \left|\mathcal{P}_p\right| = \left(\mathcal{P}_p^2\right)^{1/2} \tag{11.101}$$

is a scalar operator. Subsequent introduction of the operator $Q = P_p^{-1} \sigma\cdot\mathcal{P}_p\, R$ and frequent application of Eq. (12.59) yields the equation

$$\begin{aligned} E_p Q + Q E_p &= P_p\, A_p\, V A_p - P_p^{-1} A_p\, \sigma\cdot\mathcal{P}_p\, V \sigma\cdot\mathcal{P}_p\, A_p - Q A_p\, V A_p \\ &\quad + P_p^{-1} A_p\, \sigma\cdot\mathcal{P}_p\, V \sigma\cdot\mathcal{P}_p\, A_p P_p^{-1} Q + A_p P_p\, V P_p\, A_p Q \\ &\quad - Q A_p\, \sigma\cdot\mathcal{P}_p\, V \sigma\cdot\mathcal{P}_p\, A_p + Q A_p\, \sigma\cdot\mathcal{P}_p\, V \sigma\cdot\mathcal{P}_p\, A_p P_p^{-1} Q \\ &\quad - Q A_p\, V A_p\, P_p\, Q \end{aligned} \tag{11.102}$$

for the (2×2)-operator $Q = Q(R)$.

Though Eq. (11.102) seems to be unnecessarily complicated, it can be solved by purely numerical iterative techniques, and the matrix representation of the operator Q is obtained [516]. This result appears to be the best representation of the operator Q that can be achieved within a given basis and is only limited by machine accuracy. Note that all expressions occurring in Eq. (11.102) and hence the matrix representation of the operator Q depend only on the squared momentum $\boldsymbol{p}^2$ rather than on the momentum variable itself, which is the key feature of the computational feasibility of this approach. This is an essential trick for actual calculations of transformed two-component operators first noticed by Hess [521] (compare also section 12.5.1). Eq. (11.102) is still nonlinear and, thus, bears the possibility of negative-energy solutions for the operator Q. The choice toward the positive-energy branch has to be implemented via the boundary conditions imposed on the numerical iterative technique. Essentially, Q and hence R have to be 'small' operators with operator norms much smaller than unity.

Once the matrix representation of the operator Q is known, it can be directly used to determine the desired matrix representation of the two-component Hamiltonian f_+, which reads

$$\begin{aligned} f_+ &= \frac{1}{\sqrt{1+R^\dagger R}} \left\{ f_1^{LL} + f_1^{LS} R + R^\dagger f_1^{SL} + R^\dagger f_1^{SS} R \right\} \frac{1}{\sqrt{1+R^\dagger R}} \\ &= \frac{1}{\sqrt{1+Q^\dagger Q}} \left\{ E_p - mc^2 + A_p V A_p + A_p \boldsymbol{\sigma}\cdot\boldsymbol{P}_p V \boldsymbol{\sigma}\cdot\boldsymbol{P}_p A_p \left(1 + P_p^{-1} Q\right) \right. \\ &\qquad - A_p V A_p P_p Q - Q^\dagger P_p A_p V A_p + Q^\dagger P_p^{-1} A_p \boldsymbol{\sigma}\cdot\boldsymbol{P}_p V \boldsymbol{\sigma}\cdot\boldsymbol{P}_p A_p \\ &\qquad + Q^\dagger \Big(- E_p - mc^2 + P_p^{-1} A_p \boldsymbol{\sigma}\cdot\boldsymbol{P}_p V \boldsymbol{\sigma}\cdot\boldsymbol{P}_p A_p P_p^{-1} \\ &\qquad \left. + A_p P_p V P_p A_p \Big) Q \right\} \frac{1}{\sqrt{1+Q^\dagger Q}} \end{aligned} \tag{11.103}$$

This Hamiltonian can then be used variationally in quantum chemical calculations, since because of its derivation no negative energy states can occur. It should be anticipated that this Hamiltonian is conceptually equivalent to the infinite-order Douglas–Kroll–Hess Hamiltonian to be discussed in section 12.3, because both schemes do not apply any expansion in $1/c$. Also the expressions for E_p and A_p are strictly evaluated in closed form within both approaches. However, whereas Douglas–Kroll–Hess theory yields analytic exressions for each order in V, the infinite-order two-component method summarizes all powers of V in the final matrix representation of f_+.

11.7 Toward Well-Defined Analytic Block-Diagonal Hamiltonians

The failure of the infinite-order Foldy–Wouthuysen expansion in terms of $1/c$ does not reflect a major fundamental problem of block-diagonalization procedures applied to the Dirac Hamiltonian. It rather demonstrates that neither these unitary transformations nor the Hamiltonian must be expanded naively in $1/c$, i.e., transgressing the domain of convergence of the resulting series. At this point we may draw attention to a misconception of a type that often occurs if elements of different theories are mixed up. One might ask whether this simple convergence issue might be avoided for light elements, for which a classical picture would assign small velocities and hence small momenta to the electrons in atoms or molecules composed from these elements. The picture of classically moving electrons on trajectories with well-defined momentum is misleading here as it is in all other (quantum mechanical) cases when it is invoked. We have already explained that the transformation is to be carried out in momentum space. Consequently, all momenta play a role, and one must not restrict the discussion to arbitrarily selected low momenta. This statement holds for the general operator expression in momentum space but it can, of course, also be expressed in terms of a matrix representation of the operator for a given one-particle basis introduced in chapter 10. In the latter case one may say that the matrix represenations always involve basis functions of high momentum, if the finite dimension of this representation is to be meaningful at all (compare also the construction of p^2-basis sets in chapter 12).

If, however, an elegant *and legitimate* expansion of the decoupled Hamiltonian similar to Eq. (11.92) is to be preserved for both analytical and numerical investigations, one has to classify each term of this expansion according to a new order parameter. This is necessarily the scalar potential V (in the absence of any vector potentials $\boldsymbol{A}$), which is the only remaining possibility since all other parameters like m_e and $\boldsymbol{p}$ are already all associated with $1/c$.

The key feature of the expansion in terms of V is that the closed-form expression of the free-particle Foldy–Wouthuysen Hamiltonian f_1 given by Eq. (11.36) is to remain untouched during the whole transformation procedure, and the resulting block-diagonal Hamiltonian is well-defined for all momenta $p \in \mathbb{R}_0^+$ and features exactly the same spectrum as the original Dirac Hamiltonian. We introduce this expansion, called the Douglas–Kroll–Hess expansion, in detail in chapter 12. Of course, also the Douglas–Kroll–Hess Hamiltonian features singularities for the complex momenta $x = \pm \mathrm{i}$, i.e., $p = \pm imc$ due to the presence of terms containing E_p. As a consequence, the Douglas–Kroll–Hess Hamiltonians are only well defined on either of the sliced complex planes $\mathbb{C}_1 = \mathbb{C} \backslash \{z = \mathrm{i}y \mid y \leq -1 \vee y \geq +1\}$ or $\mathbb{C}_2 = \mathbb{C} \backslash \{z = \mathrm{i}y \mid -1 \leq y \leq +1\}$, but not simultaneously on both connected regions [510]. This situation is illustrated in Figure 11.2. Restricting

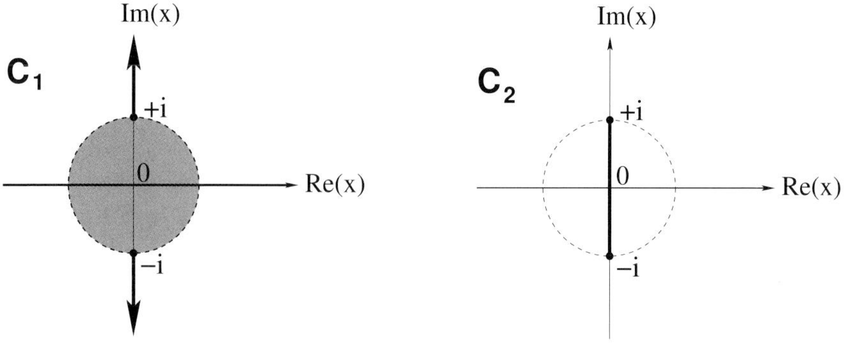

Figure 11.2 Schematic representation of the sliced complex planes $\mathbb{C}_1$ and $\mathbb{C}_2$. After removal of suitable parts of the imaginary axis, indicated by thick lines, both regions are connected domains for E_p [510]. Series expansions around $x = 0$ are only legitimate within the shaded disc of radius 1 of $\mathbb{C}_1$. Only $\mathbb{C}_1$ contains the whole real axis, i.e., all physically relevant values for the momentum $x = p/mc$.

the domain for E_p to $\mathbb{C}_1$, which contains the *whole* real axis, i.e., *all* physically relevant values for the momentum p, there is no doubt about the well-defined and unique existence of the Douglas–Kroll–Hess Hamiltonians. Furthermore, the nonrelativistic limit $x \rightarrow 0$ will also certainly not cause any problems on $\mathbb{C}_1$, since all terms contributing to the Douglas–Kroll–Hess Hamiltonians are holomorphic on the whole disc around $x = 0$ with radius 1. Hence, they may be expanded in a Taylor series, which can subsequently be truncated after the leading term yielding the nonrelativistic limit [510]. Clearly, an extension of this expansion to the region with $|x| > 1$, which would be equivalent to the Foldy–Wouthuysen expansion, is not justified.

An expansion in terms of V, i.e., the Douglas–Kroll–Hess expansion, is the *only* valid analytic expansion technique for the Dirac Hamiltonian, where the final block-diagonal Hamiltonian is represented as a series of regular even terms of well-defined order in V, which are all given in closed form. For the derivation, the initial transformation step has *necessarily* to be chosen as the closed-form, analytical free-particle Foldy–Wouthuysen transformation defined by Eq. (11.35) in order to provide an odd term depending on the external potential that can then be diminished. We now address these issues in the next chapter.

12
Douglas–Kroll–Hess Theory

Chapter 11 introduced the basic principles for elimination-of-the-small-component protocols and noted that the Foldy–Wouthuysen scheme applied to one-electron operators including scalar potentials yield ill-defined 1/c-expansions of the desired block-diagonal Hamiltonian. In contrast, the Douglas–Kroll–Hess transformation represents a unique and valid decoupling protocol for such Hamiltonians and is therefore investigated in detail in this chapter.

12.1
Sequential Unitary Decoupling Transformations

Since the energy-independent operator X is not known in closed form, a decomposition of the overall unitary transformation U of Eq. (11.15) into a sequence of tailored unitary transformations, as in Eq. (11.91),

$$U = \cdots U_4 U_3 U_2 U_1 U_0 = \prod_{i=0}^{\infty} U_i \tag{12.1}$$

is beneficial. Tailored means that each of these transformations U_i has to be parametrized such that the odd term of lowest order in a *pre-determined* expansion parameter occurring in the Hamiltonian is eliminated stepwise. As long as all infinitely many matrices U_i are taken into account, these nested unitary transformations will exactly decouple the Dirac Hamiltonian and no approximation is introduced at all. As discussed in the preceding chapter, there remains only one valid choice for the expansion parameter of the one-electron operator, which is the scalar potential V. The resulting decoupling scheme is called Douglas–Kroll–Hess (DKH) transformation.

The original idea of this procedure dates back to 1974 and is due to Douglas and Kroll [186] who mention it in the appendix of their work on the lowest ^{3}P state of He, which they study with the Bethe–Salpeter equation. More than a decade later the paper by Douglas and Kroll was rediscovered by Hess [522] who at first had to struggle with the huge problem to transform the idea into a method that allows actual calculations on molecular systems. He invoked the idea of producing a basis set that diagonalizes the matrix representation

Relativistic Quantum Chemistry. Markus Reiher and Alexander Wolf

ISBN: 978-3-527-31292-4

of p^2 — or equivalently of the kinetic energy — which he applied earlier for the square-root operator [521]. In the years to follow, the second-order DKH method was established as a very valuable tool for relativistic quantum chemistry (see Refs. [523–529] and references cited therein), which has been implemented in many quantum chemistry programs — too many to be listed separately. Surprisingly, higher-order terms, which would also shed light on the convergence of the DKH expansion, were only considered much later, when Nakajima and Hirao presented the third-order Hamiltonian [530]. The method was than revisited [511] where the generalized parametrization was introduced as well as the fourth- and fifth-order Hamiltonians. Afterwards, the sixth-order expression was also given [531], and then all explicitly derived and implemented DKH Hamiltonians became in a way superfluous when an arbitrary-order scheme [532] for the automatic symbolic derivation of these Hamiltonians became available and found its way into quantum chemistry computer programs [533, 534]. In this chapter, we shall now derive and discuss all essential ingredients of the DKH method.

In the Douglas–Kroll–Hess expansion, the block-diagonal Hamiltonian of Eq. (11.15) can be formally expressed as a series of even terms of well-defined order in the external scalar potential V,

$$h_{\mathrm{DKH}\infty} = \sum_{k=0}^{\infty} \mathcal{E}_k = \sum_{k=0}^{\infty} \begin{pmatrix} \mathcal{E}_{k+} & 0 \\ 0 & \mathcal{E}_{k-} \end{pmatrix} \tag{12.2}$$

where the subscript indicates the order in the external potential of the corresponding term. The explicit form of these even operators will be derived in the next sections. We will see that the two lowest-order even terms $\mathcal{E}_0$ and $\mathcal{E}_1$ are identical to those which we encountered in the case of the free-particle Foldy–Wouthuysen transformation in Eqs. (11.37) and (11.38). We will also see that each of these even operators can be decomposed into spin-free (sf) and spin-dependent (sd) terms,

$$\mathcal{E}_k = \begin{pmatrix} \mathcal{E}_{k+} & 0 \\ 0 & \mathcal{E}_{k-} \end{pmatrix} = \begin{pmatrix} \mathcal{E}^{\mathrm{sf}}_{k+} + \mathcal{E}^{\mathrm{sd}}_{k+} & 0 \\ 0 & \mathcal{E}^{\mathrm{sf}}_{k-} + \mathcal{E}^{\mathrm{sd}}_{k-} \end{pmatrix} \tag{12.3}$$

by exploiting Dirac's relation given in Eq. (D.11).

It should be recalled that, because of the presence of the external potential and the nonlocal form of E_p given by Eq. (11.11), all operators resulting from these unitary transformations are well defined only in momentum space (compare the discussion of the square-root operator in the context of the Klein–Gordon equation in chapter 5 and the momentum-space formulation of the Dirac equation in section 6.10). Whereas $\mathcal{E}_0$ acts as a simple multiplicative operator, all higher-order terms containing the potential V are integral operators and completely described by specifying their kernel. For example, the

(2×2)-kernel of $\mathcal{E}_{1+}$ is given by

$$\mathcal{E}_{1+}(i,j) = A_i V_{ij} A_j + A_i \boldsymbol{\sigma} \cdot \boldsymbol{P}_i \cdot V_{ij} \boldsymbol{\sigma} \cdot \boldsymbol{P}_j A_j \tag{12.4}$$

$$= \underbrace{A_i V_{ij} A_j + A_i \boldsymbol{P}_i \cdot V_{ij} \boldsymbol{P}_j A_j}_{\mathcal{E}^{\mathrm{sf}}_{1+}(i,j)} + \underbrace{\mathrm{i}\, A_i \boldsymbol{\sigma} \cdot (\boldsymbol{P}_i \times V_{ij} \boldsymbol{P}_j) A_j}_{\mathcal{E}^{\mathrm{sd}}_{1+}(i,j)} \tag{12.5}$$

where the symbolic variables i and j symbolize momentum operators $\boldsymbol{p}_i$ and $\boldsymbol{p}_j$, respectively. The action of $\mathcal{E}^{\mathrm{sf}}_{1+}$ on an arbitrary 2-spinor ψ^L is then defined by the integral expression

$$\mathcal{E}^{\mathrm{sf}}_{1+}\, \psi^L(\boldsymbol{p}_i) = \int \frac{\mathrm{d}^3 p_j}{(2\pi\hbar)^3}\, \mathcal{E}^{\mathrm{sf}}_{1+}(\boldsymbol{p}_i, \boldsymbol{p}_j)\, \psi^L(\boldsymbol{p}_j) \tag{12.6}$$

12.2 Explicit Form of the DKH Hamiltonians

This section demonstrates how the first three unitary matrices are explicitly constructed and applied to the one-electron operator f (or to some of its parts like $h^D + V_{nuc}$). The first transformation has necessarily to be the free-particle Foldy–Wouthuysen transformation U_0, which is followed by the transformation U_1. The third transformation U_2 turns out to produce even operators that depend on the parametrization chosen for U_2. Afterwards the infinite-order, coefficient-dependence-free scheme is discussed.

12.2.1 First Unitary Transformation

In order to eliminate the odd operator $\mathcal{O}_0$ of the Dirac Hamiltonian as written in Eq. (11.40) order by order in the scalar potential V, an odd operator that depends on V must be generated. As discussed in section 11.5, only the special closed-form free-particle Foldy–Wouthuysen transformation produces an operator $\mathcal{O}_1$ linear in V as indicated by the subscript. This is the mandatory starting point for subsequent transformation steps.

For convenience we recall the result of the closed-form free-particle Foldy–Wouthuysen transformation of Eq. (11.35):

$$f_1 = U_0 \left(h^D + V \right) U_0^\dagger = \mathcal{E}_0 + \mathcal{E}_1 + \mathcal{O}_1 \tag{12.7}$$

with

$$\mathcal{E}_0 = \beta E_p - m_e c^2 \tag{12.8}$$

$$\mathcal{E}_1 = A_p V A_p + A_p R_p V R_p A_p \tag{12.9}$$

$$\mathcal{O}_1 = \beta A_p \left[R_p , V \right] A_p \tag{12.10}$$

The kinematic factors E_p, A_p and R_p have been defined in Eqs. (5.4) and (11.28).

12.2.2 Second Unitary Transformation

After the initial transformation $\mathcal{U}_0$, the new odd part $\mathcal{O}_1$ in f_1 needs to be deleted by the next transformation $\mathcal{U}_1$, and all that we require is that new odd terms are of higher order in the scalar potential V. Now, we do not restrict any of the subsequent unitary transformations to any specific choice of unitary transformation and employ the most general form of Eq. (11.57) without specifying the expansion coefficients $a_{i,k}$ that satisfy the unitarity conditions Eqs. (11.60)–(11.66). Because the first term of each expansion of the unitary matrices according to Eq. (11.57) is the unit operator $\mathbf{1}$, a transformation of any operator yields exactly this operator plus correction terms. Hence, the first free-particle transformation already produced the final zeroth- and first-order even terms. So, we already have all even terms up to 1st order for the DKH Hamiltonian collected,

$$h_{\mathrm{DKH1}} = \mathcal{E}_0 + \mathcal{E}_1 = \sum_{k=0}^{1} \mathcal{E}_k \tag{12.11}$$

For later convenience the odd and antihermitean expansion parameter is denoted by W_i' instead of W_i, and shall be of exactly i-th order in V.

The transformation of f_1 with $\mathcal{U}_1$ yields

$$\begin{aligned} f_2 &= \mathcal{U}_1 f_1 \mathcal{U}_1^\dagger \\ &= \left[a_{1,0}\mathbf{1} + \sum_{k=1}^{\infty} a_{1,k} W_1'^{k} \right] \left(\mathcal{E}_0 + \mathcal{E}_1 + \mathcal{O}_1^{(1)} \right) \left[a_{1,0}\mathbf{1} + \sum_{k=1}^{\infty} (-1)^k a_{1,k} W_1'^{k} \right] \\ &= \mathcal{E}_0 + \mathcal{E}_1 + \mathcal{O}_1^{(2)} + \mathcal{E}_2 + \mathcal{O}_2^{(2)} + \mathcal{E}_3 + \mathcal{O}_3^{(2)} + \sum_{k=4}^{\infty} \left(\mathcal{E}_k^{(2)} + \mathcal{O}_k^{(2)} \right) \end{aligned} \tag{12.12}$$

with

$$\mathcal{O}_1^{(2)} = \mathcal{O}_1^{(1)} + a_{1,0} a_{1,1} \left[W_1', \mathcal{E}_0 \right], \tag{12.13}$$

$$\mathcal{E}_2 = a_{1,0} a_{1,1} \left[W_1', \mathcal{O}_1^{(1)} \right] + \frac{1}{2} a_{1,1}^2 \left[W_1', \left[W_1', \mathcal{E}_0 \right] \right], \tag{12.14}$$

$$\mathcal{E}_3 = \frac{1}{2} a_{1,1}^2 \left[W_1', \left[W_1', \mathcal{E}_1 \right] \right] \tag{12.15}$$

Note that $\mathcal{E}_0$, $\mathcal{E}_1$, and $\mathcal{O}_1^{(1)}$ are independent of W_1' and thus completely determined from the very beginning. Again, the subscript attached to each term of the Hamiltonian denotes its order in the external potential, whereas the superscript in parentheses indicates that such a term belongs to the intermediate, partially transformed Hamiltonian relevant only for the higher-order terms to be finalized by subsequent transformations. Only those even terms which will not be affected by the succeeding unitary transformations U_i, $(i = 2, 3, \dots)$ bear no superscript.

W_1' is chosen to guarantee $\mathcal{O}_1^{(2)} = 0$, and thus the following condition for W_1' is obtained,

$$\left[W_1', \mathcal{E}_0\right] = -\frac{a_{1,0}}{a_{1,1}} \mathcal{O}_1^{(1)}, \tag{12.16}$$

which is satisfied if and only if the kernel of W_1' is given by

$$W_1'(i,j) = \frac{a_{1,0}}{a_{1,1}} \beta \frac{\mathcal{O}_1^{(1)}(i,j)}{E_i + E_j} \tag{12.17}$$

It is clear that W_1' is then an odd operator. As a consequence, this operator alone cannot be represented in any truly two-component theory, but only its products with other odd operators which are then even and hence representable in a two-component theory.

It is a consequence of the so-called $(2n+1)$-rule that $\mathcal{E}_2$ and $\mathcal{E}_3$ are already completely determined after the first unitary DK transformation U_1. Hence, the block-diagonalized Hamiltonian is already defined up to third order in the external potential, although the second-order term $\mathcal{O}_2^{(2)}$ is still present and will be eliminated by the next transformation step. In general, the first $2n+1$ even terms of the *final* block-diagonal Hamiltonian depend only on the n lowest-order matrices $W_1', W_2', \dots, W_n'$. In particular, they are independent of all succeeding unitary transformations. It is exactly this feature of the DKH Hamiltonians which is responsible for the fact that *no error at all* is introduced for the first $2n + 1$ even terms of the block-diagonal Hamiltonian if the power series expansions for U_i are truncated after terms of order n. This remarkable property of the even terms originates from the central idea of the DKH method to choose the latest odd operator W_i' always in such a way that the lowest of the remaining odd terms is eliminated.

For the third transformation U_2 in the next section, the following explicit expression which are known already at this stage are needed:

$$\mathcal{O}_2^{(2)} = a_{1,0} a_{1,1} \left[W_1', \mathcal{E}_1\right] \tag{12.18}$$

$$\mathcal{O}_3^{(2)} = \frac{1}{2}a_{1,1}^2 \left[W_1', \left[W_1', \mathcal{O}_1^{(1)}\right]\right] - \frac{1}{2}a_{1,0}a_{1,1}^3\, W_1'\left[W_1', \mathcal{E}_0\right]W_1' + a_{1,0}a_{1,3}\left[W_1'^3, \mathcal{E}_0\right] \tag{12.19}$$

$$\mathcal{E}_4^{(2)} = a_{1,1}a_{1,3}\left[W_1'^3, \left[W_1', \mathcal{E}_0\right]\right] - \frac{1}{8}a_{1,1}^4\left[W_1'^2, \left[W_1'^2, \mathcal{E}_0\right]\right] + a_{1,0}a_{1,3}\left[W_1'^3, \mathcal{O}_1^{(1)}\right] - \frac{1}{2}a_{1,0}a_{1,1}^3\, W_1'\left[W_1', \mathcal{O}_1^{(1)}\right]W_1' \tag{12.20}$$

$$\mathcal{E}_5^{(2)} = -\frac{1}{8}a_{1,1}^4\left[W_1'^2, \left[W_1'^2, \mathcal{E}_1\right]\right] + a_{1,1}a_{1,3}\left[W_1'^3, \left[W_1', \mathcal{E}_1\right]\right] \tag{12.21}$$

The choice of W_1' in Eq. (12.16) satisfies all constraints, namely that it is an antihermitean operator of first order in V. W_1' depends on the chosen coefficients $a_{1,0}$ and $a_{1,1}$, i.e., it is linear in $a_{1,0}/a_{1,1}$. We therefore introduce the modified operator W_1 defined by

$$W_1(i,j) = a_{1,0}a_{1,1}\, W_1'(i,j) = \beta\,\frac{\mathcal{O}_1^{(1)}(i,j)}{E_i + E_j} \tag{12.22}$$

which is independent of the coefficients $a_{1,k}$ and still odd. With this choice of W_1 and by utilizing Eq. (12.16) the above results may be simplified to a large extent,

$$\mathcal{E}_2 = \frac{1}{2}\left[W_1, \mathcal{O}_1\right] \tag{12.23}$$

$$\mathcal{E}_3 = \frac{1}{2}\left[W_1, \left[W_1, \mathcal{E}_1\right]\right] \tag{12.24}$$

as well as the operators required for the third transformation, explicitly

$$\mathcal{O}_2^{(2)} = \left[W_1, \mathcal{E}_1\right] \tag{12.25}$$

$$\mathcal{O}_3^{(2)} = \frac{1}{2}\left[W_1, \left[W_1, \mathcal{O}_1\right]\right] + \frac{1}{2}W_1\mathcal{O}_1W_1 + \frac{a_{1,3}}{a_{1,1}^3}\left[W_1^3, \mathcal{E}_0\right] \tag{12.26}$$

$$\mathcal{E}_4^{(2)} = \frac{1}{8}\left[W_1, \left[W_1, \left[W_1, \mathcal{O}_1\right]\right]\right] \tag{12.27}$$

$$\mathcal{E}_5^{(2)} = -\frac{1}{8}\left[W_1^2, \left[W_1^2, \mathcal{E}_1\right]\right] + \frac{a_{1,3}}{a_{1,1}^3}\left[W_1^3, \left[W_1, \mathcal{E}_1\right]\right] \tag{12.28}$$

We now have all even terms up to third order for the DKH Hamiltonian collected,

$$h_{\text{DKH3}} = \mathcal{E}_0 + \mathcal{E}_1 + \mathcal{E}_2 + \mathcal{E}_3 = \sum_{k=0}^{3}\mathcal{E}_k \tag{12.29}$$

12.2.3
Third Unitary Transformation

The next unitary transformation U_2 is applied in order to eliminate the odd term of second order of f_2 given by Eq. (12.12),

$$f_3 = U_2 f_2 U_2^\dagger = \left[a_{2,0}\mathbf{1} + \sum_{k=1}^{\infty} a_{2,k} W_2'^k\right] f_2 \left[a_{2,0}\mathbf{1} + \sum_{k=1}^{\infty} (-1)^k a_{2,k} W_2'^k\right]$$

$$= \sum_{k=0}^{5} \mathcal{E}_k + \sum_{k=6}^{\infty} \mathcal{E}_k^{(3)} + \sum_{k=2}^{\infty} \mathcal{O}_k^{(3)} \tag{12.30}$$

with

$$\mathcal{O}_2^{(3)} = \mathcal{O}_2^{(2)} + a_{2,0}a_{2,1}\left[W_2', \mathcal{E}_0\right] \tag{12.31}$$

$$\mathcal{E}_4 = \mathcal{E}_4^{(2)} + a_{2,0}a_{2,1}\left[W_2', \mathcal{O}_2^{(2)}\right] + \frac{1}{2}a_{2,1}^2\left[W_2', \left[W_2', \mathcal{E}_0\right]\right] \tag{12.32}$$

$$\mathcal{E}_5 = \mathcal{E}_5^{(2)} + a_{2,0}a_{2,1}\left[W_2', \mathcal{O}_3^{(2)}\right] + \frac{1}{2}a_{2,1}^2\left[W_2', \left[W_2', \mathcal{E}_1\right]\right] \tag{12.33}$$

Again, W_2 is conveniently and uniquely chosen to eliminate the second-order odd term,

$$\mathcal{O}_2^{(3)} \stackrel{!}{=} 0 \tag{12.34}$$

i.e., it has to satisfy the condition

$$\left[W_2', \mathcal{E}_0\right] = -\frac{a_{2,0}}{a_{2,1}}\mathcal{O}_2^{(2)} = -\frac{a_{2,0}}{a_{2,1}}\left[W_1, \mathcal{E}_1\right] \tag{12.35}$$

After introduction of the modified operator $W_2 = a_{2,0}a_{2,1}W_2'$, this is guaranteed if and only if the kernel of W_2 is given by

$$W_2(i,j,k) = \beta\,\frac{W_1(i,j)\mathcal{E}_1(j,k) - \mathcal{E}_1(i,j)W_1(j,k)}{E_i + E_k} \tag{12.36}$$

Since even and odd operators obey the same multiplication rules as natural numbers, i.e., even times odd is odd, etc., this is obviously an odd and antihermitean operator of second order in the scalar potential, which is independent of the chosen parametrizations of the unitary transformations. W_2 is thus a second-order integral operator in momentum space, whose action on a 4-spinor ψ is defined by

$$W_2\,\psi(\boldsymbol{p}_i) = \int \frac{d^3p_j d^3p_k}{(2\pi\hbar)^6}\, W_2(\boldsymbol{p}_i, \boldsymbol{p}_j, \boldsymbol{p}_k)\,\psi(\boldsymbol{p}_k) \tag{12.37}$$

With this choice of W_2 the final results for the fourth- and fifth-order even terms are given by

$$\mathcal{E}_4 = \frac{1}{8}\left[W_1,\left[W_1,\left[W_1,\mathcal{O}_1^{(1)}\right]\right]\right] + \frac{1}{2}\left[W_2,\left[W_1,\mathcal{E}_1\right]\right] \tag{12.38}$$

$$\begin{aligned}\mathcal{E}_5 = & \frac{1}{2}\left[W_2,\left[W_2,\mathcal{E}_1\right]\right] + \frac{1}{2}\left[W_2,\left[W_1,\left[W_1,\mathcal{O}_1^{(1)}\right]\right]\right] + \frac{1}{2}\left[W_2,W_1\mathcal{O}_1^{(1)}W_1\right] \\ & - \frac{1}{8}\left[W_1^2,\left[W_1^2,\mathcal{E}_1\right]\right] + \frac{a_{1,3}}{a_{1,1}^3}\left[\left[W_2,W_1^3\right],\mathcal{E}_0\right]\end{aligned} \tag{12.39}$$

where we have extensively taken advantage of Eq. (12.35) in order to simplify the expressions.

We have now all even terms up to third order for the DKH Hamiltonian collected,

$$h_{\text{DKH5}} = \mathcal{E}_0 + \mathcal{E}_1 + \mathcal{E}_2 + \mathcal{E}_3 + \mathcal{E}_4 + \mathcal{E}_5 = \sum_{k=0}^{5} \mathcal{E}_k \tag{12.40}$$

All terms contributing to the fourth-order Hamiltonian h_{DKH4} are independent of the coefficients $a_{1,k}$, i.e., they are invariant under an arbitrary change of the parametrization of U_1. Terms contributing to the fifth- and higher-order DKH corrections, however, depend on the coefficients of the parametrization of U_1.

An investigation of the eigenvalues of the DKH1 to DKH5 Hamiltonians derived so far demonstrates the efficiency of the DKH expansion and shows an interesting convergence feature, which is the oscillatory convergence. All even-order energies are above the Dirac reference energy, while all odd-order energies are below the Dirac reference value but still variationally stable. This is all illustrated in Fig. 12.1, which plots the lowest electronic bound state, the $1s_{1/2}$ state, for all hydrogen-like atoms up to $Z = 130$.

12.3 Infinite-Order DKH Hamiltonians and the Arbitrary-Order DKH Method

All higher transformations produce coefficient-dependent higher-order even terms and hence a coefficient-dependent spectrum. We will later see that the dependence is negligible. Only at infinite order does the coefficient dependence of the spectrum of the transformed one-electron operator vanish exactly because of (exact) unitarity.

The higher transformations can hardly be carried out manually and require an automatic, symbolic derivation. For practical purposes it would be desirable to calculate any high-order DKH Hamiltonian which approximates the

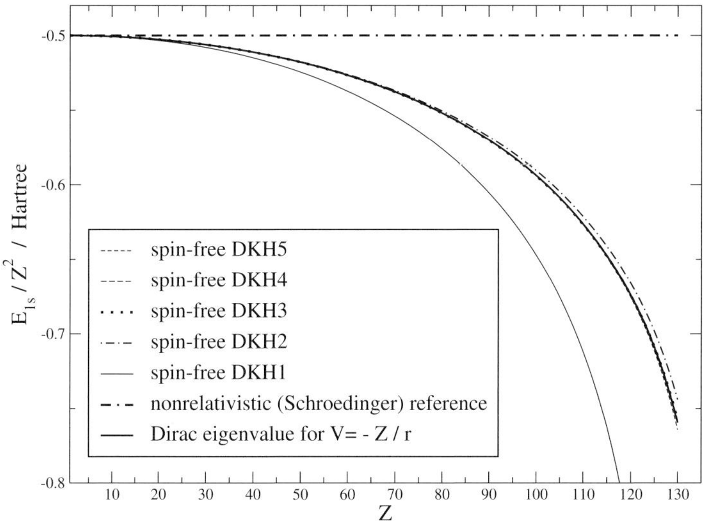

Figure 12.1 Hydrogen-like atoms in the limiting Schrödinger and Dirac pictures. Ground state eigenvalues of the scalar-relativistic DKH Hamiltonians up to fifth order are also plotted. The fourth- and fifth-order DKH ground states can hardly be distinguished from the analytic Dirac result on this scale of presentation.

exact block-diagonal Hamiltonian sufficiently closely. This is, however, not directly possible because of the nested character of the sequence of unitary transformations. Because of the linear dependence of W_1 on V, an n-th order decoupled DKH Hamiltonian $h_{\text{DKH}n}$ requires an expansion of the inner U_1 transformation up to the n-th power of W_1. That is, the higher the final order of the Hamiltonian shall be the further must the inner transformations have to be expanded.

Consequently, a high- or even infinite-order DKH Hamiltonian requires a predefinition of the order in V which determines the Taylor expansion length of all individual unitary matrices in the sequence. Before we come back to this issue, the convergence of the DKH series needs to be discussed.

12.3.1 Convergence of DKH Energies and Variational Stability

It has already been stated in chapter 11 that an expansion in the formal parameter $1/c$ is invalid because of the linear momentum associated with this formal expansion parameter. The only other remaining parameter V is the basis for the DKH expansion, and we shall now investigate how convergence can be guaranteed in terms of this parameter. The components of a DKH Hamiltonian sum *all* terms up to a given order in V, irrespective of their order in $1/c$.

Thus, even the third-order Hamiltonian h_{DKH3} contains *all* sixth-order terms plus infinitely many additional terms. This situation may be summarized by the statement that the DKH Hamiltonians contain 'implicit partial summations up to infinite order in $1/c$'.

Regarding the formal convergence behavior of the DKH series, one might argue that it might not converge at all because of increasingly higher powers of the singular Coulomb potential, which reads in the case of a single atom under consideration $-Ze^2/r$, contained in V in the case of point-like atomic nuclei. For this reason the DKH scheme, normally referred to as an expansion in the external potential V, is sometimes denoted as an expansion in the coupling strength Ze^2. However, to replace the rather general expansion parameter V by Z is an unnecessary restriction on the special class of scalar Coulomb-type electron–nucleus potentials and holds only for a single atom, i.e., it is also not suitable for the discussion of convergence in the molecular case.

Inspection of the explicit expressions for the different DKH orders reveals the reason for a rapid convergence behavior of the DKH series. The true expansion parameter is the energy-damped potential $\tilde{V}$ defined by Eq. (12.58), which is at least suppressed by a factor of $1/(2mc^2)$ as compared to the bare Coulomb potential V. Each order $\mathcal{E}_k$ ($k \geq 1$) of the DKH Hamiltonian contains exactly $(k-1)$ factors of $\tilde{V}$. Since also the kinematic factor $\mathcal{R}_p$ of Eq. (11.28) is of leading order $1/c$ — i.e., its lowest-order term occurring in a (virtual) series expansion in this parameter would be of order $\mathcal{O}(1/c)$ — each term $\mathcal{E}_k$ is of leading order Z^{k-1}/c^{2k} (for $k \geq 2$). Therefore, each higher-order term $\mathcal{E}_k$ of the Hamiltonian is formally suppressed by Z/c^2 as compared to the previous approximation $h_{\mathrm{DKH}(k-1)}$, which is the basis for the rapid convergence properties of the DKH protocol. Furthermore, this analysis demonstrates that even the Coulomb singularity near a point-like nucleus does not cause any problems within the DKH scheme. It is strongly damped in the higher-order DKH terms rather than giving rise to non-integrable and singular $1/r^k$ expressions. These peculiarities can only be resolved by a momentum-space analysis, as it is presented here, since the kinematic terms A_p and $\mathcal{R}_p$ contain square roots of momentum operators, which would lead to nonlocal terms with arbitrarily high derivative operators in a position-space formulation as already discussed in chapter 5 in the context of the Klein–Gordon equation.

The variational stability of the DKH Hamiltonians — a boundary condition for the convergence of the DKH series — has been proven for DKH2 [535] and results for the higher-order expansion have been obtained also [536]. Despite these analytical findings, which may be regarded as a first step toward an even deeper understanding of DKH theory, first numerical results obtained for the eigenvalues of higher-order DKH Hamiltonians created doubts about the method. Nakajima and Hirao [530] obtained a smooth convergence even for the third-order DKH energies. A closer inspection up to fifth order [511] and later up to 14th order [532] revealed an oscillating but convergent behavior, which is, of course, paralleled by the magnitude and sign of the truncation operator defined in Ref. [510]; the odd orders DKH1, DKH3, DKH5, DKH7, ... always yield energies below the four-component reference energy. The first result by Nakajima and Hirao turned out to be an artifact of a too small basis set. Later, additional doubts about the DKH methods arose from results obtained to approximate the degenerate Dirac energies $s_{1/2}$ and $p_{1/2}$, which turned out to be nondegenerate at low-order DKH theory [517, 518, 537]. However, all these discrepancies vanish at higher order — provided the basis set is properly chosen. The different accuracy of the low-order $s_{1/2}$ and $p_{1/2}$ energies has been rationalized by van Wüllen [537].

12.3.2 Infinite-Order Protocol

In an implementation of the DKH protocol for actual calculations, 'exact' decoupling is reached if the truncation error $\mathcal{O}\left(\tilde{V}^{n+1}\right)$ is of the order of the arithmetic precision of the computer, or at least negligible either with respect to the physical or chemical issues under investigation or with respect to additional methodological approximations like the finite size of a one-particle basis set. Given a molecular system and a desired numerical accuracy τ, we may determine a cut-off order n_{opt} of the DKH expansion prior to the calculation, beyond which no further numerical improvement (relevant for τ) will be achieved.

The necessary order n_{opt} for exact decoupling — depending on the system under investigation — can be determined prior to any calculation [532]: each term contributing to the even term $\mathcal{E}_k$ of the DKH Hamiltonian features exactly $k-1$ huge energy denominators due to the damped external potential

$$\tilde{V}_{ij} = \frac{V_{ij}}{E_i + E_j} \tag{12.41}$$

with $E_{p_i} = \sqrt{p_i^2 c^2 + m^2 c^4}$ being the relativistic energy–momentum relation. In order to keep the notation as simple as possible, we will continue to denote $\mathcal{E}_k$ as being of $(k-1)$-th order in $\tilde{V}$, though obviously it cannot always be decomposed into $k-1$ separate factors of $\tilde{V}$, but might also contain terms with nested energy denominators like W_2, for example.

Because of the structure of the free-particle Foldy–Wouthuysen transformed Hamiltonian

$$f_1 = \beta E_p - mc^2 + A_p(V + R_p V R_p)A_p + \beta A_p\left[R_p, V\right]A_p \tag{12.42}$$

each even term $\mathcal{E}_k$ comprises at least a factor of $A_p \mathcal{R}_p V \mathcal{R}_p A_p \cdot (A_p \tilde{V} A_p)^{k-1}$. Within a semi-classical analysis this factor assumes its maximum value for vanishing momentum p, and is thus always guaranteed to have a smaller operator norm than the truncation estimate operator

$$\mathcal{V}_k = (4m^2c^2)^{-1} \cdot A_p V A_p \cdot \left(A_p \tilde{V} A_p\right)^{k-1} \tag{12.43}$$

The order of magnitude of the term $\mathcal{E}_k$ can thus be estimated by investigating the eigenvalues of this truncation estimate operator $\mathcal{V}_k$. They will, of course, depend on the system under investigation, i.e., the largest nuclear charge Z of the nuclei occurring in the molecule and the basis set chosen. As soon as the absolute value of the largest eigenvalue of $\mathcal{V}_k$ is below the chosen truncation threshold τ, $|\alpha^{(k)}_{\max}| < \tau$, the higher-order even terms $\mathcal{E}_l$ with $l > k$ will only affect the total energy of the system under investigation beyond the desired accuracy established by τ.

Note that the discussion of series truncation focused on the one-electron *Hamiltonian* expanded in terms of the scalar potential. A discussion of the truncation of the *unitary matrices* has been given by Kedziera [538].

Although the mathematical complexity of the even terms $\mathcal{E}_k$ is not significantly increased by transition to higher orders, the number of contributing terms to each order is rapidly increasing. The zeroth- and first-order terms $\mathcal{E}_0$ and $\mathcal{E}_1$ consist of only one and two terms, respectively, whereas already the second-order term comprises eight basic terms, as is explicitly shown by Eq. (12.56). This number increases to 42, 160, and 432 terms for the third-, fourth-, and fifth-order Hamiltonian. It would, thus, be extremely tedious and challenging to derive and evaluate sixth- and higher-order corrections manually on a sheet of paper.

Let us assume the intermediate Hamiltonian of the DKH procedure after n transformation steps were given. From the $(2n+1)$-rule [511, 530] it may be written as

$$f_n = \sum_{k=0}^{2n-1} \mathcal{E}_k + \sum_{k=2n}^{\infty} \mathcal{E}_k^{(n)} + \sum_{k=n}^{\infty} \mathcal{O}_k^{(n)} \tag{12.44}$$

The following unitary transformation U_n is established by the odd operator W_n, which is of n-th order in V, via a general power series ansatz of the type given by Eq. (11.57) and yields

$$\begin{aligned} f_{n+1} &= U_n h_n U_n^\dagger \\ &= \sum_{k=0}^{2n+1} \mathcal{E}_k + \sum_{k=2n+2}^{\infty} \mathcal{E}_k^{(n+1)} + \underbrace{\mathcal{O}_n^{(n)} + a_{n,0}a_{n,1}\left[W_n, \mathcal{E}_0\right]}_{\mathcal{O}_n^{(n+1)}} + \sum_{k=n+1}^{\infty} \mathcal{O}_k^{(n+1)} \end{aligned} \tag{12.45}$$

W_n is determined uniquely by the requirement that it has to account for the elimination of the term $\mathcal{O}_n^{(n+1)}$. Being an n-th order integral operator, this is guaranteed if the kernel of W_n is given by

$$W_n(i,j,\ldots,n) = \frac{a_{n,0}}{a_{n,1}} \beta \frac{\mathcal{O}_n^{(n)}(i,j,\ldots,n)}{E_i + E_n} \tag{12.46}$$

Because of the nested unitary transformations, the DKH$n_{\rm opt}$ Hamiltonian can be derived once the optimum order for sufficiently accurate decoupling has been pre-determined. Since the derivation is automatic in a sense that the steps to be carried out are always the same, it can be automatized in a computer program [532].

12.3.3 Coefficient Dependence

The dependence of the DKH Hamiltonians of fifth and higher order on the expansion coefficients $a_{i,k}$ in the parametrization of the unitary transformation in Eq. (11.57) clearly shows that the DKH expansion in Eq. (12.2) cannot be unique, in the sense that it is possible to have different expansions of the block-diagonal Hamiltonian in terms of the external potential. At first sight, this seems to be odd because one would expect to obtain a unique block-diagonal and unique Hamiltonian. At infinite order, all infinitely many different expansions with respect to the external potential V formally written in Eq. (11.57) as a single unique one necessarily yield the same diagonal Hamiltonian, i.e., the same spectrum. The individual block-diagonal operators may, however, still be different in accord with what has been said in section 4.3.3.

Obviously, the different expansions are related to one another. Let us assume two different sequences of unitary transformations,

$$U = \cdots U_2 U_1 U_0 \tag{12.47}$$

and

$$U' = \cdots U_2' U_1' U_0' \tag{12.48}$$

which differ in the choice of expansion coefficients $a_{m,j}$ in Eq. (11.57). Then, we can formally write the relation between both resulting expansions of f_{bd},

$$f_{bd}^{(U)} = \cdots U_2 U_1 U_0 \, f \, U_0^\dagger U_1^\dagger U_2^\dagger \cdots = U \, f \, U^\dagger \tag{12.49}$$

and

$$f_{bd}^{(U')} = \cdots U_2' U_1' U_0' \, f \, U_0'^\dagger U_1'^\dagger U_2'^\dagger \cdots = U' \, f \, U'^\dagger \tag{12.50}$$

as a linear mapping S

$$S = \cdots U_2' U_1' U_0' U_0^\dagger U_1^\dagger U_2^\dagger \cdots = U' U^\dagger \tag{12.51}$$

which transforms the expansion $f_{bd}^{(U)} = \sum_i \mathcal{E}_i^{(U)}$ to the one $f_{bd}^{(U')} = \sum_i \mathcal{E}_i^{(U')}$,

$$f_{bd}^{(U)} = U\,U'^{\dagger}\, f_{bd}^{(U')}\, U'\,U^{\dagger} = S^{\dagger}\, f'_{bd}\, S \tag{12.52}$$

However, this formal trick to describe the relation between both expansions does not explain explicitly how the difference between the Hamiltonians (12.49) and (12.50) at a given power $n > 4$ of the expansion parameter V is compensated at some higher power $m > n$.

The dependence of the even terms $\mathcal{E}_{k\geq 5}$ on the coefficients $a_{i,k}$ does not pose a problem to the infinite-order DKH method without any truncation of the Hamiltonian. For finite-order approximations, however, application of the optimum unitary transformations U_i^{opt} has been suggested in section 11.4.3. Still, a coefficient dependence remains, but it is, for all meaningful parametrizations, orders of magnitude smaller than the effect of the DKH order. Also, the effect on the eigenvalues is diminished with increasing DKH order as it should be in order to converge to the same infinite-order result [539].

From a purist point of view, one should use DKH Hamiltonians either only up to fourth order or up to infinite order (within machine precision) in order to avoid ambiguities in the method. However, a discussion of the coefficient dependence on results obtained with fifth or higher order DKH Hamiltonians is not relevant as the numerical effect is tiny (see Fig. 12.2 for an example) and always smaller than any other approximation made (like the finite size of the one-particle basis set). To have a well-defined model Hamiltonian at hand one may always restrict a calculation to the highest coefficient-free DKH Hamiltonian, which is the fourth-order DKH4 Hamiltonian.

Interestingly, for a long time it was believed that the original square-root parametrization of Douglas and Kroll is particularly useful since the mandatory Taylor expansion of the square root according to Eq. (11.57) contains only even terms (all odd terms possess zero coefficients). Nakajima and Hirao [530] decided to use a different parametrization, namely the exponential parametrization already employed by Foldy and Wouthuysen, compare also section 11.4.1. And indeed, the square-root parametrization leads to the smallest number of terms in the DKH Hamiltonian [532], but the exponential parametrization can be expected to converge faster. However, there is hardly any gain in efficiency for any of these parametrizations for high-order DKH Hamiltonians, and up to fourth order they all yield the *same* even terms anyway. The only formal choice for an optimum parametrization is thus one which requires the expansion coefficients $a_{m,j}$ to fulfill the unitarity condition as closely as possible as given for $a_{m,3}$ by Eq. (11.70).

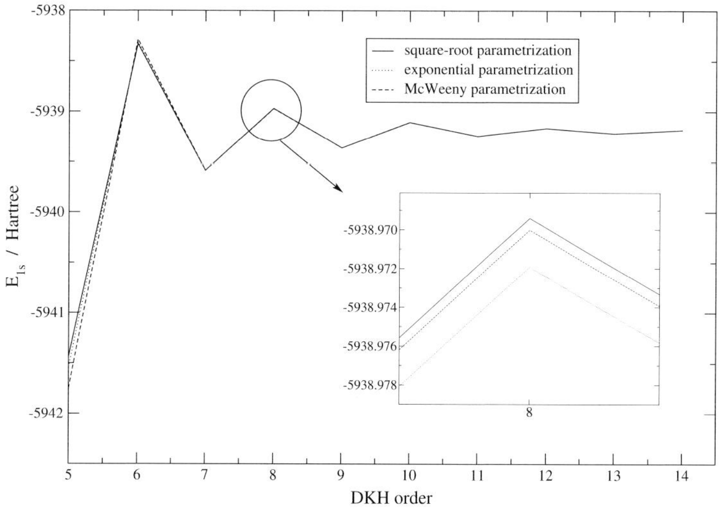

Figure 12.2 Dependence of higher than 4th-order DKH ground state energy of a one-electron atom with Z=100 on the parametrization of the unitary matrices (data adopted from Ref. [539], c=137.0359895 a.u.). To demonstrate the small effect, three closed-form parametrizations presented in chapter 11 have been chosen. Note the oscillatory convergence behavior of the DKH eigenvalue.

12.3.4
Explicit Expressions of the Positive-Energy Hamiltonians

Up to this point all Hamiltonians that have been derived still belong to the four-component world and thus possess a 2×2 super-structure. Moreover, they have been presented in a compact commutator notation, which still involves the antihermitean operators W_i. These operators cannot be calculated in a two-component framework because they would require the eliminated two-component contributions. Hence, the whole DKH idea — apart from its convergent series — only works in a two-component framework if it is possible to calculate the Hamiltonians without requiring any operators with LS, SL and SS blocks — only LL blocks are known in a truly two-component theory.

Since we are only interested in the positive-energy, i.e., electronic spectrum, the $\mathcal{E}_{k+}$ components of $\mathcal{E}_k$ need to be derived. They are obtained from the commutator expressions given above by commuting all Dirac β matrices with all other operators until they all vanish because of $\beta^2 = 1$ or until only one β survives, which is then simply neglected. The corresponding negative energy operators $\mathcal{E}_{k-}$ are derived in the same way, but the finally surviving β operators are replaced by a negative 2×2 unit matrix $-\mathbf{1}$.

For the sake of brevity, we stick to the notation $h_{\mathrm{DKH}n}$ also for the (2×2)-block diagonal operators (up to this point, these operators denoted sums of block-diagonal (4×4) even operators). Up to fourth-order (n=4), the (2×2)-dimensional DKH Hamiltonians

$$h_{\text{DKH}n} = \sum_{k=0}^{n} \mathcal{E}_{k+} \tag{12.53}$$

are independent of the chosen coefficients for the parametrizations of the unitary transformations U_i. The fifth- and all higher-order terms depend on the coefficients $a_{i,k}$, which is actually not a drawback of the DKH approach. The spectrum and the eigenfunctions of the exact, i.e., untruncated, and block-diagonal Hamiltonian $h_{\text{DKH}\infty}$ do *not* depend on the choice of the coefficients $a_{i,k}$, since all transformations applied have been unitary, i.e., invertible. The spectrum of an operator A is invariant under arbitrary similarity transformations, $\tilde{A} = SAS^{-1}$, where S is an invertible (i.e., nonsingular) operator, as it is evident by inspection of the eigenvalue equation $Av = \lambda v$. Hence unitary transformations do not change the spectrum. We have thus found an infinite family of completely equivalent Hamiltonians, which do all describe the Dirac electron perfectly and which are related via unitary transformations in Hilbert space.

The electronic, i.e., upper-left (2 × 2)-blocks of the kernels of the lowest-order terms may be explicitly given as

$$\mathcal{E}_{0+}(i) = E_i - m_e c^2 = \sqrt{\boldsymbol{p}^2 c^2 + m_e^2 c^4} - m_e c^2 \tag{12.54}$$

$$\mathcal{E}_{1+}(i,j) = A_i V_{ij} A_j + A_i \boldsymbol{\sigma} \cdot \boldsymbol{P}_i \cdot V_{ij} \boldsymbol{\sigma} \cdot \boldsymbol{P}_j A_j \tag{12.55}$$

$$\begin{aligned}
\mathcal{E}_{2+}(i,j,k) = \frac{1}{2}\Big\{ & - A_i \boldsymbol{\sigma}\cdot\boldsymbol{P}_i \tilde{V}_{ij} \boldsymbol{\sigma}\cdot\boldsymbol{P}_j A_j A_j V_{jk} A_k + A_i \boldsymbol{\sigma}\cdot\boldsymbol{P}_i \tilde{V}_{ij} A_j A_j V_{jk} \boldsymbol{\sigma}\cdot\boldsymbol{P}_k A_k \\
& + A_i \tilde{V}_{ij} A_j \boldsymbol{P}_j^2 A_j V_{jk} A_k - A_i \tilde{V}_{ij} A_j A_j \boldsymbol{\sigma}\cdot\boldsymbol{P}_j V_{jk} \boldsymbol{\sigma}\cdot\boldsymbol{P}_k A_k \\
& - A_i \boldsymbol{\sigma}\cdot\boldsymbol{P}_i V_{ij} \boldsymbol{\sigma}\cdot\boldsymbol{P}_j A_j A_j \tilde{V}_{jk} A_k + A_i \boldsymbol{\sigma}\cdot\boldsymbol{P}_i V_{ij} A_j A_j \tilde{V}_{jk} \boldsymbol{\sigma}\cdot\boldsymbol{P}_k A_k \\
& + A_i V_{ij} A_j \boldsymbol{P}_j^2 A_j \tilde{V}_{jk} A_k - A_i V_{ij} A_j A_j \boldsymbol{\sigma}\cdot\boldsymbol{P}_j \tilde{V}_{jk} \boldsymbol{\sigma}\cdot\boldsymbol{P}_k A_k \Big\}
\end{aligned} \tag{12.56}$$

$$\mathcal{E}_{3+}(i,j,k,l) = \frac{1}{2}\Big\{ A_i \boldsymbol{\sigma}\cdot\boldsymbol{P}_i \tilde{V}_{ij} \boldsymbol{\sigma}\cdot\boldsymbol{P}_j A_j A_j \tilde{V}_{jk} A_k \mathcal{E}_{1+}(k,l) + \cdots \tag{12.57}$$

with the abbreviations

$$\tilde{V}_{ij} = \frac{V_{ij}}{E_i + E_j} = \frac{V(\boldsymbol{p}_i, \boldsymbol{p}_j)}{\sqrt{\boldsymbol{p}_i^2 c^2 + m_e^2 c^4} + \sqrt{\boldsymbol{p}_j^2 c^2 + m_e^2 c^4}} \tag{12.58}$$

(compare also Eq. (11.28) in section 11.3). Furthermore, Dirac's relation given in Eq. (D.11) for Pauli spin matrices has frequently been used, which simplifies to

$$\sigma \cdot P_j \, \sigma \cdot P_j = P_j^2 \tag{12.59}$$

if no potential is involved. The expressions for the higher-order kernels are hardly more complicated but very lengthy and thus ideally suited for a automatic computer-driven derivation as described in Ref. [532].

12.3.5 Additional Peculiarities of DKH Theory

Up to this point, we have mostly discussed the analytical features of the DKH transformation scheme that allow us to expand the Dirac Hamiltonian in a well-defined, convergent series. However, it will also become clear that the DKH transformation can be turned into a viable protocol for actual calculations on atomic and molecular systems (and, of course, also on solids [451, 453, 540]); see next section. In this section, some useful facts that are worth mentioning about the DKH transformation are collected.

First of all, the DKH approach to atomic and molecular calculations spawned many different variants. Apart from the choice of the truncation order, two- and one-component variants are available. Basically, they emerge when proceeding from four-component DKH operators for negative- and positive-energy states to two-component DKH operators for the positive-energy states only to, finally, one-component DKH operators that neglect spin–orbit coupling. Accordingly, the four-dimensional operator product $(\boldsymbol{\alpha} \cdot \boldsymbol{p})$ becomes the two-component product $(\boldsymbol{\sigma} \cdot \boldsymbol{p})$ and finally through Dirac's relation, the one-component operator $\boldsymbol{p}$ in the corresponding equations.

Basically, DKH theory starts from the standard representation of the Dirac Hamiltonian introduced in section 5.2. The fact that the Dirac matrices $\boldsymbol{\alpha}$ and β are not uniquely defined suggests that a different representation of the Dirac operator might be more advantageous. In order to shed light on this question, we consider the Weyl representation of Eq. (5.38), which produces a Dirac Hamiltonian of a very different structure, namely the following

$$\begin{pmatrix} c\boldsymbol{\sigma} \cdot \boldsymbol{p} & m_e c^2 + V \\ m_e c^2 + V & -c\boldsymbol{\sigma} \cdot \boldsymbol{p} \end{pmatrix}_{\text{Weyl}} = \tilde{\mathcal{E}}_0 + \tilde{\mathcal{O}}_0 + \tilde{\mathcal{O}}_1 \tag{12.60}$$

where the tildes have been introduced to distinguish these operators from the ones of Eq. (11.75). The scalar potential enters on the off-diagonal according to the minimal coupling scheme. Hence, *two* odd operators of different order in the external potential are already present from the very beginning. Instead of the free-particle Foldy–Wouthuysen transformation applied in Eq. (12.7) to the Dirac Hamiltonian in standard representation, we would require a transformation that eliminates $\tilde{\mathcal{O}}_0$ in a first step. It is, however, clear from the structure of the Dirac–Weyl Hamiltonian in Eq. (12.60) that more simplified expressions compared to standard DKH theory are not at hand.

12.3.5.1 **Two-Component Electron Density Distribution**

For a one-electron system, the electron density is obtained as the square of the orbital in four-component as well as in nonrelativistic theory. One might be led to believe that this also holds for transformed DKH orbitals. This, however, is not the case, and the situation is much more involved than one might at first think. For the detailed discussion of this problem we need to refer to section 15.3.5 later.

The electron density ρ calculated from four-component methods does not possess any nodes because large and small components feature nodes at different positions; see section 6.8 for details. If, however, only one product, i.e., $\psi_L^\dagger \psi_L$ or $\psi_S^\dagger \psi_S$, survives (because the other is zero after DKH transformation), the density might then be zero if the product is zero for a given $\boldsymbol{r}$. This is the case in (one-component) nonrelativistic Schrödinger quantum mechanics. By contrast to the Dirac hydrogen atom, the radial electron density distribution of a *nonrelativistic* electron in a hydrogen-like atom is zero if the radial function $R_{nl}(r)$ is zero (there exist $(n - l - 1)$ radial nodes with n being the principal quantum number).

One may now ask [541] whether the two-component density $\rho_{bd} = \phi_L^\dagger \phi_L$ reproduces the mutual annihilation of nodes described for the four-component density in section 6.8 (see also Figure 6.3). At first sight, one is tempted to claim that this is not possible as there is no small component whose squared value may add a finite number to compensate for a vanishing large component as in the radial density given in Eq. (6.148). However, the two-component transformed orbitals in the product $\phi_L^\dagger \phi_L$ are in general complex-valued so that $\phi_L^\dagger \phi_L = |\phi_L|^2 = \phi_{L,\text{Re}}^2 + \phi_{L,\text{Im}}^2$, which is also non-zero if real and imaginary parts, $\phi_{L,\text{Re}}$ and $\phi_{L,\text{Im}}$, possess their nodes at different values of the variable $\boldsymbol{r}$. If ϕ_L were a real-valued function, we would have to have two different radial parts for the two components of the 2-spinor ϕ_L in order to obtain a nodeless radial density distribution as in the four-component case. In a one-component approximation, which neglects all spin-dependent terms and which thus does not resemble the four-component density, additional nodes may occur as in the Schrödinger case.

12.3.5.2 **Transformation in the Presence of Off-Diagonal Potential Operators**

So far, we have solely considered the DKH transformation of the Dirac Hamiltonian in the presence of an external block-diagonal, i.e., even potential. This restriction was introduced and discussed in section 11.1.1. Although this is certainly the most important case that allows us to describe one-electron systems exactly and atoms, molecules and solids with excellent approximation by neglecting all off-diagonal potential energy terms, this is not satisfactory. External magnetic fields, the Breit interaction and certain exchange contributions — all being odd operators — cannot be disregarded in view of their paramount importance for spectroscopy. Because of the relevance of these terms to molecular property calculations (e.g., all resonance spectroscopic techniques employ external magnetic fields), we address this issue later in a whole chapter devoted to this issue: these odd potential energy operators can be considered via perturbation theory or variationally and both options are considered in chapter 15. Of course, in the case of strong magnetic fields these cannot be treated via perturbation theory but must be considered in a variational approach.

12.3.5.3 **Nonrelativistic Limit of Approximate Two-Component Hamiltonians**

Every well-defined two-component method like the DKH transformation must have a well-defined nonrelativistic limit. In fact, in the DKH case the nonrelativistic limit is solely determined by the even terms of the free-particle Foldy–Wouthuysen transformation given in Eq. (11.37),

$$\lim_{c\to\infty} \mathcal{E}_0 = \lim_{c\to\infty} \left[\beta E_p - m_e c^2\right] = \frac{p^2}{2m_e} \tag{12.61}$$

$$\lim_{c\to\infty} \mathcal{E}_1 = \lim_{c\to\infty} \left[A_p V A_p + A_p R_p V R_p\right] = V \tag{12.62}$$

while all higher-order terms contain more and more R_p factors of order $O(1/c)$, which vanish in the limit $c \to \infty$.

12.3.5.4 **Rigorous Analytic Results**

The DKH transformation has been subjected to rigorous mathematical analysis. In particular, Siedentop and collaborators presented many analytical results [542, 543]. The variational stability of the DKH2 Hamiltonian was analyzed [535] as well as the convergence of the DKH series [536].

12.4 Many-Electron DKH Hamiltonians

The discussion so far has focused on the block-diagonalization of the one-electron Dirac-like Fock operator f of Eq. (11.1). One may argue that the transformed Fock operator finally provides transformed molecular spinors to enter Slater determinants. Then, the total electronic wave function is constructed from two-component spinors only. Nonetheless, we need to investigate how the total unitary transformation U^{tot} valid for N molecular spinors,

$$E_{el} = \langle \Psi | H_{el} | \Psi \rangle = \langle U^{tot} \Psi | U^{tot} H_{el} U^{tot\dagger} | U^{tot} \Psi \rangle \tag{12.63}$$

relates to the one-electron transformations in order to show that the V-dependence of the Fock operator does not impose a restriction for the U^{tot} that block-diagonalizes the total many-electron Hamiltonian H_{el}.

12.4.1 DKH Transformation of One-Electron Terms

To consider the effect of the DKH transformation on the many-electron Hamiltonian of Eq. (12.63), the Hilbert space formulation of many-particle quantum mechanics presented in section 8.4 needs to be recalled.

The innermost or initial transformation of the sequence of Eqs. (11.15) and (12.1), i.e., the free-particle Foldy–Wouthuysen transformation, reads in the N-electron case

$$\begin{aligned} U_0^{tot} &= U_0(1) \otimes U_0(2) \otimes \cdots \otimes U_0(n) = \bigotimes_{i=1}^{N} U_0(i) \\ &= \bigotimes_{i=1}^{N} A_p(i) \left[\mathbf{1}(i) + \beta(i) R_p(i) \right] = \bigotimes_{i=1}^{N} A_i \left(\mathbf{1}_i + \beta_i R_i \right) \end{aligned} \tag{12.64}$$

with the same A_p and R_p factors already introduced in Eq. (11.28). According to the general multiplication rule of the tensor product,

$$[A(i) \otimes B(j)] \cdot [C(i) \otimes D(j)] = A(i)C(i) \otimes B(j)D(j) \tag{12.65}$$

the action of U_0^{tot} on the sum of one-electron terms of Eq. (8.66) yields

$$\begin{aligned} H_1^{1e} &= U_0^{tot} \left(\sum_{i=1}^{N} \left[h^D(i) + V_{nuc}(i) \right] \right) U_0^{tot\dagger} \\ &= \sum_{i=1}^{N} U_0(i) \left[h^D(i) + V_{nuc}(i) \right] U_0^{\dagger}(i) \\ &= \sum_{i=1}^{N} \left(\mathcal{E}_0(i) + \mathcal{E}_1(i) + \mathcal{O}_1^{(1)}(i) \right) \end{aligned} \tag{12.66}$$

where we write the scalar electron–nucleus potential $V_{nuc}(i)$ explicitly separated from h^D as before.

Similarly, the subsequent DKH transformations

$$U_m^{tot} = U_m(1) \otimes U_m(2) \otimes \cdots \otimes U_m(N) = \bigotimes_{i=1}^{N} U_m(i) \tag{12.67}$$

to the intermediate Hamiltonian H_1^{1e}, where each (4×4)-block $U_i(j) = f(W_i(j))$ is given as the most general power series parametrization of Eq. (11.57). We arrive at the nth-order DKH approximation for the transformation of the one-electron terms,

$$H_{\text{DKH}n}^{1e} = \sum_{k=1}^{n} \sum_{i=1}^{N} \mathcal{E}_k(i) \tag{12.68}$$

where the action of each term $\mathcal{E}_k(i)$ for electron i is described by the same one-particle expression $\mathcal{E}_k$ discussed in the last sections.

12.4.2
DKH Transformation of Two-Electron Terms

The DKH transformation of the two-electron terms requires more effort, which is the reason why this is neglected in standard DKH calculations that sacrifice this transformation for the sake of efficiency. For this transformation it is vital to appreciate the double-even structure of the Coulomb operator and the double-odd structure of the Breit operator as highlighted in Eq. (8.89). Then, following the protocol for the transformation of the one-electron operators in the preceding section, the innermost unitary transformation of the two-electron part of the Hamiltonian of Eq. (8.66) reads

$$\begin{aligned} H_1^{2e} &= U_0^{\text{tot}} \sum_{i<j}^{N} g(i,j)\, U_0^{\text{tot}\dagger} \\ &= \sum_{i<j}^{N} \left[U_0(i) \otimes U_0(j)\right] g(i,j) \left[U_0^\dagger(i) \otimes U_0^\dagger(j)\right] \\ &= \sum_{i<j}^{N} A_i A_j \Big\{ \left(1+\beta R_i\right) \left(1+\beta R_j\right) g(i,j) \left(1-\beta R_i\right) \left(1-\beta R_j\right) \Big\} A_i A_j \end{aligned} \tag{12.69}$$

This expression can formally be evaluated without assuming any further simplification for the electron–electron interaction potential energy operator $g(i,j)$, i.e., $g(i,j)$ is still defined by Eq. (8.69) as the sum of the Coulomb and frequency-independent Breit interactions. As in the case of the one-electron DKH transformation, $g(i,j)$ acts as an integral operator in momentum space and hence does not commute with any expression containing momentum operators like A_i or R_i, for example. With the abbreviation $g_{ij} = g(i,j)$ the total two-electron interaction operator reads

$$H_1^{2e} = \sum_{i<j}^{N} A_i A_j \Big\{ g_{ij} - \beta R_i\, g_{ij}\, \beta R_i - \beta R_j\, g_{ij}\, \beta R_j + \beta R_i \beta R_j\, g_{ij}\, \beta R_i \beta R_j$$

$$+ \beta R_i \beta R_j g_{ij} - \beta R_i g_{ij} \beta R_j - \beta R_j g_{ij} \beta R_i + g_{ij} \beta R_i \beta R_j$$
$$- [g_{ij}, \beta R_i] - [g_{ij}, \beta R_j] + \beta R_i g_{ij} \beta R_i \beta R_j + \beta R_j g_{ij} \beta R_i \beta R_j$$
$$- \beta R_i \beta R_j g_{ij} \beta R_j - \beta R_i \beta R_j g_{ij} \beta R_i \Big\} A_i A_j \tag{12.70}$$

where the special ordering of the sixteen terms of Eq. (12.70) will become clear in the following. It is important not to forget the tensor structure of each term. All quantities bearing the subscript i refer to electron i living in a completely different Hilbert space than that of electron j. This has importance consequences for the assignment of odd and even to the individual operators. For example, the first term in the second line of Eq. (12.70), $\beta R_i \beta R_j g_{ij}$, is not an even term as one might think at first sight because of the two $\boldsymbol{\alpha}$ operators hidden in the R_i and R_j operators. Instead, it is odd in both spaces, ${}^1\mathcal{H}(i)$ and ${}^1\mathcal{H}(j)$, and *cannot* be simplified according to $\beta R_i \beta R_j g_{ij} = -R_i R_j g_{ij}$.

The DKH transformations U_m^{tot} ($m = 0, 1, \dots$) are designed to decouple, i.e., block-diagonalize, the total one-electron terms step by step. In practice, the decoupling can be done most efficiently in an approximate way if one restricts all transformations to the nuclear potential V_{nuc}. Then, the untransformed electron–electron interaction operators can be considered in some sort of perturbation theory by employing the same unitary transformations to obtain transformed operators of which the upper left block is then chosen for the two-component theory. This approximation has the advantage that the Breit interaction can be treated on the same footing. Having already introduced this approximation, we can even go on and restrict the transformation to the free-particle Foldy–Wouthuysen transformation only.

For the Coulomb-only electron–electron interaction only the four terms of the first line of Eq. (12.70) yield even expressions for both electrons under consideration. All other terms of Eq. (12.70) have at least one odd component. The result may be written as

$$H_1^{2e,\mathrm{C}} = \sum_{i<j}^{N} A_i A_j \left\{ \frac{e^2}{r_{ij}} + R_i \frac{e^2}{r_{ij}} R_i + R_j \frac{e^2}{r_{ij}} R_j + R_i R_j \frac{e^2}{r_{ij}} R_i R_j \right\} A_i A_j \tag{12.71}$$

This free-particle Foldy–Wouthuysen-transformed two-electron part has been investigated by various authors [176, 521, 525, 537, 544, 545]. Still, these are expressions of the four-component framework. Similarly to the procedure for one-electron operators the restriction to the electronic, i.e., upper-left part of these operators yields the desired two-component operators,

$$H_{1,+}^{2e,\mathrm{C}} = \sum_{i<j}^{N} A_i A_j \Big\{ \frac{e^2}{r_{ij}} + (\sigma_i \cdot \boldsymbol{P}_i) \frac{e^2}{r_{ij}} (\sigma_i \cdot \boldsymbol{P}_i) + (\sigma_j \cdot \boldsymbol{P}_j) \frac{e^2}{r_{ij}} (\sigma_j \cdot \boldsymbol{P}_j)$$
$$+ (\sigma_i \cdot \boldsymbol{P}_i)(\sigma_j \cdot \boldsymbol{P}_j) \frac{e^2}{r_{ij}} (\sigma_i \cdot \boldsymbol{P}_i)(\sigma_j \cdot \boldsymbol{P}_j) \Big\} A_i A_j \tag{12.72}$$

After a subsequent separation of spin-dependent terms for each electron by application of Dirac's relation one again arrives at a scalar-relativistic expression

$$H_{1,+}^{2e,\mathrm{sf},C} = \sum_{i<j} A_i A_j \left\{ \frac{e^2}{r_{ij}} + \boldsymbol{P}_i \frac{e^2}{r_{ij}} \boldsymbol{P}_i + \boldsymbol{P}_j \frac{e^2}{r_{ij}} \boldsymbol{P}_j + \boldsymbol{P}_i\boldsymbol{P}_j \frac{e^2}{r_{ij}} \boldsymbol{P}_i\boldsymbol{P}_j \right\} A_i A_j \tag{12.73}$$

The same discussion applies equally well for the Breit interaction. Only the four terms of the second line of Eq. (12.70) yield even operators for $g(i,j) = B(i,j)$. The corresponding two-component form of the free-particle Foldy–Wouthuysen-transformed frequency-independent Breit interaction B_0 reads

$$\begin{aligned} H_{1,+}^{2e,\mathrm{B}} &= \sum_{i<j}^{N} A_i A_j \Big\{ (\sigma_i \cdot \boldsymbol{P}_i)(\sigma_j \cdot \boldsymbol{P}_j)\tilde{B}_{0,ij} + (\sigma_i \cdot \boldsymbol{P}_i)\tilde{B}_{0,ij}(\sigma_j \cdot \boldsymbol{P}_j) \\ &\quad + (\sigma_j \cdot \boldsymbol{P}_j)\tilde{B}_{0,ij}(\sigma_i \cdot \boldsymbol{P}_i) + \tilde{B}_{0,ij}(\sigma_i \cdot \boldsymbol{P}_i)(\sigma_j \cdot \boldsymbol{P}_j) \Big\} A_i A_j \end{aligned} \tag{12.74}$$

where the two-component analog $\tilde{B}_0(i,j)$ is given by

$$\tilde{B}_0(i,j) = -\frac{e^2}{2r_{ij}} \left(\sigma_i \cdot \sigma_j + \frac{(\sigma_i \cdot \boldsymbol{r}_{ij})(\sigma_j \cdot \boldsymbol{r}_{ij})}{r_{ij}^2} \right) \tag{12.75}$$

This operator is the two-component analog of the Breit operator derived in section 8.1. The reduction has already been considered by Breit [68] and discussed subsequently by various authors (in this context see Refs. [176, 524]). We come back to this discussion when we derive the Breit–Pauli Hamiltonian in section 13.2 in exactly the same way from the free-particle Foldy–Wouthuysen transformation. However, there are two decisive differences from the DKH terms: (i) in the Breit–Pauli case the momentum operators in Eq. (12.74) are explicitly resolved by their action on all right-hand side terms, whereas in the DKH case they are taken to operate on basis functions in the bra and ket of matrix elements instead, and (ii) the Breit–Pauli Hamiltonian then results after (ill-defined) expansion in terms of $1/c$ (remember the last chapter for a discussion of this issue).

So far we have applied only the first step of the Douglas–Kroll–Hess transformation sequence, i.e., the innermost free-particle transformation U_0^{tot} to the two-electron terms. The higher-order expressions may be found by applying the subsequent unitary transformation $U_{m\geq 1}^{\mathrm{tot}}$ to the intermediate total two-electron Hamiltonian H_1^{2e} including any of the non-diagonal terms among the sixteen contributions of Eq. (12.70). Though these expressions are very lengthy, no new difficulties arise and the procedure is straightforward. However, since $\boldsymbol{P}_i$ is suppressed by large energy denominators and all DKH corrections feature an increasing number of $\boldsymbol{P}_i$-factors with increasing order, these terms provide only very small corrections as compared to the first order.

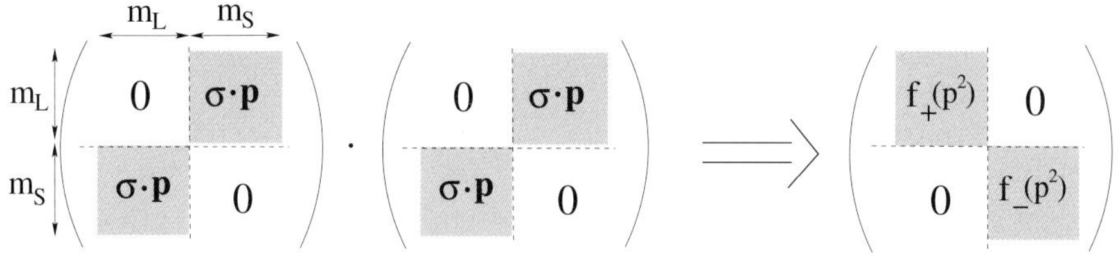

Figure 12.3 Scheme that demonstrates how even operators are generated from products of odd operators. In cases where no off-diagonal scalar and vector potential contributions occur or can be neglected, the off-diagonal blocks contain linear momentum operators that produce squared momenta in the blocn-diagonal form. The upper left block f_+ of the Hamiltonian describes the positive-energy solutions and features an $(m_L \times m_L)$-dimensional matrix representation. The representation of the negative-energy block f_- need not be of the same size but can be much more demanding to compute because of the kinetic balance restriction (compare also chapter 10).

12.5
Computational Aspects of DKH Calculations

Considering the derivation of DKH Hamiltonians so far, we are facing the problem to express all operators in momemtum space, which is somewhat unpleasant for most molecular quantum chemical calculations which employ atom-centered position-space basis functions of the Gaussian type as explained in section 10.3. The origin of the momentum-space presentation of the DKH method is traced back to the square-root operator in $\mathcal{E}_0$ of Eq. (12.54). This square root requires the evaluation of the square root of the momentum operator as already discussed in the context of the Klein–Gordon equation in chapter 5. Such a square-root expression can hardly be evaluated in a position-space formulation with linear momentum operators as differential operators. In a momentum-space formulation, however, the momentum operator takes a simple multiplicative form. We thus have to Fourier-transform all operators, which possess in *position* space a simple multiplicative form like the scalar potential V, and obtain integral operators defined by their operator kernels. Hence, the W_i operators are also to be calculated in momentum space, which is usually written for the corresponding operator kernels as in Eq. (12.46).

Even worse, none of the W_i operators can be calculated in a two-component framework as they all involve odd blocks that require a small component not existing in a truly two-component theory. The solution to this problem turns out to be simple. Since they are multiplied by other odd operators, all odd operators can be grouped and the odd-times-odd products are even and thus computable (compare Figure 12.3).

Hess realized [521] that the calculation of the square root of a momentum operator $\boldsymbol{p}$ only requires a basis, which yields a diagonal representation for $\boldsymbol{p}^2 = p^2$. Since the nonrelativistic one-electron kinetic energy matrix, $(t) = p^2/2m_e$, is calculated in all quantum chemistry program packages in a given basis set $\{\phi_i\}$, diagonalization of (t) can easily be performed. The eigenvectors then provide the transformation matrix from the original position-space basis set to a linearly transformed position-space basis, which yields a diagonal representation for p^2. After calculation and addition of all even terms in this p^2-basis by standard matrix operations, the resulting matrix representation of the DKH Hamiltonian can be transformed back to the original basis with the inverse transformation. In this new p^2-basis, the operator kernels in momentum space as given in Eqs. (12.54)–(12.57) simply translate to matrix multiplications. The indices for the different momenta in Eqs. (12.54)–(12.57) thus directly relate to the indices of the matrix representations of the operators involved. The calculation of a DKH Hamiltonian thus reduces to a couple of matrix multiplications whose actual number increases with increasing DKH order. The question now is, how many different matrix representations of operators are actually needed in order to compute a DKH Hamiltonian of arbitrary order. An ingenious trick by Hess [522] allows us to keep this number restricted to less than a dozen.

12.5.1 Exploiting a Resolution of the Identity

The peculiarity that is involved in the calculation of the DKH Hamiltonians derives from the fact that some terms in the Hamiltonians are of the form $\boldsymbol{p}V\dots V\boldsymbol{p}$ [compare Eq. (12.56)]. Hence, no momentum operators occur between the potential energy operators, and a new matrix representation would be needed for such terms. Even worse, the higher the order, the more complicated are the terms that arise. Hess' solution to this problem was the introduction of a resolution of the identity (RI),

$$\mathbf{1} = \frac{(\boldsymbol{\sigma}\cdot\boldsymbol{p})(\boldsymbol{\sigma}\cdot\boldsymbol{p})}{p^2} = \frac{\boldsymbol{p}\cdot\boldsymbol{p}}{p^2} \tag{12.76}$$

which generates $\boldsymbol{p}V\boldsymbol{p}$ or $\boldsymbol{\sigma}\cdot\boldsymbol{p}V\boldsymbol{\sigma}\cdot\boldsymbol{p}$ matrix operators, respectively, (depending on whether a spin-free or the full matrix elements are sought), which can be seen after insertion between the two external potential operators and rearranging the $\boldsymbol{p}$ operators:

$$\boldsymbol{p}V\dots V\boldsymbol{p} = \boldsymbol{p}V\dots\frac{\boldsymbol{p}\cdot\boldsymbol{p}}{p^2}\dots V\boldsymbol{p} = \boldsymbol{p}V\boldsymbol{p}\dots\frac{1}{p^2}\dots\boldsymbol{p}V\boldsymbol{p} \tag{12.77}$$

This would be necessary, for instance, for the second term of $\mathcal{E}_{2+}(i,j,k)$ in Eq. (12.56),

$$A_i\,\boldsymbol{\sigma}\cdot\boldsymbol{P}_i\,\tilde{V}_{ij}A_j\,A_jV_{jk}\,\boldsymbol{\sigma}\cdot\boldsymbol{P}_k\,A_k = A_i\,\boldsymbol{\sigma}\cdot\boldsymbol{P}_i\,\tilde{V}_{ij}\boldsymbol{\sigma}\cdot\boldsymbol{P}_jA_j\frac{1}{P_j^2}\,A_j\boldsymbol{\sigma}\cdot\boldsymbol{P}_jV_{jk}\,\boldsymbol{\sigma}\cdot\boldsymbol{P}_k\,A_k \tag{12.78}$$

Formally, we have *not* yet introduced any approximation. But the evaluation of the operator sequence on the right hand side of Eq. (12.77) requires a translation into products of operator matrices,

$$\boldsymbol{p}V\boldsymbol{p}\ldots\frac{1}{p^2}\ldots\boldsymbol{p}V\boldsymbol{p} \quad\longrightarrow\quad (\boldsymbol{p}V\boldsymbol{p})\,(\ldots)\left(1/p^2\right)(\ldots)\,(\boldsymbol{p}V\boldsymbol{p})\,, \tag{12.79}$$

where $(\ldots)$ symbolizes the matrix form. It is this last translation step which makes the RI step introduced approximate and which requires (locally) a large basis set as Eq. (12.79) is only exact in a complete basis set.

Note that this also implicitly implies that we never encounter a (single) linear momentum between two V operators and that there is always a (single) linear momentum on the right and left hand side of the right and left V operators, respectively, in Eq. (12.77) if the RI is to be inserted between two V operators. This implication is always fulfilled, as can be understood from the construction of the DKH Hamiltonian according to, e.g., Eq. (12.49), and can be systematically incorporated in the derivation of the infinite-order DKH Hamiltonian [532].

An important aspect of DKH theory is that only standard operator matrices for the nonrelativistic kinetic energy and external potential are required for the evaluation of the DKH Hamiltonian, which is after all quite remarkable. Only one non-standard matrix is needed, which can, however, be calculated with little additional effort. This is the matrix representation of $\boldsymbol{p}V\boldsymbol{p}$,

$$\begin{aligned}\langle\phi_\mu|\boldsymbol{p}\cdot V\boldsymbol{p}|\phi_\nu\rangle &= \langle p_x\phi_\mu|V|p_x\phi_\nu\rangle+\langle p_y\phi_\mu|V|p_y\phi_\nu\rangle+\langle p_z\phi_\mu|V|p_z\phi_\nu\rangle\\ &\longrightarrow (p_xVp_x)+(p_yVp_y)+(p_zVp_z)\end{aligned} \tag{12.80}$$

which is evaluated by operation of the momentum operator on the basis functions and integration of the resulting matrix elements. From diagonalization of the kinetic energy we obtain the eigenvalues to calculate the relativistic energy of a freely moving particle and the eigenvector matrix for the transformation of the V and $\boldsymbol{p}V\boldsymbol{p}$ matrices in the p^2-basis. Diagonalizing the matrix representation of the kinetic energy

$$\boldsymbol{\Omega}^\dagger\boldsymbol{T}\boldsymbol{\Omega}=\boldsymbol{t} \tag{12.81}$$

yields the diagonal matrix $\boldsymbol{t}$ with eigenvalues

$$t_i=\frac{p_i^2}{2m_e}\longrightarrow p_i^2=2m_et_i \tag{12.82}$$

Latin indices i and j refer to basis functions in p^2-space, whereas the Greek indices μ and ν denote basis functions in the original position space as before. This notation will also be adopted in section 12.5.4. The matrix representation of the DKH kinetic energy $\mathcal{E}_0$ is now easily obtained,

$$\left(\boldsymbol{\mathcal{E}}_0^{p^2\text{-space}}\right)_{ij} = \begin{cases} \sqrt{2m_e t_i c^2 + m_e^2 c^4} - m_e c^2 & i = j \\ 0 & i \neq j \end{cases} \tag{12.83}$$

The reversed transformation yields the first term of the DKH Hamiltonian in position space

$$\boldsymbol{T}^{(r)} = \boldsymbol{\Omega}\, \boldsymbol{\mathcal{E}}_0^{p^2\text{-space}}\, \boldsymbol{\Omega}^\dagger \tag{12.84}$$

Higher order corrections to the one-electron integrals involving the kinematic factors and the nuclear potential may be evaluated analogously. Thus, the DKH Hamiltonian is calculated as a matrix representation of the relativistic energy of a freely moving particle to which correction terms are added that include order-by-order the action of the scalar potential.

It should be noted that what has just been said about the very small number of basic matrix operators holds strictly for the so-called scalar-relativistic variant of DKH theory. The full two-component scheme, which is seldom used in practice, requires only a few more matrices, as is clear in view of the spin-dependent terms like the second in Eq. (12.5) that arise after application of Dirac's relation. These spin-dependent terms finally require matrices of mixed linear momentum contributions like $(p_x V p_y)$, which, however, do not pose a problem and can be computed like their 'diagonal' counterparts of Eq. (12.80).

12.5.2
Advantages of Scalar-Relativistic DKH Hamiltonians

The pleasant features of DKH Hamiltonians, which is the basis for their success in relativistic quantum chemistry, are first of all that they can be given in analytic form. Then, they are strictly variationally stable by construction, since the only source of unbounded behavior, i.e., the part of the Dirac Hamiltonian belonging to the small component, is no longer coupled to the electronic Hamiltonian f_+. The odd-order DKH Hamiltonians may not be variational, but may overestimate binding energies for electrons in external potentials, but there exists a ground state for every DKH Hamiltonian, and hence no variational collapse can occur.

But the most important feature for practical purposes comes with a certain approximation, which is the scalar-relativistic variant of DKH. This one-component DKH approximation, in which all spin-dependent operators are separated by Dirac's relation and then simply omitted, are particularly easy to implement in widely available standard nonrelativistic quantum chemistry program packages, as Figure 12.4 demonstrates. Only the one-electron operators in matrix representation are modified to account for the kinematic

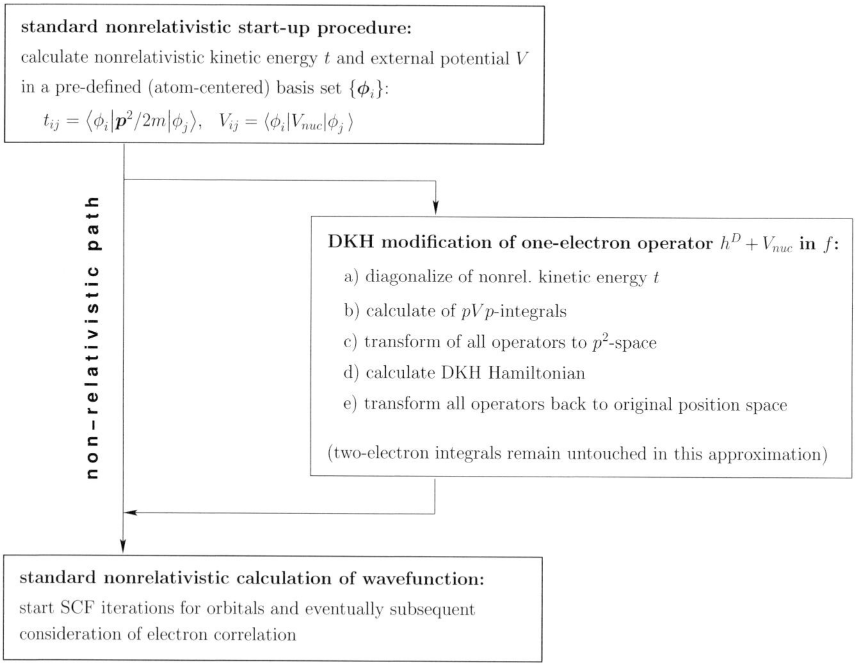

Figure 12.4 Data flow and calculations in a scalar-relativistic DKH implementation that can include kinematic relativistic contributions to arbitrary order in the external electron–nucleus potential. The flow chart indicates how a standard nonrelativistic computer program has to be modified to include the scalar-relativistic DKH one-electron operator.

or (synonymously) scalar-relativistic effects. The inclusion of the spin–orbit terms requires a two-component infrastructure of the computer program. The consequences of the neglect of spin–orbit effects have been investigate in pilot studies like Refs. [524, 546] and, naturally, the accuracy depends on the systems under consideration (see also chapter 16 for some examples and further explanation).

The terms 'scalar-relativistic', 'spin-free', 'spin-averaged' and 'one-component' all refer to this very same procedure of eliminating all Pauli spin matrices from the Hamiltonian by simple deletion after employment of Dirac's relation.

But to be more precise, most implementations of the DKH Hamiltonian transform the one-electron terms of the Fock operator f only. Hence, the general scalar potential V is reduced to the complete electron–nucleus potential V_{nuc} of an atom or a molecule. Only this additional step allows us to calculate the DKH Hamiltonian prior to any SCF iterations, as indicated in Figure 12.4, because the one-electron terms do not depend on the solution functions, i.e., they do not depend on the molecular orbitals ψ. The two-electron interaction terms are thus not considered in standard DKH calculations. They are incorporated into the Fock operator of the SCF equation in the standard nonrelativistic way. Since they are not transformed, they are affected by an error, the so-called picture change error or picture change effect, which judged on total electronic energy expectation values property integrals, is usually very small [547–549], although it may become nonnegligible in certain cases [537]. We will come back to this picture change error in chapter 16.

By inspection of the explicit expressions for the even terms $\mathcal{E}_k$ given above it appears that all even terms depend only quadratically on the momentum operator rather than on the linear operator $\boldsymbol{p}$. It is easy to realize that this is not an accidental feature of the six lowest-order terms of the DKH series but a fundamental property of *all* even terms occurring within *any* expansion of the block-diagonal Hamiltonian. The origin of this peculiarity lies, of course, in the structure of the original Dirac Hamiltonian, whose only odd component is the kinetic term $c\,\boldsymbol{\alpha}\cdot\boldsymbol{p}$, which is linear in $\boldsymbol{p}$. Since there is no other odd term available and since only the product of an even number of odd terms yields an even term, all terms contributing to a block-diagonal Hamiltonian are necessarily functions of $\boldsymbol{p}^2$ rather than $\boldsymbol{p}$. This situation is graphically illustrated in Figure 12.3.

It should also be noted that most DKH implementations invoke a prediagonalization of the atom-centered molecular basis set for numerical stability reasons (i.e., to avoid linear dependencies). Different orthogonalization schemes (like Gram–Schmidt or Löwdin's symmetric orthogonalization) have been employed to diagonalize the overlap matrix $\boldsymbol{S}$ of basis functions.

In section 12.3.1 it was explained that the true expansion parameter in Eq. (12.2) is the energy-damped potential $\tilde{V}$. The origin of this damping is the energy of the freely moving relativistic electron in E_p by Eq. (5.4) [510]. In the matrix operations needed for the evaluation of the DKH Hamiltonian the energy damping occurs automatically via Eq. (12.58), which determines how the energy-damped matrix is to be calculated in the p^2-basis. The $\boldsymbol{p}_i^2 = p_i^2$ values are simply the eigenvalues of the nonrelativistic kinetic energy matrix multiplied by two. Since we are working in a p^2-basis, we have associated with each basis vector a value for the squared momentum.

12.5.3
Efficient Approximations for the Transformation of Complicated Terms

Despite the accuracy of the (low-order) scalar-relativistic DKH expansion in powers of the total electron-nuclei potential V_{nuc} only, a consistent transformation also requires the proper inclusion of the spin–orbit and electron–electron interaction terms.

12.5.3.1 **Spin–Orbit Operators**

Of course, the inclusion of the spin–orbit terms increases the computational demand for DKH calculations because the number of terms is enlarged, and, when included in a variational procedure, the one-electron orbitals must be two-component functions that would require significant extensions to a non-relativistic one-component computer program. For the efficient calculation of two-electron spin–orbit terms, Boettger proposed an efficient screened-nuclear spin–orbit approximation [550]. Shortly after its proposal, this model has been extended and investigated further [551,552].

12.5.3.2 **Two-Electron Terms**

A more severe challenge is the efficient treatment of the DKH transformation of the two-electron terms. Here, the problem is actually two fold. Firstly, the electron density or the orbital densities which enter the electron–electron interaction integrals are affected by the picture-change effect on the density. Then, these integrals contribute to the potential into which the Fock-type operator is expanded. Moreover, the interaction integrals are not necessarily even, block-diagonal operators, which has led to the restrictions made in section 11.1.

For the case of the two-electron terms, efficient approximations have been developed. Also here, a model potential approach that takes advantage of the 'atomic structure' of molecules proved to be beneficial requiring only atomic data to make up for the neglected DKH transformation of the density in the two-electron integrals, i.e., for the neglect of the picture change effect [537]. In turn, this solution avoids the transformation in each SCF iteration. Moreover, for the Coulomb interaction energy of two electrons — the Hartree 'potential' — a charge-density fitting scheme has been devised [553, 554]. Another approach successfully adopts the effective one-electron potential energy operator of the SCF equations [555], but bear in mind the restrictions made in section 11.1 with respect to the choice of the potential of Fock-type operators to serve as an expansion parameter.

12.5.3.3 **Large One-Electron Basis Sets**

When the one-electron basis set becomes large because of increasing size of the molecule, the DKH transformation becomes computationally expensive if not prohibitive because of the size of the matrices involved. It is straightforward and also quite efficient to do the transformation only for diagonal blocks of the one-electron Hamiltonian that comprise basis functions located at one and the same center. However, early work included a more extensive approach considering also nearest-neighbor centers [556].

Especially in periodic systems, basis-set sizes prohibit a full DKH transformation. To a good approximation, this can be easily dealt with by using a short-ranged Coulomb operator, which naturally occurs in the Gaussian-augmented plane-wave scheme [557,558], in which the transformation can be performed in the atomic basin [540,559]. More elaborate local schemes have also been developed [560]

12.5.4
DKH Gradients

The importance of analytic derivatives of the total electronic energy for structure optimization purposes or molecular properties was emphasized in section 10.5. Obviously, this will be complicated in the case of DKH theory because of the momentum space or the p^2-basis formulation, because the derivative of the transformation matrices $\boldsymbol{\Omega}^{\dagger}$ and $\boldsymbol{\Omega}$ of Eq. (12.81). Let the superscript $\partial_{A,\alpha}$ indicate the derivative of a certain quantity with respect to one Cartesian component α of the coordinates of nucleus A. Differentiating an arbitrary matrix $\boldsymbol{X}$ then reads

$$\frac{\partial}{\partial R_{A,\alpha}}\boldsymbol{X} = \left\{\frac{\partial}{\partial R_{A,\alpha}}X_{ij}\right\} = \boldsymbol{X}^{\partial_{A,\alpha}} \tag{12.85}$$

corresponding to a differentiation of each element X_{ij} of the matrix. Recalling Eq. (12.84) and applying the product rule one obtains the derivative of the matrix representation of the $\mathcal{E}_0$ operator in position space,

$$(\boldsymbol{T}^{(r)})^{\partial_{A,\alpha}} = \boldsymbol{\Omega}^{\partial_{A,\alpha}}\,\boldsymbol{\mathcal{E}}_0\,\boldsymbol{\Omega}^{\dagger} + \boldsymbol{\Omega}\,\boldsymbol{\mathcal{E}}_0^{\partial_{A,\alpha}}\,\boldsymbol{\Omega}^{\dagger} + \boldsymbol{\Omega}\,\boldsymbol{\mathcal{E}}_0\,(\boldsymbol{\Omega}^{\dagger})^{\partial_{A,\alpha}} \tag{12.86}$$

Already for the lowest-order term in the Hamiltonian we see that the gradient will not be easy to evaluate. Despite these difficulties, an analytic gradient for the second-order scalar-relativistic DKH2 variant has been developed by Nasluzov and Rösch [561], and this was later generalized to a general response formalism for first- and second-order derivatives with respect to the nuclear coordinates starting from the so-called modified Dirac equation [562]; cf. section 14.1. Such a low-order approximation is certainly sufficient because the geometric structure gradient can be reliably corrected for low-order relativistic effects as demonstrated in Ref. [563].

The complexity of the analytic gradient demanded a simpler and more feasible procedure. Following the rather simple though very efficient idea of de Jong, Harrison and Dixon [564], a numerical approximation to the analytical evaluation of the scalar-relativistic DKH gradient can be devised, which only affects the derivative of the DKH-modified one-electron integrals,

$$\frac{\partial h_{\mu\nu}^{\mathrm{DKH}}}{\partial R_{A,\alpha}} = \frac{\partial h_{\mu\nu}}{\partial R_{A,\alpha}} - \frac{\partial}{\partial R_{A,\alpha}}\left(h_{\mu\nu} - h_{\mu\nu}^{\mathrm{DKH}}\right) \tag{12.87}$$

The trick is that the derivative of the difference between the nonrelativistic one-electron integrals and the one-electron DKH integrals is calculated numerically. Since one-electron integrals are easy to calculate and require negligible computing time, they can easily be calculated for distorted structures, so that the derivative can be evaluated using finite-difference techniques. An n-point Bickley central-difference formula with sufficiently high n may guarantee sufficient accuracy even in cases where relativistic effects on the gradient are negligible.

As already noted in section 12.5.2, a pre-diagonalization of the atom-centered basis is usually performed. Nasluzov and Rösch [561] discuss this issue for the derivative of the symmetric orthogonalization.

13
Elimination Techniques

A straightforward and easy reduction of the four-component one-electron equations to two-component form is the elimination of the two small components. This is achieved by expressing the small component through the large component via rearrangement of the lower two coupled (integro-)differential equations and substituting the result into the remaining upper two equations. Main issues to be considered are then from which formulation of the four-component one-electron equation one should start and how the arising energy and potential in the kinetic operator of the Hamiltonian — leading to equations which are no longer eigenvalue equations — should be treated.

In the preceding two chapters, we dealt with general unitary transformation schemes to produce a one-electron Hamiltonian valid for only the positive-energy part of the Dirac spectrum that governs the electronic bound and continuum states. Evidently, these unitary transformation schemes are elegant but quite involved. Developments in quantum chemistry always focus on efficient approximations in a sense that the main numerical contribution of some physical effect is reliably captured for any class of molecule or molecular aggregate. The so-called elimination techniques have been very successful in this sense and are therefore discussed in the present chapter.

13.1
Naive Reduction: Pauli Elimination

Historically, the first derivations of approximate relativistic operators of value in molecular science have become known as the Pauli approximation. Still, the best-known operators to capture relativistic corrections originate from those developments which provided well-known operators like the spin–orbit or the mass–velocity or the Darwin operators. Not all of these operators are variationally stable, and thus they can only be employed within the framework of perturbation theory. Nowadays, these difficulties have been overcome by, for instance, the Douglas–Kroll–Hess hierarchy of approximate Hamiltonians and the regular approximations to be introduced in a later section, so that operators like the mass–velocity and Darwin terms are no longer needed. Nev-

Relativistic Quantum Chemistry. Markus Reiher and Alexander Wolf

ISBN: 978-3-527-31292-4

ertheless, we should present their derivation in order to present a complete picture of relativistic quantum chemistry.

The central idea of all elimination methods for the small components is to employ Eq. (11.2) and to substitute ψ^S in Eq. (11.3) by an expression for the large component only. This yields a two-component eigenvalue equation for the latter only, whose most general form may be expressed as

$$\left(V + c\,\boldsymbol{\sigma}\cdot\boldsymbol{p}\,X\right)\psi^L \;=\; \epsilon\,\psi^L \tag{13.1}$$

where X may either be the energy-dependent or energy-independent operator defined by Eqs. (11.4) or (11.6), respectively. For all practical calculations based on elimination techniques for the small components the former has been employed, which yields the two-component equation

$$(V-\epsilon)\,\psi^L + \frac{1}{2m_ec^2}\left[(c\,\boldsymbol{\sigma}\cdot\boldsymbol{p})\,\omega\,(c\,\boldsymbol{\sigma}\cdot\boldsymbol{p})\right]\psi^L \;=\; 0 \tag{13.2}$$

with

$$\omega \;=\; \omega(\epsilon) \;=\; \left[1 - \frac{V-\epsilon}{2m_ec^2}\right]^{-1} \tag{13.3}$$

So far no approximation has been introduced at all, and Eq. (13.2) still yields, of course, the exact Dirac eigenvalues and large components. Also the nonrelativistic limit ($V \ll 2m_ec^2$, $\epsilon \approx 0$) of Eq. (13.2) is well defined and recovers the Schrödinger equation by setting $\omega = 1$. Since ω is a scalar operator, even this simple energy-dependent elimination of the small component based on Eqs. (13.2) and (13.3) permits an exact separation of the spin-free and spin-dependent terms of the Dirac Hamiltonian by application of Dirac's relation

$$(\boldsymbol{\sigma}\cdot\boldsymbol{p})\,\omega\,(\boldsymbol{\sigma}\cdot\boldsymbol{p}) \;=\; \boldsymbol{p}\cdot\omega\,\boldsymbol{p} + \mathrm{i}\,\boldsymbol{\sigma}\cdot(\boldsymbol{p}\times\omega\,\boldsymbol{p}) \tag{13.4}$$

(see appendix D.2).

The historically first reduction of the Dirac equation to two-component form is the so-called Pauli approximation, which utilizes the properties of the geometric series in order to convert Eq. (13.3) into the apparently simpler form

$$\omega \;=\; \sum_{k=0}^{\infty}\left(\frac{V-\epsilon}{2m_ec^2}\right)^k \tag{13.5}$$

that is subsequently truncated after the first two terms ($k = 0, 1$). The energy dependence is eliminated by substituting ϵ by its semiclassical approximation correct through order $O(c^{-2})$.

Inserting the first two terms with $k = 0, 1$ into Eq. (13.2) yields a kinetic energy operator of the form

$$\frac{1}{2m_ec^2}\left[(c\,\boldsymbol{\sigma}\cdot\boldsymbol{p})\left(1 + \frac{V-\epsilon}{2m_ec^2}\right)(c\,\boldsymbol{\sigma}\cdot\boldsymbol{p})\right] = \frac{\boldsymbol{p}^2}{2m_e} + \frac{\boldsymbol{\sigma}\cdot\boldsymbol{p}V\boldsymbol{\sigma}\cdot\boldsymbol{p}}{4m_e^2c^2} - \frac{\epsilon\,\boldsymbol{p}^2}{4m_e^2c^2} \tag{13.6}$$

For the operator in the middle we can again apply Dirac's relation,

$$\sigma \cdot pV\sigma \cdot p = pVp + \mathrm{i}\sigma \cdot (pV \times p) \tag{13.7}$$

By noting that

$$p^2V = ppV = p(pV) + p(Vp) = (p^2V) + 2(pV)p + Vp^2 \tag{13.8}$$

(the parentheses denote compound operators such that the momentum operator acts only on what is within the parentheses and not on what is still to come from the right hand side) we can write for the first term of the right hand side of Eq. (13.7)

$$\begin{aligned} pVp &= \frac{1}{2}[pVp + pVp] \\ &= \frac{1}{2}[(pV)p + Vp^2 + (pV)p + Vp^2 \;\; - (p^2V) + (p^2V)] \\ &\overset{(13.8)}{=} \frac{1}{2}[Vp^2 + p^2V - (p^2V)] \overset{(4.49)}{=} \frac{1}{2}[Vp^2 + p^2V] + \frac{\hbar^2}{2}(\Delta V) \end{aligned} \tag{13.9}$$

The next step requires us to find a more appropriate expression for the term in brackets in Eq. (13.9). For this, we recall the nonrelativistic Schrödinger equation for a hydrogen-like atom, Eq. (6.9), and write

$$V\Psi = \left(\epsilon^{nr} - \frac{p^2}{2m_e}\right)\Psi \tag{13.10}$$

which is fulfilled here only in the framework of first-order perturbation theory if the Hamiltonian to be devised enters energy expectation values that are evaluated with the nonrelativistic eigenfunction Ψ. With this approximation in mind, we may now write the anticommutator as

$$\begin{aligned} \frac{1}{2}[Vp^2 + p^2V] &\approx \frac{1}{2}\left[\left(\epsilon^{nr} - \frac{p^2}{2m_e}\right)p^2 + p^2\left(\epsilon^{nr} - \frac{p^2}{2m_e}\right)\right] \\ &= \epsilon^{nr}p^2 - \frac{p^4}{2m_e} \end{aligned} \tag{13.11}$$

Combining all individual results yields the Pauli Hamiltonian already encountered in section 11.5.2,

$$h^{\text{Pauli}} = \frac{p^2}{2m_e} + V - \frac{p^4}{8m_e^3c^2} + \frac{\hbar^2}{8m_e^2c^2}(\Delta V) + \frac{\hbar}{4m_e^2c^2}\sigma\cdot[(\nabla V)\times p] \tag{13.12}$$

where we assumed that $\epsilon^{nr} \approx \epsilon$. We then recover the nonrelativistic Hamiltonian $p^2/2m_e + V$ and find as correction terms the mass–velocity operator $-p^4/8m_e^3c^2$, the Darwin term $\hbar^2\Delta V/8m_e^2c^2$, and the spin–orbit coupling term proportional to $\boldsymbol{\sigma}\cdot\left[(\nabla V)\times\boldsymbol{p}\right]$. These additional operators describe all relativistic corrections correct up to $\mathcal{O}(c^{-2})$. Again, we emphasize the analogy to the classical energy expression derived in Eq. (3.121).

The spin–orbit term can be cast in a more convenient form for a hydrogen-like atom with a point-like nucleus by noting that

$$\nabla V = \nabla\left(-\frac{Ze^2}{r}\right) = \frac{Ze^2}{r^2}\nabla r = \frac{Ze^2\boldsymbol{r}}{r^3} \tag{13.13}$$

(see Eq. (A.23) in the appendix), where $r = \sqrt{x^2+y^2+z^2}$ if the nucleus is placed at the origin of the coordinate system, so that the spin–orbit term h^{SO} reads

$$h^{\mathrm{SO}} = \frac{\hbar}{4m_e^2c^2}\boldsymbol{\sigma}\cdot\left[(\nabla V)\times\boldsymbol{p}\right] = \frac{\hbar}{4m_e^2c^2}\frac{Ze^2\boldsymbol{\sigma}\cdot\left[\boldsymbol{r}\times\boldsymbol{p}\right]}{r^3} = \frac{Ze^2}{2m_e^2c^2}\frac{\boldsymbol{s}\cdot\boldsymbol{l}}{r^3} \tag{13.14}$$

where we utilized the definition of the angular momentum operator $\boldsymbol{l}$ from Eq. (4.87) and the definition of the spin operator $\boldsymbol{s}$ from Eq. (4.139).

When we studied the electron moving in a central field in chapter 6, we only noted the formal consequences of spin–orbit coupling, namely the fact that the Dirac Hamiltonian commutes only with the squared *total* angular momentum operator rather than with the squared *orbital* and *spin* angular momentum operators individually. For the latter no eigenvalue equations are obtained with the Dirac spinor and, hence, spin and orbital angular momentum are no longer good quantum numbers. Here, we now explicitly see how the spin–orbit interaction can be derived as an *approximation* to the full Dirac Hamiltonian. Although it is often believed that spin is an important symmetry in quantum mechanics, we see that this is only valid as an approximation, namely when spin–orbit angular momentum coupling turns out to be small for a given system. Dirac writes in Ref. [66] that all *the same there is a great deal of truth in the spinning electron model, at least as a first approximation. The most important failure of the model seems to be that the magnitude of the resultant orbital angular momentum of an electron moving in an orbit in a central field of force is not a constant, as the model leads one to expect.*

The Darwin operator for hydrogen-like atoms is often written in a different form. The Coulombic potential of a point nucleus, $\phi_{nuc} = V_{nuc}/q_e$, can be replaced by Ze/r, so that we obtain from Poisson's equation

$$\Delta\phi_{nuc} = -4\pi\rho_{nuc}(r) \qquad \Longrightarrow \qquad \Delta\left(\frac{Ze}{r}\right) \overset{(A.24)}{=} -4\pi\, Ze\,\delta^{(3)}(\boldsymbol{r}) \tag{13.15}$$

(cf. appendix A.1.3), where the δ-function is centered at the atomic nucleus, which is chosen to be in the origin of the coordinate system, and represents a so-called *contact term*. After multiplication by the electron charge $(-e)$, $V =$

$-e\phi_{nuc}$, the Darwin term h^{Darwin} can be written as

$$h^{\text{Darwin}} = \frac{\hbar^2}{8m_e^2c^2}(\Delta V) = \frac{\hbar^2 \pi Z e^2}{2m_e^2c^2}\delta^{(3)}(\boldsymbol{r}) \tag{13.16}$$

However, there is one major problem connected with the Pauli Hamiltonian that strictly prohibits its use within a variational procedure. The minus sign of the mass–velocity term leads to a strongly attractive potential for states with high momentum, and would thus lead to variational collapse, since it is not bounded from below. This insurmountable problem cannot be remedied by going to higher orders [565], since the geometric series expansion of ω given by Eq. (13.5) is only valid for $|V - \epsilon| < 2m_ec^2$, which is certainly violated in regions close to the nucleus contributing to *every* bound state. Therefore the Pauli Hamiltonian and all other operators based on simple expansions of ω in powers of c^{-2} according to Eq. (13.5) are in general singular and ill-defined, and must not be used for variational procedures (see the discussion in section 11.5.3). Also the Darwin term produces difficulties as it degenerates to the singular Delta distribution term in the case of a point-like nucleus, as shown above. However, the Pauli Hamiltonian and the next three relativistic corrections, i.e., terms up to $k = 4$ in Eq. (13.5), have been successfully used within the framework of perturbation theory [565]. But even in lowest-order perturbation theory divergent terms arise which can only be dealt with by highly sophisticated renormalization techniques. The Pauli Hamiltonian itself applied perturbatively is better behaved and yields satisfactory relativistic corrections to the energy up to the first and second transition metal row.

13.2
Breit–Pauli Theory

Pseudo-relativistic many-electron Hamiltonians have a long history beginning with the discussion of the spin–spin interaction of two electrons by Heisenberg, Pauli, Gaunt and many others as early as in the late 1920s. They are included in an essentially nonrelativistic theory as additive potential energy terms. In section 8.1 we saw that they can be rigorously derived from Dirac's principle of minimal coupling and, hence, from the Breit interaction. Therefore, we may here start from the Breit interaction to derive the pseudo-relativistic Hamiltonians instead of following the somewhat meandering historical path from 1926 to about 1932, which was mainly based on classical considerations. As usual, a quantum-electrodynamical derivation is also possible and has been presented by Itoh [566], but the sound basis of our semi-classical theory, which we pursue throughout this book, is necessarily the Breit equation. Needless to say, the rigorous transformation approach to the Dirac–

Coulomb–Breit Hamiltonian yields results identical to those from the QED-based derivation.

The most convenient way to reduce the Breit equation to an essentially two-dimensional Pauli-like equation is to apply a unitary transformation of the form discussed in chapters 11 and 12. Of course, we then meet again the one-electron Pauli Hamiltonian derived in the preceding section and in chapter 11 via a free-particle Foldy–Wouthuysen transformation carried out for a single electron. *All pathologies of the Foldy–Wouthuysen transformation hold here as well, and the resulting Hamiltonian can only be employed in perturbation theory and not in any variational calculations.* Usually, this issue is discussed for the individual terms of the one-electron Pauli Hamiltonian of the previous section. However, we may refer to the more general reasons given in chapter 11 for the general failure of Foldy–Wouthuysen-like expansions in terms of $1/c$.

The resulting Hamiltonian is the *Breit–Pauli Hamiltonian* and is still in current use in perturbation theory calculations of especially spin–orbit effects with sophisticated CI-type wave functions (see also section 14.1).

13.2.1
Foldy–Wouthuysen Transformation of the Breit Equation

The derivation of the Breit–Pauli Hamiltonian is quite tedious. It is nowadays customary to follow the Foldy–Wouthuysen approach first given by Chraplyvy [567–569], which has, for instance, been sketched by Harriman [26]. Still, many presentations of this derivation lack significant details. In the spirit of this book, we shall give an explicit derivation which is as detailed and compact as possible. A review of the same expression derived quite differently was provided by Bethe in 1933 [39]. Compared to the DKH treatment of the two-electron term to lowest order as described in section 12.4.2 we now only consider the lowest-order terms in $1/c$. We transform the Breit operator in Eq. (8.19) by the free-particle Foldy–Wouthuysen transformation,

$$\left\{ U_0 \left(h_1^D + h_2^D + \frac{q_1 q_2}{r_{12}} - \frac{q_1 q_2}{2} \left[\frac{\boldsymbol{\alpha}_1 \cdot \boldsymbol{\alpha}_2}{r_{12}} + \frac{(\boldsymbol{r}_{12} \cdot \boldsymbol{\alpha}_1)(\boldsymbol{r}_{12} \cdot \boldsymbol{\alpha}_2)}{r_{12}^3} \right] \right) U_0^\dagger \right\} \left(U_0 \Psi(\boldsymbol{r}_1, \boldsymbol{r}_2) \right) = E \left(U_0 \Psi(\boldsymbol{r}_1, \boldsymbol{r}_2) \right) \quad (13.17)$$

with the unitary transformation defined in Eq. (11.77). For this transformation we have for two particles according to Eq. (12.64),

$$U_0 = U_0(1) \otimes U_0(2) \quad (13.18)$$

For the sake of simplicity, we have chosen $a_{0,0} = a_{0,1} = 1$ in the expansion of the unitary matrix above with the odd and antihermitean parameters, $W_{[0]}(1)$

and $W_{[0]}(2)$, defined by Eq. (11.80). Its structure is most easily understood in terms of the exponential parametrization of Eq. (11.49),

$$U_0 = \exp\left[W_{[0]}(1)\right] \otimes \exp\left[W_{[0]}(2)\right] \tag{13.19}$$

These $W_{[0]}$ operators are to be undestood to operate only on the corresponding coordinates. Hence,

$$U_0 h_1^D U_0^\dagger = U_0(1) h_1^D U_0^\dagger(1) \otimes \underbrace{U_0(2) U_0^\dagger(2)}_{=\mathbf{1}}$$

$$\longrightarrow \left[U_0(1) h_1^D U_0^\dagger(1)\right]^{LL} \approx h^{\text{Pauli}}(1) \tag{13.20}$$

$$U_0 h_2^D U_0^\dagger = \underbrace{U_0(1) U_0^\dagger(1)}_{=\mathbf{1}} \otimes U_0(2) h_2^D U_0^\dagger(2)$$

$$\longrightarrow \left[U_0(2) h_2^D U_0^\dagger(2)\right]^{LL} \approx h^{\text{Pauli}}(2) \tag{13.21}$$

produce to lowest orders in $1/c$ the one-electron Pauli Hamiltonian for each of the particles as deduced in section 11.5.2 and re-visited in the previous section. The (LL) superscript denotes the upper-left part — the large–large block — of the transformed Dirac operator, which is first obtained in (2×2)-dimensional super-structure. Since we skipped the direct-product notation already in chapter 8 and, for the sake of convenience, wrote simple sums instead, we may write the total unitary transformation for two particles as

$$U_0 = \mathbf{1} + W_{[0]}(1) + W_{[0]}(2) + W_{[0]}(1)W_{[0]}(2) + \frac{1}{2}W_{[0]}^2(1) + \frac{1}{2}W_{[0]}^2(2) + O(c^{-3}) \tag{13.22}$$

where we exploited Eq. (11.61), i.e., $a_{0,2} = \frac{1}{2}\, a_{0,0}\, a_{0,1}^2 = \frac{1}{2}$. This simply results from the product of two unitary matrices in the most general representation [511] (but with the same expansion coefficients) as discussed in chapter 11.

13.2.2
Transformation of the Two-Electron Interaction

The only additional steps to make now are the reduction of the two-electron operator $g(1,2) = V_C(1,2) + B_0(1,2)$ of section 8.2.2. Its unitary transformation with U_0 reads

$$\begin{aligned} U_0\, g(1,2)\, U_0^\dagger \approx \Big(&\mathbf{1} + W_{[0]}(1) + W_{[0]}(2) + W_{[0]}(1)W_{[0]}(2) \\ &+ \frac{1}{2}W_{[0]}^2(1) + \frac{1}{2}W_{[0]}^2(2)\Big) \left[V_C(1,2) + B_0(1,2)\right] \Big(\mathbf{1} - W_{[0]}(1) \\ &- W_{[0]}(2) + W_{[0]}(1)W_{[0]}(2) + \frac{1}{2}W_{[0]}^2(1) + \frac{1}{2}W_{[0]}^2(2)\Big) \end{aligned} \tag{13.23}$$

which produces the following terms after explicit expansion of the products

$$
\begin{aligned}
U_0 g(1,2) U_0^\dagger \approx\ & \underbrace{V_C(1,2)}_{\equiv\mathcal{E}(1)\mathcal{E}(2)} + \underbrace{B_0(1,2)}_{\equiv\mathcal{O}(1)\mathcal{O}(2)} + \underbrace{\left[W_{[0]}(1) + W_{[0]}(2), V_C(1,2)\right]}_{\equiv\mathcal{O}(1)\mathcal{E}(2)+\mathcal{E}(1)\mathcal{O}(2)} \\
& + \underbrace{\left[W_{[0]}(1) + W_{[0]}(2), B_0(1,2)\right]}_{\equiv\mathcal{E}(1)\mathcal{O}(2)+\mathcal{O}(1)\mathcal{E}(2)} + \frac{1}{2}\underbrace{\left\{W_{[0]}^2(1) + W_{[0]}^2(2), V_C(1,2)\right\}}_{\equiv\mathcal{E}(1)\mathcal{E}(2)} \\
& + \frac{1}{2}\underbrace{\left\{W_{[0]}^2(1) + W_{[0]}^2(2), B_0(1,2)\right\}}_{\equiv\mathcal{O}(1)\mathcal{O}(2)} + \underbrace{\left\{W_{[0]}(1) W_{[0]}(2), V_C(1,2)\right\}}_{\equiv\mathcal{O}(1)\mathcal{O}(2)} \\
& + \underbrace{\left\{W_{[0]}(1) W_{[0]}(2), B_0(1,2)\right\}}_{\equiv\mathcal{E}(1)\mathcal{E}(2)} \\
& - \underbrace{\left(W_{[0]}(1) + W_{[0]}(2)\right) V_C(1,2) \left(W_{[0]}(1) + W_{[0]}(2)\right)}_{\equiv\mathcal{E}(1)\mathcal{E}(2)+\mathcal{O}(1)\mathcal{O}(2)} \\
& - \underbrace{\left(W_{[0]}(1) + W_{[0]}(2)\right) B_0(1,2) \left(W_{[0]}(1) + W_{[0]}(2)\right)}_{\equiv\mathcal{O}(1)\mathcal{O}(2)+\mathcal{E}(1)\mathcal{E}(2)} \qquad (13.24)
\end{aligned}
$$

where we have already omitted all terms that are of order $O(c^{-3})$ or higher and denoted the even/odd structure in underbraces (note the anticommutators in curly brackets). Here, $\mathcal{E}(1)\mathcal{O}(2)$ denotes, for example, that the corresponding term features an even structure with respect to particle 1 and an odd structure with respect to particle 2. For this assignment it is important to consider the structure of the Coulomb and Breit operators as indicated in Eq. (8.89).

13.2.2.1 **All-Even Operators**

We now identify *all-even* operators from Eq. (13.24)

$$
\begin{aligned}
\left[U_0\, g(1,2)\, U_0^\dagger\right]_{\text{all-}\mathcal{E}} =\ & V_C(1,2) + \frac{1}{2}\left\{W_{[0]}^2(1) + W_{[0]}^2(2), V_C(1,2)\right\} \\
& + \left\{W_{[0]}(1) W_{[0]}(2), B_0(1,2)\right\} - W_{[0]}(1) V_C(1,2) W_{[0]}(1) \\
& - W_{[0]}(2) V_C(1,2) W_{[0]}(2) - W_{[0]}(1) B_0(1,2) W_{[0]}(2) \\
& - W_{[0]}(2) B_0(1,2) W_{[0]}(1) + O(c^{-3}) \qquad (13.25)
\end{aligned}
$$

This can be written in a more compact form as

$$
\left[U_0\, g(1,2)\, U_0^\dagger\right]_{\text{all-}\mathcal{E}} = V_C(1,2) + \frac{1}{2}\left\{W_{[0]}^2(1) + W_{[0]}^2(2), V_C(1,2)\right\}
$$

$$-W_{[0]}(1)V_C(1,2)W_{[0]}(1) - W_{[0]}(2)V_C(1,2)W_{[0]}(2)$$

$$+ \underbrace{\left[W_{[0]}(1), \left[W_{[0]}(2), B_0(1,2)\right]\right]}_{=\left[W_{[0]}(2),\left[W_{[0]}(1),B_0(1,2)\right]\right]} + O(c^{-3}) \qquad (13.26)$$

$$= V_C(1,2) + \frac{1}{2}\left[W_{[0]}(1), \left[W_{[0]}(1), V_C(1,2)\right]\right]$$
$$+\frac{1}{2}\left[W_{[0]}(2), \left[W_{[0]}(2), V_C(1,2)\right]\right]$$
$$+ \left[W_{[0]}(1), \left[W_{[0]}(2), B_0(1,2)\right]\right] + O(c^{-3}) \qquad (13.27)$$

$$\equiv V_C(1,2) + \widetilde{V}_C(1,2) + \widetilde{B}_0(1,2) \qquad (13.28)$$

where we have introduced the abbreviations $\widetilde{V}_C(1,2)$ and $\widetilde{B}_0(1,2)$ to conveniently summarize all commutators containing $V_C(1,2)$ and $B_0(1,2)$, respectively, up to second order in $1/c$. In the sections to follow we explicitly resolve these commutators and finally arrive at the Breit–Pauli Hamiltonian. For the explicit evaluation of scalar-product sequences of Dirac's $\boldsymbol{\alpha}_i$ with momentum and distance operators, which emerge according to Eqs. (8.18) and (11.80), two guiding rules must be kept in mind:
(1) Dirac's relation, Eq. (D.24), must be employed for the resolution of the scalar products; note, however, that it requires two such neighboring products in an operator sequence to contain the $\boldsymbol{\alpha}_i$-vector for the *same* electron i.
(2) The momentum operators $\boldsymbol{p}_i$ do not commute with the distance vector $\boldsymbol{r}_{12}$. We have according to the product rule for the case $\boldsymbol{r}_1 \neq \boldsymbol{r}_2$

$$\boldsymbol{p}_1 \cdot \boldsymbol{r}_{12} = \left(\boldsymbol{p}_1 \cdot \boldsymbol{r}_{12}\right) + \boldsymbol{r}_{12} \cdot \boldsymbol{p}_1 \overset{\boldsymbol{r}_1 \neq \boldsymbol{r}_2}{=} -3i\hbar + \boldsymbol{r}_{12} \cdot \boldsymbol{p}_1 \qquad (13.29)$$

$$\boldsymbol{p}_2 \cdot \boldsymbol{r}_{12} = \left(\boldsymbol{p}_2 \cdot \boldsymbol{r}_{12}\right) + \boldsymbol{r}_{12} \cdot \boldsymbol{p}_2 \overset{\boldsymbol{r}_1 \neq \boldsymbol{r}_2}{=} +3i\hbar + \boldsymbol{r}_{12} \cdot \boldsymbol{p}_2 \qquad (13.30)$$

where the parentheses explicitly indicates that only the distance vector needs to be differentiated so that the result is simply a multiplicative operator. The explicit differentiation reads in the case $\boldsymbol{r}_1 \neq \boldsymbol{r}_2$

$$\left(\boldsymbol{p}_1 \cdot \boldsymbol{r}_{12}\right) = -i\hbar \left(\frac{\partial}{\partial x_1}(x_1 - x_2) + \frac{\partial}{\partial y_1}(y_1 - y_2) + \frac{\partial}{\partial z_1}(z_1 - z_2)\right) \overset{\boldsymbol{r}_1 \neq \boldsymbol{r}_2}{=} -3i\hbar \qquad (13.31)$$

and we have

$$\left(\boldsymbol{p}_2 \cdot \boldsymbol{r}_{12}\right) = -\left(\boldsymbol{p}_1 \cdot \boldsymbol{r}_{12}\right) \qquad (13.32)$$

The last equation also holds in the case $\boldsymbol{r}_1 = \boldsymbol{r}_2$ because then $\left(\boldsymbol{p}_2 \cdot \boldsymbol{r}_{12}\right) = 0$.

13.2.2.2 **Transformed Coulomb Contribution**

For the evaluation of the contribution of the transformed Coulomb operator,

$$\widetilde{V}_C = \frac{1}{2}\left[W_{[0]}(1), \left[W_{[0]}(1), V_C(1,2)\right]\right] + \frac{1}{2}\left[W_{[0]}(2), \left[W_{[0]}(2), V_C(1,2)\right]\right] \quad (13.33)$$

to the Breit–Pauli operator we first consider the commutators of the antihermitean operators $W_{[0]}(1)$ and $W_{[0]}(2)$ with the Coulomb operator V_C

$$\left[W_{[0]}(1), V_C(1,2)\right] = \frac{q_1 q_2}{2m_e c}\beta_1\left[\boldsymbol{\alpha}_1\cdot\boldsymbol{p}_1, \frac{1}{r_{12}}\right] \overset{(A.23)}{=} +\frac{q_1 q_2}{2m_e c}(\mathrm{i}\hbar)\beta_1\frac{\boldsymbol{\alpha}_1\cdot\boldsymbol{r}_{12}}{r_{12}^3} \quad (13.34)$$

$$\left[W_{[0]}(2), V_C(1,2)\right] = \frac{q_1 q_2}{2m_e c}\beta_2\left[\boldsymbol{\alpha}_2\cdot\boldsymbol{p}_2, \frac{1}{r_{12}}\right] \overset{(A.23)}{=} -\frac{q_1 q_2}{2m_e c}(\mathrm{i}\hbar)\beta_2\frac{\boldsymbol{\alpha}_2\cdot\boldsymbol{r}_{12}}{r_{12}^3} \quad (13.35)$$

and so the first double commutator turns out to be

$$\begin{aligned}\frac{1}{2}\left[W_{[0]}(1), \left[W_{[0]}(1), V_C(1,2)\right]\right] &= \frac{q_1 q_2}{8m_e^2c^2}(\mathrm{i}\hbar)(-\beta_1^2)\left[\boldsymbol{\alpha}_1\cdot\boldsymbol{p}_1, \frac{\boldsymbol{\alpha}_1\cdot\boldsymbol{r}_{12}}{r_{12}^3}\right]\\
&= \frac{q_1 q_2}{8m_e^2c^2}(-\mathrm{i}\hbar)\left\{(\boldsymbol{\alpha}_1\cdot\boldsymbol{p}_1)(\boldsymbol{\alpha}_1\cdot\boldsymbol{r}_{12})\frac{1}{r_{12}^3} - \frac{1}{r_{12}^3}(\boldsymbol{\alpha}_1\cdot\boldsymbol{r}_{12})(\boldsymbol{\alpha}_1\cdot\boldsymbol{p}_1)\right\}\\
&\overset{(D.24)}{=} \frac{q_1 q_2}{8m_e^2c^2}(-\mathrm{i}\hbar)\left\{(\boldsymbol{p}_1\cdot\boldsymbol{r}_{12})\frac{1}{r_{12}^3}\mathbf{1}_4 + \mathrm{i}\boldsymbol{\Sigma}_1\cdot(\boldsymbol{p}_1\times\boldsymbol{r}_{12})\frac{1}{r_{12}^3}\right.\\
&\quad\left. - \frac{1}{r_{12}^3}(\boldsymbol{r}_{12}\cdot\boldsymbol{p}_1)\mathbf{1}_4 - \frac{1}{r_{12}^3}\mathrm{i}\boldsymbol{\Sigma}_1\cdot(\boldsymbol{r}_{12}\times\boldsymbol{p}_1)\right\}\end{aligned} \quad (13.36)$$

where we employed the fact that β_1 in $W_{[0]}(1)$ commutes with even operators, while it anticommutes with odd operators. The last equation can be simplified by noting that

$$\begin{aligned}\boldsymbol{p}_1\cdot\frac{\boldsymbol{r}_{12}}{r_{12}^3} - \frac{1}{r_{12}^3}(\boldsymbol{r}_{12}\cdot\boldsymbol{p}_1) &= \left(\boldsymbol{p}_1\cdot\frac{\boldsymbol{r}_{12}}{r_{12}^3}\right) + \frac{1}{r_{12}^3}(\boldsymbol{r}_{12}\cdot\boldsymbol{p}_1) - \frac{1}{r_{12}^3}(\boldsymbol{r}_{12}\cdot\boldsymbol{p}_1)\\
&= -4\pi\mathrm{i}\hbar\delta^{(3)}(\boldsymbol{r}_{12})\end{aligned} \quad (13.37)$$

The big parentheses denote a multiplicative operator as before in Eq. (13.30). The derivative of the momentum operator, $\boldsymbol{p}_1 = -\mathrm{i}\hbar\nabla_1$, operating on the fraction $\boldsymbol{r}_{12}/r_{12}^3$ has been evaluated according to

$$-\mathrm{i}\hbar\nabla_1\cdot\frac{\boldsymbol{r}_{12}}{r_{12}^3} \overset{(A.23)}{=} -\mathrm{i}\hbar\nabla_1\cdot\left(-\nabla_1\frac{1}{r_{12}}\right) = \mathrm{i}\hbar\Delta_1\frac{1}{r_{12}} \overset{(A.24)}{=} -4\pi\mathrm{i}\hbar\,\delta^{(3)}(\boldsymbol{r}_{12}) \quad (13.38)$$

For Eq. (13.36) we thus find

$$\begin{aligned}\frac{1}{2}\left[W_{[0]}(1), \left[W_{[0]}(1), V_C(1,2)\right]\right] &= -\frac{q_1 q_2}{8m_e^2c^2}\hbar^2\mathbf{1}_4\left\{4\pi\,\delta^{(3)}(\boldsymbol{r}_{12})\right\} + \frac{q_1 q_2}{8m_e^2c^2}\hbar\\
&\quad\times\left\{\boldsymbol{\Sigma}_1\cdot(\boldsymbol{p}_1\times\boldsymbol{r}_{12})\frac{1}{r_{12}^3} - \frac{1}{r_{12}^3}\boldsymbol{\Sigma}_1\cdot(\boldsymbol{r}_{12}\times\boldsymbol{p}_1)\right\}\end{aligned} \quad (13.39)$$

For the other double commutator we have accordingly

$$\frac{1}{2}\left[W_{[0]}(2),\left[W_{[0]}(2),V_{\mathrm{C}}(1,2)\right]\right]=\frac{q_1q_2}{8m_e^2c^2}(\mathrm{i}\hbar)\beta_2^2\left[\boldsymbol{\alpha}_2\cdot\boldsymbol{p}_2,\frac{\boldsymbol{\alpha}_2\cdot\boldsymbol{r}_{12}}{r_{12}^3}\right] \tag{13.40}$$

which, after the manipulations considered before now carried out for the momentum operator of the second electron,

$$-\mathrm{i}\hbar\nabla_2\cdot\frac{\boldsymbol{r}_{12}}{r_{12}^3}=-\mathrm{i}\hbar\Delta_2\frac{1}{r_{12}}=+4\pi\mathrm{i}\hbar\,\delta^{(3)}(\boldsymbol{r}_{12}) \tag{13.41}$$

and

$$\boldsymbol{p}_2\cdot\frac{\boldsymbol{r}_{12}}{r_{12}^3}-\frac{1}{r_{12}^3}(\boldsymbol{r}_{12}\cdot\boldsymbol{p}_2)=+4\pi\mathrm{i}\hbar\,\delta^{(3)}(\boldsymbol{r}_{12}) \tag{13.42}$$

is now reduced to

$$\begin{aligned}\frac{1}{2}\left[W_{[0]}(2),\left[W_{[0]}(2),V_{\mathrm{C}}(1,2)\right]\right]=&-\frac{q_1q_2}{8m_e^2c^2}\hbar^2\mathbf{1}_4\left\{4\pi\,\delta^{(3)}(\boldsymbol{r}_{12})\right\}+\frac{q_1q_2}{8m_e^2c^2}\hbar\\&\times\left\{-\boldsymbol{\Sigma}_2\cdot(\boldsymbol{p}_2\times\boldsymbol{r}_{12})\frac{1}{r_{12}^3}+\frac{1}{r_{12}^3}\boldsymbol{\Sigma}_2\cdot(\boldsymbol{r}_{12}\times\boldsymbol{p}_2)\right\}\end{aligned} \tag{13.43}$$

Together, both double commutators yield the correction to the transformed Coulomb contribution

$$\begin{aligned}\widetilde{V}_{\mathrm{C}}(1,2)=&-\frac{q_1q_2}{4m_e^2c^2}\hbar^2\mathbf{1}_4\left\{4\pi\,\delta^{(3)}(\boldsymbol{r}_{12})\right\}+\frac{q_1q_2}{8m_e^2c^2}\hbar\\&\times\left\{\boldsymbol{\Sigma}_1\cdot(\boldsymbol{p}_1\times\boldsymbol{r}_{12})\frac{1}{r_{12}^3}-\frac{1}{r_{12}^3}\boldsymbol{\Sigma}_1\cdot(\boldsymbol{r}_{12}\times\boldsymbol{p}_1)\right.\\&\left.-\boldsymbol{\Sigma}_2\cdot(\boldsymbol{p}_2\times\boldsymbol{r}_{12})\frac{1}{r_{12}^3}+\frac{1}{r_{12}^3}\boldsymbol{\Sigma}_2\cdot(\boldsymbol{r}_{12}\times\boldsymbol{p}_2)\right\}\end{aligned} \tag{13.44}$$

13.2.2.3 **Transformed Breit Contribution**

Analogously to the transformed Coulomb interaction, the anticommutator with the Breit operator in Eq. (13.25),

$$\begin{aligned}\left\{W_{[0]}(1)W_{[0]}(2),B_0(1,2)\right\}=&\frac{1}{4m_e^2c^2}\beta_1\beta_2\Big[(\boldsymbol{\alpha}_1\cdot\boldsymbol{p}_1)(\boldsymbol{\alpha}_2\cdot\boldsymbol{p}_2)B_0(1,2)\\&+B_0(1,2)(\boldsymbol{\alpha}_1\cdot\boldsymbol{p}_1)(\boldsymbol{\alpha}_2\cdot\boldsymbol{p}_2)\Big]\end{aligned} \tag{13.45}$$

can be resolved. Here, no sign change occurs because of the anticommutation of β_i with odd operators of particle i ($i=1,2$). Since the α-matrices in the Gaunt-type operator yield

$$(\boldsymbol{\alpha}_1\cdot\boldsymbol{p}_1)(\boldsymbol{\alpha}_2\cdot\boldsymbol{p}_2)\boldsymbol{\alpha}_1\boldsymbol{\alpha}_2=[(\boldsymbol{\alpha}_1\cdot\boldsymbol{p}_1)\boldsymbol{\alpha}_1][(\boldsymbol{\alpha}_2\cdot\boldsymbol{p}_2)\boldsymbol{\alpha}_2]=(\boldsymbol{p}_1\cdot\boldsymbol{p}_2)\mathbf{1}_4 \tag{13.46}$$

where the four-dimensional unit matrix $\mathbf{1}_4$ will be left out below as usual, the transformed (negative) Gaunt-type term becomes

$$\boldsymbol{p}_1 \cdot \boldsymbol{p}_2 \frac{1}{2r_{12}} + \frac{1}{2r_{12}} \boldsymbol{p}_1 \cdot \boldsymbol{p}_2 = \frac{\boldsymbol{p}_1 \cdot \boldsymbol{p}_2}{r_{12}} + \left(\boldsymbol{p}_1 \frac{1}{2r_{12}} \right) \cdot \boldsymbol{p}_2 + \boldsymbol{p}_1 \cdot \left(\boldsymbol{p}_2 \frac{1}{2r_{12}} \right) \tag{13.47}$$

where the product rule still has to be applied to the last term on the right hand side (to be carried out below). By noting that

$$\nabla_1 \frac{1}{r_{12}} = -r_{12}^{-2} \nabla_1 r_{12} \overset{(A.21)}{=} -\frac{\boldsymbol{r}_{12}}{r_{12}^3} \tag{13.48}$$

$$\nabla_2 \frac{1}{r_{12}} = -r_{12}^{-2} \nabla_2 r_{12} \overset{(A.22)}{=} +\frac{\boldsymbol{r}_{12}}{r_{12}^3} \tag{13.49}$$

we can rewrite Eq. (13.47) to obtain

$$\begin{aligned} \boldsymbol{p}_1 \cdot \boldsymbol{p}_2 \frac{1}{2r_{12}} + \frac{1}{2r_{12}} \boldsymbol{p}_1 \cdot \boldsymbol{p}_2 &= \frac{\boldsymbol{p}_1 \cdot \boldsymbol{p}_2}{r_{12}} + \left(\frac{i\hbar}{2} \frac{\boldsymbol{r}_{12}}{r_{12}^3} \right) \cdot \boldsymbol{p}_2 + \boldsymbol{p}_1 \cdot \left(-\frac{i\hbar}{2} \frac{\boldsymbol{r}_{12}}{r_{12}^3} \right) \\ &= \frac{\boldsymbol{p}_1 \cdot \boldsymbol{p}_2}{r_{12}} + \frac{i\hbar}{2} \frac{\boldsymbol{r}_{12} \cdot \boldsymbol{p}_2}{r_{12}^3} - \frac{i\hbar}{2} \frac{\boldsymbol{r}_{12} \cdot \boldsymbol{p}_1}{r_{12}^3} + \frac{\hbar^2}{2} \left(\nabla_1 \frac{\boldsymbol{r}_{12}}{r_{12}^3} \right) \\ &\overset{(13.38)}{=} \frac{\boldsymbol{p}_1 \cdot \boldsymbol{p}_2}{r_{12}} + \frac{i\hbar}{2} \frac{\boldsymbol{r}_{12} \cdot \boldsymbol{p}_2}{r_{12}^3} - \frac{i\hbar}{2} \frac{\boldsymbol{r}_{12} \cdot \boldsymbol{p}_1}{r_{12}^3} + \frac{\hbar^2}{2} \left(4\pi \delta^{(3)}(\boldsymbol{r}_{12}) \right) \end{aligned} \tag{13.50}$$

It must be emphasized that the momentum operators in Eq. (13.50) (as in most equations to follow) only operate on the wave function and not on any other position-dependent quantity in the operators. Thus, they must be read as, e.g., $(\boldsymbol{p}_1 \cdot \boldsymbol{p}_2)/r_{12} = 1/r_{12}\,(\boldsymbol{p}_1 \cdot \boldsymbol{p}_2)$. The second term in the Breit operator is more complicated to transform. The transformation requires us to study

$$\begin{aligned} (\boldsymbol{\alpha}_1 \cdot \boldsymbol{p}_1)(\boldsymbol{\alpha}_2 \cdot \boldsymbol{p}_2) \frac{(\boldsymbol{\alpha}_1 \cdot \boldsymbol{r}_{12})(\boldsymbol{\alpha}_2 \cdot \boldsymbol{r}_{12})}{r_{12}^3} &= (\boldsymbol{\alpha}_1 \cdot \boldsymbol{p}_1)(\boldsymbol{\alpha}_2 \cdot \boldsymbol{p}_2) \frac{1}{r_{12}^3} (\boldsymbol{\alpha}_1 \cdot \boldsymbol{r}_{12})(\boldsymbol{\alpha}_2 \cdot \boldsymbol{r}_{12}) \\ &= \left\{ (\boldsymbol{\alpha}_1 \cdot \boldsymbol{\alpha}_2) \left(\boldsymbol{p}_1 \cdot \left[\boldsymbol{p}_2 \frac{1}{r_{12}^3} \right] \right) + \left(\boldsymbol{\alpha}_2 \cdot \left[\boldsymbol{p}_2 \frac{1}{r_{12}^3} \right] \right) (\boldsymbol{\alpha}_1 \cdot \boldsymbol{p}_1) \right. \\ &\quad \left. + \left(\boldsymbol{\alpha}_1 \cdot \left[\boldsymbol{p}_1 \frac{1}{r_{12}^3} \right] \right) (\boldsymbol{\alpha}_2 \cdot \boldsymbol{p}_2) + \frac{1}{r_{12}^3} (\boldsymbol{\alpha}_1 \cdot \boldsymbol{p}_1)(\boldsymbol{\alpha}_2 \cdot \boldsymbol{p}_2) \right\} \\ &\quad \times (\boldsymbol{\alpha}_1 \cdot \boldsymbol{r}_{12})(\boldsymbol{\alpha}_2 \cdot \boldsymbol{r}_{12}) \end{aligned} \tag{13.51}$$

which we can simplify with

$$\nabla_1 \frac{1}{r_{12}^3} = -3r_{12}^{-4} \nabla_1 r_{12} \overset{(A.21)}{=} -3\frac{\boldsymbol{r}_{12}}{r_{12}^5} \tag{13.52}$$

$$\nabla_2 \frac{1}{r_{12}^3} = -3r_{12}^{-4}\, \nabla_2 r_{12} \overset{(A.22)}{=} +3\frac{\boldsymbol{r}_{12}}{r_{12}^5} \tag{13.53}$$

to obtain

$$\begin{aligned}
(\boldsymbol{\alpha}_1 \cdot \boldsymbol{p}_1)(\boldsymbol{\alpha}_2 \cdot \boldsymbol{p}_2)\frac{(\boldsymbol{\alpha}_1 \cdot \boldsymbol{r}_{12})(\boldsymbol{\alpha}_2 \cdot \boldsymbol{r}_{12})}{r_{12}^3} &= \Bigg\{ (\boldsymbol{\alpha}_1 \cdot \boldsymbol{\alpha}_2)\left(\hbar^2 4\pi\delta^{(3)}(\boldsymbol{r}_{12})\right) \\
&-3i\hbar\left(\boldsymbol{\alpha}_2 \cdot \frac{\boldsymbol{r}_{12}}{r_{12}^5}\right)(\boldsymbol{\alpha}_1 \cdot \boldsymbol{p}_1) + 3i\hbar\left(\boldsymbol{\alpha}_1 \cdot \frac{\boldsymbol{r}_{12}}{r_{12}^5}\right)(\boldsymbol{\alpha}_2 \cdot \boldsymbol{p}_2) \\
&+\frac{1}{r_{12}^3}(\boldsymbol{\alpha}_1 \cdot \boldsymbol{p}_1)(\boldsymbol{\alpha}_2 \cdot \boldsymbol{p}_2)\Bigg\}(\boldsymbol{\alpha}_1 \cdot \boldsymbol{r}_{12})(\boldsymbol{\alpha}_2 \cdot \boldsymbol{r}_{12})
\end{aligned} \tag{13.54}$$

The transformed second operator then reads

$$\begin{aligned}
-\frac{1}{2}\Bigg\{ (\boldsymbol{\alpha}_1 \cdot \boldsymbol{p}_1)(\boldsymbol{\alpha}_2 \cdot \boldsymbol{p}_2)\frac{(\boldsymbol{\alpha}_1 \cdot \boldsymbol{r}_{12})(\boldsymbol{\alpha}_2 \cdot \boldsymbol{r}_{12})}{r_{12}^3} &+ \frac{(\boldsymbol{\alpha}_1 \cdot \boldsymbol{r}_{12})(\boldsymbol{\alpha}_2 \cdot \boldsymbol{r}_{12})}{r_{12}^3} \\
\times(\boldsymbol{\alpha}_1 \cdot \boldsymbol{p}_1)(\boldsymbol{\alpha}_2 \cdot \boldsymbol{p}_2)\Bigg\} &= -2\pi\hbar^2\delta^{(3)}(\boldsymbol{r}_{12})(\boldsymbol{\alpha}_1 \cdot \boldsymbol{\alpha}_2) \\
&-\frac{1}{r_{12}^3}(\boldsymbol{\alpha}_1 \cdot \boldsymbol{p}_1)(\boldsymbol{\alpha}_2 \cdot \boldsymbol{p}_2)(\boldsymbol{\alpha}_1 \cdot \boldsymbol{r}_{12})(\boldsymbol{\alpha}_2 \cdot \boldsymbol{r}_{12}) \\
&+\frac{3i\hbar}{2}\left(\boldsymbol{\alpha}_2 \cdot \frac{\boldsymbol{r}_{12}}{r_{12}^5}\right)(\boldsymbol{\alpha}_1 \cdot \boldsymbol{p}_1)(\boldsymbol{\alpha}_2 \cdot \boldsymbol{r}_{12})(\boldsymbol{\alpha}_1 \cdot \boldsymbol{r}_{12}) \\
&-\frac{3i\hbar}{2}\left(\boldsymbol{\alpha}_1 \cdot \frac{\boldsymbol{r}_{12}}{r_{12}^5}\right)(\boldsymbol{\alpha}_2 \cdot \boldsymbol{p}_2)(\boldsymbol{\alpha}_1 \cdot \boldsymbol{r}_{12})(\boldsymbol{\alpha}_2 \cdot \boldsymbol{r}_{12})
\end{aligned} \tag{13.55}$$

where we have commuted the last two scalar products in the second last term on the right hand side. In order to group those pairs of scalar products which belong to the same coordinate, we note that

$$\begin{aligned}
(\boldsymbol{\alpha}_2 \cdot \boldsymbol{p}_2)(\boldsymbol{\alpha}_1 \cdot \boldsymbol{r}_{12}) &= \alpha_{2,x}\alpha_{1,x}p_{2,x}r_{12,x} + \alpha_{2,x}\alpha_{1,y}p_{2,x}r_{12,y} + \cdots + \alpha_{2,z}\alpha_{1,z}p_{2,z}r_{12,z} \\
&= \alpha_{2,x}\alpha_{1,x}(p_{2,x}r_{12,x}) + \alpha_{2,x}\alpha_{1,x}r_{12,x}p_{2,x} + \alpha_{2,x}\alpha_{1,y}(p_{2,x}r_{12,y}) \\
&\quad +\alpha_{2,x}\alpha_{1,y}r_{12,y}p_{2,x} + \cdots \alpha_{2,z}\alpha_{1,z}(p_{2,z}r_{12,z}) + \alpha_{2,z}\alpha_{1,z}r_{12,z}p_{2,z} \\
&= \boldsymbol{\alpha}_1\boldsymbol{\alpha}_2\,(\boldsymbol{p}_2 \cdot \boldsymbol{r}_{12}) + (\boldsymbol{\alpha}_1 \cdot \boldsymbol{r}_{12})(\boldsymbol{\alpha}_2 \cdot \boldsymbol{p}_2)
\end{aligned} \tag{13.56}$$

where again terms $(p_{2,i}r_{12,j})$ denote multiplicative operators, which vanish if $i \neq j$. Accordingly, the remaining term in parentheses $\sum_i(p_{2,i}r_{12,i}) = (\boldsymbol{p}_2 \cdot \boldsymbol{r}_{12})$ can be written as

$$(\boldsymbol{p}_2 \cdot \boldsymbol{r}_{12}) = \begin{cases} +3i\hbar & , \quad \text{for } \boldsymbol{r}_1 \neq \boldsymbol{r}_2 \quad \text{[see Eq. (13.30)]} \\ 0 & , \quad \text{for } \boldsymbol{r}_1 = \boldsymbol{r}_2 \end{cases} \tag{13.57}$$

In the following, we omit the special case for which the two coordinates coincide, $\boldsymbol{r}_1 = \boldsymbol{r}_2$. Accordingly, we find for the product of scalar products with exchanged indices,

$$(\boldsymbol{\alpha}_1 \cdot \boldsymbol{p}_1)(\boldsymbol{\alpha}_2 \cdot \boldsymbol{r}_{12}) = \boldsymbol{\alpha}_1\boldsymbol{\alpha}_2 \left(-3i\hbar\right) + (\boldsymbol{\alpha}_2 \cdot \boldsymbol{r}_{12})(\boldsymbol{\alpha}_1 \cdot \boldsymbol{p}_1) \tag{13.58}$$

Moreover, we note the commutation of scalar products involving only distance operators, i.e., $(\boldsymbol{\alpha}_1 \cdot \boldsymbol{r}_{12})(\boldsymbol{\alpha}_2 \cdot \boldsymbol{r}_{12}) = (\boldsymbol{\alpha}_2 \cdot \boldsymbol{r}_{12})(\boldsymbol{\alpha}_1 \cdot \boldsymbol{r}_{12})$. Now, we may write

$$\begin{aligned}\left(\boldsymbol{\alpha}_1 \cdot \frac{\boldsymbol{r}_{12}}{r_{12}^5}\right)(\boldsymbol{\alpha}_2 \cdot \boldsymbol{p}_2)(\boldsymbol{\alpha}_1 \cdot \boldsymbol{r}_{12})(\boldsymbol{\alpha}_2 \cdot \boldsymbol{r}_{12}) &= \left(\boldsymbol{\alpha}_1 \cdot \frac{\boldsymbol{r}_{12}}{r_{12}^5}\right)\boldsymbol{\alpha}_1\boldsymbol{\alpha}_2 \left(3i\hbar\right)(\boldsymbol{\alpha}_2 \cdot \boldsymbol{r}_{12}) \\ &\quad + \left(\boldsymbol{\alpha}_1 \cdot \frac{\boldsymbol{r}_{12}}{r_{12}^5}\right)(\boldsymbol{\alpha}_1 \cdot \boldsymbol{r}_{12})(\boldsymbol{\alpha}_2 \cdot \boldsymbol{p}_2)(\boldsymbol{\alpha}_2 \cdot \boldsymbol{r}_{12}) \\ &= \underbrace{\frac{\boldsymbol{r}_{12} \cdot \boldsymbol{r}_{12}}{r_{12}^5}}_{1/r_{12}^3}(3i\hbar) + \underbrace{\frac{\boldsymbol{r}_{12} \cdot \boldsymbol{r}_{12}}{r_{12}^5}}_{1/r_{12}^3}(\boldsymbol{\alpha}_2 \cdot \boldsymbol{p}_2)(\boldsymbol{\alpha}_2 \cdot \boldsymbol{r}_{12})\end{aligned} \tag{13.59}$$

and obtain then for the right-hand side of Eq. (13.55)

$$\begin{aligned}-\frac{1}{2}\left\{\cdots\right\} &= -2\pi\hbar^2\delta^{(3)}(\boldsymbol{r}_{12})(\boldsymbol{\alpha}_1 \cdot \boldsymbol{\alpha}_2) - \frac{3i\hbar}{r_{12}^3}(\boldsymbol{p}_1 \cdot \boldsymbol{r}_{12}) \\ &\quad - \frac{1}{r_{12}^3}(\boldsymbol{\alpha}_1 \cdot \boldsymbol{p}_1)(\boldsymbol{\alpha}_1 \cdot \boldsymbol{r}_{12})(\boldsymbol{\alpha}_2 \cdot \boldsymbol{p}_2)(\boldsymbol{\alpha}_2 \cdot \boldsymbol{r}_{12}) \\ &\quad + \frac{9\hbar^2}{2}\frac{1}{r_{12}^3} + \frac{3i\hbar}{2}\frac{1}{r_{12}^3}(\boldsymbol{\alpha}_1 \cdot \boldsymbol{p}_1)(\boldsymbol{\alpha}_1 \cdot \boldsymbol{r}_{12}) \\ &\quad + \frac{9\hbar^2}{2}\frac{1}{r_{12}^3} - \frac{3i\hbar}{2}\frac{1}{r_{12}^3}(\boldsymbol{\alpha}_2 \cdot \boldsymbol{p}_2)(\boldsymbol{\alpha}_2 \cdot \boldsymbol{r}_{12})\end{aligned} \tag{13.60}$$

The remaining terms for the transformed Breit operator in Eq. (13.25) can be resolved to give

$$-W_{[0]}(1)B_0(1,2)W_{[0]}(2) = \frac{1}{4m_e^2c^2}\beta_1\beta_2(\boldsymbol{\alpha}_1 \cdot \boldsymbol{p}_1)B_0(1,2)(\boldsymbol{\alpha}_2 \cdot \boldsymbol{p}_2) \tag{13.61}$$

$$-W_{[0]}(2)B_0(1,2)W_{[0]}(1) = \frac{1}{4m_e^2c^2}\beta_1\beta_2(\boldsymbol{\alpha}_2 \cdot \boldsymbol{p}_2)B_0(1,2)(\boldsymbol{\alpha}_1 \cdot \boldsymbol{p}_1) \tag{13.62}$$

where we again took care of sign changes since β_1 and β_2 anticommute with the odd operators of electrons 1 and 2, respectively, in $B_0(1,2)$. In order to resolve the operator sequences we note for the first term containing the Gaunt-type operator

$$-(\boldsymbol{\alpha}_1 \cdot \boldsymbol{p}_1)\frac{\boldsymbol{\alpha}_1\boldsymbol{\alpha}_2}{2r_{12}}(\boldsymbol{\alpha}_2 \cdot \boldsymbol{p}_2) = +i\hbar\left(\nabla_1\frac{1}{2r_{12}}\right)\cdot \boldsymbol{p}_2 - \frac{\boldsymbol{p}_1 \cdot \boldsymbol{p}_2}{2r_{12}} \stackrel{(A.23)}{=} -i\hbar\frac{\boldsymbol{r}_{12} \cdot \boldsymbol{p}_2}{2r_{12}^3} - \frac{\boldsymbol{p}_1 \cdot \boldsymbol{p}_2}{2r_{12}} \quad (13.63)$$

and accordingly for the Gaunt-type operator in Eq. (13.62)

$$-(\boldsymbol{\alpha}_2 \cdot \boldsymbol{p}_2)\frac{\boldsymbol{\alpha}_1\boldsymbol{\alpha}_2}{2r_{12}}(\boldsymbol{\alpha}_1 \cdot \boldsymbol{p}_1) \stackrel{(A.23)}{=} +i\hbar\frac{\boldsymbol{r}_{12} \cdot \boldsymbol{p}_1}{2r_{12}^3} - \frac{\boldsymbol{p}_1 \cdot \boldsymbol{p}_2}{2r_{12}} \quad (13.64)$$

Again, the second term of the Breit operator is more difficult to transform. We first consider the action of the momentum operator for which we need to evaluate

$$(\boldsymbol{\alpha}_1 \cdot \boldsymbol{p}_1)(\boldsymbol{\alpha}_1 \cdot \boldsymbol{r}_{12})\frac{1}{r_{12}^3} = (\boldsymbol{\alpha}_1 \cdot \boldsymbol{p}_1)\frac{1}{r_{12}^3}(\boldsymbol{\alpha}_1 \cdot \boldsymbol{r}_{12})$$

$$= \frac{1}{r_{12}^3}(\boldsymbol{\alpha}_1 \cdot \boldsymbol{p}_1)(\boldsymbol{\alpha}_1 \cdot \boldsymbol{r}_{12}) + \left(\boldsymbol{\alpha}_1 \cdot \left[\boldsymbol{p}_1\frac{1}{r_{12}^3}\right]\right)(\boldsymbol{\alpha}_1 \cdot \boldsymbol{r}_{12}) \quad (13.65)$$

and then simplify with Eq. (13.52) to

$$(\boldsymbol{\alpha}_1 \cdot \boldsymbol{p}_1)(\boldsymbol{\alpha}_1 \cdot \boldsymbol{r}_{12})\frac{1}{r_{12}^3} = \frac{(\boldsymbol{\alpha}_1 \cdot \boldsymbol{p}_1)(\boldsymbol{\alpha}_1 \cdot \boldsymbol{r}_{12})}{r_{12}^3} - 3i\hbar\frac{1}{r_{12}^5}(\boldsymbol{\alpha}_1 \cdot \boldsymbol{r}_{12})(\boldsymbol{\alpha}_1 \cdot \boldsymbol{r}_{12})$$

$$= \frac{(\boldsymbol{\alpha}_1 \cdot \boldsymbol{p}_1)(\boldsymbol{\alpha}_1 \cdot \boldsymbol{r}_{12})}{r_{12}^3} - 3i\hbar\underbrace{\frac{\boldsymbol{r}_{12} \cdot \boldsymbol{r}_{12}}{r_{12}^5}}_{1/r_{12}^3} \quad (13.66)$$

so that the complete transformed second term of (negative) $B_0(1,2)$ becomes

$$(\boldsymbol{\alpha}_1 \cdot \boldsymbol{p}_1)(\boldsymbol{\alpha}_1 \cdot \boldsymbol{r}_{12})\frac{1}{2r_{12}^3}(\boldsymbol{\alpha}_2 \cdot \boldsymbol{r}_{12})(\boldsymbol{\alpha}_2 \cdot \boldsymbol{p}_2) = \frac{(\boldsymbol{\alpha}_1 \cdot \boldsymbol{p}_1)(\boldsymbol{\alpha}_1 \cdot \boldsymbol{r}_{12})(\boldsymbol{\alpha}_2 \cdot \boldsymbol{r}_{12})(\boldsymbol{\alpha}_2 \cdot \boldsymbol{p}_2)}{2r_{12}^3} - 3i\hbar\frac{1}{2r_{12}^3}(\boldsymbol{\alpha}_2 \cdot \boldsymbol{r}_{12})(\boldsymbol{\alpha}_2 \cdot \boldsymbol{p}_2) \quad (13.67)$$

and

$$(\boldsymbol{\alpha}_2 \cdot \boldsymbol{p}_2)(\boldsymbol{\alpha}_2 \cdot \boldsymbol{r}_{12})\frac{1}{2r_{12}^3}(\boldsymbol{\alpha}_1 \cdot \boldsymbol{r}_{12})(\boldsymbol{\alpha}_1 \cdot \boldsymbol{p}_1) = \frac{(\boldsymbol{\alpha}_2 \cdot \boldsymbol{p}_2)(\boldsymbol{\alpha}_2 \cdot \boldsymbol{r}_{12})(\boldsymbol{\alpha}_1 \cdot \boldsymbol{r}_{12})(\boldsymbol{\alpha}_1 \cdot \boldsymbol{p}_1)}{2r_{12}^3} + 3i\hbar\frac{1}{2r_{12}^3}(\boldsymbol{\alpha}_1 \cdot \boldsymbol{r}_{12})(\boldsymbol{\alpha}_1 \cdot \boldsymbol{p}_1) \quad (13.68)$$

respectively. The total Breit contribution to the even–even part of the transformed Dirac–Coulomb–Breit Hamiltonian can be obtained by summation of all four contributions derived in this section. We leave it to the reader to perform this collection. Some terms evidently are identical and the total number of terms is thus reduced after summation. However, still some momentum operators have not yet been commuted with distance operators so that the momentum operators are finally found at the utmost right position, where they will directly operate on the wave function. This needs to be done in close analogy to the derivation of terms discussed so far. Once the rearranging is accomplished, Dirac's relation must be exploited like in the case of the transformed Coulomb interaction discussed in the last section. All these manipulations are straightforward — though tedious — when compared to the ones discussed so far.

All that then needs to be done is to collect the individual terms to obtain the total transformed two-electron term, i.e., transformed Coulomb plus transformed Breit terms. Then, the reduction to two-component form is performed as in chapter 12, i.e., through replacing the β_i matrices by two-dimensional $\mathbf{1}_i$ for the electronic states (for the positronic states one would have to replace them by $(-\mathbf{1}_i)$ and through a transition from the (4×4)-dimensional Dirac matrices $\boldsymbol{\alpha}$ to the (2×2)-dimensional Pauli matrices $\boldsymbol{\sigma}$. Although we skip all these steps, we discuss the result in the next section.

13.2.3 The Breit–Pauli Hamiltonian

Finally, the full Breit–Pauli Hamiltonian contains the one-electron terms discussed in the beginning of this chapter as well as the magnetic interactions of the electrons that can be rewritten as 'interacting' (coupled) angular momenta. Starting from the full Dirac–Breit equation for two electrons, Eq. (8.19), we obtain the purely two-component external-field-free Breit–Pauli Hamiltonian [39, p. 377],

$$H_{\text{BP}} = H_0 + H_1 + H_2 + H_3 + H_4 + H_5 \tag{13.69}$$

to be used at most within perturbation theory with the individual operators defined as

$$H_0 = \sum_{i=1}^{2} \left[\frac{\boldsymbol{p}_i^2}{2m_e} + V_{nuc}(i) \right] + \frac{e^2}{r_{12}} = \sum_{i=1}^{2} h^S(i) + \frac{e^2}{r_{12}} \tag{13.70}$$

$$H_1 = -\sum_{i=1}^{2} \frac{\boldsymbol{p}_i^4}{8m_e^3 c^2} \tag{13.71}$$

$$H_2 = -\frac{e^2}{2m_e^2 c^2} \left[\frac{\boldsymbol{p}_1 \cdot \boldsymbol{p}_2}{r_{12}} + \frac{(\boldsymbol{r}_{12} \cdot \boldsymbol{p}_1)(\boldsymbol{r}_{12} \cdot \boldsymbol{p}_2)}{r_{12}^3} \right] \tag{13.72}$$

$$H_3 = \frac{e\hbar}{2m_e^2c^2}\left\{(\boldsymbol{E}_1 \times \boldsymbol{p}_1)\cdot \boldsymbol{s}_1 + \frac{2e}{r_{12}^3}(\boldsymbol{r}_{12}\times \boldsymbol{p}_2)\cdot \boldsymbol{s}_1\right\}$$
$$+\frac{e\hbar}{2m_e^2c^2}\left\{(\boldsymbol{E}_2 \times \boldsymbol{p}_2)\cdot \boldsymbol{s}_2 + \frac{2e}{r_{12}^3}(\boldsymbol{r}_{21}\times \boldsymbol{p}_1)\cdot \boldsymbol{s}_2\right\} \quad (13.73)$$

$$H_4 = -\frac{\mathrm{i}e\hbar}{4m_e^2c^2}\sum_{i=1}^{2}\boldsymbol{p}_i\cdot \boldsymbol{E}_i \quad (13.74)$$

$$H_5 = \frac{e^2\hbar^2}{m_e^2c^2}\left[-\frac{8\pi}{3}(\boldsymbol{s}_1\cdot \boldsymbol{s}_2)\delta^{(3)}(\boldsymbol{r}_{12}) + \frac{\boldsymbol{s}_1\cdot \boldsymbol{s}_2}{r_{12}^3}\delta^{(3)}(\boldsymbol{r}_{12})\right.$$
$$\left.-3\frac{(\boldsymbol{r}_{12}\cdot \boldsymbol{s}_1)(\boldsymbol{r}_{12}\cdot \boldsymbol{s}_2)}{r_{12}^5}\right] \quad (13.75)$$

where H_0 defines the unperturbed nonrelativistic many-electron Hamiltonian that can be employed to generate the unperturbed wave function from the variational principle. The Breit–Pauli Hamiltonian can be written in many different forms (see, for instance, a couple of options presented by Balasubramanian [493]), which is the reason why the explicit though lengthy derivation in this section is so important.

In the Breit–Pauli Hamiltonian, the one-electron Pauli Hamiltonians and the two-dimensional Breit operator can be clearly identified. The factor $e\hbar/(2m_ec)$ in these operators represents the Bohr magneton μ_B, which we have already encountered in section 5.4.3, and the gradient of the nuclear potential is the electric field strength

$$\boldsymbol{E}_i = -\nabla_i\left(\frac{V_{nuc}(i)}{q_e}\right) \quad (13.76)$$

— recall also Eq. (11.89), but now the potential energy operator of the two-electron interaction must also be considered. The operator for electrons 1 and 2 is easily generalizable to arbitrary particle numbers by extension of the summation boundaries in case of the one-electron terms and by summation over all nonredundant two-electron terms along the lines described in section 8.2. However, remember the *caveats* made with respect to the use of the one-electron terms above. Witness also that the one-electron spin–orbit terms are hidden in the vector product of the electric field with the momentum operator in the H_3-operator above. The H_4-operator produces the Laplacian of the potential, i.e., the one-electron Darwin term as derived previously. However, the potential due to the other electron(s) also produces terms of the Darwin-form

$$\boldsymbol{p}_i\cdot \boldsymbol{E}_i = -\mathrm{i}\hbar\Delta_i\left[V_{nuc}(i)+V_{ee}\right] = \mathrm{i}\hbar\, 4\pi\sum_{I=1}^{M} Z_I e\, \delta^{(3)}(\boldsymbol{r}_{iI}) - \mathrm{i}\hbar\, 4\pi e\, \delta^{(3)}(\boldsymbol{r}_{ij}) \quad (13.77)$$

where a summation over all atomic nuclei M is now taken into account and j is either 1 or 2 depending on the value of i, whereas a summation must be carried out if there are more than two electrons in the system.

Note that the operator H_2 exactly resembles the classical Darwin energy given in Eq. (3.250), while no such expression occurred in the Breit operator, which required only velocity rather than momentum operators. The Gaunt-like spin–spin interaction, which can be considered a part of Breit's operator according to Eq. (8.20), is, however, also present, namely in H_5. The spin–spin coupling results in a splitting of different M_S components of the unperturbed wave function even in the absence of an external magnetic field. In the original Dirac–Breit formulation this splitting would be automatically included in a formalism that can be subjected to a variational approach (under the well-known precautions discussed in chapter 8). However, the Breit–Pauli Hamiltonian is widely applied in the molecular response theory of properties; e.g., for magnetic resonance spectroscopy (see Refs. [570–572] and the references therein). The splitting of states due to the interaction of all internal angular momenta is called zero-field splitting. For electron spin resonance spectroscopy a phenomenological spin Hamiltonian is employed to describe these splittings of states; see section 15.4.4 for a discussion of this Hamiltonian in the presence of external magnetic fields. The spin–spin coupling is then described by a contribution

$$H_{\text{eff}}^{(0)} = \boldsymbol{S} \cdot \boldsymbol{D} \cdot \boldsymbol{S} \tag{13.78}$$

where $\boldsymbol{S}$ denotes the total (effective) spin operator, while the zero-field splitting tensor $\boldsymbol{D}$ represents the coupling strength. An explicit expression for $\boldsymbol{D}$ in the elementary two-spin problem can be deduced, of course, from the explicit expression of H_5.

In the Breit–Pauli Hamiltonian, we identify 'iso'-coupling terms, namely the orbit–orbit — also called orbit–other-orbit — and spin–spin — also called spin–other-spin — terms. The latter is the two-component Gaunt interaction derived and discussed in sections 8.1.2 and 8.1.3. On the other hand, we also found various spin–orbit coupling terms. The spin–orbit part of the Breit–Pauli Hamiltonian can be subdivided into a sum of one-electron terms

$$h_{\text{BP}}^{\text{SO}} = \frac{1}{2m^2c^2} \sum_i \sum_I \frac{Z_I (\boldsymbol{r}_{iI} \times \boldsymbol{p}_i) \cdot \boldsymbol{s}_i}{r_{iI}^3} \tag{13.79}$$

and a sum of two-electron terms stemming from the Breit interaction

$$g_{\text{BP}}^{\text{SO}} = \underbrace{-\frac{1}{2m^2c^2} \sum_{i,j\neq i} \frac{(\boldsymbol{r}_{ij} \times \boldsymbol{p}_i) \cdot \boldsymbol{s}_i}{r_{ij}^3}}_{g_{\text{BP}}^{\text{SSO}}} + \underbrace{\frac{1}{m^2c^2} \sum_{i,j\neq i} \frac{(\boldsymbol{r}_{ij} \times \boldsymbol{p}_i) \cdot \boldsymbol{s}_j}{r_{ij}^3}}_{g_{\text{BP}}^{\text{SOO}}} \tag{13.80}$$

which all result, when operator H_3 is evaluated. The details of this derivation should be easy to figure out considering that we have presented this for a single nucleus at the beginning of this chapter. Hence, the nuclear contribution to $\boldsymbol{E}_i$ just needs to be generalized for all M nuclei and the same needs to be performed for the electron–electron Coulomb operator in order to derive the two-electron terms $g_{\rm BP}^{\rm SO}$. The one-electron term $h_{\rm BP}$ represents the interaction of the spin of a particular electron i with the magnetic moment, generated by the very same electron by its interaction with the electric field of the nuclei. This is evident by comparison with Eq. (13.14) from the previous section in the case of the central field potential from a single atomic nucleus. However, in the presence of many nuclei, we may recover the angular momentum operator according to

$$(\boldsymbol{r}_{iA} \times \boldsymbol{p}_i) \cdot \boldsymbol{s}_i = (\boldsymbol{r}_i \times \boldsymbol{p}_i) \cdot \boldsymbol{s}_i - (\boldsymbol{R}_A \times \boldsymbol{p}_i) \cdot \boldsymbol{s}_i = \boldsymbol{l}_i \cdot \boldsymbol{s}_i - (\boldsymbol{R}_A \times \boldsymbol{p}_i) \cdot \boldsymbol{s}_i \quad (13.81)$$

Accordingly, we observe

$$(\boldsymbol{r}_{ij} \times \boldsymbol{p}_i) \cdot \boldsymbol{s}_i = \boldsymbol{l}_i \cdot \boldsymbol{s}_i - (\boldsymbol{r}_j \times \boldsymbol{p}_i) \cdot \boldsymbol{s}_i \quad (13.82)$$

$$(\boldsymbol{r}_{ij} \times \boldsymbol{p}_i) \cdot \boldsymbol{s}_j = \boldsymbol{l}_i \cdot \boldsymbol{s}_j - (\boldsymbol{r}_j \times \boldsymbol{p}_i) \cdot \boldsymbol{s}_j \quad (13.83)$$

which is the reason why the two-electron operators in $g_{\rm BP}$ have been separated into a spin–same-orbit (SSO) and spin–other-orbit (SOO) part, respectively. In analogy to the one-electron term, the spin–same-orbit contribution describes the interaction of the spin of electron i with the magnetic moment of electron i, which results from interaction with the electric field generated by electron j. The spin–other-orbit contribution can thus be understood as the interaction of the spin of electron i with the orbital magnetic moment of electron j. Because of its dependence on the nuclear charge number Z_A the one-electron term yields a dominant contribution to spin–orbit coupling when heavy elements are involved. Recall that a structurally closely related spin–orbit Hamiltonian was also derived in the framework of DKH theory as a no-pair spin–orbit Hamiltonian [524]. The only difference between Eqs. (13.79)–(13.80) and the no-pair expression appears in the form of so-called kinematic factors which embrace the operator expression in Eqs. (13.79)–(13.80).

As a further complication, the addition of externally applied scalar and vector fields leads to new operators, which can be found, e.g., in Bethe's review [39, p. 377] and which we will introduce in chapter 15 dealing with external electromagnetic fields (see, in particular, section 15.2.4). It is important to note that there are *no* additional scalar and vector potentials generated by electron 1 and felt by electron 2 (and vice versa) as in the four-component case in section 8.1. Recall from chapter 8 that the nuclear system is truly external to the electronic system. Only the dominating electron–nucleus monopole interactions are included. However, the nuclei are also moving, i.e., rotating and vibrating electric charges, and hence they also generate magnetic fields. To include them in the Breit–Pauli Hamiltonian generates more and more terms which are difficult to keep track of [573, 574]. Only in the original, four-component framework, the number of such terms is comparatively small.

13.3 The Cowan–Griffin and Wood–Boring Approach

If the basic idea of the Pauli elimination is to be employed for variational calculations, the correct expression for ω given by Eq. (13.3) has to be used. This yields an improved and well-defined version of the original Pauli approximation with elimination of the small component developed by Cowan and Griffin in Dirac–Fock-type atomic structure calculations and then adopted by Wood and Boring for atomic DFT calculations [448, 575–581]. The rationale for the approach is again the observation that, even in the case of the uranium atom, 99.2% of the radial density (as calculated by DFT with the Slater exchange functional) stems from the large-component radial function $P_i(r)$, while $Q_i(r)$ contributes only 0.8%, so that it is desirable to completely eliminate the small component from the one-electron SCF-type equations. The basic idea of this approach is to re-arrange the two coupled radial equations of atoms, see Eqs. (9.105), for their most general form that also covers the DFT case for vanishing inhomogeneities, so that they read in the homogeneous case (i.e., $X_i^P(r) = X_i^Q(r) = 0$)

$$P_i(r) = -\frac{1}{V_i^P(r) - \epsilon_{ii}} A_i^\dagger(r) Q_i(r) \tag{13.84}$$

$$Q_i(r) = -\frac{1}{V_i^Q(r) - \epsilon_{ii}} A_i(r) P_i(r) \tag{13.85}$$

(for the definition of all quantities see section 9.5). Now, the second of these equations can be used to eliminate the small-component radial function $Q_i(r)$ in the upper equation. Then, a second-order differential equation for $P_i(r)$ emerges (compare the analogous derivations for the nonrelativistic limit in section 6.6). This procedure has not yet involved any approximation, and the positive and negative eigenvalues of the Dirac Hamiltonian are, of course, reproduced. It has successfully been applied in atomic many-body calculations [582, 583]. However, all operators of this kind are non-hermitean and energy-dependent, and thus plagued by orbitals that are not mutually orthogonal. Nevertheless, a similar approach based on Kohn–Sham density functional theory has been successfully implemented and tested for calculations on molecules [584].

13.4
Elimination for Different Representations of Dirac Matrices

As a side remark, we may note that elimination procedures like the one just discussed depend on the choice of the Dirac matrices; see section 5.2. All that has been said so far holds for the standard representation, for which we have in split notation,

$$(m_e c^2 + V)\psi^L + c\boldsymbol{\sigma}\cdot\boldsymbol{p}\psi^S = \epsilon\psi^L \tag{13.86}$$
$$c\boldsymbol{\sigma}\cdot\boldsymbol{p}\psi^L - (m_e c^2 - V)\psi^S = \epsilon\psi^S \tag{13.87}$$

while we have in the Weyl representation of Eq. (5.38)

$$(c\boldsymbol{\sigma}\cdot\boldsymbol{p} + V)\psi^L + m_e c^2\psi^S = \epsilon\psi^L \tag{13.88}$$
$$m_e c^2\psi^L - (c\boldsymbol{\sigma}\cdot\boldsymbol{p} + V)\psi^S = \epsilon\psi^S \tag{13.89}$$

If we now seek to eliminate the small component, we must always solve that equation which does not contain the derivative of ψ^S. In the case of the standard representation we therefore rearrange the second one, Eq. (13.87),

$$\psi^S = \frac{c\boldsymbol{\sigma}\cdot\boldsymbol{p}}{\epsilon + m_e c^2 - V}\psi^L \tag{13.90}$$

which can be used to eliminate the small component in Eq. (13.86) and yields the elimination procedures discussed explicitly in this chapter. In the Weyl representation, however, we solve for the first of the set of equations, namely Eq. (13.88),

$$\psi^S = \frac{\epsilon - c\boldsymbol{\sigma}\cdot\boldsymbol{p} - V}{m_e c^2}\psi^L \tag{13.91}$$

and then substitute in the second one, Eq. (13.89). Now it is interesting to see that resolving

$$m_e c^2\psi^L - (c\boldsymbol{\sigma}\cdot\boldsymbol{p} + V)\,\frac{\epsilon - c\boldsymbol{\sigma}\cdot\boldsymbol{p} - V}{m_e c^2}\psi^L = \epsilon\,\frac{\epsilon - c\boldsymbol{\sigma}\cdot\boldsymbol{p} - V}{m_e c^2}\psi^L \tag{13.92}$$

then yields the Klein–Gordon equation of section 5.1 for the large component,

$$m_e^2 c^4\psi^L + \underbrace{\left[c(\boldsymbol{\sigma}\cdot\boldsymbol{p})^2 + c\boldsymbol{\sigma}\cdot\boldsymbol{p}V + Vc\boldsymbol{\sigma}\cdot\boldsymbol{p} + V^2\right]}_{(c\boldsymbol{\sigma}\cdot\boldsymbol{p}+V)^2}\psi^L = \epsilon^2\psi^L \tag{13.93}$$

Hence, we do not gain anything from the Weyl representation when subjected to an elimination procedure.

13.5
Regular Approximations

In the mid-1980s an approach to eliminating the small component was developed in order to arrive at regular expansions for the Hamiltonian that facilitate their use in variational approaches [503, 585]. These regular approximations are based on the general theory of effective Hamiltonians [322, 586]. In the case of the Dirac Hamiltonian the basic idea is to rewrite the expression for ω of Eq. (13.3) in the form

$$\omega = \frac{2m_e c^2}{2m_e c^2 - V}\left[1 + \frac{\epsilon}{2m_e c^2 - V}\right]^{-1} = \frac{2m_e c^2}{2m_e c^2 - V}\sum_{k=0}^{\infty}\left(\frac{\epsilon}{V - 2m_e c^2}\right)^k \quad (13.94)$$

and to choose the new expansion parameter $\epsilon/(V - 2m_e c^2)$, which is the starting point for the so called regular approximations developed to a working quantum chemical method by van Lenthe, Baerends, Snijders and collaborators [507, 587–590]. Note that this new expansion parameter is always strictly smaller than unity. All electronic solutions for point-like Coulombic potentials feature energies larger than $-m_e c^2$, and the potential V is strictly negative. The region close to the nucleus is effectively damped, and the geometric series expansion thus well defined. A truncation of this expansion for ω defines the zero- and first-order regular approximation abbreviated as ZORA and FORA, respectively [587]. The ZORA Hamiltonian is also called the Chang–Pelissier–Durand Hamiltonian owing to the earlier paper [585]. Note that the infinite-order regular (IORA) approximation [508] is an improvement upon ZORA, but does not yield exact Dirac eigenvalues.

A particular noteworthy feature of ZORA is that even in this zeroth order approximation there is an efficient relativistic correction for the region close to the nucleus, where the main relativistic effects come from. Excellent agreement of orbital energies and other valence shell properties with the results from the Dirac equation is obtained, and can even be improved within the so-called *scaled* ZORA variant [507], which takes the renormalization to the transformed large component approximately into account. The main disadvantage of the method is its gauge dependence, i.e., a constant shift of the electrostatic potential does not lead to a constant shift in the energy, because the potential enters nonlinearly in the denominator of the Hamiltonian. This deficiency can, however, be approximately remedied by suitable means [507, 591]. Of course, the construction of the ZORA kinetic energy operator requires the knowledge of the potential V, which is not so straightforward to include in Fock-type operators if off-diagonal exchange contributions show up, which we have already discussed as a general complication for all reduction methods in section 11.1.1. However, one may either use the full potential energy operator [592], which is to be updated in each SCF cycle, or fall back on some efficient approximation like a frozen-core potential [593]. The ZORA method comes in a scalar relativistic variant, in which all spin-dependent terms have been separated, and in the fully fledged spin–orbit version [594, 595]. ZORA is certainly one of the most extensively used efficient quasi-relativistic approaches in quantum chemistry — mostly because of its implementation into the Amsterdam Density Functional (ADF) program suite [596, 597], which allows one to tackle sophisticated problems in molecular science from the first-principles perspective.

Part V

Chemistry with Relativistic Hamiltonians

14
Special Computational Techniques

Smart algorithms and efficient approximations that do not sacrifice accuracy are key building blocks of tractable computational schemes. While this has already been an issue in the last chapters, a couple of important methods have not yet been discussed. It should be obvious that the options to combine the ingredients developed in chapters 8–13 lead to numerous 'methods'. However, we shall now present the most important techniques not mentioned so far. Moreover, effects of relativistic quantum chemistry affect the region close to a heavy atom or to its nucleus, respectively. Especially, molecules composed of mostly light atoms can be described efficiently by one-component Schrödinger quantum mechanics. Hence, efficient approximations can be introduced for actual numerical calculations, which exploit the local nature of relativistic effects.

In this chapter, we present further computationally efficient methods for relativistic calculations of the electronic structures for extended systems like molecules and molecular aggregates. While the theory has been developed in detail in the preceding chapters, we now ask the question how it can be transformed to computationally feasible methods. Though we have already encountered many issues in chapters 10, 12, and 13 that have involved consideration of the computational effort, we shall now discuss more of these efficient approximations.

In principle, one may distinguish two types of approximations, namely those which leave the accuracy of a given solution approach unchanged and those which introduce more drastic approximations. For instance, the number of two-electron integrals is so large that their calculation benefits significantly from efficient approximations (in one-component methods it formally scales with the fourth power of the number of basis functions). Efficient integral pre-screening techniques based on a Schwartz-type inequality reduce the number of the two-electron integrals that are mandatory for a pre-defined accuracy. With so-called density-fitting approaches, which are widely employed in actual calculations, one may expand an orbital product or density in a two-electron Coulomb integral into a new (auxiliary) basis set — reducing the formal scaling of the number of integrals to third power in the number of basis functions — or even in the already chosen one-electron basis set introduced for the representation of the orbitals [598] by hardly compromising the accuracy

Relativistic Quantum Chemistry. Markus Reiher and Alexander Wolf

ISBN: 978-3-527-31292-4

of the total electronic energy expectation value. On the other hand, we may introduce approximations that have a significant effect on the core orbital energies and hence on the total electronic energy, although this error will often drop out as soon as relative valence-shell-dominated energies like chemical reaction energies are considered.

Hence, in this chapter we proceed further on our way from the fundamental theory to different presentations of first-quantized relativistic quantum chemistry — now guided mostly by questions of algorithmic technique and feasibility. For the sake of compactness, the focus in this chapter must be on techniques that are specific to the relativistic realm. In nonrelativistic theory numerous approaches have been devised to reduce the computational effort of quantum chemical calculations. Apart from the just mentioned density-fitting approach, which also goes by the name resolution-of-the-identity approach, specific linear-scaling techniques have been devised [599, 600] that ensure a linear increase of the computational effort with system size (measured by the atom or electron number or directly by the number of basis functions). These employ, for example, localized orbitals or sparse-matrix operations. All these techniques also apply instantaneously to the relativistic analogs.

The preceding three chapters have already introduced Hamiltonians of reduced dimension. Particularly successful in variational calculations are the DKH and ZORA approaches. In their scalar-relativistic variant, they can easily be implemented in a computer program for nonrelativistic quantum chemistry so that spin remains a good quantum number leading to great computational advantages (if this approximation is justifiable). Already for these methods we have seen that numerous approximations can be made in order to increase their computational efficiency with little or even no loss of accuracy compared to a four-component reference calculation with the same type of total wave function.

However, as soon as spin–orbit coupling cannot be neglected, the computational effort increases, and a four-component framework can be quite efficient. Therefore, we discuss some efficient ways of technical importance to rewrite four-component one-electron equations in the next section. Subsequently, section 14.2 discusses approaches that consider spin–orbit effects mainly as some sort of perturbation, and hence we tie in with section 13.2. Finally, we enter the realm of chemistry at the end of this chapter and consider the locality of relativistic contributions.

14.1 The Modified Dirac Equation

Many attempts have been undertaken to rewrite the one-electron Dirac equation — of hydrogen-like atoms as well as of the mean-field SCF type derived in chapter 8 and in matrix form in chapter 10 — to obtain a form that is most suitable for numerical computations. Historically, the transformation and elimination techniques first emerged from such endeavors and were only later studied from a formal point of view as an essential part of the complete picture of relativistic many-electron theory. For instance, the DKH theory was first developed as an efficient low-order approximation to the Dirac equation. Only in the new millennium, was it studied as an exact protocol with well-defined properties sharply distinguished from the much older Foldy–Wouthuysen transformation as detailed in chapters 11 and 12.

Full four-component electronic structure methods do not need to be considered as unavoidably expensive, as was believed in the early days of relativistic quantum chemistry. Four-component correlation methods are not significantly more expensive in computational terms once the four-index transformation from four-component atomic basis functions to molecular spinors has been efficiently accomplished. And even the SCF step can be performed very efficiently with quite a small prefactor as compared to the nonrelativistic situation. Nevertheless, one may wonder what might be the most appropriate representation of a Fock-type operator in such one-electron equations for computational purposes.

Interestingly, all these efforts allow us now to finally understand all relativistic matrix techniques as different flavors of the same brand. For instance, if a matrix representation of a four-dimensional one-electron operator is available, a block-diagonalization, i.e., a DKH transformation, can be accomplished in one shot, though one would, of course, instead aim at a straightforward complete diagonalization to obtain spinors and spinor energies. Such a procedure [601] is also much like the IOTC protocol of Barysz and Sadlej described in chapter 11 with the important difference that the latter aims at the two-dimensional upper-left block of the Hamiltonian without ever computing the full four-dimensional operators. The connection can also be made to the regular approximation methods and to the trick of not explicitly evaluating the kinetic balance condition for the small-component basis functions discussed in section 10.1. The close relations are, of course, no surprise, as all attempts to arrive at a smart computational approach start from the same four-component theory as reference. It is somewhat like the situation described by Lindh for the different approaches of the molecular integrals [190], where a truly efficient method inevitably depends on its smart algorithmic realization and implementation, which may level out formal advantages or disadvantages of different methods.

In order to understand all these connections the best point to start with is the so-called *modified Dirac equation*. The modified Dirac equation [602,603] is the basis of the so-called normalized elimination of the small component (NESC) approach worked out by Dyall [508,604–606]. Here, the small component ψ^S

of the 4-spinor ψ is replaced by a pseudo-large component ϕ^L defined by the relation

$$c\,\sigma\cdot\boldsymbol{p}\,\phi^L = 2m_e c^2 \psi^S \tag{14.1}$$

Insertion of this relation into the split form of the Dirac equation, as given by Eq. (5.128), yields the modified Dirac equation

$$(V-\epsilon)\,\psi^L + T\,\phi^L = 0 \tag{14.2}$$

$$T\,\psi^L + \left[\frac{1}{4m_e^2c^2}(\sigma\cdot\boldsymbol{p})(V-\epsilon)(\sigma\cdot\boldsymbol{p}) - T\right]\phi^L = 0 \tag{14.3}$$

where we have neglected the vector potential. $T = \boldsymbol{p}^2/2m_e$ is the nonrelativistic kinetic energy operator of a single electron. Note that we have assumed a block-diagonal potential — we may even consider this as an equation for a one-electron system rather than as a Fock-type equation for the time being — and, hence, make the same assumption as in section 11.1. Such assumptions can always be generalized afterwards, but the main idea can best be demonstrated for this simple case. The many-electron generalization follows then from the same principles but can easily become crowded and cumbersome.

The modified Dirac equation is a pair of two coupled second-order equations relating the two components of the modified Dirac spinor $\psi' = (\psi^L, \phi^L)$. Since the operator $(\sigma\cdot\boldsymbol{p})/2m_ec$ was extracted from the small component in the definition of the pseudo-large component, the original large component ψ^L and this new *small* component ϕ^L have the same symmetry properties and can thus be expanded in the same basis set. By exploiting this special feature of the matrix representation of the modified Dirac equation it is possible to preserve the proper normalization of the large component during the elimination of the small component ϕ^L. This normalized elimination procedure results in energy eigenvalues which deviate only in the order c^{-4} from the correct Dirac eigenvalues, whereas the standard (un-normalized) elimination techniques (UESC) are only correct up to the order c^{-2}. In addition, the NESC method is free from the singularities which plague the UESC methods, and can be simplified systematically by a sequence of approximations [465, 605, 606], which can reduce the computational cost significantly.

Further improvements were presented by Filatov and Cremer, who succeeded in expanding the exact relativistic Hamiltonian for electronic states by means of linear energy-independent operators [607]. The resulting effective relativistic Hamiltonian is used to obtain a perturbational expansion of the exact relativistic electronic energy, which in zeroth order approximation leads to the infinite-order regular approximation (IORA) formalism [508]. Higher-order corrections to the IORA energy turned out to converge fast, rendering this method a promising alternative to traditional approaches within the framework of regular approximations.

The development of approximations to the NESC technique [465, 605, 606] has been the subject of continuous work by Dyall, and has helped to reduce the computational effort of the original approach. Moreover, promising results for helium-like ions were presented by Filatov and Cremer by introducing regular approximations to the exact NESC equations [607, 608]. An elimination technique closely related to NESC has also been developed by Kutzelnigg and Liu [609–615]. Filatov has emphasized the close similarities of these approaches [612, 613].

It is now common to call exact two-component methods *X2C methods* though all such schemes are not exactly identical (often because of approximations that have been introduced to increase the efficiency with almost no loss of accuracy).

14.2 Efficient Calculation of Spin–Orbit Coupling Effects

A major advantage of spin-dependent relativistic quantum-chemical methods is that electronic transitions, which are spin-forbidden in nonrelativistic theory but nevertheless observed in experiment, can be properly studied [616]. In particular, this is important for electronic transitions from an excited triplet state to a singlet ground state. In nonrelativistic theory, spin symmetry would lead to a vanishing transition matrix element (note that we cannot elaborate the intensity theory of spectroscopic techniques, which largely rests on time-dependent perturbation theory and thus on Fermi's golden rule [617]; see also chapter 12). Accordingly, such an excited state can be rather long lived, leading eventually to the emission of light, which is called *phosphorescence* (see Ref. [618] for a sample study). The spin symmetry is lifted because of spin–orbit interactions, which are consistently taken into account in relativistic methods that go beyond the kinematic or scalar regime. However, the quantum chemical methods that can be employed must facilitate the description of excited states, and hence CI- or MCSCF-like methods are most appropriate. Since a fully four-component calculation can be computationally very demanding, various approximate schemes that still take spin–orbit coupling into account have been devised, and we consider these in the following sections.

We have already noted in chapter 13 that the Pauli approximation produces a spin–orbit coupling operator that may be employed in essentially one-component, i.e., nonrelativistic or scalar-relativistic methods via perturbation theory. Of course, this is an approximation compared to fully fledged four-component methods, but it can be a very efficient one that requires less computational effort without significant loss of accuracy.

The spin–orbit part of the Breit–Pauli Hamiltonian (see section 13.1) proved to yield quite accurate spin–orbit-corrected electronic energies. The full Breit–Pauli Hamiltonian is computationally demanding because of the large number of integrals to be calculated. Therefore, further approximations have been devised and tested. Neglecting the two-electron contribution leads to an overestimation of spin–orbit effects because two-center one- and two-electron terms tend to cancel to some extent [526]. A remedy to this shortcoming has been proposed by Hess *et al.* in the form of a mean-field approximation [526]. Their spin–orbit mean-field approach is based on an effective one-electron Hamiltonian. The general structure of this Hamiltonian features operators which closely resemble Coulomb and exchange operators in Hartree–Fock theory. This idea has been applied with success in the so-called *atomic mean-field integral (AMFI) method*, where all multi-center integrals are neglected [619–622]. The reliability of the spin–orbit mean-field approach has also been analyzed for molecular g-tensors [623–625].

In molecules containing heavy atoms the effect of spin–orbit coupling on spectroscopic parameters and on transition energies tends to be of the same order of magnitude as those of electron correlation. Consequently, spin–orbit-coupling and electron-correlation effects must be properly taken care of — especially since they cannot be expected to be additive. Instead of a configuration-interaction ansatz based on 4-spinors, which is computationally very demanding because of the number of integrals and the size of the N-electron basis, it is straightforward to include spin–orbit effects as a perturbation *a posteriori* to a one-component-based configuration-interaction calculation. Then, the configuration space for the spin–orbit calculation can be chosen to be smaller as compared to the one for the calculation of the correlated wave function. Typically, a CI calculation exploiting spin symmetry is performed first with a large configuration-interaction expansion of the total electronic state. Then, a subset of key states from the full set of eigenfunctions of the CI calculation is selected for the construction of the many-electron spin–orbit perturbation matrix. Since the unperturbed Hamiltonian has been diagonalized in the preceding CI calculation, the resulting energies of the key states can be placed on the diagonal of the spin–orbit matrix. In the case of a truncated CI expansion the diagonal elements of the spin–orbit matrix may also be replaced by extrapolated energy eigenvalues in order to correcting for contributions from discarded configurations [222, 626–628]. The corresponding spin–orbit matrix elements are consequently located on off-diagonal positions. The perturbation calculation is carried out by diagonalization of the spin–orbit matrix. Since the spin–orbit-coupling operator is thus represented in a truncated basis of N-electron states whose contribution to the desired spin–orbit-coupled states is expected to be dominant, this approach is also called 'contracted' spin–orbit configuration interaction [629–633]. The spin–orbit-coupled states are obtained by quasi-degenerate perturbation theory [634,635].

An improved approach to correct for spin-polarization effects by Vallet *et al.* explicitly selects configurations that significantly contribute to electron-correlation effects or spin–orbit-coupling effects in a pre-defined reference space [636]. Of course, the reduced configuration space in the spin–orbit-coupling calculation introduces an error. Teichteil *et al.* [637] proposed a configuration interaction approach with perturbation including spin–orbit coupling (CIPSO) where a perturbative correction of second order is added to matrix elements of the contracted model space in order to account for the contribution of discarded configurations [637].

Alternative 'uncontracted' spin-orbit configuration interaction methods have been developed by Christiansen *et al.* [638,639], by Esser [236], by Balasubramanian [640], and by Alekseyev *et al.* [641]. In the recent approach by Marian and coworkers [642,643] a standard MRCI calculation without spin–orbit coupling is performed first in order to obtain LS-coupled multi-reference states. Then, a matrix is set up that consists of the matrix representation of the spin-free and spin–orbit Hamiltonians involving the full uncontracted electronic configuration space. By diagonalizing a spin–orbit sub-matrix constructed from the most important LS-contracted states, solutions corresponding to the 'contracted' approach are obtained. These eigenvectors now serve as a starting point for the iterative diagonalization of the full spin–orbit MRCI matrix. Needless to say, the uncontracted approach generally yields better results compared to the contracted one because of the much larger N-electron basis set employed. However, the contracted approach still represents a convenient feasible alternative for the lowest-lying spin–orbit eigenstates.

For more details on configuration interaction methods that include spin–orbit coupling we refer to the reviews by Marian [644] and by Hess, Marian and Peyerimhoff [616]. Finally, we also mention that the four-component coupled-cluster approaches discussed in section 8.9 have two-component relatives (see Refs. [645,646] for examples).

14.3
Locality in Four-Component Methods

The main bottleneck of four-component calculations has its origin in the kinetic balance condition that generates a very large basis set for the small component, which contributes only a little to expectation values for moderately large nuclear charge numbers Z. Of course, the size of this effect changes for super-heavy atoms in molecules, where the 'small' component becomes large so that its contribution to expectation values and integrals is increased.

In chapter 10 we have already discussed how the size of the small-component basis set can be made equal to that of the large-component basis set by absorbing the kinetic-balance operator into the one-electron Hamiltonian. However, what has just been said still holds and produces a large pre-factor when we consider the scaling of a four-component SCF calculation compared to a corresponding nonrelativistic one. In chapter 9 we saw plots of large- and small-component radial functions for Kr-like atoms. From these plots it was clear that *all* small component functions are rather localized in the vicinity of the nucleus. This effect transfers to molecules.

Accordingly, the simplest approach to reducing the computational effort of the evaluation of two-electron integrals for small components only is to simply neglect them by increasing thresholds in pre-screening techniques for the integrals to be evaluated or by neglecting these integrals entirely. Of course, this would change the relativistic Roothaan equations in a somewhat uncontrollable manner, although the small magnitude of the effect will guarantee that no inaccuracies occur especially when only a few heavy atoms are bound to mostly light atoms [162, 647–649].

A more consistent way to avoid these integrals has been suggested by Visscher [650], who found that a simple point charge model (the so-called Simple Coulombic Correction) can be used very efficiently to correct for the complete neglect of integrals which contain solely small-component 2-spinors. The reason for this is that the molecular small-component density can be approximated by a superposition of atomic small-component densities. The working procedure of this approximation is surprisingly simple: calculate the potential energy curve without all-small-component integrals and correct *a posteriori* the energy by adding half of the point-charge interaction of the small component's atomic charges, which can be obtained from numerical atomic Dirac–Fock calculations, at a given molecular structure plus the difference in total energy of the atoms calculated with and without all-small-component integrals. Results for I_2 and At_2 demonstrate that results of 'full' DHF and CCSD(T) calculations of bond distances, harmonic frequencies, total energies, and dissociation energies are accurately recovered. The main advantage of Visscher's trick is that it works independently of the particular electronic structure method employed.

If we now consider a large molecule with many light atoms and only one heavy atom, we would like to have a method at hand that treats only the heavy atom with elaborate relativistic methods. For instance, if the DKH transformation is only performed in the atomic basis set of the heavy atom, one may easily see how this could eventually be accomplished (see section 12.5 for more details). The effective-core potential approach is another such method, and is discussed in the following section.

14.4
Relativistic Effective Core Potentials

Based on experimental results in chemistry, atoms in molecules are thought of as governed by their valence shell, while the core shells are little affected when molecular structures and reaction energies are to be determined — though chemical effects on the latter are still spectroscopically detectable via photoelectron spectroscopy. We may therefore call such properties valence-shell properties. In fact, the whole idea of homologous elements in one group of the periodic table is based on this picture. On the other hand, we already know that mostly core shells are affected by relativistic effects. Naturally, one may wonder how this can be exploited. One option would be to consider a frozen core of atomic orbitals, but we may also introduce surrogate potentials [651] that simply substitute the atomic core shells. If these potentials are designed such that they resemble the effect of 'relativistic' core electrons [652], they can be efficiently applied in nonrelativistic quantum chemical calculations. Even spin–orbit effects can be included, as we shall see later.

All relativistic Hamiltonians discussed so far were designed to treat all electrons of a system explicitly. This, however, results in a significant computational effort since quantum mechanical calculations scale with the number of electrons (or more precisely with the number of basis functions). One efficient and consistent approach that replaces the core electrons by a potential acting on the valence electrons only is the effective-core-potential (ECP) ansatz — although a decomposition of the molecular many-particle Hamiltonian into atomic valence and core regions is not strictly possible. Nevertheless, ECPs are derived for isolated atoms and then applied in molecular calculations. Apart from the fact that they reduce the computational effort directly as fewer electrons are treated explicitly, they have the advantage that the modified one-electron Hamiltonians combine easily with any kind of ansatz for the many-electron wave function, be it of configuration-interaction, coupled-cluster or any other orbital-based type. Of course, in the case of DFT, subtleties arise from the fact that the electron density can only be constructed from a valence-only density since the core density is completely missing while the approximate exchange–correlation density functional has been obtained for all-electron calculations. Pragmatic tests have, however, shown that there is little reason for worries about actual calculations — at least as long as approximate functionals still lack sufficient accuracy.

Qualitatively speaking, effects of the core electrons on the valence orbitals due to relativistically contracted core shells must be exerted by the surrogate potential. If an appropriate analytical form for this potential has been chosen, its parameters can be adjusted in four-component atomic structure calculations. Hence, the most important step is the choice of the analytical representation of the ECP. Formally we replace the many-electron Hamiltonian H_{el} of

Eq. (8.75) by an effective valence-only Hamiltonian

$$H_v = \sum_i^{N_v} h_v(i) + \sum_i^{N_v} \sum_{j>i}^{N_v} g(i,j) + V_{cc} + V_{cpp} \tag{14.4}$$

where N_v denotes the number of valence electrons. V_{cc} comprises the Coulomb repulsion energy of all core electrons and all nuclei of a particular molecule, and V_{cpp} is the so-called core polarization potential [653]. The one-electron Hamiltonian can then be chosen to be the nonrelativistic Schrödinger one-electron Hamiltonian

$$h_v(i) = \frac{p^2(i)}{2m_e} + V_{cv}(i) \tag{14.5}$$

where $V_{cv}(i)$ is the ECP. In the case of a molecule, $V_{cv}(i)$ becomes a superposition of all atomic ECPs

$$V_{cv}(i) = \sum_I^M \left[-\frac{Q_I}{r_{Ii}} + \Delta V_{cv}^I(i,I) \right] \tag{14.6}$$

The ECPs contain projection operators to ensure that the valence orbitals are kept orthogonal to and pushed out of the core. They can be expanded in Gaussian basis functions and enter molecular integrals to be solved with the standard techniques introduced in section 10.3. For explicit forms of the ECPs we may refer to review articles listed below (e.g., Ref. [653]).

The first term on the right-hand side of Eq. (14.6) describes the electrostatic attraction potential of the atomic core with the charge $Q_I = (Z_I - N_I^{\text{core}})e$, where Z_I and N_I^{core} are the nuclear charge number of atom I and the number of electrons in atom I to be attributed to the core region, respectively. The distance between nucleus I and electron i is denoted as usual $r_{Ii} = |\boldsymbol{R}_I - \boldsymbol{r}_i|$.

A structurally similar term can be specified for the repulsive interaction of all core electrons and nuclei

$$V_{cc} = \sum_I^M \sum_{I<J}^M \left[\frac{Q_I Q_J}{R_{IJ}} + \Delta V_{cc}^{IJ}(I,J) \right] \tag{14.7}$$

where $R_{IJ} = |\boldsymbol{R}_I - \boldsymbol{R}_J|$ is the distance between nucleus I and nucleus J as before (compare chapter 8). Additional correction terms, ΔV_{cv}^I and ΔV_{cc}^{IJ}, are needed to account for the errors introduced by the simplification of the original all-electron Hamiltonian. The core polarization potential V_{cpp} [654, 655] models the polarization of the core by the electric field from the interacting cores, nuclei and valence electrons. It is a sophisticated correction within the valence-only approximation. Its use is often avoided by the choice of a rather small core to be replaced by the ECP yielding a *small-core ECP*. Large-core

ECPs which substitute many electrons are much more prone to core polarization effects [656–658].

Different ECP approaches can be classified according to various criteria. If the original radial-node structure of the atomic valence orbitals is preserved, a *model potential* is produced [659–663]. If the nodal structure is not conserved, the ECP is called *pseudo potential* [664–667]. While *shape-consistent* pseudo potentials [668–671] are optimized to obtain a maximum resemblance in the shape of pseudo-valence orbitals and original valence orbitals, *energy-consistent* pseudo potentials [672–677] reproduce the experimental atomic spectrum very accurately.

Because of the various approximations involved in the valence-only ansatz, one should keep in mind that extensive testing and validation are needed to obtain reliable predictions for relative energies and spectroscopic parameters. Excellent reviews on ECPs and their application are available of which we may include Refs. [529,653,678–680] here.

Spin–orbit coupling has also been introduced into relativistic ECPs [681, 682]. A spin–orbit pseudo-operator has also been employed by Teichteil *et al.* to reproduce the results from an all-electron approach [637]. Effective spin–orbit Hamiltonians derived from the difference between l- and j-dependent relativistic ECPs have been proposed by, amongst others, Christiansen, Ross, Ermler and coworkers [669,671,683–688], by Dolg, Stoll, Preuss and coworkers [672,673,689,690] and are under constant development [691,692]).

15
External Electromagnetic Fields and Molecular Properties

The relativistic calculation of observables and molecular properties follows the lines elaborated first in nonrelativistic quantum chemistry, in which the small electromagnetic perturbation is considered in a Taylor expansion of the electronic energy. Molecular properties are then defined as the derivatives in this expansion. However, the relativistic minimal coupling principle allows for a rigorous inclusion of electromagnetic fields. Thus, four-component theory of response properties derived from this perturbation is naturally a consistent framework. Such a theory, however, requires the perturbed wave function, which is usually expanded in terms of the set of eigenfunctions of the unperturbed Hamiltonian so that the role of the negative-energy states has to be clarified. In particular, they become decisive for the explanation of the diamagnetic contribution of magnetic-response parameters. Again, two-component schemes are free of negative-energy states, but this benefit needs to be analyzed carefully. Also, they suffer from the so-called picture-change error.

In this chapter, we shall now come back to the question how physical observables are associated with proper operator descriptions, which has already been addressed in section 4.3. All preceding chapters dealt with the proper construction of Hamiltonians for the calculation of energies and wave functions of many electron systems. Here, we shall now transfer this knowledge to the construction of relativistic expressions for *first-principles* calculations of molecular properties for many-electron systems. The basic guideline for this is the fact that all molecular properties can be expressed as total electronic energy derivatives.

Any electromagnetic property of a molecular system is probed by an external electromagnetic field. The external electromagnetic field produced by an incident beam of light is completely characterized by the classical picture of an electromagnetic wave, and all optical (spectroscopic) techniques are to be derived from the very same theoretical framework [693]. We may treat the incoming oscillating electromagnetic wave within the framework of time-dependent perturbation theory from which the intensity theory and Fermi's golden rule also follow [617]. Electronic excitations are mediated by visible or ultraviolet light and give rise to UV/vis absorption, fluorescence and phosphorescence spectroscopy. Externally applied magnetic fields interact with

Relativistic Quantum Chemistry. Markus Reiher and Alexander Wolf

ISBN: 978-3-527-31292-4

the nuclear or electron spin and give rise to a state splitting that is modulated by the electronic structure of the particular molecular system. After the discussion of locality in relativistic electronic structure theory in chapter 14, it is clear that the nonrelativistic theory may fail if properties in close spatial proximity to a heavy atomic nucleus are probed. Examples of this kind are the quadrupole splitting and the isomer shift of Mößbauer spectroscopy, which directly probes the electron density at the nucleus.

Before we consider explicit spectroscopic techniques, we address the question of how, in general, external scalar and vector potentials are to be accounted for if they give rise to the definition of molecular properties. The key to this issue is a series expansion of the electronic energy with respect to the strength of the components of the electromagnetic fields. In principle, we may discuss this issue for the one-electron case first. We must distinguish between electromagnetic fields that are internal or external to the electronic system. Many weak electromagnetic effects affect the electronic structure of a molecule. Typical effects on the energy levels are the hyperfine splitting caused by the interaction of the electrons with the electric and magnetic multipole moments of the nucleus.

In general, they can all be properly dealt with in the framework of perturbation (response) theory. According to the discussion in section 5.4, we may add external electromagnetic fields acting on individual electrons to the one-electron terms in the Hamiltonian of Eq. (8.66). Fields produced by other electrons, so that contributions to the one- and two-electron interaction operators in Eq. (8.66) arise, are not of this kind as they are considered to be internal and are properly accounted for in the Breit (section 8.1) or Breit–Pauli Hamiltonians (section 13.2). Though the external-field-free Breit–Pauli Hamiltonian comprises all internal interactions, like spin–spin and spin–other-orbit terms, they may nevertheless also be considered as a perturbation in molecular property calculations. While our derivation of the Breit–Pauli Hamiltonian did not include additional external fields (like the magnetic field applied in magnetic resonance spectroscopies), we now need to consider these fields as well.

However, other non-electromagnetic effects also influence the electronic structure. Examples are isotope shifts or electro-weak interactions, that lead to parity nonconservation. The latter result from the motion of a nucleus (in an atom: relative to the center of mass) and yield a dependence of energy levels on the nuclear mass (and on the finite size of the nucleus, as we have already discussed in chapters 6 and 9). The energy levels are thus *shifted*, while the electromagnetic perturbations result in a *splitting* of these levels.

The capabilities of molecular property calculations depend not only on the proper relativistic Hamiltonians, but also crucially on the consideration of electron-correlation effects due to the use of the mean-field orbital model. The achievements in the first-principles prediction of molecular properties are very impressive and the state of the art is under constant review (see, e.g., Refs. [694–697]).

15.1 Four-Component Perturbation and Response Theory

Any electromagnetic perturbation of the set of N electrons in a molecule finally leads to sums of operators that may be classified according to the number of electronic coordinates involved as one- or two-electron operators. Of course, if an electron experiences a truly external field a one-electron operator will describe this interaction. If two electrons interact via retarded magnetic fields of the moving electrons as expressed in the frequency-independent Breit interaction and, hence, in the Breit–Pauli Hamiltonian, two-electron interaction operators arise.

For the sake of simplicity, we may consider one-electron operators and incorporate them in the many-electron Hamiltonian via

$$H_{el}(\lambda) = H_{el} + \lambda \mathcal{X} = H_{el} + \lambda \sum_{i=1}^{N} X(i) \tag{15.1}$$

If $\mathcal{X}$ contained two-electron operators, the subsequent analysis given for the one-electron pertubations would be analogous. However, two-electron effects necessarily refer to the description of the electron–electron interaction, while all external potentials affect the potential energy of each electron separately and hence take the form of one-electron operators $X(i)$. Note that these one-electron perturbation operators X must not be confused with the X-operator of decoupling schemes introduced in chapter 11.

The reason why $\mathcal{X}$ can be considered as a simple additive operator is the fact that external vector and scalar potentials hidden in $X(i)$ are simply added to the field-free one-electron Dirac Hamiltonian by the principle of minimal coupling discussed in section 5.4. Therefore, they can easily be separated from the field-free many-electron Hamiltonian H_{el} discussed so far.

The exact field-dependent Hamiltonian $H_{el}(\lambda)$ of Eq. (15.1) is characterized by the parameter λ having a value of one. The influence of the external field described by $\mathcal{X}$ turns out to be small for molecular systems for many relevant cases (for instance, if it stems from incident light or from the magnetic fields of magnetic resonance spectrometer). Actually, this was the reason why certain interactions have been neglected right away for the construction of the field-free many-electron Hamiltonian H_{el} in chapter 8. It is convenient to have a coupling strength parameter like λ that will allow us to make extensive use of series expansions around the field-free case for $\lambda = 0$. In this way, the resulting quantum mechanical equations need not be solved variationally but can be treated as perturbative extensions of the already solved field-free problem. Note that the dependence of any quantity on the perturbation $\mathcal{X}$ will consequently be denoted by explicitly writing down the functional dependence on λ in this chapter. Quantities without explicit dependence on λ do thus not depend on the perturbation $\mathcal{X}$.

The full stationary many-electron equation now reads

$$H_{el}(\lambda)\,\Psi_A(\lambda) \;=\; E_A(\lambda)\,\Psi_A(\lambda) \tag{15.2}$$

where the index $A = 0, 1, \ldots$ shall uniquely label the different electronic states. For the sake of brevity and in contrast to chapter 8 we have now dropped the subscript *'el'* for the electronic wave function and for the electronic energy. Solving the full Dirac Eq. (15.2) can in general be accomplished either variationally or perturbatively.

15.1.1 Variational Treatment

External electromagnetic fields (e.g., intense laser fields) may be so strong that a perturbation approach would not be sensible. Then, the exact energy $E_A(\lambda)$ is sought, and a direct solution of Eq. (15.2) is required, which is usually obtained by means of the variational principle. Both the energy and the wave function depend on the external field. Having determined an exact eigenfunction or a variational approximation to Eq. (15.2), the corresponding energy can always be expressed as the expectation value of the perturbed Hamiltonian with regard to the state under consideration,

$$E_A(\lambda) \;=\; \langle \Psi_A(\lambda) | H_{el}(\lambda) | \Psi_A(\lambda) \rangle \tag{15.3}$$

15.1.2 Perturbation Theory

Often, the exact energy of Eq. (15.3) is not particularly important. On the contrary, a systematic expansion of the energy $E_A(\lambda)$ in terms of the perturbation strength λ is more desirable because this allows us to define properties like the polarizability as an (n-dimensional) coefficient in this series expansion. If such an approximation does not agree with experimental observations, we may simply define further properties as higher-derivative (n'-dimensional) coefficients — like hyperpolarizabilities to stick to this example — in the expansion.

Often, only the linear or quadratic *response* of the unperturbed system to the perturbation $\mathcal{X}$ is required (the perturbative approach is often also denoted as response theory). In other words, we have to calculate the lowest-order corrections of stationary Rayleigh–Schrödinger perturbation theory for the energy and the wave function. Therefore it is assumed that both quantities are analytic functions of λ and may thus be expanded in a power series in λ around the unperturbed solution,

$$E_A(\lambda) = E_A + \lambda E_A^{(1)} + \lambda^2 E_A^{(2)} + \ldots \tag{15.4}$$

$$\Psi_A(\lambda) = \Psi_A + \lambda \Psi_A^{(1)} + \lambda^2 \Psi_A^{(2)} + \ldots \tag{15.5}$$

The expansion coefficients $E_A^{(k)}$ as well as the functions $\Psi_A^{(k)}$ uniquely depend on the perturbation but do *not* feature any λ-dependence. For a meaningful solution of the eigenvalue problem defined by Eq. (15.2) the radius of convergence of this expansion must be larger than unity, i.e., the full perturbed problem with $\lambda = 1$ must be covered by this ansatz (which is by no means a trivial condition).

The perturbed Hamiltonian $H_{el}(\lambda)$ need *not* to be expanded, since it exactly depends linearly on λ and thus does not feature any higher-order corrections. The perturbed energy $E_A(\lambda)$ can also be expanded into a Taylor series around $\lambda = 0$,

$$E_A(\lambda) = \sum_{k=0}^{\infty} \frac{1}{k!} \left[\frac{\mathrm{d}^k E_A(\lambda)}{\mathrm{d}\lambda^k} \right]_{\lambda=0} \lambda^k = \sum_{k=0}^{\infty} \underbrace{\frac{1}{k!} \left[E_A^{(k)}(\lambda) \right]_{\lambda=0}}_{\equiv E_A^{(k)}} \lambda^k \tag{15.6}$$

where a self-explanatory notation for higher-order derivatives has been introduced in the last step. The underbrace in the last expression establishes the connection between the k-th derivative of the perturbed energy $E_A(\lambda)$ taken at $\lambda = 0$ and the λ-independent expansion coefficient $E_A^{(k)}$ defined by Eq. (15.4). For weak perturbations one may assume rapid convergence of the expansions for energy and wave function, and the series can be truncated after the leading orders in λ.

Insertion of the series ansätze given by Eqs. (15.4) and (15.5) into Eq. (15.2) and reordering of terms with respect to their order in λ yields the lowest-order corrections for the energy

$$E_A^{(1)} = \left[\frac{\mathrm{d}E_A(\lambda)}{\mathrm{d}\lambda} \right]_{\lambda=0} = \langle \Psi_A | \mathcal{X} | \Psi_A \rangle \tag{15.7}$$

$$E_A^{(2)} = \frac{1}{2} \left[\frac{\mathrm{d}^2 E_A(\lambda)}{\mathrm{d}\lambda^2} \right]_{\lambda=0} = \langle \Psi_A | \mathcal{X} | \Psi_A^{(1)} \rangle \tag{15.8}$$

Note that the correction $\Psi_A^{(1)}$ to the wave function does not contribute to the first-order energy since it can always be chosen to be orthogonal to the unperturbed wave functions,

$$\langle \Psi_A^{(1)} | H_{el} | \Psi_A \rangle = E_A \langle \Psi_A^{(1)} | \Psi_A \rangle = 0 \tag{15.9}$$

and the same holds for its complex conjugate. This is the reason why all first-order corrections to the wave function $\Psi_A^{(1)}$ (the so-called linear response of the wave function) do not contribute to the first-order energy expression, which is also known as the Hellmann–Feynman theorem [34, 36] and thus follows trivially from this analysis.

The first-order correction to the unperturbed wave function turns out to be

$$\Psi_A^{(1)} = \sum_{\substack{B=0 \\ B \neq A}}^{\infty} \frac{\langle \Psi_B | \mathcal{X} | \Psi_A \rangle}{E_A - E_B} \Psi_B \tag{15.10}$$

With this perturbative correction to the wave function, we can rewrite the second-order energy expression in the familiar form as [279]

$$E_A^{(2)} = \sum_{\substack{B=0 \\ B \neq A}}^{\infty} \frac{\langle \Psi_A | \mathcal{X} | \Psi_B \rangle \langle \Psi_B | \mathcal{X} | \Psi_A \rangle}{E_A - E_B} \tag{15.11}$$

The above analysis is restricted to the framework of *nondegenerate* perturbation theory here in order to focus on the essentials of our discussion of the transition from four- to two-component formulations, but emphasizes that all findings hold equally for the degenerate case.

Depending on the perturbation $\mathcal{X}$, the resulting energy derivatives $E_A^{(k)}$ of Eq. (15.6) determine the molecular property. Of course, there could be more than one perturbation, and the scheme is easily generalized to account also for such combinations by the introduction of additional coupling parameters, say κ for a second perturbation. Hence, we may have pure and mixed derivatives with respect to $\lambda, \kappa, \ldots$ The following list illustrates how this is formally achieved, though the explicit expressions for the perturbation $\mathcal{X}$ will be introduced at a later stage in this chapter,

1) $\sigma_{ij}^{(A)} = \left(\frac{\partial^2 E_A}{\partial B_i \partial \mu_j^{(A)}} \right)_{B_i = \mu_j^{(A)} = 0}$ NMR shift tensor

2) $J_{ij}^{(AB)} = \left(\frac{\partial^2 E_A}{\partial \mu_i^{(A)} \partial \mu_j^{(B)}} \right)_{\mu_i^{(A)} = \mu_j^{(B)} = 0}$ NMR spin–spin coupling

3) $g_{ij} = \left(\frac{\partial^2 E_A}{\partial B_i \partial S_j} \right)_{B_i = S_j = 0}$ ESR g-tensor

4) $A_{ij}^{(A)} = \left(\frac{\partial^2 E_A}{\partial I_i^{(A)} \partial S_j} \right)_{I_i^{(A)} = S_j = 0}$ ESR hyperfine A-tensor

5) $\xi_{ij} = -\left(\frac{\partial^2 E_A}{\partial B_i \partial B_j'} \right)_{B_i = B_j' = 0}$ magnetizability

(where we did not denote the explicit dependence of the target energy on the perturbation $E_A(\lambda, \kappa, ...)$; i, j denote Cartesian coordinates). The first two properties are important parameters of nuclear magnetic resonance (NMR) spectroscopy and the second two those of electron spin resonance (ESR) spectroscopy (also known as electron paramagnetic resonance; EPR). Note that we include all relevant quantities in the full perturbation energy operators $\mathcal{X}$, and thus in X, so that the perturbation parameters λ and κ are still dimensionless coupling strengths. The components of an external magnetic field $\boldsymbol{B}$ are denoted as B_i and those of the spin operator as S_i. The quantities $\boldsymbol{\mu}^{(A)}$ and $\boldsymbol{\mu}^{(B)}$ are the magnetic moments of the nuclei A and B generated by the nuclear spins $\boldsymbol{I}_A$ and $\boldsymbol{I}_B$ according to $\boldsymbol{\mu} = \gamma \boldsymbol{I}$ (γ is the nuclear gyromagnetic ratio). The NMR spin–spin coupling constant requires, of course, the inclusion of these nucleus-specific properties γ_A and γ_B in each one-electron perturbation-energy operator X, as already noted. Like many such material-specific constants, however, they are not required in the explicit evaluation of the derivative but must then be multiplied by the derivative afterwards. Additionally, further manipulations may be necessary to derive the observable quantity. For instance, the isotropic shift can be calculated from the trace of the NMR shift tensor and can then be related to the isotropic shift of a reference molecule to finally produce the chemical shift for the nucleus under consideration.

The energy derivatives can be evaluated numerically in the limit of vanishing perturbation (finite-field methods) or analytically, which is more accurate and more efficient. In practice, a quite straightforward analytic approach produces response equations that are suitable for computational implementation. The single-determinant approaches *Random Phase Approximation* (also known as Time-Dependent Hartree–Fock) and the analogous Time-Dependent Density Functional Theory (TD-DFT) are of this kind and rest on time-dependent perturbation theory [279, 698, 699]. These latter methods are in wide use for the calculation of wave function (or density) response and properties like polarizabilities or NMR chemical shifts for large molecular aggregates. As usual with computational methods, various realizations of response calculations are possible depending on the relativistic Hamiltonian chosen [700–707].

15.1.3
The Dirac-Like One-Electron Picture

An external perturbation that interacts with all electrons, as introduced in Eq. (15.1), modifies the one-electron operators by recalling Eq. (8.73),

$$H_{el}(\lambda) = \sum_{i}^{N} \left[h^D(i) + \lambda X(i) \right] + \sum_{i<j}^{N} g(i,j) \qquad (15.12)$$

(the nucleus–nucleus repulsion term is suppressed; cf. chapter 8). Consequently, the first-order energy perturbation can be reduced to one-electron integrals over the spinors, $\langle \psi_i | X | \psi_j \rangle$,

$$E_A^{(1)} = \langle \Psi_A | \mathcal{X} | \Psi_A \rangle = \sum_{ij}^{N} \gamma_{ij} \langle \psi_i | X | \psi_j \rangle \tag{15.13}$$

weighted by structure factors γ_{ij} according to the rules for the calculation of such matrix elements developed for the one-electron part in the many-electron Hamiltonian in chapter 8. Therefore, we may simplify our discussion by proceeding to an analysis of the one-electron problem only.

Since the electrons of a molecular system are indistinguishable with respect to their properties (i.e., charge and mass) we may drop the electron label i and write the perturbed Dirac Hamiltonian as

$$h^D(\lambda) = h^D + \lambda X \tag{15.14}$$

The unperturbed operator h^D — in this context then also called *zeroth-order* operator — is the one-electron Dirac Hamiltonian of Eq. (8.64)

$$h^D = h^D(\lambda = 0) = c\,\boldsymbol{\alpha} \cdot \boldsymbol{p} + (\beta - 1) m_e c^2 + V \tag{15.15}$$

but still including a scalar potential energy operator V. This potential energy operator V must not be confused with the scalar part of the perturbing electromagnetic potential X. By contrast, it is convenient to incorporate certain potential terms in V so that the zeroth-order operator includes all dominating potential terms while X represents a weak perturbing potential. The electron–nucleus monopole interaction V_{nuc} is of paramount importance and must be included in the zeroth-order Hamiltonian; usually it is the only potential of such magnitude. Note also that we have chosen to use the energy-shifted one-electron operator of section 6.7, which is, however, of no relevance to the analysis.

For later convenience we introduce the *perturbed potential*

$$V(\lambda) = V + \lambda X \tag{15.16}$$

which is not necessarily an even, i.e., block-diagonal operator — in contrast to the electron–nucleus interaction V_{nuc}, which is even diagonal (see section 15.1.4).

Since the first-order energy, i.e., the energy correct up to terms linear in λ, is just the energy expectation value of the perturbed system with respect to the unperturbed wave function Ψ_A,

$$E_A(\lambda) \approx \left\langle \Psi_A \left| \sum_i^N \left[h^D(i) + \lambda X(i) \right] + \sum_{i<j}^N g(i,j) \right| \Psi_A \right\rangle + \mathcal{O}(\lambda^2) \tag{15.17}$$

the *linear response* of the system with respect to the perturbation X is given by the expectation value requiring only the *unperturbed* wave function $\langle \Psi_A | \mathcal{X} | \Psi_A \rangle$. For the evaluation of linear response quantities, knowledge of the unperturbed wave function is thus sufficient.

15.1.4
Two Types of Properties

There are only two basic types of properties that can consistently occur within a four-component (electromagnetic) theory. *Electric* or *electric-field-like* properties are in general described by an even, i.e., block-diagonal operator $X_{\mathcal{E}}$,

$$X_{\mathcal{E}} = \begin{pmatrix} X^{LL} & 0 \\ 0 & X^{SS} \end{pmatrix} \tag{15.18}$$

which can always be combined with the external electron–nucleus potential V to yield the even perturbed potential $V_{\mathcal{E}}(\lambda)$. In almost all cases of practical interest, the upper and lower blocks of the property operator X, i.e., the positive- and negative-energy (2×2) components, respectively, are identical, $X^{LL} = X^{SS} = X_{(2\times 2)}$. For the sake of convenience this puristic notation with subscripts denoting the dimensionality of the corresponding quantity will be suppressed in the following if the dimensionality is obvious from the context. This general form for electric properties is further simplified for one-electron operators. Because of the time-like component of the minimal-coupling procedure of section 5.4,

$$E_A = i\hbar \frac{\partial}{\partial t} \longrightarrow E_A - e\phi = i\hbar \frac{\partial}{\partial t} - e\phi \tag{15.19}$$

(with the one-component scalar potential ϕ) *all* one-electron electric property operators not only feature block-diagonal but even truly diagonal form. That is, the four diagonal entries of the four-dimensional operator are just given by the familiar one-dimensional nonrelativistic expression X_{nr},

$$X = diag(X_{\text{nr}}, X_{\text{nr}}, X_{\text{nr}}, X_{\text{nr}}) = X_{\text{nr}} \mathbf{1}_4 \tag{15.20}$$

with $X_{\text{nr}} = e\phi$. Many important properties, which have been the subject of careful studies employing both four- and two-component techniques over the last decade, are exactly of this diagonal type, e.g., the electric field gradient (EFG), the electric dipole moment $\boldsymbol{\mu}$, various radial moments (i.e., radial expectation values) $\langle r^k \rangle$, etc.

In contrast to even electric properties, so-called *magnetic* or *magnetic-field-like* properties are described by an odd, i.e., off-diagonal operator $X_{\mathcal{O}}$

$$X_{\mathcal{O}} = \begin{pmatrix} 0 & X^{LS} \\ X^{SL} & 0 \end{pmatrix} \tag{15.21}$$

In order to represent an observable, any operator $X_{\mathcal{O}}$ necessarily has to satisfy the hermiticity constraint $X^{LS^{\dagger}} = X^{SL}$. For one-electron magnetic properties this form can further be restricted to expressions with identical and hermitean (2×2)-blocks on the off-diagonal, which can always be cast into the form

$$X^{LS} = X^{SL} = X_{(2\times 2)} = -q_e\, \boldsymbol{\sigma}\cdot\boldsymbol{A} \tag{15.22}$$

because of the principle of minimal coupling (compare section 5.4). We emphasize that it is exactly the minimal coupling procedure that guarantees Lorentz and gauge invariance of electromagnetic properties and of the Hamiltonian. Recall from section 5.4 that the spatial components of minimal coupling are given by

$$\boldsymbol{p} \longrightarrow \boldsymbol{\pi} = \boldsymbol{p} - \frac{q_e}{c}\boldsymbol{A} \tag{15.23}$$

The general form of electric and magnetic properties can, of course, be simultaneously derived from a covariant formulation of the minimal coupling procedure in terms of the 4-momentum $p^{\mu} = (E/c, \boldsymbol{p})$ and the electromagnetic 4-potential $A^{\mu} = (\phi, \boldsymbol{A})$ covering both temporal and spatial components discussed above. If s different sources of electromagnetic fields perturb a molecule or molecular aggregate, the minimal coupling procedure is easily extended as all these sources add up to a total vector potential and a total scalar potential simply obtained as

$$\boldsymbol{A} = \sum_{I=1}^{s} \boldsymbol{A}_I \quad \text{and} \quad \phi = \sum_{I=1}^{s} \phi_I \tag{15.24}$$

where we introduced capital-letter indices to clearly distinguish these sums from those running over electron coordinates. Important examples of magnetic properties investigated by means of four- and two-component approaches are NMR chemical shifts, g-tensors, various phenomena involving electron spin, hyperfine coupling tensors, and magnetic shielding constants (compare the next sections and references given in Refs. [5,6,694,708]).

The most general electromagnetic (4×4)-property X can always be written as a linear combination of an electric and a magnetic contribution,

$$X = X_{\mathcal{E}} + X_{\mathcal{O}} \tag{15.25}$$

15.2
Reduction to Two-Component Form and Picture Change Artifacts

The unitary transformation of the Dirac Hamiltonian to two-component form is accompanied by a corresponding reduction of the wave function. As discussed in detail in chapters 11 and 12, the four-component Dirac spinor ψ will have only two nonvanishing components as soon as complete decoupling of the Dirac Hamiltonian is achieved and can thus be used as a two-component spinor. This feature can be exploited to calculate expectation values of hermitean operators representing observables. However, the procedure requires some precautions to be taken care of with respect to the representation of the operators, i.e., their transition from the original (4×4)-matrix representation (often referred to as the Dirac picture) to a suitable two-component representation (the DKH picture) has to be taken into account.

The neglect of this mandatory transition has been referred to as the 'picture change effect' or 'picture change error' (PCE) [709, 710]. It should be emphasized that this phrase can easily cause misconceptions and has been a source of confusion over the years. As a matter of fact, the PCE is not a physical effect at all but a pure artifact due to an incomplete and thus incorrect formulation of the property expression under consideration within the two-component framework.

Neglecting this obvious necessity of the latter transformation is just wrong and will yield erroneous results if expectation values of the untransformed operator X are to be evaluated with transformed wave functions. Though being formally very simple, the consistent application of the unitary transformation U to the property X can be very challenging in actual calculations and is thus often (at least partially) neglected. Also in the case of higher- or even infinite-order DKH protocols, avoiding these picture change artifacts becomes more and more cumbersome. However, their magnitude and thus influence on numerical results is rapidly diminished with increasing number of unitary decoupling transformations. It is clear that the transformation protocols developed for the decoupling of the large and small components in chapters 11 and 12 which affected the Hamiltonian are the same for the property transformation. Their purpose is not to decouple the property operator X but to properly account for the transformed one-electron states from which, in combination with the correct representation of X, an observable is calculated.

In the following sections, we introduce the basic principles of the correct calculation of expectation values for a *single* electron moving in external fields. A generalization to the many-electron case and the relation to response theory is straighforward along the principles highlighted in the preceding sections.

15.2.1
Origin of Picture Change Errors

The expectation value of the hermitean (4×4)-operator X can be written as

$$\begin{aligned} \langle X \rangle &= \langle \psi | X | \psi \rangle = tr(DX) \\ &= \langle \psi^L | X^{LL} | \psi^L \rangle + \langle \psi^L | X^{LS} | \psi^S \rangle + \langle \psi^S | X^{SL} | \psi^L \rangle + \langle \psi^S | X^{SS} | \psi^S \rangle \end{aligned} \quad (15.26)$$

The (4×4) *density matrix* or *density operator* $D = D(\boldsymbol{r})$ is defined by

$$D \equiv \psi\psi^\dagger = \begin{pmatrix} D^{LL} & D^{LS} \\ D^{SL} & D^{SS} \end{pmatrix} = \begin{pmatrix} \psi^L\psi^{L,\dagger} & \psi^L\psi^{S,\dagger} \\ \psi^S\psi^{L,\dagger} & \psi^S\psi^{S,\dagger} \end{pmatrix} \tag{15.27}$$

with each block of this density operator being a (2×2)-matrix, e.g.,

$$D^{LL} = |\psi^L\rangle\langle\psi^L| = \begin{pmatrix} \psi_1^L \\ \psi_2^L \end{pmatrix} \otimes (\psi_1^{L,*}, \psi_2^{L,*}) = \begin{pmatrix} \psi_1^L\psi_1^{L,*} & \psi_1^L\psi_2^{L,*} \\ \psi_2^L\psi_1^{L,*} & \psi_2^L\psi_2^{L,*} \end{pmatrix} \tag{15.28}$$

The trace of this one-electron density matrix D gives the one-electron (probability) density ρ,

$$trD = trD^{LL} + trD^{SS} = \rho^L + \rho^S = \rho \tag{15.29}$$

This representation is called the *Dirac picture*. Alternatively, this expectation value may be reformulated within the two-component DKH framework employing the large component of the transformed spinor $\tilde{\psi}^L$ only,

$$\begin{aligned} \langle X \rangle &= \langle \psi \,|\, U^\dagger U X U^\dagger U \,|\, \psi \rangle = \langle \tilde{\psi} \,|\, U X U^\dagger \,|\, \tilde{\psi} \rangle = \langle \tilde{\psi}^L | \tilde{X}^{LL} | \tilde{\psi}^L \rangle \\ &= tr(\tilde{D}\tilde{X}) = tr(\tilde{D}^{LL}\tilde{X}^{LL}) \end{aligned} \tag{15.30}$$

It must be noted that the unitary transformation U and its adjoint do *not* commute with the property operator X even if it was a multiplicative electric-field-like operator in position space. The reason for this is the fact that the transformation U always contains momentum operators.

Solely due to the vanishing small component of the transformed (decoupled) spinor $\tilde{\psi}$ only the upper left block $\tilde{X}^{LL}$

$$\tilde{X}^{LL} = U^{LL}X^{LL}U^{LL^\dagger} + U^{LL}X^{LS}U^{LS^\dagger} + U^{LS}X^{SL}U^{LL^\dagger} + U^{LS}X^{SS}U^{LS^\dagger} \tag{15.31}$$

of the transformed operator

$$\tilde{X} = UXU^\dagger = \begin{pmatrix} \tilde{X}^{LL} & \tilde{X}^{LS} \\ \tilde{X}^{SL} & \tilde{X}^{SS} \end{pmatrix} \tag{15.32}$$

is required in order to evaluate expectation values *exactly*. This decoupled, i.e., two-component formulation of relativistic quantum chemistry is denoted here as the *DKH picture*. These two alternatives of calculating expectation values are completely equivalent and no approximation has yet been introduced. However, in two-component quantum chemical calculations of expectation values the unitary transformation of the operator X is often not taken into account. Instead, the quantity

$$\bar{X} \equiv \langle \tilde{\psi}^L \,|\, X^{LL} \,|\, \tilde{\psi}^L \rangle = tr(\tilde{D}^{LL}X^{LL}) \tag{15.33}$$

is calculated, i.e., the change of picture for the operator X is neglected completely. As a consequence of this obviously erroneous proceeding, the artifical difference

$$\mathrm{PCE}(X) = \langle X \rangle - \bar{X} \tag{15.34}$$

is the *picture change error* of the transition from the four-component to the two-component representation.

From these considerations it should be unambiguously clear that the change of picture is an artificial effect resulting from basic properties of the theory of unitary linear transformations rather than a complicated feature of two-component theories. It has already been discussed by Newton and Wigner [711] and Foldy and Wouthuysen [509].

According to Eq. (15.30) the PCE will vanish if one is able to calculate the matrix representation of $\tilde{X}^{LL}$. The expectation value of the operator X may then be evaluated exactly within the two-component framework. The most general form of $\tilde{X}^{LL}$, which is valid for any operator X, is given by Eq. (15.31) and features serious conceptual problems. Apart from being lengthy and complicated at first glance, these expressions — at least for some operators — may *in principle* not be available within a decoupled two-component approach, i.e., strictly without reference to the small components of the Dirac spinor and, hence, without the introduction of a small component basis. As explained in section 12.5, the even blocks U^{LL} and U^{SS} of the unitary transformation U are always functions of $\boldsymbol{p}^2$, whereas the odd blocks U^{LS} and U^{SL} contain an odd number of momentum operators, i.e., they depend on $\boldsymbol{p}$ rather than $\boldsymbol{p}^2$ [510]. It is the key feature of any decoupled two-component theory that all even expressions may be efficiently evaluated in p^2-space by means of the large component only. However, if an expression contained an odd number of momentum operators, it could not be evaluated by means of the large component solely according to the standard DKH procedure, since the proper basis set representation of the (linear) derivative operator $\boldsymbol{p} = -i\hbar\nabla$ requires kinetically balanced basis sets for both the large and the small components.

Usually, X^{LL} and X^{SS} does in fact contain an even number of momentum operators, and the odd terms X^{LS} and X^{SL} are then functions of $\boldsymbol{p}$. However, a vector potential $\boldsymbol{A}$ introduced through minimal coupling does not depend on $\boldsymbol{p}$ at all and may lead to linear momentum operators $\boldsymbol{p}$ in the final even operators, which somewhat destroys the elegant structure of the vector-potential-free DKH transformation and requires additional computational effort. $\tilde{X}^{LL}$ is thus only efficiently accessible within a standard two-component approach if all four terms occurring in Eq. (15.31) are functions of $\boldsymbol{p}^2$. However, especially in the case of external magnetic fields, the emerging vector potentials produce terms that are of odd order in the linear momentum $\boldsymbol{p}$ and then require many additional terms to be calculated, leveling out the advantages of two-component compared to four-component methods.

A plethora of studies have been conducted where the magnitude of the picture change error for various systems of chemical interest was analyzed numerically [527, 548, 549, 710, 712–720]. According to these studies, the picture change error seems to be significant for operators which assume large values in the vicinity of the nuclei, e.g., electric field gradients at heavy nuclei. Since calculated electric field gradients are used to determine the nuclear quadrupole moments from molecular quadrupole coupling constants, the so-called molecular values of nuclear quadrupole moments are very sensitive to the picture change error (see section 15.5 for more details).

15.2.2
Picture-Change-Free Transformed Properties

The PCE vanishes if the upper left block of X is not affected by the unitary transformation U applied to the four-component one-electron perturbation X,

$$\tilde{X} = UXU^{\dagger} \stackrel{!}{=} X \quad \Longleftrightarrow \quad [U, X] \stackrel{!}{=} 0 \tag{15.35}$$

as this can only be fulfilled for a constant operator,

$$X = \text{constant} \tag{15.36}$$

which is of no physical significance. Any dependence on the spatial coordinate $\boldsymbol{r}$ automatically leads to non-commutation with U, so that Eq. (15.35) can never be fulfilled. However, this does not exclude the possibility that the PCE may be small for certain $\boldsymbol{r}$-dependent properties. The latter will be the case if spatial regions far away from atomic cores are probed, i.e., for operators that depend on some positive power of the position operator like the electric dipole moment operator [719].

15.2.3
Free-Particle Foldy–Wouthuysen Transformation of Properties

Since all unitary decoupling transformations have necessarily to start with the free-particle Foldy–Wouthuysen transformation U_0, it is convenient to introduce the free-particle Foldy–Wouthuysen-transformed operator

$$X_0 = U_0 X U_0^{\dagger} = \begin{pmatrix} X_0^{LL} & X_0^{LS} \\ X_0^{SL} & X_0^{SS} \end{pmatrix} \tag{15.37}$$

For *block-diagonal* operators X, i.e., $X^{LS} = X^{SL} = 0$, as it is the case for electric field gradients, for example, all expressions are simplified to a large extent, and the free-particle Foldy–Wouthuysen transformed observable is then given by

$$X_0 = A_p \begin{pmatrix} X^{LL} + \sigma\cdot P_p X^{SS} \sigma\cdot P_p & -X^{LL}\sigma\cdot P_p + \sigma\cdot P_p X^{SS} \\ -\sigma\cdot P_p X^{LL} + X^{SS}\sigma\cdot P_p & X^{SS} + \sigma\cdot P_p X^{LL}\sigma\cdot P_p \end{pmatrix} A_p \quad (15.38)$$

Because of the increasing suppression of higher-order corrections within the Douglas–Kroll–Hess sequence to be discussed in the following sections as a way to approach the exactly picture-change-free property, the initial free-particle Foldy–Wouthuysen transformation U_0 has by far the largest numerical effect on the expectation value of any operator X. Therefore, it is at least necessary to account for the effects of U_0 on the operator X if expectation values are to be calculated within the DKH protocol. According to Eqs. (15.38) this correction is given by

$$\tilde{X}_0^{LL} = A_p X^{LL} A_p + A_p \frac{c\,\sigma\cdot p}{E_p + mc^2} X^{SS} \frac{c\,\sigma\cdot p}{E_p + mc^2} A_p \quad (15.39)$$

A formal analysis of the size of picture change errors is, however, by no means simple. Though one could immediately try to write down (for electric properties) the seemingly leading terms in $1/c$ of the PCE as

$$\text{PCE}(X_\mathcal{E}) = \left\langle \psi_i^L \middle| -\frac{p^2 X_\mathcal{E} + X_\mathcal{E} p^2}{8m^2c^2} + \frac{\sigma\cdot p X_\mathcal{E} \sigma\cdot p}{4m^2c^2} + \mathcal{O}\left(\frac{1}{c^3}\right) \middle| \psi_i^L \right\rangle \quad (15.40)$$

this analysis would be completely misleading, since any $1/c$-expansion of the quantities E_p, A_p, and R_p given by Eq. (11.28) is not allowed [721] — at least not in variational schemes. The formal investigation of the magnitude of the PCE thus requires a completely new strategy for estimating the influence of each term which does not rely on a series expansion in the inverse speed of light. This analysis is presented in the next section.

15.2.4
Breit–Pauli Hamiltonian with External Electromagnetic Fields

At this stage we should add the missing external-field-dependent operators to the Breit–Pauli Hamiltonian reviewed already by Bethe [39]. By contrast to what follows, these terms are also derived in the spirit of the ill-defined Foldy–Wouthuysen expansion in powers of $1/c$. However, since the molecular property calculation is carried out in a perturbation theory anyhow, we may utilize the complete field-dependent Breit–Pauli Hamiltonian in such calculations.

Naturally, the field-dependent Breit–Pauli Hamiltonian automatically results as the low-order limit of the Foldy–Wouthuysen-transformed field-dependent Dirac–Breit Hamiltonian. Once this has been carried out along the lines of the field-free case given explicitly in section 13.2, we obtain a number of additional terms. While all scalar electrostatic contributions can be directly absorbed in the electric field strength $\boldsymbol{E}_i$ of Eq. (13.76), the external magnetic fields give rise to a sixth *two-electron* Hamiltonian,

$$H_6 = \frac{e\hbar}{m_e c} \sum_i \boldsymbol{B}_i \cdot \boldsymbol{s}_i + \frac{e}{m_e c} \sum_i \boldsymbol{A}_i \cdot \boldsymbol{p}_i \tag{15.41}$$

to be added to that of Eq. (13.69). Here, $\boldsymbol{B}_i$ is the magnetic field while $\boldsymbol{A}_i$, $\boldsymbol{s}_i$, and $\boldsymbol{p}_i$ are as usual the spin and momentum operator of the i-th electron and the vector potential acting on it, respectively. Note that the external electromagnetic field also contributes to the *one-electron* terms of the Breit–Pauli Hamiltonian as sketched for arbitrary scalar and vector fields in Eq. (5.134). Hence, in combination with Eq. (15.21) the external-field free Breit Pauli Hamiltonian produces a plethora of operators that may all be assigned a specific type of interaction (see the appendix in Harriman's monograph for an exhaustive list [26]). In a perturbation theory of molecular properties all these operators contribute and may be interpreted accordingly [722, 723].

15.3
Douglas–Kroll–Hess Property Transformation

For the sake of simplicity the DKH property transformation has often been restricted to zeroth or first order, but higher-order property calculations are now routinely available [549, 719, 721]. The dilemma is that the DKH decoupling protocol of the one-particle states is involved and lengthy expressions emerge that have rather small numerical effects on the total observable.

An important aspect of the DKH approach to molecular properties is to understand the necessity to start at the four-component Dirac framework with a Hamiltonian containing the property X under investigation. The evaluation of X within this four-component picture may then be accomplished either variationally or by means of perturbation theory up to some well-defined order as discussed in section 15.1. The reduction to two-component formulations can be realized by suitably chosen DKH transformations for both the variational and the perturbative treatment of X. However, the unitary transformations to be applied are different for both schemes [721], which is to be shown in the following. Of course, this distinction holds irrespective of the specific features of X. The differences will only vanish for infinite-order perturbation theory.

15.3.1
The Variational DKH Scheme for Perturbing Potentials

As before, we stick to the one-electron case and leave the generalization to N electrons to the reader. The proper choice of the DKH expansion parameter, V or $V(\lambda)$, for the DKH transformation is a decisive question in the sequential unitary transformation scheme as it produces a series expansion of the block-diagonal operator with each term to be classified according to a well-defined order in the expansion parameter. In the case of a one-step decoupling scheme (for instance, in a purely numerical fashion as suggested by Barysz and Sadlej for the Hamiltonian; see section 11.6) all derivations of this section are also valid.

If the property is incorporated variationally into the four-component Dirac picture (as in section 15.1), the λ-dependent energy is the reference value which has to be reproduced by the DKH calculation. In order to evaluate this energy within a two-component framework, the perturbed Hamiltonian $h^D(\lambda)$ has to be decoupled by a suitably chosen unitary transformation $U(\lambda)$,

$$h_{bd}(\lambda) = U(\lambda)\, h^D(\lambda)\, U(\lambda)^\dagger = \sum_{k=0}^{\infty} \mathcal{E}_k(\lambda) = \begin{pmatrix} h_+(\lambda) & 0 \\ 0 & h_-(\lambda) \end{pmatrix} \tag{15.42}$$

(bd denotes the $b$lock-$d$iagonal form of the transformed Hamiltonian). The corresponding transformed wave function for positive-energy solutions reads

$$\tilde{\psi}_i(\lambda) = U(\lambda)\, \psi_i(\lambda) = \begin{pmatrix} \tilde{\psi}_i^L(\lambda) \\ 0 \end{pmatrix} \tag{15.43}$$

and no longer couples with the negative-energy components of the Dirac Hamiltonian. The energy defined by Eq. (15.3) can thus be calculated exactly by means of the two-component expectation value

$$\epsilon_i(\lambda) = \langle \tilde{\psi}_i^L(\lambda) | h_+(\lambda) | \tilde{\psi}_i^L(\lambda) \rangle \tag{15.44}$$

This one-electron energy $\epsilon_i(\lambda)$ may be generalized to the many-electron case to yield $E_i(\lambda)$. According to the basic philosophy of the generalized DKH theory [511], the global unitary transformation $U(\lambda)$ is constructed step by step as the product

$$U(\lambda) = \ldots U_2(\lambda)\, U_1(\lambda)\, U_0 \tag{15.45}$$

of unitary transformations $U_i(\lambda)$, $(i = 1, 2, \ldots)$, which are parametrized by a suitably chosen antihermitean operator $W_i(\lambda)$ of exactly i-th order in $V(\lambda)$. It is crucial to realize that the perturbed potential $V(\lambda)$ is the *only* valid expansion parameter for this variational treatment of X within the DKH framework, since each term $\mathcal{E}_k(\lambda)$ contributing to $h_{bd}(\lambda)$ has a well-defined order in this parameter. However, because of Eq. (15.16) these terms cannot be assigned a unique order either in the unperturbed potential V or in the perturbation X. In other words, neither V nor X are valid expansion parameters when taken separately.

15.3.2
Most General Electromagnetic Property

For arbitrary electromagnetic properties defined by Eq. (15.25) the most general form of the expansion parameter is given by $V(\lambda) = V + \lambda X_{\mathcal{E}} + \lambda' X_{\mathcal{O}}$, and features both even and odd components. Suitable choices of λ and λ' activate (1) or deactivate (0) the electric and magnetic interactions, respectively. For the sake of brevity we will restrict this study to one common perturbation parameter λ, so that the perturbed potential is then given by

$$V(\lambda) = V + \lambda X_{\mathcal{E}} + \lambda X_{\mathcal{O}} \equiv V_{\mathcal{E}}(\lambda) + V_{\mathcal{O}}(\lambda) \tag{15.46}$$

In the following, explicit expressions for the lowest-order terms of the decoupled Hamiltonian $h_{bd}(\lambda)$, including most general electromagnetic property operators of the form given by Eq. (15.46), will be derived [721].

Alternatively, one could try to include the magnetic perturbation $V_{\mathcal{O}}(\lambda)$ in the odd kinematic term $(c\,\boldsymbol{\alpha}\cdot\boldsymbol{p})$ of the Dirac Hamiltonian before starting the DKH procedure. That is, the canonical momentum $\boldsymbol{p}$ has to be replaced by the kinematic momentum $\boldsymbol{\pi}$ defined by Eq. (15.23). For purely magnetic properties this was first suggested by Fukui and Baba [724] and was later on worked out in detail by Dyall [701] for properties which may be expanded in solid spherical harmonics. However, this approach is technically very demanding since the perturbing vector potential already enters the zeroth-order square-root term $E_\pi = \sqrt{(c\,\boldsymbol{\sigma}\cdot\boldsymbol{\pi})^2 + m^2c^4}$ via the kinematic momentum, which — in contrast to the canonical momentum — in quantum mechanics does not feature a vanishing vector product with itself, i.e., $\boldsymbol{\pi}\times\boldsymbol{\pi} \neq \mathbf{0}$. Furthermore, even terms arising by such a scheme can no longer be uniquely classified according to their order in the perturbed potential $V = V(\lambda)$.

The initial transformation U_0 is the familiar free-particle Foldy–Wouthuysen transformation [509] and thus independent of the perturbation. Its application to $h^D(\lambda)$ yields the perturbed free-particle Foldy–Wouthuysen Hamiltonian

$$h_1(\lambda) = U_0 h^D(\lambda) U_0^\dagger = \mathcal{E}_0 + \mathcal{E}_1(\lambda) + \mathcal{O}_1(\lambda) \tag{15.47}$$

with

$$\mathcal{E}_0 = \beta E_p - mc^2 \tag{15.48}$$

being the standard zeroth-order term of DKH theory [511]. It is, of course, not affected by the perturbation. Because of the presence of the property operator

X, however, the first-order terms are modified and read

$$\mathcal{E}_1(\lambda) = A_p\big(V_{\mathcal{E}}(\lambda) + R_p V_{\mathcal{E}}(\lambda) R_p\big) A_p + \beta A_p \big\{R_p, V_{\mathcal{O}}(\lambda)\big\} A_p \tag{15.49}$$

$$\mathcal{O}_1(\lambda) = \beta A_p \big[R_p, V_{\mathcal{E}}(\lambda)\big] A_p + A_p\big(V_{\mathcal{O}}(\lambda) - R_p V_{\mathcal{O}}(\lambda) R_p\big) A_p \tag{15.50}$$

where, as before, '$[,]$' and '$\{,\}$' denote the commutator and anticommutator, respectively. The positive and negative signs in these expressions are due to the commuting and anticommuting behavior of $V_{\mathcal{E}}(\lambda)$ and $V_{\mathcal{O}}(\lambda)$ with β, respectively. The energy-momentum relation E_p, the kinematic factor A_p, and the operator R_p have been defined in section 11.3. Both the even and odd terms $\mathcal{E}_1(\lambda)$ and $\mathcal{O}_1(\lambda)$ are exactly of first order (i.e., linear) in the expansion parameter $V(\lambda)$, but neither is strictly linear in V or X. It is important to emphasize that because of this result the vector-potential-independent free-particle Foldy–Wouthuysen transformation U_0 is also the mandatory first step for the DKH theory of properties.

Within the framework of the generalized DKH transformation the next unitary transformation $U_1(\lambda)$ (see section 11.4 and the entire chapter 12) is given by

$$U_1(\lambda) = a_{1,0}\mathbf{1} + \sum_{k=1}^{\infty} a_{1,k}\big[W_1(\lambda)\big]^k \tag{15.51}$$

where the odd and antihermitean first-order operator $W_1(\lambda)$ is uniquely determined by the requirement that it has to account for the elimination of the first-order odd term,

$$\big[W_1(\lambda), \mathcal{E}_0\big] = -\mathcal{O}_1(\lambda) \tag{15.52}$$

It is thus given by

$$W_1(\lambda) = A_p\big[R_p, \tilde{V}_{\mathcal{E}}(\lambda)\big] A_p + \beta A_p\big(\tilde{V}_{\mathcal{O}}(\lambda) - R_p \tilde{V}_{\mathcal{O}}(\lambda) R_p\big) A_p \tag{15.53}$$

where $\tilde{V}$ is the energy-damped quantity, $\tilde{V}_{ij} = V_{ij}/(E_i + E_j)$ (see chapter 12).

In accordance with the $(2n+1)$-rule the application of $U_1(\lambda)$ yields the final second- and third-order even terms, whose most compact form seems to be identical to the familiar unperturbed expressions,

$$\mathcal{E}_2(\lambda) = \tfrac{1}{2}\big[W_1(\lambda), \mathcal{O}_1(\lambda)\big] \tag{15.54}$$

$$\mathcal{E}_3(\lambda) = \tfrac{1}{2}\big[W_1(\lambda), [W_1(\lambda), \mathcal{E}_1(\lambda)]\big] \tag{15.55}$$

However, because of the odd components of the perturbed potential [cf. Eq. (15.46)] these expressions are much more involved. As an example we give the second-order even term [510]

$$\begin{aligned}\mathcal{E}_2(\lambda) \;=\; & \tfrac{1}{2}\Big\{ -\beta A_p\big[R_p,\tilde{V}_{\mathcal{E}}(\lambda)\big]A_p^2\big[R_p,V_{\mathcal{E}}(\lambda)\big]A_p \\ & -\beta A_p\big[R_p,V_{\mathcal{E}}(\lambda)\big]A_p^2\big[R_p,\tilde{V}_{\mathcal{E}}(\lambda)\big]A_p \\ & +A_p\big[R_p,\tilde{V}_{\mathcal{E}}(\lambda)\big]A_p^2\big(V_{\mathcal{O}}(\lambda)-R_pV_{\mathcal{O}}(\lambda)R_p\big)A_p \\ & +A_p\big[R_p,V_{\mathcal{E}}(\lambda)\big]A_p^2\big(\tilde{V}_{\mathcal{O}}(\lambda)-R_p\tilde{V}_{\mathcal{O}}(\lambda)R_p\big)A_p \\ & -A_p\big(\tilde{V}_{\mathcal{O}}(\lambda)-R_p\tilde{V}_{\mathcal{O}}(\lambda)R_p\big)A_p^2\big[R_p,V_{\mathcal{E}}(\lambda)\big]A_p \\ & -A_p\big(V_{\mathcal{O}}(\lambda)-R_pV_{\mathcal{O}}(\lambda)R_p\big)A_p^2\big[R_p,\tilde{V}_{\mathcal{E}}(\lambda)\big]A_p \\ & +\beta A_p\big(\tilde{V}_{\mathcal{O}}(\lambda)-R_p\tilde{V}_{\mathcal{O}}(\lambda)R_p\big)A_p^2\big(V_{\mathcal{O}}(\lambda)-R_pV_{\mathcal{O}}(\lambda)R_p\big)A_p \\ & +\beta A_p\big(V_{\mathcal{O}}(\lambda)-R_pV_{\mathcal{O}}(\lambda)R_p\big)A_p^2\big(\tilde{V}_{\mathcal{O}}(\lambda)-R_p\tilde{V}_{\mathcal{O}}(\lambda)R_p\big)A_p\Big\}\end{aligned} \tag{15.56}$$

The first two terms correspond to the familiar unperturbed second-order expressions, and the following six terms are due to the magnetic perturbation. Accordingly, subsequent application of the remaining unitary transformations $U_i(\lambda)$ will yield all even terms $\mathcal{E}_k(\lambda)$ up to the desired order n of decoupling. The corresponding DKHn approximation of the energy $E_i(\lambda)$ may then be evaluated using Eq. (15.44). In the limit of exact decoupling, $n \to \infty$, it exactly reproduces the four-component expression given by Eq. (15.3), of course, since no approximation has been introduced by the DKH protocol. However, the number of terms contributing to higher-order DKH corrections and thus the computational effort are rather large.

For purely electric properties ($X_{\mathcal{O}} = 0$) the perturbed potential $V(\lambda)$ is strictly even, and the resulting DKH expressions are significantly simplified. The familiar DKH expressions for the unperturbed Dirac Hamiltonian h^D can be directly transferred to the full system just by replacing the electron–nucleus interaction V by the even perturbed potential

$$V_{\mathcal{E}}(\lambda) \;=\; V + \lambda X_{\mathcal{E}} \tag{15.57}$$

Depending on both the molecular system and the specific form of the property under investigation, the necessary DKH order n_{opt} for exact decoupling, i.e., decoupling up to machine precision, can be predicted prior to any quantum chemical calculation by means of the recently proposed truncation error analysis, which has been described in detail in Ref. [532]. The truncation estimate operator for the variational treatment of electric properties reads

$$\mathcal{V}_k(\lambda) = (4m^2c^2)^{-1}\cdot A_p\,V_{\mathcal{E}}(\lambda)A_p\cdot\big(A_p\,\tilde{V}_{\mathcal{E}}(\lambda)A_p\big)^{k-1} \tag{15.58}$$

15.3.3
Perturbative Approach

In most cases of chemical and physical interest the property X is incorporated perturbatively by means of the linear response expression given by Eq. (15.7). This four-component reference value has thus to be reproduced by the infinite-order two-component DKH scheme, and any finite-order DKH approximation should converge toward this reference with increasing order of decoupling. We emphasize that it is *not* the exact variational energy $\epsilon_i(\lambda)$ discussed above which has to be reproduced by the DKH linear response treatment but the first-order correction $\epsilon_i^{(1)}$.

For the proper choice of the unitary DKH transformations the question of whether or not these unitary transformations shall depend on the perturbation needs to be answered. For example, in a study on hyperfine coupling tensors Malkin *et al.* [725] suggested using a unitary DKH transformation which only depends on the electron–nucleus interaction V and not on the hyperfine operator, i.e., $U = U(V)$. In other investigations on the DKH calculation of electric field gradients [548, 726], dipole moments [450], magnetic shielding constants [727], however, unitary transformations have been applied which also explicitly depend on the property operator X, i.e., $U = U(V(\lambda)) = U(V, X)$. Though both approaches obviously yield different *formal* expressions for the DKH-transformed property operator, both schemes would be numerically equivalent in the limit of exact decoupling. However, the numerical differences for finite-order approximations are comparatively small. These different choices in the literature are accompanied by the question about the proper choice of the expansion parameter [V or $V(\lambda)$] for a perturbative treatment. But one may give rational arguments in favor of the much simpler direct approach, which is independent of the perturbing potential [721]. In the following we shed light on these issues and rederive explicit DKH expressions for the expectation value given by Eq. (15.7) up to third order [721].

15.3.3.1 Direct DKH Transformation of First-Order Energy

Assume the unperturbed Dirac Hamiltonian h^D given by Eq. (15.15) has been decoupled by the perturbation-independent unitary transformation U to yield the standard (i.e., perturbation-free) block-diagonal infinite-order DKH Hamiltonian h_{bd} with the unitary transformation U decomposed into a sequence of unitary transformations U_i to be applied stepwise as discussed in depth in chapter 12. The transformed unperturbed wave function for positive-energy solutions is — according to Eq. (11.16) — given by

$$\tilde{\psi}_i = U\psi_i = \begin{pmatrix} \tilde{\psi}_i^L \\ 0 \end{pmatrix} \tag{15.59}$$

Then, the individual unitary transformations do *not* depend on the perturbation. Since it is the *unperturbed* wave function ψ_i that enters the linear-response energy correction of Eq. (15.7), the two-component DKH expression for $\epsilon_i^{(1)}$ is most easily derived by smart insertion of 'unperturbed' identity operators ($\mathbf{1} = U^\dagger U$) according to

$$\begin{aligned}\epsilon_i^{(1)} &\equiv \langle X \rangle = \left\langle U\psi_i \middle| UXU^\dagger \middle| U\psi_i \right\rangle \\ &= \left\langle \tilde{\psi}_i \middle| X_{\mathrm{DKH}\infty} \middle| \tilde{\psi}_i \right\rangle = \left\langle \tilde{\psi}_i^L \middle| X_{\mathrm{DKH}\infty}^{LL} \middle| \tilde{\psi}_i^L \right\rangle \end{aligned} \tag{15.60}$$

In contrast to the decoupled DKH Hamiltonian h_{bd}, the transformed property operator $X_{\mathrm{DKH}\infty} = UXU^\dagger$ is *not* block-diagonal, but its off-diagonal components do not affect $\epsilon_i^{(1)}$ because of the vanishing lower components of $\tilde{\psi}_i$. In practical calculations this is even the case for any finite-order DKH scheme because the lower two components are never calculated and are thus set equal to zero. Since the first-order energy correction $\epsilon_i^{(1)}$ is given as the unperturbed expectation value of the property operator X, a compact yet slightly imprecise notation $\langle X \rangle$ has been introduced in Eq. (15.60) which suppresses the information on the state for which the expectation value is evaluated. It is crucial to realize that all terms contributing to the transformed property

$$X_{\mathrm{DKH}\infty} = \ldots U_2 U_1 \underbrace{U_0 X U_0^\dagger}_{X_1} U_1^\dagger U_2^\dagger \ldots = \sum_{k=0}^{\infty} \left(X_{\mathcal{E},k} + X_{\mathcal{O},k} \right) \tag{15.61}$$

are exactly first-order (i.e., linear) in X and can be uniquely classified according to their order in V, which is represented by the subscript attached to each term. For instance, the term $X_{\mathcal{E},2}$ is even and of second order in the external potential V. Here the notations $X_{i+1} = U_i X_i U_i^\dagger$ ($i \in \mathbb{N}_0$ and $X_0 = X$) for the intermediate, partially transformed property operator and $X_{\mathrm{DKH}\infty}$ (instead of X_∞) for the final transformed property operator have been employed. Since the antihermitean operator W_k parametrizing the unitary transformation U_k is of k-th order in V, the intermediate, partially transformed property operator can always be written as

$$X_{i+1} = U_i X_i U_i^\dagger = \sum_{k=0}^{i} \left(X_{\mathcal{E},k} + X_{\mathcal{O},k} \right) + \sum_{k=i+1}^{\infty} \left(X_{\mathcal{E},k}^{(i+1)} + X_{\mathcal{O},k}^{(i+1)} \right) \tag{15.62}$$

where superscripts in parentheses indicate terms which are modified by subsequent unitary transformations. In other words, if the DKH property operator up to n-th order in V is sought, one has to apply all unitary transformations up to U_n. This rule is denoted as the *n-rule* of the DKH calculation of molecular properties [721]. It also holds for purely electric properties ($X_{\mathcal{O}} = 0$)

since the free-particle Foldy–Wouthuysen transformation creates a nonvanishing zeroth-order odd term $X_{\mathcal{O},0}$ (cf. Eq. (15.68)) in any case. As a consequence, it is the n-rule instead of the familiar $(2n+1)$-rule which has to be taken into account for the calculation of DKH property operators, a fact that has not fully been recognized in earlier studies.

In the limit of exact decoupling (DKH∞) no approximation has been introduced by Eq. (15.60). In most applications, however, both the DKH Hamiltonian and thus the wave function and the DKH property are obtained only approximately up to a specific order in V. Though it seems most natural at first sight to choose the same order n of decoupling for both the wave function and the property, it is legitimate to determine the DKH wave function ψ_i with the help of the unperturbed DKHn Hamiltonian and to evaluate the expectation value given by Eq. (15.60) with the DKH-transformed property operator correct up to m-th order in V. As a matter of fact, depending on the specific form of the property X, this approach of employing different DKH orders for wave function and property $(n \neq m)$ can be very sensible for an efficient convergence toward exact decoupling in practice. Such approximations will be denoted as $\langle X\rangle(n,m)$ in the following,

$$\langle X\rangle(n,m) = \left\langle \tilde{\psi}_i^{\mathrm{DKH}n} \middle| \sum_{k=0}^{m} X_{\mathcal{E},k} \middle| \tilde{\psi}_i^{\mathrm{DKH}n} \right\rangle \tag{15.63}$$

15.3.3.2 **Explicit Expressions up to Third Order in the Unperturbed Potential**

For any arbitrary electromagnetic property operator of the form given by Eq. (15.25) the even components up to third order in V read [721]

$$X_{\mathcal{E},0} = A_p\left(X_{\mathcal{E}} + R_p X_{\mathcal{E}} R_p\right)A_p + \beta A_p\left\{R_p, X_{\mathcal{O}}\right\}A_p \tag{15.64}$$

$$X_{\mathcal{E},1} = \left[W_1, X_{\mathcal{O},0}\right] \tag{15.65}$$

$$X_{\mathcal{E},2} = \tfrac{1}{2}\left[W_1,\left[W_1, X_{\mathcal{E},0}\right]\right] + \left[W_2, X_{\mathcal{O},0}\right] \tag{15.66}$$

$$X_{\mathcal{E},3} = \left[a_3 W_1^3 + W_3, X_{\mathcal{O},0}\right] - \tfrac{1}{2}W_1\left[W_1, X_{\mathcal{O},0}\right]W_1 + \left[W_2,\left[W_1, X_{\mathcal{E},0}\right]\right] \tag{15.67}$$

with the zeroth-order odd term given by

$$X_{\mathcal{O},0} = \beta A_p\left[R_p, X_{\mathcal{E}}\right]A_p + A_p\left(X_{\mathcal{O}} - R_p X_{\mathcal{O}} R_p\right)A_p \tag{15.68}$$

The transformed *third-order* operator X_{DKH3} is then given by

$$\begin{aligned} X_{\mathrm{DKH3}} = {} & X_0 + \left[W_1, X_0\right] + \left[W_2, X_0\right] + \left[W_3, X_0\right] + \frac{1}{2}\left[W_1,\left[W_1, X_0\right]\right] \\ & + \left[W_2,\left[W_1, X_0\right]\right] + a_{1,3}\left[W_1^3, X_0\right] + \frac{1}{2}\left[W_1 X_0 W_1, W_1\right] + O(V^4) \end{aligned} \tag{15.69}$$

Though lengthy, these third-order expressions can be evaluated easily within any two-component theory. The expectation value of an odd term like $X_{\mathcal{O},0}$

would vanish because of the vanishing small components in a two-component framework (independently of the DKH order).

For simplicity it has been assumed that all unitary transformations U_i feature the same parametrization with $a_{i,1} = 1$, i.e., the cubic coefficient a_3 is the first coefficient to be chosen freely. The odd operators $W_i = W_i(\lambda = 0)$ are the familiar perturbation-independent operators parametrizing the standard unitary DKH transformations U_i; see chapter 12. As an important consequence, no terms containing the energy-damped property $(\tilde{X})$ occur within $X_{\text{DKH}\infty}$. In order to emphasize this general structure and to avoid any misconceptions, we give the explicit expression of the first-order term of the transformed property [721],

$$X_{\mathcal{E},1} = -\beta A_p\left[R_p,\tilde{V}\right]A_p^2\left[R_p,X_{\mathcal{E}}\right]A_p + A_p\left[R_p,\tilde{V}\right]A_p^2\left(X_{\mathcal{O}} - R_pX_{\mathcal{O}}R_p\right)A_p$$
$$-\beta A_p\left[R_p,X_{\mathcal{E}}\right]A_p^2\left[R_p,\tilde{V}\right]A_p - A_p\left(X_{\mathcal{O}} - R_pX_{\mathcal{O}}R_p\right)A_p^2\left[R_p,\tilde{V}\right]A_p \quad (15.70)$$

The third-order term $X_{\mathcal{E},3}$ seems to depend on the parametrization of the unitary transformations since it explicitly contains the coefficient a_3. However, the operator W_3 also depends on the chosen parametrization

$$\left[W_3,\mathcal{E}_0\right] = -\left[W_2,\mathcal{E}_1\right] + \left(a_3 - \tfrac{1}{2}\right)\left[W_1,\left[W_1,\mathcal{O}_1\right]\right] + \left(3a_3 - \tfrac{1}{2}\right)W_1\mathcal{O}_1W_1 \quad (15.71)$$

and the a_3-dependence of $X_{\mathcal{E},3}$ will thus exactly cancel. In other words, up to third-order in V the transformed property operator is independent of the parametrization of the unitary transformations. Fourth- and higher-order terms, however, depend on the chosen parametrization. This situation is comparable to the DKH Hamiltonian, where even terms up to fourth order are invariant on changing the parametrization of the unitary transformations [511]. Depending on the specific form of the property X, this dependence will be more or less pronounced [539,549,719]. Furthermore, the derivation and evaluation of higher-order terms $X_{\mathcal{E},k}$ $(k \geq 4)$ becomes computationally more and more demanding with increasing order in V and needs to be accomplished fully automatically [719].

15.3.3.3 **Alternative Transformation for First-Order Energy**

So far the direct and simplest calculation of the linear response correction $\epsilon_i^{(1)}$ has been presented [721]. Starting directly from the expectation value $\langle\psi_i|X|\psi_i\rangle$, i.e., the right hand side of Eq. (15.7), this approach employs λ-*independent* unitary transformations U_i only and is based on the transformed property $X_{\text{DKH}\infty}$ given by Eq. (15.61). This result can, of course, equally well be obtained starting with the left hand equality of Eq. (15.7), i.e., employing the first derivative of the exact energy $\epsilon_i(\lambda)$ with respect to λ taken at $\lambda = 0$. By insertion of perturbation-independent identity operators ($\mathbf{1} = U^\dagger U$), application of the Hellmann–Feynman theorem, and taking advantage of Eqs. (15.59) and (15.61), the same result as that given by Eq. (15.60) is recovered.

In the literature, however, it has most often been suggested to employ perturbation-*dependent* identity operators $\left(\mathbf{1} = U^\dagger(\lambda)U(\lambda)\right)$ for the calculation of $\epsilon_i^{(1)}$ within the two-component DKH scheme based on the left hand equality of Eq. (15.7). This alternative scheme for calculating $\epsilon_i^{(1)}$ will yield formally different and more involved expressions for each order $X_{\mathcal{E},k}$ of the transformed property operator as compared to the direct approach. In the limit of infinite order, however, both schemes will yield identical results since only exact unity operators have been inserted and thus no approximations have been introduced at all.

The alternative approach relies on the insertion of perturbation-*dependent* identity operators, which will cast the linear response expression given by Eq. (15.7) into the form

$$E_i^{(1)} = \left[\frac{\mathrm{d}}{\mathrm{d}\lambda}\left\langle \underbrace{U(\lambda)\psi_i(\lambda)}_{\tilde{\psi}_i(\lambda)} \middle| U(\lambda)h^D(\lambda)U^\dagger(\lambda) \middle| \underbrace{U(\lambda)\psi_i(\lambda)}_{\tilde{\psi}_i(\lambda)} \right\rangle\right]_{\lambda=0}$$

$$= \left\langle \tilde{\psi}_i \middle| \left[\frac{\mathrm{d}}{\mathrm{d}\lambda} U(\lambda)h^D(\lambda)U^\dagger(\lambda)\right]_{\lambda=0} \middle| \tilde{\psi}_i \right\rangle \tag{15.72}$$

$$= \left\langle \tilde{\psi}_i \middle| \left[\frac{\mathrm{d}U(\lambda)}{\mathrm{d}\lambda}\right]_{\lambda=0} h^D U^\dagger + U h^D \left[\frac{\mathrm{d}U^\dagger(\lambda)}{\mathrm{d}\lambda}\right]_{\lambda=0} + X_{\mathrm{DKH}\infty} \middle| \tilde{\psi}_i \right\rangle \tag{15.73}$$

where again the Hellmann–Feynman theorem and $\lim_{\lambda\to 0}\tilde{\psi}_i(\lambda) = \tilde{\psi}_i$ have been employed. In addition to the term containing the transformed property $X_{\mathrm{DKH}\infty}$, which has already been present in the direct approach, this alternative procedure yields two additional terms containing derivatives of $U(\lambda)$, which are very tedious to evaluate. Reconsidering Eq. (15.72) and taking advantage of Eq. (15.42), however, this expression for $\epsilon_i^{(1)}$ can be converted into the equivalent, yet slightly simpler expression

$$\epsilon_i^{(1)} = \left\langle \tilde{\psi}_i \middle| \left[\frac{\mathrm{d}h_{bd}(\lambda)}{\mathrm{d}\lambda}\right]_{\lambda=0} \middle| \tilde{\psi}_i \right\rangle = \sum_{k=1}^{\infty} \left\langle \tilde{\psi}_i \middle| \underbrace{\left[\frac{\mathrm{d}\mathcal{E}_k(\lambda)}{\mathrm{d}\lambda}\right]_{\lambda=0}}_{X'_{\mathcal{E},k-1}} \middle| \tilde{\psi}_i \right\rangle$$

$$= \sum_{k=0}^{\infty} \left\langle \tilde{\psi}_i \middle| X'_{\mathcal{E},k} \middle| \tilde{\psi}_i \right\rangle \tag{15.74}$$

which defines $X'_{\mathcal{E},k}$. Eq. (15.74) features the advantages that no derivatives of the unitary transformations $U_i(\lambda)$ have to be evaluated and that all odd terms have already cancelled. Apart from these technical aspects, however, the expressions given by Eqs. (15.73) and (15.74) are completely equivalent, since no approximation has been introduced. In the following we will compare the most important features of this alternative approach to the direct scheme presented above.

In contrast to the direct DKH transformation of section 15.3.3.1, the alternative transformation with $U = U(\lambda)$ gives rise to additional terms which are not present in Eq. (15.60). In particular, terms containing energy-damped property operators $\tilde{X}$ are present in Eqs. (15.73) and (15.74) and lead to much more complicated expressions compared to the direct transformation with $U = U(V)$. For example, already the first-order even term of the transformed property operator is modified as compared to Eq. (15.65) and reads

$$X'_{\mathcal{E},1} = \frac{1}{2}\Big(\lfloor W_1, X_{\mathcal{O},0} \rfloor + \lfloor W_1^X, \mathcal{O}_1 \rfloor \Big) \tag{15.75}$$

where W_1^X denotes the odd and antihermitean W_1-operator given by Eq. (15.53), with $V(\lambda)$ being replaced by the property X, i.e.,

$$W_1^X = A_p\big[R_p, \tilde{X}_\mathcal{E}\big]A_p + \beta A_p\big(\tilde{X}_\mathcal{O} - R_p\tilde{X}_\mathcal{O}R_p\big)A_p \tag{15.76}$$

Also all other higher-order terms $X'_{\mathcal{E},k}$ with $k \geq 1$ are modified as compared to the direct approach. Finite-order approximations to this exact alternative scheme have been proposed by several authors, cf. Refs. [450, 519, 548, 726–728]. However, labeling of terms and counting of orders within the alternative approach have not always been consistent. One has to realize that — similarly to $X_{\text{DKH}\infty}$ discussed above — each term contributing to Eq. (15.73) is exactly linear in X. Terms of higher order in X are eliminated by evaluating the derivatives at $\lambda = 0$. That is, these terms can only be sensibly classified according to their order in the unperturbed potential V. From Eq. (15.46) and the fact that $\mathcal{E}_i(\lambda)$ is exactly of i-th order in $V(\lambda)$,

$$\mathcal{E}_i(\lambda) \sim V^i(\lambda) = V^i + \lambda\big(V^{i-1}X + V^{i-2}XV + \cdots + XV^{i-1}\big) + \mathcal{O}\big(\lambda^2\big) \tag{15.77}$$

one has to determine all terms up to $\mathcal{E}_{n+1}(\lambda)$ if $\epsilon_i^{(1)}$ given by Eq. (15.74) is to be correct up to n-th order in V.

15.3.4
Automated Generation of DKH Property Operators

The evaluation of DKH property operators up to any predefined order in the external potential requires a computational scheme, which cannot rely on a direct noniterative numerical method since odd operators have no representation in a truly two-component formalism (which includes the one-component approximation). Only the product of two odd operators may be evaluated, and therefore one needs to derive the operator expression first. This fact has already been discussed for the DKH Hamiltonian in chapter 12.

Each order of the upper left block $\tilde{X}^{LL}$ of the transformed property operator in the decoupled DKH picture has to be determined purely algebraically, i.e., by symbolical evaluation of the required unitary transformations. Because of the increasing complexity of the higher-order terms, their determination, however, is only possible if the algebraic operations are evaluated automatically by a suitable parser routine yielding analytic formulae [719]. Subsequently, the program translates the resulting closed-form operator expressions into corresponding matrix multiplications. Compared to the DKH Hamiltonian, the property parser requires significant changes since, for instance, the $(2n+1)$-rule for the Hamiltonian does not hold for the property operators, which obey the n-rule [721].

15.3.5
Consequences for the Electron Density Distribution

A hideous feature of transformation schemes is the fact that the density is no longer calculated from the squared transformed orbitals. Especially close to atomic nuclei this leads to significant deviations of squared transformed orbitals from squared spinors, while the effect becomes negligible at larger distances [729, 730]. Already in the case of a one-electron system, the absolute square of a four-component spinor $\psi(\boldsymbol{r})$ is related to the square of two- (or one)-component wave functions $\phi_L(\boldsymbol{r})$ only by introducing a distance-dependent error $\Delta\rho(\boldsymbol{r})$

$$\rho(\boldsymbol{r}) = \psi^\dagger(\boldsymbol{r}) \cdot \psi(\boldsymbol{r}) = |\tilde{\psi}^L(\boldsymbol{r})|^2 + \Delta\rho(\boldsymbol{r}) \tag{15.78}$$

to yield the electron density distribution $\rho(\boldsymbol{r})$. In order to get rid of the $\Delta\rho(\boldsymbol{r})$ error, a picture change transformation must be performed. The correct picture-change-transformed two-component density is then equal to the four-component density, $\rho^{(4c)}(\boldsymbol{r}) = \rho^{(2c)}(\boldsymbol{r})$.

Hence, the DKH density cannot simply be obtained from a sum of squared DKH orbitals, though this can be a very good approximation for spatial regions that do not penetrate the vicinity of the nucleus. We must DKH-transform the corresponding operator for the density at the position $\boldsymbol{r}$

$$\hat{\rho}_{\boldsymbol{r}} = \sum_{i=1}^{N} \delta^{(3)}(\boldsymbol{r}_i - \boldsymbol{r}) \tag{15.79}$$

where $\boldsymbol{r}$ may be any position in space as introduced already in Eq. (4.74). As usual, N is the number of electrons and $\boldsymbol{r}_i$ denotes the coordinate of the ith electron, whereas $\delta^{(3)}(\boldsymbol{r}_i - \boldsymbol{r})$ represents Dirac's delta distribution in three dimensions. The electron density is $\rho(\boldsymbol{r}) = \langle\hat{\rho}_{\boldsymbol{r}}\rangle$ as introduced in section 4.3.9.

The orbital contribution to the expectation value for the density then reads

$$\rho_{ii}(\boldsymbol{s}) = \left\langle U\psi_i^{(\lambda=0)} \middle| U\delta^{(3)}(\boldsymbol{s}-\boldsymbol{r})U^{\dagger} \middle| U\psi_i^{(\lambda=0)} \right\rangle_{\boldsymbol{r}} = \left\langle \tilde{\psi}_i^L \middle| \delta_{\text{DKH}\infty}^{LL}(\boldsymbol{s}-\boldsymbol{r}) \middle| \tilde{\psi}_i^L \right\rangle \tag{15.80}$$

For s-type Gaussian basis functions, $\phi_\mu(\boldsymbol{r}_A) = (2\zeta_\mu/\pi)^{3/4} \exp(-\zeta_\mu|\boldsymbol{r}-\boldsymbol{R}_A|^2)$, the matrix representation of the δ-distribution simply reduces to a matrix with entries $(4\zeta_\mu\zeta_\nu/\pi^2)^{3/4}$, which can then be subjected to a DKH property transformation as described above for general one-electron properties. Within an IOTC protocol, the exact transformed density can be obtained in one shot. Unfortunately, the transformation must be done for any position $\boldsymbol{s}$ desired. Of course, smart algorithms may be implemented that allow one to newly calculate only those quantities of the expectation value that explicitly depend on the given position $\boldsymbol{s}$ of a numerical grid, while the rest are computed only once.

If the unitary matrix U is represented by the sequence of infinitely many unitary matrices U_m, this sequence may be truncated with respect to a predefined DKH order yielding an approximate decoupling scheme

$$\rho_{ii}^{(n,m)}(\boldsymbol{r}) = \left\langle \tilde{\psi}_i^{L,\text{DKH}n} \middle| \sum_{k=0}^{m} \delta_{\mathcal{E},k}(\boldsymbol{r}) \middle| \tilde{\psi}_i^{L,\text{DKH}n} \right\rangle \tag{15.81}$$

which can be classified according to the DKH order of the Hamiltonian n and of the property operator m, respectively. The DKH density at an atomic nucleus is particularly prone to picture change artifacts, and a rather large number of unitary transformations is required to accurately reproduce the four-component reference density at that nucleus [731]. Interestingly, this only holds true for the operator transformation, while the DKH orbitals that enter the expectation value can be of low order; e.g., DKH2 orbitals are sufficient.

Finally, we may note that the continuity equation, which we used to define the electron density in chapter 8 (compare also chapter 4), reads

$$\frac{\partial}{\partial t}\left\langle \Psi \middle| U^{\dagger}U\hat{\rho}_{\boldsymbol{r}}U^{\dagger}U \middle| \Psi \right\rangle = -\nabla\cdot cN\left\langle \Psi \middle| U^{\dagger}U\boldsymbol{\alpha}\delta(\boldsymbol{r}_1-\boldsymbol{r})U^{\dagger}U \middle| \Psi \right\rangle \tag{15.82}$$

which can be rewritten with the transformed total wave function $\tilde{\Psi} = U\Psi$ as

$$\frac{\partial}{\partial t}\left\langle \tilde{\Psi} \middle| U\hat{\rho}_{\boldsymbol{r}}U^{\dagger} \middle| \tilde{\Psi} \right\rangle = -\nabla\cdot cN\left\langle \tilde{\Psi} \middle| U\boldsymbol{\alpha}\delta(\boldsymbol{r}_1-\boldsymbol{r})U^{\dagger} \middle| \tilde{\Psi} \right\rangle \tag{15.83}$$

15.3.6
DKH Perturbation Theory with Magnetic Fields

Magnetic vector potentials introduce additional complications to the DKH transformation. The key to properly transformed corrections of the unperturbed wave function is, of course, the equation used in four-component theory. This aspect can hardly be overemphasized. Once it is clear how the proper four-component equations look, these may *then* be transfered to the two-component picture using the very same principles discussed in this work for the first-order property (expectation value) case. It is, however, not guaranteed that the resulting expressions are easier to handle from a computational point of view. The principles introduced in the preceding sections can be transfered to the magnetic case and we refer to the detailed presentation of this matter in Ref. [707] for the sake of brevity.

15.4 Magnetic Fields in Resonance Spectroscopies

Magnetic resonance spectroscopies like NMR and ESR belong to the most important spectroscopic techniques for the elucidation of molecular structures. In order to be able to interpret or even predict the spectra of molecules, calculations are essential. We have seen earlier in this chapter how particular interactions are obtained in the Breit–Pauli theory including external electromagnetic fields when electron spin operators interact with the external magnetic field generated by nuclear spins. Many difficulties need to be overcome, from the derivation of equations to implementations in computer programs designed for calculations of ESR and NMR parameters such as g-tensors [584,623,732–738], hyperfine coupling tensors [443,584,725,739–741], NMR shielding tensors [163,724,727,728,742–756] and spin–spin coupling constants [757–772].

15.4.1 The Notorious Diamagnetic Term

It is known from Pauli theory that two one-electron terms should emerge in the nonrelativistic limit as is evident from Eq. (5.137). One of these terms is quadratic (bilinear) with respect to the vector potentials,

$$\frac{q_e^2}{2m_ec^2}\boldsymbol{A}^2 \stackrel{(15.24)}{=} \frac{q_e^2}{2m_ec^2}\left(\sum_I \boldsymbol{A}_I^2 + \sum_{IJ}\boldsymbol{A}_I\boldsymbol{A}_J\right) \tag{15.84}$$

This is called the diamagnetic term. Since a term bilinear in the vector potentials is not present in the fully relativistic four-component formulation, the question arises how it emerges when the speed of light is set equal to infinity, i.e., in the nonrelativistic limit. It was shown early by Sternheim [773] that the in a first-quantized scheme artificial negative-energy states are responsible for the diamagnetic term [774]. However, this creates new difficulties as it should be possible to represent a property like the shielding tensor on the basis of positive-energy (i.e., electronic) states only. An interesting suggestion by Kutzelnigg [775] solves this problem on the operator level by invoking a unitary transformation of the electromagnetic-field-containing one-electron Dirac Hamiltonian, which depends explicitly on the vector potential and generates a bilinear vector potential term in the pre-transformed four-component Dirac Hamiltonian. This transformation does not decouple upper and lower components of the one-electron spinors — which could only be achieved in a DKH scheme, though Foldy and Wouthuysen made a first attempt in their original approach [509] — but is designed such that exact expressions are obtained in a perturbation theory framework. In a variational treatment they would be useless, as the transformation also generates an infinite sum of operators out of the Dirac operator; in molecular-property response theory it is, however, well defined. Especially in the case of the unitary-transformation schemes, the many options for different starting points and routes make the different approaches difficult to understand. Instead of presenting all these options, we refer to the analysis in Ref. [707].

15.4.2 Gauge Origin and London Orbitals

Property operators which include the interaction with the magnetic part of the electromagnetic field, $\boldsymbol{B}$, induce a gauge-origin dependence in the calculations. Translation of the molecule in space yields thus different results before and after the translation, which is physically not sensible.

If we define the vector potential of a homogeneous magnetic field $\boldsymbol{B}$ as $\boldsymbol{A}(\boldsymbol{r}) = \boldsymbol{B} \times (\boldsymbol{r} - \boldsymbol{R}_O)/2$, then the *gauge origin* $\boldsymbol{R}_O$ dependence can be understood as a gauge transformation with the gauge function defined by $\chi = -(\boldsymbol{B} \times \boldsymbol{R}_O) \cdot \boldsymbol{r}/2$ (compare section 2.4). This vector potential then produces terms that depend on the arbitrary position $\boldsymbol{R}_O$. Only for exact wave functions, these terms vanish (complete basis set), while they cannot be neglected in any (small) finite one-electron basis set.

To avoid these artifacts, London atomic orbitals $\chi_\mu(\boldsymbol{B}, \boldsymbol{R}_A, \boldsymbol{r})$ are often employed [776,777],

$$\chi_\mu(\boldsymbol{B}, \boldsymbol{R}_A, \boldsymbol{r}) = \exp\left(-\frac{\mathrm{i}}{c}\boldsymbol{A} \cdot \boldsymbol{r}\right) \phi_\mu(\boldsymbol{R}_A, \boldsymbol{r}) \tag{15.85}$$

given in Hartree atomic units. These functions are also known as Gauge Including Atomic Orbitals (GIAO) [778]. $\boldsymbol{R}_A$ denotes the center of the basis function $\phi_\mu(\boldsymbol{R}_A, \boldsymbol{r})$. The effect of the vector-field-dependent prefactor is that matrix elements of the kinetic energy become independent of the origin of $\boldsymbol{A}$. Other choices of orbitals are also possible, and we may refer to Ref. [708] for further details.

15.4.3
Explicit Form of Perturbation Operators

The explicit form of the property operators is derived in two steps. First, the scalar and vector fields are determined and, second, they are inserted into the Dirac–Breit [779, 780] or any other quasi-relativistic operator derived from it like the external-field-containing DKH operator, the external-field-containing ZORA Hamiltonian or the similar Breit–Pauli Hamiltonian. The theory of NMR parameters has been derived by Ramsey [781–783] from the nonrelativistic perspective and by Pyykkö from the relativistic perspective [784,785].

As an example of the explicit expression we give the vector potential produced by the nuclear spin of a *point* nucleus

$$\boldsymbol{A}_I = \frac{\boldsymbol{\mu}^{(I)} \times \boldsymbol{r}_I}{r_I^3} \tag{15.86}$$

and the magnetic field is then

$$\boldsymbol{B}_I = -\frac{\boldsymbol{\mu}^{(I)}}{r_I^3} + 3\frac{\boldsymbol{r}_I(\boldsymbol{\mu}^{(I)} \cdot \boldsymbol{r}_I)}{r_I^5} + \frac{8\pi}{3}\delta^{(3)}(\boldsymbol{r}_I)\,\boldsymbol{\mu}^{(I)} \tag{15.87}$$

according to $\boldsymbol{B}_I = \nabla \times \boldsymbol{A}_I$, with the difference vector $\boldsymbol{r}_I$ and its length defined as $\boldsymbol{r}_I = \boldsymbol{r} - \boldsymbol{R}_I$ and $r_I = |\boldsymbol{r} - \boldsymbol{R}_I|$, respectively. $\boldsymbol{R}_I$ denotes the position of the nucleus. Note that different vector potentials and magnetic fields result for a finite-size nucleus, where the singular Coulomb potential is to be replaced by some model potential as introduced in section 6.9. These quantities can now be employed in every one-electron term of Eq. (5.134) to produce the perturbation operators $X(i)$ in the Hamiltonian — and, in a more accurate and complete treatment, also in the two-electron terms given in Eq. (15.41) to produce also two-electron perturbation operators $X(i,j)$. Of course, all fields are additive as indicated in Eq. (15.24). From the many terms of the Breit–Pauli Hamiltonian we already understand that a large number of individual terms can contribute to a certain property. Such terms may be individually analyzed in order to gain deeper understanding of how a net result emerges for a given molecule. As an example we may refer to the work by Autschbach and coworkers on NMR parameters [722,723,786].

15.4.4
Spin Hamiltonian

The discussion so far has rested on the first-principles idea of quantum mechanics. All terms in the external-field-containing Breit–Pauli Hamiltonian or, for more accurate considerations, in the external-field-containing Dirac–Coulomb–Breit Hamiltonian give rise to specific quantum mechanical states. Magnetic resonance spectroscopy probes the transitions between these states.

The probed energy gaps of these experiments can therefore be related to calculated energy differences between the quantum mechanical states. We have already seen in Eq. (13.78) that magnetic-field-mediated angular-momentum interactions that are truly internal to an electronic system, i.e., those that arise already for the external-field-free Dirac–Coulomb–Breit or external-field-free Breit–Pauli Hamiltonian, contribute to this splitting. However, external magnetic fields produced in a laboratory set-up or generated by the spins of nuclei make additional contributions. For this purpose, the phenomenological spin Hamiltonian of Eq. (13.78) must be extended by the following two terms [26,787]

$$H_{\text{eff}}^{\text{ESR}} = H_Z + H_{\text{hf}} = \mu_B \boldsymbol{B} \cdot \boldsymbol{g} \cdot \boldsymbol{S} + \sum_{I=1}^{M} \boldsymbol{S} \cdot \boldsymbol{A}^{(I)} \cdot \boldsymbol{I}_I \tag{15.88}$$

which are of paramount importance to ESR spectroscopy. This Hamiltonian covers the electronic Zeeman (Z) interaction of an *effective spin* operator $\boldsymbol{S}$ with the external magnetic field as well as the hyperfine (hf) interaction of the magnetic field produced by a nuclear spin $\boldsymbol{I}_I$ with this effective spin operator. The (3×3)-matrix $\boldsymbol{g}$ is the so-called g-tensor, which accounts for the dependence of the interaction on the orientation of the molecule with respect to the external magnetic field (the eigenvalues are taken to measure the strength of the Zeeman interaction). Accordingly, the hyperfine interaction tensor $\boldsymbol{A}^{(I)}$ takes care of the orientational dependence with respect to the magnetic field generated by nucleus I. The latter is usually split into an isotropic, orientation-independent part and an anisotropic remainder, $\boldsymbol{A}^{(I)} = A_{\text{iso}}^{(I)} + \boldsymbol{A}_{\text{noniso}}^{(I)}$. While the former is the portion measured in solution, because of the spatial averaging, the latter is a traceless operator. One should not confuse the A-tensor with a vector potential, which we would also denote by a bold-face $\boldsymbol{A}$. The principal values of $\boldsymbol{A}^{(I)}$ are also called hyperfine coupling constants. $A_{\text{iso}}^{(I)}$ arises from the contact interaction at nucleus I and is also called the *Fermi contact term*. We may note that neither $\boldsymbol{g}$ nor $\boldsymbol{A}$ are true tensors, i.e., their components do not fulfill the required transformation properties. The spin Hamiltonian operates on a direct product basis of normalized electron and nuclear spin functions.

Naturally, additional (phenomenological) interaction terms can be formulated relevant to other spectroscopies. The nuclear spin–spin coupling term arises through the polarization of the electronic structure by the magnetic moment of one nuclear spin, a change which is then experienced by the magnetic moment of the second spin. Consequently, it is written as

$$H_{\text{eff}}^{\text{NMR},J} = \boldsymbol{I}_1 \cdot \boldsymbol{J} \cdot \boldsymbol{I}_2 \tag{15.89}$$

where $\boldsymbol{I}_i$ are nuclear spins as before. The coupling tensor $\boldsymbol{J}$ is characteristic for the electronic structure. The spin Hamiltonian relevant to NMR spectroscopy contains further interaction operators, namely the nuclear Zeeman term and the (nuclear) quadrupole interaction term for nuclear spins larger than 1/2. These terms were introduced by Abragam and Pryce [788] who base their analysis on the Breit–Pauli Hamiltonian including the magnetic field of the nuclear spins and with the external magnetic field. Higher magnetic and electric poles than the magnetic moment of the nuclei (linear in the nuclear spin $\boldsymbol{I}$) and the electric quadrupole moment $\boldsymbol{Q}$, respectively, are omitted because they are considered to be negligibly small.

15.5 Electric Field Gradient and Nuclear Quadrupole Moment

In contrast to magnetic properties, the theory of electric-field-like properties is much easier to cast into a set of working equations. One of them has attracted particular interest, and that is the electric field gradient (EFG). This property is of decisive importance to Mössbauer spectroscopy, i.e., to the spectroscopy of excited *nuclear* states whose energies are modulated by the molecular structure (the 'chemical environment'). In order to see how this property arises, we study the electrostatic electron–nucleus interaction of extended, not spherically symmetric charge distributions. For this we apply a multipole expansion in order to generate the 'properties' term by term.

The classical electrostatic electron–nucleus interaction energy $E_{e,nuc}$ of a positive nuclear charge distribution $\rho_{\mathrm{N}}(\boldsymbol{r})$ with the surrounding electronic density $\rho_e(\boldsymbol{r})$ is given by the Poisson integral [cf. Eq. (2.139)],

$$\begin{aligned} E_{e,nuc} &= \int \mathrm{d}^3r \int \mathrm{d}^3s \frac{\rho_{nuc}(\boldsymbol{r})\rho_e(\boldsymbol{s})}{|\boldsymbol{r}-\boldsymbol{s}|} \\ &= \int \mathrm{d}^3s\, \rho_e(\boldsymbol{s}) V_{nuc}(\boldsymbol{s}) = \int \mathrm{d}^3r\, \rho_{nuc}(\boldsymbol{r}) V_e(\boldsymbol{r}) \end{aligned} \tag{15.90}$$

in which V_{nuc} is the electrostatic potential generated by the total nuclear charge distribution and V_e is the corresponding potential generated by the electronic charge distribution. If we are interested in the interaction energy of the individual nuclei with the electronic system, we employ the obvious decomposition of the total nuclear charge distribution in terms of the atomic contributions,

$$\rho_{nuc}(\boldsymbol{r}) = \sum_{I=1}^{M} \rho_{nuc}^{(I)}(\boldsymbol{R}_I, \boldsymbol{r}) \tag{15.91}$$

and obtain

$$E_{e,nuc} = \sum_{I=1}^{M} \int \mathrm{d}^3r\, \rho_{nuc}^{(I)}(\boldsymbol{R}_I, \boldsymbol{r}) V_e(\boldsymbol{r}) \equiv \sum_{I=1}^{M} E_{e,nuc}^{(I)} \tag{15.92}$$

For the determination of the electronic wave function in standard electronic structure calculations, one has to include V_{nuc} in the electronic Hamiltonian operator to account for the electron–nucleus interaction. However, this would require knowledge of the exact nuclear charge density distribution $\rho_{nuc}(\boldsymbol{r})$, which can presently neither be calculated from quantum field theory of the nucleus nor unambiguously deduced from experiment (cf. the discussion in section 6.9). On the other hand, inclusion of the *exact* $\rho_{nuc}^{(A)}$ or $V_{nuc}^{(A)}$ for a given nucleus A into the electronic Hamiltonian in a variational procedure yields results which would depend on the particular isotope chosen for a given atom. Although the effect would be very small, it is not desirable to obtain (slightly) different electronic wave functions for the same molecule due to different isotopes.

Therefore, one usually employs perturbation theory as discussed throughout this chapter and sticks to the point-charge model for atomic nuclei (or to some simplified spherically symmetric model density distribution) in quantum chemistry. The tiny effects of multipole moments of the nuclear charge density are then not included in the variational procedure for the determination of the electronic wave function.

Since it is possible to calculate an accurate electronic density ρ_e by quantum chemical methods, we may use V_e instead in order to obtain information on ρ_{nuc} according to Eq. (15.90). In this case, we have to use perturbation theory because the terms depending on the electronic density can only be obtained with the unperturbed electronic wave function calculated for an external potential of point-like (or spherically symmetric) nuclei. It can then be assumed that V_e varies only a little within the small spatial extension of the nucleus A, i.e., of $\rho_{nuc}^{(A)}$ so that we can use a truncated Taylor expansion of V_e around the center of the nucleus A

$$\begin{aligned} V_e(\boldsymbol{r}) &= V_e(\boldsymbol{R}_A) + \sum_{i=1}^{3}(r_i - R_{A,i}) \left(\frac{\partial V_e(\boldsymbol{r})}{\partial r_i}\right)_{\boldsymbol{r}=\boldsymbol{R}_A} \\ &\quad + \frac{1}{2!}\sum_{i,j=1}^{3}(r_i - R_{A,i})\,(r_j - R_{A,j}) \left(\frac{\partial^2 V_e(\boldsymbol{r})}{\partial r_i \partial r_j}\right)_{\boldsymbol{r}=\boldsymbol{R}_A} + \cdots \end{aligned} \tag{15.93}$$

The third term in this expansion contains the EFG components, $X_{ij}^{(A)}$, as expansion coefficient. Since the first derivatives of V_e are the components of the electric field, E_j, at nucleus A, the second derivatives can be understood as the gradient of the electric field components,

$$X_{ij}^{(A)} \equiv \left(\frac{\partial^2 V_e(\boldsymbol{r})}{\partial r_i \partial r_j}\right)_{\boldsymbol{r}=\boldsymbol{R}_A} = -\left(\frac{\partial}{\partial r_i}E_j\right)_{\boldsymbol{r}=\boldsymbol{R}_A} = \left(\frac{\partial^2}{\partial r_i \partial r_j}\int d^3s\,\frac{\rho_e(\boldsymbol{s})}{|\boldsymbol{r}-\boldsymbol{s}|}\right)_{\boldsymbol{r}=\boldsymbol{R}_A} \tag{15.94}$$

hence the name *electric field gradient*. Substituting this expansion into the atomic contribution $E_{e,nuc}^{(A)}$ as defined in Eq. (15.92) to $E_{e,nuc}$ of Eq. (15.90) yields

$$\begin{aligned} E_{e,nuc}^{(A)} &= V_e(\boldsymbol{R}_A) \int d^3r \rho_{nuc}^{(A)}(\boldsymbol{R}_A, \boldsymbol{r}) \\ &+ \sum_{i=1}^{3} \left[\left(\frac{\partial V_e(\boldsymbol{r})}{\partial r_i} \right)_{\boldsymbol{r}=\boldsymbol{R}_A} \underbrace{\int d^3r \, (r_i - R_{A,i}) \, \rho_{nuc}^{(A)}(\boldsymbol{R}_A, \boldsymbol{r})}_{=0 \text{ for sym. reasons}} \right] \\ &+ \frac{1}{2!} \sum_{i,j=1}^{3} \left[\left(\frac{\partial^2 V_e(\boldsymbol{r})}{\partial r_i \partial r_j} \right)_{\boldsymbol{r}=\boldsymbol{R}_A} \right. \\ &\quad \times \left. \int d^3r \, (r_i - R_{A,i}) \, (r_j - R_{A,j}) \, \rho_{nuc}^{(A)}(\boldsymbol{R}_A, \boldsymbol{r}) \right] \cdots \\ &\equiv \sum_{k=0}^{\infty} E_{e,nuc;k}^{(A)} \end{aligned} \tag{15.95}$$

which represents a multipole expansion of the electron–nucleus interaction energy. The first term is the standard monopole interaction energy, which is the dominating term and therefore always included in the zeroth-order Hamiltonian and wave function. Note that the dipole moment of the nucleus in the second term vanishes for symmetry reasons. Only multipole moments which feature inversion symmetry are nonvanishing. The third term in this expansion is the *electric* quadrupole term

$$E_{e,nuc;2}^{(A)} = \frac{1}{2} \sum_{i,j=1}^{3} \left[\left(\frac{\partial^2 V_e(\boldsymbol{r})}{\partial r_i \partial r_j} \right)_{\boldsymbol{r}=\boldsymbol{R}_A} \int d^3r \, (r_i - R_{A,i}) \, (r_j - R_{A,j}) \, \rho_{nuc}^{(A)}(\boldsymbol{R}_A, \boldsymbol{r}) \right] \tag{15.96}$$

which contains the electric nuclear quadrupole moment (NQM), $\boldsymbol{Q}^{(A)}$,

$$Q_{ij}^{(A)} = \int d^3r \, (r_i - R_{A,i}) \, (r_j - R_{A,j}) \, \rho_{nuc}^{(A)}(\boldsymbol{R}_A, \boldsymbol{r}) \tag{15.97}$$

This energy $E_{e,nuc;2}^{(A)}$ can be measured experimentally. Since the tensor components of the EFG, X_{ij}, evaluated at the position of the nucleus, can be calculated with high accuracy in electronic structure calculations, it is used for the determination of the NQM. An up-to-date collection of NQMs is maintained by Pyykkö [789].

From the definition of V_e in Eq. (15.90) we understand that the inverse-distance-weighted density integral needs to be rewritten as a matrix element containing the bra and ket of the wave function (instead of the electronic density) for quantum chemical, orbital-based calculations. The inverse distance is then to be differentiated *twice* as demanded by Eq. (15.94), and we obtain the usual type of formula,

$$X_{ij}^{(I)}(\boldsymbol{r}, \boldsymbol{R}_I) = -e\frac{3(\boldsymbol{r}-\boldsymbol{R}_I)_i(\boldsymbol{r}-\boldsymbol{R}_I)_j - |\boldsymbol{r}-\boldsymbol{R}_I|^2\delta_{ij}}{|\boldsymbol{r}-\boldsymbol{R}_I|^5} \tag{15.98}$$

which we have encountered throughout the book (e.g., in section 13.2 or in the appendix). Note, however, that in addition to the electronic part, the full EFG contribution comprises also a nuclear contribution,

$$X_{ij}^{(I),nuc}(\boldsymbol{R}_I) = \sum_{\substack{J=1\\J\neq I}}^{M} Z_J e\frac{3(\boldsymbol{R}_J-\boldsymbol{R})_i(\boldsymbol{r}-\boldsymbol{R})_j - |\boldsymbol{R}_J-\boldsymbol{R}|^2\delta_{ij}}{|\boldsymbol{R}_J-\boldsymbol{R}_I|^5} \tag{15.99}$$

which arises from the electrostatic interaction of nuclear charge densities,

$$E_{nuc,nuc} = \frac{1}{2}\int d^3r \int d^3s \frac{\rho_{nuc}(\boldsymbol{r})\rho_{nuc}(\boldsymbol{s})}{|\boldsymbol{r}-\boldsymbol{s}|} = \frac{1}{2}\sum_{IJ,I\neq J}^{M} \rho_{nuc}^{(I)}(\boldsymbol{R}_I,\boldsymbol{r})V_{nuc}^{(J)}(\boldsymbol{s}) \tag{15.100}$$

along the same lines indicated here for the electron–nucleus charge interaction (note that the factor of one half avoids double counting of interactions, as usual). We may refer to Ref. [549] for a comparative study on the calculation of EFGs from relativistic wave functions and to Refs. [720,790–792] for prime examples of the NQM calculation for ^{197}Au, ^{139}La and ^{119}Sn, respectively. For further references we may redirect the reader to references provided by these papers.

15.6
Parity Violation and Electro-Weak Chemistry

In chapter 8 we abandoned radiative corrections of QED from our semi-classical theory for molecular science. This was justified by the fact that we aim at a theory that describes the energy and properties of molecular systems at 'ordinary' (usually ambient) temperature. However, high-resolution molecular spectroscopy may very well aim at the detection of tiny effects in order to complete our picture of physical effects in molecular systems. Then, our custom-made theory needs to be put again into the broader perspective of a quantum field theory. This is especially true for effects beyond electromagnetic interactions. Parity nonconservation effects are a famous example from electro-weak theory, which unites electrodynamical and weak forces in a single quantum field theory. While such effects were investigated some time ago for atomic systems, studies of them in molecules have a quite young history (see the recent review by Quack and collaborators [793] and references therein).

Parity nonconservation (PNC) effects are produced by electro-weak interactions and lead to asymmetries on a macroscopic level. The parity-violation processes in nature, which were discovered by Lee and Yang [794, 795], attribute a very small energy difference to two enantiomeric molecules [796, 797]. This energy difference has been considered (and controversely discussed) as a possible reason for the existence of homochirality, i.e., the natural existence of L-amino acids and D-monosaccharides in our biosphere [798–800]. The behavior of a system under inversion (reflection through the origin of the coordinate system) is denoted by its parity. The wave functions of two enantiomeric molecules are not eigenfunctions of the parity operator because the parity operator converts one into the other. The energy difference, which can be attributed to both isomers through a parity-odd interaction operator, has been given and discussed, for instance, in Ref. [440]. Although this difference is extremely small, it may well be attacked within a perturbation-theory framework so that numerical accuracy of the calculations is not the predominant issue, while the smallness is a true challenge to experiment [793].

The electric dipole transition from shell ns to $(n+1)s$, for example, is parity forbidden; only the magnetic-dipole and electric-quadrupole transitions are allowed. However, taking account of a 'model' interaction Hamiltonian for the exchange of an intermediate Z^0 vector boson between the nucleus and an electron introduces (odd) matrix operators that exchange upper and lower components, which have different parity. Through PNC interaction the spinors become parity mixed such that a non-zero transition amplitude results [801]. This PNC effect was first discussed for the electronic structure of atoms (see, e.g., for reviews Refs. [801–805], for recent experimental work Refs. [806, 807] and for theoretical work Refs. [808–810]). The number of studies of the PNC effect in molecules with electronic structure calculations has grown rapidly since the end of the 1990s, first with nonrelativistic wave functions [811–814]. Various molecules as optimum candidates for high-resolution spectroscopy have been proposed as two examples may demonstrate [815, 816].

Probably one of the most appealing features of four-component methods is the possibility to test fundamental physical symmetries through accurate electronic structure calculations of molecules. First four-component studies on molecules relied on single-determinantal Dirac–Hartree–Fock calculations for TlF [817, 818], YbF [819], H_2X_2 (with X=O, S, Se, Te, Po) [820], and chiral halogenides of methane and its higher homologs [821]. More advanced four-component coupled-cluster results have been presented for H_2O_2 and H_2S_2 demonstrating that electronic correlation contributions are small in these cases but depend critically on the molecular structure [822]. Consequently, density functional theory can be beneficial for the calculation of PNC effects [823, 824]. However, other effects can become non-negligible like the inclusion of the Breit interaction [825, 826]. Apart from PNC effects on vibrational spectroscopy, nuclear magnetic resonance has been considered a fruitful target [827–830].

16
Relativistic Effects in Chemistry

We have developed the relativistic theory of molecular science from the first principles of fundamental physics, namely from quantum mechanics and from the special theory of relativity. In principle, we are now able to study any molecular system using quantum chemical methods of controllable accuracy. Comparisons with purely nonrelativistic calculations highlight so-called relativistic effects. Prominent macroscopic examples are the yellowish color of elemental gold and the fluidity of mercury at ambient temperature. This final chapter comprises some important examples for which relativity is of paramount importance.

The importance of relativistic quantum mechanics to molecular science was realized in the 1970s [831–835], which came somewhat as a surprise since it was not expected that low-energy molecular physics requires a quantum mechanical description beyond Schrödinger quantum mechanics.

The notion *relativistic effect* denotes the difference between relativistic and nonrelativistic expectation values. Though this is a seemingly straightforward definition, the difficulties are hidden in the details. On the one hand, the nonrelativistic observable cannot be measured and, on the other hand, the nonrelativistic limit is difficult to define in a meaningful manner. Of course, the limit of an infinitely large speed of light is a rigorous choice, but if electromagnetic fields from the moving electrons and nuclei are to be considered, the $1/c$ prefactor of vector potentials in the minimal coupling procedure makes all magnetic fields vanish in the limit $c \rightarrow \infty$. Such a strict nonrelativistic limit would not be a sensible reference. Another example is the separation of scalar and spin–orbit relativistic effects, which can be defined rigorously in different relativistic approaches though the results obtained do not necessarily match, and hence different numerical results in spin-free calculations depend on the quasi-relativistic scheme employed [836].

Since relativistic calculations are not significantly more expensive than nonrelativistic Schrödinger-theory-based calculations (except for a small prefactor at most), the relativistic calculation is always to be preferred as should have become evident especially in the last couple of chapters. Hence, we may be a bit sloppy in discussing *qualitative* relativistic effects and use the term mostly to emphasize that certain effects (like spin–orbit coupling), which nat-

Relativistic Quantum Chemistry. Markus Reiher and Alexander Wolf

ISBN: 978-3-527-31292-4

urally arise in the relativistic theory, may lead to important differences if they were neglected in the calculation.

We may discuss relativistic effects at the example of reaction energies. Imagine the decomposition of an ether which, for illustration purposes, we may formally write as

$$\text{R–O–R}' + \text{H}_2\text{O} \longrightarrow \text{R–OH} + \text{R}'\text{–OH}$$

with R and R′ being organic residues. Temperature effects, which could be included based on the solution of the nuclear Schrödinger equation of section 8.3, are certainly negligible, and we may calculate the nonrelativistic reaction energy ΔE_{el}^{nr} at 0 Kelvin for 1 mol from the energy expectation values of the individual species weighted by Avogadro's constant, N_A,

$$\Delta E_{el}^{nr} = N_A \left[E_{el}^{nr}(\text{R–OH}) + E_{el}^{nr}(\text{R}'\text{–OH}) - E_{el}^{nr}(\text{R–O–R}') - E_{el}^{nr}(\text{H}_2\text{O}) \right] \quad (16.1)$$

and accordingly the relativistic value

$$\Delta E_{el}^{rel} = N_A \left[E_{el}^{rel}(\text{R–OH}) + E_{el}^{rel}(\text{R}'\text{–OH}) - E_{el}^{rel}(\text{R–O–R}') - E_{el}^{rel}(\text{H}_2\text{O}) \right] \quad (16.2)$$

Now, it turns out that these reaction energies are almost identical,

$$\Delta E_{el}^{nr} \approx \Delta E_{el}^{rel} \quad (16.3)$$

and certainly equal on an energy range of about 1 kJ/mol relevant to chemistry [usually, the Avogadro constant is hidden in the unit conversion factor like 2625.5 kJ/(mol Hartree). For molecules containing heavy atoms whose bonds are broken or rearranged, these energy differences can deviate a lot — if, for example, the oxygen atom in the example above is replaced by a heavier analog like Te to yield a telluro-ether. And exactly for such cases one would like to assign the label *relativistic effect* in the sense of a chemical concept and, of course, the relativistic result is to be preferred (provided the calculations are sound).

There exist obvious cases where relativistic methods are mandatory like reactions in which the educts and products feature electronic ground states of different spin. This implies that the spin symmetry is changed on the reaction path (mediated by spin–orbit coupling). Such events occur quite often in transition metal chemistry (for an example see the recombination of carbon monoxide with iron tetra-carbonyl [837, 838]), so often that even a new chemical concept has been coined: *two-state reactivity* [839–842]. This denotes the fact that a reaction path of lowest energy is not necessarily the one on a potential energy surface E_{el} of fixed spin (say, the total spin state of the educts) [843–846].

Besides the energies, relativity also affects molecular structures [847–852] like bond angles and bond distances as well as any molecular property [694–

697]. Some properties may even not be calculable within a nonrelativistic framework at all. For instance, the g-tensor of ESR spectroscopy may not be looked at without the consideration of spin–orbit coupling, while the ESR isotropic hyperfine coupling constant may come out quite accurately in a purely scalar-relativistic framework. Innumerable theoretical studies on relativistic effects have been carried out during the past decade from which we select some illustrative examples. Prototypical elements for which significant relativistic effects have been observed are heavy transition metals like Au, Pd, Pt, heavy main group elements like Tl, Pb, Bi as well as lanthanides and actinides. Since quantum chemical studies on molecules require sufficient computational resources, the most approximate methods were among the first to be applied. Most important was the effective-core-potential approach (see chapter 14) in combination with single- and multi-determinant methods (see chapter 8). Such theoretical studies of isolated molecules, i.e., with complete neglect of any environmental effects (often misleadingly denoted as 'calculations in the gas phase') demand experiments which match these models best and mass spectrometry is one of them. In the mid 1990s, beautiful studies by Schwarz and collaborators [853–857] demonstrated the reliability and usefulness of calculations on elementary pre-catalytic systems (namely on bare metal ions with small ligands) for which no structural information other than that from calculations could be obtained.

An extraordinary interplay between experiment and theory took also place in the development of the chemistry of gold [858–861]. In particular, the surprising and unexpected structures and their interactions that have been discovered by Schmidbaur and collaborators [862, 863], like the six-coordinate carbon atom in a gold octahedron [864–866], stimulated theoretical work [867–869]. Pyykkö and collaborators carried out a number of studies on gold complexes and clusters [870–873] in order to understand the peculiar features that have been referred to as *aurophilic effects*. Especially the interacting closed-shell gold complexes demanded theoretical elucidation. Pyykkö tackled this problem in an interesting fashion. As closed-shell molecules, their interaction should be dominated by dispersion forces. Such van der Waals forces are usually weak, though this is not true in the case of gold clusters. A dispersion-dominated interaction energy can only be reliably calculated from an accurate electronic energy including electron-correlation effects. Hence, single-determinant Hartree–Fock theory, which lacks any such correlation effects by definition, is uncapable of describing systems bound by van der Waals forces. For example, all dimers/clusters of rare gas atoms or alkanes, which are found to be stable experimentally, come out (erroneously) unbound in Hartree–Fock calculations. Also all contemporary density functionals are unable to capture such dispersion effects, and DFT can therefore also not be employed (except if empirical corrections are added that include an *ad hoc* attractive $-1/d^6$ term with d being the distance between the interacting fragments). Møller–Plesset perturbation theory, however, can be successfully employed (although it often yields too deep potential wells, i.e., it 'overbinds'). Pyykkö turned the deficiencies of Hartree–Fock theory into a feature and compared Møller–Plesset second-order calculations with Hartree–Fock results. This allows one to isolate the dispersion contribution in the interaction. Since also relativistic effects can be 'switched' on and off by employing the nonrelativistic many-electron Hamiltonian or a relativistic Hamiltonian, all tools have become available to decompose the interaction energy of gold clusters.

16.1
Effects in Atoms with Consequences for Chemical Bonding

The importance of relativity to chemistry was first noticed on the basis of atomic calculations [832,833], because effects on bonding may be qualitatively anticipated in terms of the LCAO model. Moreover, atoms and their ions can be studied with the utmost accuracy (in the case of one-electron ions even exactly), and we are therefore advised to start with a comparison of Schrödinger and Dirac atoms. Such comparisons have been made several times throughout this book on a formal basis. Now, we recall some typical effects on one-electron states and orbital energies.

From the short-range behavior of the radial shell functions in chapter 9 (compare also chapter 6) we understand that the $ns_{1/2}$ and $np_{1/2}$ shells, i.e., the shells with $|\kappa| = 1$ are affected most by relativity as they assign a non-negligible probability to finding the electron in the close vicinity of the nucleus. Qualitatively speaking, these shells 'experience' the atomic nucleus directly and are not pushed away from it by a centrifugal barrier (cf. section 6.6). As a consequence, the $ns_{1/2}$ and $np_{1/2}$ shells are found to contract compared to their nonrelativistic analogs. Other shells can then (indirectly) be affected by this contraction, although their behavior cannot be uniformly predicted compared to the nonrelativistic case. Especially the valence shells with $|\kappa| > 1$ may expand, because of the increased shielding of the nucleus due to contracted core-penetrating $|\kappa| = 1$ shells, with effects on chemical bonding. Figure 16.1 demonstrates these effects for the $1s$, $4s$, and $3d$ shells in the Kr-like Rn ion already discussed in section 9.8.2. It is interesting to observe the more pronounced contraction of the $4s$ shell compared to the $1s$ shell, while the $3d$ shell does not show any noticeable change. Such comparisons were provided by Burke and Grant as early as in 1967 [874].

Figure 16.2 shows the experimental energy splitting of the excited states in the carbon isoelectronic series. The ground state is taken to be the zero-energy reference for each atom. The figure clearly shows the increasing spin–orbit

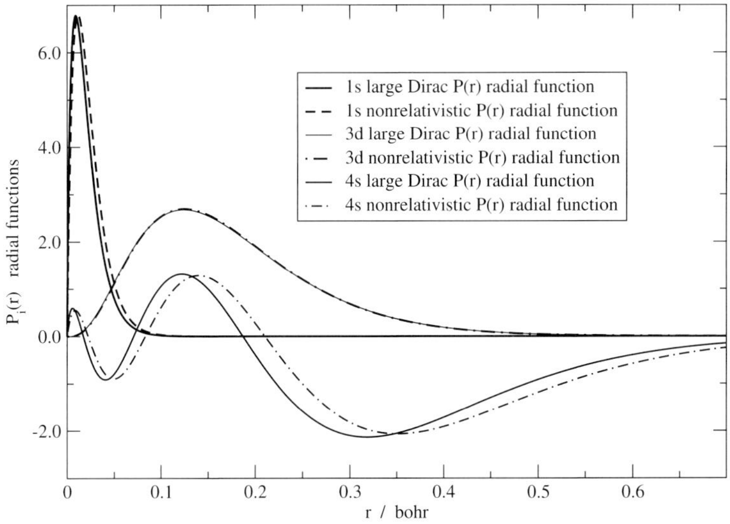

Figure 16.1 Relativistic Dirac–Hartree–Fock calculation for Kr-like Rn of Figure 9.3 compared to a nonrelativistic calculation where the speed of light has been set to $c = 10^5$. Depicted are the large-component radial functions $P_{n\kappa}(r)$ only. While the small-component radial function is negligible in the nonrelativistic calculation, it is of non-negligible size in the relativistic calculation, which is important for considerations based on the electronic density where the small component contributes. Note the relativistic contraction of the $1s$ and $4s$ shells and the negligible effect on the $3d$ valence shell.

splitting of the ${}^3P$ ground state with nuclear charge number Z. While the spin–orbit splitting is negligible in the case of carbon, the energy difference $|{}^3P_2 - {}^3P_0|$ amounts to 10650.5 cm^{-1} in the case of lead. This large spin–orbit splitting corresponds to the energy difference between the ground ${}^3P$ and the excited ${}^1D_2$ state of carbon, which is 10193.7 cm^{-1}. Clearly, the quantum chemical calculation of the excited states of carbon does not require us to consider spin–orbit splitting, while it is essential to the electronic spectrum of lead.

The observation that large spin–orbit splittings of total states occur for the heavy p-block elements also has consequences for the qualitative one-particle picture which is commonly used in chemistry to understand electronic structure. In other words, a molecular orbital diagram for a molecule containing such elements should take into account the spin–orbit splitting even for the one-particle spinors — which is, of course, naturally incorporated if the di-

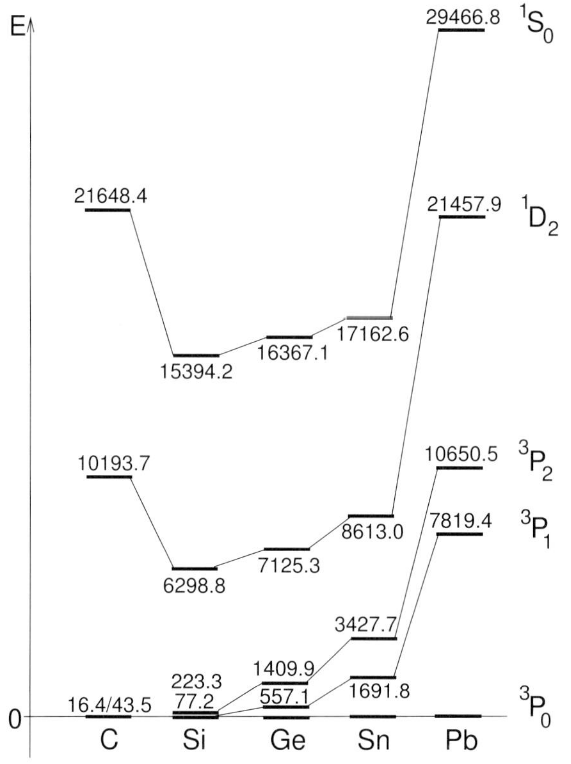

Figure 16.2 Experimental energy splitting in cm^{-1} of the lowest lying terms of neutral atoms from the 4th main group of the periodic table of the elements. Note the increasing spin–orbit splitting of the ^3P ground state with nuclear charge Ze. The plot was produced from atomic spectroscopy data of Ref. [875].

agram is constructed from four- or two-component spinors. Most important for the qualitative understanding of chemical bonding is the fact that the spin–orbit splitting of the molecular orbitals (spinors) can be so large that the $np_{1/2}$ and $np_{3/2}$ valence atomic orbitals lead to different types of bonds. Thus, hybridization of s- and p-type atomic spinors changes if the p-valence shell splits significantly. Of course, for the more close-lying pairs of spin–orbit split d- and f-shells it becomes increasingly difficult to clearly distinguish between the various effects that shift the one-particle energies [876].

Typical examples of such qualitatively different types of bonds are found, for instance, in the molecules TlH (see below) and HAt [877]. TlH can be regarded as a truly closed-shell molecule since the $6p_{1/2}$ shell is completely filled while the $6p_{3/2}$ is energetically well separated so that it may be neglected in the qualitative molecular orbital picture. By contrast, the valence-isoelectronic molecule BH is an open-shell system because of the (near-)degeneracy of the boron (s- and) p-shell. Moreover, the wealth of experimental data has led in comparative studies to the concept of the *inert pair effect* which expresses the fact that the occupied valence s-shell in the third main group of the periodic table is so much contracted for the high-Z elements that it hardly contributes to hybridization and bonding. Consequently, lower oxidation numbers are found for the heavy elements when compared to their lighter homologs.

Since especially diatomics built from atoms with large nuclear charge number Z and an open p shell are significantly affected by scalar and spin–orbit effects, molecules containing thallium became perfect test cases for relativistic electronic structure methods. Some representative results for TlH are collected in a later section.

Prominent bulk properties are also linked to relativistic effects though their theoretical decription is not straightforward and quite involved. These are the yellowish color of bulk gold and the liquid state of mercury at ambient temperature. It must be emphasized that these are true condensed-phase systems requiring solid-state calculations and molecular dynamics, respectively. Moreover, a relativistic calculation must be compared to a nonrelativistic analog in order to clearly separate the effects that would simply show up because of the high nuclear charge and large number of electrons — especially when related to lighter homologous elements in the periodic table. In the case of gold the identification of the relativistic effect on its color may be achieved in a one-particle model within a periodic-boundaries framework, in which the energetical position of the partially occupied valence band is compared to the energy range of the conduction band in a relativistic and a nonrelativistic calculation. The fluidity of mercury is a much more difficult problem to tackle. While the weakly bound mercury dimer — bound via attractive van-der-Waals-type closed-shell intercations like the heavy rare gas dimers — can reliably be described by correlated relativistic calculations [878, 879], polynuclear clusters are difficult to assess. Computationally less demanding approaches like DFT are hampered by the presently existing difficulties of density functionals to treat dispersion interactions. Then, the description of the melting behavior of mercury by molecular dynamics requires reliable many-body potentials, which are, however, a tough task to generate for the highly polarizable mercury atoms as Schwerdtfeger and coworkers demonstrated in pioneering studies [880–882].

16.2
Is Spin a Relativistic Effect?

When Dirac set out to derive his fundamental quantum mechanical equation for the electron, he aimed at solving two problems at once, namely the lack of a relativistic quantum mechanical equation and the fact that electron spin was merely an *ad hoc* assumption. Dirac noted [66] that *the incompleteness of the previous theories* lies *in their disagreement with relativity [...]. It appears that the simplest Hamiltonian for a point-charge electron satisfying the requirements of both relativity and the general transformation theory leads to an explanation of all duplexity phenomena without further assumption.* Spin enters Dirac's equation through the $\boldsymbol{\alpha}$ matrices required to fulfill the Lorentz covariance condition. Clearly, this connection leads one directly to think of spin as a 'relativistic effect', but the situation is more involved as can be seen already from the fact that the Klein–Gordon equation is a *relativistic equation* but for particles *without spin*. Moreover, quite the opposite question is interesting, namely why can spin play such a prominent role as an add-on to nonrelativistic Schrödinger theory although it is not a good quantum number in fundamental relativistic theory (recall also section 13.1).

The question of spin as a relativistic effect got quite an interesting twist in the late 1960s when Lévy-Leblond [883, 884] showed that spin 1/2 also emerges in a *non*relativistic theory from irreducible representations of the Galilei group, which was shortly afterwards generalized for arbitrary spin [885,886]. Hence, the Lévy-Leblond equation is a nonrelativistic equation including spin, which can be written for the stationary states of a single electron in a scalar potential as

$$\begin{pmatrix} V & c\boldsymbol{\sigma}\cdot\boldsymbol{p} \\ c\boldsymbol{\sigma}\cdot\boldsymbol{p} & -2m_ec^2 \end{pmatrix} \begin{pmatrix} \tilde{\Psi}^L \\ \tilde{\Psi}^S \end{pmatrix} = E^{nr} \begin{pmatrix} 1 & 0 \\ 0 & 0 \end{pmatrix} \begin{pmatrix} \tilde{\Psi}^L \\ \tilde{\Psi}^S \end{pmatrix} \tag{16.4}$$

After what has been derived in chapter 13, it is clear how this equation 'works'. All difficulties with the elimination of the small component in chapter 13 stem from the fact that the relation between large and small component contains the potential energy operator V as well as the energy. In the Lévy-Leblond equation this is avoided because the lower part of the matrix equation does not depend on either of them. This is most easily seen in split notation,

$$V\tilde{\Psi}^L + c\boldsymbol{\sigma}\cdot\boldsymbol{p}\tilde{\Psi}^S = E^{nr}\tilde{\Psi}^L \tag{16.5}$$

$$c\boldsymbol{\sigma}\cdot\boldsymbol{p}\tilde{\Psi}^L - 2m_ec^2\tilde{\Psi}^S = 0 \tag{16.6}$$

From the second equation we obtain

$$\tilde{\Psi}^S = \frac{c\boldsymbol{\sigma}\cdot\boldsymbol{p}}{2m_ec^2}\tilde{\Psi}^L \tag{16.7}$$

which yields the Schrödinger equation once inserted in the first equation,

$$\left[V + c\boldsymbol{\sigma} \cdot \boldsymbol{p} \frac{c\boldsymbol{\sigma} \cdot \boldsymbol{p}}{2m_e c^2}\right] \tilde{\Psi}^L \overset{(D.18)}{=} \left[V + \frac{\boldsymbol{p}^2}{2m_e c^2}\right] \tilde{\Psi}^L = E\tilde{\Psi}^L \tag{16.8}$$

where we exploited Dirac's relation yet another time for the squared scalar product $(\boldsymbol{\sigma} \cdot \boldsymbol{p})^2 = \boldsymbol{p}^2$.

16.3 Z-Dependence of Relativistic Effects: Perturbation Theory

As should have become clear at various places throughout this book, the magnitude of relativistic effects depends on the nuclear charge expressed by the parameter Z. In general, these effects will be the more pronounced the larger Z is. The actual Z-dependence of the relativistic effect of a given property depends on this property (and possibly also but to a smaller extent on how it has been assessed). The analysis of relativistic effects depending on the nuclear charge number has a long history (see, for example, the early study of atomic properties provided by Desclaux [351]).

Perturbation theory allows us to deduce definite Z-dependences of relativistic effects on atomic and molecular energies and properties [887]. Perturbative expansions in powers of the parameters Z and $1/c$ have a long history — as long as relativistic quantum mechanics itself, as illustrated by the Pauli Hamiltonian. A most sophisticated formulation is the so-called *direct perturbation theory* [512,888–896]. Especially Kutzelnigg has worked out this perturbation theory from all perspectives and discussed the results in the light of how one might rigorously define relativistic effects depending on what equation is considered the nonrelativistic limit [887]. As a suitable starting point for this perturbation theory, the Dirac equation with rest-energy shift $(-m_e c^2)$ is written such that a c pre-factor is absorbed in the small component,

$$\begin{pmatrix} V & \boldsymbol{\sigma} \cdot \boldsymbol{p} \\ c^2 \boldsymbol{\sigma} \cdot \boldsymbol{p} & V - 2m_e c^2 \end{pmatrix} \begin{pmatrix} \Psi^L \\ c\Psi^S \end{pmatrix} = E \begin{pmatrix} \Psi^L \\ c\Psi^S \end{pmatrix} \tag{16.9}$$

Now, the second part of the Dirac equation can be divided by $1/c^2$,

$$\begin{pmatrix} V & \boldsymbol{\sigma} \cdot \boldsymbol{p} \\ \boldsymbol{\sigma} \cdot \boldsymbol{p} & \frac{V}{c^2} - 2m_e \end{pmatrix} \begin{pmatrix} \Psi^L \\ c\Psi^S \end{pmatrix} = E \underbrace{\begin{pmatrix} 1 & 0 \\ 0 & c^{-2} \end{pmatrix}}_{\equiv S} \begin{pmatrix} \Psi^L \\ c\Psi^S \end{pmatrix} \tag{16.10}$$

where now a metric S emerges so that the ordinary normalization condition of the spinor is kept. A similar equation can be obtained for the negative-energy states if the first part of the Dirac equation is divided by c [891]. The unperturbed reference equation is obtained for $c \to \infty$, which is immediately identified as the Lévy-Leblond equation [Eq. (16.4)].

We have already discussed in chapters 12 and 13 that low-order scalar-relativistic operators like DKH2 or ZORA provide very efficient *variational* schemes, which comprise all effects for which the (non-variational) Pauli Hamiltonian could account for (as is clear from the derivations in chapters 11 and 13). It is for this reason that the significance of especially scalar relativistic corrections, which can only be considered perturbatively (like the mass-velocity and Darwin terms in the Pauli approximation of section 13.1), faded away. There is also no further need to develop new pseudo-relativistic one- and two-electron operators. This is very beneficial in view of the desired comparability of computational studies. In other words, if there were very many pseudo-relativistic Hamiltonians available, computational studies with different operators of this sort on similar molecular systems would hardly be comparable.

16.4
Potential Energy Surfaces and Spectroscopic Parameters

A reliable prediction of spectroscopic phenomena in heavy-element compounds requires a balanced description of scalar-relativistic, spin–orbit and electron-correlation effects. In some cases one or more of these effects can be dominant, requiring an elaborate method to take this into account, whereas the others may be treated in a more approximate way or can even be completely neglected. The choice of the Hamiltonian is a crucial issue in relativistic calculations of spectroscopic quantities. Four-component methods employing the Dirac–Coulomb or Dirac–Coulomb–Breit Hamiltonian offer the most consistent theoretical framework for relativistic calculations, treating scalar-relativistic effects and spin–orbit coupling in the one- and two-electron terms on the same footing.

The calculation of molecular spectra of large molecules inevitably demands approximations to both relativistic and electron-correlation effects. Consideration of the size, the electronic structure of a particular molecule under investigation, and the spectroscopic quantity to be predicted often leads to the adoption of a custom-made approach for feasibility reasons. In the following, different relativistic approaches applied to model systems involving main group as well as transition metal atoms are compared. In order to demonstrate the effect of a relativistic versus a nonrelativistic quantum chemical method for the calculation of relative energies, it is most appropriate to study the potential energy curve of diatomic molecules. For such small molecules, very accurate quantum chemical calculations can be carried out, reliable total electronic wave functions are obtained, and the results for dissociation energies and equilibrium distances may be extrapolated to reaction energies and molecular structures of general molecules, respectively.

As an example, we refer to a study [897] in which the four-component Dirac–Coulomb, a spin-free and an infinite-order two-component Hamiltonian were applied in a multi-reference configuration interaction and a multi-reference coupled-cluster formalism [898] for the prediction of the double photoionization spectrum of molecular bromine. The results exemplify the fact that accurate vertical transition energies require a sophisticated treatment of electron-correlation and relativistic effects within a multi-reference configuration interaction scheme. A balanced description of electron-correlation and spin–orbit effects turned out to be crucial in regions with a high density of total electronic states, because the energetical separation of states and the spin–orbit splitting are then of the same order of magnitude.

While we select some diatomics in more detail in the following discussion, we may refer to the book by Balasubramanian [899] for an exhaustive account on diatomics and triatomics from the whole periodic table.

The shape and curvature of the potential energy $E'_{el,B}(\{\boldsymbol{R}_I\})$ of Eq. (8.84) can be viewed as a continuous measure of the changing total electronic energy. In theoretical vibrational spectroscopy, the curvature of this potential energy function finally enters the expression for the harmonic vibrational frequency ω_e and its anharmonic correction $\omega_e x_e$ [205–207]. Other important spectroscopic parameters are, of course, the equilibrium bond length R_e, which is to be calculated from the set of nuclear coordinates $\{\boldsymbol{R}_I\}$ for which $E'_{el,B}$ assumes a minimal value, and the electronic contribution to the dissociation energy D_e of the diatomic molecule. Since R_e and D_e refer to the solution of the *electronic* Schrödinger equation given in Eq. (8.76), they are not observables. Only when combined with the solutions of the nuclear complement in Eq. (8.84), i.e., when corrected for molecular vibrations, are R_0 and D_0 obtained, so that they can be directly compared with suitable experimental data. However, in the tables of this section, R_e and D_e have been extracted from the experimental raw data, and therefore no vibrational effects are considered.

16.4.1
Thallium Hydride

Many relativistic electronic structure calculations have been carried out on TlH [528], and we select some material from the collection in Ref. [900]. Due to the fact that spin–orbit coupling is essential for the correct description of the ground state of TlH, only those results are included in Table 16.1 that treat spin–orbit coupling either variationally or perturbatively.

Rakowitz and Marian [642] have presented results of spin–orbit-free MRCI calculations. Scalar-relativistic effects were included via the scalar-relativistic second-order DKH2 Hamiltonian in the ΛS-coupled wave function. The spin–orbit coupling was included *a posteriori* with three different methods, of which two are reported in Table 16.1, namely, quasi-degenerate perturbation theory and a spin–orbit configuration interaction (SOCI). The quasi-degenerate perturbation theory overestimates the equilibrium bond length R_e and underestimates the harmonic vibrational frequency ω_e when compared to experimental data. When spin-polarization effects are taken into account via SOCI, R_e decreases by 3 pm and ω_e increases by about 80 cm^{-1}, closely approaching the experimental value. Later, Seth *et al.* [901] performed all-electron four-component MP2 and CCSD(T) calculations on the basis of the Dirac–Coulomb (DC) Hamiltonian denoted as DC-CCSD(T) in Table 16.1. The CCSD(T) results are in good agreement with the experimental data; R_e is overestimated only by about 2 pm. The MP2 result, however, underestimates R_e as well as D_e and overshoots ω_e considerably by 46 cm^{-1}. This effect has later been qualitatively confirmed by Abe *et al.* [437] although the predicted spectroscopic parameters differ from the data by Seth *et al.*, which is likely because of the different one-electron basis sets used.

Table 16.1 Equilibrium distance (R_e) in pm, harmonic vibrational frequency (ω_e) in cm^{-1} and potential energy well depth (D_e) in eV of the ground state of TlH obtained from various relativistic approaches. The abbreviation "var." indicates a variational and "pert." perturbational treatment of spin–orbit coupling (SOC).

Method	SOC	Correlated e^-	R_e	ω_e	D_e	Ref.
MRCI	pert.	14	190	1309	2.08	[642]
MRCI	SOCI	14	187	1388	2.15	
DC-MP2	var.		186.2	1437	1.83	[901]
DC-CCSD(T)	var.		188.5	1376	2.07	
DC-CCSD(T)	var.	14	187.6	1385	2.00	[902]
DC-CCSD(T)	var.	20	187.4	1378	1.98	
DC-CCSD(T)	var.	36	187.4	1371	1.98	
DCG-CCSD(T)	var.	36	187.7	1376	2.06	
DC-MP2	var.	14	184.6	1454	1.83	[437]
DC-MP2	var.	36	183.8	1467	1.84	
DC-CCSD	var.	36	186.2	1410	2.06	
DC-CCSD(T)	var.	36	186.2	1404	2.10	
DC-CASPT2	var.	14	189.3	1351	1.87	[438]
Exp.			186.6	1390.7	2.06	[903]
			186.8	1391.3	—	[904]

The work required for the calculations is significantly reduced if a frozen core of electrons is employed so that only a few electrons are treated explicitly in the excitation scheme used for setting up the wave function (compare section 8.5). Faegri and Visscher [902] carried out DC-CCSD(T) calculations with large uncontracted basis sets. They systematically investigated the effect of core correlation and could demonstrate that increasing the number of correlated electrons from 14 to 20 and finally 36 leads to minor effects on the equilibrium bond length and that ω_e tends to decrease slightly upon increased core correlation. The same applies to the dissociation energy D_e. Moreover, the effect of the Gaunt term on spectroscopic parameters was investigated in Ref. [902]. These DCG-CCSD(T) results suggest that the Gaunt interaction somewhat counteracts the core-correlation effects, yielding slightly elongated equilibrium bond lengths and increased ω_e and D_e results.

Abe *et al.* [438] report Dirac–Coulomb CASPT2 calculations on TlH. The reference for the CASPT2 treatment is constructed via a complete active space configuration interaction (CASCI) ansatz. The results exhibit less good agreement with the experimental spectroscopic parameters, as one would expect considering the limited size of the active space feasible in any CAS approach. The equilibrium bond length is overestimated by 2.5 pm and ω_e by 40 cm^{-1}, which cannot challenge the accuracy of the highly correlated DC-CCSD(T) results.

16.4.2
The Gold Dimer

Another prototypical molecule, which has been extensively investigated in numerous studies with relativistic electronic structure methods is the gold dimer Au_2 [507, 544, 591, 594, 847, 848, 905–927]. Spin–orbit coupling may be neglected owing to its closed-shell electronic structure. Electron-correlation effects, however, play a decisive role for obtaining reliable spectroscopic parameters. Table 16.2 provides results obtained with various methods.

The work of Lee *et al.* [923] compares results obtained with relativistic effective core potentials (ECP). The ECP calculations have been carried out with a Kramers-restricted (KR) two-component formalism as described in Ref. [682]. These results demonstrate the accuracy of the ECP-CCSD(T) approach when compared with experiment. Tsuchiya *et al.* [926] investigated Au_2 with MP2, CCSD and CCSD(T) employing a scalar-relativistic third-order DKH3 Hamiltonian. Again CCSD(T) yields results that reproduce all spectroscopic parameters remarkably well.

Hess and Kaldor [925] conducted the most comprehensive study on the gold dimer investigating the effects of the size of the one-electron basis set as well as of the choice of the approximation for the total wave function. All results were obtained with the scalar-relativistic second-order DKH2 Hamiltonian and demonstrate that a proper approximation of the many-electron wave function is crucial for accurate results. Correlating the $5p$-shell by switching from 22 to 34 correlated electrons (i.e., including the corresponding orbitals in the excitation process for the construction of the wave-function approximation) yields a shortening of the equilibrium bond length by about 1 pm and an increase in D_e by about 0.05 eV for the very accurate CCSD(T) calculations. A fully satisfactory convergence of electron-correlation effects can only be achieved if at least one-electron basis functions of i-angular momentum (i.e., $l = 6$) are included in the one-particle basis set.

Table 16.2 Equilibrium distance (R_e) in pm, harmonic vibrational frequency (ω_e) in cm^{-1} and potential energy well depth (D_e) in eV of the ground state of Au_2 obtained from various relativistic approaches (selection from the review in Ref. [900]). The abbreviation "var." indicates a variational treatment of spin–orbit coupling (SOC).

Method	SOC	correlated e^-	R_e	ω_e	D_e	Ref.
ECP-KRMP2	var.	34	244.4	204	2.44	[923]
ECP-KRCCSD	var.	34	249.6	186	2.01	
ECP-KRCCSD(T)	var.	34	249.0	187	2.23	
DKH3-MP2	—	34	244.1	208.3	2.32	[926]
DKH3-CCSD	—	34	250.0	187.0	1.96	
DKH3-CCSD(T)	—	34	249.4	188.6	2.17	
DKH2-MP2	—	34	241.8	212.4	—	[925]
DKH2-MP3	—	34	250.1	185.4	—	
DKH2-MP4	—	34	244.2	202.2	—	
DKH2-CCSD	—	34	249.4	186.6	1.98	
DKH2-CCSD(T)	—	34	248.8	186.9	2.19	
SFDC-CISD	—	34	249.8	185.3	5.03	[927]
SFDC-MR(2 in 2)CISD	—	34	250.8	180.2	1.95	
DC-CCSD	var.	34	248.4	190.4	2.06	
DC-CCSD(T)	var.	34	247.7	192.1	2.28	
ZORA-DFT	—	—	250	176	2.30	[591]
ZORA-DFT	—	—	252.1	175	2.25	[924]
4comp. DFT	var.	—	251.3	183	2.22	
Exp.	—	—	247.2	191	2.29	[928–930]

Hess and Kaldor have further shown that even MP4 with 34 correlated electrons is not able to reproduce the experimental results with sufficient accuracy. Their results have been challenged by Fleig and Visscher [927], who performed four-component but spin-free calculations where the spin-dependent part has been separated from the spin-free terms. This approach represents a proper reference for calculations with truncated scalar-relativistic DKH Hamiltonians. The calculations were carried out within a general active space approach [243, 931] with single and double excitations. A single reference ansatz for the CI expansion shows that D_e is overestimated. Turning to a multi-reference approach significantly improves the result for the dissociation energy. But Fleig and Visscher have also investigated the effects of spin–orbit coupling on the spectroscopic parameters with Dirac–Coulomb CCSD(T) calculations. Their results suggest that spin–orbit coupling tends to reduce the equilibrium bond length by only 0.4 pm as anticipated.

Table 16.3 Comparison of spectroscopic parameters for the ground states of SnO and CsH: equilibrium bond length R_e in pm, dissociation energies D_e in eV, harmonic frequencies ω_e in cm^{-1}, and anharmonic vibrational constants $\omega_e x_e$ in cm^{-1}. This table demonstrates the effect of the approximation of the total electronic wave function (represented by the methods CI, MRCI, CCSD(T), and DMRG) compared to the choice of the one-electron Hamiltonian (DKH and effective core potential) with a special emphasis on the order of the DKH Hamiltonian. The calculated results are compared to data from experiment. The effect of the one-particle basis set can be seen for CsH from the entries for the contracted bases of quality $[11s9p8d]$ and $[11s9p8d2f1g]$, respectively. Note also the importance of the approximation for the total wave function (as, for example, highlighted in the case of SnO for uncorrelated Dirac–Hartree–Fock or even nonrelativistic Hartree–Fock compared to the coupled-cluster results).

Method	R_e	D_e	ω_e	$\omega_e x_e$	Ref.
			SnO		
CCSD(T)/nonrelativistic	184.0	5.53	824.6	7.08	[547]
CCSD(T)/DKH2	183.6	5.45	812.4	3.88	
CCSD(T)/DKH3	183.6	5.45	812.3	3.88	
CCSD(T)/DKH4	183.6	5.45	812.3	3.88	
CCSD(T)/DKH5	183.6	5.45	812.3	3.88	
Hartree–Fock (nonrel.)	180.8		955		[932]
Dirac–Hartree–Fock	180.1		946		
Exp.			810±10		[933]
	183.3	5.45	815	3.73	[934]
			CsH		
DMRG/DKH10/$[11s9p8d]$	259.0	1.658	856	11.6	[935]
MRCI/DKH10/$[11s9p8d]$	258.1	1.639	856	14.5	
CCSD(T)/DKH10/$[11s9p8d]$	257.8	—	832	11.2	
CCSD(T)/DKH10/$[11s9p8d2f1g]$	251.6	—	882	13.0	
MRCI/DKH2	254.8	1.752	896	—	[936]
CI/ECP+CPP	244.8	1.847	885	—	[937]
MP2/DKH2	250.7	—	888	—	
Exp.	249.4	1.834	891	12.9	[934,938]

As an example of a pseudo-relativistic all-electron DFT method, Table 16.2 lists results from van Wüllen's scalar-relativistic ZORA-DFT study [591]. Leaving the comparatively large error in ω_e aside, we note that the remaining spectroscopic constants are quite well reproduced. A subsequent relativistic

DFT study on the gold dimer employed a four-component Dirac–Kohn–Sham method and the scalar-relativistic ZORA approach [924]. Table 16.2 also shows that the scalar-relativistic ZORA approach yields quite good results compared to the four-component reference. Recall that the accuracy of such DFT calculations is determined by the approximate nature of the exchange–correlation functional employed (see section 8.8).

16.4.3 Tin Oxide and Cesium Hydride

To demonstrate the effect of different orders in DKH calculations, Table 16.3 presents results obtained for SnO and CsH (different ansätze for the electronic wave function were employed). As can be seen from the table, all spectroscopic parameters converge fast with increasing DKH order. The accuracy is mostly determined by the quality of the wave-function approximation. Note that DKHn denotes the scalar-relativistic variant, which has also been called the spin-averaged (i.e., the spin-free) DKH approach. Also, the two-electron terms have not been transformed. Since the nuclear charge numbers of Sn and Cs are not very high, these elements are not 'ultrarelativistic' cases. However, also for Au_2 it was found that the spectroscopic parameters are already converged with the second-order DKH2 Hamiltonian [939], which is typical for such valence-shell dominated properties.

16.5 Lanthanides and Actinides

One of the most interesting targets for relativistic quantum chemistry is provided by molecules containing heavy elements, like the lanthanides and actinides [940]. Especially radioactive heavy elements and trans-actinides are a target of computational research aimed at improving experimental procedures for extraction and isolation of these radioactive species from nuclear waste. A concrete example is the optimization of chelate ligands that can be employed in extraction processes of uranium and plutonium. Accordingly, numerous uranium-containing molecules have been studied with relativistic quantum chemical methods. Typical molecules are uranium oxides, in particular UO_2^{2+}, and uranium fluoride, of which UF_6 is the most important example for technical processes; see for some of these studies Refs. [941–944].

Gagliardi and Roos conducted a series of studies on actinide compounds. They follow a combined approach with DKH/AMFI Hamiltonians combined with CASSCF/CASPT2 for the energy calculation and an *a posteriori* added spin–orbit perturbation expanded in the space of nonrelativistic CSFs. This strategy aims to establish a balance of sufficiently accurate wave function and Hamiltonian approximations. Since the CASSCF wave function provides chemically reasonable but not highly accurate results (as witnessed, for instance, in the preceding section), it is combined with a quasi-relativistic Hamiltonian, namely the scalar-relativistic DKH one-electron Hamiltonian. Additional effects — dynamic correlation and spin–orbit coupling — are then considered via perturbation theory.

The calculation of spectroscopic parameters of the uranium dimer, U_2, received considerable attention [945]. The bonding situation of the U_2-molecule implies a large number of nearly-degenerate electronic states, which require the multi-configurational approach for the calculation of the electronic wave function. This approach was also successfully applied to U_2^{2+} [946], to the actinide dimers Th_2, Ac_2 and Pa_2 [947], and to the dinuclear clusters PhU-UPh [948], U_2Cl_6 and $U_2(OCHO)_6$ [949].

16.5.1 Lanthanide and Actinide Contraction

The 'filling' of inner f shells in lanthanide and actinide atoms is accompanied by a steady decrease of the size of the atom (measured by suitable means like the first radial moment). As a consequence, the size of atoms from the same group in the second and third transition metal series of the periodic table turns out to be similar. That is, the increased number of electrons with increasing nuclear charge number Z does not lead to an increased size of the atom under consideration. This reduced-size effect is called the *lanthanide contraction*. It is important to understand that the effect is also witnessed in purely nonrelativistic calculations — the effect is simply amplified in a relativistic description.

The main difficulty here is to clearly separate effects that can hardly be separated, namely relativistic and electron-correlation effects. Nevertheless, pioneering studies of this effect date back to the mid 1970s [950]. Four-component methods have been employed to determine the contribution which is solely due to relativity [951]. The four-component approach, for which Dirac–Hartree–Fock and — to also account for correlation effects — relativistic MP2 calculations have been utilized, confirms results first obtained with relativistic effective core potential methods [952, 953]. It has been found [951] that between 10% and 30% of the lanthanide contraction and 40% to 50% of the actinide contraction are caused by relativity in monohydrides, trihydrides, and monofluorides of La, Lu and Ac, Lr, respectively.

16.5.2
Electronic Spectra of Actinide Compounds

The calculation of electronic spectra of lanthanide or actinide compounds poses a considerable challenge to relativistic electronic structure methods. Apart from the mandatory relativistic Hamiltonians, the small energetic separation of valence *f* and *d* orbitals in lanthanide and actinide atoms gives rise to a prominent multi-reference nature of the N-electron wave function. Even the most accurate methods of calculating multiple spin–orbit-coupled states like MRCI+SO and CASPT2+SO suffer from the limited number of configurations or the restricted size of the active space to be included in a calculation, restricting the range of application of these models. Four-component Fock-space coupled cluster methods, which include jj-coupled spinors already in the SCF procedure, are therefore required to challenge the former standard methods. We select PuO_2^{2+} as a representative example to compare excitation energies obtained with MRCISD+Q+SO [954], with CASPT2+SO [955] and with the intermediate Hamiltonian Fock-space coupled cluster method iHFSCCSD [956] in Table 16.4.

Table 16.4 Vertical excitation energies T_e of PuO_2^{2+} in cm^{-1} calculated at different distances R_e with three different relativistic quantum chemical approaches (the data have been reproduced from the collection in Ref. [900]).

MRCISD+Q+SO Ref. [954] R_e=169.9 pm		CASPT2+SO Ref. [955] R_e=167.7 pm		iHFSCCSD Ref. [956] R_e=164.5 pm		Exp. Ref. [957]	
State	T_e	State	T_e	State	T_e	State	T_e
4_g	0	4_g	0	4_g	0	$^3H_4$	0
0_g	4295	0_g	4190	0_g	2530	Σ_0	—
5_g	6593	1_g	6065	1_g	4870	Π_1	—
1_g	7044	5_g	8034	5_g	6700	$^3H_5$	—
0_g	7393	0_g	12874	0_g	10334	$^3\Pi_0$	10185
6_g	7848	1_g	12906	1_g	10983	Σ_1	10500
0_g	9415	6_g	14326	0_g	11225	$^3\Pi_0$	10700
1_g	12874	0_g	14606	6_g	11651	$^3H_6$	—
2_g	14168	2_g	14910	0_g	12326	$^3\Pi_2$	12037
5_g	16984	—	—	0_g	16713	$^1\Gamma_4$	15420
4_g	23091	—	—	1_g	17737	Σ_0	16075
1_g	27005	—	—	4_g	18565	Σ_1	17800
6_g	30254	—	—	0_g	20029	$^3\Gamma_3$	19100
3_g	33164	—	—	1_g	22703	Σ_0	19810
0_g	33314	—	—	6_g	22889	$^3\Phi_2$	22200
4_g	33318	—	—	5_g	23022	$^1H_5$	21840

Maron *et al.* [954] used a small-core ECP with 34 valence electrons for the Pu atom and 6 valence electrons for the O atom. The active space for the spin–

orbit calculations was set up to reproduce the $5f$-manifold of the PuO_2^{2+} electronic spectrum by distributing two electrons in the f valence orbitals. Atomic mean-field integrals were used to calculate the spin–orbit matrix elements. The diagonal elements of the spin–orbit matrix were replaced by the energies obtained from a large Davidson-corrected MRCI calculation with single and double excitations involving 26 electrons (denoted as MRCISD+Q), and finally the spin–orbit matrix was diagonalized in a QDPT procedure to obtain the spin–orbit states. The Davidson correction was used to obtain an estimate of the contribution from higher order excitations to the electronic energy.

Clavaguera-Sarrio *et al.* [955] also used an ECP-based approach with 34 valence electrons for Pu and a similar electron configuration to generate the reference space. Electron correlation was taken into account either by a CASPT2 approach or by the so-called difference-dedicated configuration interaction method [958], which was custom made to reproduce the energy difference between electronic states correctly. Spin–orbit coupling was calculated by applying an SOCI method to the reference space and by taking into account the spin-free energy obtained from the CASPT2 or difference-dedicated configuration interaction calculations.

The intermediate Hamiltonian Fock-space coupled-cluster calculations of Infante *et al.* [956] were performed in an all-electron approach explicitly correlating 26 electrons and using large de-contracted basis sets.

A comparison of the data given in Table 16.4 shows that the iHFSCC model features lower excitation energies from the ground state to the first three excited states. Moreover, the order of the six states of lowest energy given in Ref. [955] is reproduced by the iHFSCC approach. However, a direct comparison to experiment is not possible for the first three low-lying states, due to the fact that these transitions are dipole forbidden and therefore feature only very low intensities. For the subsequent excited states a remarkable agreement with experimental data within a few hundred cm^{-1} is found in case of the iHFSCC calculations. The data provided by Refs. [955] and [954] features larger deviations of far more than 1000 cm^{-1}. The iHFSCC approach proves to be the best but also the most expensive choice in order to reproduce the experimental excitation energies sufficiently well.

16.6
Electron Density of Transition Metal Complexes

In previous sections we have discussed the relativistic effect on the electron density of one-electron atoms and on atomic orbitals, about which many research papers have appeared (see Refs. [874, 959] for a quite old and a recent example). We shall now consider the case of molecules where polarization ef-

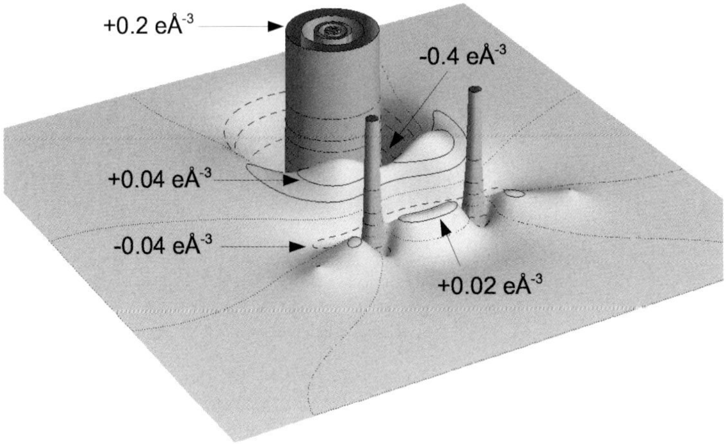

Figure 16.3 Relativistic effect on the electron density of $Pt(C_2H_2)$. The difference between the electron densities from a four-component and a nonrelativistic Hartree–Fock calculation is plotted for the plane in which all atomic nuclei lie. The figure is based on data from Ref. [730].

fects play a role. The relativistic effect on the electron density of an extended molecule can be conveniently demonstrated by inspection of electron density differences. Figure 16.3 depicts such a density difference for the plane of the atomic nuclei of the mono-acetylene complex $Pt(C_2H_2)$ [730]. The densities have been calculated from simple Hartree–Fock determinants according to Eq. (8.206) as the sum of squared molecular spinors. The relativistic contraction of the electron density in the core of the platinum atom appears as a huge tower of the difference density in the middle of the plot. Naturally, the effect is smaller at the core of the lighter carbon atoms and almost vanishes at the hydrogen atom cores. Also the valence shell is affected by relativistic effects, which are, in the case of molecules, structured — in contrast to spherically symmetric atoms — owing to the different polarization effects within a molecule.

To illustrate the magnitude of relativistic effects — kinematic as well as spin–orbit effects — electron density differences are depicted in Figure 16.4 for $Ni(C_2H_2)$ and $Pt(C_2H_2)$. For these plots various approximate electron densities have been subtracted from the four-component reference result.

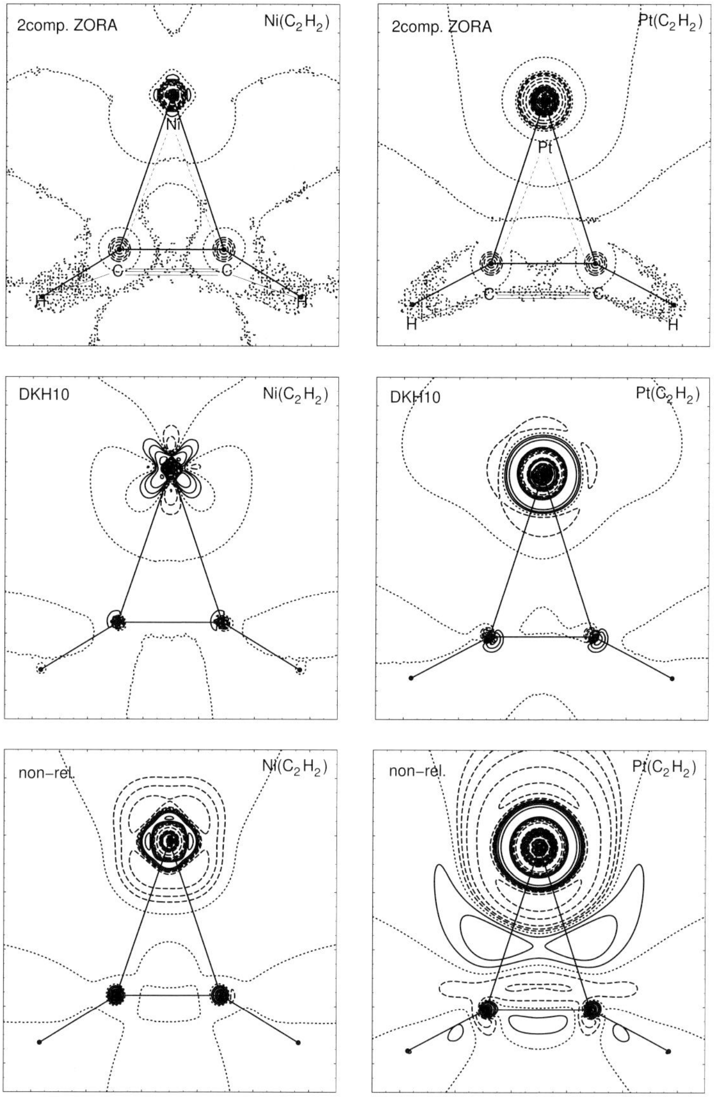

Figure 16.4 Total electronic densities of $M(C_2H_2)$ with M=Ni,Pt from Hartree–Fock calculations with two-component ZORA, scalar-relativistic DKH10, and nonrelativistic Schrödinger one-electron operators subtracted from the four-component Dirac–Hartree–Fock reference densities (data taken from Ref. [730]). The molecular structure of the complexes is indicated by element symbols and lines positioned just below the atomic nuclei (top panel). Asymmetries in the plot are due to the discretization of the density on a cubic grid of points. The DKH densities have not been corrected for the picture-change effect and, hence, deviate from the four-component reference density in the closest proximity to the nuclei. But these effects can hardly be resolved on the numerical grid employed to represent the densities.

The upper two graphs in Figure 16.4 illustrate the difference between the four-component Dirac–Hartree–Fock reference density and the two-component ZORA density, which is only prominent in the inner core of the atoms in both complexes. The two plots in the middle of Figure 16.4 display the difference between the four-component density and the scalar-relativistic DKH density. The DKH density differs from the reference only close to the nucleus of the heavy atom Pt — spin–orbit effects are thus negligible for such a sixth-row atom. The lighter Ni analogue features differences only very close to the atomic nucleus.

In the case of the Ni compound, Figure 16.4 shows in detail four maxima of +0.02 $e\text{Å}^{-3}$ at the nickel center in the four-component–ZORA difference plot. All other differences are negligible, which is indicated by the noisy lines that fail to resolve the tiny differences. This is also visible in the case of the Pt complex. Of course, close to the platinum nucleus, the difference is more pronounced, and a circular minimum of −0.2 $e\text{Å}^{-3}$ dominates. The more approximate one-component DKH10 calculations yield more features in the difference map. In case of the Ni system, four maxima of +0.2 $e\text{Å}^{-3}$, minima in *cis* and *trans* position to the acetylene ligand, and maxima at the carbon atoms are found. The Pt homolog features a circular maximum at +0.12 $e\text{Å}^{-3}$ and four additional minima at −0.02 $e\text{Å}^{-3}$. Of course, the largest number of features in the difference map shows up in the case of the nonrelativistic density as compared to the four-component Dirac–Hartree–Fock reference. In this case, the difference density map for the Ni complex depicts three maxima of +0.2 $e\text{Å}^{-3}$ *trans* and parallel to the ligand, one maximum of +0.08 $e\text{Å}^{-3}$ in *cis*-position to the ligand, as well as four minima at the Ni atom of −0.08 $e\text{Å}^{-3}$ and maxima of +2.0 $e\text{Å}^{-3}$ at the carbon atoms. The situation is even more dramatic in the Pt case. Here, we discover one maximum of +1.4 $e\text{Å}^{-3}$ and a minimum of −0.4 $e\text{Å}^{-3}$ in *trans* position to the ligand. Another minimum of −0.4 $e\text{Å}^{-3}$ between Pt and the ligand as well as two maxima on the bond path of the Pt–C bond of +0.02 $e\text{Å}^{-3}$ and maxima of +4 $e\text{Å}^{-3}$ at the carbon atoms are found.

Hence, for the Pt complex, the scalar-relativistic DKH Hamiltonian and the spin–orbit ZORA-Hamiltonian recover relativistic effects on the electron density of the valence region found in four-component calculations. This is particularly remarkable since the DKH density has not been corrected for the picture-change error in these calculations. The difference density maps for the Pt complex clearly show the importance of the scalar-relativistic contraction of the inner shells, while spin–orbit effects can be neglected for this closed-shell complex (otherwise, the DKH results would deviate much more from the four-component and ZORA data). In the case of the Ni system the spin–orbit-coupling-induced splitting of the *d* shell is small but noticeable.

Knowledge of the accurate electron density is decisive especially for the development of chemical concepts that are based on the analysis of this observable. Such concepts are Gillespie's valence-shell electron-pair repulsion model [960] or the ligand-induced charge concentrations [730, 961–963] that are designed to predict molecular structures and even chemical reactivity. Both approaches can be related to Bader's theory of atoms in molecules [964, 965].

16.7
Relativistic Quantum Chemical Calculations in Practice

The case studies discussed in the preceding sections show that the accuracy of molecular spectra of heavy element compounds is, qualitatively speaking, governed by relativistic and electron-correlation effects. The numerous methods, which are in practical use in quantum chemistry and whose foundations have been elaborated in chapter 8, require a lot of experience in order to produce sound results. Fortunately, there exist two excellent books, which present the wealth of computational methods with a thorough analysis of their capabilities, to which we may refer the reader [966, 967].

Relativity affects the N-electron wave function through the proper choice of the many-electron Hamiltonian. On the other hand, electron correlation is considered through a multi-determinantal ansatz for the N-electron wave function (or through a suitable choice of an approximate exchange–correlation functional in a DFT framework). It is desirable to somehow systematize the many different options from which one can choose in actual calculations. Mainly three different methodological branches emerge in a somewhat over-simplified picture:

(1) The most straightforward and accurate approaches to treating electron correlation and relativistic effects are all-electron four-component methods employing the Dirac–Coulomb–Breit or Dirac–Coulomb Hamiltonians within a (multi-reference) coupled-cluster or configuration interaction ansatz. The major drawback of four-component methods is the explicit coupling of the large and small component of the Dirac-spinor requiring larger computational resources in actual calculations. However, when high accuracy is sought for, highly correlated four-component methods yield the most accurate results. Of course, one may also use the four-component approach in connection with present-day density functionals, although this represents a rather unbalanced combination of the accuracy achieved for relativistic and correlation effects. Nevertheless, orbitals are then subjected to spin–orbit coupling and split, which is advantageous especially for heavy p-block as well as for super-heavy elements (say, $Z>92$). An efficient alternative of balanced accuracy would be an approach based on a two-component model Hamiltonian, which includes spin–orbit effects.

(2) Since the accuracy required to solve chemical problems — like clarifying a reaction mechanism — is smaller than the accuracy that required for high-resolution spectroscopy, tailor-made methods have been devised. One example is the relativistic DKH-CASSCF/CASPT2+SO approach [968,969]. In view of the limited reliability of the CASSCF wave function (due to the limited size of the active space), we may well combine this ansatz with a one-component DKH Hamiltonian and sacrifice, to a certain extent, accuracy for feasibility. Dynamic correlation effects can be considered in a subsequent CASPT2 calculation. Spin–orbit interaction, when necessary, can be included *a posteriori* as a perturbation. This intrinsically one-component formalism allows one to treat molecules of extended size compared to the four-component approach. As an alternative to CASSCF/CASPT2, truncated MRCI calculations can be envisaged. Limitations arise at the horizon when spin–orbit splitting of atomic one-electron shells starts to become large. Then, a spin-averaged orbital picture will no longer be sufficient. It can be expected that this will be the case for the heavy open-shell *p*-block elements Bi, Po, and At.

Of course, this second branch also accommodates DFT methods with approximate contemporary functionals. Again, approximate relativistic Hamiltonians like the DKH or the ZORA Hamiltonian will do a good job. This holds true for the second-order DKH2 Hamiltonian, but one may always choose the fourth-order DKH4 Hamiltonian, which is still a uniquely defined operator (i.e., it is independent of the parametrization chosen for the unitary transformation). The fact that the lowest-order Hamiltonians in the regular approximation and in the DKH scheme work so well is also the reason why ZORA and DKH2 Hamiltonians yield similarly good results [970]. Both Hamiltonians are now well established and heavily used in actual calculations. Needless to say, these all-electron methods are mandatory when effects of the core electrons become decisive (as, for instance, in Mößbauer spectroscopy).

(3) The next step in the hierarchy of approximations abandons the explicit consideration of all electrons. Instead, the core electrons of the heavy atoms are replaced by a suitably parametrized effective core potential. Since the core-relativistic effects are now caught by the ECP, the valence electrons are affected by this surrogate potential and can be treated in a formally nonrelativistic scheme with a one-component valence-only Hamiltonian (see section 14.4). Electron correlation can be captured with standard nonrelativistic approaches like coupled-cluster or configuration interaction methods, for example. Spin–orbit coupling is then either neglected or considered as a perturbation *a posteriori* or included via an effective spin–orbit ECP Hamiltonian. Usually, the ECP can be chosen as a small-core ECP so that core correlation effects do not come into play. The relativistic ECP approach is very accurate for all properties and relative energies dominated by the valence shell. Moreover, it has the salient feature that larger and larger molecular aggregates can be studied, as

the computational effort is significantly reduced after introduction of an ECP into the self-consistent field equations.

Although *computational methods* are under constant development in relativistic quantum chemistry, it can be expected that the principles given above will endure for all time. This holds even more true for the *theory* of relativistic quantum chemistry derived in this book.

Appendices

A
Vector and Tensor Calculus

In relativistic theory one often encounters vector and tensor expressions in both three- and four-dimensional form. The most important of these expressions are briefly summarized in this section, where Einstein's summation convention (cf. section 3.1.2) is strictly applied for four-dimensional objects.

A.1
Three-Dimensional Expressions

A.1.1
Algebraic Vector and Tensor Operations

In chapter 2 the Kronecker delta δ_{ij} and the totally antisymmetric Levi-Cività symbol ε_{ijk} were introduced by Eqs. (2.24) and (2.9), respectively. Their relation to each other is given by

$$\sum_{i=1}^{3} \varepsilon_{ijk}\varepsilon_{ilm} = \delta_{jl}\delta_{km} - \delta_{jm}\delta_{kl} \tag{A.1}$$

This immediately yields

$$\sum_{i,j=1}^{3} \varepsilon_{ijk}\varepsilon_{ijm} = \sum_{j=1}^{3} \left(\delta_{jj}\delta_{km} - \delta_{jm}\delta_{jk}\right) = 2\delta_{km} \tag{A.2}$$

since the trace of the Kronecker symbol is given by

$$\sum_{i=1}^{3} \delta_{ii} = 3 \tag{A.3}$$

The Euclidean scalar product between two arbitrary three-dimensional vectors $\boldsymbol{A}$ and $\boldsymbol{B}$ is given by

$$\boldsymbol{A} \cdot \boldsymbol{B} = \sum_{i=1}^{3} A_i B_i = AB \cos\gamma \tag{A.4}$$

Relativistic Quantum Chemistry. Markus Reiher and Alexander Wolf

ISBN: 978-3-527-31292-4

with A and B being the length of the vectors $\boldsymbol{A}$ and $\boldsymbol{B}$, respectively, and γ being the angle spanned by these vectors, $\gamma = \angle(\boldsymbol{A}, \boldsymbol{B})$. The *vector product* or *cross product* between these vectors yields again a three-vector $\boldsymbol{C}$ defined as

$$\boldsymbol{C} = \boldsymbol{A} \times \boldsymbol{B} = \begin{pmatrix} A_y B_z - A_z B_y \\ A_z B_x - A_x B_z \\ A_x B_y - A_y B_x \end{pmatrix} \tag{A.5}$$

and its ith component might thus be written as

$$C_i = (\boldsymbol{A} \times \boldsymbol{B})_i = \sum_{j,k=1}^{3} \varepsilon_{ijk} A_j B_k \tag{A.6}$$

A.1.2
Differential Vector Operations

The *gradient* of a scalar function $\phi = \phi(\boldsymbol{r})$ is defined as the three-dimensional vector of its Cartesian partial derivatives,

$$\text{grad}\, \phi(\boldsymbol{r}) = \nabla \phi(\boldsymbol{r}) = \left(\frac{\partial \phi}{\partial x}, \frac{\partial \phi}{\partial y}, \frac{\partial \phi}{\partial z} \right)^T \in \mathbb{R}^3 \tag{A.7}$$

which is consequently interpreted as a column-vector in this book. The *divergence* of a vector field $\boldsymbol{A} = \boldsymbol{A}(\boldsymbol{r})$ is a measure of the sources defined by

$$\text{div} \boldsymbol{A} = \nabla \cdot \boldsymbol{A} = \sum_{i=1}^{3} \frac{\partial A_i}{\partial x_i} = \sum_{i=1}^{3} \partial_i A_i \in \mathbb{R} \tag{A.8}$$

whereas the *curl* of this vector field is itself a vector field defined by

$$\text{curl}\, \boldsymbol{A} = \nabla \times \boldsymbol{A} = \begin{pmatrix} \partial_y A_z - \partial_z A_y \\ \partial_z A_x - \partial_x A_z \\ \partial_x A_y - \partial_y A_x \end{pmatrix} \in \mathbb{R}^3 \tag{A.9}$$

(in the German literature, it is also common to use 'rot' instead of 'curl'). Finally, the *Laplacian* of a scalar function ϕ might be considered as a generalization of the familiar one-dimensional second derivative as is given by

$$\Delta \phi(\boldsymbol{r}) = \text{div grad}\, \phi(\boldsymbol{r}) = \sum_{i=1}^{3} \frac{\partial^2 \phi(\boldsymbol{r})}{\partial x_i^2} = \sum_{i=1}^{3} \partial_i^2 \phi(\boldsymbol{r}) \in \mathbb{R} \tag{A.10}$$

Application of the Laplacian to a vector field $\boldsymbol{A}$ has to be understood as acting separately on each component. The result is then again a three-vector, of course. We note some important identities for the above vector operations.

For any differentiable vector fields $\boldsymbol{A}$, $\boldsymbol{B}$ and scalar functions ϕ, ψ the following relations hold:

$$\text{grad}\,(\phi\psi) = \phi\,\text{grad}\,\psi + \psi\,\text{grad}\,\phi \tag{A.11}$$

$$\text{div}\,(\phi\boldsymbol{A}) = \boldsymbol{A}\cdot\text{grad}\,\phi + \phi\,\text{div}\boldsymbol{A} \tag{A.12}$$

$$\text{curl}\,(\phi\boldsymbol{A}) = (\text{grad}\,\phi)\times\boldsymbol{A} + \phi\,\text{curl}\,\boldsymbol{A} \tag{A.13}$$

$$\text{div}\,(\boldsymbol{A}\times\boldsymbol{B}) = \boldsymbol{B}\cdot\text{curl}\,\boldsymbol{A} - \boldsymbol{A}\cdot\text{curl}\,\boldsymbol{B} \tag{A.14}$$

$$\text{curl}\,(\boldsymbol{A}\times\boldsymbol{B}) = \boldsymbol{A}\,\text{div}\boldsymbol{B} - \boldsymbol{B}\,\text{div}\boldsymbol{A} + (\boldsymbol{B}\cdot\nabla)\boldsymbol{A} - (\boldsymbol{A}\cdot\nabla)\boldsymbol{B} \tag{A.15}$$

$$\Delta\boldsymbol{A} = \text{grad}\,(\text{div}\boldsymbol{A}) - \text{curl}\,(\text{curl}\,\boldsymbol{A}) \tag{A.16}$$

$$\text{div}\,\text{curl}\,\boldsymbol{A} = 0 \tag{A.17}$$

$$\text{curl}\,\text{grad}\,\phi = 0 \tag{A.18}$$

The proof of all these relations is straightforward and thus omitted here. Eq. (A.17) states that any rotational vector field has no sources, and Eq. (A.18) summarizes the fact that the curl of any gradient field is zero.

A.1.3
Integral Theorems and Distributions

In classical mechanics and electrodynamics the integral theorems of Gauss and Stokes may often be employed beneficially. Given a sufficiently smooth (i.e., differentiable) and well-behaved vector field $\boldsymbol{A}$, *Gauss' theorem* may be expressed in its most elementary form as

$$\int_V \text{div}\boldsymbol{A}\,\text{d}^3r = \int_{\partial V}\boldsymbol{A}\cdot\text{d}\boldsymbol{\sigma} \tag{A.19}$$

where ∂V denotes the closed surface of the volume V and σ is the outer normal unit vector perpendicular to the plane ∂V. Eq. (A.19) is also denoted as the *divergence theorem* and relates the volume integral over all sinks and sources (i.e., the divergence) of a vector field $\boldsymbol{A}$ to the net flow of this vector field through the volume's boundary. Similarly, in its simplest form *Stokes' theorem* relates the vortices or curls of a vector field $\boldsymbol{A}$ within a given plane or surface S to the line integral of this vector field along the surface's boundary ∂S,

$$\int_S (\text{curl}\,\boldsymbol{A})\cdot\text{d}\boldsymbol{\sigma} = \oint_{\partial S}\boldsymbol{A}\cdot\text{d}\boldsymbol{r} \tag{A.20}$$

Here σ again denotes the outer normal unit vector perpendicular to the surface S. Eq. (A.20) is sometimes also referred to as the *curl theorem* .

In discussions of the Coulomb potential the following relations are often useful. The gradients of a distance between two positions, $\boldsymbol{r}$ and $\boldsymbol{r}'$, read

$$\nabla_{\boldsymbol{r}} |\boldsymbol{r}-\boldsymbol{r}'| = \nabla_{\boldsymbol{r}} \sqrt{(\boldsymbol{r}-\boldsymbol{r}')^2} = \frac{1}{2}\frac{1}{\sqrt{(\boldsymbol{r}-\boldsymbol{r}')^2}}\left[2(\boldsymbol{r}-\boldsymbol{r}')\right] = \frac{\boldsymbol{r}-\boldsymbol{r}'}{|\boldsymbol{r}-\boldsymbol{r}'|} \quad \text{(A.21)}$$

$$\nabla_{\boldsymbol{r}'} |\boldsymbol{r}-\boldsymbol{r}'| = \frac{1}{2}\frac{1}{\sqrt{(\boldsymbol{r}-\boldsymbol{r}')^2}}\left[2(\boldsymbol{r}-\boldsymbol{r}')\right](-1) = -\frac{\boldsymbol{r}-\boldsymbol{r}'}{|\boldsymbol{r}-\boldsymbol{r}'|} \quad \text{(A.22)}$$

and the gradient and Laplacian of the inverse distance become

$$\nabla_{\boldsymbol{r}} \frac{1}{|\boldsymbol{r}-\boldsymbol{r}'|} = -\frac{\boldsymbol{r}-\boldsymbol{r}'}{|\boldsymbol{r}-\boldsymbol{r}'|^3} = -\nabla_{\boldsymbol{r}'}\frac{1}{|\boldsymbol{r}-\boldsymbol{r}'|} \quad \text{(A.23)}$$

$$\Delta \frac{1}{|\boldsymbol{r}-\boldsymbol{r}'|} = -4\pi\,\delta^{(3)}(\boldsymbol{r}-\boldsymbol{r}') \quad \text{(A.24)}$$

The indices in Eq. (A.21) label the variable with respect to which the derivative is taken. We note that the proof of Eq. (A.24) is only trivial for $\boldsymbol{r}-\boldsymbol{r}' \neq 0$ for which the right-hand side becomes zero. In this case, where $\boldsymbol{r} \neq \boldsymbol{r}'$, we may simply differentiate the inverse distance and obtain this result right away,

$$\begin{aligned}\left(\Delta\frac{1}{|\boldsymbol{r}-\boldsymbol{r}'|}\right)_{\boldsymbol{r}\neq\boldsymbol{r}'} &= \nabla_{\boldsymbol{r}}\cdot\left(\nabla_{\boldsymbol{r}}\frac{1}{|\boldsymbol{r}-\boldsymbol{r}'|}\right) \overset{(A.23)}{=} \nabla_{\boldsymbol{r}}\cdot\left(-\frac{\boldsymbol{r}-\boldsymbol{r}'}{|\boldsymbol{r}-\boldsymbol{r}'|^3}\right)\\ &= -\left(\frac{1}{|\boldsymbol{r}-\boldsymbol{r}'|^3}\underbrace{\nabla_{\boldsymbol{r}}\cdot(\boldsymbol{r}-\boldsymbol{r}')}_{3} + \underbrace{(\boldsymbol{r}-\boldsymbol{r}')\cdot\nabla_{\boldsymbol{r}}\frac{1}{|\boldsymbol{r}-\boldsymbol{r}'|^3}}_{-3(\boldsymbol{r}-\boldsymbol{r}')\cdot(\boldsymbol{r}-\boldsymbol{r}')/|\boldsymbol{r}-\boldsymbol{r}'|^5}\right) = 0\end{aligned} \quad \text{(A.25)}$$

(Δ may equally well be resolved by a differentiation with respect to $\boldsymbol{r}'$). Eq. (A.24) states that the function

$$G(\boldsymbol{r},\boldsymbol{r}') = -\frac{1}{4\pi\,|\boldsymbol{r}-\boldsymbol{r}'|} \quad \text{(A.26)}$$

is the so-called *Green's function* of the Laplacian, since application of the differential operator Δ to G yields the three-dimensional *delta distribution* $\delta^{(3)}(\boldsymbol{r}-\boldsymbol{r}')$ [971, p. 22ff.]. The three-dimensional delta distribution is the product of three one-dimensional delta distributions,

$$\delta^{(3)}(\boldsymbol{r}-\boldsymbol{r}') = \delta(x-x')\,\delta(y-y')\,\delta(z-z') \quad \text{(A.27)}$$

which may sloppily be thought of as being zero everywhere except at $x = x'$, where it is appropriately infinite such that

$$\int_{-\infty}^{\infty} \mathrm{d}x\,\delta(x-x')f(x) = f(x') \quad \text{(A.28)}$$

for *all* continuous and integrable test functions f. A very useful and convenient integral representation of the delta distribution is given by

$$\delta(x - x') = \frac{1}{2\pi} \int_{-\infty}^{\infty} dk\, e^{ik(x-x')} \tag{A.29}$$

which will become important in appendix E in the discussion of Fourier transforms.

A.1.4
Total Differentials and Time Derivatives

In both classical mechanics and quantum mechanics one often has to deal with *total* time derivatives of functions depending on many time-dependent arguments. For example, consider a real-valued function f depending on N time-dependent arguments $x_i(t)$ and explicitly on the time t itself,

$$f = f(x_1(t), x_2(t), \ldots, x_N(t), t) \tag{A.30}$$

The partial time derivative of the function f is simply given by

$$\partial_t f = \frac{\partial f}{\partial t} \tag{A.31}$$

i.e., the time-dependence of the arguments x_i does not matter at all for the calculation of the partial time derivative of f. Only the explicit time dependence is taken into account. For the calculation of the *total* time derivative, however, both explicit and implicit dependences on the time t have to be taken into account, i.e., the total time derivative is given by application of the chain rule of multi-dimensional calculus,

$$\dot{f} = \frac{df}{dt} = \sum_{i=1}^{N} \frac{\partial f}{\partial x_i} \frac{dx_i}{dt} + \frac{\partial f}{\partial t} \tag{A.32}$$

This rule has, for example, been extensively applied in the discussion of Hamiltonian mechanics in section 2.3. By formal multiplication of Eq. (A.32) by dt the total differential of the function f is recovered,

$$df = \sum_{i=1}^{N} \frac{\partial f}{\partial x_i} dx_i + \frac{\partial f}{\partial t} dt \tag{A.33}$$

A.2
Four-Dimensional Expressions

A.2.1
Algebraic Vector and Tensor Operations

As a generalization of the three-dimensional Levi-Cività symbol defined in chapter 2 by Eq. (2.9) we have introduced the four-dimensional totally anti-symmetric (pseudo-)tensor $\varepsilon^{\alpha\beta\gamma\delta}$, whose contravariant components have been defined by

$$\varepsilon^{\alpha\beta\gamma\delta} = \left\{ \begin{array}{lll} +1 & & (\alpha\beta\gamma\delta) \text{ is an even permutation of } (0123) \\ -1 & \text{if} & (\alpha\beta\gamma\delta) \text{ is an odd permutation of } (0123) \\ 0 & & \text{else} \end{array} \right\} \tag{A.34}$$

We have further introduced the four-dimensional generalization of the scalar product between any two 4-vectors a and b by

$$a \cdot b = a^T g\, b = a^\mu b_\mu = a^0 b^0 - a^i b^i = a^0 b^0 - \boldsymbol{a} \cdot \boldsymbol{b} \tag{A.35}$$

As a consequence, the four-dimensional distance between 2 infinitesimal neighboring events can now be expressed as

$$ds^2 = dx \cdot dx = dx^T g\, dx = g_{\mu\nu}\, dx^\mu\, dx^\nu \tag{A.36}$$

A.2.2
Differential Vector Operations

Similarly to the nonrelativistic situation [cf. Eq. (2.28)], the components of the Lorentz transformation matrix Λ may be expressed as derivatives of the new coordinates with respect to the old ones or vice versa. However, since we have to distinguish contra- and covariant components of vectors in the relativistic framework, there are now four different possibilities to express these derivatives:

$$x'^\mu = \Lambda^\mu{}_\nu x^\nu \longrightarrow \Lambda^\mu{}_\nu = \frac{\partial x'^\mu}{\partial x^\nu} = \partial_\nu x'^\mu \tag{A.37}$$

$$x^\mu = \Lambda_\nu{}^\mu x'^\nu \longrightarrow \Lambda_\mu{}^\nu = \frac{\partial x^\nu}{\partial x'^\mu} = \partial'_\mu x^\nu \tag{A.38}$$

$$x'_\mu = \Lambda_\mu{}^\nu x'_\nu \longrightarrow \Lambda_\mu{}^\nu = \frac{\partial x'_\mu}{\partial x_\nu} = \partial^\nu x'_\mu \tag{A.39}$$

$$x_\nu = \Lambda^\mu{}_\nu x'_\mu \longrightarrow \Lambda^\mu{}_\nu = \frac{\partial x_\nu}{\partial x'_\mu} = \partial'^\mu x_\nu \tag{A.40}$$

where we have employed a shorthand notation for the 4-gradient defined by

$$\partial_\mu \equiv \frac{\partial}{\partial x^\mu} = \left(\frac{1}{c}\frac{\partial}{\partial t}, \nabla \right) = \left(\frac{1}{c}\frac{\partial}{\partial t}, \frac{\partial}{\partial x}, \frac{\partial}{\partial y}, \frac{\partial}{\partial z} \right) \tag{A.41}$$

The 4-gradient has been written as a row vector above solely for our convenience; it still is to be interpreted mathematically as a column vector, of course. Being defined as the derivative with respect to the contravariant components x^μ, the 4-gradient ∂_μ is *naturally* a covariant vector whose contravariant components may be obtained by application of the metric g and read

$$\partial^\mu = g^{\mu\nu}\partial_\nu = \frac{\partial}{\partial x_\mu} = \left(\frac{1}{c}\frac{\partial}{\partial t}, -\nabla\right) \tag{A.42}$$

The four-dimensional generalization of the Laplacian has been identified to be the d'Alembert operator

$$\Box = \partial_\mu \partial^\mu = \frac{1}{c^2}\frac{\partial^2}{\partial t^2} - \Delta \tag{A.43}$$

B
Kinetic Energy in Generalized Coordinates

For a system of N point particles in three spatial dimensions the kinetic energy expressed in Cartesian coordinates is always given by

$$T = T(\dot{\boldsymbol{r}}) = \frac{1}{2}\sum_{i=1}^{N} m_i \dot{\boldsymbol{r}}_i^2 = \frac{1}{2}\sum_{i=1}^{3N} m_i \dot{x}_i^2 \tag{B.1}$$

with x_i ($i = 1,\ldots,3N$) being the ith Cartesian component. For the masses m_i an obvious notation has been employed. If there are R constraints on the system, the Cartesian coordinates may be replaced by $f = 3N - R$ suitably chosen generalized coordinates q_k according to Eq. (2.44), or in a more compact notation

$$x_i = x_i(q,t), \qquad \forall i = 1,2,\ldots,3N \tag{B.2}$$

This coordinate transformation does *not* in general depend on the velocities $\dot{q}$. Accordingly the total time derivatives of the Cartesian coordinates read

$$\dot{x}_i = \frac{\mathrm{d}x_i}{\mathrm{d}t} = \sum_{k=1}^{f} \frac{\partial x_i}{\partial q_k}\dot{q}_k + \frac{\partial x_i}{\partial t} = \dot{x}_i(q,\dot{q},t), \qquad \forall i = 1,2,\ldots,3N \tag{B.3}$$

and depend on both the generalized coordinates and velocities and the time. Inserting this expression into (B.1) one obtains

$$\begin{aligned} T = {} & \frac{1}{2}\sum_{k,l=1}^{f} \underbrace{\sum_{i=1}^{3N} m_i \left(\frac{\partial x_i}{\partial q_k}\right)\left(\frac{\partial x_i}{\partial q_l}\right)}_{m_{kl}(q,t)} \dot{q}_k \dot{q}_l + \sum_{k=1}^{f} \underbrace{\sum_{i=1}^{3N} m_i \left(\frac{\partial x_i}{\partial q_k}\right)\left(\frac{\partial x_i}{\partial t}\right)}_{a_k(q,t)} \dot{q}_k \\ & + \underbrace{\frac{1}{2}\sum_{i=1}^{3N} m_i \left(\frac{\partial x_i}{\partial t}\right)^2}_{b(q,t)} = T(q,\dot{q},t) \end{aligned} \tag{B.4}$$

This is the most general expression for the kinetic energy T in terms of the generalized coordinates. As soon as other than Cartesian coordinates are em-

Relativistic Quantum Chemistry. Markus Reiher and Alexander Wolf

ISBN: 978-3-527-31292-4

ployed, T may no longer depend only on the velocities but also on the coordinates itself. Because of its definition by Eq. (B.4) the *mass matrix* is symmetric, $m_{kl} = m_{lk}$.

Equation (B.4) is simplified to a large extent if the constraints do not explicitly depend on time, i.e., the coordinate transformation given by Eq. (B.2) itself does not explicitly depend on time,

$$x_i = x_i(q(t)), \qquad \forall i = 1, 2, \ldots, 3N \tag{B.5}$$

Then the partial derivatives $\partial x_i / \partial t$ vanish and the kinetic energy is given by

$$T(q, \dot{q}) = \frac{1}{2} \sum_{k,l=1}^{f} m_{kl}(q)\, \dot{q}_k \dot{q}_l \tag{B.6}$$

If furthermore the potential U does *not* depend on the velocities, $U = U(q, t)$, one finds

$$\sum_{k=1}^{f} \frac{\partial L}{\partial \dot{q}_k} \dot{q}_k = \sum_{k=1}^{f} \frac{\partial T}{\partial \dot{q}_k} \dot{q}_k \overset{(B.6)}{=} \sum_{k,l=1}^{f} m_{kl} \dot{q}_k \dot{q}_l = 2T(q, \dot{q}) \tag{B.7}$$

Under those circumstances the energy of the system is thus given by

$$\sum_{k=1}^{f} \frac{\partial L}{\partial \dot{q}_k} \dot{q}_k - L \overset{(B.7)}{=} 2T - T + U = T + U = E \tag{B.8}$$

C
Technical Proofs for Special Relativity

C.1
Invariance of Space-Time Interval

In section 3.1.2 we found the invariance under Lorentz transformations of the squared space-time interval s_{12}^2 between two events connected by a light signal being solely based on the relativity principle of Einstein, i.e., the constant speed of light in all inertial frames, cf. Eq. (3.5),

$$s_{12}^2 = s_{12}'^2 = 0\,, \qquad \forall\ \text{IS}' \qquad \text{(for light signal)} \tag{C.1}$$

As a consequence of Eq. (C.1) and the homogeneity of space and time and the isotropy of space, we now formally prove the invariance of the space-time interval for *any* two events E_1 and E_2, cf. Eq. (3.6),

$$s_{12}^2 = s_{12}'^2\,, \qquad \forall\ \text{IS}' \qquad \text{(for any two events)} \tag{C.2}$$

Proof: We thus consider these two arbitrary events with reference to two inertial frames IS and IS′ moving with velocity v_1 relative to each other. We can always express the relationship between the four-dimensional distances s_{12}^2 and $s_{12}'^2$ between these two events as

$$s_{12}'^2 = F_{v_1}\left(s_{12}^2\right) \tag{C.3}$$

where $F_{v_1}(s)$ shall be a sufficiently smooth function permitting its Taylor expansion around $s = 0$, i.e.,

$$s_{12}'^2 \approx A(v_1) + B(v_1)\, s_{12}^2 + \frac{1}{2} C(v_1) \left(s_{12}^2\right)^2 + \ldots \tag{C.4}$$

Eq. (C.4) must hold for *any* two events and thus also for events connected by a light signal. From Eq. (C.1) we thus find $A(v_1) = 0$. Furthermore, Eq. (C.4) must also hold for events in the infinitesimal neighborhood of each other where $s_{12}^2 = \mathrm{d}s^2$ is an infinitesimal, i.e., very small quantity. For those events we thus arrive at

$$\mathrm{d}s'^2 = B(v_1)\, \mathrm{d}s^2 \tag{C.5}$$

Relativistic Quantum Chemistry. Markus Reiher and Alexander Wolf

ISBN: 978-3-527-31292-4

Because of the assumption of a homogeneous space-time the function $B(\boldsymbol{v}_1)$ cannot depend on the space-time coordinates t and $\boldsymbol{r}$, and because of the assumption of spatial rotational invariance (isotropy of space) the function $B(\boldsymbol{v}_1)$ must *not* depend on the direction of $\boldsymbol{v}_1$ but only on its magnitude $v_1 = |\boldsymbol{v}_1|$, i.e.,

$$ds'^2 = B(v_1)\, ds^2 \tag{C.6}$$

The last ingredient for this proof is the group structure of Lorentz transformations, i.e., the possibility to consider a further inertial frame IS″ which moves with velocity $\boldsymbol{v}_2$ relative to IS. Its relative velocity against IS′ shall be $\boldsymbol{v}_{12}$. We thus arrive at the following relations for the squared space-time intervals between our two infinitesimally neighboring events,

$$ds'^2 = B(v_1)\, ds^2 \tag{C.7}$$

$$ds''^2 = B(v_2)\, ds^2 \tag{C.8}$$

$$ds''^2 = B(v_{12})\, ds'^2 \tag{C.9}$$

A simple rearrangement of Eqs. (C.7)–(C.9) yields

$$B(v_{12}) = \frac{B(v_2)}{B(v_1)} \tag{C.10}$$

The term on the left hand side of this equation is a function of the *angle* between $\boldsymbol{v}_1$ and $\boldsymbol{v}_2$, cf. section 3.2.3, whereas the term on the right hand side is not. Eq. (C.10) can thus only be valid in general if $B(v_{12})$ is also *not* a function of the angle between $\boldsymbol{v}_1$ and $\boldsymbol{v}_2$. Since we have made no further assumptions for both the size and the orientation of the velocities $\boldsymbol{v}_1$ and $\boldsymbol{v}_2$ this immediately implies

$$B(v) = const. = 1\,, \qquad \forall \, \text{velocities } v \tag{C.11}$$

This finally yields the desired result for infinitesimally neighboring events,

$$ds'^2 = ds^2 \tag{C.12}$$

and since Eq. (C.12) holds for *any* infinitesimally neighboring events within the whole space-time, it holds for any two events and thus Eq. (C.2) is proven.

C.2
Uniqueness of Lorentz Transformations

In section 3.1.3 we mentioned that Lorentz transformations of the form as given by Eq. (3.12) are the only *nonsingular* coordinate transformations from

IS to IS′, i.e., $x \to x'(x)$, that leave the four-dimensional space-time interval $\mathrm{d}s^2$ invariant. Nonsingular in this context means that both $x' = x'(x)$ and $x = x(x')$ are sufficiently smooth and well-behaved functions that feature a well-defined inverse.

Proof: We consider an arbitrary nonsingular coordinate transformation $x \to x'(x)$ and calculate the four-dimensional distance $\mathrm{d}s'^2$ between two infinitesimally neighboring events,

$$\begin{aligned} \mathrm{d}s'^2 &= g_{\alpha\beta}\,\mathrm{d}x'^\alpha\,\mathrm{d}x'^\beta = g_{\alpha\beta}\,\frac{\partial x'^\alpha}{\partial x^\mu}\,\frac{\partial x'^\beta}{\partial x^\nu}\,\mathrm{d}x^\mu\,\mathrm{d}x^\nu \\ &= g_{\alpha\beta}\,(\partial_\mu x'^\alpha)\,(\partial_\nu x'^\beta)\,\mathrm{d}x^\mu\mathrm{d}x^\nu \end{aligned} \tag{C.13}$$

where we have taken advantage of the relation $\mathrm{d}x'^\alpha = (\partial_\mu x'^\alpha)\mathrm{d}x^\mu$ for infinitesimal space-time distances (at the second equality). Since we are looking for transformations that leave the space-time interval invariant for *any* two events,

$$\mathrm{d}s'^2 \overset{!}{=} \mathrm{d}s^2 = g_{\mu\nu}\,\mathrm{d}x^\mu\,\mathrm{d}x^\nu \tag{C.14}$$

this yields the condition

$$g_{\mu\nu} = g_{\alpha\beta}\,(\partial_\mu x'^\alpha)\,(\partial_\nu x'^\beta) \tag{C.15}$$

We now differentiate this equation with reference to the arbitrary space-time component x^σ and, by the product rule, arrive at

$$0 = g_{\alpha\beta}\,(\partial_\mu\partial_\sigma x'^\alpha)\,(\partial_\nu x'^\beta) + g_{\alpha\beta}\,(\partial_\mu x'^\alpha)\,(\partial_\nu\partial_\sigma x'^\beta) \tag{C.16}$$

In order to solve for the second derivatives we write down Eq. (C.16) with indices μ and σ interchanged,

$$0 = g_{\alpha\beta}\,(\partial_\mu\partial_\sigma x'^\alpha)\,(\partial_\nu x'^\beta) + g_{\alpha\beta}\,(\partial_\sigma x'^\alpha)\,(\partial_\nu\partial_\mu x'^\beta) \tag{C.17}$$

and again with indices ν and σ interchanged,

$$0 = g_{\alpha\beta}\,(\partial_\mu\partial_\nu x'^\alpha)\,(\partial_\sigma x'^\beta) + g_{\alpha\beta}\,(\partial_\mu x'^\alpha)\,(\partial_\nu\partial_\sigma x'^\beta) \tag{C.18}$$

We now calculate (C.16) + (C.17) − (C.18) and, bearing in mind that α and β are just dummy indices which are summed over, we find

$$0 = 2\,g_{\alpha\beta}\,(\partial_\mu\partial_\sigma x'^\alpha)\,(\partial_\nu x'^\beta) \tag{C.19}$$

Since we have assumed a nonsingular coordinate transformation, the last term $(\partial_\nu x'^\beta)$ *cannot* vanish identically, and thus Eq. (C.19) can only hold in general if the second derivatives identically vanish, i.e.,

$$\partial_\mu\partial_\sigma x'^\alpha = 0 \tag{C.20}$$

This immediately implies a linear coordinate transformation of the form as given by Eq. (3.12), i.e.,

$$x' = \Lambda x + a \qquad \Longleftrightarrow \qquad x'^{\mu} = \Lambda^{\mu}{}_{\nu}\, x^{\nu} + a^{\mu} \tag{C.21}$$

where the entries of the matrix Λ and the 4-vector a must be constants which do *not* depend on x. Insertion of Eq. (C.21) in Eq. (C.15) yields,

$$g_{\mu\nu} = g_{\alpha\beta}\,(\partial_\mu x'^{\alpha})\,(\partial_\nu x'^{\beta}) = g_{\alpha\beta}\,\Lambda^{\alpha}{}_{\mu}\,\Lambda^{\beta}{}_{\nu} \tag{C.22}$$

i.e., the fundamental property of Lorentz transformations as given by Eq. (3.17) is recoverd. This completes the proof of the uniqueness of Lorentz transformations as the natural symmetry transformations within the four-dimensional Minkowski space equipped with the metric g.

C.3
Useful Trigonometric and Hyperbolic Formulae for Lorentz Transformations

The trigonometric functions may be related to the exponential function via Euler's relation

$$e^{\mathrm{i}x} = \cos x + \mathrm{i}\sin x\,, \quad \forall x \in \mathbb{R} \tag{C.23}$$

which may be inverted to yield

$$\cos x = \tfrac{1}{2}\left(e^{\mathrm{i}x} + e^{-\mathrm{i}x}\right), \quad \sin x = \tfrac{1}{2\mathrm{i}}\left(e^{\mathrm{i}x} - e^{-\mathrm{i}x}\right), \quad \forall x \in \mathbb{R} \tag{C.24}$$

The tangent tan and cotangent cot are defined as

$$\tan x = \frac{\sin x}{\cos x}\,, \quad \forall x \in \mathbb{R} \setminus \left\{(2k+1)\tfrac{\pi}{2}\,,\ k \in \mathbb{Z}\right\} \tag{C.25}$$

$$\cot x = \frac{\cos x}{\sin x} = \frac{1}{\tan x}\,, \quad \forall x \in \mathbb{R} \setminus \left\{k\pi\,,\ k \in \mathbb{Z}\right\} \tag{C.26}$$

The inverse functions of the trigonometric functions (on suitably chosen domains) are denoted as $\arccos x$, $\arcsin x$, $\arctan x$, and $\operatorname{arccot} x$. Important relations between the trigonometric functions and/or their inverse functions are

$$\cos^2 x + \sin^2 x = 1\,, \quad \forall x \in \mathbb{R} \tag{C.27}$$

$$\arcsin x = \arctan \frac{x}{\sqrt{1-x^2}}\,, \quad \forall x \in]-1,1[\tag{C.28}$$

$$\arccos x = \operatorname{arccot} \frac{x}{\sqrt{1-x^2}}\,, \quad \forall x \in]-1,1[\tag{C.29}$$

$$\arctan x + \arctan y = \arctan\left(\frac{x+y}{1-xy}\right), \quad \forall x, y \in \mathbb{R} \text{ with } xy < 1 \tag{C.30}$$

Similarly to the definition of the usual trigonometric functions, the hyperbolic functions are given by

$$\cosh x = \tfrac{1}{2}\left(e^x + e^{-x}\right), \quad \sinh x = \tfrac{1}{2}\left(e^x - e^{-x}\right), \quad \forall x \in \mathbb{R} \tag{C.31}$$

and

$$\tanh x = \frac{\sinh x}{\cosh x} = \frac{e^x - e^{-x}}{e^x + e^{-x}}, \quad \forall x \in \mathbb{R} \tag{C.32}$$

As a direct consequence it holds that

$$\cosh^2 x - \sinh^2 x = 1, \quad \forall x \in \mathbb{R} \tag{C.33}$$

$$\cosh(x+y) = \cosh x \cosh y + \sinh x \sinh y, \quad \forall x \in \mathbb{R} \tag{C.34}$$

$$\sinh(x+y) = \sinh x \cosh y + \cosh x \sinh y, \quad \forall x \in \mathbb{R} \tag{C.35}$$

$$\tanh(x+y) = \frac{\tanh x + \tanh y}{1 + \tanh x \tanh y}, \quad \forall x \in \mathbb{R} \tag{C.36}$$

The inverse hyperbolic functions (on suitably chosen domains) are the *area-functions* arcosh, arsinh, and artanh. They satisfy

$$\operatorname{artanh} x = \operatorname{arcosh}\left(\frac{1}{\sqrt{1-x^2}}\right), \quad \forall x \in]-1,1[\tag{C.37}$$

$$\operatorname{artanh} x \pm \operatorname{artanh} y = \operatorname{artanh}\left(\frac{x \pm y}{1 \pm xy}\right), \quad \forall x, y \in \mathbb{R} \text{ with } xy < 1 \tag{C.38}$$

D
Relations for Pauli and Dirac Matrices

D.1
Pauli Spin Matrices

The Pauli spin matrices introduced in Eq. (4.140) fulfill some important relations. First of all, the squared matrices yield the (2×2) unit matrix $\mathbf{1}_2$,

$$\sigma_x^2 = \sigma_y^2 = \sigma_z^2 = \begin{pmatrix} 1 & 0 \\ 0 & 1 \end{pmatrix} = \mathbf{1}_2 \tag{D.1}$$

which is an essential property when calculating the square of the spin operator. Next, multiplication of two different Pauli spin matrices yields the third one multiplied by the (positive or negative) imaginary unit,

$$\sigma_x\sigma_y = \mathrm{i}\sigma_z \quad , \quad \sigma_x\sigma_z = -\mathrm{i}\sigma_y \quad , \quad \sigma_y\sigma_z = \mathrm{i}\sigma_x \tag{D.2}$$

$$\sigma_y\sigma_x = -\mathrm{i}\sigma_z \quad , \quad \sigma_z\sigma_x = \mathrm{i}\sigma_y \quad , \quad \sigma_z\sigma_y = -\mathrm{i}\sigma_x \tag{D.3}$$

This may be expressed in more compact form for all cyclic permutations of $i, j, k \in \{1, 2, 3\}$ as

$$\sigma_i\sigma_j = \delta_{ij}\mathbf{1}_2 + \mathrm{i}\sum_{k=1}^{3} \varepsilon_{ijk}\sigma_k \tag{D.4}$$

where $\{1, 2, 3\}$ and $\{x, y, z\}$ are used synonymously. As a direct consequence of Eq. (D.4) the commutation and anticommutation relations for Pauli spin matrices are given by

$$\left[\sigma_i\,,\,\sigma_j\right] = 2\mathrm{i}\sum_{k=1}^{3} \varepsilon_{ijk}\sigma_k \qquad \text{and} \qquad \left\{\sigma_i\,,\,\sigma_j\right\} = 2\delta_{ij}\mathbf{1}_2 \tag{D.5}$$

These relations may be generalized to the four-component case if we consider the even matrix $\mathbf{\Sigma}$ and the Dirac matrices $\boldsymbol{\alpha}$ and β; cf. chapter 5, for which we have

$$\alpha_x^2 = \alpha_y^2 = \alpha_z^2 = \beta^2 = \mathbf{1}_4 \tag{D.6}$$

Relativistic Quantum Chemistry. Markus Reiher and Alexander Wolf

ISBN: 978-3-527-31292-4

$$\alpha_i \alpha_j = \mathbf{1}_2 \otimes \sigma_i \sigma_j = \begin{pmatrix} \sigma_i\sigma_j & 0 \\ 0 & \sigma_i\sigma_j \end{pmatrix} \tag{D.7}$$

so that commutators and anticommutators read

$$\left[\alpha_i , \alpha_j\right] = 2\mathrm{i} \sum_{k=1}^{3} \varepsilon_{ijk} \Sigma_k \tag{D.8}$$

$$\left\{\alpha_i , \alpha_j\right\} = 2\delta_{ij} \mathbf{1}_4 \qquad \text{and} \qquad \left\{\alpha_i , \beta\right\} = 0 \tag{D.9}$$

The tensor product denoted by '$\otimes$' is to be evaluated according to the general prescription

$$\begin{pmatrix} a_{11} & a_{12} \\ a_{21} & a_{22} \end{pmatrix} \otimes \begin{pmatrix} b_{11} & b_{12} \\ b_{21} & b_{22} \end{pmatrix} = \begin{pmatrix} a_{11}b_{11} & a_{11}b_{12} & a_{12}b_{11} & a_{12}b_{12} \\ a_{11}b_{21} & a_{11}b_{22} & a_{12}b_{21} & a_{12}b_{22} \\ a_{21}b_{11} & a_{21}b_{12} & a_{12}b_{11} & a_{22}b_{12} \\ a_{21}b_{21} & a_{21}b_{22} & a_{12}b_{21} & a_{22}b_{22} \end{pmatrix} \tag{D.10}$$

D.2
Dirac's Relation

A relation that is often exploited in the book is Dirac's relation [66], which for two arbitrary vector operators $\boldsymbol{A}$ and $\boldsymbol{B}$ reads

$$(\boldsymbol{\sigma} \cdot \boldsymbol{A})(\boldsymbol{\sigma} \cdot \boldsymbol{B}) = \boldsymbol{A} \cdot \boldsymbol{B}\, \mathbf{1}_2 + \mathrm{i}\boldsymbol{\sigma} \cdot (\boldsymbol{A} \times \boldsymbol{B}) \tag{D.11}$$

where the (2×2) unit matrix $\mathbf{1}_2$ is usually omitted. This relation can be verified by evaluating the scalar products on the left hand side of the relation,

$$\begin{aligned} (\boldsymbol{\sigma} \cdot \boldsymbol{A})(\boldsymbol{\sigma} \cdot \boldsymbol{B}) &= (\sigma_x A_x + \sigma_y A_y + \sigma_z A_z)(\sigma_x B_x + \sigma_y B_y + \sigma_z B_z) \\ &= \sigma_x^2 A_x B_x + \sigma_x \sigma_y A_x B_y + \sigma_x \sigma_z A_x B_z + \sigma_y \sigma_x A_y B_x + \sigma_y^2 A_y B_y \\ &\quad + \sigma_y \sigma_z A_y B_z + \sigma_z \sigma_x A_z B_x + \sigma_z \sigma_y A_z B_y + \sigma_z^2 A_z B_z \\ &= A_x B_x + A_y B_y + A_z B_z + \mathrm{i}\sigma_z A_x B_y - \mathrm{i}\sigma_y A_x B_z \\ &\quad - \mathrm{i}\sigma_z A_y B_x + \mathrm{i}\sigma_x A_y B_z + \mathrm{i}\sigma_y A_z B_x - \mathrm{i}\sigma_x A_z B_y \\ &= \boldsymbol{A} \cdot \boldsymbol{B} + \mathrm{i}\boldsymbol{\sigma} \cdot (\boldsymbol{A} \times \boldsymbol{B}) \end{aligned} \tag{D.12}$$

if we use the relations of the Pauli spin matrices given in appendix D.1. This proof can also be given in more compact form as

$$(\boldsymbol{\sigma} \cdot \boldsymbol{A})(\boldsymbol{\sigma} \cdot \boldsymbol{B}) = \sum_{i,j=1}^{3} \sigma_i A_i \sigma_j B_j \tag{D.13}$$

$$\stackrel{(D.4)}{=} \sum_{i,j=1}^{3} \left(\delta_{ij} \mathbf{1}_2 + \mathrm{i} \sum_{k=1}^{3} \varepsilon_{ijk} \sigma_k \right) A_i B_j \tag{D.14}$$

$$\stackrel{(A.6)}{=} \sum_{i=1}^{3} A_i B_i + \mathrm{i} \sum_{k=1}^{3} \sigma_k (\boldsymbol{A} \times \boldsymbol{B})_k \tag{D.15}$$

$$= \boldsymbol{A} \cdot \boldsymbol{B} + \mathrm{i}\boldsymbol{\sigma} \cdot (\boldsymbol{A} \times \boldsymbol{B}) \tag{D.16}$$

Obviously, if $\boldsymbol{A} = \boldsymbol{B}$ the Dirac relation simplifies to

$$(\boldsymbol{\sigma} \cdot \boldsymbol{A})^2 = \boldsymbol{A}^2 + \mathrm{i}\boldsymbol{\sigma} \cdot (\boldsymbol{A} \times \boldsymbol{A}) \tag{D.17}$$

which reads in the case of $\boldsymbol{A} = \boldsymbol{p}$

$$(\boldsymbol{\sigma} \cdot \boldsymbol{p})^2 = \boldsymbol{p}^2 \tag{D.18}$$

because $\boldsymbol{p} \times \boldsymbol{p} = 0$.

D.2.1
Momenta and Vector Fields

The situation is more complicated in the presence of vector potentials, where we have

$$(\boldsymbol{\sigma} \cdot \boldsymbol{\pi})(\boldsymbol{\sigma} \cdot \boldsymbol{\pi}) = \boldsymbol{\pi}^2 + \mathrm{i}\boldsymbol{\sigma} \cdot (\boldsymbol{\pi} \times \boldsymbol{\pi}) \tag{D.19}$$

By noting that $\boldsymbol{\pi} = \boldsymbol{p} - \frac{q_e}{c}\boldsymbol{A}$ the vector product of the kinematical momentum operator with itself can be simplified to

$$\begin{aligned} \boldsymbol{\pi} \times \boldsymbol{\pi} &= -\frac{q_e}{c} (\boldsymbol{p} \times \boldsymbol{A} + \boldsymbol{A} \times \boldsymbol{p}) \\ &= -\mathrm{i}\hbar \frac{q_e}{c} (\nabla \times \boldsymbol{A} + \boldsymbol{A} \times \nabla) \end{aligned} \tag{D.20}$$

since the vector products of the canonical momentum $\boldsymbol{p}$ and the vector potential $\boldsymbol{A}$ with themselves vanish, respectively. We now consider the action of the i-th component of the operator $\nabla \times \boldsymbol{A}$ on a two-component spinor ψ^{L},

$$\begin{aligned} \left(\nabla \times \boldsymbol{A}\psi^{\mathrm{L}}\right)_i &= \sum_{j,k=1}^{3} \varepsilon_{ijk} \partial_j (A_k \psi^{\mathrm{L}}) \\ &= \sum_{j,k=1}^{3} \varepsilon_{ijk} (\partial_j A_k) \psi^{\mathrm{L}} + \sum_{j,k=1}^{3} \varepsilon_{ijk} A_k (\partial_j \psi^{\mathrm{L}}) \\ &= B_i \psi^{\mathrm{L}} - \left(\boldsymbol{A} \times \nabla \psi^{\mathrm{L}}\right)_i \end{aligned} \tag{D.21}$$

where we have employed $\boldsymbol{B} = \nabla \times \boldsymbol{A}$ in the last step (where now $\boldsymbol{A}$ is *not* some general vector but the electromagnetic vector potential, of course.) This immediately implies for the vector product of the kinematical momentum with itself

$$\boldsymbol{\pi} \times \boldsymbol{\pi} = \frac{\mathrm{i}\hbar q_e}{c} \boldsymbol{B} \tag{D.22}$$

and yields thus the final result

$$(\sigma \cdot \pi)(\sigma \cdot \pi) = \pi^2 - \frac{q_e \hbar}{c} \sigma \cdot B \tag{D.23}$$

D.2.2
Four-Dimensional Generalization

Dirac's relation can also be generalized to the four-component framework if σ is substituted by α,

$$\begin{aligned} (\alpha \cdot A)(\alpha \cdot B) &= \mathbf{1}_2 \otimes [(A \cdot B)\, \mathbf{1}_2 + \mathrm{i}\sigma \cdot (A \times B)] && \text{(D.24)} \\ &= (A \cdot B)\, \mathbf{1}_4 + \mathrm{i}\, \Sigma \cdot (A \times B) && \text{(D.25)} \end{aligned}$$

because

$$(\alpha \cdot A)(\alpha \cdot B) = \begin{pmatrix} (\sigma \cdot A)(\sigma \cdot B) & \mathbf{0} \\ \mathbf{0} & (\sigma \cdot A)(\sigma \cdot B) \end{pmatrix} \tag{D.26}$$

which is a (4×4)-matrix.

E
Fourier Transformations

E.1
Definition and General Properties

Consider a sufficiently smooth function f of one real (or complex) variable x,

$$f : \begin{cases} \mathbb{R} \to \mathbb{R} \\ x \mapsto f(x) \end{cases} \tag{E.1}$$

In many situations in mathematics, physics and chemistry it is advantageous to consider not the function f itself but a somehow transformed variant $\tilde{f}$ of the function f. The so-called *Fourier transform* or *Fourier transformation* (FT) of the function f is defined as

$$\tilde{f}(k) = \int_{-\infty}^{\infty} dx\, e^{-ikx} f(x) = \int dx\, e^{-ikx} f(x) \tag{E.2}$$

It is essential to note that f and $\tilde{f}$ are two different functions and not merely the same function depending on two different variables. For the sake of simplicity this distinction is not always reflected by the notation; however, we will explicitly distinguish these two functions by the $\tilde{f}$-notation in this appendix. Furthermore, since all integrals in this appendix extend over the whole real line it is convenient to not explicitly write down the limits of integration, which has been done in the second step of Eq. (E.2). Given the transformed function $\tilde{f}$ it is always possible to extract the original function f by a so-called *Fourier back transformation* (FBT) defined by

$$f(x) = \frac{1}{2\pi} \int dk\, e^{+ikx} \tilde{f}(k) \tag{E.3}$$

It is easy to see that the expressions for the two transformations (FT and FBT) are perfectly consistent with each other if and only if the integral representa-

Relativistic Quantum Chemistry. Markus Reiher and Alexander Wolf

ISBN: 978-3-527-31292-4

tion of Dirac's delta-function as given by Eq. (A.29) holds,

$$f(x) \stackrel{(E.3)}{=} \frac{1}{2\pi}\int dk\, e^{+ikx}\tilde{f}(k) \stackrel{(E.2)}{=} \int dx' \underbrace{\frac{1}{2\pi}\int dk\, e^{ik(x-x')}}_{\delta(x-x')} f(x')$$

$$\stackrel{(A.28)}{=} f(x) \tag{E.4}$$

Because of this feature Eq. (E.3) is also known as the *Fourier reciprocity theorem.*

We now investigate the Fourier transformation of the derivative $f' = df/dx$ of a function f. According to Eq. (E.2) it is given by

$$\begin{aligned}\widetilde{f'}(k) &= \int dx\, e^{-ikx}\left[\frac{d}{dx}f(x)\right] \stackrel{(p.I.)}{=} -\int dx \left[\frac{d}{dx}e^{-ikx}\right] f(x) \\ &= ik\int dx\, e^{-ikx}f(x) = ik\tilde{f}(k)\end{aligned} \tag{E.5}$$

where integration by parts (p.I.) has been employed at the second equality. The surface term necessarily has to vanish for any integrable function f and has thus been neglected. We have derived the very important result that a derivative operator in normal (position) space reduces to a simple multiplicative operator in Fourier (momentum) space. Especially in quantum mechanics (cf. chapter 4) this feature is often conveniently exploited.

For the discussion of the Douglas–Kroll–Hess transformation in chapter 12 the Fourier transformation of a product of two functions, $h(x) = f(x)g(x)$, has been employed. In one dimension it is given by the convolution integral of the Fourier transformations of f and g,

$$\begin{aligned}\tilde{h}(k) &= \int dx\, e^{-ikx}f(x)g(x) = \int dx \int dx'\, e^{-ikx}\,\delta(x-x')\, f(x)g(x') \\ &\stackrel{(A.29)}{=} \int dx \int dx'\, e^{-ikx}\frac{1}{2\pi}\int dk' e^{ik'(x-x')}f(x)g(x') \\ &= \frac{1}{2\pi}\int dk' \int dx\, e^{-i(k-k')x}f(x)\int dx' e^{-ik'x'}g(x') \\ &= \frac{1}{2\pi}\int dk' \tilde{f}(k-k')\tilde{g}(k')\end{aligned} \tag{E.6}$$

The formulae for Fourier transformations in three (or more) dimensions are straightforward generalizations of the one-dimensional formulae presented above and are thus not explicitly given in this appendix.

E.2
Fourier Transformation of the Coulomb Potential

The Coulomb potential, or more precisely, the Coulomb potential energy V between two charged particles with charges q_1 and q_2 in three dimensions in position-space (**r**-space) representation is given by

$$V(\boldsymbol{r}) = \frac{q_1 q_2}{|\boldsymbol{r}|} = \frac{q_1 q_2}{r} \tag{E.7}$$

where $r = |\boldsymbol{r}|$ denotes the spatial distance between the two charges. If q_1 and q_2 feature the same sign, i.e., either both charges are positive or negative, the potential will be repulsive, otherwise it will be negative and thus attractive.

For the discussion of the Douglas–Kroll–Hess transformation in chapter 12 it has been advantageous to consider the momentum-space representation of the Coulomb potential, which may be obtained via a Fourier transformation of $V(\boldsymbol{r})$. It is given by

$$\tilde{V}(\boldsymbol{k}) = \int d^3 r \, e^{-i\boldsymbol{k}\cdot\boldsymbol{r}} V(\boldsymbol{r}) \tag{E.8}$$

However, Eq. (E.8) cannot be evaluated directly in a straightforward manner. We thus introduce a suitable *cutoff* which damps the Coulomb potential sufficiently and define the family of cutoff potentials

$$V_\mu(\boldsymbol{r}) = \frac{q_1 q_2}{r} e^{-\mu r}, \qquad \forall \mu > 0 \tag{E.9}$$

Obviously $V_\mu(\boldsymbol{r}) \to V(\boldsymbol{r})$ for $\mu \to 0$.

Now we can try to calculate the Fourier transformation of $V_\mu(\boldsymbol{r})$. It is given by

$$\begin{aligned} \widetilde{V_\mu}(\boldsymbol{k}) &= q_1 q_2 \int d^3 r \, e^{-i\boldsymbol{k}\cdot\boldsymbol{r}} \, V_\mu(\boldsymbol{r}) = q_1 q_2 \int d^3 r \, \frac{e^{-i\boldsymbol{k}\cdot\boldsymbol{r}} \, e^{-\mu r}}{r} \\ &= 2\pi \, q_1 q_2 \int_0^\infty dr \int_0^\pi d\theta \, r^2 \sin\theta \, \frac{e^{-ikr\cos\theta} \, e^{-\mu r}}{r} \end{aligned} \tag{E.10}$$

where we have introduced spherical polar coordinates in the last step with $\theta = \angle(\boldsymbol{k}, \boldsymbol{r})$. If we now introduce the substitution

$$x = \cos\theta \quad \Longrightarrow \quad dx = -\sin\theta \, d\theta \tag{E.11}$$

with $x(\theta = 0) = +1$ and $x(\theta = \pi) = -1$, we arrive at

$$\begin{aligned} \widetilde{V_\mu}(\boldsymbol{k}) &= 2\pi \, q_1 q_2 \int_0^\infty dr \, r e^{-\mu r} \int_{-1}^{+1} dx \, e^{-ikrx} \\ &= \frac{2\pi \, q_1 q_2}{-ik} \int_0^\infty dr \, e^{-\mu r} \left(e^{-ikr} - e^{ikr} \right) \\ &= \frac{2\pi i \, q_1 q_2}{k} \int_0^\infty dr \left(e^{-(\mu+ik)r} - e^{-(\mu-ik)r} \right) \\ &= \frac{2\pi i \, q_1 q_2}{k} \left(\frac{1}{\mu + ik} - \frac{1}{\mu - ik} \right) = \frac{4\pi \, q_1 q_2}{\mu^2 + k^2} \end{aligned} \tag{E.12}$$

For $\mu \to 0$ we finally arrive at the momentum-space representation of the Coulomb potential,

$$\tilde{V}(\boldsymbol{k}) = \lim_{\mu \to 0} \widetilde{V_\mu}(\boldsymbol{k}) = \frac{4\pi q_1 q_2}{k^2} \tag{E.13}$$

The Coulomb potential thus features a $1/k^2$-dependence in momentum space.

F
Discretization and Quadrature Schemes

As a first-order differential equation, the radial Dirac equation, which is the central equation to be solved in the case of atoms, requires a discretization scheme, and several options are at hand of which some should be presented here for the sake of completeness. The first one is analogous to the Numerov procedure for second-order differential equations without first derivatives [404,405]. The derivation in terms of Taylor series expansions provides a derivation which is easier to understand. However, using operator techniques is the most elegant way for this particular task.

F.1
Numerov Approach toward Second-Order Differential Equations

A general second-order differential equation of the form

$$\chi''(s) + F(\epsilon, s)\chi(s) = G(s) \tag{F.1}$$

can be discretized on an equidistant grid of step size h, for instance, by the so-called Numerov method [406, 407, 409, 412, 972, 973]. The special feature of this method is that the truncation error is of comparatively high order in the step size h while only three points are required to discretize the second derivative. This is achieved through the explicit use of Eq. (F.1) as we shall see in the following.

For this, we expand the function $\chi(s_p) = \chi_p$ at grid point s_p and its second derivative χ_p'' at that grid point into a Taylor series

$$\chi_{p\pm1} = \chi(s_p \pm h) = \chi_p \pm \frac{1}{1!}\chi_p' h + \frac{1}{2!}\chi_p'' h^2 \pm \frac{1}{3!}\chi_p''' h^3 + \cdots \tag{F.2}$$

$$\chi_{p\pm1}'' = \chi(s_p \pm h)'' = \chi_p'' \pm \frac{1}{1!}\chi_p''' h + \frac{1}{2!}\chi_p^{(4)} h^2 \pm \frac{1}{3!}\chi_p^{(5)} h^3 + \cdots \tag{F.3}$$

Addition of Eq. (F.2) — and similarly for Eq. (F.3) — leads to

$$\chi_{p-1} + \chi_{p+1} = 2\chi_p + \chi_p'' h^2 + \frac{1}{12}\chi_p^{(4)} h^4 + O(h^6) \tag{F.4}$$

Relativistic Quantum Chemistry. Markus Reiher and Alexander Wolf

ISBN: 978-3-527-31292-4

and

$$\chi''_{p-1} + \chi''_{p+1} = 2\chi''_p + \chi_p^{(4)} h^2 + O(h^4) \tag{F.5}$$

If the second derivative of χ in Eq. (F.5) is replaced by the right-hand side of the re-arranged Eq. (F.1)

$$\chi'' = G - F\chi \tag{F.6}$$

and if we solve for $\chi_p^{(4)}$, we obtain

$$h^2 \chi_p^{(4)} = G_{p-1} - 2G_p + G_{p+1} - F_{p-1}\chi_{p-1} + 2F_p\chi_p - F_{p+1}\chi_{p+1} + O(h^4) \tag{F.7}$$

Now, Eq. (F.7) is applied to delete $\chi_p^{(4)}$ from Eq. (F.4). Solving for χ''_p then yields

$$\begin{aligned} \chi''_p &= \left(\frac{1}{h^2} + \frac{F_{p-1}}{12}\right)\chi_{p-1} - 2\left(\frac{1}{h^2} + \frac{F_p}{12}\right)\chi_p + \left(\frac{1}{h^2} + \frac{F_{p+1}}{12}\right)\chi_{p+1} \\ &\quad - \frac{1}{12}\left(G_{p-1} - 2G_p + G_{p+1}\right) + O(h^4) \end{aligned} \tag{F.8}$$

Insertion of this result in the original differential equation yields a set of linear equations whose p-th equation reads

$$\begin{aligned} \left(\frac{1}{h^2} + \frac{F_{p-1}}{12}\right)\chi_{p-1} - \left(\frac{2}{h^2} - \frac{10F_p}{12}\right)\chi_p + \left(\frac{1}{h^2} + \frac{F_{p+1}}{12}\right)\chi_{p+1} \\ = \frac{1}{12}\left(G_{p-1} + 10G_p + G_{p+1}\right) + O(h^4) \end{aligned} \tag{F.9}$$

Multiplication by h^2 and introduction of the definitions

$$b_p \equiv 1 + \frac{h^2}{12}F_p\,, \qquad a_p \equiv -2 + \frac{10h^2}{12}F_p = 10b_p - 12 \tag{F.10}$$

and

$$d_p \equiv \frac{h^2}{12}\left(G_{p-1} + 10G_p + G_{p+1}\right) \tag{F.11}$$

finally yields

$$a_{p-1}\chi_{p-1} + b_p\chi_p + a_{p+1}\chi_{p+1} = b_p \tag{F.12}$$

All such equations can be collected in a single matrix equation that reads

$$\begin{pmatrix} a_1 & b_2 & & & & 0 \\ b_1 & a_2 & b_3 & & & \\ & b_2 & a_3 & b_4 & & \\ & & \ddots & \ddots & \ddots & \\ & & & b_{n-2} & a_{n-1} & b_n \\ 0 & & & & b_{n-1} & a_n \end{pmatrix} \begin{pmatrix} \chi_1 \\ \chi_2 \\ \chi_3 \\ \vdots \\ \chi_{n-1} \\ \chi_n \end{pmatrix} = \begin{pmatrix} d_1 \\ d_2 \\ d_3 \\ \vdots \\ d_{n-1} \\ d_n \end{pmatrix} \tag{F.13}$$

Note that the boundaries would have to be corrected via simple extrapolation in order to preserve the numerical accuracy wanted. The determinant of the matrix in Eq. (F.13) can be calculated by a Sturm chain and the eigenvalue is then solved for by bisection [408,974].

F.2
Numerov Approach for First-Order Differential Equations

We now turn to the discretization of the coupled first-order differential equations as they occur in the solution of the Dirac radial equation for atoms (see chapter 9). While the Numerov scheme is well established for second-order differential equations — and, hence, for the solution of the radial Schrödinger equation for atoms — this is not the case for first-order differential equations. Indeed, it was long believed that the Numerov scheme cannot be used at all in this context [973].

We again denote as $\chi(s)$ the function to be discretized on an equidistant grid in the variable s. The differential equation to be solved reads

$$\frac{1}{a}\chi'(s) + F(s)\chi(s) = G^{\star}(s) \tag{F.14}$$

where a can take the values ± 1 to account for the different signs of the first derivatives in the coupled Dirac–Hartree–Fock equations. h is the step size of the equidistant grid and the index p denotes the p-th grid point at s_p. To keep the equations clearly and simply arranged, $\chi(s_p)$ will be abbreviated as χ_p.

Note that the derivations will be held as general as possible. Thus, the coefficient functions of the coupled Dirac–Hartree–Fock equations, Eqs. (9.237) and (9.238), connected at a time with the other radial function are implicitly introduced via the inhomogeneity,

$$G^{\star}(s) \quad = \quad G(s) - F_{UL}(s)\, y_L(s) \tag{F.15}$$

$$H^{\star}(s) \quad = \quad H(s) - aF_{LU}(s)\, y_U(s) \tag{F.16}$$

The function $\chi(s)$ is assumed to be analytic. We should, however, note that the requirement of infinite differentiability is the reason for numerical problems with the untransformed non-analytic radial functions in calculations with point nuclei (cf. the non-integral exponent in the short-range series expansions derived in section 9.6.1). An analytic function $\chi(s)$ can be expanded into a Taylor series around grid point p,

$$\chi_{p\pm 1} = \chi_p \pm \frac{1}{1!}\chi'_p h + \frac{1}{2!}\chi''_p h^2 \pm \frac{1}{3!}\chi'''_p h^3 + \cdots \tag{F.17}$$

which may be subtracted from each other to become

$$-\chi_{p-1} + \chi_{p+1} = 2\chi'_p h + \frac{1}{3}\chi'''_p h^3 + O(\chi_p^{(5)} h^5) \tag{F.18}$$

where we use E. Landau's O-symbolism to indicate that the main term of the truncated series is of fifth order in h and contains a fifth derivative of $\chi(s)$ at grid point p. Note that every second term in the series expansion vanishes. The same procedure has to be repeated for the first derivative $\chi'(s)$, which yields the series expansions

$$\chi'_{p\pm1} = \chi'_p \pm \frac{1}{1!}\chi''_p h + \frac{1}{2!}\chi'''_p h^2 \pm \frac{1}{3!}\chi_p^{(4)} h^3 + \cdots \tag{F.19}$$

which can be added to obtain

$$\chi'_{p-1} + \chi'_{p+1} = 2\chi'_p + \chi'''_p h^2 + O(\chi_p^{(5)} h^4) \tag{F.20}$$

This equation may be rearranged to

$$\chi'''_p h^2 = \chi'_{p-1} - 2\chi'_p + \chi'_{p+1} + O(\chi_p^{(5)} h^4) \tag{F.21}$$

which yields with the discretized differential equation (cf. Eq. (F.14))

$$\chi'_p = a(-F_p \chi_p + G_p^\star) \tag{F.22}$$

the expression

$$\begin{aligned} \chi'''_p h^2 &= a\Big[(G^\star_{p-1} - F_{p-1}\chi_{p-1} - 2(G^\star_p - F_p\chi_p) \\ &\quad + G^\star_{p+1} - F_{p+1}\chi_{p+1})\Big] + O(\chi_p^{(5)} h^4) \end{aligned} \tag{F.23}$$

for the unknown term $h/3\ (\chi'''_p h^2)$ needed in Eq. (F.18). This equation then becomes, after division by $(2h)$, the discretized first derivative

$$\begin{aligned} \chi'_p &= \left(-\frac{1}{2h} + \frac{a}{6}F_{p-1}\right)\chi_{p-1} - \frac{a}{3}F_p\chi_p + \left(\frac{1}{2h} + \frac{a}{6}F_{p+1}\right)\chi_{p+1} \\ &\quad - \frac{a}{6}G^\star_{p-1} + \frac{a}{3}G^\star_p - \frac{a}{6}G^\star_{p+1} + O(\chi_p^{(5)} h^4) \end{aligned} \tag{F.24}$$

where, if we use Eq. (F.22) for a second time and remove the division by $(2h)$, the final result (i.e., the discretized differential equation) will be

$$\begin{aligned} &\left(-1 + \frac{ah}{3}F_{p-1}\right)\chi_{p-1} + \frac{4ah}{3}F_p\chi_p + \left(1 + \frac{ah}{3}F_{p+1}\right)\chi_{p+1} \\ &\qquad = \frac{ah}{3}G^\star_{p-1} + \frac{4ah}{3}G^\star_p + \frac{ah}{3}G^\star_{p+1} + O(\chi_p^{(5)} h^5) \end{aligned} \tag{F.25}$$

Remember that a can take the values ±1 only! Obviously, without using the differential equation, we obtain from Eqs. (F.18) and (F.21), the general expression

$$\begin{aligned} \chi_{p+1} - \chi_{p-1} &= 2\chi'_p h + \frac{h}{3}(\chi'_{p-1} - 2\chi'_p + \chi'_{p+1}) + O(\chi_p^{(5)} h^5) \\ &= \frac{h}{3}(\chi'_{p-1} + 4\chi'_p + \chi'_{p+1}) + O(\chi_p^{(5)} h^5) \end{aligned} \tag{F.26}$$

F.3
Simpson's Quadrature Formula

For the numerical evaluation of expectation values, numerical quadrature schemes are needed. In the case of atoms (see chapter 9), these are one-dimensional and, hence, particularly simple. As an example, we discuss Simpson's rule in the following.

The famous Simpson rule for numerical quadrature can be written as [392, p.102]

$$\begin{aligned}\int_{s_{p-1}}^{s_{p+1}} f(s)\mathrm{d}s &= \frac{h}{3}\left[f(s_{p+1}) + 4f(s_p) + f(s_{p-1})\right] \\ &\quad - \frac{h}{90}\delta^4 f(s_p) + O(h\delta^6 f(s))\end{aligned} \tag{F.27}$$

where the operator for a central difference, δ, is defined as $\delta f(x) = f(x + h/2) - f(x - h/2)$. If we rewrite this equation in terms of the integrated function $\chi(s) = \int f(s)\,\mathrm{d}s$ we obtain

$$\begin{aligned}\chi(s_{p+1}) - \chi(s_{p-1}) &= h/3\left[\chi'(s_{p+1}) + 4\chi'(s_p) + \chi'(s_{p-1})\right] \\ &\quad - \frac{h}{90}\delta^4\chi'(s_p) + O\left[h\delta^6\chi'(s)\right]\end{aligned} \tag{F.28}$$

and get Eq. (F.26) — which will lead to Eq. (F.25) if the differential Eq. (F.22) is introduced to remove the derivatives of $\chi(s)$.

F.4
Bickley's Central-Difference Formulae

Simple discretization schemes use derivatives of Lagrangian interpolation polynomials that approximate the function known only at the grid points $\{s_i\}$. These schemes consist of tabulated numbers multiplied by the function's values at m contiguous grid points and are referred to as "m-point-formulae" by Bickley [411] (cf. [381, p. 914]). The accuracy of the numerical differentiation increases with the number of neighboring grid points needed. For an acceptable truncation error $O(h^t)$, $t = 4$ or higher, m is larger than t, which leads to an extended amount of computation. The general structure and usage of Bickley's central-difference formulae can be demonstrated for the simplest approximation of a first-order derivative,

$$\frac{\mathrm{d}f(s)}{\mathrm{d}s} = \lim_{\Delta s \to 0} \frac{\Delta f(s)}{\Delta s} \tag{F.29}$$

which becomes for an equidistant grid $h = s_{k+1} - s_k$ at a given position s_k

$$\lim_{\Delta s \to 0} \left. \frac{\Delta f(s)}{\Delta s} \right|_{s=s_k} \approx \frac{f(s_{k+1}) - f(s_k)}{s_{k+1} - s_k} = \frac{1}{h}(-1, 1) \cdot \begin{pmatrix} f(s_k) \\ f(s_{k+1}) \end{pmatrix} \tag{F.30}$$

The last step carried out for all grid points yields the combined result

$$\frac{\mathrm{d}f(s)}{\mathrm{d}s} \approx \frac{1}{h} \begin{pmatrix} -1 & 1 & 0 & 0 & \cdots & 0 & 0 \\ 0 & 1 & 1 & 0 & \cdots & 0 & 0 \\ \vdots & & & \ddots & & & \vdots \\ 0 & 0 & 0 & 0 & \cdots & -1 & 1 \\ 0 & 0 & 0 & 0 & \cdots & 0 & -1 \end{pmatrix} \cdot \begin{pmatrix} f(s_1) \\ f(s_2) \\ \vdots \\ f(s_{n-1}) \\ f(s_n) \end{pmatrix} \tag{F.31}$$

As a result, we obtain a matrix representation for the differential operator. Note that the differentiation of $f(s_n)$ at the upper boundary would require an asymptotic correction in order to preserve the overall numerical accuracy.

G
List of Abbreviations and Acronyms

nr, non-rel.	nonrelativistic
AIMP	ab initio model potential
AMFI	atomic mean-field integrals
B3LYP	Becke's three-parameter hybrid density functional
CASSCF	complete-active-space self-consistent field
CASPT2	complete-active-space 2nd-order perturbation theory
CC	coupled cluster
CCSD	coupled cluster with single and double excitations
CCSD(T)	CCSD with perturbative treatment of triple excitations
CI	configuration interaction
CI(SD)	CI (with single and double excitations)
CPP	core polarization potential
CSF	configuration state function
DC	Dirac–Coulomb
DCB	Dirac–Coulomb–Breit
DFC	Dirac–Fock–Coulomb
DFT	density functional theory
DHF	Dirac–Hartree–Fock
DKHn	Douglas–Kroll–Hess (n-th order Hamiltonian)
DKH(n,m)	DKH with nth-order Hamiltonian, m-th order property
DMRG	density matrix renormalization group
ECP	effective core potential
ESR	electron spin resonance
FSCC	Fock-space coupled cluster
FSCCSD	FSCC with single and double excitations
GIAO	gauge including atomic orbitals

Relativistic Quantum Chemistry. Markus Reiher and Alexander Wolf

ISBN: 978-3-527-31292-4

HF	Hartree–Fock
iHFSCC	intermediate-Hamiltonian FSCC
IORA	infinite-order regular approximation
IOTC	infinite-order two-components
KR	Kramers-restricted
MCSCF	multi-configuration self-consistent field
MP (MPn)	Møller–Plesset perturbation theory (of n-th order)
MRCI	multi-reference CI
MRCISD+Q	MRCI up to double excitations plus Davidson correction
NMR	nuclear magnetic resonance
PCE	picture change effect
QED	quantum electrodynamics
SCF	self-consistent field
SO	spin–orbit
ZORA	zeroth-order regular approximation

H
List of Symbols

α	fine structure constant (if explicitly noted as such)
$\boldsymbol{\alpha}$	3-vector of Dirac's (4×4) α-matrices
β	Dirac's (4×4) β-matrix
c	speed of light
∂_μ	covariant 4-gradient
$\not{\partial}$	Feynman's slash-notation for 4-derivatives
e	elementary charge (i.e., charge of the proton)
γ^ν	Dirac's (4×4) γ-matrices
$g(i,j)$, g_{ij}	two-electron interaction term
$\hbar$	reduced Planck constant
j^μ	charge–current 4-density
$\boldsymbol{j}$	electric current density
m_e	rest mass of the electron
$\boldsymbol{p}$	momentum 3-vector or momentum operator
ψ	four-component Dirac matter field
$\bar{\psi}$	adjoint Dirac field
Ψ	four-component Dirac spinor
$\Psi^\dagger$	hermitean conjugate Dirac spinor
$\Psi_{el,A}$	total electronic wave function for state A
ψ_i	Dirac one-electron 4-spinor
$\boldsymbol{r}$	position 3-vector
ϱ	electric charge density
$\boldsymbol{\sigma}$	3-vector of Pauli's (2×2) spin matrices
x^μ	space–time position 4-vector
A^μ	4-potential
$\boldsymbol{A}$	vector potential
A_p	kinematic A-factor

Relativistic Quantum Chemistry. Markus Reiher and Alexander Wolf

ISBN: 978-3-527-31292-4

$\mathbf{B}$	magnetic field
$\mathbf{E}$	electric field
$\mathcal{E}_k$	even term of k-th order in V
$\mathcal{E}_{[2k]}$	even term of $2k$-th order in $1/c$
E_p	relativistic energy–momentum relation
$F^{\mu\nu}$	electromagnetic field tensor
${}^1\mathcal{H}$	one-particle Hilbert space
$\mathcal{H}$	Hamiltonian density
h_{bd}	block-diagonal Dirac Hamiltonian
h^D	four-component, one-electron Dirac Hamiltonian
	(note that V may be included or not as in chapters 6 and 11, resp.)
h^S	one-electron Schrödinger Hamiltonian
h_+	electronic part of H_{bd}
$\mathcal{L}$	Lagrangian density
$\Lambda^\mu{}_\nu$	Lorentz transformation
m	number of one-particle basis functions
m_L	number of large component's basis functions
m_S	number of small component's basis functions
M	number of atomic nuclei
N	number of electrons
Ω	unitary transformation to p^2-space
$\boldsymbol{P}_p$	generalized momentum variable
ϕ	scalar potential
Q	total electric charge
S	action
$\boldsymbol{\Sigma}$	Dirac's (4×4) $\boldsymbol{\Sigma}$-matrix
U_0	free-particle Foldy–Wouthuysen transformation
U_i	i-th unitary transformation of a sequence
V	external electron–nucleus interaction potential
$\tilde{V}$	damped external potential
W_k	odd and antihermitean operator, which is k-th order in V
X	(2×2) X-operator relating large and small components
Z	charge number of an atomic nucleus
$\mathbf{1}$	unity (identity) operator
$:\psi^\dagger\psi:$	normal-ordered product of field operators
$\otimes$	tensor product, direct product

References

1 P. Pyykkö. *Relativistic Theory of Atoms and Molecules I – A Bibliography 1916–1985.* Lecture Notes in Chemistry. Springer-Verlag, Berlin, 1986.

2 P. Pyykkö. *Relativistic Theory of Atoms and Molecules II – A Bibliography 1986-1992,* Volume 60 of *Lecture Notes in Chemistry.* Springer-Verlag, Berlin, 1993.

3 P. Pyykkö. *Relativistic Theory of Atoms and Molecules III – A Bibliography 1993-1999,* Volume 76 of *Lecture Notes in Chemistry.* Springer-Verlag, Berlin, 2000.

4 P. Pyykkö. Database 'RTAM' (relativistic quantum chemistry database 1915–1998; http://www.csc.fi/lul/rtam/rtamquery.html), 2006.

5 P. Schwerdtfeger (Ed.). *Relativistic Electronic Structure Theory — Part I. Fundamentals.* Theoretical and Computational Chemistry. Elsevier, Amsterdam, 2002.

6 P. Schwerdtfeger (Ed.). *Relativistic Electronic Structure Theory — Part II. Applications.* Theoretical and Computational Chemistry. Elsevier, Amsterdam, 2004.

7 B. A. Hess. *Relativistic Effects in Heavy Element Chemistry and Physics.* Wiley Series in Theoretical Chemistry. John Wiley & Sons Ltd., Chichester, 2003.

8 P. J. Mohr, B. N. Taylor. CODATA recommended values of the fundamental physical constants: 1998. *Rev. Mod. Phys.*, **72**(2) (2000) 351–495; http://www.codata.org/.

9 Y. Lau. An easy method for converting equations between SI and Gaussian units. *Am. J. Phys.*, **56**(2) (1988) 135–137.

10 I. Newton. *Philosophiae Naturalis Principia Mathematica.* 1687.

11 J. C. Maxwell. A Dynamical Theory of the Electromagnetic Field. *Philosophical Transactions of the Royal Society of London,* **155** (1865) 459–512.

12 J. D. Jackson. *Classical Electrodynamics.* Wiley, New York, 1975.

13 W. Ehrenberg, R. E. Siday. The refractive index in electron optics and the principle of dynamics. *Proc. Phys. Soc. B*, **62**(1) (1949) 8–21.

14 Y. Aharonov, D. Bohm. Significance of electromagnetic potentials in the quantum theory. *Phys. Rev.*, **115**(3) (1959) 485–491.

15 R. P. Feynman, R. B. Leighton, M. Sands. *The Feynman lectures on physics — Volume II.* Addison-Wesley, Reading, Massachusetts, 1964.

16 J. J. Sakurai. *Modern Quantum Mechanics.* Addison-Wesley, Reading, Massachusetts, rev. ed., 1995.

17 R. P. Feynman, R. B. Leighton, M. Sands. *The Feynman lectures on physics — Volume I.* Addison-Wesley, Reading, Massachusetts, 1964.

18 H. Goldstein. *Classical Mechanics.* Addison-Wesley, Reading, Massachusetts, 1950.

19 L. D. Landau, E. M. Lifshitz. *Course of Theoretical Physics. Volume 1: Mechanics.* Butterworth Heinemann, Oxford, Burlington (MA), 3rd ed., 1995.

20 L. D. Landau, E. M. Lifshitz. *Course of Theoretical Physics. Volume 2: The classical theory of fields.* Butterworth Heinemann, Oxford, 4th ed., 1995.

21 V. I. Arnold. *Mathematical Methods of Classical Mechanics.* Springer, Berlin, New York, 2nd ed., 1989.

22 O. Jahn, V. V. Sreedhar. The maximal invariance group of Newton's equations for a free point particle. *Am. J. Phys.*, **69**(10) (2001) 1039–1043.

23 C. G. Darwin. The Dynamical Motions of Charged Particles. *Phil. Mag.*, **39**(233) (1920) 537–550.

24 L. Szasz. *The Electronic Structure of Atoms.* John Wiley & Sons, New York, 1992.

25 O. D. Jefimenko. Direct calculation of the electric and magnetic fields of an electric point charge moving with constant velocity. *Am. J. Phys.*, **62** (1994) 79–85.

26 J. E. Harriman. *Theoretical Foundations of Electron Spin Resonance*, Volume 37 of *Physical Chemistry — A Series of Monographs.* Academic Press, New York, 1978.

27 S. Weinberg. *Gravitation and Cosmology: Principles and Applications of the General Theory of Relativity.* John Wiley & Sons, New York, 1972.

28 D. Bohm. *The special theory of relativity.* The Benjamin/Cummings Publishing Company, Inc., Reading, Massachusetts, 1965.

29 W. Pauli. *Relativitätstheorie.* Springer, Berlin, 2000.

30 D. E. Mook, T. Vargish. *Inside Relativity.* University Presses of CA, 1991.

31 S. Parrott. *Relativistic Electrodynamics and Differential Geometry.* Springer, Berlin, 1987.

32 F. Gross. *Relativistic Quantum Mechanics and Field Theory.* John Wiley & Sons, Inc., New York, 1993.

33 R. B. Lindsay, H. Margenau. *Foundations of Physics.* Ox Bow Press, Woodbridge, 1981.

34 H. Hellmann. Zur Rolle der kinetischen Elektronenenergie für die zwischenatomaren Kräfte. *Z. Phys.*, **85** (1933) 180–190.

35 H. Hellmann. *Einführung in die Quantenchemie.* Deuticke, Leipzig, 1937.

36 R. P. Feynman. Forces in Molecules. *Phys. Rev.*, **56** (1939) 340–343.

37 A. R. Edmonds. *Angular Momentum in Quantum Mechanics.* Princeton Landmarks in Physics. Princeton University Press, Princeton NJ, 1996.

38 E. U. Condon, G. H. Shortley. *The Theory of Atomic Spectra.* Cambridge University Press, Cambridge, 1970.

39 H. A. Bethe. *Quantenmechanik der Ein- und Zwei-Elektronenprobleme*, Volume 24, Part 1 of *Handbuch der Physik*, p. 273–560. Springer-Verlag, Berlin, 1933.

40 L. Schiff. *Quantum Mechanics.* McGraw Hill, 1949.

41 B. R. Judd. *Operator Techniques in Atomic Spectroscopy.* Princeton Landmarks in Physics. Princeton University Press, Princeton NJ, 1998.

42 W. R. Johnson. *Atomic Structure Theory — Lectures on Atomic Physics.* Springer, Berlin, 2007.

43 T. E. Phipps, J. B. Taylor. The Magnetic Moment of the Hydrogen Atom. *Phys. Rev.*, **29** (1927) 309–320.

44 G. E. Uhlenbeck, S. Goudsmit. Ersetzung der Hypothese vom unmechanischen Zwang durch eine Forderung bezüglich des inneren Verhaltens jedes einzelnen Elektrons. *Naturwissenschaften*, **47** (1925) 953–954.

45 G. Herzberg. *Molecular Spectra and Molecular Structure — I. Spectra of Diatomic Molecules.* Structure of Matter Series. D. van Nostrand, New York, 2nd ed., 1950.

46 F. Schwabl. *Advanced Quantum Mechanics.* Springer-Verlag, Berlin, 3rd ed., 2005.

47 I. P. Grant. *Relativistic Quantum Theory of Atoms and Molecules — Theory and Computation.* Springer Series on Atomic, Optical, and Plasma Physics. Springer, New York, 2007.

48 H. A. Bethe, E. E. Salpeter. *Quantum Mechanics of One- and Two-Electron Atoms.* Plenum, New York, 1977.

49 R. P. Feynman, R. B. Leighton, M. Sands. *The Feynman lectures on physics — Volume III.* Addison-Wesley, Reading, Massachusetts, 1964.

50 A. Messiah. *Quantum Mechanics.* Dover Publications, 1999.

51 A. S. Davydov. *Quantum Mechanics.* Pergamon Press, 2nd ed., 1976.

52 L. D. Landau, E. M. Lifshitz. *Course of Theoretical Physics. Volume 3: Quantum Mechanics.* Butterworth Heinemann, Oxford, Burlington (MA), 3rd ed., 1981.

53 E. Merzbacher. *Quantum Mechanics.* John Wiley & Sons, Inc., New York, 3rd ed., 1998.

54 C. Cohen-Tannoudji, B. Diu, F. Laloe. *Quantum Mechanics.* Wiley-Interscience, 2006.

55 P. A. M. Dirac. *The Principles of Quantum Mechanics.* Oxford University Press, Oxford, 4th ed., 1958.

56 A. Bohm. *Quantum mechanics: foundations and applications.* Springer, New York, 3rd ed., 1993.

57 C. J. Isham. *Lectures on Quantum Theory — Mathematical and Structural Foundations.* Imperial College Press, London, 1995.

58 M. Weissbluth. *Atoms and Molecules.* Academic Press, New York, 1978.

59 J. von Neumann. *Mathematical Foundations of Quantum Mechanics.* Princeton University Press, Princeton, New Jersey, 1955.

60 Y. Aharonov, D. Rohrlich. *Quantum Paradoxes — Quantum Theory for the Perplexed*. Wiley-VCH, Berlin, 2005.

61 R. Omnès. *The Interpretation of Quantum Mechanics*. Princeton University Press, Princeton, New Jersey, 1994.

62 R. Omnès. *Understanding Quantum Mechanics*. Princeton University Press, Princeton, New Jersey, 1999.

63 F. Strocchi. *An Introduction to the Mathematical Structure of Quantum Mechanics*, Volume 27 of *Advanced Series in Mathematical Physics*. World Scientific, Singapore, 2005.

64 D. Giulini, E. Joos, C. Kiefer, J. Kupsch, I.-O. Stamatescu, H. D. Zeh. *Decoherence and the Appearance of a Classical World in Quantum Theory*. Springer-Verlag, Berlin, Heidelberg, 1996.

65 J. D. Bjorken, S. D. Drell. *Relativistic Quantum Mechanics*. McGraw-Hill, New York, 1964.

66 P. A. M. Dirac. The Quantum Theory of the Electron. *Proc. Roy. Soc. London A*, **117** (1928) 610–624.

67 P. A. M. Dirac. The Quantum Theory of the Electron (Part II). *Proc. Roy. Soc. London A*, **118** (1928) 351–361.

68 G. Breit. The Effect of Retardation on the Interaction of Two Electrons. *Phys. Rev.*, **34**(4 (2nd Series)) (1929) 553–573.

69 G. Breit. On the Interpretation of Dirac's α Matrices. *Proc. Nat. Acad. Sci. USA*, **17** (1931) 70–73.

70 C. G. Darwin. The Wave Equations of the Electron. *Proc. Roy. Soc. London A*, **118** (1928) 654–680.

71 J. J. Sakurai. *Advanced Quantum Mechanics*. Addison-Wesley, Reading, Massachusetts, rev. ed., 1967.

72 P. A. M. Dirac. A theory of electrons and protons. *Proc. Roy. Soc. London A*, **126** (1930) 360–365.

73 F. A. B. Coutinho, Y. Nogami, L. Tomio. Reply to "Comment on 'Validity of Feynman's prescription of disregarding the Pauli principle in intermediate states' ". *Phys. Rev. A*, **62**(1) (2000) 016102.

74 F. A. B. Coutinho, D. Kiang, Y. Nogami, L. Tomio. Dirac's hole theory versus quantum field theory. *Can. J. Phys.*, **80** (2002) 837–840.

75 P. Strange. *Relativistic Quantum Mechanics — with applications in condensed matter and atomic physics*. Cambridge University Press, Cambridge, 1998.

76 B. Thaller. *The Dirac Equation*. Texts and Monographs in Physics. Springer-Verlag, New York, 1992.

77 B. Thaller. *The Dirac Operator*, p. 23–106. Theoretical and Computational Chemistry. Elsevier, Amsterdam, 2002.

78 M. E. Rose. *Relativistic Electron Theory*. John Wiles & Sons, New York, 1961.

79 G. W. Erickson, D. R. Yennie. Radiative Level Shifts I: Formulation and Lowest Order Lamb Shift. *Ann. Phys. (NY)*, **35** (1965) 271–313.

80 G. W. Erickson, D. R. Yennie. Radiative Level Shifts II: Higher Order Contributions ot the Lamb Shift. *Ann. Phys. (NY)*, **35** (1965) 447–510.

81 H. Grotch, D. R. Yennie. Effective Potential Model for Calculating Nuclear Corrections to the Energy Levels of Hydrogen. *Rev. Mod. Phys.*, **41** (1969) 350–374.

82 W. Gordon. Die Energieniveaus des Wasserstoffatoms nach der Diracschen Theorie des Elektrons. *Z. Phys.*, **48** (1928) 11–14.

83 H. Margenau, G. M. Murphy. *The Mathematics of Physics and Chemistry*. D. van Nostrand Co., Princeton, 2nd ed., 1956.

84 B. Swirles. The Relativistic Self-Consistent Field. *Proc. Roy. Soc. London A*, **152** (1935) 625–649.

85 T. P. Das. *Relativistic Quantum Mechanics of Electrons*. Harper & Row, New York, 1973.

86 M. E. Rose. *Relativistische Elektronentheorie II*, Volume 554a of *BI Hochschultaschenbücher*. Bibliographisches Institut, Mannheim, Wien, Zürich, 1971.

87 H. E. White. Pictorial Representations of the Dirac Electron Cloud for Hydrogen-Like Atoms. *Phys. Rev.*, **38** (1931) 513–520.

88 M. Reiher. *Development and Implementation of Numerical Algorithms for the Solution of Multi-Configuration Self-Consistent Field Equations for Relativistic Atomic Structure Calculations*. PhD thesis, Fakultät für

Chemie, University of Bielefeld, Germany, 1998.

89 I. P. Grant. Relativistic Calculations of Atomic Structures. *Adv. Phys.*, **19** (1970) 747–811.

90 M. Bander, C. Itzykson. Group Theory and the Hydrogen Atom (I). *Rev. Mod. Phys.*, **38**(2) (1966) 330–345.

91 M. Bander, C. Itzykson. Group Theory and the Hydrogen Atom (II). *Rev. Mod. Phys.*, **38**(2) (1966) 346–358.

92 P. G. Nelson. How Do Electrons Get Across Nodes? *J. Chem. Educ.*, **67**(8) (1990) 643–647.

93 D. Andrae. Recursive evaluation of expectation values $\langle r^k \rangle$ for arbitrary states of the relativistic one-electron atom. *J. Phys. B: At. Mol. Opt. Phys.*, **30** (1997) 4435–4451.

94 D. Andrae. Finite nuclear charge density distributions in electronic structure calculations for atoms and molecules. *Phys. Rep.*, **336**(6) (2000) 413–525.

95 H. A. Bethe, E. E. Salpeter. *Quantum Mechanics of One- and Two-Electron Systems*, Volume 35 of *Handbuch der Physik*, p. 88–436. Springer-Verlag, Berlin, 1957.

96 L. Bergmann, C. Schaefer. *Constituents of Matter: Atoms, Molecules, Nuclei and Particles*. W. de Gruyter, New York, 1997.

97 R. Abtamowicz, P. Lounesto. *Clifford Algebras and Spinor Structures*. Kluwer Academic, Dordrecht, 1995.

98 E. Fermi. Quantum Theory of Radiation. *Rev. Mod. Phys.*, **4** (1932) 87–132.

99 S. Weinberg. *The Quantum Theory of Fields. Volume I. Foundations*. Cambridge University Press, Cambridge, 1995.

100 S. N. Gupta. Theory of longitudinal photons in quantum electrodynamics. *Proc. Phys. Soc. (London)*, **A63** (1950) 681–691.

101 K. Bleuler. Eine neue Methode zur Behandlung der longitudinalen und skalaren Photonen. *Helv. Phys. Acta*, **23** (1950) 567–586.

102 C. Itzykson, J.-B. Zuber. *Quantum Field Theory*. McGraw-Hill, New York, 1980.

103 S.-I. Tomonaga. On a Relativistically Invariant Formulation of the Quantum Theory of Wave Fields. *Prog. Theor. Phys. (Kyoto)*, **1** (1946) 27–42.

104 S.-I. Tomonaga, J. R. Oppenheimer. On Infinite Field Reactions in Quantum Field Theory. *Phys. Rev.*, **74**(2) (1948) 224–225.

105 J. Schwinger. On Quantum-Electrodynamics and the magnetic moment of the electron. *Phys. Rev.*, **73**(4) (1948) 416–417.

106 J. Schwinger. Quantum Electrodynamics. I. A Covariant Formulation. *Phys. Rev.*, **74**(10) (1948) 1439–1461.

107 J. Schwinger. Quantum Electrodynamics. II. Vacuum Polarization and Self-Energy. *Phys. Rev.*, **75**(4) (1949) 651–679.

108 R. P. Feynman. Space-Time Approach to Non-Relativistic Quantum Mechanics. *Rev. Mod. Phys.*, **20**(2) (1948) 367–387.

109 R. P. Feynman. A Relativistic Cut-Off for Classical Electrodynamics. *Phys. Rev.*, **74**(8) (1948) 939–946.

110 R. P. Feynman. Relativistic Cut-Off for Quantum Electrodynamics. *Phys. Rev.*, **74**(10) (1948) 1430–1438.

111 R. P. Feynman. The Theory of Positrons. *Phys. Rev.*, **76**(6) (1949) 749–759.

112 R. P. Feynman. Space-Time Approach to Quantum Electrodynamics. *Phys. Rev.*, **76**(6) (1949) 769–789.

113 F. J. Dyson. The Radiation Theories of Tomonaga, Schwinger, and Feynman. *Phys. Rev.*, **75**(3) (1949) 486–502.

114 F. J. Dyson. The S Matrix in Quantum Electrodynamics. *Phys. Rev.*, **75**(11) (1949) 1736–1755.

115 N. M. Kroll, W. E. Lamb, Jr. On the self-energy of a bound electron. *Phys. Rev.*, **75**(3) (1949) 388–398.

116 P. J. Mohr, G. Plunien, G. Soff. QED Corrections in Heavy Atoms. *Phys. Rep.*, **293** (1998) 227–369.

117 P. Indelicato, P. J. Mohr. Coordinate-space approach to the bound-electron self-energy: Self-energy screening calculation. *Phys. Rev. A*, **63**(5) (2001) 052507.

118 B. Åsén, S. Salomonson, I. Lindgren. Two-photon-exchange QED effects in the 1s2s S-1 and S-3 states of heliumlike ions. *Phys. Rev. A*, **65**(3) (2002) 032516.

119 P. Teller. *An Interpretive Introduction to Quantum Field Theory*. Princeton University Press, Princeton, New Jersey, 1995.

120 R. P. Feynman. *QED: The Strange Theory of Light and Matter*. Princeton Science Library. Princeton University Press, Princeton, New Jersey, 2006.

121 R. P. Feynman. *Quantum Electrodynamics*. W. A. Benjamin New York, 1961.

122 S. S. Schweber. *An Introduction to Relativistic Quantum Field Theory*. Harper & Row Publishers, New York, 2nd printing ed., 1966.

123 J. D. Bjorken, S. D. Drell. *Relativistic Quantum Fields*. McGraw-Hill, New York, 1965.

124 A. I. Akhiezer, V. B. Berestetskii. *Quantum Electrodynamics*, Volume XI of *Interscience Monographs and Texts in Physics and Astronomy*. Interscience Publishers, New York, 1965.

125 L. D. Landau, E. M. Lifshitz. *Course of Theoretical Physics. Volume 4: Quantum Electrodynamics*. Pergamon Press, Oxford, 1982.

126 N. F. Mott, I. N. Sneddon. *Wave Mechanics and Its Applications*. Dover Publications, 1963.

127 E. E. Salpeter, H. A. Bethe. A Relativistic Equation for Bound-State Problems. *Phys. Rev.*, **84**(6) (1951) 1232–1242.

128 E. E. Salpeter. Mass Corrections to the Fine Structure of Hydrogen-Like Atoms. *Phys. Rev.*, **87**(2) (1952) 328–343.

129 A. A. Broyles. Relativistic equation for the multielectron atom. *Phys. Rev. A*, **38**(3) (1988) 1137–1148.

130 A. A. Broyles. The Derivation of the Relativistic Hamiltonian for Molecules. *Int. J. Quantum Chem., Quantum Chem. Symp.*, **29** (1995) 257–275.

131 G. Breit. The Fine Structure of He As a Test of the Spin Interactions of Two Electrons. *Phys. Rev.*, **36**(3) (1930) 383–397.

132 G. Breit. Dirac's Equation and the Spin-Spin Interactions of Two Electrons. *Phys. Rev.*, **39** (1932) 616–624.

133 J. A. Gaunt. The Triplets of Helium. *Proc. Roy. Soc. London A*, **122** (1929) 513–532.

134 J. A. Gaunt. IV. The Triplets of Helium. *Phil. Trans. Roy. Soc. (London)*, **A288** (1929) 151–196.

135 H. Bethe, E. Fermi. Über die Wechselwirkung von zwei Elektronen. *Z. Phys.*, **77** (1932) 296–306.

136 J. B. Mann, W. R. Johnson. Breit Interaction in Multielectron Atoms. *Phys. Rev. A*, **4**(1) (1971) 41–51.

137 T. Saue. Spin-Interactions and the Nonrelativistic Limit of Electrodynamics. *Adv. Quantum Chem.*, **48** (2005) 383–405.

138 I. P. Grant, N. C. Pyper. Breit interaction in multi-configuration relativistic atomic calculations. *J. Phys. B: Atom. Molec. Phys.*, **9**(5) (1976) 761–774.

139 I. P. Grant, B. J. McKenzie. The transverse electron-electron interaction in atomic-structure calculations. *J. Phys. B: At. Molec. Phys.*, **13**(14) (1980) 2671–2681.

140 J. Hata, I. P. Grant. Comments on relativistic 2-body interaction in atoms. *J. Phys. B: At. Mol. Phys.*, **17**(5) (1984) L107–L113.

141 B. P. Das, I. P. Grant. Multiconfiguration Dirac-Fock approach to the fine-structure splitting in the 3s3p configuration of the magnesium sequence. *J. Phys. B: At. Mol. Phys.*, **19** (1986) L7–L11.

142 H. M. Quiney, I. P. Grant, S. Wilson. The Dirac equation in the algebraic approximation: V. Self-consistent field studies including the Breit interaction. *J. Phys. B: At. Mol. Phys.*, **20** (1987) 1413–1422.

143 E. Lindroth, A.-M. Mårtensson-Pendrill. Further analysis of the complete Breit interaction. *Phys. Rev. A*, **39**(8) (1989) 3794–3802.

144 E. Lindroth, A.-M. Mårtensson-Pendrill, A. Ynnerman, P. Öster. Self-consistent treatment of the Breit interaction, with application to the electric dipole moment of thallium. *J. Phys. B: At. Mol. Opt. Phys.*, **22** (1989) 2447–2464.

145 Y. Ishikawa. Atomic Dirac–Fock–Breit Self-Consistent Field Calculations. *Int. J. Quantum Chem., Quantum Chem. Symp.*, **24** (1990) 383–391.

146 Y. Ishikawa. Dirac–Fock Gaussian Basis Calculations: Inclusion of the Breit Interaction in the Self-Consistent Field Procedure. *Chem. Phys. Lett.*, **166**(3) (1990) 321–325.

147 S. Okada, M. Shinada, O. Matsuoka. Relativistic well-tempered Gaussian basis sets for helium through mercury. Breit interaction included. *J. Chem. Phys.*, **93**(7) (1990) 5013–5019.

148 Y. Ishikawa, H. M. Quiney, G. L. Malli. Dirac–Fock–Breit self-consistent-field method: Gaussian basis-set calculations on many-electron atoms. *Phys. Rev. A*, **43**(7) (1991) 3270–3278.

149 Y. Ishikawa, G. L. Malli, A. J. Stacey. Matrix Dirac–Fock–Breit SCF calculations on heavy atoms using geometric basis sets of Gaussian functions. *Chem. Phys. Lett.*, **188**(1,2) (1992) 145–148.

150 A. K. Mohanty. A Dirac–Fock Self-Consistent Field Method for Closed-Shell Molecules Including Breit Interaction. *Int. J. Quantum Chem.*, **42** (1992) 627–662.

151 F. A. Parpia, A. K. Mohanty, E. Clementi. Relativistic calculations for atoms: self-consistent treatment of Breit interaction and nuclear volume effect. *J. Phys. B: At. Mol. Opt. Phys.*, **25** (1992) 1–16.

152 Y. Ishikawa, H. M. Quiney. Relativistic many-body perturbation-theory calculations based on Dirac–Fock–Breit wave equations. *Phys. Rev. A*, **47** (1993) 1732–1739.

153 Y. Ishikawa, K. Koc. Relativistic many-body perturbation theory based on the no-pair Dirac–Coulomb–Breit Hamiltonian: Relativistic correlation energies for the noble-gas sequence through Rn (Z=86), the group-IIB atoms through Hg, and the ions of Ne isoelectronic sequence. *Phys. Rev. A*, **50**(6) (1994) 4733–4742.

154 E. Eliav, U. Kaldor, Y. Ishikawa. Open-shell relativistic coupled-cluster method with Dirac–Fock–Breit wave functions : Energies of the gold atom and its cation. *Phys. Rev. A*, **49**(3) (1994) 1724–1729.

155 E. Eliav (Ilyabaev), U. Kaldor, Y. Ishikawa. Relativistic coupled cluster method based on Dirac–Coulomb–Breit wavefunctions. Ground state energies of atoms with two to five electrons. *Chem. Phys. Lett.*, **222** (1994) 82–87.

156 F. E. Jorge, A. B. F. da Silva. On the inclusion of the Breit interaction term in the closed-shell generator coordinate Dirac–Fock formalism. *J. Chem. Phys.*, **105**(13) (1996) 5503–5509.

157 G. L. Malli, J. Styszynski. Ab initio all-electron fully relativistic Dirac–Fock–Breit calculations for molecules of the superheavy transactinide elements: Rutherfordium tetrachloride. *J. Chem. Phys.*, **109**(11) (1998) 4448–4455.

158 M. Reiher, J. Hinze. Self-consistent treatment of the frequency-independent Breit interaction in Dirac–Fock and MCSCF calculations of atomic structures. I. Theoretical considerations. *J. Phys. B: At. Mol. Opt. Phys.*, **32** (1999) 5489–5505.

159 M. Reiher, C. Kind. Self-consistent treatment of the frequency-independent Breit interaction in Dirac–Fock calculations of atomic structures. II. He- and Be-like ions. *J. Phys. B: At. Mol. Opt. Phys.*, **34**(15) (2001) 3133–3156.

160 O. Visser, L. Visscher, P. J. Aerts, W. C. Nieuwpoort. Relativistic all-electron molecular Hartree–Fock–Dirac–(Breit) calculations on CH_4, SiH_4, GeH_4, SnH_4, PbH_4. *Theor. Chim. Acta*, **81** (1992) 405–416.

161 G. L. Malli, J. Styszynski. *Ab initio* all-electron Dirac–Fock–Breit calculations for UF_6. *J. Chem. Phys.*, **104**(3) (1996) 1012–1017.

162 J. Styszyński, X. Cao, G. L. Malli, L. Visscher. Relativistic All-Electron Dirac–Fock–Breit Calculations on Xenon Fluorides (XeF_n, $n = 1,2,4,6$). *Int. J. Quantum Chem.*, **18**(5) (1997) 601–608.

163 H. M. Quiney, H. Skaane, I. P. Grant. Relativistic, quantum electrodynamic and many-body effects in the water molecule. *Chem. Phys. Lett.*, **290** (1998) 473–480.

164 L. Rosenfeld. Bemerkung zur korrespondenzmäßigen Behandlung des relativistischen Mehrkörperproblems. *Z. Phys.*, **73** (1931) 253–259.

165 K. Nikolsky. The Interaction of Charges in Dirac's Theory. *Phys. Z. Sowjetunion*, **2** (1932-3) 447–452.

166 C. Møller. Über den Stoß zweier Teilchen unter Berücksichtigung der Retardierung der Kräfte. *Z. Phys.*, **70** (1931) 786–795.

167 C. Møller. Zur Theorie des Durchgangs schneller Elektronen durch Materie. *Ann. Phys. (Berlin; 5. Folge)*, **14** (1931) 531–585.

168 Y. Nambu. Force Potentials in Quantum Field Theory. *Prog. Theor. Phys.*, **5**(4) (1950) 614–633.

169 H. M. Quiney, I. P. Grant, S. Wilson. On the Relativistic Many-Body Perturbation

Theory of Atomic and Molecular Electronic Structure. p. 307–344, 1989.

170 G. E. Brown. XLII. Electron–Electron Interaction in Heavy Atoms. *Phil. Mag.*, **43** (1952) 467–471.

171 J. Sucher. On the choice of the electron–electron potential in relativistic atomic physics. *J. Phys. B: At. Mol. Opt. Phys.*, **21** (1988) L585–L591.

172 W. H. Furry. On Bound States and Scattering in Positron Theory. *Phys. Rev.*, **81**(1) (1951) 115–124.

173 W. H. E. Schwarz, E. Wechsel-Trakowski. The Two Problems Connected with Dirac–Breit–Roothaan Calculations. *Chem. Phys. Lett.*, **85**(1) (1982) 94–97.

174 G. E. Brown, D. G. Ravenhall. On the interaction of two electrons. *Proc. Roy. Soc. London A*, **208** (1951) 552–559.

175 M. H. Mittleman. Configuration-Space Hamiltonian for Heavy Atoms and Correction to the Breit Interaction. *Phys. Rev. A*.

176 J. Sucher. Foundations of the relativistic theory of many-electron atoms. *Phys. Rev. A*, **22**(2) (1980) 348–362.

177 M. H. Mittleman. Theory of relativistic effects on atoms: Configuration-space Hamiltonian. *Phys. Rev. A*.

178 J. Sucher. Relativistic Many-Electron Hamiltonians. *Phys. Scr.*, **36** (1987) 271–281.

179 P. Indelicato, J. P. Desclaux. Projection Operator in the Multiconfiguration Dirac–Fock Method. *Phys. Scr.*, **T46** (1993) 110–114.

180 P. Indelicato. Projection operators in multiconfiguration Dirac–Fock calculations: Application to the ground state of heliumlike ions. *Phys. Rev. A*, **51**(2) (1995) 1132–1145.

181 K. Pachucki. Simple derivation of helium Lamb shift. *J. Phys. B.: At. Mol. Opt. Phys.*, **31** (1998) 5123–5133.

182 G. Soff, P. J. Mohr. Vacuum polarization in a strong external-field. *Phys. Rev. A*, **38**(10) (1988) 5066–5075.

183 E. A. Uehling. Polarization Effects in the Positron Theory. *Phys. Rev.*, **48** (1935) 55–63.

184 E. H. Wichmann, N. H. Kroll. Vacuum Polarization in a Strong Coulomb Field. *Phys. Rev.*, **101**(2) (1956) 843–859.

185 L. W. Fullerton, G. A. Rinker, Jr. Accurate and efficient methods for the evaluation of vacuum-polarization potentials. *Phys. Rev. A*, **13**(3) (1976) 1283–1287.

186 M. Douglas, N. M. Kroll. Quantum Electrodynamical Corrections to the Fine Structure of Helium. *Ann. Phys.*, **82** (1974) 89–155.

187 W. R. Johnson, G. Soff. The Lamb Shift in Hydrogen-Like Atoms. *At. Data Nucl. Data Tables*, **33** (1985) 405–446.

188 W. R. Johnson, K. T. Cheng. *Relativistic and Quantum Electrodynamic Effects on Atomic Inner Shells (in: Atomic Inner-Shell Physics)*, p. 3–30. Physics of Atoms and Molecules. Plenum Press, New York, London, 1985.

189 P. Indelicato, E. Lindroth. Relativistic effects, correlation, and QED corrections on $K\alpha$ transitions in medium to very heavy atoms. *Phys. Rev. A*, **46**(5) (1992) 2426–2436.

190 I. Lindgren. Electron correlation and quantum electrodynamics. *Mol. Phys.*, **94**(1) (1998) 19–28.

191 I. Lindgren. A Relativistic Coupled-Cluster Approach with Radiative Corrections. In U. Kaldor, Ed., *Many-Body Methods in Quantum Chemistry*, Volume 52 of *Lecture Notes in Chemistry*, p. 293–306, Heidelberg, 1989. Springer-Verlag.

192 P. Indelicato, O. Gorceix, J. P. Desclaux. Multiconfigurational Dirac-Fock studies of two-electron ions: II. Radiative corrections and comparison with experiment. *J. Phys. B: At. Mol. Phys.*, **20** (1987) 651–663.

193 P. Indelicato, J. P. Desclaux. Multiconfiguration Dirac-Fock calculations of transition energies with QED corrections in three-electron ions. *Phys. Rev. A*, **42** (1990) 5139–5149.

194 P. Indelicato. Correlation and Negative Continuum Effects for the Relativistic *M*1 Transition in Two-Electron Ions using the Multiconfiguration Dirac–Fock Method. *Phys. Rev. Lett.*, **77**(16) (1996) 3323–3326.

195 L. Labzowsky, A. Nefiodov, G. Plunien, G. Soff, P. Pyykkö. Vacuum-polarization corrections to the hyperfine-structure splitting of highly charged $^{209}_{83}$Bi ions. *Phys. Rev. A*, **56**(6) (1997) 4508–4516.

196 T. Beier, P. J. Mohr, H. Persson, G. Plunien, M. Greiner, G. Soff. Current status of Lamb shift predictions for heavy hydrogen-like ions. *Phys. Lett. A*, **236**(4) (1997) 329–338.

197 P. Pyykkö, M. Tokman, L. N. Labzowsky. Estimated valence-level Lamb shifts for group 1 and group 11 metal atoms. *Phys. Rev. A*, **57**(7) (1998) R689–R692.

198 L. Labzowsky, I. Goidenko, M. Tokman, P. Pyykkö. Calculated self-energy contributions for an *ns* valence electron using the multiple-commutator method. *Phys. Rev. A*, **59**(4) (1999) 2707–2711.

199 I. Goidenko, L. Labzowsky, E. Eliav, U. Kaldor, P. Pyykkö. QED corrections to the binding energy of the eka-radon (Z=118) negative ion. *Phys. Rev. A*, **67** (2003) 020102(R).

200 T. H. Dinh, V. A. Dzuba, V. V. Flambaum, J. S. M. Ginges. Calculations of the spectra of superheavy elements Z=119 and Z=120$^+$. *Phys. Rev. A*, **78** (2008) 022507.

201 K. Pachucki. Quantum electrodynamics of weakly bound systems. *Hyperfine Int.*, **114** (1998) 55–70.

202 P. Pyykkö, K. G. Dyall, A. G. Császár, G. Tarczay, O. L. Polyansky, J. Tennyson. Estimation of Lamb-shift effects for molecules: Application to the rotation-vibration spectra of water. *Phys. Rev. A*, **63** (2001) 024502.

203 H. Köppel, W. Domcke, L. S. Cederbaum. Multimode molecular dynamics beyond the Born–Oppenheimer approximation. Volume 57 of *Adv. Chem. Phys.*, p. 59–246. John Wiley & Sons Ltd., Chichester, 1984.

204 W. Domcke, D. R. Yarkony, H. Köppel. *Conical Intersections: Electronic Structure, Dynamics & Spectroscopy*, Volume 15 of *Advanced Series in Physical Chemistry*. World Scientific, Singapore, 2004.

205 E. B. Wilson, J. C. Decius, P. C. Cross. *Molecular Vibrations*. McGraw-Hill, New York, 1955.

206 J. Neugebauer, M. Reiher, C. Kind, B. A. Hess. Quantum Chemical Calculation of Vibrational Spectra of Large Molecules — Raman and IR Spectra for Buckminsterfullerene. *J. Comput. Chem.*, **23** (2002) 895–910.

207 C. Herrmann, M. Reiher. First Principles Approach to Vibrational Spectroscopy of Biomolecules. *Top. Curr. Chem.*, **268** (2007) 85–132.

208 D. Marx, J. Hutter. *Ab initio* Molecular Dynamics: Theory and Implementation. In J. Grotendorst, Ed., *Modern Methods and Algorithms of Quantum Chemistry*, Volume 3 of *NIC Series*, p. 329–477. 2nd ed., 2000.

209 J. Thar, W. Reckien, B. Kirchner. Car–Parrinello molecular dynamics simulations and biological systems. *Top. Curr. Chem.*, **268** (2007) 133–171.

210 L. S. Cederbaum. The multistate vibronic coupling problem. *J. Chem. Phys.*, **78** (1983) 5714–5728.

211 L. Zülicke. *Quantenchemie, ein Lehrgang — Atombau, chemische Bindung und molekulare Wechselwirkungen*, Volume 2. Dr. Alfred Hüthig Verlag, Heidelberg, 1985.

212 S. R. White. Density Matrix Formulation for Quantum Renormalization Groups. *Phys. Rev. Lett.*, **69**(19) (1992) 2863–2866.

213 S. R. White. Density-matrix algorithms for quantum renormalization groups. *Phys. Rev. B*, **48**(14) (1993) 10345–10356.

214 G. K.-L. Chan, M. Head-Gordon. Highly correlated calculations with a polynomial cost algorithm: A study of the density matrix renormalizaton group. *J. Chem. Phys.*, **116**(11) (2002) 4462–4476.

215 U. Schollwöck. The density-matrix renormalization group. *Rev. Mod. Phys.*, **77** (2005) 259–315.

216 G. Moritz, M. Reiher. Decomposition of Density Matrix Renormalization Group States into a Slater Determinant Basis. *J. Chem. Phys.*, **126** (2007) 244109.

217 T. Helgaker, P. Jørgensen, J. Olsen. *Molecular Electronic-Structure Theory*. John Wiley & Sons, Chichester, England, 2000.

218 R. Pauncz. *Spin Eigenfunctions: Construction and use*. Plenum, New York, 1979.

219 R. Pauncz. *The Construction of Spin Eigen-Functions — An Exercise Book*. Springer, 2000.

220 S. R. Langhoff, E. R. Davidson. Configuration interaction calculations on the nitrogen molecule. *Int. J. Quantum Chem.*, **8** (1974) 61–72.

221 R. J. Buenker, S. D. Peyerimhoff. CI Method for the Study of General Molecular Potentials. *Theoret. Chim. Acta* (Berl.), **12** (1968) 183–199.

222 R. J. Buenker, S. D. Peyerimhoff. Individualized Configuration Selection in CI Calculations with Subsequent Energy Extrapolation. *Theoret. Chim. Acta* (Berl.), **35** (1974) 33–58.

223 J. Hinze. MC-SCF. I. The multi-configuration self-consistent-field method. *J. Chem. Phys.*, **59**(12) (1973) 6424–6432.

224 H.-J. Werner. Matrix-Formulated Direct Multiconfiguration Self-Consistent Field and Multiconfiguration Reference Configuration-Interaction Methods. *Adv. Chem. Phys.*, **69** (1987) 1–62.

225 R. Shepard. The Multiconfiguration Self-Consistent Field Method. *Adv. Chem. Phys.*, **69** (1987) 63–200.

226 B. O. Roos. The Complete Active Space Self-Consistent Field Method and Its Applications in Electronic Structure Calculations. *Adv. Chem. Phys.*, **69** (1987) 399–445.

227 K. Andersson, B. O. Roos. in: Modern Electronic Structure Theory, D. Yarkony (Ed.), p. 55. World Scientific, Singapore, 1995.

228 K. Andersson, P.-A. Malmqvist, B. O. Roos, A. J. Sadlej, K. J. Wolinski. Second order perturbation theory with a CASSCF reference function. *J. Phys. Chem.*, **94** (1990) 5483–5488.

229 K. Andersson, P.-Å. Malmqvist, B. O. Roos. Second-order perturbation theory with a complete active space self-consistent field reference function. *J. Chem. Phys.*, **96** (1992) 1218.

230 C. Møller, M. S. Plesset. Note on an Approximation Treatment for Many-Electron Systems. *Phys. Rev.*, **46** (1934) 618–622.

231 J. S. Binkley, J. A. Pople. Møller–Plesset Theory for Atomic Ground State Energies. *Int. J. Quantum Chem.*, **9** (1975) 229.

232 P. J. Knowles, K. Somasundram, N. C. Handy, K. Hirao. The Calculation of Higher-Order Energies in the Many-Body Perturbation Theory Series. *Chem. Phys. Lett.*, **113**(1) (1985) 8–12.

233 N. C. Handy, P. J. Knowles, K. Somasundram. On the convergence of the Møller-Plesset perturbation series. *Theor. Chim. Acta*, **68** (1985) 87–100.

234 L. R. N. C. H. R. H. Nobes, J. A. Pople, P. J. Knowles. Slow Convergence of the Møller–Plesset Perturbation Series: The Dissociation Energy of Hydrogen Cyanide and the Electron Affinity of the Cyano Radical. *Chem. Phys. Lett.*, **138** (1987) 481.

235 M. Esser. Role of time-reversal symmetry in the graphical representation of Gelfand basis sets within the relativistic CI approach. *Chem. Phys. Lett.*, **111**(1,2) (1984) 58–63.

236 M. Esser. Direct MRCI Method for the Calculation of Relativistic Many-Electron Wavefunction. I. General Formalism. *Int. J. Quantum Chem.*, **26** (1984) 313–338.

237 W. C. Nieuwpoort, P. J. C. Aerts, L. Visscher. Molecular Electronic Structure Calculations based on the Dirac–Coulomb–(Breit) Hamiltonian. p. 59–70, 1994.

238 L. Visscher, O. Visser, P. J. C. Aerts, H. Merenga, W. C. Nieuwpoort. Relativistic quantum chemistry: the MOLFDIR program package. *Comp. Phys. Comm.*, **81** (1994) 120–144.

239 L. Visscher, W. A. de Jong, O. Visser, P. J. C. Aerts, H. Merenga, W. C. Nieuwpoort. Relativistic Quantum Chemistry: the MOLFDIR program package. p. 169–218, 1995.

240 P. J. C. Aerts. *Towards Relativistic Quantum Chemistry — On the ab initio calculation of relativistic electron wave functions for molecules in the Hartree–Fock–Dirac approximation*. PhD thesis, Rijksuniversiteit te Groningen, Netherlands, 1986.

241 P. J. C. Aerts, W. C. Nieuwpoort. On the Use of Gaussian Basis Sets To Solve the Hartree–Fock–Dirac Equation. II. Application to Many-Electron Atomic and Molecular Systems. *Int. J. Quantum Chem., Quantum Chem. Symp.*, **19** (1986) 267–277.

242 O. Visser, P. J. C. Aerts, L. Visscher. Open Shell Relativistic Molecular Dirac–Hartree–Fock SCF-Program. In Wilson et al. [975], p. 185–195.

243 T. Fleig, J. Olsen, C. M. Marian. The Generalized Active Space Concept for the

Relativistic Treatment of Electron Correlation I. Kramers-restricted two-component configuration interaction. *J. Chem. Phys.*, **114**(11) (2001) 4775–4790.

244 J. Thyssen. *Development and Application of Methods for Correlated Relativistic Calculations of Molecular Properties*. PhD thesis, Department of Chemistry, University of Southern Denmark, Odense, Denmark, 2001.

245 T. Saue, K. Fægri, T. Helgaker, O. Gropen. Principles of direct 4-component relativistic SCF: application to caesium auride. *Mol. Phys.*, **91**(5) (1997) 937–950.

246 T. Saue, H. J. A. Jensen. Quaternion symmetry in relativistic molecular calculations: The Dirac–Hartree–Fock method. *J. Chem. Phys.*, **111**(14) (1999) 6211–6222.

247 L. Visscher, T. Saue. Approximate relativistic electronic structure methods based on the quaternion modified Dirac equation. *J. Chem. Phys.*, **113**(10) (2000) 3996–4002.

248 E. A. Hylleraas. Über den Grundzustand des Heliumatoms. *Z. Phys.*, **48** (1928) 469–494.

249 E. A. Hylleraas. Neue Berechnung der Energie des Heliums im Grundzustande, sowie des tiefsten Terms von Ortho-Helium. *Z. Phys.*, **54** (1929) 347–366.

250 E. A. Hylleraas. Über den Grundterm der Zweielektronenprobleme von H^-, He, Li^+, Be^{++} usw. *Zeitschr. f. Phys.*, **65** (1930) 209–225.

251 R. Jastrow. Many-Body Problem with Strong Forces. *Phys. Rev.*, **98**(5) (1955) 1479–1484.

252 W. Klopper, W. Kutzelnigg. Møller-Plesset Calculations Taking Care of the Correlation Cusp. *Chem. Phys. Lett.*, **134**(1) (1987) 17–22.

253 W. Kutzelnigg, W. Klopper. Wave functions with terms linear in the interelectronic coordinates to take care of the correlation cusp. I. General theory. *J. Chem. Phys.*, **94** (1991) 1985–2001.

254 A. J. May, E. Valeev, R. Polly, F. R. Manby. Analysis of the errors in explicitly correlated electronic structure theory. *Phys. Chem. Chem. Phys.*, **7** (2005) 2710–2713.

255 V. Termath, W. Klopper, W. Kutzelnigg. Wave functions with terms linear in the interelectronic coordinates to take care of the correlation cusp. III. Second-order Møller-Plesset (MP2-R12) calculations on closed-shell atoms. *J. Chem. Phys.*, **94** (1991) 2002–2019.

256 W. Klopper, W. Kutzelnigg. Wave functions with terms linear in the interelectronic coordinates to take care of the correlation cusp. III. Second-order Møller-Plesset (MP2-R12) calculations on molecules of first two atoms. *J. Chem. Phys.*, **94** (1991) 2020–2030.

257 J. Noga, W. Kutzelnigg, W. Klopper. CC-R12, a correlation cusp corrected coupled-cluster method with a pilot application to the Be_2 potential curve. *Chem. Phys. Lett.*, **199**(5) (1992) 497–504.

258 R. Röhse, W. Klopper, W. Kutzelnigg. Configuration interaction calculations with terms linear in the interelectronic coordinate for the ground state of H_3^+. A benchmark study. *J. Chem. Phys.*, **99**(11) (1993) 8830–8839.

259 J. Noga, W. Kutzelnigg. Coupled cluster theory that takes care of the correlation cusp by inclusion of linear terms in the interelectronic coordinates. *J. Chem. Phys.*, **101**(9) (1994) 7738–7762.

260 S. Ten-no. Initiation of explicitly correlated Slater-type geminal theory. *Chem. Phys. Lett.*, **398**(1-3) (2004) 56–61.

261 D. P. Tew, W. Klopper. New correlation factors for explicitly correlated electronic wave functions. *J. Chem. Phys.*, **123** (2005) 074101.

262 D. P. Tew, W. Klopper, C. Neiss, C. Hättig. Quintuple-ζ quality coupled-cluster correlation energies with triple-ζ basis sets. *Phys. Chem. Chem. Phys.*, **9** (2007) 1921–1930.

263 H.-J. Werner, T. B. Adler, F. R. Manby. General orbital invariant MP2-F12 theory. *J. Chem. Phys.*, **126** (2007) 164102.

264 S. Bubin, L. Adamowicz. Matrix elements of N-particle explicitly correlated Gaussian basis functions with complex exponential parameters. *J. Chem. Phys.*, **124** (2006) 224317.

265 S. Bubin, L. Adamowicz. Energy and energy gradient matrix elements with N-particle explicitly correlated Gaussian basis functions with $L = 1$. *J. Chem. Phys.*, **128** (2008) 114107.

266 Y. Suzuki, K. Varga. *Stochastic Variational Approach to Quantum-Mechanical Few-Body Problems*. Lecture Notes in Physics. Springer-Verlag, Berlin, 1998.

267 M. Cafiero, S. Bubin, L. Adamowicz. Non-Born-Oppenheimer calculations of atoms and molecules. *Phys. Chem. Chem. Phys.*, **5** (2003) 1491–1501.

268 A. D. Bochevarov, E. F. Valeev, C. D. Sherrill. The electron and nuclear orbitals model: current challenges and future prospects. *Mol. Phys.*, **102** (2004) 111–123.

269 S. P. Webb, T. Iordanov, S. Hammes-Schiffer. Multiconfigurational nuclear-electronic orbital approach: Incorporation of nuclear quantum effects in electronic structure calculations. *J. Chem. Phys.*, **117**(9) (2002) 4106–4118.

270 A. Reyes, M. V. Pak, S. Hammes-Schiffer. Investigation of isotope effects with the nuclear-electronic orbital approach. *J. Chem. Phys.*, **123** (2005) 064104.

271 C. Swalina, M. V. Pak, A. Chakraborty, S. Hammes-Schiffer. Explicit Dynamical Electron–Proton Correlation in the Nuclear–Electronic Orbital Framework. *J. Phys. Chem. A*, **110** (2006) 9983–9987.

272 M. V. Pak, A. Chakraborty, S. Hammes-Schiffer. Density Functional Theory Treatment of Electron Correlation in the Nuclear–Electronic Orbital Approach. *J. Phys. Chem. A*, **111** (2007) 4522–4526.

273 S. Bubin, M. Cafiero, L. Adamowicz. Matrix elements of N-particle explicitly correlated Gaussian basis functions with complex exponential parameters. *Adv. Chem. Phys.*, **131** (2005) 377–475.

274 F. L. Pilar. *Elementary Quantum Chemistry*. Chemistry Series. McGraw-Hill International Editions, Singapore, 2nd ed., 1990.

275 F. A. Berezin. *The Method of Second Quantization*, Volume 24 of *Pure and Applied Physics*. Academic Press, 1966.

276 H. C. Longuet-Higgins. Second Quantization in the Electronic Theory of Molecules. p. 105–121, New York, 1966. Academic Press.

277 J. Hinze and F. Biegler-König. *Numerical Relativistic and Non-Relativistic MCSCF for Atoms and Molecules*, p. 405–446. Elsevier Scientific Publishing Company, Amsterdam, 1990.

278 P. Jørgensen, J. Simons. *Second Quantization-based Methods in Quantum Chemistry*. Academic Press, 1982.

279 R. McWeeny. *Methods of Molecular Quantum Mechanics*. Academic Press, New York, 2nd ed., 1992.

280 J. D. Talman. Minimax Principle for the Dirac Equation. *Phys. Rev. Lett.*, **57**(9) (1986) 1091–1094.

281 A. Kołakowska. Application of the minimax principle to the Dirac–Coulomb problem. *J. Phys. B: At. Mol. Opt. Phys.*, **29** (1996) 4515–4527.

282 M. Griesemer, H. Siedentop. A Minimax Principle for the Eigenvalues in Spectral Gaps. *J. London Math. Soc.*, **60**(2) (1999) 490–500.

283 M. Griesemer, R. T. Lewis, H. Siedentop. A Minimax Principle for Eigenvalues in Spectral Gaps: Dirac Operators with Coulomb Potentials. *Doc. Math.*, **4** (1999) 275–283.

284 M. J. Esteban, E. Séré. Solutions of the Dirac–Fock Equations for Atoms and Molecules. *Comm. Math. Phys.*, **203**(3) (1999) 499–530.

285 J. Dolbeault, M. J. Esteban, E. Séré. Variational methods in relativistic quantum mechanics: new approach to the computation of Dirac eigenvalues. p. 211–226, 2000.

286 J. Dolbeaut, M. J. Esteban, E. Séré, M. Vanbreugel. Minimization Methods for the One-Particle Dirac Equation. *Phys. Rev. Lett.*, **85**(19) (2000) 4020–4023.

287 J. Dolbeault, M. J. Esteban, E. Séré. On the eigenvalues of operators with gaps. Application to Dirac operators. *J. Funct. Anal.*, **174** (2000) 208–226.

288 J. Dolbeault, M. J. Esteban, E. Séré. Variational characterization for eigenvalues of Dirac operators. *Cal. Var.*, **10** (2000) 321–347.

289 M. J. Esteban, E. Séré. A max-min principle for the ground state of the Dirac–Fock functional. *Contemp. mathem.*, **307** (2002) 135–139.

290 M. Huber, H. Siedentop. Solutions of the Dirac-Fock equations and the energy of the electron-positron field. *Arch. Rat. Mech. Anal.*, **184**(1) (2007) 1–22.

291 M. Moshinsky, A. Sharma. Variational energy spectra of relativistic Hamiltonians. *J. Phys. A: Math. Gen.*, **31** (1998) 397–408.

292 P. Falsaperla, G. Fonte, J. Z. Chen. Two methods for solving the Dirac equation without variational collapse. *Phys. Rev. A*, **56**(2) (1997) 1240–1248.

293 R. Szmytkowski. The Dirac–Coulomb Sturmians and the Series Expansion of the Dirac–Coulomb Green Functions: Application to the Relativistic Polarizability of the Hydrogen Like Atoms. *J. Phys. B: At. Mol. Opt. Phys.*, **30** (1997) 825–861.

294 R. Szmytkowski. Variational R-Matrix Methods for Many-Electron Systems: Unified Relativistic Theory. *Phys. Rev. A*, **63** (2001) 062704–1–14.

295 C. Lanczos. An Iteration Method for the Solution of the Eigenvalue Problem of Linear Differential and Integral Operators (Research Paper 2133). *J. Res. Nat. Bur. Stand.*, **45**(4) (1950) 255.

296 E. R. Davidson. The Iterative Calculation of a Few of the Lowest Eigenvalues and Corresponding Eigenvectors of Large Real-Symmetric Matrices. *J. Comput. Phys.*, **17** (1975) 87–94.

297 C. W. Murray, S. C. Racine, E. R. Davidson. Improved Algorithms for the Lowest Few Eigenvalues and Associated Eigenvectors of Large Matrices. *J. Comp. Phys.*, **103** (1992) 382–389.

298 H. J. J. Van Dam, J. H. Van Lenthe, G. L. G. Sleijpen, H. A. van der Vorst. An improvement of Davidson's iteration method: Applications to MRCI and MRCEPA calculations. *J. Comput. Chem.*, **17**(3) (1996) 267–272.

299 H. A. van der Vorst, G. L. G. Sleijpen. *A parallelizable and fast algorithm for very large generalized eigenproblems (in: Applied Parallel Computing, Proceedings of PARA'96)*, Volume 1184 of *Lect. Notes Comp. Science*, p. 686–696. Springer-Verlag, Berlin, 1996.

300 H. A. V. der Vorst, T. F. Chan. *Linear System Solvers: Sparse Iterative Methods (in: Parallel Numerical Algorithms)*, Volume 4 of *ICASE/LaRC Interdisc. Ser. in Science and Engineering*, p. 167–202. Kluwer Academic, Dordrecht, 1997.

301 G. H. Golub, H. A. van der Vorst. *Closer to the solution: Iterative linear solvers (in: The State of the Art in Numerical Analysis)*, p. 63–92. Clarendon Press, Oxford, 1997.

302 H. A. van der Vorst, G. H. Golub. *150 Years old and still alive: eigenproblems (in: The State of the Art in Numerical Analysis)*, p. 93–119. Clarendon Press, Oxford, 1997.

303 H. Primas, U. Müller-Herold. *Elementare Quantenchemie*. Teubner Studienbücher. B. G. Teubner, Stuttgart, 1984.

304 P. Pulay. Convergence Acceleration of Iterative Sequences. The Case of SCF Iteration. *Chem. Phys. Lett.*, **73**(2) (19) 393–398.

305 P. Pulay. Improved SCF Convergence Acceleration. *J. Comput. Chem.*, **3** (1982) 556–560.

306 T. H. Fischer, J. Almlöf. General Methods for Geometry and Wave Function Optimization. *J. Phys. Chem.*, **96** (1992) 9768–9774.

307 R. G. Parr, W. Yang. *Density-Functional Theory of Atoms and Molecules*, Volume 16 of *International Series of Monographs on Chemistry*. Oxford Science Publications, New York, 1989.

308 R. M. Dreizler, E. K. U. Gross. *Density Functional Theory — An Approach to the Quantum Many-Body Problem*. Springer-Verlag, Heidelberg, 1990.

309 H. Eschrig. *The Fundamentals of Density Functional Theory*, Volume 32 of *Teubner-Texte zur Physik*. B. G. Teubner Verlagsgesellschaft, Stuttgart, Leipzig, 1996.

310 P. Hohenberg, W. Kohn. Inhomogeneous Electron Gas. *Phys. Rev.*, **136**(3B) (1964) B864–B871.

311 W. Koch, M. C. Holthausen. *A Chemist's Guide to Density Functional Theory*. Wiley-VCH, Weinheim, 2000.

312 E. Engel, S. Keller, R. M. Dreizler. Generalized gradient approximation for the relativistic exchange-only energy functional. *Phys. Rev. A*, **53**(3) (1996) 1367–1374.

313 E. Engel. *Relativistic Density Functional Theory: Foundations and Basic Formalism*, p. 523–621. Theoretical and Computational Chemistry. Elsevier, Amsterdam, 2002.

314 J. J. Ladik. Viewpoint: Four-component Dirac–Hartree–Fock equations for solids; generalization of the relativistic Hartree–Fock equations. *J. Mol. Struc. (THEOCHEM)*, **391** (1997) 1–14.

315 V. Theileis, H. Bross. Relativistic modified augmented plane wave method and its application to the electronic structure of gold and platinum. *Phys. Rev. B*, **62**(20) (2000) 13338–13346.

316 A. K. Rajagopal, J. Callaway. Inhomogeneous Electron Gas. *Phys. Rev. B*, **7** (1973) 1912–1919.

317 W. Kohn, L. J. Sham. Self-Consistent Equations Including Exchange and Correlation Effects. *Phys. Rev.*, **140**(4A) (1965) A1133–A1138.

318 A. H. MacDonald, S. H. Vosko. A relativistic density functional formalism. *J. Phys. C: Solid State Phys.*, **12** (1979) 2977–2990.

319 T. Saue, T. Helgaker. Four-Component Relativistic Kohn–Sham Theory. *J. Comput. Chem.*, **23** (2002) 814–823.

320 C. van Wüllen. Spin densities in two-component relativistic density functional calculations: Noncollinear versus collinear approach. *J. Comput. Chem.*, **23** (2002) 779–785.

321 J. Cizek. On the Correlation Problem in Atomic and Molecular Systems. Calculation of Wavefunction Components in Ursell-Type Expansion Using Quantum-Field Theoretical Methods. *J. Chem. Phys.*, **45** (1966) 4256.

322 I. Lindgren, J. Morrison. *Atomic Many-Body Theory*, Volume 3 of *Springer Series on Atoms and Plasmas*. Springer-Verlag, Berlin, Heidelberg, 2nd ed., 1986.

323 I. Lindgren. The Coupled-Cluster Approach to Non-relativistic and Relativistic Many-Body Calculations. *Phys. Scr.*, **36** (1987) 591–601.

324 P. R. Taylor. *Coupled-cluster Methods in Quantum Chemistry*, Volume 64 of *Lecture Notes in Chemistry*, Chapter 3, p. 125–202. Springer-Verlag, Berlin, 1994.

325 L. Visscher, K. G. Dyall, T. J. Lee. Kramers-Restricted Closed-Shell CCSD Theory. *Int. J. Quantum Chem. Quantum Chem. Symp.*, **29** (1995) 411–419.

326 L. Visscher, T. J. Lee, K. G. Dyall. Formulation and implementation of a relativistic unrestricted coupled-cluster method including noniterative connected triples. *J. Chem. Phys.*, **105**(19) (1996) 8769–8776.

327 D. Mukherjee, S. Pal. Use of Cluster expansion methods in the open shell correlation problem. *Adv. Quantum Chem.*, **20** (1989) 291–373.

328 U. Kaldor. The Fock Space Coupled Cluster Method: Theory and Application. *Theor. Chim. Acta*, **80** (1991) 427–439.

329 U. Kaldor, E. Eliav, A. Landau. in: Relativistic Quantum Chemistry — Part II. Applications, P. Schwerdtfeger (Ed.), p. 81. Elsevier, Amsterdam, 2004.

330 E. Ilyabaev, U. Kaldor. The relativistic open shell coupled cluster method: Direct calculation of excitation energies in the Ne atom. *J. Chem. Phys.*, **97**(11) (1992) 8455–8458.

331 E. Ilyabaev, U. Kaldor. Relativistic coupled cluster calculations for closed-shell atoms. *Chem. Phys. Lett.*, **194**(1-2) (1992) 95–98.

332 E. Eliav, U. Kaldor, Y. Ishikawa. Ground State Electron Configuration of Rutherfordium: Role of Dynamic Correlation. *Phys. Rev. Lett.*, **74**(7) (1995) 1079–1082.

333 E. Eliav, U. Kaldor, Y. Ishikawa. Transition energies of barium and radium by the relativistic coupled-cluster method. *Phys. Rev. A*, **53**(5) (1996) 3050–3056.

334 E. Eliav, U. Kaldor. The relativistic four-component coupled cluster method for molecules: spectroscopic constants of SnH_4. *Chem. Phys. Lett.*, **248** (1996) 405–408.

335 E. Eliav, U. Kaldor, B. A. Hess. The relativistic Fock-space coupled-cluster method for molecules: CdH and its ions. *J. Chem. Phys.*, **108**(9) (1998) 3409–3415.

336 A. Landau, E. Eliav, U. Kaldor. Intermediate Hamiltonian Fock-space coupled-cluster method. *Chem. Phys. Lett.*, **313** (1999) 399–403.

337 A. Landau, E. Eliav, U. Kaldor. Intermediate Hamiltonian Fock-space coupled-cluster method. *Adv. Quantum Chem.*, **39** (2001) 171–188.

338 A. L. Fetter, J. D. Walecka. *Quantum Theory of Many-Particle Systems.* McGraw-Hill, New York, 1971.

339 A. Hibbert. Developments in atomic structure calculations. *Rep. Prog. Phys.*, **38** (1975) 1217–1338.

340 I. P. Grant. Relativistic Electronic Structure of Atoms and Molecules. *Adv. At. Mol. Opt. Phys.*, **32** (1994) 169–186.

341 G. Racah. Theory of Complex Spectra. I. *Phys. Rev.*, **61** (1942) 186–197.

342 G. Racah. Theory of Complex Spectra. II. *Phys. Rev.*, **62** (1942) 438–462.

343 G. Racah. Theory of Complex Spectra. III. *Phys. Rev.*, **63** (1943) 367–382.

344 G. Racah. Theory of Complex Spectra. IV. *Phys. Rev.*, **76**(9) (1949) 1352–1365.

345 I. P. Grant. Relativistic self-consistent fields. *Proc. Roy. Soc. London A*, **262** (1961) 555–576.

346 I. P. Grant. Relativistic self-consistent fields. *Proc. Phys. Soc.*, **86** (1965) 523–527.

347 F. C. Smith, W. R. Johnson. Relativistic Self-Consistent Fields with Exchange. *Phys. Rev.*, **160**(1) (1967) 136–142.

348 J. T. Waber, D. T. Cromer, D. Liberman. SCF Dirac-Slater Calculations of the Translawrencium Elements. *J. Chem. Phys.*, **51**(2) (1969) 664–668.

349 J. B. Mann, J. T. Waber. SCF Relativistic Hartree-Fock Calculations on the Superheavy Elements 118-131. *J. Chem. Phys.*, **53**(6) (1970) 2397–2406.

350 J. B. Mann, J. T. Waber. Self-consistent Relativistic Dirac-Hartree-Fock Calculations of Lanthanide Atoms. *Atomic Data*, **5**(2) (1973) 201–229.

351 J. P. Desclaux. Relativistic Dirac–Fock expectation values for atoms with Z=1 to Z=120. *At. Data Nucl. Data Tables*, **12**(4) (1973) 311–406.

352 L. Visscher, K. G. Dyall. Dirac-Fock Atomic Electronic Structure Calculations Using Different Nuclear Charge Distributions. *At. Data Nucl. Data Tables*, **67** (1997) 207–224.

353 J. P. Desclaux. A Multiconfiguration Relativistic Dirac–Fock Program. *Comp. Phys. Commun.*, **9** (1975) 31–45.

354 J. P. Desclaux. *Relativistic Multiconfiguration Dirac–Fock Package*, Volume A – Small Systems, p. 253–274. STEF, Cagliari, 1993.

355 J. P. Desclaux, P. Indelicato. *The relativistic atomic program MCDFGME V 2005.10.* Published at http://dirac.spectro.jussieu.fr/mcdf/ on August 17, 2005.

356 I. P. Grant, B. J. McKenzie, P. H. Norrington, D. F. Mayers, N. C. Pyper. An Atomic Multiconfigurational Dirac-Fock Package. *Comp. Phys. Commun.*, **21** (1980) 207–231.

357 B. J. McKenzie, I. P. Grant, P. H. Norrington. A Program to Calculate Transverse Breit and QED Corrections to Energy Levels in a Multiconfiguration Dirac-Fock Environment. *Comp. Phys. Commun.*, **21** (1980) 233–246.

358 K. G. Dyall, I. P. Grant, C. T. Johnson, F. A. Parpia, E. P. Plummer. GRASP: A general-purpose relativistic atomic structure program. *Comp. Phys. Commun.*, **55** (1989) 425–456.

359 J. P. Desclaux, D. F. Mayers, F. O'Brien. Relativistic atomic wave functions. *J. Phys. B: Atom. Molec. Phys.*, **4** (1971) 631–642.

360 F. A. Parpia, C. F. Fischer, I. P. Grant. GRASP92: a package for large-scale relativistic atomic structure calculations. *Comp. Phys. Comm.*, **94** (1996) 249–271.

361 H. M. Quiney, I. P. Grant, S. Wilson. Relativistic many-body perturbation theory using analytic basis functions. *J. Phys. B: At. Mol. Opt. Phys.*, **23** (1990) L271–L278.

362 F. A. Parpia, A. K. Mohanty. Relativistic basis-set calculations for atoms with Fermi nuclei. *Phys. Rev. A*, **46**(7) (1992) 3735–3745.

363 E. Ilyabaev, U. Kaldor. Relativistic coupled-cluster calculations for open-shell atoms. *Phys. Rev. A*, **47**(1) (1993) 137–142.

364 E. Eliav, U. Kaldor, P. Schwerdtfeger, B. A. Hess, Y. Ishikawa. Ground State Electron Configuration of Element 111. *Phys. Rev. Lett.*, **73**(24) (1994) 3203–3206.

365 E. Eliav, U. Kaldor, Y. Ishikawa. Relativistic Coupled Cluster Theory Based on the No-Pair Dirac–Coulomb–Breit Hamiltonian: Relativistic Pair Correlation Energies of the Xe Atom. *Int. J. Quantum Chem. Quantum Chem. Symp.*, **28** (1994) 205–214.

366 D. Sundholm, P. Pyykkö, L. Laaksonen. Two-Dimensional Fully Numerical Solutions of Second-order Dirac Equations for Diatomic Molecules, Part 3. *Phys. Scr.*, **36** (1987) 400–402.

367 D. Sundholm. Two-Dimensional, Fully Numerical Solution of Molecular Dirac Equations. Dirac–Slater Calculations on LiH, Li_2, BH and CH^+. *Chem. Phys. Lett.*, **149**(3) (1987) 251–256.

368 L. Yang, D. Heinemann, D. Kolb. Relativistic self-consistent calculations for small diatomic molecules by the finite element method. *Chem. Phys. Lett.*, **192**(5,6) (1992) 499–502.

369 D. Sundholm. Fully numerical solutions of molecular Dirac equations for highly charged one-electron homonuclear diatomic molecules. *Chem. Phys. Lett.*, **223** (1994) 469–473.

370 C. Düsterhöft, L. Yang, D. Heinemann, D. Kolb. Solution of the one-electron Dirac equation for the heavy diatomic quasi-molecule NiPb(109+) by the finite element method. *Chem. Phys. Lett.*, **229** (1994) 667–670.

371 C. Düsterhöft, D. Heinemann, D. Kolb. Dirac–Fock–Slater calculations for diatomic molecules with a finite element defect correction method (FEM-DKM). *Chem. Phys. Lett.*, **296** (1998) 77–83.

372 A. v. Kopylow, D. Heinemann, D. Kolb. Kohn-Sham density functionals accurately solved by a Finite Element Multigrid Method (FEM-MG) for lighter atoms and diatomic molecules. *J. Phys. B: At. Mol. Opt. Phys.*, **31** (1998) 4743–4754.

373 A. v. Kopylow, D. Kolb. Accurate finite-element multi-grid (FEM-MG) description for angular momentum and spin dependencies of Kohn-Sham density functionals for axially restricted calculations on first-row atoms and dimers. *Chem. Phys. Lett.*, **295** (1998) 439–446.

374 O. Kullie, C. Duesterhoeft, D. Kolb. Dirac–Fock Finite-Element-Method (FEM) calculations for some diatomic molecules. *Chem. Phys. Lett.*, **314** (1999) 307–310.

375 A. D. Becke, R. M. Dickson. Numerical solution of Poisson's equation in polyatomic molecules. *J. Chem. Phys.*, **89**(5) (1988) 2993–2997.

376 A. D. Becke, R. M. Dickson. Numerical solution of Schrödinger's equation in polyatomic molecules. *J. Chem. Phys.*, **92**(6) (1990) 3610–3612.

377 J. D. Talman. Variationally Optimized Numerical Orbitals for Molecular Calculations. *Phys. Rev. Lett.*, **84**(5) (2000) 855–858.

378 D. Andrae. Numerical self-consistent field method for polyatomic molecules. *Mol. Phys.*, **99**(4) (2001) 327–334.

379 T. Shiozaki, S. Hirata. Grid-based numerical Hartree-Fock solutions of polyatomic molecules. *Phys. Rev. A*, **76**(4) (2007) 040503.

380 R. Courant, D. Hilbert. *Methods of Mathematical Physics*. Wiley-Interscience, New York, 1989.

381 M. Abramowitz, I. A. Stegun. *Handbook of Mathematical Functions*. Dover Publications, New York, 9th ed., 1972.

382 R. D. Cowan. *The Theory of Atomic Structure and Spectra*. Los Alamos Series in Basic and Applied Sciences. University of California Press, Berkeley, Los Angeles, London, 1981.

383 B. L. Silver. *Irreducible Tensor Methods — An Introduction for Chemists*, Volume 36 of *Physical Chemistry*. Academic Press, New York, San Francisco, London, 1976.

384 U. Fano, G. Racah. *Irreducible Tensorial Sets*, Volume 4 of *Pure and Applied Physics*. Academic Press Inc., New York, 1959.

385 Z. Rudzikas. *Theoretical Atomic Spectroscopy*, Volume 7 of *Cambridge Monographs on Atomic, Molecular and Chemical Physics*. Cambridge University Press, Cambridge UK, 1997.

386 Y.-K. Kim, F. Parente, J. P. Marques, P. Indelicato, J. P. Desclaux. Failure of multiconfiguration Dirac–Fock wave functions in the nonrelativistic limit. *Phys. Rev. A*, **58**(3) (1998) 1885–1888.

387 P. Indelicato, E. Lindroth, J. P. Desclaux. Nonrelativistic limit of Dirac-Fock codes: The role of Brillouin configurations. *Phys. Rev. Lett.*, **94** (2005) 013002.

388 C. Froese Fischer. *The Hartree–Fock Method for Atoms*. John Wiley & Sons, New York, 1st ed., 1977.

389 C. Froese Fischer, T. Brage, P. Jönsson. *Computational Atomic Structure — An MCHF Approach*. Institute of Physics Publishing, Bristol, Philadelphia, 1997.

390 F. V. Atkinson. *Multiparameter Eigenvalue Problems*. Academic Press, New York, 1982.

391 H. Volkmer. *Multiparameter Eigenvalue Problems and Expansion Theorems*. Springer-Verlag, Berlin, 1988.

392 D. R. Hartree. Representation of the Exchange Terms in Fock's Equations by a Quasi-Potential. *Phys. Rev.*, **109**(3) (1958) 840–841.

393 J. C. Slater. *Quantum Theorie of Matter*. McGraw-Hill, New York, 1968.

394 G. Eickerling, M. Reiher. The shell structure of atoms. *J. Chem. Theory Comput.*, **4** (2008) 286–296.

395 T. Kato. On the Eigenfunctions of Many-Particle Systems in Quantum Mechanics. *Comm. Pure Appl. Math.*, **10** (1957) 151–177.

396 W. Kutzelnigg. Generalization of Kato's Cusp Conditions to the Relativistic Case. In D. Mukherjee, Ed., *Aspects of Many-Body Effects in Molecules and Extended Systems*, Volume 50 of *Lecture Notes in Chemistry*, p. 353–366, Berlin, Heidelberg, 1989. Springer-Verlag.

397 J. Neugebauer, M. Reiher, J. Hinze. Analysis of the asymptotic and short-range behavior of quasilocal Hartree–Fock and Dirac–Fock–Coulomb electron–electron interaction potentials. *Phys. Rev. A*, **65** (2002) 032518.

398 R. Latter. Atomic Energy Levels for the Thomas-Fermi and Thomas-Fermi-Dirac Potential. *Phys. Rev.*, **99**(2) (1955) 510–519.

399 J. D. Talman, W. F. Shadwick. Optimized effective atomic central potential. *Phys. Rev. A*, **14**(1) (1976) 36–40.

400 C.-O. Almbladh, U. von Barth. Exact results for the charge and spin densities, exchange-correlation potentials, and density-functional eigenvalues. *Phys. Rev. B*, **31**(6) (1985) 3231–3244.

401 J. Chen, J. B. Krieger, R. O. Esquivel, M. J. Stott, G. J. Iafrate. Kohn-Sham effective potentials for spin-polarized atomic systems. *Phys. Rev. A*, **54**(3) (1996) 1910–1921.

402 G. J. Laming, V. Termath, N. C. Handy. A general purpose exchange–correlation energy functional. *J. Chem. Phys.*, **99**(11) (1993) 8765–8773.

403 R. Neumann, N. C. Handy. Investigations using the Becke–Roussel exchange functional. *Chem. Phys. Lett.*, **246** (1995) 381–386.

404 D. Andrae, J. Hinze. Numerical electronic structure calculations for atoms. I. Generalized variable transformation and nonrelativistic calculations. *Int. J. Quantum Chem.*, **63** (1997) 65–91.

405 D. Andrae, M. Reiher, J. Hinze. Numerical electronic structure calculations for atoms. II. The generalized variable transformation in relativistic calculations. *Int. J. Quantum Chem.*, **76** (2000) 473–499.

406 D. R. Hartree. *The Calculation of Atomic Structures*. Structure of Matter Series. John Wiley & Sons, New York, 1st ed., 1957.

407 C.-E. Fröberg. *Numerical Mathematics - Theory and Computer Applications*. The Benjamin/Cummings Publishing Company, Menlo Park, 1st ed., 1985.

408 W. H. Press, S. A. Teukolsky, W. T. Vetterling, B. P. Flannery. *Numerical Recipes in Fortran - The Art of Scientific Computing*. Cambridge University Press, Cambridge, 2nd ed., 1992.

409 L. Collatz. *The Numerical Treatment of Differential Equations*, Volume 60 of *Die Grundlehren der mathematischen Wissenschaften in Einzeldarstellungen*. Springer-Verlag, Berlin, 3rd ed., 1966.

410 L. F. Richardson. The Approximate Arithmetical Solution by Finite Differences of Physical Problems involving Differential Equations, with an Application to the Stresses in a Masonry Dam. *Phil. Trans. Roy. Soc. London A*, **210** (1910) 307–357.

411 W. G. Bickley. Formulae for Numerical Differentiation. *Mathematical Gazette*, **25** (1941) 19–27.

412 B. Noumeroff. Méthode nouvelle de la détermination des orbites et le calcul des éphémérides en tenant compte des perturbations. *Publications de L'Observatoire Astrophysique Central de Russie*, **II** (1923) 188–288.

413 T. T. Nguyen-Dang, S. Durocher, O. Atabek. Direct Numerical Integrations of Coupled Equations With Non-Adiabatic Interactions. *Chem. Phys.*, **129** (1989) 451–462.

414 J. Stoer, R. Bulirsch. *Einführung in die Numerische Mathematik II*, Volume 114 of *Heidelberger Taschenbücher*. Springer-Verlag, Heidelberg, 1973.

415 P. Deuflhard, A. Hohmann. *Numerische Mathematik — Eine algorithmisch orientierte Einführung*. Walter de Gruyter, Berlin, 1991.

416 J. Neugebauer, M. Reiher, J. Hinze. Analytical local electron–electron interaction model potentials for atoms. *Phys. Rev. A*, **66** (2002) 022717.

417 U. I. Safronova, W. R. Johnson, H. G. Berry. Excitation energies and transition rates in magnesiumlike ions. *Phys. Rev. A*, **61** (2000) 052503.

418 L. J. Curtis, P. S. Ramanujam. Isoelectronic wavelength predictions for magnetic-dipole, electric-quadrupole, and intercombination transitions in the Mg sequence. *J. Opt. Soc. Am.*, **73**(8) (1983) 979–984.

419 J. O. Ekberg, U. Feldman, J. F. Seely, C. M. Brown. Transitions and Energy Levels in Mg-like Ge XXI-Zr XXIX Observed in Laser-Produced Linear Plasmas. *Phys. Scr.*, **40** (1989) 643–651.

420 J. Migdalek, M. Stanek. Comparison of "optimal-level" and "average-level" multiconfigurational Dirac–Fock as well as of relativistic configuration-interaction calculations for the $ns^2\,{}^1S_0 - nsnp\,{}^3P_1,\,{}^1P_1$ transitions. *Phys. Rev. A*, **41**(5) (1990) 2869–2872.

421 D. Andrae, M. Reiher, J. Hinze. A comparative study of finite nucleus models for low-lying states of few-electron high-Z atoms. *Chem. Phys. Lett.*, **320** (2000) 457–468.

422 B. W. Shore, D. H. Menzel. *Principles of Atomic Spectra*. John Wiley & Sons, New York, Chichester, 1968.

423 Y.-K. Kim. Relativistic Self-Consistent-Field Theory for Closed-Shell Atoms (Erratum: *Phys. Rev.* **159**, (1967), 190). *Phys. Rev.*, **154**(1) (1967) 17–39.

424 C. C. J. Roothaan. New Developments in Molecular Orbital Theory. *Rev. Mod. Phys.*, **23** (1951) 69–89.

425 G. G. Hall. The molecular orbital theory of chemical valency. VIII. A method of calculating ionization potentials. *Proc. Roy. Soc. London A*, **205** (1951) 541–552.

426 F. R. Franz Mark, Bans Lischka. Variational solution of the Dirac equation within a multicentre basis set of Gaussian functions. *Chem. Phys. Lett.*, **71**(3) (1980) 507–512.

427 G. Malli, J. Oreg. *Ab initio* relativistic self-consistent-field (RSCF) wavefunctions for the diatomics Li_2 and Be_2. *Chem. Phys. Lett.*, **69**(2) (1980) 313–314.

428 Y. S. Lee, A. D. McLean. Relativistic effects on R_e and D_e in AgH and AuH from all-electron Dirac–Hartree–Fock calculations. *J. Chem. Phys.*, **76**(1) (1982) 735–736.

429 R. E. Stanton, S. Havriliak. Kinetic balance: A partial solution to the problem of variational safety in Dirac calculations. *J. Chem. Phys.*, **81**(4) (1984) 1910–1918.

430 Dirac, a relativistic ab initio electronic structure program. Release DIRAC08 (2008), written by H. J. Aa. Jensen, T. Saue, and L. Visscher *et al.*; http://dirac.chem.sdu.dk.

431 J. K. Laerdahl, T. Saue, K. Faegri Jr. Direct relativistic MP2: properties of ground state CuF, AgF and AuF. *Theor. Chem. Acc.*, **97**(1-4) (1997) 177–184.

432 H. J. A. Jensen, K. G. Dyall, T. Saue, K. Fægri, Jr. Relativistic four-component multiconfigurational self-consistent-field theory for molecules: Formalism. *J. Chem. Phys.*, **104**(11) (1996) 4083–4097.

433 T. Yanai, T. Nakajima, Y. Ishikawa, K. Hirao. A new computational scheme for the Dirac–Hartree–Fock method employing an efficient integral algorithm. *J. Chem. Phys.*, **114**(15) (2001) 6526–6538.

434 T. Yanai, H. Iikura, T. Nakajima, Y. Ishikawa, K. Hirao. A new implementation of four-component relativistic density functional method for heavy-atom polyatomic systems. *J. Chem. Phys.*, **115**(18) (2001) 8267–8273.

435 T. Yanai, H. Nakano, T. Nakajima, T. Tsuneda, S. Hirata, Y. Kawashima, Y. Nakao, M. Kamiya, H. Sekino, K. Hirao. *UTChem — A Program for ab initio*

Quantum Chemistry, Volume 2660 of *Lecture Notes in Computer Science*, p. 84–95. Springer-Verlag, Berlin, Heidelberg, 1995.

436 T. Nakajima, K. Hirao. Pseudospectral approach to relativistic molecular theory. *J. Chem. Phys.*, **121** (2004) 3438.

437 M. Abe, T. Yanai, T. Nakajima, K. Hirao. A four-index transformation in Dirac's four-component relativistic theory. *Chem. Phys. Lett.*, **388** (2004) 68–73.

438 M. Abe, T. Nakajima, K. Hirao. The relativistic complete active-space second-order perturbation theory with the four-component Dirac Hamiltonian. *J. Chem. Phys.*, **125** (2006) 234110.

439 M. Abe, G. Gopakmar, T. Nakajima, K. Hirao. *Relativistic Multireference Perturbation Theory: Complete Active-Space Second-Order Perturbation Theory (CASPT2) With The Four-Component Dirac Hamiltonian*, Volume 4 of *Challenges and Advances in Computational Chemistry and Physics*, p. 157–177. Springer, 2008.

440 H. M. Quiney, H. Skaane, I. P. Grant. Ab initio relativistic quantum chemistry: four-components good, two-components bad. *Adv. Quantum Chem.*, **32** (1998) 1–49.

441 I. P. Grant, H. M. Quiney. Application of relativistic theories and quantum electrodynamics to chemical problems. *Int. J. Quantum Chem.*, **80**(3) (2000) 283–297.

442 H. M. Quiney, H. Skaane, I. P. Grant. Relativistic calculation of electromagnetic interactions in molecules. *J. Phys. B: At. Mol. Opt. Phys.*, **30** (1997) L829–L834.

443 H. M. Quiney, P. Belanzoni. Relativistic calculation of hyperfine and electron spin resonance parameters in diatomic molecules. *Chem. Phys. Lett.*, **353** (2002) 253–258.

444 L. Belpassi, F. Tarantelli, A. Sgamellotti, H. M. Quiney. Computational strategies for a four-component Dirac–Kohn–Sham program: Implementation and first applications. *J. Chem. Phys.*, **122** (2005) 184109.

445 L. Belpassi, F. Tarantelli, A. Sgamellotti, H. M. Quiney. Electron density fitting for the Coulomb problem in relativistic density-functional theory. *J. Chem. Phys.*, **124** (2006) 124104.

446 L. Belpassi, F. Tarantelli, A. Sgamellotti, H. M. Quiney. Poisson-transformed density fitting in relativistic four-component Dirac–Kohn–Sham theory. *J. Chem. Phys.*, **128** (2008) 124108.

447 O. K. Andersen. Linear methods in band theory. *Phys. Rev. B*, **12**(8) (1975) 3060–3083.

448 J. H. Wood, A. M. Boring. Improved Pauli Hamiltonian for local-potential problems. *Phys. Rev. B*, **18**(6) (1978) 2701–2711.

449 X. Wang, X.-G. Zhang, W. H. Butler, G. M. Stocks, B. N. Harmon. Relativistic-multiple-scattering theory for space-filling potentials. *Phys. Rev. B*, **46**(15) (1992) 9352–9358.

450 S. bei der Kellen, A. J. Freeman. Self-consistent relativistic full-potential Korringa-Kohn-Rostocker total-energy method and applications. *Phys. Rev. B*, **54**(16) (1996) 11187–11198.

451 N. J. M. Geipel, B. A. Heß. Scalar-relativistic effects in solids in the framework of a Douglas–Kroll transformed Dirac–Coulomb Hamiltonian. *Chem. Phys. Lett.*, **273** (1997) 62–70.

452 A. B. Shick, D. L. Novikov, A. J. Freeman. Relativistic spin-polarized theory of magnetoelastic coupling and magnetic anisotropy strain dependence: Application to Co/Cu(001). *Phys. Rev. B*, **56**(22) (1997) R14259–R14262.

453 J. C. Boettger. Scalar-relativistic linear combination of Gaussian-type-orbitals technique for crystalline solids. *Phys. Rev. B*, **57**(15) (1998) 8743–8746.

454 A. B. Shick, Y. N. Gornostyrev, A. J. Freeman. Magnetoelastic mechanism of spin-reorientation transitions at step edges. *Phys. Rev. B*, **60**(5) (1999) 3029–3032.

455 W. Kutzelnigg. Relativistic one-electron Hamiltonians 'for electrons only' and the variational treatment of the Dirac equation. *Chem. Phys.*, **225** (1997) 203–222.

456 J. Bieroń, C. Froese Fischer, A. Ynnerman. Note on MCDF Correlation Calculations for High-Z Ions. *J. Phys. B: At. Mol. Opt. Phys.*, **27** (1994) 4829–4834.

457 H. M. Quiney, I. P. Grant, S. Wilson. On the accuracy of Dirac–Hartree–Fock calculations using analytic basis sets. *J. Phys. B: At. Mol. Opt. Phys.*, **22** (1989) L15–L19.

458 G. L. Malli, A. B. F. Da Silva, Y. Ishikawa. Universal Gaussian basis functions in relativistic quantum chemistry: atomic Dirac–Fock–Coulomb and Dirac–Fock–Breit calculations. *Can. J. Chem.*, **70** (1992) 1822–1826.

459 G. L. Malli, A. B. F. Da Silva, Y. Ishikawa. Universal Gaussian basis set for accurate *ab initio* relativistic Dirac–Fock calculations. *Phys. Rev. A*, **47**(1) (1993) 143–146.

460 A. B. F. Da Silva, G. L. Malli, Y. Ishikawa. Relativistic universal Gaussian basis set for Dirac–Fock–Coulomb and Dirac–Fock–Breit SCF calculations on heavy atoms. *Chem. Phys. Lett.*, **203**(2,3) (1993) 201–203.

461 G. L. Malli, A. B. F. Da Silva, Y. Ishikawa. Highly accurate relativistic universal Gaussian basis set: Dirac–Fock–Coulomb calculations for atomic systems up to nobelium. *J. Chem. Phys.*, **101**(8) (1994) 6829–6833.

462 K. G. Dyall, K. Fægri Jr. Optimization of Gaussian basis sets for Dirac–Hartree–Fock calculations. *Theor. Chim. Acta*, **94**(1) (1996) 39–51.

463 Y. Ishikawa, K. Koc, W. H. E. Schwarz. The use of Gaussian spinors in relativistic electronic structure calculations: the effect of the boundary of the finite nucleus of uniform proton charge distribution. *Chem. Phys.*, **225** (1997) 239–246.

464 F. E. Jorge, A. B. F. da Silva. A Segmented Contraction Methodology for Gaussian Basis Sets to be Used in Dirac–Fock Atomic and Molecular Calculations. *Chem. Phys. Lett.*, **289** (1998) 469–472.

465 K. G. Dyall, T. Enevoldsen. Interfacing relativistic and nonrelativistic methods. III. Atomic 4-spinor expansions and integral approximations. *J. Chem. Phys.*, **111**(22) (1999) 10000–10007.

466 A. Hu, P. Otto, J. Ladik. Relativistic Gaussian functions for atoms by fitting numerical results with adaptive nonlinear least-square algorithm. *J. Comp. Chem.*, **20**(7) (1999) 655–664.

467 K. Faegri, Jr., K. G. Dyall. in: Relativistic Quantum Chemistry — Part I. Fundamentals, P. Schwerdtfeger (Ed.). p. 259–290. Elsevier, Amsterdam, 2002.

468 K. G. Dyall, K. Fægri Jr., P. R. Taylor. Polyatomic Molecular Dirac–Hartree–Fock Calculations with Gaussian Basis Sets. In Wilson et al. [975], p. 167–184.

469 L. Visscher, P. J. C. Aerts, O. Visser. General contraction in four-component relativistic Hartree–Fock calculations. In Wilson et al. [975], p. 197–205.

470 L. Visscher, P. J. C. Aerts, O. Visser, W. C. Nieuwpoort. Kinetic Balance in Contracted Basis Sets for Relativistic Calculations. *Int. J. Quantum Chem., Quantum Chem. Symp.*, **25** (1991) 131–139.

471 I. P. Grant, H. M. Quiney. Foundations of the Relativistic Theory of Atomic and Molecular Structure. *Adv. At. Mol. Phys.*, **23** (1988) 37–86.

472 L. E. McMurchie, E. R. Davidson. One- and Two-Electron Integrals over Cartesian Gaussian Functions. *J. Comput. Phys.*, **26** (1978) 218–231.

473 S. Obara, A. Saika. Efficient recursive computation of molecular integrals over Cartesian Gaussian functions. *J. Chem. Phys.*, **84**(7) (1986) 3963–3974.

474 S. Obara, A. Saika. General recurrence formulas for molecular integrals over Cartesian Gaussian functions. *J. Chem. Phys.*, **89**(3) (1988) 1540–1559.

475 T. Yanai, T. Nakajima, Y. Ishikawa, K. Hirao. A highly efficient algorithm for electron repulsion integrals over relativistic four-component Gaussian-type spinors. *J. Chem. Phys.*, **116**(23) (2002) 10122–10128.

476 A. D. Becke. A new mixing of Hartree-Fock and local density-functional theories. *J. Chem. Phys.*, **98**(2) (1993) 1372–1377.

477 A. D. Becke. Density-functional thermochemistry. III. The role of exact exchange. *J. Chem. Phys.*, **98**(7) (1993) 5648–5652.

478 W.-D. Sepp, D. Kolb, W. Sengler, H. Hartung, B. Fricke. Relativistic Dirac–Fock–Slater program to calculate potential-energy curves for diatomic molecules. *Phys. Rev. A*, **33**(6) (1986) 3679–3687.

479 W. Liu, G. Hong, D. Dai, L. Li, M. Dolg. The Beijing four-component density functional program package (BDF) and its application to EuO, EuS, YbO and YbS. *Theor. Chem. Acc.*, **96** (1997) 75–83.

480 S. Varga, B. Fricke, H. Nakamatsu, T. Mukoyama, J. Anton, D. Geschke,

A. Heitmann, E. Engel, T. Baştuğ. Four-component relativistic density functional calculations of heavy diatomic molecules. *J. Chem. Phys.*, **112**(8) (2000) 3499–3506.

481 H. M. Quiney, P. Belanzoni. Relativistic density functional theory using Gaussian basis sets. *J. Chem. Phys.*, **117** (2002) 5550.

482 M. Tinkham. *Group Theory and Quantum Mechanics*. Dover Publications, 2003.

483 M. Hamermesh. *Group Theory and Its Application to Physical Problems*. Dover Publications, 1989.

484 D. M. Bishop. *Group Theory and Chemistry*. Dover Publications, 1993.

485 H. A. Kramers. Théorie générale de la rotation paramagnétique dans les cristaux. *Proc. Amsterdam*, **XXXIII** (1930) 959–972.

486 E. Wigner. Über die Operation der Zeitumkehr in der Quantenmechanik. *Göttinger Nachr.*, **31** (1932) 546–559.

487 L. Visscher. On the construction of double groups molecular symmetry functions. *Chem. Phys. Lett.*, **253** (1996) 20–26.

488 N. Rösch. Time-Reversal Symmetry, Kramers' Degeneracy and the Algebraic Eigenvalue Problem. *Chem. Phys. Lett.*, **80** (1983) 1–5.

489 H. Bethe. Termaufspaltung in Kristallen. *Ann. Phys. (Serie 5)*, **3**(2) (1929) 133–208.

490 S. L. Altmann. *Rotations, Quarternions, and Double Groups*. Dover Publications, Mineola, New York, 2005.

491 P. J. C. Aerts. Use of Molecular Symmetry in Hartree–Fock–Dirac SCF Calculations. *Chem. Phys. Lett.*, **104**(1) (1984) 28–30.

492 T. Yanai, R. J. Harrison, T. Nakajima, Y. Ishikawa, K. Hirao. New implementation of molecular double point-group symmetry in four-component relativistic Gaussian-type spinors. *Int. J. Quantum Chem.*, **107**(6) (2006) 1382–1389.

493 K. Balasubramanian. *Relativistic Effects in Chemistry, Part A, Theory and Techniques*. John Wiley & Sons, New York, Chichester, 1997.

494 Y. Yamaguchi, Y. Osamura, J. D. Goddard, H. F. S. III. *A New Dimension to Quantum Chemistry — Analytic Derivative Methods in Ab Initio Molecular Electronic Structure Theory*, Volume 29 of *The International Series of Monographs on Chemistry*. Oxford University Press, New York, Oxford, 1994.

495 K. G. Dyall. Second-order Møller–Plesset perturbation theory for molecular Dirac–Hartree–Fock wavefunctions. Theory for up to two open-shell electrons. *Chem. Phys. Lett.*, **224** (1994) 186–194.

496 M. Esser, W. Butscher, W. H. E. Schwarz. Complex two- and four-index transformation over two-component Kramers-degenerate spinors. *Chem. Phys. Lett.*, **77**(2) (1981) 359–364.

497 T. Fleig, C. M. Marian, J. Olsen. Spinor optimization for a relativistic spin-dependent CASSCF program. *Theor. Chem. Acc.*, **97** (1997) 125–135.

498 A. Szabo, N. S. Ostlund. *Modern Quantum Chemistry — Introduction to Advanced Electronic Structure Theory*. Dover Publications, 1989.

499 B. O. Roos, Ed. *Lecture Notes in Quantum Chemistry — European Summer School in Quantum Chemistry*, Volume 58 of *Lecture Notes in Chemistry*. Springer-Verlag, Berlin, 1992.

500 B. O. Roos, Ed. *Lecture Notes in Quantum Chemistry II — European Summer School in Quantum Chemistry*, Volume 64 of *Lecture Notes in Chemistry*. Springer-Verlag, Berlin, 1994.

501 D. B. Cook. *Handbook of Computational Quantum Chemistry*. Dover Publications, 2005.

502 P. von Ragué Schleyer. *Encyclopedia of Computational Chemistry*. John Wiley & Sons, Chichester, 1998.

503 J.-L. Heully, I. Lindgren, E. Lindroth, S. Lundquist, A.-M. Mårtensen-Pendrill. Diagonalisation of the Dirac Hamiltonian as a basis for a relativistic many-body procedure. *J. Phys. B: At. Mol. Phys.*, **19** (1986) 2799–2815.

504 R. E. Moss, A. Okniński. Diagonal forms of the Dirac Hamiltonian. *Phys. Rev. D*, **14**(12) (1976) 3358–3361.

505 A. G. Nikitin. On exact Foldy-Wouthuysen transformation. *J. Phys. A*, **31**(14) (1998) 3297–3300.

506 A. Wolf, M. Reiher, B. A. Hess. *Two-component methods and the generalized Douglas–Kroll transformation*, p. 622–663.

Theoretical and Computational Chemistry. Elsevier, Amsterdam, 2002.

507 E. van Lenthe, E. J. Baerends, J. G. Snijders. Relativistic total energy using regular approximations. *J. Chem. Phys.*, **101** (1994) 9783–9792.

508 K. G. Dyall, E. van Lenthe. Relativistic regular approximations revisited: An infinite-order relativistic approximation. *J. Chem. Phys.*, **111** (1999) 1366–1372.

509 L. L. Foldy, S. A. Wouthuysen. On the Dirac Theory of Spin 1/2 Particles and Its Non-Relativistic Limit. *Phys. Rev.*, **78**(1) (1950) 29–36.

510 M. Reiher, A. Wolf. Exact decoupling of the Dirac Hamiltonian. I. General Theory. *J. Chem. Phys.*, **121**(5) (2004) 2037–2047.

511 A. Wolf, M. Reiher, B. A. Hess. The generalized Douglas–Kroll transformation. *J. Chem. Phys.*, **117**(20) (2002) 9215–9226.

512 W. Kutzelnigg. Perturbation theory of relativistic corrections 2. Analysis and classification of known and other possible methods. *Z. Phys. D — At. Mol. Clust.*, **15** (1990) 27–50.

513 R. McWeeny. Some recent advances in density matrix theory. *Rev. Mod. Phys.*, **32**(2) (1960) 335–369.

514 M. Barysz, A. J. Sadlej. Two-component methods of relativistic quantum chemistry: from the Douglas–Kroll approximation to the exact two-component formalism. *J. Mol. Struct. (THEOCHEM)*, **573**(1-3) (2001) 181–200.

515 M. Barysz. Systematic treatment of relativistic effects accurate through arbitrarily high order in α^2. *J. Chem. Phys.*, **114** (2001) 9315–9324.

516 M. Barysz, A. J. Sadlej. Infinite-order two-component theory for relativistic quantum chemistry. *J. Chem. Phys.*, **116** (2002) 2696–2704.

517 D. Kędziera, M. Barysz. Two-component relativistic methods for the heaviest elements. *J. Chem. Phys.*, **121** (2004) 6719–6727.

518 D. Kędziera, M. Barysz. Spin-orbit interactions and supersymmetry in two-component relativistic methods. *Chem. Phys. Lett.*, **393** (2004) 521–527.

519 D. Kędziera, M. Barysz, A. J. Sadlej. Expectation values in spin-averaged Douglas–Kroll and Infinite-order relativistic methods. *Struct. Chem.*, **15**(5) (2004) 369–377.

520 M. Barysz, A. J. Sadlej, J. G. Snijders. Nonsingular two/one-component relativistic Hamiltonians accurate through arbitrary high order in α^2. *Int. J. Quantum Chem.*, **65**(3) (1997) 225–239.

521 B. A. Hess. Applicability of the no-pair equation with free-particle projection operators to atomic and molecular structure calculations. *Phys. Rev. A*, **32** (1985) 756–763.

522 B. A. Hess. Relativistic electronic-structure calculations employing a two-component no-pair formalism with external-field projection operators. *Phys. Rev. A*, **33** (1986) 3742–3748.

523 G. Jansen, B. A. Hess. Revision of the Douglas–Kroll transformation. *Phys. Rev. A*, **39**(11) (1989) 6016–6017.

524 R. Samzow, B. A. Hess. Spin–orbit effects in the Br atom in the framework of the no-pair theory. *Chem. Phys. Lett.*, **184** (1991) 491–495.

525 R. Samzow, B. A. Hess, G. Jansen. The two-electron terms of the no-pair Hamiltonian. *J. Chem. Phys.*, **96**(2) (1992) 1227–1231.

526 B. A. Heß, C. M. Marian, U. Wahlgren, O. Gropen. A mean-field spin–orbit method applicable to correlated wavefunctions. *Chem. Phys. Lett.*, **251** (1996) 365–371.

527 V. Kellö, A. J. Sadlej, B. A. Hess. Relativistic effects on electric properties of many-electron systems in spin-averaged Douglas-Kroll and Pauli approximations. *J. Chem. Phys.*, **105**(5) (1996) 1995–2003.

528 B. A. Hess, C. M. Marian. in: Computational Molecular Spectroscopy, P. Jensen, P. R. Bunker (Ed.), p. 169. John Wiley & Sons Ltd., Chichester, 2000.

529 B. A. Hess, M. Dolg. *Relativistic Quantum Chemistry with Pseudopotentials and Transformed Hamiltonians*, Volume 57 of *Wiley Series in Theoretical Chemistry*, p. 89. John Wiley & Sons Ltd., Chichester, 2002.

530 T. Nakajima, K. Hirao. The higher-order Douglas–Kroll transformation. *J. Chem. Phys.*, **113** (2000) 7786–7789.

531 C. van Wüllen. Relation between different variants of the generalized Douglas–Kroll transformation through sixth order. *J. Chem. Phys.*, **120** (2004) 7307–7313.

532 M. Reiher, A. Wolf. Exact decoupling of the Dirac Hamiltonian. II. The generalized Douglas–Kroll–Hess transformation up to arbitrary order. *J. Chem. Phys.*, **121** (2004) 10945–10956.

533 H.-J. Werner, P. J. Knowles, R. Lindh, M. Schütz *et al.* MOLPRO 2006.2, a package of *ab initio* programs; http://www.molpro.net.

534 G. Karlström, R. Lindh, P.-A. Malmqvist, B. O. Roos, U. Ryde, V. Veryazov, P.-O. Widmark, M. Cossi, B. Schimmelpfennig, P. Neogrady, L. Seijo. MOLCAS: a program package for computational chemistry. *Comput. Mat. Sci.*, **28** (2003) 222–239.

535 R. Brummelhuis, H. Siedentop, E. Stockmeyer. The ground state energy of relativistic one-electron atoms according to Hess and Jansen. *Documenta Mathematica*, **7** (2002) 167–182.

536 H. Siedentop, E. Stockmeyer. An Analytic Douglas–Kroll–Heß Method. *Phys. Lett. A*, **341** (2005) 473–478.

537 C. van Wüllen, C. Michauk. Accurate and efficient treatment of two-electron contributions in quasirelativistic high-order Douglas–Kroll density-functional calculations. *J. Chem. Phys.*, **123** (2005) 204113.

538 D. Kędziera. Convergence of Approximate Two-Component Hamiltonians: How far is the Dirac Limit? *J. Chem. Phys.*, **123** (2005) 074109.

539 M. Reiher, A. Wolf. Regular no-pair Dirac operators: Numerical study of the convergence of high-order Douglas–Kroll–Hess transformations. *Phys. Lett. A*, **360** (2007) 603–607.

540 J. E. Peralta, J. Uddin, G. E. Scuseria. Scalar relativistic all-electron density functional calculations on periodic systems. *J. Chem. Phys.*, **122** (2005) 084108.

541 M. Reiher. Douglas–Kroll–Hess Theory: a relativistic electrons-only theory for chemistry. *Theor. Chem. Acc.*, **116** (2006) 241–252.

542 H. Siedentop, E. Stockmeyer. The Douglas–Kroll–Heß Method: Convergence and Block-Diagonalization of Dirac Operators. *Ann. Henri Poincaré*, **7** (2006) 45–58.

543 D. H. Jakubaßa-Amundsen. Pseudorelativistic Operator for a Two-Electron Ion. *Phys. Rev. A*, **71** (2005) 032105.

544 C. Park, J. E. Almlöf. Two-electron relativistic effects in molecules. *Chem. Phys. Lett.*, **231** (1994) 269–276.

545 J. Seino, M. Hada. Examination of accuracy of electron–electron Coulomb interactions in two-component relativistic methods. *Chem. Phys. Lett.*, **461** (2008) 327–331.

546 M. Mayer, S. Krüger, N. Rösch. A two-component variant of the Douglas–Kroll relativistic linear combination of Gaussian-type orbitals density-functional method: Spin–orbit effects in atoms and diatomics. *J. Chem. Phys.*, **115**(10) (2001) 4411–4423.

547 A. Wolf, M. Reiher, B. A. Hess. Correlated ab initio calculations of spectroscopic parameters of SnO within the framework of the higher-order generalized Douglas–Kroll transformation. *J. Chem. Phys.*, **120** (2004) 8624–8631.

548 F. Neese, A. Wolf, T. Fleig, M. Reiher, B. A. Hess. Calculation of electric-field gradients based on higher-order generalized Douglas–Kroll transformations. *J. Chem. Phys.*, **122** (2005) 204107.

549 R. Mastalerz, G. Barone, R. Lindh, M. Reiher. Analytic High-Order Douglas–Kroll–Hess Electric-Field Gradients. *J. Chem. Phys.*, **127** (2007) 074105.

550 J. C. Boettger. Approximate two-electron spin-orbit coupling term for density-functional-theory DFT calculations using the Douglas–Kroll–Hess transformation. *Phys. Rev. B*, **62**(12) (2000) 7809–7815.

551 S. Majumder, A. V. Matveev, N. Rösch. Spin-orbit interaction in the Douglas–Kroll approach to relativistic density functional theory: the screened nuclear potential approximation for molecules. *Chem. Phys. Lett.*, **382** (2003) 186–193.

552 J. E. Peralta, G. E. Scuseria. Relativistic all-electron two-component self-consistent density functional calculations including one-electron scalar and spin-orbit effects. *J. Chem. Phys.*, **120**(13) (2004) 5875–5881.

553 A. Matveev, N. Rösch. The electron–electron interaction in the Douglas–Kroll–Hess approach to the Dirac–Kohn–Sham problem. *J. Chem. Phys.*, **118**(9) (2003) 3997–4012.

554 A. V. Matveev, S. Majumder, N. Rösch. Efficient treatment of the Hartree interaction in the relativistic Kohn–Sham problem. *J. Chem. Phys.*, **123** (2005) 164104.

555 T. Nakajima, K. Hirao. Extended Douglas–Kroll transformations applied to the relativistic many-electron Hamiltonian. *J. Chem. Phys.*, **119**(8) (2003) 4105–4111.

556 L. Gagliardi, N. C. Handy, A. G. Ioannou, C.-K. Skylaris, S. Spencer, A. Willetts, A. M. Simper. A two-centre implementation of the Douglas–Kroll transformation in relativistic calculations. *Chem. Phys. Lett.*, **283** (1998) 187–193.

557 G. Lippert, J. Hutter, M. Parrinello. The Gaussian and augmented-plane-wave density functional method for *ab initio* molecular dynamics simulations. *Theor. Chem. Acc.*, **103** (1999) 124–140.

558 M. Krack, M. Parrinello. All-electron *ab-initio* molecular dynamics. *Phys. Chem. Chem. Phys.*, **2**(10) (2000) 2105–2112.

559 J. Thar, B. Kirchner. Relativistic all-electron molecular dynamics simulations. *J. Chem. Phys.*, (2008) submitted.

560 A. V. Matveev, N. Rösch. Atomic approximation to the projection on electronic states in the Douglas–Kroll–Hess approach to the relativistic Kohn–Sham method. *J. Chem. Phys.*, **128** (2008) 244102.

561 V. A. Nasluzov, N. Rösch. Density functional based structure optimization for molecules containing heavy elements: analytical energy gradients for the Douglas–Kroll–Hess scalar relativistic approach to the LCGTO-DF method. *Chem. Phys.*, **210** (1996) 413–425.

562 A. V. Matveev, V. A. Nasluzov, N. Rösch. Linear response formalism for the Douglas–Kroll–Hess approach to the Dirac–Kohn–Sham problem: First- and second-order nuclear displacement derivatives of the energy. *Int. J. Quantum Chem.*, **107** (2007) 3236–3249.

563 J. H. van Lenthe, S. Faas, J. G. Snijders. Non-Relativistic Gradients in the *Ab Initio* Scalar ZORA Approach. *Chem. Phys. Lett.*, **328** (2000) 107–112.

564 W. A. de Jong, R. J. Harrison, D. A. Dixon. Parallel Douglas–Kroll energy and gradient in NWChem: Estimating scalar relativistic effects using Douglas–Kroll contracted basis sets. *J. Chem. Phys.*, **114** (2001) 48–53.

565 J. D. Morrison, R. E. Moss. Approximate solution of the Dirac-equation using the Foldy-Wouthuysen Hamiltonian. *Mol. Phys.*, **41**(3) (1980) 491–507.

566 T. Itoh. Derivation of Nonrelativistic Hamiltonian for Electrons from Quantum Electrodynamics. *Rev. Mod. Phys.*, **37**(1) (1965) 159–165.

567 Z. V. Chraplyvy. Reduction of Relativistic Two-Particle Wave Equations to Approximate Forms. I. *Phys. Rev.*, **91**(2) (1953) 388–391.

568 Z. V. Chraplyvy. Reduction of Relativistic Two-Particle Wave Equations to Approximate Forms. II. *Phys. Rev.*, **92**(5) (1953) 1310–1315.

569 W. A. Barker, F. N. Glover. Reduction of Relativistic Two-Particle Wave Equations to Approximate Forms. III. *Phys. Rev.*, **99**(1) (1955) 317–324.

570 O. Vahtras, O. Loboda, B. Minaev, H. Ågren, K. Ruud. Ab initio calculations of zero-field splitting parameters. *Chem. Phys.*, **279**(2-3) (2002) 133–142.

571 S. Sinnecker, F. Neese. Spin–Spin Contributions to the Zero-Field Splitting Tensor in Organic Triplets, Carbenes and Biradicals — A Density Functional and Ab Initio Study. *J. Phys. Chem. A*, **110** (2006) 12267–12275.

572 N. Gilka, P. R. Taylor, C. M. Marian. Electron spin–spin coupling from multireference configuration interaction wave functions. *J. Chem. Phys.*, **129** (2008) 044102.

573 B. J. Howard, R. E. Moss. The molecular hamiltonian. I. Non-linear molecules. *Mol. Phys.*, **19**(4) (1970) 433–450.

574 B. J. Howard, R. E. Moss. The molecular hamiltonian. II. Linear molecules. *Mol. Phys.*, **20**(1) (1971) 147–159.

575 P.-O. Löwdin. Studies in Perturbation Theory Part VIII. Separation of Dirac equation + Study of Spin-Orbit Coupling

+ Fermi Contact Terms. *J. Mol. Spectrosc.*, **14**(2) (1964) 131–144.

576 R. D. Cowan, D. C. Griffin. Approximate relativistic corrections to atomic radial wave-functions. *J. Opt. Soc. Am.*, **66**(10) (1976) 1010–1014.

577 J. C. Barthelat, M. Pelissier, P. Durand. Analytical relativistic self-consistent-field calculations for atoms. *Phys. Rev. A*, **21**(6) (1980) 1773–1785.

578 J. Karwowski, J. Kobus. An effective quasirelativistic hamiltonian. *Chem. Phys.*, **55** (1981) 361–369.

579 J. Karwowski, M. Szulkin. Relativistic calculations on the alkali atoms by a modified Hartree-Fock method. *J. Phys. B*, **14** (1981) 1915–1927.

580 J. Karwowski, J. Kobus. Quasirelativistic methods. *Int. J. Quantum Chem.*, **28** (1985) 741–756.

581 J. Wood, I. P. Grant, S. J. Wilson. The Dirac equation in the algebraic approximation. IV. Application of the partitioning technique. *J. Phys. B: At. Mol. Phys.*, **18** (1985) 3027–3041.

582 J.-L. Heully. Approximate relativistic treatment of the interaction between external fields and atoms. *J. Phys. B: At. Mol. Phys.*, **15** (1982) 4079–4091.

583 J.-L. Heully, S. Salomonson. Approximate relativistic many-body calculations of the hyperfine interaction in excited s states of the rubidium atom. *J. Phys. B: At. Mol. Phys.*, **15** (1982) 4093–4101.

584 S. Komorovský, M. Repiský, O. L. Malkina, V. G. Malkin, I. Malkin, M. Kaupp. Resolution of identity Dirac–Kohn–Sham method using the large component only: Calculations of g-tensor and hyperfine tensor. *J. Chem. Phys.*, **124** (2006) 084108.

585 C. Chang, M. Pelissier, P. Durand. Regular Two-Component Pauli-Like Effective Hamiltonians in Dirac Theory. *Phys. Scr.*, **34** (1986) 394–404.

586 P. Durand. Direct determination of effective Hamiltonians by wave-operator methods. I. General formalism. *Phys. Rev. A*, **28** (1983) 3184–3192.

587 E. van Lenthe, E. J. Baerends, J. G. Snijders. Relativistic regular two-component Hamiltonians. *J. Chem. Phys.*, **99** (1993) 4597–4610.

588 R. van Leeuwen, E. van Lenthe, E. J. Baerends, J. G. Snijders. Exact solutions of regular approximate relativistic wave equations for hydrogen-like atoms. *J. Chem. Phys.*, **101** (1994) 1272–1281.

589 E. van Lenthe, R. van Leeuwen, E. J. Baerends, J. G. Snijders. Relativistic regular two-component Hamiltonians. *Int. J. Quantum Chem.*, **57** (1996) 281–293.

590 A. J. Sadlej, J. G. Snijders, E. van Lenthe, E. J. Baerends. Four component regular relativistic Hamiltonians and the perturbational treatment of Dirac's equation. *J. Chem. Phys.*, **102**(4) (1995) 1758–1766.

591 C. van Wüllen. Molecular density functional calculations in the regular relativistic approximation: Method, application to coinage metal diatomics, hydrides, fluorides and chlorides, and comparison with first-order relativistic calculations. *J. Chem. Phys*, **109**(2) (1998) 392–400.

592 W. Klopper. R12 methods, gaussian geminals. p. 181–229, 2000.

593 E. van Lenthe. *The ZORA Equation*. PhD thesis, Vrije Universiteit te Amsterdam, The Netherlands, 1996.

594 E. van Lenthe, J. G. Snijders, E. J. Baerends. The zero-order regular approximation for relativistic effects: The effect of spin-orbit coupling in closed shell molecules. *J. Chem. Phys.*, **105** (1996) 6505–6516.

595 S. Faas, J. H. van Lenthe, A. C. Hennum, J. G. Snijders. An *ab initio* two-component relativistic method including spin–orbit coupling using the regular approximation. *J. Chem. Phys.*, **113**(10) (2000) 4052–4059.

596 ADF2006.01. SCM, Theoretical Chemistry, Vrije Universiteit, Amsterdam, The Netherlands, http://www.scm.com.

597 G. te Velde, F. M. Bickelhaupt, S. J. A. van Gisbergen, C. F. Guerra, E. J. Baerends, J. G. Snijders, T. Ziegler. Chemistry with ADF. *J. Comput. Chem.*, **22** (2001) 931–967.

598 F. Aquilante, T. B. Pedersen, R. Lindh. Low-cost evaluation of the exchange Fock matrix from Cholesky and density fitting representations of the electron repulsion integrals. *J. Chem. Phys.*, **126** (2007) 194106.

599 S. Goedecker, G. E. Scuseria. Linear Scaling Electronic Structure Methods in Chemistry and Physics. *Comp. Sci. Engin.*, **5**(4) (2003) 14–21.

600 C. Ochsenfeld, J. Kussmann, D. S. Lambrecht. Linear-Scaling Methods in Quantum Chemistry. *Rev. Comput. Chem.*, **23** (2007) 1–82.

601 M. Ilias, T. Saue. An infinite-order two-component relativistic Hamiltonian by a simple one-step transformation. *J. Chem. Phys.*, **126** (2007) 064102.

602 W. Kutzelnigg. Basis set expansion of the Dirac operator without variational collapse. *Int. J. Quantum Chem.*, **25** (1984) 107–129.

603 K. G. Dyall. An exact separation of the spin-free and spin-dependent terms of the Dirac–Coulomb–Breit Hamiltonian. *J. Chem. Phys.*, **100**(3) (1994) 2118–2127.

604 K. G. Dyall. Interfacing relativistic and nonrelativistic methods. I. Normalized elimination of the small component in the modified Dirac equation. *J. Chem. Phys.*, **106**(23) (1997) 9618–9626.

605 K. G. Dyall. Interfacing relativistic and nonrelativistic methods. II. Investigation of a low-order approximation. *J. Chem. Phys.*, **109** (1998) 4201–4208.

606 K. G. Dyall. Interfacing relativistic and nonrelativistic methods. IV. One- and two-electron scalar approximations. *J. Chem. Phys.*, **115** (2001) 9136–9143.

607 M. Filatov, D. Cremer. Representation of the exact relativistic electronic Hamiltonian within the regular approximation. *J. Chem. Phys.*, **119** (2003) 11526–11540.

608 M. Filatov, D. Cremer. Connection between the regular approximation and the normalized elimination of the small component in relativistic quantum chemistry. *J. Chem. Phys.*, **122** (2005) 064104.

609 W. Kutzelnigg, W. Liu. Quasirelativistic theory equivalent to fully relativistic theory. *J. Chem. Phys.*, **123** (2005) 241102.

610 W. Kutzelnigg, W. Liu. Quasirelativistic theory I. Theory in terms of a quasi-relativistic operator. *Mol. Phys.*, **104**(13-14) (2006) 2225–2240.

611 W. Liu, D. Peng. Infinite-order quasirelativistic density functional method based on the exact matrix quasirelativistic theory. *J. Chem. Phys.*, **125** (2006) 044102.

612 M. Filatov. Comment on "Quasirelativistic theory equivalent to fully relativistic theory" [J. Chem. Phys. 123, 241102 **2005**]. *J. Chem. Phys.*, **125** (2006) 107101.

613 W. Kutzelnigg, W. Liu. Response to "Comment on 'Quasirelativistic theory equivalent to fully relativistic theory' [J. Chem. Phys. 123, 241102 **2005**]". *J. Chem. Phys.*, **125** (2006) 107102.

614 W. Liu, W. Kutzelnigg. Quasirelativistic theory. II. Theory at matrix level. *J. Chem. Phys.*, **126** (2007) 114107.

615 D. Peng, W. Liu, Y. Xiao, L. Cheng. Making four- and two-component relativistic density functional methods fully equivalent based on the idea of "from atoms to molecule". *J. Chem. Phys.*, **127** (2007) 104106.

616 B. A. Heß, C. M. Marian, S. D. Peyerimhoff. *Ab Initio Calculations of Spin–Orbit Effects in Molecules Including Electron Correlation*, p. 152–278. Advanced Series in Physical Chemistry. World Scientific, Singapore, 1995.

617 G. C. Schatz, M. A. Ratner. *Quantum Mechanics in Chemistry*. Dover Publications, Mineola, New York, 2002.

618 K. Ruud, B. Schimmelpfennig, H. Ågren. Internal and external heavy-atom effects on phosphorescence radiative lifetimes calculated using a mean-field spin–orbit Hamiltonian. *Chem. Phys. Lett.*, **310**(1-2) (1999) 215–221.

619 B. Schimmelpfennig, L. Maron, U. Wahlgren, C. Teichteil, H. Fagerli, O. Gropen. On the combination of ECP-based CI calculations with all-electron spin-orbit mean-field integrals. *Chem. Phys. Lett.*, **286**(3-4) (1998) 267–271.

620 B. Schimmelpfennig, L. Maron, U. Wahlgren, C. Teichteil, H. Fagerli, O. Gropen. On the efficiency of an effective Hamiltonian in spin-orbit CI calculations. *Chem. Phys. Lett.*, **286**(3-4) (1998) 261–266.

621 L. Gagliardi, B. Schimmelpfennig, L. Maron, U. Wahlgren, A. Willetts. Spin–orbit coupling within a two-component density functional theory approach: theory, implementation and first applications. *Chem. Phys. Lett.*, **344**(1-2) (2001) 207–212.

622 P. Å. Malmqvist, B. O. Roos, B. Schimmelpfennig. The restricted active space (RAS) state interaction approach with spin–orbit coupling. *Chem. Phys. Lett.*, **357**(3-4) (2002) 230–240.

623 F. Neese. Efficient and Accurate Approximations to the Molecular Spin-Orbit Coupling Operator and their use in Molecular *g*-Tensor Calculations. *J. Chem. Phys.*, **122** (2005) 034107.

624 D. Ganyushin, F. Neese. First-principles calculations of zero-field splitting parameters. *J. Chem. Phys.*, **125** (2006) 024103.

625 F. Neese. Calculation of the zero-field splitting tensor on the basis of hybrid density functional and Hartree–Fock theory. *J. Chem. Phys.*, **127** (2007) 164112.

626 S. R. Langhoff, E. R. Davidson. Configuration interaction calculations on the nitrogen molecule. *Int. J. Quantum Chem.*, **8** (1974) 61–72.

627 B. Huron, J. Malrieu, P. Rancurel. Iterative perturbation calculations of ground and excited state energies from multiconfigurational zeroth-order wavefunctions. *J. Chem. Phys.*, **58** (1973) 5745–5759.

628 J. A. Pople, R. Seeger, R. Krishnan. Variational configuration interaction methods and comparison with perturbation-theory. *Int. J. Quantum Chem.*, **Suppl. Y-11** (1977) 149–163.

629 B. A. Hess, R. J. Buenker, C. M. Marian, S. D. Peyerimhoff. Ab Initio Calculation of the Zero-Fiels Splittings of the $X\,^3\Sigma_g^-$ and $B\,^3\Pi_g$ States of the S_2 Molecule. *Chem. Phys.*, **71** (1982) 79–85.

630 S. J. Havriliak, D. R. Yarkony. On the use of the Breit–Pauli approximation for evaluating line strengths for spin-forbidden transitions: Application to NF. *J. Chem. Phys.*, **83**(3) (1985) 1168–1172.

631 D. R. Yarkony. On the use of the breit–pauli approximation for evaluating line strengths for spin forbidden transitions. ii. the symbolic element method.

632 R. J. Buenker, A. B. Alekseyev, H.-P. Liebermann, R. Lingott, G. Hirsch. Comparison of spin–orbit configuration interaction methods employing relativistic effective core potentials for the calculation of zero-field splittings of heavy atoms with a $^2P^o$ ground state. *J. Chem. Phys.*, **108**(9) (1998) 3400–3408.

633 A. Berning, M. Schweizer, H.-J. Werner, P. J. Knowles, P. Palmieri. Spin–orbit matrix elements for internally contracted multireference configuration interaction wavefunctions. *Mol. Phys.*, **98**(21) (2000) 1823–1833.

634 B. H. Brandow. Linked-Cluster Expansions for the Nuclear Many-Body Problem. *Rev. Mod. Phys.*, **39** (1967) 771–828.

635 P. S. Epstein. The Stark Effect from the Point of View of Schroedinger's Quantum Theory. *Phys. Rev.*, **28** (1926) 695–710.

636 V. Vallet, L. Maron, C. Teichteil, J.-P. Flament. A two-step uncontracted determinantal effective Hamiltonian-based SO–CI method. *J. Chem. Phys.*, **113**(4) (2000) 1391–1402.

637 C. Teichteil, M. Pellisier, F. Spiegelmann. Ab Initio Molecular Calculations Including Spin–Orbit Coupling. I. Method and Atomic Tests. *Chem. Phys.*, **81** (1983) 273–282.

638 P. A. Christiansen, K. Balasubramanian, K. S. Pitzer. Relativistic *ab initio* molecular structure calculations including configuration interaction with application to six states of TlH. *J. Chem. Phys.*, **76**(10) (1982) 5087–5092.

639 G. A. DiLabio, P. A. Christiansen. Low-lying 0^+ states of bismuth hydride. *Chem. Phys. Lett.*, **277** (1997) 473–477.

640 K. Balasubramanian. Relativistic configuration interaction calculations for polyatomics: Applications to PbH_2, SnH_2, and GeH_2. *J. Chem. Phys.*, **89** (1988) 5731–5738.

641 A. B. Alekseyev, R. J. Buenker, H.-P. Liebermann, G. Hirsch. Spin–orbit configuration interaction study of the potential energy curves and radiative lifetimes of the low-lying states of bismuth hydride. *J. Chem. Phys.*, **100** (1994) 2989–3001.

642 F. Rakowitz, C. M. Marian. An extrapolation scheme for the spin–orbit configuration interaction energies applied to the ground and excited electronic states of thallium hydride. *Chem. Phys.*, **225** (1997) 223–238.

643 M. Kleinschmidt, J. Tatchen, C. M. Marian. SPOCK.CI: A multireference spin-orbit configuration interaction method for large molecules. *J. Chem. Phys.*, **124** (2006) 124101.

644 C. M. Marian. Spin–Orbit Coupling in Molecules. In K. B. Lipkowitz, D. B. Boyd, Ed., *Reviews in Computational Chemistry*, Volume 17 of *Reviews in Computational Chemistry*, p. 99–204. John Wiley & Sons Ltd., Chichester, 2001.

645 O. Christiansen, J. Gauss, B. Schimmelpfennig. Spin-orbit coupling constants from coupled-cluster response theory. *Phys. Chem. Chem. Phys.*, **2**(5) (2000) 965–971.

646 F. Wang, J. Gauss, C. van Wüllen. Closed-shell coupled-cluster theory with spin-orbit coupling. *J. Chem. Phys.*, **129** (2008) 064113.

647 K. G. Dyall, P. R. Taylor, K. Fægri Jr., H. Partridge. All-electron molecular Dirac–Hartree–Fock calculations: The group IV tetrahydrides CH_4, SiH_4, GeH_4, SnH_4, and PbH_4. *J. Chem. Phys.*, **95**(4) (1991) 2583–2594.

648 K. G. Dyall. All-electron molecular Dirac–Hartree–Fock calculations: Properties of the XH_4 and XH_2 molecules and the reaction energy $XH_4 \rightarrow XH_2+H_2$, X=Si, Ge, Sn, Pb. *J. Chem. Phys.*, **96**(2) (1992) 1210–1217.

649 L. Pisani, E. Clementi. Relativistic Dirac–Fock Calculations for Closed-Shell Molecules. *J. Comp. Chem.*, **15** (1994) 466–474.

650 L. Visscher. Approximate molecular relativistic Dirac–Coulomb calculations using a simple Coulombic correction. *Theor. Chem. Acc.*, **98**(2/3) (1997) 68–70.

651 L. Szasz. *Pseudopotential Theory of Atoms and Molecules*. Wiley Interscience, New York, 1985.

652 Y. S. Lee, W. C. Ermler, K. S. Pitzer. *Ab initio* effective core potentials including relativistic effects. I. Formalism and applications to the Xe and Au atoms. *J. Chem. Phys.*, **67** (1977) 5861.

653 M. Dolg. Effective Core Potentials. In J. Grotendorst, Ed., *Modern Methods and Algorithms of Quantum Chemistry*, Volume 3 of *NIC series*, p. 507. John von Neumann Institute for Computing, Jülich, 2000.

654 J. Migdalek, W. E. Baylis. Influence of relativistic valence-core correlation on p-state fine-structure in some univalent systems. *Can. J. Phys.*, **59**(6) (1981) 769–774.

655 W. Müller, J. Flesch, W. Meyer. Treatment of intershell correlation effects in ab initio calculations by use of core polarization potentials. Method and application to alkali and alkaline earth atoms. *J. Chem. Phys.*, **80** (1984) 3297–3310.

656 P. Schwerdtfeger, T. Fischer, M. Dolg, G. Igel-Mann, A. Nicklass, H. Stoll. The accuracy of the pseudopotential approximation. I. An analysis of the spectroscopic constants for the electronic ground states of InCl and nCl_3 using various three valence electron pseudopotentials for indium. *J. Chem. Phys.*, **102**(5) (1995) 2050–2062.

657 T. Leininger, A. Nicklas, H. Stoll, M. Dolg, P. Schwerdtfeger. The accuracy of the pseudopotential approximation. II. A comparison of various core sizes for indium pseudopotentials in calculations for spectroscopic constants of InH, InF, and InCl. *J. Chem. Phys.*, **105** (1996) 1052–1059.

658 P. Schwerdtfeger, J. R. Brown, J. K. Laerdahl, H. Stoll. The accuracy of the pseudopotential approximation. III. A comparison between pseudopotential and all-electron methods for Au and AuH. *J. Chem. Phys.*, **113**(17) (2000) 7110–7118.

659 Z. Barandiáran, L. Seijo. The Ab Initio Model Potential method. Cowan-Griffin relativistic core potentials and valence basis sets from Li (Z=3) to La (Z=57). *Can. J. Chem.*, **70** (1992) 409–415.

660 Z. Barandiáran, L. Seijo. Quasirelativistic *ab initio* model potential calculations on the group IV hydrides (XH_2, XH_4; X=Si,Ge,Sn,Pb) and oxides (XO; X=Ge,Sn,Pb). *J. Chem. Phys.*, **101** (1994) 4049–4054.

661 L. Seijo. Relativistic ab initio model potential calculations including spin–orbit effects through the Wood–Boring Hamiltonian. *J. Chem. Phys.*, **102** (1995) 8078–8088.

662 M. Casarrubios, L. Seijo. The ab initio model potential method. Relativistic

Wood-Boring valence spin-orbit potentials and spin-orbit-corrected basis sets from B(Z= 5) to Ba(Z=56). *J. Mol. Struc. (THEOCHEM)*, **426** (1998) 59–74.

663 M. Casarrubios, L. Seijo. The ab initio model potential method: Third-series transition metal elements. *J. Chem. Phys.*, **110** (1999) 784–796.

664 H. Hellmann. A New Approximation Method in the Problem of Many Electrons. *J. Chem. Phys.*, **3** (1935) 61.

665 P. Gombás. Über die metallische Bindung. *Z. Phys.*, **94** (1935) 473–488.

666 H. Preuss. Untersuchungen zum kombinierten Näherungsverfahren. *Z. Naturf.*, **10a**(5) (1955) 365–373.

667 J. C. Phillips, L. Kleinmann. New Method for Calculating Wave Functions in Crystals and Molecules. *Phys. Rev.*, **116** (1959) 287–294.

668 P. Durand, J. C. Barthelat. Theoretical method to determine atomic pseudopotentials for electronic-structure calculations of molecules and solids. *Theor. Chim. Acta*, **38**(4) (1975) 283–302.

669 L. F. Pacios, P. A. Christiansen. *Ab initio* relativistic effective potentials with spin-orbit operators. I. Li through Ar. *J. Chem. Phys.*, **82** (1985) 2664–2671.

670 P. J. Hay, W. R. Wadt. *Ab initio* effective core potentials for molecular calculations. Potentials for the transition metal atoms Sc to Hg. *J. Chem. Phys.*, **82** (1985) 270.

671 C. S. Nash, B. E. Bursten, W. C. Ermler. Ab Initio Relativistic Effective Potentials with Spin-Orbit Operators. VII. Am through Element 118. *J. Chem. Phys.*, **106** (1997) 5133–5142.

672 M. Dolg, H. Stoll, H. Preuss. Energy-adjusted *ab initio* pseudopotentials for the rare earth elements. *J. Chem. Phys.*, **90** (1989) 1730–1734.

673 W. Küchle, M. Dolg, H. Stoll, H. Preuss. Ab initio pseudopotentials for Hg through Rn. I. Parameter sets and atomic calculations. *Mol. Phys.*, **74**(6) (1991) 1245–1263.

674 W. Küchle, M. Dolg, H. Stoll, H. Preuss. Energy-adjusted pseudopotentials for the actinides. Parameter sets and test calculations for thorium and thorium monoxide. *J. Chem. Phys.*, **100** (1994) 7535–7542.

675 M. Dolg. Improved relativistic energy-consistent pseudopotentials for 3d-transition metals. *Theor. Chem. Acc.*, **114**(4-5) (2005) 297–304.

676 K. A. Peterson, D. Figgen, M. Dolg, H. Stoll. Energy-consistent relativistic pseudopotentials and correlation consistent basis sets for the 4*d* elements Y–Pd. *J. Chem. Phys.*, **126** (2007) 124101.

677 A. Moritz, X. Cao, M. Dolg. Quasirelativistic energy-consistent 5f-in-core pseudopotentials for divalent and tetravalent actinide elements. *Theor. Chem. Acc.*, **118** (2007) 845–854.

678 K. Balasubramanian. in: Encyclopedia of Computational Chemistry, P. v. R. Schleyer (Ed.). p. 2471. John Wiley & Sons Ltd., Chichester, 1998.

679 L. Seijo, Z. Barandiarán. . In J. Leszczynski, Ed., *Computational Chemistry: Reviews of Current Trends*, Volume 4, p. 55. World Scientific, Singapore, 1999.

680 M. Dolg. in: Relativistic Quantum Chemistry — Part I. Fundamentals, P. Schwerdtfeger (Ed.). p. 793. Elsevier, Amsterdam, 2002.

681 P. A. Christiansen, Y. S. Lee, K. S. Pitzer. Improved ab initio effective core potentials for molecular calculations. *J. Chem. Phys.*, **71** (1979) 4445–4450.

682 W. C. Ermler, Y. S. Lee, P. A. Christiansen, K. S. Pitzer. Ab Initio Effective Core Potentials Including Relativistic Effects. A Procedure for the Inclusion of Spin-Orbit Coupling in Molecular Wavefunctions. *Chem. Phys. Lett.*, **81** (1981) 70–74.

683 M. M. Hurley, L. F. Pacios, P. A. Christiansen, R. B. Ross, W. C. Ermler. Ab Initio Relativistic Effective Potentials with Spin-Orbit Operators. II. K through Kr. *J. Chem. Phys.*, **84** (1986) 6840–6853.

684 L. A. LaJohn, P. A. Christiansen, R. B. Ross, T. Atashroo, W. C. Ermler. *Ab initio* relativistic effective potentials with spin–orbit operators. III. Rb through Xe. *J. Chem. Phys.*, **87** (1988) 2812–2824.

685 R. B. Ross, J. M. Powers, T. Atashroo, W. C. Ermler, L. A. LaJohn, P. A. Christiansen. Ab initio relativistic effective potentials with spin–orbit operators. IV.

Cs through Rn. *J. Chem. Phys.*, **93** (1990) 6654–6670.

686 R. B. Ross, S. Gayen, W. C. Ermler. Ab initio relativistic effective potentials with spin–orbit operators. V. Ce through Lu. *J. Chem. Phys.*, **100** (1994) 8145–8155.

687 W. C. Ermler, R. B. Ross, P. A. Christiansen. Ab Initio Relativistic Effective Potentials with Spin-Orbit Operators. VI. Fr through Pu. *Int. J. Quantum Chem.*, **40** (1991) 829–846.

688 S. A. Wildman, G. A. DiLabio, P. A. Christiansen. Accurate relativistic effective potentials for the sixth-row main group elements. *J. Chem. Phys.*, **107** (1997) 9975–9979.

689 D. Andrae, U. Haussermann, M. Dolg, H. Stoll, H. Preuss. Energy-adjusted ab initio pseudopotentials for the 2nd and 3rd row transition-elements. *Theor. Chim. Acta*, **77**(2) (1990) 123–141.

690 T. Leininger, A. Berning, A. Nicklass, H. Stoll, H.-J. Werner, H.-J. Flad. Spin-orbit interaction in heavy group 13 atoms and TlAr. *Chem. Phys.*, **217**(1) (1997) 19–27.

691 Y. S. Kim, Y. S. Lee. The Kramers' restricted complete active space self-consistent-field method for two-component molecular spinors and relativistic effective core potentials including spin-orbit interactions. *J. Chem. Phys.*, **119** (2003) 12169.

692 E. Fromager, C. Teichteil, L. Maron. Atomic spin-orbit pseudopotential definition and its relation to the different relativistic approximations. *J. Chem. Phys.*, **123** (2005) 034106.

693 D. P. Craig, T. Thirunamachandran. *Molecular Quantum Electrodynamics: An Introduction to Radiation-Molecule Interactions*. Theoretical Chemistry; a Series of Monographs. Academic Press, New York, 1984.

694 J. Autschbach, T. Ziegler. Double perturbation theory: a powerful tool in computational coordination chemistry. *Coord. Chem. Rev.*, **238-239** (2003) 83–126.

695 F. Neese. Quantum Chemical Calculations of Spectroscopic Properties of Metalloproteins and Model Compounds: EPR and Mössbauer Properties. *Curr. Opin. Chem. Biol.*, **7** (2003) 125–135.

696 J. Autschbach. Density functional theory applied to calculating optical and spectroscopic properties of metal complexes: NMR and optical activity. *Coord. Chem. Rev.*, **251**(13-14) (2007) 1796–1821.

697 S. Koßmann, B. Kirchner, F. Neese. Performance of modern density functional theory for the prediction of hyperfine structure: meta-GGA and double hybrid functionals. *Mol. Phys.*, **105** (2007) 2049–2071.

698 D. J. Thouless. *The Quantum Mechanics of Many-Body Systems*. Pure & Applied Physics. Academic Press, 1972.

699 M. E. Casida. *Time-dependent density functional response theory for molecules*, Volume I, Chapter 5, p. 155–192. World Scientific, Singapore, 1995.

700 W. Kutzelnigg. Relativistic Corrections to Magnetic Properties. *J. Comput. Chem.*, **20**(12) (1999) 1199–1219.

701 K. G. Dyall. Relativistic Electric and Magnetic Property Operators for Two-Component Transformed Hamiltonians. *Int. J. Quantum Chem.*, **78**(6) (2000) 412–421.

702 J. Autschbach, T. Ziegler. Calculating molecular electric and magnetic properties from time-dependent density functional response theory. *J. Chem. Phys.*, **116**(3) (2002) 891–896.

703 P. G. Roura, J. I. Melo, M. C. Ruiz de Azúa, C. G. Giribet. Mean field linear response within the elimination of the small component formalism to evaluate relativistic effects on magnetic properties. *J. Chem. Phys.*, **125** (2006) 064107.

704 T. Helgaker, A. C. Hennum, W. Klopper. A second-quantization framework for the unified treatment of relativistic and nonrelativistic molecular perturbations by response theory. *J. Chem. Phys.*, **125** (2006) 024102.

705 D. Zaccari, M. C. Ruiz de Azúa, J. I. Melo, C. G. Giribet. Formal relations connecting different approaches to calculate relativistic effects on molecular magnetic properties. *J. Chem. Phys.*, **124** (2006) 054103.

706 J. Henriksson, T. Saue, P. Norman. Quadratic response functions in the relativistic four-component Kohn–Sham approximation. *J. Chem. Phys.*, **128** (2008) 024105.

707 S. Luber, I. M. Ondik, M. Reiher. Electromagnetic fields in relativistic one-particle equations. *Chem. Phys.*, (2008) doi:10.1016/j.chemphys.2008.10.021.

708 M. Kaupp, M. Bühl, V. G. Malkin (Eds.). *Calculation of NMR and EPR Parameters.* Wiley-VCH, Weinheim, 2004.

709 E. J. Baerends, W. H. E. Schwarz, P. Schwerdtfeger, J. G. Snijders. Relativistic atomic orbital contractions and expansions: magnitudes and explanations. *J. Phys. B: At. Mol. Phys.*, **23** (1990) 3225–3240.

710 V. Kellö, A. J. Sadlej. Picture Change and Calculations of Expectation Values in Approximate Relativistic Theories. *Int. J. Quantum Chem.*, **68** (1998) 159–174.

711 T. D. Newton, E. P. Wigner. Localized States for Elementary Systems. *Rev. Mod. Phys.*, **21** (1948) 400–406.

712 M. Pernpointner, M. Seth, P. Schwerdtfeger. A point-charge model for the nuclear quadrupole moment: Coupled-cluster, Dirac–Fock, Douglas–Kroll, and nonrelativistic Hartree–Fock calculations for the Cu and F electric field gradients in CuF. *J. Chem. Phys.*, **108**(16) (1998) 6722–6738.

713 M. Pernpointner, P. Schwerdtfeger, B. A. Hess. The nuclear quadrupole moment of ^{133}Cs: Accurate relativistic coupled cluster calculations for CsF within the point-charge model for nuclear quadrupole moments. *J. Chem. Phys.*, **108** (1998) 6739–6747.

714 V. Kellö, A. J. Sadlej. The point charge model of nuclear quadrupoles: How and why does it work. *J. Chem. Phys.*, **112** (2000) 522–526.

715 V. Kellö, P. Pyykkö, A. J. Sadlej, P. Schwerdtfeger, J. Thyssen. The nuclear quadrupole moment of ^{91}Zr from molecular data for ZrO and ZrS. *Chem. Phys. Lett.*, **318**(1-3) (2000) 222–231.

716 V. Kellö, A. J. Sadlej, P. Pyykkö, D. Sundholm, M. Tokman. Electric quadrupole moment of the ^{27}Al nucleus: Converging results from the AlF and AlCl molecules and the Al atom. *Chem. Phys. Lett.*, **304**(5-6) (1999) 414–422.

717 V. Kellö, A. J. Sadlej, P. Pyykkö. The nuclear quadrupole moment of ^{45}Sc. *Chem. Phys. Lett.*, **329** (2000) 112–118.

718 L. Visscher, T. Enevoldsen, T. Saue, J. Oddershede. Molecular relativistic calculations of the electric field gradients at the nuclei in the hydrogen halides. *J. Chem. Phys.*, **109**(22) (1998) 9677–9684.

719 A. Wolf, M. Reiher. Exact decoupling of the Dirac Hamiltonian. IV. Automated evaluation of molecular properties within the Douglas–Kroll–Hess theory up to arbitrary order. *J. Chem. Phys.*, **124** (2006) 064103.

720 G. Barone, R. Mastalerz, R. Lindh, M. Reiher. Nuclear Quadrupole Moment of ^{119}Sn. *J. Phys. Chem. A*, **112** (2008) 1666–1672.

721 A. Wolf, M. Reiher. Exact decoupling of the Dirac Hamiltonian. III. Molecular properties. *J. Chem. Phys.*, **124** (2006) 064102.

722 J. Autschbach. Analyzing molecular properties calculated with two-component relativistic methods using spin-free Natural Bond Orbitals: NMR spin-spin coupling constants. *J. Chem. Phys.*, **127** (2007) 124106.

723 J. Autschbach, B. Le Guennic. Analyzing and interpreting NMR spin-spin coupling constants from molecular orbital calculations. *J. Chem. Educ.*, **84** (2007) 156–171.

724 H. Fukui, T. Baba. Calculation of nuclear magnetic shieldings. XII. Relativistic no-pair equation. *J. Chem. Phys.*, **108**(10) (1998) 3854–3862.

725 I. Malkin, O. L. Malkina, V. G. Malkin, M. Kaupp. Scalar relativistic calculations of hyperfine coupling tensors using the Douglas–Kroll–Hess method. *Chem. Phys. Lett.*, **396** (2004) 268–276.

726 I. Malkin, O. L. Malkina, V. G. Malkin. Relativistic calculations of electric field gradients using the Douglas–Kroll method. *Chem. Phys. Lett.*, **361** (2002) 231–236.

727 R. Fukuda, M. Hada, H. Nakatsuji. Quasirelativistic theory for the magnetic shielding constant. I. Formulation of Douglas–Kroll–Hess transformation for the magnetic field and its application to atomic systems. *J. Chem. Phys.*, **118**(3) (2003) 1015–1026.

728 R. Fukuda, M. Hada, H. Nakatsuji. Quasirelativistic theory for the magnetic shielding constant. II. Gauge-including atomic orbitals and applications to molecules. *J. Chem. Phys.*, **118**(3) (2003) 1027–1035.

729 M. Reiher. On the definition of local spin in relativistic and nonrelativistic quantum chemistry. *Faraday Discuss.*, **135** (2007) 97–124.

730 G. Eickerling, R. Mastalerz, V. Herz, H.-J. Himmel, W. Scherer, M. Reiher. Relativistic Effects on the Topology of the Electron Density. *J. Chem. Theory Comput.*, **3** (2007) 2182–2197.

731 R. Mastalerz, R. Lindh, M. Reiher. The Douglas–Kroll–Hess Electron Density at an Atomic Nucleus. *Chem. Phys. Lett.*, **465** (2008) 157–164.

732 E. van Lenthe, P. E. S. Wormer, A. van der Avoird. Density functional calculations of molecular *g*-tensors in the zero order regular approximation for relativistic effects. *J. Chem. Phys.*, **107** (1997) 2488–2498.

733 F. Neese. Prediction of electron paramagnetic resonance *g* values using coupled perturbed Hartree–Fock and Kohn–Sham theory. *J. Chem. Phys.*, **115** (2001) 11080–11096.

734 O. Vahtras, M. Engström, B. Schimmelpfennig. Electronic *g*-tensors obtained with the mean-field spin–orbit Hamiltonian. *Chem. Phys. Lett.*, **351**(5-6) (2002) 424–430.

735 F. Neese. Correlated ab initio calculation of electronic g-tensors using a sum over states formulation. *Chem. Phys. Lett.*, **380** (2003) 721–728.

736 K. M. Neyman, D. I. Ganyushin, A. V. Matveev, V. A. Nasluzov. Calculation of Electronic g-Tensors Using a Relativistic Density Functional Douglas–Kroll Method. *J. Phys. Chem. A*, **106** (2002) 5022–5030.

737 F. Neese. Analytic Derivative Calculation of Electronic g-Tensors. *Mol. Phys.*, **105** (2007) 2507–2514.

738 N. Gilka, J. Tatchen, C. M. Marian. The g-tensor of AlO: Principal problems and first approaches. *Chem. Phys.*, **343**(2-3) (2008) 258–269.

739 E. van Lenthe, A. van der Avoird, P. Wormer. Density functional calculations of molecular hyperfine interactions in the zero order regular approximation for relativistic effects. *J. Chem. Phys.*, **108** (1998) 4783–4796.

740 F. Neese. Metal and ligand hyperfine couplings in transition metal complexes: The effect of spin–orbit coupling as studied by coupled perturbed Kohn–Sham theory. *J. Chem. Phys.*, **118**(9) (2003) 3939–3948.

741 I. M. Ondik. *Development, validation, and application of new relativistic methods for all-electron unrestricted two-component calculations of EPR parameters*. PhD thesis, University of Würzburg, Germany, 2006.

742 H. Fukui, T. Baba, H. Inomata. Erratum: Calculation of nuclear magnetic shieldings X. Relativistic effects [J. Chem. Phys. 105,3175 (1996)]. *J. Chem. Phys.*, **106** (1997) 2987.

743 O. L. Malkina, B. Schimmelpfennig, M. Kaupp, B. A. Hess, P. Chandra, U. Wahlgren, V. G. Malkin. Spin–orbit corrections to NMR shielding constants from density functional theory. How important are the two-electron terms? *Chem. Phys. Lett.*, **296**(1-2) (1998) 93–104.

744 J. Vaara, K. Ruud, O. Vahtras. Correlated Response Calculations of the Spin–Orbit Interaction Contribution to Nuclear Spin–Spin Couplings. *J. Comput. Chem.*, **20** (1999) 1314–1327.

745 S. K. Wolff, T. Ziegler, E. van Lenthe, E. J. Baerends. Density functional calculations of nuclear magnetic shieldings using the zeroth-order regular approximation (ZORA) for relativistic effects: ZORA nuclear magnetic resonance. *J. Chem. Phys.*, **110** (1999) 7689–7698.

746 J. I. Melo, M. C. R. de Azua, C. G. Giribet, G. A. Aucar, R. H. Romero. Relativistic effects on the nuclear magnetic shielding tensor. *J. Chem. Phys.*, **118** (2003) 471–486.

747 H. Fukui, T. Baba, Y. Shiraishi, S. Imanishi, K. Kudo, K. Mori, M. Shimoji. Calculation of nuclear magnetic shieldings: infinite-order Foldy–Wouthuysen transformation. *Mol. Phys.*, **102** (2004) 641–648.

748 S. A. Perera, R. J. Bartlett. A Reinvestigation of Ramsey's Theory of NMR Coupling. *Adv. Quantum Chem.*, **48** (2005) 435–467.

749 R. Fukuda, H. Nakatsuji. Quasirelativistic theory for the magnetic shielding constant. III. Quasirelativistic second-order Møller–Plesset perturbation theory and its application to tellurium compounds. *J. Chem. Phys.*, **123** (2005) 044101.

750 Y. Ootani, H. Yamaguti, H. Maeda, H. Fukui. Relativistic calculation of nuclear magnetic shielding tensor including two-electron spin-orbit interactions. *J. Chem. Phys.*, **125** (2006) 164106.

751 P. Lantto, R. H. Romero, S. S. Gómez, G. A. Aucar. Relativistic heavy-atom effects on heavy-atom nuclear shieldings. *J. Chem. Phys.*, **125** (2006) 184113.

752 Y. Xiao, W. Liu, L. Cheng, D. Peng. Four-component relativistic theory for nuclear magnetic shielding constants: Critical assessments of different approaches. *J. Chem. Phys.*, **126** (2007) 214101.

753 Y. Xiao, D. Peng, W. Liu. Four-component relativistic theory for nuclear magnetic shielding constants: The orbital decomposition approach. *J. Chem. Phys.*, **126** (2007) 081101.

754 M. Hanni, P. Lantto, M. Ilias, H. J. A. Jensen, J. Vaara. Relativistic effects in the intermolecular interaction-induced nuclear magnetic resonance parameters of xenon dimer. *J. Chem. Phys.*, **127** (2007) 164313.

755 S. Komorovský, M. Repiský, O. L. Malkina, V. G. Malkin, I. Malkin, M. Kaupp. A fully relativistic method for calculation of nuclear magnetic shielding tensors with a restricted magnetically balanced basis in the framework of the matrix Dirac–Kohn–Sham equation. *J. Chem. Phys.*, **128** (2008) 104101.

756 L. B. Casabianca, A. C. de Dios. *Ab initio* calculations of NMR chemical shifts. *J. Chem. Phys.*, **128** (2008) 052201.

757 V. G. Malkin, O. L. Malkina, D. R. Salahub. Spin–orbit correction to NMR shielding constants from density functional theory. *Chem. Phys. Lett.*, **261** (1996) 335–345.

758 B. F. Minaev, J. Vaara, K. Ruud, O. Vahtras, H. Ågren. Internuclear distance dependence of the spin–orbit coupling contributions to proton NMR chemical shifts. *Chem. Phys. Lett.*, **295** (1998) 455–461.

759 L. Visscher, T. Enevoldsen, T. Saue, H. J. A. Jensen, J. Oddershede. Full four-component relativistic calculations of NMR shielding and indirect spin-spin coupling tensors in hydrogen halides. *J. Comp. Chem.*, **20**(12) (1999) 1262–1273.

760 T. Baba, H. Fukui. Calculation of nuclear magnetic shieldings XIII. Gauge-origin independent relativistic effects. *J. Chem. Phys.*, **110** (1999) 131–137.

761 J. Autschbach, T. Ziegler. Nuclear spin–spin coupling constants from regular approximate relativistic density functional calculations. I. Formalism and scalar relativistic results for heavy metal compounds. *J. Chem. Phys.*, **113** (2000) 936–947.

762 J. Autschbach, T. Ziegler. Nuclear spin–spin coupling constants from regular approximate relativistic density functional calculations. II. Spin–orbit coupling effects and anisotropies. *J. Chem. Phys.*, **113** (2000) 9410–9418.

763 T. Enevoldsen, L. Visscher, T. Saue, H. J. A. Jensen, J. Oddershede. Relativistic four-component calculations of indirect nuclear spin-spin couplings in MH_4 (M=C, Si, Ge, Sn, Pb) and $Pb(CH_3)_3H$. *J. Chem. Phys.*, **112**(8) (2000) 3493–3498.

764 T. Baba, H. Fukui. Calculation of nuclear magnetic shieldings XIV. Relativistic mass-velocity corrected perturbation Hamiltonians. *Mol. Phys.*, **100** (2002) 623–633.

765 J. Vaara, P. Pyykkö. Relativistic, nearly basis-set-limit nuclear magnetic shielding constants of the rare gases He–Rh: A way to absolute nuclear magnetic resonance shielding scales. *J. Chem. Phys.*, **118** (2003) 2973–2976.

766 P. Manninen, P. Lantto, J. Vaara, K. Ruud. Perturbational *ab initio* calculations of relativistic contributions to nuclear magnetic resonance shielding tensors. *J. Chem. Phys.*, **119** (2003) 2623–2637.

767 J. I. Melo, M. C. Ruiz de Azua, C. G. Giribet, G. A. Aucar, P. F. Provasi. Relativistic effects on nuclear magnetic shielding constants in HX and CH_3X (X=Br,I) based on the linear response within the elimination of small component approach. *J. Chem. Phys.*, **121** (2004) 6798–6808.

768 J. I. Melo, M. C. R. de Azúa, J. E. Peralta, G. E. Scuseria. Relativistic calculation of indirect NMR spin-spin couplings using the Douglas–Kroll–Hess approximation. *J. Chem. Phys.*, **123** (2005) 204112.

769 L. Visscher. Magnetic Balance and Explicit Diamagnetic Expressions for Nuclear Magnetic Resonance Shielding Tensors. *Adv. Quantum Chem.*, **48** (2005) 369–381.

770 K. Kudo, H. Fukui. Calculation of nuclear magnetic shieldings using an analytically differentiated relativistic shielding formula. *J. Chem. Phys.*, **123** (2005) 114102.

771 Y. Ootani, H. Maeda, H. Fukui. Decoupling of the Dirac equation correct to the third order for the magnetic perturbation. *J. Chem. Phys.*, **127** (2007) 084117.

772 J. Autschbach. Two-component relativistic hybrid density functional computations of nuclear spin-spin coupling tensors using Slater-type basis sets and density-fitting techniques. *J. Chem. Phys.*, **129** (2008) 094105.

773 M. M. Sternheim. Second-Order Effects of Nuclear Magnetic Fields. *Phys. Rev.*, **128** (1962) 676–677.

774 G. A. Aucar, T. Saue, L. Visscher, H. J. A. Jensen. On the origin and contribution of the diamagnetic term in four-component relativistic calculations of magnetic properties. *J. Chem. Phys.*, **110**(13) (1999) 6208–6218.

775 W. Kutzelnigg. Diamagnetism in relativistic theory. *Phys. Rev. A*, **67** (2003) 032109.

776 F. London. Quantum theory of interatomic currents in aromatic compounds. *J. Phys. Radium*, **8** (1937) 397–409.

777 G. G. Hall. Gauge Invariant Gaussian Orbitals and the Ab Initio Calculation of Diamagnetic Susceptibility for Molecules. *Int. J. Quantum Chem.*, **VII** (1973) 15–25.

778 R. Ditchfield. Self-consistent perturbation theory of diamagnetism: I. A gauge-invariant LCAO method for N.M.R. chemical shifts. *Mol. Phys.*, **27** (1974) 789–807.

779 N. C. Pyper. The Breit interaction in external magnetic fields. *Chem. Phys. Lett.*, **96**(2) (1983) 211–217.

780 N. C. Pyper, Z. C. Zhang. Exact relativistic analogues of the non-relativistic hyperfine structure operators. *Mol. Phys.*, **64** (1988) 933–961.

781 N. F. Ramsey. Magnetic Shielding of Nuclei in Molecules. *Phys. Rev.*, **78** (1950) 699–703.

782 N. F. Ramsey. Dependence of Magnetic Shielding of Nuclei upon Molecular Orientation. *Phys. Rev.*, **83**(3) (1951) 540–541.

783 N. F. Ramsey. Chemical Effects in Nuclear Magnetic Resonance and in Diamagnetic Susceptibility. *Phys. Rev.*, **86**(2) (1952) 243–246.

784 P. Pyykkö. Relativistic Theory Of Nuclear Spin–Spin Coupling In Molecules. *Chem. Phys.*, **22** (1977) 289–296.

785 P. Pyykko. On the relativistic theory of NMR chemical shifts. *Chem. Phys.*, **74** (1983) 1–7.

786 J. Autschbach. Analyzing NMR shielding tensors calculated with two-component relativistic methods using spin-free localized molecular orbitals. *J. Chem. Phys.*, **128** (2008) 164112.

787 A. Schweiger, G. Jeschke. *Principles of pulse electron paramagnetic resonance*. Oxford University Press, Oxford, 2001.

788 A. Abragam, M. H. L. Pryce. Theory of the nuclear hyperfine structure of paramagnetic resonance spectra in crystals. *Proc. Roy. Soc. London A*, **205** (1951) 135–153.

789 P. Pyykkö. Spectroscopic nuclear quadrupole moments. *Mol. Phys.*, **99**(19) (2001) 1617–1629.

790 P. Schwerdtfeger, R. Bast, M. C. L. Gerry, C. R. Jacob, M. Jansen, V. Kellö, A. V. Mudring, A. J. Sadlej, T. Saue, T. Söhnel, F. E. Wagner. The quadrupole moment of the $3/2^+$ nuclear ground state of ^{197}Au from electric field gradient relativistic coupled cluster and density-functional theory of small molecules and the solid state. *J. Chem. Phys.*, **122** (2005) 124317.

791 L. Belpassi, F. Tarantelli, A. Sgamellotti, H. M. Quiney, J. N. P. van Stralen, L. Visscher. Nuclear electric quadrupole moment of gold. *J. Chem. Phys.*, **126** (2007) 064314.

792 C. R. Jacob, L. Visscher, C. Thierfelder, P. Schwerdtfeger. Nuclear quadrupole moment of ^{139}La from relativistic electronic structure calculations of the electric field gradients in LaF, LaCl, LaBr, and LaI. *J. Chem. Phys.*, **127** (2007) 204303.

793 M. Quack, J. Stohner, M. Willeke. High-Resolution Spectroscopic Studies and Theory of Parity Violation in Chiral Molecules. *Annu. Rev. Phys. Chem.*, **59** (2008) 741–769.

794 T. D. Lee, C. N. Yang. Question of Parity Conservation in Weak Interactions. *Phys. Rev.*, **104**(1) (1956) 254–258.

795 C. Wu, E. Ambler, R. W. Hayward, D. D. Hoppes, R. P. Hudson. Experimental Test of Parity Conservation in Beta Decay. *Phys. Rev.*, **105** (1957) 1413–1415.

796 M. Quack. On the measurement of the parity violating energy difference between enantiomers. *Chem. Phys. Lett.*, **132** (1986) 147–153.

797 M. Quack. Structure and dynamics of chiral molecules. *Angew. Chem. Int. Ed. Engl.*, **28** (1989) 571–586.

798 R. Berger, M. Quack, G. S. Tschumper. Electroweak Quantum Chemistry for Possible Precursor Molecules in the Evolution of Biomolecular Homochirality. *Helv. Chim. Acta*, **83** (2000) 1919–1950.

799 W. A. Bonner. Parity violation and the evolution of biomolecular homochirality. *Chirality*, **12**(3) (2000) 114–126.

800 M. Quack. How important is parity violation for molecular and biomolecular chirality? *Angew. Chem. Int. Ed. Engl.*, **41** (2002) 4618–4630.

801 A. C. Hartley, P. G. H. Sandars. Relativistic Calculations of Parity Non-Conserving Effects in Atoms. In Wilson et al. [975], p. 67–81.

802 I. B. Chriplovic. *Parity nonconservation in atomic phenomena*. Gordon and Breach Science Publ., Philadelphia, 1991.

803 A.-M. Mårtensson-Pendrill. *Calculation of P- and T-Violating Properties in Atoms and Molecules*, Volume 4, p. 99–156. Plenum Press, New York, 1992.

804 I. B. Khriplovich, S. K. Lamoreaux. *CP violation without strangeness: electric dipole moments of particles, atoms, and molecules*. Texts and Monographs in Physics. Springer-Verlag, Berlin, 1997.

805 J. S. M. Ginges, V. V. Flambaum. Violations of fundamental symmetries in atoms and tests of unification theories of elementary particles. *Phys. Rep.*, **397** (2004) 63–154.

806 C. S. Wood, S. C. Bennett, D. Cho, B. P. Masterson, J. L. Roberts, C. E. Tanner, C. E. Wieman. Measurement of Parity Non-conservation and an Anapole Moment in Cesium. *Science*, **275** (1997) 1759–1763.

807 S. C. Bennett, C. E. Wieman. Measurement of the $6S \longrightarrow 7S$ Transition Polarizability in Atomic Cesium and an Improved Test of the Standard Model (Erratum: *Phys. Rev. Lett.*, **82**, (1999), 4153). *Phys. Rev. Lett.*, **82**(12) (1999) 2484–2487.

808 W. R. Johnson, J. Sapirstein, S. A. Blundell. Atomic Structure Calculations Associated with PNC Experiments in Atomic Cesium. *Phys. Scr.*, **T46** (1993) 184–192.

809 V. A. Dzuba, C. Harabati, W. R. Johnson, M. S. Safronova. Breit correction to the parity-nonconservation amplitude in cesium. *Phys. Rev. A*, **63** (2001) 044103.

810 I. Bednyakov, W. R. Johnson, G. Plunien, G. Soff. Vacuum-Polarization Corrections to Parity-Nonconserving Effects in Atomic Systems. *Hyperfine Int.*, **146/147** (2003) 67–70.

811 A. Bakasov, T.-K. Ha, M. Quack. *Ab initio* calculation of molecular energies including parity violating interactions (Erratum: J. Chem. Phys., **110**, (1999), 6081). *J. Chem. Phys.*, **109**(17) (1998) 7263–7285.

812 A. Bakasov, M. Quack. Representation of parity violating potentials in molecular main chiral axes. *Chem. Phys. Lett.*, **203** (1999) 547–557.

813 R. Berger, M. Quack. Electroweak Quantum Chemistry of Alanine: Parity Violation in Gas and Condensed Phase. *Chem. Phys. Chem.*, **1**(1) (2000) 57–60.

814 R. Berger, M. Quack. Multiconfiguration linear response approach to the calculation of parity violating potentials in polyatomic molecules. *J. Chem. Phys.*, **112**(7) (2000) 3148–3158.

815 A. S. Lahamer, S. M. Mahurin, R. N. Compton, D. House, J. K. Laerdahl,

M. Lein, P. Schwerdtfeger. Search for a Parity-Violating Energy Difference between Enantiomers of a Chiral Iron Complex. *Phys. Rev. Lett.*, **85**(21) (2000) 4470–4473.

816 J. K. Laerdahl, R. Wesendrup, P. Schwerdtfeger. D- or L-Alanine: That Is the Question. *ChemPhysChem*, **1**(1) (2000) 60–62.

817 J. K. Laerdahl, T. Saue, K. Faegri Jr., H. M. Quiney. *Ab initio* Study of *PT*-Odd Interactions in Thallium Fluoride. *Phys. Rev. Lett*, **79**(9) (1997) 1642–1645.

818 H. M. Quiney, J. K. Laerdahl, K. Fægri, Jr., T. Saue. *Ab initio* Dirac–Hartree–Fock calculations of chemical properties and *PT*-odd effects in thallium fluoride. *Phys. Rev. A*, **57**(2) (1998) 920–944.

819 H. M. Quiney, H. Skaane, I. P. Grant. Hyperfine and *PT*-odd effects in YbF $^2\Sigma$. *J. Phys. B: At. Mol. Opt. Phys.*, **38** (1998) L85–L95.

820 J. K. Laerdahl, P. Schwerdtfeger. Fully relativistic *ab initio* calculations of the energies of chiral molecules including parity-violating weak interactions. *Phys. Rev. A*, **60**(6) (1999) 4439–4453.

821 J. K. Laerdahl, P. Schwerdtfeger, H. M. Quiney. Theoretical Analysis of Parity-Violating Energy Differences between the Enantiomers of Chiral Molecules. *Phys. Rev. Lett.*, **84**(17) (2000) 3811–3814.

822 J. Thyssen, J. K. Laerdahl, P. Schwerdtfeger. Fully Relativistic Coupled Cluster Treatment for Parity-Violating Energy Differences in Molecules. *Phys. Rev. Lett*, **85**(15) (2000) 3105–3108.

823 R. Berger, C. van Wüllen. Density functional calculations of molecular parity-violating effects within the zeroth-order regular approximation. *J. Chem. Phys.*, **122** (2005) 134316.

824 R. Berger, J. L. Stuber. Electroweak interactions in chiral molecules: two-component density functional theory study of vibrational frequency shifts in polyhalomethanes. *Mol. Phys.*, **105**(1) (2007) 41–49.

825 R. Berger, J. L. Stuber. *Electroweak quantum chemistry: Do it Breit!*, Volume 7B of *Recent Progress in Computational Science and Engineering*, p. 858–864. Brill Academic Publishers, Leiden, 2006.

826 R. Berger. Breit interaction contribution to parity violating potentials in chiral molecules containing light nuclei. *submitted*, 2008.

827 G. Laubender, R. Berger. Ab initio Calculation of Parity-Violating Chemical Shifts in NMR Spectra of Chiral Molecules. *ChemPhysChem*, (4) (2003) 395–399.

828 R. Bast, P. Schwerdtfeger, T. Saue. Parity nonconservation contribution to the nuclear magnetic resonance shielding constants of chiral molecules: A four-component relativistic study. *J. Chem. Phys.*, **125** (2006) 064504.

829 V. Weijo, R. Bast, P. Manninen, T. Saue, J. Vaara. Methodological aspects in the calculation of parity-violating effects in nuclear magnetic resonance parameters. *J. Chem. Phys.*, **126** (2007) 074107.

830 S. Nahrwold, R. Berger. Zeroth Order Regular Approximation Approach to Parity Violating Nuclear Magnetic Resonance Shielding Tensors. *submitted*, 2008.

831 R. E. Powell. Relativistic Quantum Chemistry. *J. Chem. Educ.*, **45**(9) (1968) 558–563.

832 K. S. Pitzer. Relativistic Effects on Chemical Properties. *Acc. Chem. Res.*, **12**(8) (1979) 271–276.

833 P. Pyykkö, J.-P. Desclaux. Relativity and the Periodic System of Elements. *Acc. Chem. Res.*, **12**(8) (1979) 276–281.

834 P. Pyykkö. On the Interpretation of 'Secondary Periodicity' in the Periodic System. *J. Chem. Res. (S)*, 1979) 380–381.

835 L. J. Norrby. Why Is Mercury Liquid? — Or, Why Do Relativistic Effects Not Get into Chemistry Textbooks? *J. Chem. Educ.*, **68**(2) (1991) 110–113.

836 L. Visscher, E. van Lenthe. On the distinction between scalar and spin-orbit relativistic effects. *Chem. Phys. Lett.*, **306** (1999) 357–365.

837 J. N. Harvey, M. Aschi. Modelling Spin-Forbidden Reactions: Recombination of Carbon Monoxide with Iron Tetracarbonyl. *Faraday Disc.*, **124** (2003) 129–143.

838 T. Tsuchiya, B. O. Roos. A theoretical study of the spin-forbidden reaction

$Fe(CO)_4 + CO \rightarrow Fe(CO)_5$. *Mol. Phys.*, **104**(6-7) (2006) 1123–1131.

839 M. Filatov, S. Shaik. Theoretical Investigation of Two-State-Reactivity Pathways of H–H Activation by FeO^+: Addition–Elimination, "Rebound", and Oxene-Insertion Mechanisms. *J. Phys. Chem. A*, **102** (1998) 3835–3846.

840 N. Harris, S. Shaik, D. Schröder, H. Schwarz. Single- and Two-State Reactivity in the Gas-Phase C–H Bond Activation of Norbornane by 'Bare' FeO^+. *Helv. Chim. Acta*, **82** (1999) 1784–1797.

841 N. Harris, S. Cohen, M. Filatov, F. Ogliaro, S. Shaik. Two-State Reactivity in the Rebound Step of Alkane Hydroxylation by Cytochrome P-450: Origins of Free Radicals with Finite Lifetimes. *Angew. Chem. Int. Ed.*, **39**(11) (2000) 2003–2007.

842 D. Schröder, S. Shaik, H. Schwarz. Two-State Reactivity as a New Concept in Organometallic Chemistry. *Acc. Chem. Res.*, **33** (2000) 139–145.

843 B. F. Minaev, H. Ågren. Spin-catalysis phenomena. *Int. J. Quantum Chem.*, **57**(3) (1996) 519–532.

844 J. C. Green, J. N. Harvey, R. Poli. Theoretical investigation of the spin crossover transition states of the addition of methane to a series of Group 6 metallocenes using minimum energy crossing points. *J. Chem. Soc., Dalton Trans.*, (2002) 1861–1866.

845 R. Poli, J. N. Harvey. Spin forbidden chemical reactions of transition metal compounds. New ideas and new computational challenges. *Chem. Soc. Rev.*, **32**(1) (2003) 1–8.

846 J. N. Harvey. DFT Computation of Relative Spin-State Energetics of Transition Metal Compounds. *Struct. Bonding*, **112** (2004) 151–183.

847 T. Ziegler, J. G. Snijders, E. J. Baerends. On the Origin of Relativistic Bond Contraction. *Chem. Phys. Lett.*, **75**(1) (1980) 1–4.

848 T. Ziegler, J. G. Snijders, E. J. Baerends. Relativistic effects on bonding. *J. Chem. Phys.*, **74**(2) (1981) 1271–1284.

849 P. A. Christiansen, W. C. Ermler, K. S. Pitzer. Relativistic Effects in Chemical Systems. *Annu. Rev. Phys. Chem.*, **36** (1985) 407–432.

850 W. H. E. Schwarz. Relativistic Calculations of Molecules — Relativity and Bond Lengths. *Phys. Scr.*, **36** (1987) 403–411.

851 P. Pyykkö. Relativistic Effects in Structural Chemistry. *Chem. Rev.*, **88** (1988) 563–594.

852 W. H. E. Schwarz, A. Rutkowski, S. G. Wang. Understanding Relativistic Effects of Chemical Bonding. *Int. J. Quantum Chem.*, **57** (1996) 641–653.

853 D. Schröder, J. Hrušàk, R. H. Hertwig, W. Koch, P. Schwerdtfeger, H. Schwarz. Experimental and Theoretical Studies of Gold(I) Complexes $Au(L)^+$ (L = H_2O, CO, NH_3, C_2H_4, C_3H_6, C_4H_6, C_6H_6, C_6F_6). *Organometallics*, **14** (1995) 312–316.

854 J. Hrušàk, R. H. Hertwig, D. Schröder, P. Schwerdtfeger, W. Koch, H. Schwarz. Relativistic Effects in Cationic Gold(I) Complexes: A Comparative Study of ab Initio Pseudopotential and Density Functional Methods. *Organometallics*, **14** (1995) 1284–1291.

855 R. H. Hertwig, W. Koch, D. Schröder, H. Schwarz, J. Hrušàk, P. Schwerdtfeger. A Comparative Computational Study of Cationic Coinage Metal-Ethylene Complexes $(C_2H_4)M^+$ (M=Cu, Ag, and Au). *J. Phys. Chem.*, **100** (1996) 12253–12260.

856 J. Schwarz, D. Schröder, H. Schwarz, C. Heinemann, J. Hrušàk. Theory-enforced Re-investigation of the Origin of the Large Metal-Carbon Bond Strength in $PdCH_2I^+$ and Its Reactions with Unsaturated Hydrocarbons. *Helv. Chim. Acta*, **79** (1996) 1110–1120.

857 D. Schröder, M. Diefenbach, H. Schwarz, A. Schier, H. Schmidbaur. Experimental Probes of Relativistic Effects in the Chemistry of Heavy *d* and *f* Elements. In B. A. Hess, Ed., *Relativistic Effects in Heavy Element Chemistry and Physics*, Wiley Series in Theoretical Chemistry, p. 245–258. John Wiley & Sons Ltd., Chichester, 2003.

858 P. Pyykkö. Strong closed-shell interactions in inorganic chemistry. *Chem. Rev.*, **97** (1997) 597–636.

859 P. Pyykkö. Theoretical Chemistry of Gold. *Angew. Chem. Int. Ed.*, **43** (2004) 4412–4456.

860 D. J. Gorin, F. D. Troste. Relativistic effects in homogeneous gold catalysis. *Nature*, **446** (2007) 395–403.

861 P. Pyykkö. Theoretical Chemistry of Gold. III. *Chem. Soc. Rev.*, **37** (2008) 1967–1997.

862 H. S. *et al.* *Strukturphänomene in der Gold-Chemie und ihre Zuordnung zu Relativistischen Effekten (in: Unkonventionelle Wechselwirkungen in der Chemie metallischer Elemente)*, Chapter C.3, p. 373–385. DFG Forschungsbericht. VCH, Weinheim, 1992.

863 H. Schmidbaur. The Aurophilicity Phenomenon: A Decade of Experimental Findings, Theoretical Concepts and Emerging Applications. *Gold Bull.*, **33**(1) (2000) 3–10.

864 F. Scherbaum, A. Grohmann, B. Huber, C. Krüger, H. Schmidbaur. "Aurophilie" als Konsequenz relativistischer Effekte: Das Hexakis(triphenylphosphanaurio)-methan-Dikation $[(Ph_3PAu)_6C]^{2+}$. *Angew. Chem.*, **100** (1988) 1602–1604.

865 N. Rösch, A. Görling, D. E. Ellis, H. Schmidbaur. Aurophilie als konzertierter Effekt: Relativistische MO-Berechnungen für Kohlenstoff-zentrierte Goldcluster. *Angew. Chem.*, **101**(10) (1989) 1410–1412.

866 H. Schmidbaur, B. Brachthäuser, O. Steigelmann, H. Beruda. Preparation and Structure of Hexakis[(trialkylphosphane)aurio(I)]methanium(2+) Salts $[(LAu)_6C]^{2+}(X^-)_2$ with L=Et_3P, $i Pr_3P$ and X=BF_4^-, $B_3O_3F_4^-$. *Chem. Ber.*, **125** (1997) 2705–2710.

867 A. Görling, N. Rösch, D. E. Ellis, H. Schmidbaur. Electronic Structure of Main-Group-Element-Centered Octahedral Gold Clusters. *Inorg. Chem.*, **30** (1991) 3986–3994.

868 J. Neugebauer, M. Reiher. Vibrational Center–Ligand Couplings in Transition Metal Complexes. *J. Comput. Chem.*, **25** (2004) 587–597.

869 B. Le Guennic, J. Neugebauer, M. Reiher, J. Autschbach. The "invisible" ^{13}C chemical shift of the central carbon atom in $[(Ph_3PAu)_6C]^{2+}$: A theoretical investigation. *Chen. Eur. J.*, **11** (2005) 1677–1686.

870 P. Pyykkö, K. Angermaier, B. Assmann, H. Schmidbaur. Calculated Structures of SAu_3^+ and $S(AuPH_3)_3^+$. *J. Chem. Soc., Chem. Comm.*, (1995) 1889–1890.

871 P. Pyykkö, N. Runeberg, F. Mendizabal. Theory of the d^{10}–d^{10} Closed-Shell Attraction: 1. Dimers Near Equilibrium. *Chem. Eur. J.*, **3** (1997) 1451–1457.

872 P. Pyykkö, T. Tamm. Theory of the $d^{10}-d^{10}$ Closed-Shell Attraction. 4. $X(AuL)_n^{m+}$ Centered Systems. *Organometallics*, **17** (1998) 4842–4852.

873 P. Pyykkö, F. Mendizabal. Theory of d^{10}–d^{10} Closed-Shell Attraction. III. Rings. *Inorg. Chem.*, **37** (1998) 3018–3025.

874 V. M. Burke, I. P. Grant. The effect of relativity on atomic wave function. *Proc. Phys. Soc.*, **90** (1967) 297–314.

875 C. E. Moore. *Atomic Energy Levels — Vols. I–III*, Volume 35 of *Nat. Stand. Ref. Data Ser.* Nat. Bur. Stand. (USA), 1971.

876 J. Autschbach, S. Siekierski, M. Seth, P. Schwerdtfeger, W. H. E. Schwarz. Dependence of relativistic effects on electronic configuration in the neutral atoms of *d*- and *f*-block elements. *J. Comput. Chem.*, **23** (2002) 804–813.

877 A. S. P. Gomes, L. Visscher. The influence of core correlation on the spectroscopic constants of HAt. *Chem. Phys. Lett.*, **399** (2004) 1–6.

878 C. Kunz, C. Hättig, B. A. Heß. Ab initio study of the individual interaction energy components in the ground state of the mercury dimer. *Mol. Phys.*, **89** (1996) 139–156.

879 M. Dolg, H.-J. Flad. Size dependent properties of Hg_n clusters. *Mol. Phys.*, **91** (1997) 815–825.

880 P. Schwerdtfeger, R. Wesendrup, G. E. Moyano, A. J. Sadlej, J. Greif, F. Hensel. The potential energy curve and dipole polarizability tensor of mercury dimer. *J. Chem. Phys.*, **115** (2001) 7401–7412.

881 G. E. Moyano, R. Wesendrup, T. Söhnel, P. Schwerdtfeger. Properties of Small- to Medium-Sized Mercury Clusters from a Combined ab initio, Density-Functional, and Simulated-Annealing Study. *Phys. Rev. Lett.*, **89**(10) (2002) 103401.

882 N. Gaston, P. Schwerdtfeger. From the van der Waals dimer to the solid state of mercury with relativistic ab initio and density functional theory. *Phys. Rev. B*, **74** (2006) 024105.

883 J.-M. Lévy-Leblond. Nonrelativistic Particles and Wave Equations. *Commun. Math. Phys.*, **6** (1967) 286–311.

884 J.-M. Lévy-Leblond. Minimal electromagnetic coupling as a consequence of Lorentz invariance. *Ann. Phys.*, **57**(2) (1970) 481–495.

885 W. J. Hurley. Nonrelativistic Quantum Mechanics for Particles with Arbitrary Spin. *Phys. Rev. D*, **3** (1970) 2339–2347.

886 W. J. Hurley. Relativistic Wave Equations for Particles with Arbitrary Spin. *Phys. Rev. D*, **4** (1971) 3605–3616.

887 W. Kutzelnigg. Perturbation Theory of Relativistic Effects. In P. Schwerdtfeger, Ed., *Relativistic Electronic Structure Theory — Part I. Fundamentals*, p. 664–757, Amsterdam, 2002. Elsevier.

888 A. Rutkowski. Relativistic perturbation theory: I. A new perturbation approach to the Dirac equation. *J. Phys. B: At. Mol. Opt. Phys.*, **19** (1986) 149–158.

889 A. Rutkowski. Relativistic perturbation theory: II. One-electron variational perturbation calculations. *J. Phys. B: At. Mol. Opt. Phys.*, **19** (1986) 3431–3441.

890 A. Rutkowski. Relativistic perturbation theory: III. A new perturbation approach to the two-electron Dirac–Coulomb equation. *J. Phys. B: At. Mol. Opt. Phys.*, **19** (1986) 3443–3455.

891 W. Kutzelnigg. Perturbation theory of relativistic corrections 1. The non-relativistic limit of the Dirac equation and a direct perturbation expansion. *Z. Phys. D — At. Mol. Clust.*, **11** (1989) 15–28.

892 A. Rutkowski, W. H. E. Schwarz. Relativistic perturbation theory of chemical properties. *Theor. Chim. Acta*, **76** (1990) 391–410.

893 W. H. E. Schwarz, A. Rutkowski, G. Collignon. *Nonsingular Relativistic Perturbation Theory and Relativistic Changes of Molecular Structure*, p. 135–147. Plenum Press, New York, 1991.

894 A. Rutkowski, D. Rutkowska, W. H. E. Schwarz. Relativistic perturbation theory of molecular structure. *Theor. Chim. Acta*, **84** (1992) 105–114.

895 A. Rutkowski. Regular perturbation theory of relativistic corrections: Basic aspects. *Phys. Rev. A*, **53**(1) (1996) 145–151.

896 W. Kutzelnigg. Effective Hamiltonians for degenerate and quasidegenerate direct perturbation theory of relativistic effects. *J. Chem. Phys.*, **110**(17) (1999) 8283–8294.

897 T. Fleig, D. Edvardsson, S. T. Banks, J. H. D. Eland. A theoretical and experimental study of the double photoionisation of molecular bromine and a new double ionisation mechanism. *Chem. Phys.*, **343** (2008) 270–280.

898 T. Fleig, L. K. Sørensen, J. Olsen. A relativistic 4-component general-order multireference coupled cluster method: initial implementation and application to HBr. *Theor. Chem. Acc.*, **118** (2007) 347–356.

899 K. Balasubramanian. *Relativistic Effects in Chemistry, Part B, Relativistic Effects in Chemistry*. John Wiley & Sons, New York, Chichester, 1997.

900 R. Mastalerz, M. Reiher. *Relativistic Electronic Structure Theory for Molecular Spectroscopy*, p. in press. Wiley, Chichester, 2008.

901 M. Seth, P. Schwerdtfeger, K. Faegri. The chemistry of superheavy elements. III. Theoretical studies on element 113 compounds. *J. Chem. Phys.*, **111**(14) (1999) 6422–6433.

902 K. Fægri Jr., L. Visscher. Relativistic calculations on thallium hydride. *Theor. Chem. Acc.*, **105**(3) (2001) 265–267.

903 B. Grundström, P. Valberg. Das Bandenspektrum des Thalliumhydrids. I. *Z. Phys.*, **108**(5-6) (1935) 293–306.

904 R.-D. Urban, A. H. Bahnmaier, U. Magg, H. Jones. The diode laser spectrum of thallium hydride (^{205}TlH and ^{203}TlH) in its ground electronic state. *Chem. Phys. Lett.*, **158** (1989) 443–446.

905 Y. S. Lee, W. C. Ermler, K. S. Pitzer, A. D. McLean. *Ab initio* effective core potentials including relativistic effects. III. Ground state Au_2 calculations. *J. Chem. Phys.*, **70** (1979) 288–292.

906 W. C. Ermler, Y. S. Lee, K. Pitzer. *Ab initio* effective core potentials including relativistic effects. IV. Potential energy curves for the ground and several excited states of Au_2. *J. Chem. Phys.*, **70** (1979) 293.

907 H. Gollisch. Density matrix study of Cs_2, Au_2 and CsAu. *J. Phys. B*, **15**(16) (1982) 2569–2578.

908 S. Rabii, C. Y. Wang. Relativistic electronic structures of the Ag_2 and Au_2 molecules. *Chem. Phys. Lett.*, **105**(5) (1984) 480–483.

909 R. B. Ross, W. C. Ermler. Ab Initio Calculations Including Relativistic Effects for Ag_2, Au_2, AgAu, AgH, and AuH. *J. Phys. Chem.*, **89** (1985) 5202–5206.

910 P. A. Christiansen, W. C. Ermler. Relativistic Bond Length and Atomic Orbital Contraction. *Mol. Phys.*, **55** (1985) 1109–1111.

911 M. H. McAdon, I. W. A. Goddard. Charge Density Waves, Spin-Density Waves, and Peierls Distortions in One-Dimensional Metals. 2. Generalized Valence Bond Studies of Cu, Ag , Au, Li, and Na. *J. Phys. Chem.*, **92**(5) (1988) 1352–1365.

912 C. W. Bauschlicher, Jr., S. R. Langhoff, H. Partridge. Theoretical study of the structures and electron affinities of the dimers and trimers of the group IB metals (Cu, Ag, and Au). *J. Chem. Phys.*, **91** (1989) 2412.

913 P. Schwerdtfeger, M. Dolg, W. H. E. Schwarz, G. A. Bowmaker, P. D. W. Boyd. Relativistic effects in gold chemistry. I. Diatomic gold compounds. *J. Chem. Phys.*, **91** (1989) 1762–1774.

914 K. Balasubramanian, P. Y. Feng. The ionization potentials of Ag_n and Au_n and binding energies of Ag_n, Au_n, Ag_n^+ and Au_n^+ (n=1–4). *Chem. Phys. Lett.*, **159** (1989) 452–458.

915 H. Partridge, C. W. Bauschlicher, Jr., S. R. Langhoff. Theoretical study of the positive ions of the dimers and trimers of the group IB metals (Cu, Ag, and Au). *Chem. Phys. Lett.*, **175**(5) (1990) 531–535.

916 D. Strömberg, U. Wahlgren. First-order relativistic calculations on Au_2 and Hg_2^{2+}. *Chem. Phys. Lett.*, **169**(1-2) (1990) 109–115.

917 D. Andrae, U. Häussermann, M. Dolg, H. Stoll, H. Preuss. Energy-adjusted *ab initio* pseudopotentials for the second and third row transition elements - molecular test for M_2 (M=Ag,Au) a nd MH (M=Ru,Os). *Theor. Chim. Acta*, **78** (1991) 247–266.

918 P. Schwerdtfeger. Relativistic and electron-correlation contributions in atomic and molecular properties: benchmark calculations on Au and Au_2. *Chem. Phys. Lett.*, **183** (1991) 457–463.

919 O. D. Häberlen, N. Rösch. A scalar-relativistic extension of the linear combination of Gaussian-type orbitals local density functional method: application to AuH, AuCl and Au_2. *Chem. Phys. Lett.*, **199** (1992) 491–496.

920 T. Baştuğ, D. Heinemann, W.-D. Sepp, D. Kolb, B. Fricke. All-electron Dirac–Fock–Slater SCF calculations of the Au_2 molecule. *Chem. Phys. Lett.*, **211** (1993) 119–124.

921 M. Mayer, O. D. Häberlen, N. Rösch. Relevance of relativistic exchange-correlation functionals and of finite nuclei in molecular density-functional calculations. *Phys. Rev. A*, **54** (1996) 4775–4782.

922 U. Wahlgren, B. Schimmelpfennig, S. Jusuf, H. Strömsnes, O. Gropen, L. Maron. A local approximation for relativistic scalar operators applied to the uranyl ion and to Au. *Chem. Phys. Lett.*, **287** (1998) 525–530.

923 H.-S. Lee, Y.-K. Han, M. C. Kim, C. Bae, Y. S. Lee. Spin–orbit effects calculated by two-component coupled-cluster methods: test calculations on AuH, Au_2, TlH and Tl_2. *Chem. Phys. Lett.*, **293** (1998) 97–102.

924 W. Liu, C. van Wüllen. Spectroscopic constants of gold and eka-gold (element 111) diatomic compounds: The importance of spin-orbit coupling. *J. Chem. Phys.*, **110**(8) (1999) 3730–3735.

925 B. A. Hess, U. Kaldor. Relativistic all-electron coupled-cluster calculations on Au_2 in the framework of the Douglas–Kroll transformation. *J. Chem. Phys.*, **112**(4) (2000) 1809–1813.

926 T. Tsuchiya, M. Abe, T. Nakajima, K. Hirao. Accurate relativistic Gaussian basis sets for H through Lr determined by atomic self-consistent field calculations with the third-order Douglas–Kroll approximation. *J. Chem. Phys.*, **115**(10) (2001) 4463–4472.

927 T. Fleig, L. Visscher. Large-scale electron correlation calculations in the framework of the spin-free Dirac formalism: the Au_2 molecule revisited. *Chem. Phys.*, **311** (2005) 113–120.

928 K. Hilpert, K. A. Gingerich. Atomization enthalpies of the molecules Cu_3, Ag_3, and Au_3. *Ber. Bunsenges. Phys. Chem.*, **84**(8) (1980) 739–745.

929 B. Simard, P. A. Hackett. High Resolution Study of the (0,0) and (1,1) Bands of the $A0_u^+$–$X0_g^+$ System of Au_2^1. *J. Mol. Spectrosc.*, **142** (1990) 310–318.

930 L. L. Ames, R. F. Barrow. Rotational analysis of bands of gaseous Au_2 molecule. *Trans. Faraday Soc.*, **63**(529P) (1967) 39–44.

931 J. Olsen, B. O. Roos, P. Jørgensen, H. J. A. Jensen. Determinant based configuration interaction algorithms for complete and restricted configuration interaction spaces. *J. Chem. Phys.*, **89** (1988) 2185–2192.

932 K. G. Dyall. All-electron molecular Dirac–Hartree–Fock calculations: Properties of the group IV monoxides GeO, SnO, and PbO. *J. Chem. Phys.*, **98**(3) (1993) 2191–2197.

933 G. E. Davico, T. M. Ramond, W. C. Lineberger. *J. Chem. Phys.*, **113**(19) (2000) 8852–8853.

934 K. P. Huber, G. Herzberg. *Constants of Diatomic Molecules*, Volume IV of *Molecular Spectra and Molecular Structure*. Van Nostrand Reinhold, New York, 1979.

935 G. Moritz, A. Wolf, M. Reiher. Relativistic DMRG calculations on the curve crossing of cesium hydride. *J. Chem. Phys.*, **123** (2005) 184105.

936 M. Carnell, S. D. Peyerimhoff, B. A. Hess. Ab initio MRD CI calculations on the cesium hydride (CsH) molecule. *Z. Phys. D*, **13** (1989) 317–333.

937 M. Dolg. Fully relativistic pseudopotentials for alkaline atoms: Dirac–Hartree–Fock and configuration interaction calculations of alkaline monohydrides. *Theor. Chim. Acta*, **93** (1996) 141–156.

938 W. C. Stwalley, W. T. Zemke, S. C. Yang. Spectroscopy and Structure of the Alkali Hydride Diatomic Molecules and Their Ions. *J. Chem. Phys. Ref. Data*, **20**(1) (1991) 153–187.

939 A. Wolf, M. Reiher, B. A. Hess. *University of Erlangen–Nuremberg*, 2002) unpublished results.

940 M. Dolg. Lanthanides and Actinides. In P. v. R. Schleyer *et al.*, Ed., *Encyclopedia of Computational Chemistry*, p. 1478–1486. John Wiley & Sons Ltd., Chichester, 1998.

941 J. S. Craw, M. A. Vincent, I. H. Hillier, A. L. Wallwork. *Ab Initio* Quantum Chemical Calculations on Uranyl UO_2^{2+}, Plutonyl PuO_2^{2+}, and Their Nitrates and Sulfates. *J. Opt. Soc. Am.*, **99** (1995) 10181–10185.

942 N. Ismail, J.-L. Heully, T. Saue, J.-P. Daudey, C. J. Marsden. Theoretical studies of the actinides: method calibration for the UO_2^{2+} and PuO_2^{2+} ions. *Chem. Phys. Lett.*, **300** (1999) 296–302.

943 K. G. Dyall. Bonding and Bending in the actinyls. *Mol. Phys.*, **96**(4) (1999) 511–518.

944 P. Pyykkö, N. Runeberg, M. Straka, K. G. Dyall. Could uranium(XII)hexoxide, $UO_6(O_h)$ exist? *Chem. Phys. Lett.*, **328** (2000) 415–419.

945 L. Gagliardi, B. O. Roos. Quantum chemical calculations show that the uranium molecule U_2 has a quintuple bond. *Nature*, **433** (2005) 848–851.

946 L. Gagliardi, P. Pyykkö, B. O. Roos. A very short uranium-uranium bond: The predicted metastable U_2^{2+}. *Phys. Chem. Chem. Phys.*, **7** (2005) 2415–2417.

947 B. O. Roos, P.-A. Malmqvist, L. Gagliardi. Exploring the Actinide–Actinide Bond: Theoretical Studies of the Chemical Bond in Ac_2, Th_2, Pa_2, and U_2. *J. Am. Chem. Soc.*, **128**(51) (2006) 17000–17006.

948 G. La Maccia, M. Brynda, L. Gagliardi. Quantum chemical calculations predict the diphenyl diuranium compound [PhU-UPh] to have a stable ^1A_g ground state. *Angew. Chem. Int. Ed.*, **45** (2006) 6210–6213.

949 B. O. Roos, L. Gagliardi. Quantum chemistry predicts multiply bonded diuranium compounds to be stable. *Inorg. Chem.*, **45**(2) (2006) 803–807.

950 P. S. Bagus, Y. S. Lee, K. S. Pitzer. Effects of Relativity and of the Lanthanide Contraction on the Atoms from Hafnium to

Bismuth. *Chem. Phys. Lett.*, **33**(3) (1975) 408–411.

951 J. K. Laerdahl, K. Fægri, Jr., L. Visscher, T. Saue. A fully relativistic Dirac–Hartree–Fock and second-order Møller–Plesset study of the lanthanide and actinide contraction. *J. Chem. Phys.*, **109**(24) (1998) 10806–10817.

952 W. Küchle, M. Dolg, H. Stoll. Ab Initio Study of the Lanthanide and Actinide Contraction. *J. Phys. Chem. A*, **101** (1997) 7128–7133.

953 M. Seth, M. Dolg, P. Fulde, P. Schwerdtfeger. Lanthanide and actinide contractions: relativistic and shell structure effects. *J. Am. Chem. Soc.*, **117** (1995) 6597–6598.

954 L. Maron, T. Leininger, B. Schimmelpfennig, V. Vallet, J.-L. Heully, C. Teichteil, O. Gropen, U. Wahlgren. Investigation of the low-lying excited states of PuO_2^{2+}. *Chem. Phys.*, **244** (1999) 195–201.

955 C. Clavaguéra-Sario, V. Vallet, D. Maynau, C. J. Marsden. Can density functional methods be used for open-shell actinide molecules? Comparison with multiconfigurational spin-orbit studies. *J. Chem. Phys.*, **121** (2004) 5312–5321.

956 I. Infante, A. S. P. Gomes, L. Visscher. On the performance of the intermediate Hamiltonian Fock-space coupled-cluster method on linear triatomic molecules: The electronic spectra of NpO_2^{+}, NpO_2^{2+}, and PuO_2^{2+}. *J. Chem. Phys.*, **125** (2006) 074301.

957 *Gmelin Handbooks of Inorganic Chemistry, Transuranium Elements A2*. Springer-Verlag, New York, 1973.

958 V. M. Garcia, O. Castell, R. Caballol, J. P. Malrieu. An iterative difference-dedicated configuration interaction. Proposal and test studies. *Chem. Phys. Lett.*, **238**(4-6) (1995) 222–229.

959 J. Autschbach, W. H. E. Schwarz. Relativistic electron densities in the four-component Dirac representation and in the two-component picture — Hydrogen-like systems. *Theor. Chem. Acc.*, **104** (2000) 82–88.

960 R. J. Gillespie, E. A. Robinson. Elektronendomänen und das VSEPR-Modell der Molekülgeometrie. *Angew. Chem.*, **108** (1996) 539–560.

961 W. Scherer, G. S. McGrady. Agostic interactions in d(0) metal alkyl complexes. *Angew. Chem. Int. Ed.*, **43**(14) (2004) 1782–1806.

962 W. Scherer, G. Eickerling, D. Shorokhov, E. Gullo, G. S. McGrady, P. Sirsch. Valence shell charge concentrations and the Dewar–Chatt–Duncanson bonding model. *New J. Chem.*, **30**(3) (2006) 309–312.

963 N. Hebben, H.-J. Himmel, G. Eickerling, C. Herrmann, M. Reiher, V. Herz, M. Presnitz, W. Scherer. The electronic structure of the tris(ethylene) complexes $M(C_2H_4)_3$ (M = Ni, Pd, and Pt): A combined experimental and theoretical study. *Chem. Eur. J.*, **13** (2007) 10078–10087.

964 R. F. W. Bader. *Atoms in Molecules - A Quantum Theory*, Volume 22 of *International Series of Monographs on Chemistry*. Oxford Science Publications, 1st ed., 1990.

965 J. Cioslowski, J. Karwowski. *Quantum-Mechanical Theory of Atoms in Molecules: A Relativistic Formulation*, p. 101. Mathematical and Computational Chemistry. Kluwer Academic, Dordrecht, 2001.

966 C. J. Cramer. *Essentials of Computational Chemistry: Theories and Models*. J. Wiley & Sons, New York, 2 ed., 2004.

967 F. Jensen. *Introduction to Computational Chemistry*. Wiley, New York, 2nd ed., 2006.

968 B. O. Roos, P.-Å. Malmqvist. On the Effects of Spin–Orbit Coupling on Molecular Properties: Dipole Moment and Polarizability of PbO and Spectroscopic Constants for the Ground and Excited States. *Adv. Quantum Chem.*, **47** (2004) 37–49.

969 L. Gagliardi, B. O. Roos. Multiconfigurational quantum chemical methods for molecular systems containing actinides. *Chem. Soc. Rev.*, **36** (2007) 893–903.

970 G. Hong, M. Dolg, L. Li. A comparison of scalar-relativistic ZORA and DKH density functional schemes: monohydrides, monooxides and monofluorides of La, Lu, Ac and Lr. *Chem. Phys. Lett.*, **334** (2001) 396–402.

971 T. Fließbach. *Elektrodynamik — Lehrbuch zur Theoretischen Physik II*. Spektrum Akademischer Verlag, Weinheim, 2nd ed., 1997.

972 B. R. Johnson. New numerical methods applied to solving the one-dimensional eigenvalue problem. *J. Chem. Phys.*, **67**(9) (1977) 4086–4093.

973 D. F. Mayers. *Computational Methods in the Differential Equations of Atomic Physics (in: New Directions in Atomic Physics)*, Volume 1 of *Yale Series in the Sciences*, Chapter 9, p. 233–244. Yale University Press, New Haven, 1972.

974 W. Barth, R. S. Martin, J. H. Wilkinson. *(in: Linear Algebra)*, p. 249. Springer-Verlag, 1st ed., 1971.

975 S. Wilson, I. P. Grant, B. Gyorffy, Ed. *The Effects of Relativity in Atoms, Molecules and the Solid State*. Plenum Press, New York, 1991.

Index

Relativistic Quantum Chemistry. Markus Reiher and Alexander Wolf

ISBN: 978-3-527-31292-4